Parasitism

The diversity and ecology of animal parasites

Parasitism in its many, sometimes subtle, guises is the most prevalent life style among organisms. Not surprisingly then, most organisms are exploited by parasites of one type or another. But what is parasitism and what are parasites? In this textbook, the authors describe parasitism as an ecological relationship. Ecology is the study of the relationship between organisms and their environment; what better definition can there be for the host–parasite relationship? The study of the host–parasite relationship differs markedly from studies on free-living organisms in one very important way – the hosts for parasites are alive and are thus capable of responding, positively or negatively, to the presence of the parasite. This feature alone weds such traditionally disparate disciplines as immunology and physiology with ecology and epidemiology!

This textbook describes the diversity of the major eukaryotic parasites of animals, who they are and how they live. With a consideration of both pathology and regulation, it considers the impact that parasites have on their hosts (and, with a discussion of immunology, vice versa). It also includes chapters on parasite populations, communities, perpetuation, biogeography, and evolution, emphasizing the ecological overtones of the host–parasite relationships.

Text boxes containing anecdotal, supplemental, and interesting observations and discussions pepper the book to bring the subject matter to life. Well illustrated and written in an easy, accessible style, *Parasitism: The Diversity and Ecology of Animal Parasites* represents a new and different approach to the study of parasitology.

ALBERT O. BUSH is Professor and Chair of the Zoology Department, Brandon University, Brandon, Manitoba, Canada; he is a member on the editorial boards of several parasitology journals, is past-President of the Parasitology Section of the Canadian Society of Zoologists, and is an internationally known and respected ecological parasitologist.

JACQUELINE C. FERNÁNDEZ is a Research Associate in the Department of Biology, Wake Forest University, Winston-Salem, North Carolina, USA and is a recognized expert on digenean–snail interactions and parasite biogeography.

GERALD W. ESCH is Editor of the *Journal of Parasitology* and recipient of the 1999 Mentor Award from the American Society of Parasitologists; he is presently Charles M. Allen Professor of Biology at Wake Forest University.

J. RICHARD SEED is a Ward Medallist and past-President of the American Society of Parasitologists; he is currently Professor of Parasitology in the Department of Epidemiology, School of Public Health, University of North Carolina at Chapel Hill, North Carolina, USA.

Dedication

We dedicate this book to our students: after all, they sanction our professional lives.

Parasitism:
The diversity and ecology of animal parasites

Albert O. Bush
Brandon University, Manitoba, Canada

Jacqueline C. Fernández
Wake Forest University, North Carolina, USA

Gerald W. Esch
Wake Forest University, North Carolina, USA

J. Richard Seed
University of North Carolina at Chapel Hill, USA

PUBLISHED BY THE PRESS SYNDICATE OF THE UNIVERSITY OF CAMBRIDGE
The Pitt Building, Trumpington Street, Cambridge, United Kingdom

CAMBRIDGE UNIVERSITY PRESS
The Edinburgh Building, Cambridge CB2 2RU, UK
40 West 20th Street, New York, NY 10011-4211, USA
10 Stamford Road, Oakleigh, VIC 3166, Australia
Ruiz de Alarcón 13, 28014 Madrid, Spain
Dock House, The Waterfront, Cape Town 8001, South Africa

http://www.cambridge.org

First published 2001

Printed in the United Kingdom at the University Press, Cambridge

Typeface Swift Regular 9.5/12.25 pt. *System* QuarkXPress™ [SE]

A catalogue record for this book is available from the British Library

ISBN 0 521 66278 8 hardback
ISBN 0 521 66447 0 paperback

Contents

Preface

The present book derives from another, entitled *A Functional Biology of Parasitism*, published by Chapman & Hall in 1993 and written by two of the present authors, GWE and JCF. Many colleagues liked the Chapman & Hall book but lamented the lack of information on the diversity of parasites, which, in turn, made that book unsuitable as a text for a first course in parasitology. To that end, several chapters in the present book might look familiar; however, all of those which were included here have been revised and some have been rewritten completely. We have also included chapters that deal with the diversity of parasites and their hosts. Moreover, no modern textbook on parasitology would be complete without consideration of the biochemical, molecular, and immunological aspects of parasitism, and we have attempted to deal with these topics as well. Our ultimate goal in the present effort was to provide an undergraduate textbook that stresses the fundamental nature of parasitism by using a decidedly ecological approach. We also hope that teaching faculty might find the text a useful supplement for a seminar course, following its use as an introductory book.

Writing any sort of a book, especially in some ways a textbook, is not an easy chore. This is also true when we know that there are several parasitology textbooks in the marketplace, all of which were authored by parasitologists for whom we have great respect as scientists. With most of these authors, we also have had very long and personal friendships. Under such circumstances, it might seem that even the thought of writing another introductory tome on parasitology would be somewhat of an audacious undertaking.

Our interest in developing and writing a new parasitology textbook was provoked, however, by our pervasive interest in ecology and, more specifically, in ecological parasitology. We believe that ecology and parasitology are, in fact, almost redundant in terms of their approach and content. We contend that the very core of all host–parasite interactions rests within an ecological framework. After all, does not ecology imply the study of relationships between organisms and their environment, and does not this completely embody the notion of the host–parasite relationship? The only possible caveat for this analogy is that, most of the time, the environment of a parasite is alive, but this should not present an obstacle to considering host–parasite interactions within the framework of ecology. Indeed, it makes the ecological approach even more exciting since this part of a parasite's environment is alive and usually can respond to the presence of the parasite. So, the decision to write a new textbook was not based on our disdain for any of the current 'models' in the marketplace, or most certainly for their authors. We simply felt we had some important ideas about parasites and wanted to share them with students.

An objective of the present book was to trim the detail as much as possible. If we have any serious criticism for some of the current parasitology textbooks, it is that several are encyclopedic in their approaches. We realize that we also may be criticized for some of the same bias, but we have made a conscious effort to be as balanced in our treatment of the various topics and groups as possible. Some will not approve of the 'mega life-cycle' format that we have incorporated in several schematics. Our intent was, whenever possible, to refrain from including a relentless tautophony of life-cycle diagrams while, at the same time, still providing an overview of the incredible diversity of strategies employed by parasites to ensure their perpetuation.

We also endeavored to use only new illustrative material, although obtaining it was not an easy task. Some of our illustrations are from the primary literature, which means they have been already published in one forum or another, but mostly just once. The published material is acknowledged in the text, giving credit to the author(s) who created it and the journal in which it was published. We thank each of these sources again here. A great deal of the material, however, came from the private collections of various colleagues throughout the world. These too are credited individually and we again thank them for sharing their personal material with us (even though space limitations prevented us from using some of it). We also thank, particularly, Lisa Esch-McCall for drafting some of the original figures and Maggie Bush for sizing, editing, and annotating virtually all of the illustrations. We thank Tom and Linda Arcure at the Wake Forest Medical School for their aid in scanning many 35-mm slides. All of the photographs lacking credits were taken by one of us, and all figures lacking credits are original.

There are a number of other people who have been most helpful, or encouraging, throughout our efforts and we want to thank each of them for their assistance. To wit, Al Bush thanks Gordon Goldsborough of the Botany Department at the University of Manitoba for his unrelenting and painstaking patience in teaching an incompetent to draw; he also thanks colleagues in the Zoology Department at Brandon University for their patience and understanding. Also, all of us thank Zella Johnson and Cindy Davis, departmental secretaries at Wake Forest, for their help; Herman Eure for his encouragement in the project; Gary Alwine for his assistance in several ways; and the Grady Britt Fund for subsuming some of the costs in preparing the illustrations. We are especially grateful to Vickie Hennings for her time from the beginning. Ward Cooper, our Commissioning Editor at Cambridge University Press, had a lot of confidence in us, and our project, so much so that we all began it together at Chapman & Hall before that company was sold. Ward then took the idea of the book with him to Cambridge University Press where he found himself and the book a new 'home'.

Finally, we want to thank our families for their patience and their sharing in so many different ways. There were many times during the book's preparation when we 'unloaded' on them our frustrations with one or another of our author colleagues, or with someone who failed to meet a commitment for one reason or another. In particular, Al Bush thanks his wife Maggie for her direct participation and tolerance, as well as his sons Jason and Jonathan for sacrificing so much time that was rightly theirs; Jackie Fernández thanks her husband Steve and sons Gabriel and Nicholas for allowing her to use so much of their family time during the preparation of the book; Jerry Esch expresses his appreciation to Ann for being such a marvelous 'listener' and for her support from the book's inception; Dick Seed would like to thank Judy for her patience and assistance during the preparation of this text.

Chapter 1

Introduction

You had no right to be born; for you make no use of life. Instead of living for, in, and with yourself, as a reasonable being ought, you seek only to fasten your feebleness on some other person's strength.

Novelist Eliza Reed to her sister, Georgiana in Charlotte Brontë's Jane Eyre

1.1 Parasitism in perspective

To most people, the word 'parasite' conjures up an image of disease and pathology, blood and guts, gross disfigurement, or even death. This notion may be based on some sort of vague imagery suggested by a newspaper article describing mortality caused by malaria or, even more sensational, a graphic television commercial asking for contributions to help victims of 'river blindness'. For the pet owner, it is because of 'parasites' that your veterinarian is likely to ask you to bring a fecal sample (a distasteful task for most!) when you take your pet for its annual checkup. If you are a world traveler, you may be immunized, or must begin taking pills, for one or more parasites, the names of which you never heard before, and may not even be able to pronounce! The physician administering the shots or the pills also is likely to warn you about not drinking the local water, about not eating fresh leafy vegetables, or about cooking meats thoroughly. In all of these instances, parasites of one sort or another are the reasons for the precaution. For the vast segment of the world's population who, live in tropical or subtropical countries, however, many of these parasites are commonplace. It is estimated that >1.4 billion people are currently infected with the roundworm *Ascaris lumbricoides* (Crompton, 1999). Therefore, approximately 20% of the world's population is infected with this one eukaryotic parasite. Three hundred and forty-two helminth parasites have been detected in humans. In addition to *Ascaris*, many of these 342 species also infect hundreds of millions of people, and multiple parasitic infections in a single individual are common. It is, therefore, safe to suggest that over one half of the world's population is infected with a wide range of these beasts! These people live mostly in Third World countries (Table 1.1). Why? For various reasons, but mostly because these folks live in abject poverty, are poorly educated, live where sanitary conditions are poorly developed (if at all) and without access to even the most basic of medicines or medical facilities. Moreover, it is in these tropical and subtropical countries that many of these parasites flourish. Again, why? In the main, it is because parasite diversity seems to be higher in tropical and subtropical areas and because the environmental conditions are conducive to transmitting the parasites that produce these diseases.

Table 1.1 Estimates of current human infections (and distributions) caused by the major parasitic organisms

Disease	Numbers (in millions)	Distribution (primary)
Hookworm	1298	cosmopolitan
Ascariasis	1472	cosmopolitan
Trichuriasis	1050	cosmopolitan
Filariasis	100	Asia; southwest Pacific Islands
Onchocerciasis	18	Central, South America; sub-Saharan Africa
Paragonimiasis	21	Asia; South Africa
Schistosomiasis	200	Asia; Africa
Strongyloidiasis	70	cosmopolitan
Malaria	300	Asia; sub-Saharan Africa; Central and South America
Leishmaniasis	80	Asia; sub-Saharan Africa; Central and South America
Chagas' disease	18	Central and South America
African trypanosomiasis	20	sub-Saharan Africa
Amoebiasis	>500	cosmopolitan
Giardiasis	200	cosmopolitan

Source:
Data from Crompton (1999) and other sources. Some of the infection data for protozoans are updated in Boxes in Chapter 3.
(Modified from Crompton [1999], with permission, *Journal of Parasitology*, **85**, 397–404.)

Parasites are not only common among humans, they are ubiquitous among all plant and animal groups. In fact, various estimates suggest that at least 50% of all plants and animals are parasitic at some stage during their life cycles. This is probably a slight exaggeration, but it is not far from being accurate. In the broadest sense, all viruses, and many bacteria and fungi are parasitic, but traditionally most parasitologists focus on **eukaryotic** animal parasites.

Parasites in humans have been known for thousands of years. Different aspects of their occurrence, cure, and transmission have been of great historical interest (Box 1.1). In the following chapters, we will attempt to provide you with descriptions of some of the most devastating parasites and the diseases they cause in humans and domesticated animals. We also want to stimulate your interest in parasites as biological entities, deserving of study in their own right. Of necessity (because we lack the space in this book), we will restrict our discussion to the more 'conventional' protozoan and metazoan parasites of animals. Throughout, however, we implore you to remember that parasitism is a way of life that transcends all phylogenetic boundaries.

Box 1.1 A brief historical perspective on parasitology and the completion of the first life cycle of a parasitic helminth

Sometime around 1500 BC, an Egyptian physician, or perhaps a group of physicians, assembled a large body of medical information regarding the diagnosis and treatment of diseases known to occur during that period. Written in hieroglyphics on papyrus and sealed in a tomb, it was discovered in 1872, then initially translated by Georg Ebers in 1873; it became known among Egyptologists as the Ebers' Papyrus, an enormously invaluable source that documented the medical profession and various cures used in the ancient world.

Based on these writings, we now know that the early Egyptian physicians were certainly aware of at least two parasitic helminths infecting their patients. One of these was a roundworm, probably *Ascaris*; the recommended treatment for infection by this apparently common parasite included such remedies as turpentine and goose fat, among others. The second parasite was a tapeworm, most likely *Taenia saginata*, for which a special poultice applied to the abdomen was the recommended treatment. Whereas the digenean *Schistosoma haematobium* was not described *per se*, the hematuria (bloody urine) produced by this parasite was well known. Moreover, eggs of this worm have been since identified in mummies from the thirteenth century BC. It is also possible that *Ancylostoma duodenale* was present based on descriptions in the Ebers' Papyrus of a 'deathly pallor' in some patients, a condition that could have been caused by hookworm anemia.

Another group of ancients was equally acquainted with a number of helminth parasites in the fertile Nile Valley at the same time. Thus, for example, consider Numbers 21:6–9, which refers to the Fiery Serpent, now recognized as the nematode *Dracunculus medinensis*. When the Israelites misbehaved during their trek out of Egypt, they were directed by God, through Moses, to 'make a serpent of brass and put it upon a pole'. And, 'when he beheld the serpent of brass, he lived'. This treatment is still used today (see Chapter 5), that is, to remove the parasite from its subcutaneous site of infection, slowly twist the parasite on a stick. Many feel the Hebrew law against eating the flesh of an 'unclean' animal, e.g., a pig, can be traced to a nematode parasite, probably the nematode *Trichinella spiralis*, and maybe even the tapeworm *Taenia solium*, although there is certainly no direct evidence for either suggestion. On the other hand, the Talmud (a sacred Jewish book), written in AD 390, referenced the hydatid cysts of the tapeworm *Echinococcus granulosus*, indicating that they were not fatal.

Periodic fevers due to malaria were mentioned in Chinese writings from around 2700 BC and in virtually every civilization since then. Hippocrates (460–377 BC) provided the earliest detailed description of these periodic fevers. Both Hippocrates and Aristotle (383–322 BC) were aware of 'worms' and refer to cucumber and melon seeds in the 'dung' of humans. Both references are probably to the gravid proglottids of *Taenia saginata*. The word *Taenia* was coined by the Greek writer Pliny (AD 23?–79), and has remained associated with the parasite as the generic name ever since. Galen (AD 130–200) actually referred to the intestinal phases of what were probably *Ascaris lumbricoides* and *Enterobius vermicularis*, saying that the former worms preferred the upper portion of the gut whereas the latter were closer to the anus. Tapeworms, he opined, were found throughout the length of the intestine (the first reference to site specificity by a parasite?).

The gap between Galen's time and the Renaissance, beginning in the thirteenth century, was not a particularly productive period for parasitology in the western world, although the Chinese around AD 200 proclaimed that parasitic worms were created when certain kinds of foods were 'coated with warm blood and nourished by the vital elements of the host'. By the twelfth century, however, the Chinese were on the right track when it was written that humans became infected with worms by 'eating fruits and vegetables or animals' viscera'.

The earliest use of the microscope, by Antony von Leeuwenhoek in the seventeenth century, provided a unique breakthrough for the biological sciences and parasitology. He actually observed, and described, the protozoan parasite *Giardia lamblia* in his own feces. Also in the seventeenth century, a number of other contributions were made through the work of such scholars as Fehr, Spigelius, and Tyson, who prepared detailed drawings of a number of parasitic helminths. The father of modern parasitology was, however, Francesco Redi (1626–1697) who not only determined that mites could make one itch, but apparently was an inveterate collector, dissecting everything in sight and describing some 108 species of parasites in the process. Perhaps Redi's greatest contribution was that he showed that parasites produce eggs, dispelling the widespread myth that parasites developed through spontaneous generation. The idea of spontaneous generation persisted for many years, however, and took Louis Pasteur's now classic experiments in nineteenth-century Paris to quash the notion once and for all.

L. Dufour in 1828 described gregarines from insects and in 1841 G. G. Valentin observed trypanosomes in the blood of fishes. The late nineteenth and early twentieth centuries were times of major discoveries dealing with some of the protozoan and helminth scourges of humans. Patrick Manson in 1878 identified *Wuchereria bancrofti* as the causative agent for elephantiasis and determined that mosquitoes were the insect vectors of the disfiguring disease. Charles Laveran was the first to find the malarial parasite *Plasmodium* sp. in human blood. Ronald Ross, while working in India in 1897, demonstrated that the mosquito was the vector for *Plasmodium*; he was subsequently knighted and won the second Nobel Prize for physiology in 1902. Griffith Evans in 1881 identified the connection between trypanosomes in horse blood and the disease called surra, and David Bruce in 1894 implicated tsetse flies as the vectors for African trypanosomiasis. At the turn of the century, Paul Erlich described the first chemotherapeutic agents for African trypanosomiasis and syphilis. With this discovery, he correctly hypothesized that it should be possible to find organic molecules with selective toxicity to parasitic organisms and, for this, is considered the father of modern chemotherapy. Between 1907 and 1912, Carlos Chagas determined the identity of trypanosomes that cause Chagas' disease and worked out the trypanosome's life cycle in the reduviid intermediate host.

The first recognizable description of the liver fluke *F. hepatica* was in a volume published by Sir Anthony Fitzherbert in 1523. The first published illustration of *F. hepatica* was made by Redi in 1668. With greater use of the microscope inexorably came the development of a radically new concept in biology, the notion of alternation of generation and, with it, the discovery of the complete life cycle of *F. hepatica* by Algernon Phillips Withiel Thomas, a graduate of Balliol College, Oxford. Thomas began his work on the parasite's life cycle in 1880 and had most of it completed by 1883. His work was paralleled by the great German parasitologist, Rudolph Leuckart, who published his version of the parasite's life cycle almost simultaneously with that of Thomas. Both are given credit for this remarkable discovery.

During the winter of 1879–80, liver rot, caused by *F. hepatica*, killed some three million sheep in Great Britain. Seeking a solution to the problem, the Royal Agricultural Society of England approached Thomas who eagerly accepted the challenge. Thomas was not without some insight with respect to the biology of the parasite. For example, cercariae and encysted metacercariae of *F. hepatica* had been described by La Valette St. George in 1855. Subsequently, the German parasitologist David F. Weinland reported in 1875 the finding of what he called 'cercaria-sacs' in the livers of the pulmonate snail *Lymnaea truncatula,* and that the cercariae had a tendency to encyst on inanimate objects. He suggested that the encysted metacercariae on blades of grass could be consumed by grazing sheep, thereby completing the life cycle.

With all of this in mind, in the summer of 1881 Thomas located an appropriate study area where sheep losses had been high the previous winter. He next found *Lymnaea truncatula* in a marsh and discovered rediae in the livers of the snails. He was on the right track for making the discovery. As luck would have it though, in the summer of 1882 he could not find any snails. He believed, however, that their disappearance was correlated with the absence of liver rot in sheep during the previous winter. The following summer brought local flooding and the return of his snails. His initial inclination was to obtain encysted metacercariae and use them to infect rabbits. Thomas, however, decided to focus his efforts on the first part of the life cycle. So, he obtained adult parasites from infected sheep and then eggs. He incubated the eggs which subsequently hatched, releasing what he termed 'embryos'. He was intensely fascinated with this part of the cycle, describing in great detail the swimming behavior of the 'embryos', their penetration into the snails, and their subsequent intramolluscan development. He made these and a number of other highly significant observations, with the almost complete description of the cycle being published in the *Quarterly Journal of Microscopical Science* in January, 1883. The one thing he did not do, however, was to expose any experimental animals to the metacercariae. Unfortunately, his work on *F. hepatica* came to a close when he moved to New Zealand where he taught and, alas, became an academic administrator. Several months before his death in 1937, at the age of 80, he was knighted for his many contributions in the field of education.

The next to the last step in solving the problem of the enigmatic life cycle was made in Hawaii by Adolpho Lutz in 1893 who succeeded in infecting several guinea pigs, a rabbit, a goat, and a brown rat with the parasite, although J. E. Alicata later asserted that Lutz was working with *Fasciola gigantica* and not *F. hepatica*. The final step was taken by D. F. Sinitsin, the famous Russian parasitologist, who early in his career worked at Shanjasky University in Moscow before being forced by the Russian revolution into fleeing to the United States in 1923. Sinitsin in 1914 proved that, upon excysting in the gut of the definitive host, the parasite actually penetrated the intestinal wall and migrated via the liver to its final resting place in the bile ducts and gall bladder rather than directly from the lumen of the intestine into the bile ducts as was believed by Leuckart.

The history of parasitology is a fascinating one, filled with mysteries solved and new mysteries created by their resolution. But this is the nature of science itself, isn't it? By answering old questions, new ones should always be discovered. Thomas and Leuckart will be remembered primarily for their solving a great scientific mystery but, in doing so, they and their pioneering contemporaries pointed the way for all those who followed, and succeeded, in resolving so many other parasitological mysteries.

1.2 Symbiotic relationships

Parasitology is the science that deals with one of several different kinds of symbiotic relationships. We would be remiss if we did not provide you with a definition of a parasite, at least to the extent that conventional wisdom dictates, i.e., the dreaded dictionary definition. According to the 2nd Edition of the *Oxford English Dictionary*, a parasite is:

> An animal or plant which lives in or upon another organism (technically called its host) and draws its nutrients directly from it. Also extended to animals or plants that live as tenants of others, but not at their expense (strictly called commensal or symbiotic); also to those which depend on others in various ways for sustenance, as the cuckoo, the skua-gull etc.

Webster's Third New International Dictionary of the English Language invokes directly the concept of harm:

> An organism living in or on another living organism obtaining from it part or all of its organic nutrient, and commonly exhibiting some degree of adaptive structural modification – such an organism that causes some degree of real damage to its host.

Perhaps useful to some, these definitions define a word, but tell us little about the concept of parasitism. We suspect that if you assemble 10 scientists and ask them to define parasitism, you would obtain 10 different answers. Our approach to parasitism in this book is decidedly ecological and we favor treating the subject as one of several, broad and often overlapping classes of symbiotic relationships. We consider symbiosis to mean, simply, organisms living together. In this case, there is no implication with respect to the length or outcome of the relationship, or the degree of adaptation.

Given such a broad definition of symbiosis, a functional separation can be made in terms of trophic relationships and then, if and how energy is transferred between symbiotic organisms. Such categories should best be viewed as a continuum, with vague boundaries (Fig. 1.1); some consider such a continuum as a broad trend in evolution.

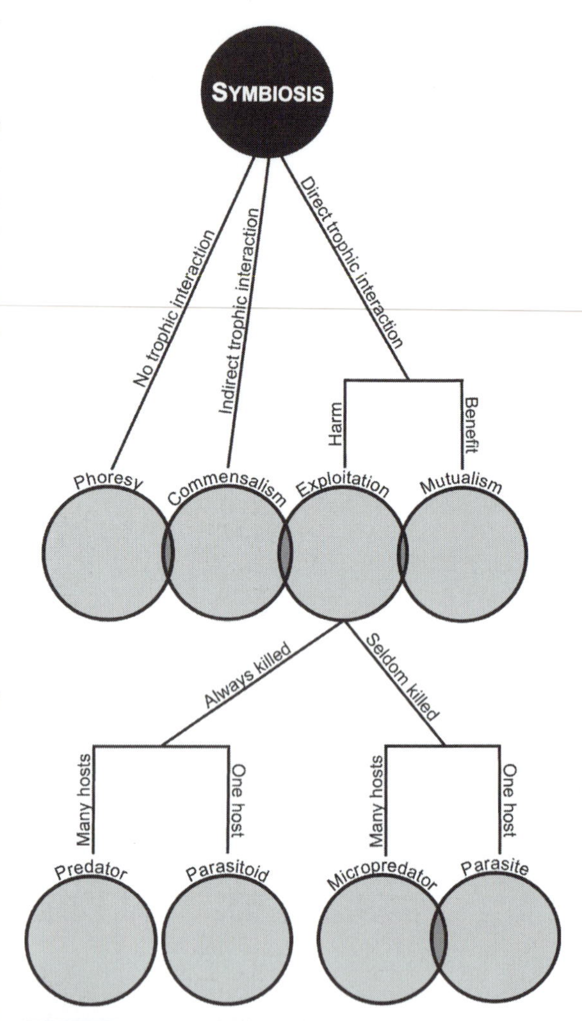

Fig. 1.1 An attempt to find 'parasitism's place' within the context of symbiotic relationships. This is only one way of looking at parasitism and it is based, initially, on trophic relationships, followed by 'harm', and finally, quantity of hosts involved. The final criterion, number of hosts attacked, is meaningful only if restricted to a single life-history stage. For example, adult parasitoids may attack many prey but their larvae live in, and consume, only a single individual. Likewise, a typical helminth parasite may have both intermediate and definitive hosts but each life-history stage will infect only a single host individual. We cannot emphasize too strongly that the overlap between many of the relationships reflects the extraordinary diversity of life styles found in nature. Seldom, if ever, can one classify a group of organisms exclusively.

As we note above, there are many views on parasites, and Fig. 1.1 should provide a point of discussion for a parasitology class. For example, one might consider separating life styles based on immunological interactions, or suggest that there are but two fundamental life styles, e.g., parasitic (a 'host' is involved) versus free-living (no 'host' is involved).

If there is no trophic interaction between the organisms in the symbiotic relationship, then the relationship is called phoresy (Fig. 1.1). The process of pollination is an excellent example. When a butterfly obtains nectar from a flower, it will become dusted with pollen and then, when it moves to the next flower, pollen is carried with it ensuring fertilization of the second flower. There is, however, no trophic interaction or transfer of energy associated with the interaction between the butterfly and the pollen.

Phoresy grades into commensalism, a symbiosis in which there is a trophic relationship and a transfer of energy between the symbionts (Fig. 1.1). The benefit gained is unidirectional, one partner benefits and the other is neither harmed nor helped. A frequently cited example is the relationship between sharks and remoras. When sharks feed on large prey, they scatter pieces of flesh. Remoras feed on these scraps, thus deriving energy from the actions of the host even though the transfer of energy is indirect.

When there is a direct transfer of energy between the partners, the interaction may be either mutualistic or exploitative (Fig. 1.1). In a mutualistic relationship, both symbionts not only obtain benefit, but neither can survive without the other. Lichens are a classic example of an obligate association between a fungus and an alga. In this case, the fungus provides protection and moisture for the alga and the alga in turn provides nutrients for the fungus. Similar relationships are thought to exist between algae and many of the coral reef-forming cnidarians. The relationship between ruminants and the microorganisms in their stomach is also mutualistic. On the one hand, the ruminant host provides an almost continuous supply of carbohydrate in the form of cellulose, plus an otherwise constant environment. On the other hand, the rumen-dwelling microorganisms secrete enzymes that convert the cellulose into glucose. The rumen-dwelling symbiotic organisms first use these glucose molecules as an energy source. Living in an **anaerobic** environment means that the intermediary carbohydrate metabolism of the microorganisms is inefficient (see Chapter 2). The end products of glucose degradation by these symbionts include mostly short-chain fatty acids. Even though these fatty acids are metabolic 'waste' products from the microorganisms, they still possess substantial levels of potential energy. The ruminant absorbs the fatty acids in its intestine, transports them to the liver, and converts them into glucose. The converted glucose is then used as an energy source by the ruminant in the same way other mammals use it. The complexity of this mutualistic arrangement is obvious and clearly is the product of a long evolutionary history.

In most exploitative interactions, however, benefit is in one direction and, moreover, some form of disadvantage, or harm, is the outcome for the other partner. Several major categories of this kind of exploitation can be recognized based primarily on the number of hosts attacked by the symbiont and the subsequent fate of the organism assaulted (Fig. 1.1). If more than one organism is attacked, but typically not killed, then the aggressor is called a micropredator. Hematophagous organisms such as mosquitoes, and some leeches and biting flies, for example, are highly successful micropredators. If more than one organism is attacked and always killed, then the aggressor is considered a **predator** (predatory relationships should need no further elaboration!). If only one host is attacked and is always killed, then the aggressor is usually referred to as a parasitoid, most of which are hymenopterans and dipterans. For example, an adult female wasp may deposit her egg(s) on, or into, an insect. On hatching, the larval parasitoid will consume the host, killing it in the process. Finally, if only one host is attacked, but typically is not killed, the aggressor is a parasite. Indeed, remember that our human attempts to categorize relationships may often be inadequate. For example, as we have suggested above, parasitism denotes some 'harm' to the host. Frankly, most of the time that is true. Interestingly, however, there are a few experimental studies that show potential benefits to being parasitized (e.g., Lincicombe, 1971; Munger & Holmes, 1988). In fact, unlikely as it may seem, many years ago *Plasmodium* infections were used as a control for syphilis and the rarity of tertiary syphilis in Africa is thought to be due to the high prevalence of malaria (Garnham, 1981).

1.3 Kinds of parasites

Endoparasites include those parasites that are confined within the host's body. They include the more familiar animal parasites such as protozoans, digeneans, cestodes, nematodes, and acanthocephalans. Many bacteria and all viruses are also endoparasitic. Parasites typically confined to the exterior of the host's body are called **ectoparasites**. Most parasitic arthropods and most monogeneans are ectoparasitic.

Another dichotomous method for classifying parasites is based on their size. Generally, **macroparasites** are large and can be viewed without the aid of a microscope. They can be endoparasitic, such as digeneans, cestodes, nematodes, and acanthocephalans, or ectoparasitic, such as arthropods and monogeneans. **Microparasites**, as their name implies, are mostly microscopic and can be ectoparasitic or endoparasitic. They may also be intracellular, or extracellular, or both. Eukaryotic microparasites are primarily protozoans.

Most **obligate parasites** are parasitic as adults. The larvae of these organisms, however, may include both obligatory parasitic forms and/or free-living stages. The adults of some species are commonly free-living but, should the opportunity be presented, their progeny may become parasitic. These organisms, mostly protozoans, and a few nematodes and isopods, are referred to as **facultative parasites**.

Parasites can have parasites too. These parasites of parasites are called **hyperparasites**. Hyperparasitism appears to be much more common than was once believed. Due to their small size, some hyperparasites may have been overlooked. Hyperparasites are usually bacteria or viruses, but some protozoans, cestodes and crustaceans have been found parasitizing other parasites.

1.4 Kinds of hosts

The organism in, or on, which a parasite reaches sexual maturity is the **definitive host**. Some parasites require only one host to complete their life cycles. These cycles are said to be **direct life cycles**. All monogeneans, and some nematodes and arthropods, have direct life cycles. Many animal parasites, however, have obligate **intermediate hosts** in which the parasites undergo some developmental and morphological change, but do not reach sexual maturity (there are several exceptions, i.e., progenesis and neoteny, but discussion of these patterns will be deferred to subsequent chapters). Life cycles in which more than one host is required are **indirect life cycles**.

Some protozoans and filarial worms employ **vectors** as hosts. Vectors are micropredators that transmit infections from one host to another. A vector may be an intermediate or a definitive host, depending on whether the sexual phase of the parasite's life cycle occurs in it or not. Being a vector implies a more active role in transmission rather than a passive one. For example, the insect vectors for species of *Plasmodium*, the causative agents of malaria, are female mosquitoes that actively inoculate infective agents of the parasite into the vertebrate host during their blood meal.

A number of parasites may employ hosts in which there is no development and that are not always obligatory for the completion of a parasite's life cycle. These are called **paratenic** or **transport hosts**. Such hosts are most frequently used to bridge an ecological, or trophic, gap. For example, a parasite may require an ostracod as a second intermediate host and a frog as the definitive host. Under normal conditions, frogs do not prey on ostracods, but they do consume odonate naiads which feed on ostracods. The parasite may be transferred from an ostracod to the naiad, then to a frog definitive host. Its chances for reaching the frog host are thereby immeasurably increased by bridging the trophic gap between the ostracod and frog hosts. Ecologically, transport or paratenic hosts are important because they may help disseminate the infective stages of the parasite, or they may aid these stages in avoiding unfavorable conditions such as the temporary absence of a definitive host.

A number of animals are normal hosts for parasites that may also infect humans. These are called **reservoir hosts**. Ecologically they are similar to transport or paratenic hosts since they may keep the parasite from becoming locally extinct when the natural host is unavailable. These parasites, because of their normal associations with animals in nature, are particularly difficult to control.

1.5 Ecology and the host–parasite relationship

As we will emphasize throughout this book, the essence of parasitism rests with the nature of host–parasite relationships. If we accept the simple definition of ecology as the study of the relationships between organisms and their environment, then it is not difficult to understand why parasitism is an ecological concept. Ecologically, however, the host–parasite relationship is a 'double-edged sword'. This is because in dealing with parasites and their hosts from an ecological perspective, one must simultaneously consider the ecology of the host(s) in a parasite's life cycle, as well as the host as a habitat for the parasite. Many of the biotic and abiotic vagaries affecting the ecology of the host will also affect the parasite. But the parasite also must deal with a host that is alive, and capable of responding physiologically and immunologically to the parasite. It must be understood that these latter interactions between the parasite and the host are as 'ecological' as those involving the host's relationships with its own environment.

The study of parasitism, whether from an ecological or physiological perspective, is a fascinating exploration of organisms that make their living at the expense of others. We hope that you will enjoy this brief exploration and that it will serve to stimulate you to learn more about these fascinating creatures. To that end, we provide Box 1.2, which, current at the time of writing this book, provides a number of web sites about parasites. If you continue on in the sciences, no matter what discipline, remember parasites. They can often prove useful for addressing a variety of questions and hypotheses. If your future is not in science, remember parasites anyway – they make extraordinary dinner conversation!

Box 1.2 Parasitology in cyberspace

Even though we believe that nothing can replace the warm feeling of printed information, the Internet is a vast and, often, very useful resource. Here, we provide a list of websites current at the time of the writing of this book. Some sites focus on specific information about parasites, others are photographic galleries with outstanding images, still others provide excellent links to other sites of parasitological interest. An exhaustive listing is nearly impossible and highly redundant. Surfing is inevitable.

http://asp.unl.edu
Official website of the American Society of Parasitologists. Offers information about the society and its activities as well as links to relevant parasitological sites.

http://www.parasitology.org.uk
Official website of the British Society for Parasitology. Offers information about the society and its activities as well as links to relevant parasitological sites.

http://dspace.dial.pipex.com/town/plaza/aan18/urls.htm
Excellent website created and maintained by Dr. David Gibson at the Natural History Museum of London. It provides an exhaustive list of parasitological URLs taken from a poster on Internet Resources presented at a meeting of the

British Society for Parasitology. This list is updated regularly and includes more than 400 URLs. This website provides links to sites with information about parasites, parasitological societies, parasitological resources, images, newsgroups, journals, books, people, courses, meetings, etc. A great surfing site.

http://www.dpd.cdc.gov/dpdx/
Useful site for the identification and diagnosis of parasites of public health concern. It provides information about life cycle, geographical distribution, clinical features, diagnosis and treatment for each of the parasites listed. It also includes a Parasite Image Gallery.

http://www.biosci.ohio-state.edu/~parasite/home.html
Very useful website maintained by Dr. Peter Pappas at Ohio State University. It is aptly called Parasites and Parasitological Resources.

http://www.ksu.edu/parasitology
Website about *Cryptosporidium* and Coccidial Research at Kansas State University. It contains relevant, up-to-date information about these parasites as well as a nice image tutorial to test our knowledge of parasites. It also provides links to the source of the images used in the tutorial.

http://www.cvm.okstate.edu/~users/jcfox/htdocs/clinpara/Index.htm
Website of Veterinary Clinical Parasitology Images created by Professor J. Carl Fox of Oklahoma State University. An excellent site with images, keys, and other interesting features about parasites.

http://www.ag.arizona.edu/tree
This website is called the Tree of Life. It is a multi-authored Internet site containing information about phylogeny and biodiversity. The information is linked together in the form of and evolutionary tree connecting all organisms to each other. The site is changing constantly as new information is added. Look for your favorite parasite!

http://www.parasitology.org
Website of Veterinary Parasitology at the University of Missouri. The site includes lecture notes from a parasitology course, as well as images, diagnosis information, and even a glossary of terms used in parasitology.

http://www.riaes.org/resources/ticklab/
Website of the Tick Research Laboratory, which is devoted to the study of various aspects of tick-borne diseases. The site provides detailed information about tick-borne diseases as well as images of ticks.

http://parasite.biology.uiowa.edu
This website contains 2320 images and information about parasites taken from Dr. Herman Zaiman's publication 'A Pictorial Presentation of Parasites'. Although the site is password protected, everyone can access it by using the user name 'guest' and the password 'visitor'.

http://parasitology.icb2.usp.br/marcelocp/
Website with many good original images of parasitic insects, ticks and mites.

http://cal.vet.upenn.edu/
Website of the University of Pennsylvania, School of Veterinary Medicine, Computer-Aided Learning Project. Follow the shortcuts to 'Diagnosis of Veterinary Endoparasitic Infections' (**http://cal.vet.upenn.edu/dxendopar/**), 'Parasitology' (**http://cal.vet.upenn.edu/parasit/P_index.html**), and 'Parasitology Course 4001: Laboratory Demonstrations' (**http://cal.vet.upenn.edu/paralab/index.html**). These sites provide lecture notes, images, and diagnostic procedures for parasites.

http://www.medicalweb.it/aumi/echinonet/
An online newsletter of the WHO Informal Working Group on Echinococcosis. The website belongs to the Tropical Diseases Web Ring, a network of websites dedicated to tropical diseases, all of which include original information and regular updates. Once the user logs into one of the sites in the Ring, he/she can go to all the other sites included in the ring just by clicking on a single icon in the web page.

http://info.dom.uab.edu/geomed/index.html
Site of the Division of Geographic Medicine of the University of Alabama at Birmingham. It offers information about traveler's medicine and links to other websites.

News:bionet.parasitology
Access to a parasitology newsgroup.

http://www.cdc.gov/
Home page of the Centers for Disease Control.

http://www.who.ch/
Home page of the World Health Organization. Of special interest is **http://www.who.int/ctd**, which is the WHO Division of Control of Tropical Diseases. It includes good updates on tropical diseases caused by parasites.

http://www.mic.ki.se/Diseases/c3.html
Website of the Karolinska Institutet, a prestigious medical facility in Sweden. Offers organized links to numerous websites of interest for parasitologists, some of which are little known but interesting.

http://www.nhm.ac.uk/
Website of the Natural History Museum, London. The search engine of the site provides access to parasitological information.

http://www.iss.it/
Istituto Superiore di Sanità. Italian counterpart of the Centers for Disease Control.

http://www.pasteur.fr/
Home Page of the Institut Pasteur in France. Offers links to areas of interest.

http://www.lshtm.ac.uk
Home page of the London School of Hygiene and Tropical Medicine.

References

Crompton, D. W. T. (1999) How much human helminthiasis is there in the world? *Journal of Parasitology*, **85**, 397–404.

Garnham, P. C. C. (1981) Multiple infections of parasites. *Mémoires du Muséum National d'Histoire Naturelle Série A, Zoologie*, **123**, 39–46.

Lincicombe, D. R. (1971) The goodness of parasitism. In *The Biology of Symbiosis*, ed. T. C. Cheng, pp. 139–227. Baltimore: University Park Press.

Munger, J. C. & Holmes, J. C. (1988) Benefits of parasitic infection: a test using a ground squirrel – trypanosome system. *Canadian Journal of Zoology*, **66**, 222–227.

Chapter 2

Immunological, pathological, and biochemical aspects of parasitism

2.1 Host resistance: immunoecology

2.1.1 Introduction

The host is a remarkable organism, surviving and reproducing in a sea of infectious organisms. Infectious agents can be found in the air we breathe, the food and water we consume, the walls, floors, and kitchen counters of our homes, or in the soil on which we walk. These agents truly surround us, yet, for the majority of our lives, most of us are completely healthy. Why? Both nature and experimental research have provided answers to this question. We naturally are resistant to many potential **pathogens** and, moreover, we have the ability to respond actively to most, if not all of the potentially pathogenic parasitic organisms. Proof of this is that, if the immune system is suppressed either because of a genetic deficiency or for medical reasons (transplant surgery), the individual becomes highly susceptible to a variety of infectious disease agents. We also know that unless we step on a nail we are resistant to the tetanus **bacillus**, suggesting the skin is a marvelous barrier to infectious organisms. Or, after treatment with oral antibiotics for a problem such as a streptococcal sore throat, a bout of diarrhea will often follow. The simple explanation for this is that after antibiotic treatment, the normal composition of the bacterial community inhabiting our intestine is altered. This allows for the growth of opportunistic antibiotic-resistant microorganisms that then cause the illness. Only after the typical gut bacterial community has been re-established does one become well. The normal microbial communities that inhabit our nose, mouth, intestinal tract, vagina, and skin are protective.

Resistance to infectious agents can be attributed to two different sets of mechanisms. One is called innate resistance (or natural immunity) because the factors involved are always present, and they are non-specific. In other words, they work against a wide variety of different potential pathogens. The previous examples of skin and normal gut microbial communities are both part of an animal's innate resistance. Innate mechanisms of resistance, however, can be observed in all animal groups. Protective coverings, anti-microbial substances, and cells involved in wound repair, occur throughout the animal kingdom. In fact, hundreds of broad-spectrum anti-microbial peptides have been discovered having activity against viruses, bacteria, fungi, and other parasitic organisms in all metazoan animal groups. These innate mechanisms of immunity appear to have arisen very early in our evolutionary history (Hoffmann *et al.*, 1999). In contrast, the second set of mechanisms is confined to the higher animal phyla, particularly the vertebrates. These responses are restricted to a particular pathogen and are inducible. In other words, in contrast to the skin, the cells involved will be specifically increased in number following invasion by a parasite and will respond only to that particular pathogen (**antigen**). In addition, if an animal is challenged a second time with the same pathogen, it appears to remember, an anamnestic

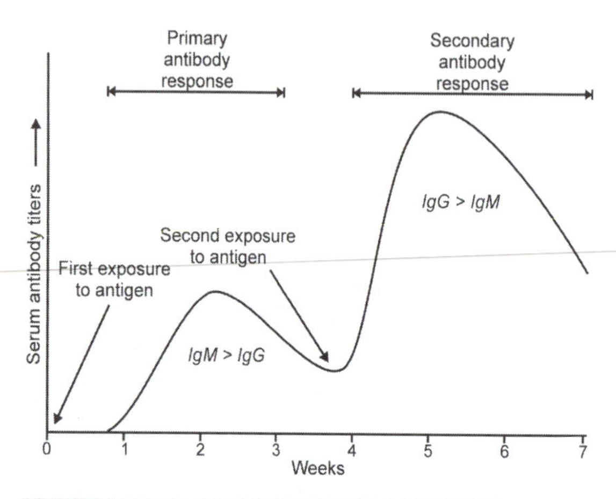

Fig. 2.1 Diagram of the primary and secondary antibody responses. Note that the secondary response is more rapid and that the plateau in antibody concentration is considerably greater. In addition, the lag time for the primary response is longer than that for the secondary response. The antibody affinity of the primary response is low compared to that of the secondary response. The predominant class of antibody produced during the primary response is IgM, whereas in the secondary response it is IgG.

response or immunological memory, the specific nature of the primary exposure and responds both faster and to a greater degree to the subsequent exposure (Fig. 2.1). This second mechanism is usually referred to as immunity or the immune response.

The immune response in mammals involves three different cell types. These include antigen-presenting cells (**macrophages** and **dendritic cells**, phagocytic white blood cells), **B-** (bone-marrow derived) and **T-** (thymus derived) **lymphocyte cells**. Following invasion by a parasite, it is these three cell types which, through an exchange of chemical messengers (**lymphokines** or **interleukins**), expand and respond specifically to a particular infectious agent. It must be obvious that in the healthy individual, both innate resistance factors and the immune system interact to protect them from that sea of potentially infectious pathogens (Wakelin, 1993; Janeway & Travers, 1994). For a detailed recent review of the immune responses to some important intracellular parasites, see the volume edited by Liew & Cox (1998).

Macrophages are also part of the innate system. They are phagocytic white blood cells that, in the normal host, are always present and capable of engulfing and killing infectious disease agents. However, it is the macrophage that processes the engulfed parasite and presents small pieces (**epitopes**) of the parasite's protein to the lymphocytes (T-cells and B-cells). They are, therefore, also part of the immune system. This transfer of information, from macrophage to lymphocyte, induces the lymphocyte to divide and expand in number. The activated lymphocyte can then specifically respond to the parasite that contained the information originally processed and transferred by the macrophage. In addition, the macrophage itself can be induced to become an activated, or angry, macrophage. The activated macrophage is a better killing machine and is an integral part of the immune system. Finally, not only are there complex direct cell-to-cell interactions between macrophages and the other cells of the immune system, there is also communication or cross-talk via chemical messengers (hormones and **cytokines**) between the various organs of the body and the immune system (Fig. 2.2). For example, macrophages release a chemical messenger called IL-1, or interleukin 1, that signals the brain to initiate a fever response (Fig. 2.2). The brain can also, through a series of hormones, stimulate the adrenals to produce corticosteroids. Corticosteroids interact with macrophages and other cell types to produce an anti-inflammatory type response (Zwilling, 1992; Adams & Boyce, 1995; DeJong-Brink, 1995; Hiramoto *et al.*, 1997). The point is that, although various individual components involved in innate resistance will be discussed here, as well as in the immune response, one may ultimately consider the overall defense system within the context of a single complex ecosetting, which has evolved to protect animals from all infectious invaders (Seed, 1993).

2.1.2 Innate resistance

We begin the discussion of the animal defense system with a brief description of various aspects of innate resistance. The external surface of an animal's body is protected by a variety of structures. In homeotherms this includes feathers or hair, in poikilotherms it might be scales. Hair, feathers, and scales can act as a physical barrier to the penetration of some ectoparasites, such as

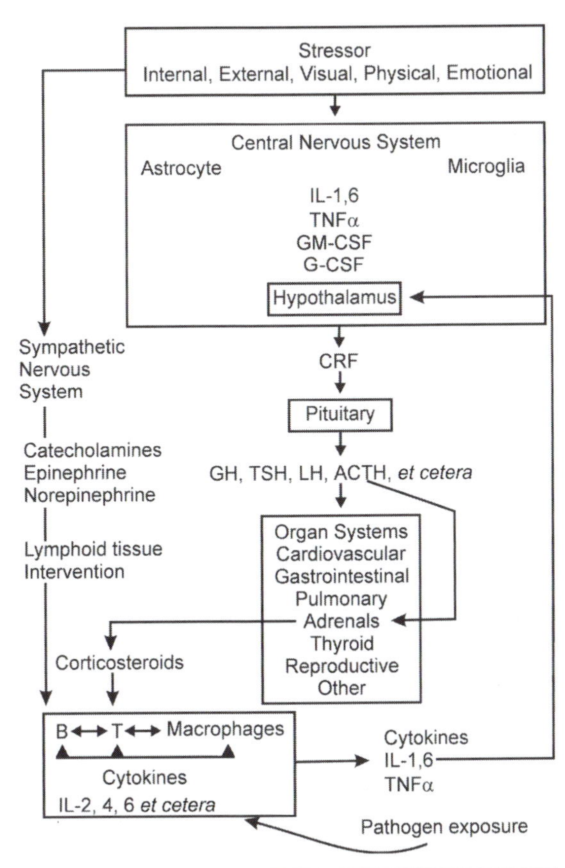

Fig. 2.2 The interaction between the central nervous system, other organ systems, and the immune system. Abbreviations: IL, interleukins; TNFα, tumor necrosis factor alpha; CRF, corticotrophin releasing factor; GH, growth hormone; TSH, thyroid stimulating hormone; ACTH, adrenocorticotrophin hormone; GM-CSF, granulocyte-macrophage-colony stimulating factor; G-CSF, granulocyte-colony stimulating factor; LH, luteinizing hormone.

insects, crustaceans, ticks, etc. If the parasite successfully penetrates the external structure, however, it then meets the skin. In mammals, the most exterior part of the skin includes compacted layers of dead cells on to which are excreted salts from perspiration, and small organic acids and large fatty acids from glands in the skin. High salt concentrations, lactic and other small organic acids released on to the skin, plus a number of fatty acids, are known to have anti-bacterial activity. In addition, other animals, such as some frog species, produce anti-microbial **peptides** that are present in skin secretions. The skin, therefore, acts both as a physical and a chemical barrier to many infectious agents. Individuals with large segments of skin removed, e.g., those with severe burns, are susceptible to many microorganisms not normally considered a problem.

Now then, assuming that the parasite has successfully penetrated the skin into deeper tissues, another set of innate defense mechanisms will become involved. Most parasitic organisms release **chemotactic** substances and will stimulate the attraction of both **neutrophils** and macrophages to the site of invasion. Many parasitic invasions may cause at least minor breaks in lymphatic and vascular beds. This leads to an infiltration, at the site of the insult, of cells in the blood and lymph, and a variety of macromolecules involved in the **clotting**, **complement**, and **kinin** pathways, plus other proteins that have antimicrobial activity. Each of these pathways includes a series of proteins in which one protein or peptide in the pathway will activate a second protein and the second set of reactions will activate a third in a cascading response. The final end product, and in some pathways intermediate products, have biological activity. The surface macromolecules of many foreign invaders are capable of interacting with proteins in the complement pathway, or activating the clotting mechanism, or both. In the clotting pathway, the end result is the formation of a clot consisting of proteins and cellular elements. A clot, in addition to the obvious control of bleeding, also forms a matrix of protein and cellular elements. This matrix not only traps some parasitic organisms, and therefore localizes them; the trapping of the parasite also permits easier **phagocytosis** by the white blood cells. The kinin and complement pathways also function in defense against parasitic organisms. The products formed during the kinin cascade produce changes in the vascular bed that allow easier movement of the white blood cells and antimicrobial substances from the blood and lymph into the sites of parasite invasion. Finally, the complement pathway plays at least two major roles in resistance to parasitic organisms. First, during the enzymatic cascade, products are formed that are chemotactic for white blood cells. Second, the end product of the complement pathway can, when present on the surface of some cells (parasites), cause them to lyse. Therefore the clotting, kinin, and complement pathways are all important in innate immunity and in the **inflammatory** response to parasitic organisms. In both blood and lymph, there are certain macromolecules that have inhibitory properties against most classes of parasitic organisms. In addition, the neutrophils that migrate to the site of invasion secrete chemotactic molecules that further attract macrophages to the site of the parasitic invasion.

Therefore, very early in the host–parasite interaction, there is a non-specific attempt by the host to surround and sequester the parasitic invader and then to kill it by antimicrobial macromolecules, or by the phagocytic activity of the white blood cells (WBCs), or both. The neutrophils and macrophages increase in number, and thereby also increase metabolism at the site. This results in an increased excretion of acid catabolites and a decrease in O_2 concentration, which ultimately causes the death of some of these phagocytic cells and some microbial parasites as well. The death of the phagocytic cells and parasites leads to the release of a variety of their hydrolytic **enzymes** including proteases, RNAase, etc., which can have a direct toxic effect on parasitic organisms, as well as on surrounding host cells.

These processes are all part of the inflammatory response that occurs whenever the external barriers are breached and a parasitic organism enters the body. These innate mechanisms constitute an animal's first line of defense. In closing this section, note that fever (high body temperature) itself can slow or inhibit microbial growth and, therefore, help protect the host against the parasite. If the host successfully inhibits growth of the invading microorganism, the inflammatory response will decrease and wound healing will occur.

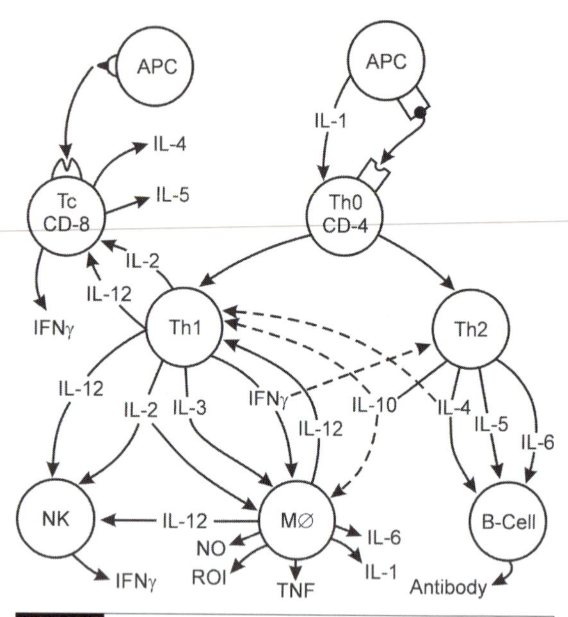

Fig. 2.3 Simplified diagram of the host cells, interleukins, and other reactive molecules involved in the host's immune response. This diagram demonstrates how dynamic the host–parasite interaction can be. For example, following the processing and presentation of parasite antigens by the antigen-presenting cells (APC) to the Th0 T-cell, the response may be directed towards either cellular immunity (Th1 T-cells) or a humoral response (Th2 T-cells and, subsequently, B-cells). Note that once stimulated, the Th1 or the Th2 cells produce interleukins such as interferon (IFNγ) or interleukin-4 (IL-4) which either up regulate (stimulate, solid lines), or down regulate (broken lines) the other cell type. For example, if the Th1 cells are stimulated during a leishmanial infection, the Th1 cells produce interleukins such as IL-2 and IL-12 which can activate (solid lines) natural killer (NK) cells, macrophages (MØ), or cytotoxic T-cells (Tc) but they also secrete IFNγ which down regulates the Th2 cell population. This insures a preferential cellular (or Th1) type response. Abbreviations: NO, nitric oxide; ROI, reactive oxygen radicals. Both NO and ROI are known to have antimicrobial activity. TNF, tumor necrosis factor (see Table 2.2); IL-2, IL-3, etc., interleukins with different functions; CD-4 and CD-8, T-cell populations having different biological functions and identified by having different surface markers.

2.1.3 Antibody response

If the parasite successfully escapes an animal's innate defense mechanisms, it will induce an immune response specific to the invader. In higher vertebrates, the immune system can be divided into two parts (Fig. 2.3). One involves the plasma cell, or B-cell, which synthesizes and secretes a large-molecular-weight **protein** called either an **immunoglobulin** or an **antibody**. An antibody is a large-molecular-weight protein, 150 000 daltons or greater, consisting of a minimum of four peptide chains (two light and two heavy chains), linked by disulfide bonds (Fig. 2.4). Individual antibody molecules include at least two antigen-combining sites per molecule. Each combining site is composed of one light and

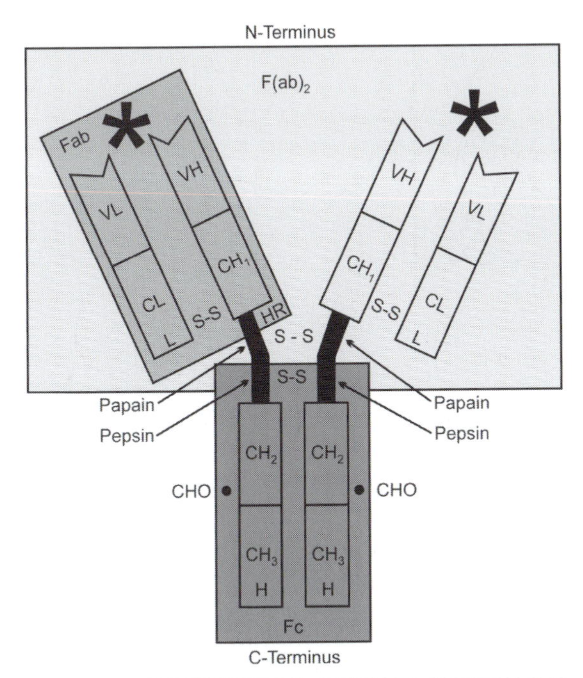

Fig. 2.4 Diagram of an antibody molecule. L, light chain; H, heavy chain; VL, variable region of the light chain; CL, constant region of the light chain; VH, variable region of the heavy chain; CH_1, CH_2, and CH_3, the three constant domains of the antibody molecule; S-S, disulfide bonds; CHO, carbohydrate moiety; *, combining sites of the L and H chains; HR, hinge region of the molecule. Papain, sites sensitive to papain digestion; following papain proteolysis, Fab and Fc portions of the antibody molecule are released. Note that there are two Fab regions. Pepsin, sites sensitive to pepsin digestion. Following pepsin treatment, the two Fab regions are still joined together by disulfide bonds at the hinge region. This portion of the molecule is referred to as the $F(ab)_2$ region. Note that during pepsin treatment the Fc region is cleaved further into smaller fragments. Several points should be apparent from the diagram: (1) the antibody molecule is made up of four peptide chains, two light and two heavy chains; (2) the antibody molecule has at least two combining sites, each combining site being composed of the variable region of a light and a heavy chain; (3) the Fc fragment is important in many of the biological properties of the antibody molecule such as in antibody binding to Fc receptors on cells.

one heavy chain molecule. Antibodies combine with a very small part (an epitope) of a foreign (parasite) macromolecule (most often a **polysaccharide** or protein). Although macromolecules usually contain a number of different epitopes, each antibody molecule can combine with only one specific epitope. Years ago, it was shown that antibody molecules are so sensitive they could distinguish between chemical isomers (Landsteiner, 1962). Therefore, antibody–antigen (epitope) interactions have high degrees of specificity, similar to those involving enzyme–substrate interactions.

Antibodies are soluble macromolecules produced by B-cells in response to specific foreign epitopes (parasites). They are present in blood, lymph, and cerebrospinal fluid. There are five classes of antibody molecules, each of which is distinguished from the other by their heavy chains (Table 2.1). The IgG immunoglobulin class can be further subdivided into a number of subclasses. These subdivisions are based on minor differences in the **amino acid** sequence of their heavy chains (Janeway & Travers, 1994). The various major classes of immunoglobulins have different biological functions based on differences in their heavy chain (Table 2.1). The immunoglobulins can be differentiated based on their capability to **agglutinate** (clump), to **neutralize** toxic or enzymatic activity, or to **precipitate** antigens out of solution. In addition, the heavy chains of the immunoglobulin E (IgE) molecule are able to combine with receptors on **mast cells** and **platelets** and, following interaction with an antigen, cause the release of small-molecular-weight mediators such as histamine. The release of these molecules produces many of the symptoms characteristic of an **allergic** response. In contrast, the heavy chains of immunoglobulin M (IgM) do not combine with mast or platelet cell receptors. The various immunoglobulin classes are also predominantly associated with different parts of the body. For example, the large immunoglobulin M is located predominantly in the vascular bed, and immunoglobulin A is found in body secretions (gut, saliva). IgG occurs in all of the internal body fluids.

Antibody formation usually involves three cell types (the macrophage, the T-cell, and the B-cell). Their individual surface markers and their biological functions distinguish these cells from each other. Macrophages will engulf and process the foreign antigen into small epitopes. These epitopes are placed on the surface of the macrophage in association with specific surface macromolecules, the major histocompatibility antigens. The

Table 2.1 Characteristics of immunoglobulin classes

	Immunoglobulin class				
Character	IgG	IgM	IgA	IgD	IgE
Heavy chain	γ	μ	α	δ	ε
Number of binding sites	2	10	2 or 4	2	2
Number of subclasses	4	1	2	1	1
Size (kda)	150[a]	970	160[b]	180	190
Serum level (mg/ml/class or subclass)[c]	1 to 9	1.5	0.5 to 3.0	0.03	0.00005
Primary site of action	intravascular, interstitial, transplacental	intravascular	secretions, breast milk	B-cell surface	subcutaneous, submucosal
Binds to:					
Macrophages	+	–	±	–	±
PMN	+	–	–	–	–
Basophils	±	–	–	–	+
Mast cells	±	–	–	–	+
Biological functions	complement activation, precipitation, neutralization, opsonization	complement activation, agglutination	neutralization, in surface secretions and intestine	not known	mast cell sensitization and eosinophil activation

Notes:

[a] The size given is an approximate average of the weights of the four subclasses of IgG.

[b] Similar to IgG. The size shown is the average of the two different subclasses of IgA. In addition, IgA varies in size depending upon whether the molecule is in the serum (160 kda) or in secretions (340 kda). In secretions, two IgA molecules are joined through a peptide referred to as the J chain.

[c] For IgG and IgA, the range of their subclass concentrations in serum is shown.

macrophage can then interact with specific receptors on a lymphocyte, e.g., a helper T-cell (also called CD-4 cells or Th-2 cells) and the B-cell (Fig. 2.3). These cellular interactions are highly specific and a B-cell is presented with one specific epitope. Once the epitope information has been passed to a B-cell, it is induced to produce antibodies to that one epitope. In addition, the T-cells produce chemical messengers (interleukins) that induce the B-cell to divide. Thus, the B-cell, in cooperation with the T-cell, is induced to produce antibody, as well as being stimulated to divide and form a colony or clone of antibody-producing cells to the one epitope. It is estimated that animals have the genetic capability to respond to over one million different antigenic epitopes. As a consequence, although each immune response to a parasitic invader is highly specific, the immune system has the enormous potential to respond to a vast array of different parasitic organisms. Note that parasitic organisms are not composed of a single antigenic epitope, but rather, their surface is a topographical mosaic of different macromolecular antigens and, therefore, epitopes. The host will recognize a number of these different epitopes and there will be a polyclonal antibody response to any parasitic organism. In addition, the animal will produce a number of different immunoglobulin classes having the same antigen specificity.

Antibody production is an important part of the immune system. Antibody molecules, by combining directly with key parasite epitopes, can neutralize the ability of the parasites to enter cells or they can inhibit toxic parasite products. In addition, antigen(parasite)–antibody interactions can activate complement and kinin pathways and, therefore, a portion of the inflammatory response. Finally parasite(antigen)–antibody complexes can be linked through the heavy chains of antibodies to the surfaces of macrophages. This interaction increases the ability of macrophages to trap and engulf parasitic microorganisms.

2.1.4 Cellular immunity

In addition to the antibody response, there is often a host cellular response to parasitic organisms (Wakelin, 1984; Janeway & Travers, 1994). The cellular response involves another set of immune cells, the CD-4, Th-1, T-cell, and the CD-8, cytotoxic T-cells (Fig. 2.3) that are lymphocytes derived from the thymus. Their surface markers also distinguish these cells from each other. The CD-4, Th-1, T-cell, and the CD-8 cells contain surface receptors that interact with specific epitopes on foreign antigens. The specificity of the CD-4 cellular response is similar to that induced with the antibody response. Each activated T-cell responds to a single epitope and, following combination with a specific antigen, there is clonal proliferation of these cells. There is also the release of regulatory (chemical messenger) interleukins. Cellular immunity, like the antibody response, involves macrophages and the presentation of antigenic epitopes in combination with specific macrophage surface molecules (the major histocompatibility proteins) to the T-cells. Specific cytotoxic T-cells are capable of combining with antigenic epitopes on the surface of parasitic cells and ultimately cause their death. Depending upon the particular interleukin released, activated cytotoxic T-cells can inhibit (down regulate) the response of the helper T-cells involved in the antibody response. Similarly, helper T-cells can also down regulate the proliferation of the cytotoxic T-cells. There is, therefore, a complex array of chemical messengers that integrate the function of the cells involved in the overall immune response (Fig. 2.3). Note that different parasites preferentially stimulate different helper T-cell subsets (the Th-1 or the Th-2 populations) (Table 2.2). Each helper T-cell subset secretes a characteristic set of cytokines, referred to as a T-cell cytokine profile (Fig. 2.3). For example, in the mouse, *Leishmania* spp. generally induce a Th-1 response with a characteristic cytokine profile of IL-2, IL-3, IL-12, and IFN. In contrast, *Plasmodium* spp. preferentially induce an antibody Th-2 response with a set of cytokines consisting of IL-4, IL-6, and IL-10.

As in any biological process, the immune system must be able to dampen its responses. This is accomplished through suppressor T-cells and their assorted chemical messengers. In addition, macrophages are also capable of causing immune suppression. This suppression can be critical to the health of an animal since much of the pathology caused by parasitic organisms can be related to the immune response of the host itself. For example, the cellular infiltration that occurs

Table 2.2 Major T-cell types involved in resistance[a]

Parasite	T-cell type
Leishmania major	Th1
Cryptosporidium parvum	Th1
Trypanosoma cruzi	Th1
Schistosoma mansoni	Th1
Trypanosoma brucei brucei	Th2, a T-independent B-cell
Plasmodium falciparum	Th2
Trichuris muris	Th2

Notes:

[a] In most parasitic infections, the factors involved in resistance depend upon a variety of host responses, not a single response. Innate factors such as host age, sex, nutritional status, and genetic background all play a role and, in many infections, cellular and humoral immunity are involved in both resistance and pathology.

around schistosome eggs in the liver parenchyma is due to a cellular immune response. The continuous influx of T-cells and macrophages around the egg ultimately leads to inflammatory changes and host cell death. If the parasite persists, however, although modulated by suppressors, the cellular infiltrate around the egg continues to grow slowly. There are then both cell death and areas of wound healing, with the replacement of normal liver tissue by fibroblasts and collagen. Ultimately, if the immune system is successful, the egg is destroyed, and then calcified.

Similar to cellular immunopathology, there are also forms of antibody-mediated pathology. Antibody, in combination with an antigen, can activate parts of the inflammatory response, e.g., kinin and complement pathways. Locally, this can produce edema, redness, and pain. Systemically, these reactions can produce a total body response referred to as **anaphylactic shock**. Therefore, although the immune system is a critical part of an animal's defense against parasitic organisms, it can also cause pathology if the parasite is not eliminated and the system is not adequately suppressed (see section 2.2).

2.1.5 Host–parasite interactions

These descriptions of the mechanisms by which animals resist parasitic organisms are simplistic and not intended to substitute for a text or course in immunology. However, they should illustrate that, following invasion, or expansion of a parasite population in a host (or both), there is a complex interaction that takes place between the parasites and different host cell populations, as well as their chemical messages, under different *in vivo* 'ecosettings'. Finally, it should be emphasized that the host–parasite interaction is continuously changing.

Initially, there is a short period following invasion by the parasite in which there is multiplication, migration, and/or maturation, depending on the particular parasite. This is followed by host recognition of the parasite, with infiltration of neutrophils and macrophages into the site of infection. There are then inflammatory changes in the vascular bed with clotting and edema, followed by chemical signals that are released by neutrophils and macrophages. There is further cellular infiltration and expansion of the activated T- and B-cells, resulting in an antibody or a cellular response, or both. Dramatic localized environmental changes occur during this time period. These include changes in temperature, pH, and O_2 tension, all of which contribute to both parasite and host cell death.

The response of the host to an infectious agent requires the expenditure of considerable energy. The synthesis of macromolecules, and movement and proliferation of cells, all involve the use of both the necessary building blocks, i.e., amino acids, as well as energy. Therefore, the nutritional status and the energy resources of the host, at the start of an infection, will play a role in determining the ultimate outcome of the host–parasite relationship. It is known, for example, that mal-

Table 2.3 Examples of metabolic functions and cytokines suggested to take part in host–parasite interactions

Physiological parameter	Action	Suggested regulatory factor[a]
Sleep patterns	altered	IL-1
Body temperature (fever)	increased	IL-1, TNFα, IFNγ
Voluntary food intake	decreased	IL-1, TNFα
Gluconeogenesis	increased	IL-1
Fatty acid synthesis in adipocytes	decreased	TNFα, TNFβ, IFNγ
Hepatic acute phase protein synthesis	increased	IL-1, IL-6, TNFα
Skeletal muscle protein degradation	increased	IL-1, TNFα
Catecholamine release (norepinephrine, dopamine)	increased	IL-1, IL-2, IL-6, TNFα
Indolamine release (serotonin)	increased	IL-1, IL-6
Corticosteroid release	increased	IL-1, IL-6
Growth hormone release	decreased	IL-1, IL-6, TNFα
Thyroxin release	decreased	IL-1
Glucagon release	increased	IL-1, TNFα
Insulin release	increased	IL-1

Notes:

[a] IL-1, 2, and 6 are interleukins, TNFα or β are tumor necrosis factors, and IFNγ is interferon.

nutrition will reduce an individual's ability to respond immunologically to a parasite. In addition, once within the host, the successful parasite will also require small molecular building blocks and energy for its own growth, or proliferation, or both. It must compete with its host for these substances. The parasite's metabolism will, therefore, also tax the host's nutritional status and energy reserves.

When considering innate factors involved in resistance and the immune system, it is also essential that the host's entire 'ecosetting' be considered (Seed, 1993; Wakelin, 1993; Wassom, 1993). It is well documented that a host's diet, physical environment, and social setting, plus the behavioral responses to these factors, may influence the immune response through neuro-endocrine and hormonal interactions with various organ systems (Fig. 2.2; Table 2.3) (Khansari *et al.*, 1990; Crompton, 1991; Müller & Ackenheil, 1998). Host aggression, hostility, and stress are, in general, known to decrease resistance to infectious diseases. Depending upon the outcome of the host–parasite interaction, there may be a continued immune response leading to host pathology, or the host may successfully eliminate the parasite allowing for healing to occur. In either case, the host–parasite interaction takes place in a complex ecological setting in which the *in vivo*, and external, environments are changing continuously.

2.1.6 Parasite escape

Since the human population numbers in the billions and is currently growing at a logarithmic rate, it is obvious that resistance to infectious diseases allows most people to reach a reproductive age. In other words, humans have evolved remarkably successful mechanisms to evade or control infectious diseases. In fact, if one considers the entire system, one must ask the questions, why do we ever become infected, and how do parasites successfully avoid the host's defense system? The host–parasite system has also evolved over a long period of time. Therefore, whereas humans and other animals have evolved successful defense mechanisms, successful parasites have evolved mechanisms to avoid them. There are at least four major parasite escape mechanisms that will be discussed. They include antigenic mimicry, or masking, living in an intracellular habitat, antigenic variation, and immunosuppression.

One mechanism is to avoid host immune detection by mimicking key host macromolecules, or by masking their foreign antigens with a coating of host macromolecules. The adult schistosomes in mammalian hosts appear to use both of these mechanisms.

A second parasite escape mechanism is to avoid detection by occupying an intracellular site. There are many different protozoan species that have intracellular habitats. Indeed, many of these protozoans are capable of surviving within the very cells of the host that are an integral part of the host's immune defense mechanism. For example, *Trypanosoma cruzi*, the American trypanosome, and various species of *Leishmania*, are capable of growing within macrophages. As noted previously, this cell is normally an important part of the host's defense mechanism. It acts by phagocytizing and then killing the invading microorganisms. The macrophage also presents antigenic epitopes to those lymphocytes responsible for both antibody production and cell-mediated immunity. The mechanisms by which these protozoans survive within the cell are not completely understood, but one is simply to invade cells that are not part of the immune system. Many sporozoan species use this approach, e.g., they invade liver cells, epithelial cells or red blood cells (RBCs). In these locations, they are both protected from the host's defense mechanisms and avoid detection by the host's immune system. The malaria parasite within the host RBC is a good example of a parasite that is partially hidden from the immune system, as well as sequestered from direct attack by either antibodies or cytotoxic lymphocytes.

The strategies of those parasites that inhabit cells of the host's immune system, i.e., macrophages, include surviving within the **phagolysosomal** vacuole in which they reside (*Leishmania*), or escaping from the phagocytic vacuole in which they entered the cell by moving into the cytoplasm of the cell (*T. cruzi*). Some intracellular parasites can also prevent the fusion of the **lysosomal** vacuole with the phagocytic vacuole in which they enter and reside (*Toxoplasma gondii*). Each of these strategies is complex and requires close coordination between the host cell and the parasite. For example, for *Leishmania* to survive within a phagolysosomal vacuole, it must be able to prevent a drop in pH as well as the reactive oxygen radicals that are formed inside the vacuole. It has to also inhibit, or survive, the various digestive enzymes within the lysosomal vacuole. It can do this by having a surface that is not sensitive to enzymatic attack, producing enzyme inhibitors that are released on to the parasite's surface, or preventing the release or activation of the lysosomal enzymes. It should be noted here that although the intracellular habitat has advantages, there are a number of difficulties that a parasite must overcome in using this habitat. First the parasite must gain entrance into the cell. This usually requires specific parasite surface receptors on the host cell. Once attached to the host cell membrane, the parasite must have a mechanism for entry into the cell's interior. This can be accomplished directly by host cell phagocytosis, or via some type of zipper mechanism by which the parasite induces the host membrane to slowly encircle it. Once inside, the parasite must survive normal host cellular defenses, as well as obtain the necessary nutrients for growth and survival. In the *Plasmodium*–RBC example, the parasites must obtain their nutrients (and transport all their end products) through not one, but three sets of membranes, i.e., the external RBC membrane, the inner RBC membrane surrounding the vacuole in which the parasite resides, and finally the parasite cell membrane itself. In the life cycle of the nematode *Trichinella spiralis*, a larval stage penetrates a striated muscle cell of the mammalian host. Once within the cell, there is a complex and coordinated interaction between host cell and the parasite. The host cell is modified to insure survival of the larval stages and is appropriately called a nurse cell. Fig. 2.5 shows this highly integrated host–parasite association. The larval stage has both nutritional and energy requirements that must be met if it is to survive. Note from Fig. 2.5 that it would appear that several mitochondria are closely associated with the parasite cuticle; presumably, this insures the parasite access to a rich energy source. It seems that many parasitic species have successfully used the intracellular habitat both as a mechanism to escape immune detection as well as to insure their growth and development.

The intracellular parasite or its progeny must also have a mechanism to exit from the host cell in order to invade new cells as well as for possible

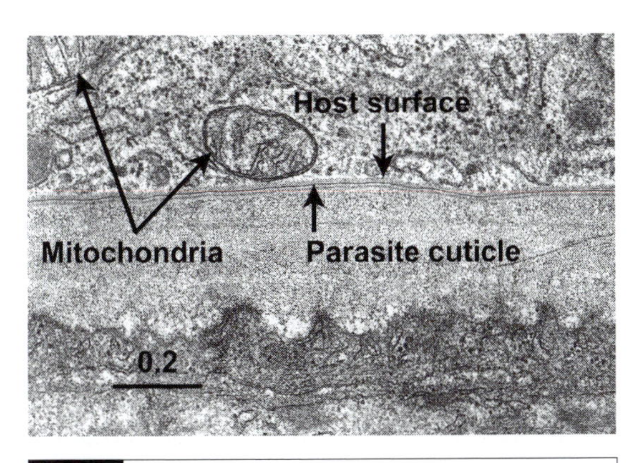

Fig. 2.5 *Trichinella spiralis* larva. The larva is intracellular in a striated muscle cell called a nurse cell (see also Fig. 5.13). The micrograph shows the interface between the mature, host nurse cell cytoplasm and the parasite cuticular surface. There appears to be a concentration of mitochondria around the parasite. (Photograph courtesy of Dickson Despommier, Columbia University, New York).

transmission to a new host. Although exit by killing and lysing the host cell may seem like a simple process, it is not. The exit step must be carefully controlled. If lysis of the host cell occurs prematurely, there could be limited parasite growth, or, immature stages, not infective to other host cells, would be released. It is, therefore, critical that the biochemical steps involved in the exit process be carefully controlled. Finally, since most intracellular protozoans cannot grow and survive extracellularly, they must have a means for the rapid invasion of new host cells. This could be through some form of motility, or by occupying an environment in which uninfected cells are both numerous and in the immediate vicinity of an infected cell. The latter would insure maximum opportunity for the released parasites to contact a new cell rapidly.

A third parasite escape mechanism would be to undergo a series of changes in a parasite's surface antigens. In this way, by the time a particular set of parasite surface antigens is recognized by the host's immune system, at least some of the parasites in the population have already changed the antigenic makeup of their surface coat. They would not, therefore, be susceptible to the host's immune response. The African trypanosomes (*Trypanosoma brucei* group) have over 1000 different genes within their genome coding for different surface antigens. They avoid the host immune response by continuously changing their antigenic coat. *Plasmodium* spp. and *Giardia lamblia* are also known to alter their surface coats.

A fourth possible escape mechanism involves the ability by some parasites to avoid the host immune response by directly suppressing it. Depending upon the particular parasite and host, this process can vary from parasite-specific immunosuppression to a rather generalized suppression of both the antibody and cellular responses to any antigen. Most parasitic organisms are known to immunosuppress their host to some extent. For example, the African trypanosomes profoundly immunosuppress both the humoral and cellular components of the immune system.

In summary, parasites can survive in an immunologically competent host by avoiding recognition, hiding from the host's immune system in an intracellular site, changing their antigenic composition, or by suppressing the host's immune system. Complicating the problem even further, there may be combinations of these avoidance mechanisms. For example, the African trypanosomes undergo antigenic change, but they also immunosuppress their host and, in the appropriate host, can adsorb host proteins onto their surface to presumably avoid immune recognition. In fact, the African trypanosomes have been shown to produce a protein (T-lymphocyte triggering factor) that activates CD-8 cells to secrete interferon (Donelson *et al.*, 1998). This important host cytokine is known to stimulate trypanosome growth. Similarly, *Leishmania mexicana* can use the host's insulin-like growth factor (IGF-1) to stimulate its own growth (Gomes *et al.*, 1997). The malarial parasite hiding within a RBC can undergo antigenic change as well as cause immunosuppression. Finally, adult schistosomes avoid the immune response by masking and mimicry, or they may undergo antigenic surface changes during parasite maturation from the schistosomula stage to the adult worms, or even immunosuppress the host's immune response.

Although the above discussion emphasizes the ways in which parasites avoid the host's immune system, it must be recognized that parasites have also evolved mechanisms to avoid the host's innate defense mechanisms. The most obvious example of this is that parasites have found ways to be able to penetrate the skin. A number of parasite forms employ insect vectors that inject the parasite into their host; others, such as hookworms or the schistosomes, have larval stages that are capable of attaching and actively burrowing through the skin. What is obvious is that parasites have evolved complex mechanisms to avoid both the host's innate and immune factors involved in resistance.

The relationship between the host and parasite is, therefore, a dynamic one. It is, moreover, a continuously evolving relationship. Parasites are being continuously selected for traits that allow the parasite to avoid the host's defense system and to successfully complete their life cycle. Simultaneously, the hosts are also continuously being selected for individuals that can avoid, control, or eliminate potentially harmful parasitic invaders. In Chapter 16, this 'evolutionary arms race' is further explored.

2.2 Pathology

Parasitic organisms can produce pathology in their hosts by a number of different mechanisms. These vary from the direct penetration of the intestinal wall by the adult *Ascaris* that produces mechanical damage and potentially a secondary bacterial invasion and peritonitis, to immunosuppression caused by the African trypanosomes that also leads to secondary opportunistic infections. There are several general observations that can be made (Table 2.4). Pathology may be due to mechanical insult. Both the growth of larval stages (*Taenia* and cysticercosis), and the migration of larval helminths (*Fasciola*) through host tissues can produce mechanical tissue damage. It may also occur by the consumption of host tissue. Hookworms, for example, feed on host blood and this can lead to severe **anemia**. However, pathology may also be indirect occurring through biochemical interventions of various sorts. *Entamoeba histolytica* releases hydrolytic enzymes that are capable of destroying host tissue. They then cleave the released macromolecules and use the small-molecular-weight cleavage products for their own nutritional needs. Many parasites consume host nutrients and can, in some cases, severely deplete the host of essential nutrients required for its survival. Some parasites (trypanosomes, plasmodia, and schistosomes) produce toxic molecules that are believed to be capable of altering the host physiology, or the host's immunological response, or both. Parasites can also alter their host's hormonal response and therefore the host's physiological, immunological, and behavioral responses to external environmental stimuli (Khansari *et al.*, 1990; Crompton, 1991; Müller & Ackenheil, 1998). Or, as will be discussed below, the pathology caused by the parasite may be due to a direct host–parasite interaction, but mediated by the host's immune responsiveness resulting in inflammatory reactions (Table 2.4). For a further description of the mechanisms of pathology in *Plasmodium* infection, see Clark *et al.* (1997).

As previously noted, there are both non-specific and specific host defense mechanisms. Activation of the complement, the clotting, and the kinin cascades all are part of the host's defense mechanism. However, they also are part of the inflammatory reaction to foreign antigens (parasites). It is an inflammatory reaction, for example, that initiates the raised, red, hard, painful **pustule** following a splinter penetrating the skin. After the initial changes in the vascular bed and edema, there is an influx of phagocytic cells, first non-specifically, but then, if the immune system is activated, mediated by a series of cytokine messengers. If the microorganisms on the splinter are eliminated, the reaction quickly subsides; however, if they persist and the host continues to respond immunologically, the small pustule may enlarge into a boil, tissue damage will follow, and scarring may result.

African trypanosomiasis illustrates a much more complex pathological syndrome. Following injection of the parasite by the vector, the parasite numbers increase stimulating both an acute phase response (fever, **malaise**, **cachexia**) by the release of the cytokines IL-1 and TNF (tumor necrosis factor), followed by a humoral B-cell response. This host defense mechanism is remarkably successful leading to a profound reduction in the parasitemia, but the parasite is nonetheless able to persist, albeit initially in very low numbers. The anti-trypanosome antibody is

Table 2.4 Some examples of pathological mechanisms in parasitic diseases

Organism	Observed pathology
	Direct mechanical factors
Plasmodium	Lysed RBCs, clog CNS microvascular bed
Hookworm	Feed on blood from intestinal wall
Ascaris	Penetrate intestinal wall
Taenia solium larvae	Cysticercosis (larval growth in confined space)
	Biochemical factors
	(1) Nutritional depletion
African trypanosomes	Hypoglycemia
Plasmodium	Anemia, Fe^{2+} deficiency
Hookworm	Anemia, Fe^{2+} deficiency
Diphyllobothrium latum	Vitamin B12 deficiency
	(2) Toxic parasite products
	(a) Small molecular weight
African trypanosomes	Aromatic amino acid catabolites
Fasciola hepatica	Proline excess, bile duct hyperplasia
	(b) Large molecular weight
Entamoeba histolytica	Secreted proteases, pore-forming peptides, lectin-mediated adherence
African trypanosomes	Hydrolytic enzymes (proteases), B-cell mitogen, T-lymphocyte triggering factor, endotoxin
	Immunological factors
	(1) Immediate hypersensitivity
African trypanosomes	Glomerular nephritis
Plasmodium	Glomerular nephritis
Schistosomes	Granuloma formation
Taenia solium larvae	Cysticercosis, allergic reaction
	(2) Delayed hypersensitivity
Leishmaniasis	Mucocutaneous, cutaneous pathology
Schistosomiasis	Egg granulomas, pipe-stem fibrosis
Toxoplasmosis	Brain pathology
	(3) Immunosuppression
Plasmodium, African trypanosomes, schistosomes, possibly most parasitic infections	Decreased ability to respond immunologically; increased secondary infections
	(4) Auto immunity
Plasmodium	Anti-RBC, anemia
Trypanosoma cruzi	Anti-nerve and/or heart tissue; nerve and heart damage

known to both lyse the trypanosome in the presence of complement, as well as enhance phagocytosis of the trypanosomes. The antigen–antibody complexes (trypanosome antigens–anti-trypanosome antibodies) will be trapped in various tissue sites, including the kidneys. These antigen–antibody complexes are known to activate complement, which attracts phagocytic cells to the site of deposition. This accumulation of phagocytes ultimately leads to a decrease in pH and oxygen tension at the site, and the release of a variety of hydrolytic enzymes that can cause host tissue destruction. In addition, the antigen–antibody complex activates the kinin cascade, producing changes in the vascular bed and resulting in edema. This is an immediate **hypersensitivity**, or an Arthus-type, response. In trypanosomiasis, the infection persists, and there is continued trapping of antigen–antibody complexes in the kidneys and other tissue sites. This continued response could ultimately lead to severe tissue damage. In *Trypanosoma cruzi* infections, there is evidence to suggest that the parasite induces the formation of antibodies to the host's own proteins (autoantibody). It is suggested that this autoantibody is, at least partially, responsible for the damage to heart tissue (Kaplan, 1997).

Another example of host immunopathology is chronic schistosomiasis in which the host responds immunologically to the eggs that are often trapped in the liver. Within weeks of a schistosome infection, the host is immunologically activated, producing both a B-cell and T-cell response to the schistosome eggs. When an egg is trapped in the liver, there is active migration of T- and B-cells, macrophages, and **eosinophils** to the site of the egg deposition. Although both antibody and B-cells can be observed in areas surrounding the egg, the most prominent response involves T-cells in a delayed-type hypersensitivity reaction. The continued influx of T-cells, macrophages, and their associated cytokines leads to a ring of host cells surrounding the egg. Histologically, this response is referred to as a **granuloma**. Since the egg often remains viable in the presence of the host defenses, the granuloma continues to enlarge. Eventually, the environment in the granuloma changes with decreased oxygen, increased acid metabolic products, etc., followed by host cell death and the release of hydrolytic enzymes. Since female schistosomes continue to produce eggs over a period of years, new granulomas are continually being formed. Therefore, as eggs are eventually walled off and killed, there is wound healing, with collagen deposition and egg calcification. Eventually, the granulomas, with scarring and calcified areas, join together, producing changes in the vascular bed. This ultimately results in severe liver damage (Kojama, 1998). Similar tissue destruction may occur in any host in which larval stages, or eggs, persist.

In African trypanosomiasis, one cause of pathology would appear to be an immediate hypersensitivity response, but it may not be the only one. For example, evidence suggests that the African trypanosomes produce an **endotoxin**-like substance that can activate the fever response and induce other physiological responses (Alafiatayo *et al.*, 1993). Moreover, animals chronically infected with the trypanosomes become immunosuppressed and secondary infections are a common cause of morbidity and mortality. Similarly, although the delayed type of hypersensitivity is a major cause of pathology in chronic schistosomiasis, there are other factors that may contribute to the pathology. These include parasite toxins involved in the acute phase of the disease, mechanical destruction due to fibrosis, and parasite-induced immunosuppression (Table 2.5). Finally, it has been suggested that schistosome catabolites (carcinogens) can induce bladder cancer. A similar mechanism of pathology may also exist for the liver fluke *Opisthorchis viverrini* (Kirby *et al.*, 1994).

In some cases, the pathological changes induced by the parasite appear to have evolved to assist the parasite in the completion of its life cycle. For example, changes in host behavior, malaise, and lethargy may increase the transmission of parasites that require capture and consumption by a predator. This is an example in which a parasite exploits food web relationships. Similarly, there is considerable evidence to suggest that granuloma formation is essential for the movement of schistosome eggs through the intestinal wall, and their subsequent release into the intestinal lumen. Finally, parasite-induced host immunosuppression could permit the parasite to survive within the host for an extended period of time increasing the opportunity for transmission to occur.

It should be apparent that the same mechanisms that are involved in the defense of the host are also involved in causing host pathology. Both the humoral and cellular arms of the defense system can lead to severe disease if the parasite is able to persist within the host. Therefore, the host–parasite relationship is very delicate (Fig.

Table 2.5 Some pathological mechanisms suggested in African trypanosomiasis and schistosomiasis[a]

Mechanism	African trypanosomes	Schistosomes
Toxin(s)	+	+
Immunosuppression	+	+
Immediate hypersensitivity	++	±
Delayed hypersensitivity	±	++
Mechanical damage	−	+ (pipe-stem fibrosis)
Metabolic rate and catabolite production	+ (hypoglycemia)	+ (malignancy)

Notes:
[a] (+) Present, may play some role in pathology; (++) suggested as a major cause of pathology; (±) believed to play a limited role; and (−) not known to play a role in pathology.

2.6). The host remains healthy as long as its immune system is intact and it is able to rapidly eliminate the parasite. In contrast, if the parasite can persist even at low levels, the immune mechanisms, which are so important in maintaining the low parasitemia, can also continue to respond and thereby lead to both localized tissue damage, e.g., schistosome–egg granuloma, or to a systemic inflammatory response, e.g., African trypanosome–glomerular nephritis.

2.3 Summary comments

In the preceding sections, our discussion was limited to several of the better-studied host–parasite systems. However, the basic mechanisms involved in resistance and pathology are similar for all infectious disease agents. In other words, the same basic principles explain host immunity, hypersensitivity, and inflammation in the protozoa, the helminths, or the ectoparasites. In other words, the mechanisms used by the protozoa to escape host defenses are similar to those used by other parasitic organisms. It is only when a comparison of the details involved in resistance or pathology of specific host–parasite relationships is made that differences are observed. For example, variations can be found in the type of toxin(s) produced, whether the host response is primarily a cellular or an antibody (T- or B-cell) response, as well as in the specific molecules that are involved in initiating one (or all) of the inflammatory cascades. Finally, the details on parasite escape mechanisms also vary. The molecular mechanisms involved in antigenic variation in the African trypanosomes would appear to differ from other parasitic protozoa.

In summary, remember that the basic principles are similar whether you are studying a virus or an ectoparasite. However, the details of the mechanisms involved in immunity, etc., will often differ even between two species in the same genus (African versus American trypanosomes). Finally, these differences can often be explained by differences in a parasite's habitat, life cycle, or both.

2.4 Biochemistry: physiological ecology

2.4.1 Introduction

All living organisms consist primarily of water and four different macromolecules: proteins, **lipids**, **nucleic acids**, and **carbohydrates**. Each of these macromolecules is synthesized by plant or animal cells from small-molecular-weight building blocks during cell repair, growth, and division. Proteins are synthesized from amino acids, nucleic acids from **purine** and **pyrimidine** nucleotides, carbohydrates or polysaccharides from simple **sugars**, and lipids from **fatty acids**. The synthesis of these macromolecules requires the expenditure of a readily available source of energy. Therefore, both the host and parasite

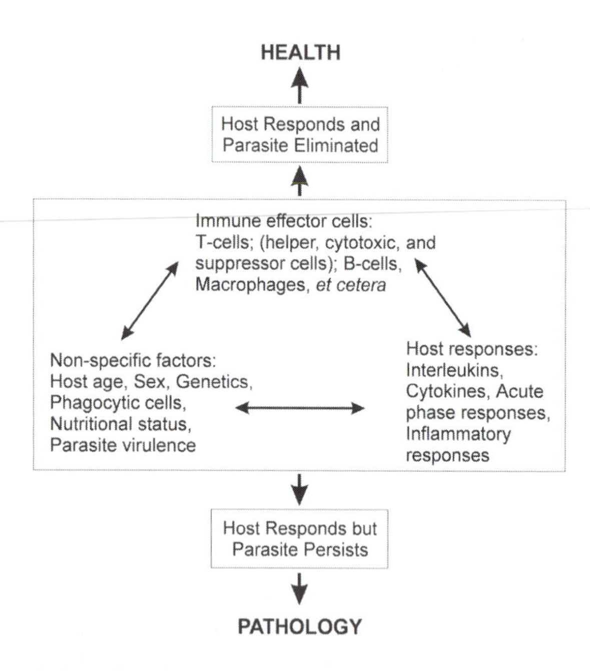

Fig. 2.6 Diagram of some factors involved in the host–parasite relationship.

must have available the necessary small-molecular-weight building blocks, and enzyme co-factors, i.e., metal ions and nicotinamide adenine dinucleotide (NAD), as well as a sufficient energy supply.

Since both the host and metazoan parasites are composed of eukaryotic cells, they will have biochemical pathways that are similar and will require identical resources. It is, therefore, obvious that when a parasite enters a host, it must compete with the host for key nutrients and energy resources. Since the host is considered a closed system containing finite resources, it is also apparent that if the parasite outcompetes the host for this common pool of nutrients, the host itself will suffer. It is reasonable to assume that the higher the parasite numbers in a setting in which the parasite can successfully compete with its host, the greater the drain on the host's resources and the more severe the parasite's impact will be. The loss of host resources can only be corrected if the host is able to resupply its nutritional needs.

Several conclusions can be drawn from this brief discussion. First, the host and parasite have similar nutritional requirements. They require essential amino acids, nucleotides, sugars, and fatty acids for both the synthetic (or **anabolic**) processes involved in the synthesis of the macromolecules needed for growth and reproduction, as well as for the breakdown (**catabolism**) processes which release the energy that is necessary to drive the synthetic machinery. Second, both host and parasite compete for a common finite pool of nutrients. Finally, the parasite must be able to successfully compete with its host for these resources if it is to complete its life cycle. We will address these conclusions by describing briefly the metabolic interactions observed in several host–parasite associations.

Mice experimentally infected with virulent African trypanosomes develop very high parasitemias (10^9 trypanosomes/ml blood). The mice will die within 7 days following the injection of a single trypanosome. At the time of host death, blood glucose levels approach zero and liver glycogen is almost totally depleted. The host is thus in an extremely hypoglycemic state. If the infected mouse is injected intraperitoneally with a small volume of a concentrated glucose solution, the mouse will survive for an additional period of hours and the trypanosome population will double in number. It is obvious that the energy resources of the host are insufficient to maintain host **homeostasis**, and that the parasite has successfully outcompeted its host for carbohydrate resources. It is estimated from these observations that the African trypanosomes consume their own weight in glucose each hour. It is also apparent that the size of the parasite population is limited by the size of the host's own carbohydrate pool. Finally, in this experimental model, it has also been shown that the blood of the host becomes depleted of O_2, and that **acidosis** occurs during the terminal stages of infection. In this biochemical process, for every molecule of glucose catabolized by the parasite, two molecules of pyruvic acid are produced and released into the host's blood. It is, therefore, assumed that the trypanosome's rapid rate of glucose utilization is responsible for the host's **hypoglycemia**, as well as for the decrease in blood pH and O_2 tension (von Brand, 1966).

A second example of metabolic interactions between hosts and parasites again involves the African trypanosomes. In an animal model with a

Table 2.6 Changes in host blood glucose and glycogen reserves during parasitic infections

Host	Parasite	Host glucose	Host liver glycogen
Rat/mouse	*T. brucei brucei*	hypoglycemic prior to death	greatly reduced prior to death
Microtus montanus	*T. brucei gambiense*	reduced	reduced
Man	*T. brucei gambiense*	normal	—
Rat	*Plasmodium berghei*	hypoglycemic	decreased
Man	*Plasmodium* species	often hypoglycemic	—
Man	*Schistosoma mansoni*	normal	—
Snail	*Schistosoma mansoni*	—	total body glycogen decreased
Man	*Ascaris*	occasionally reduced	—

chronic, or long-lasting, infection, it has been demonstrated that infected animals have dramatically decreased blood amino acid levels. In this model, tryptophan, an essential amino acid, was not detectable in the blood of the host and tyrosine, another essential amino acid, was reduced by approximately 50%.

In both of these examples, it is apparent that the host and parasite are competing for common resources and that the parasites are enormously successful in this competition. Changes in host glucose and **glycogen** are shown for several parasitic infections in Table 2.6. In many cases, but not all, the host's energy resources are reduced (von Brand, 1966).

Interactions between the host and parasite are quite complex. It has been shown, for example, that depriving a host of essential amino acids can reduce its ability to respond immunologically. In addition, in order to maintain homeostasis, the depletion of host blood glucose will initiate a set of physiological responses that result in the liver and other tissues depleting their glycogen reserves in order to maintain blood glucose levels. Similarly, it follows that there will be a transfer of key blood amino acids from the breakdown of muscle proteins in order to supply the blood and other organs.

2.4.2 General carbohydrate pathways

In the balance of this chapter some of the basic biochemical pathways involved in supplying parasites with adenosine triphosphate (ATP), a high-energy compound used by all cells in their synthetic pathways, will be explored (Box 2.1). Our discussion of these pathways will focus on their similar roles in host and parasite metabolism, as well as on the variability in the pathways between different parasites based upon a parasite's habitat.

Several different biochemical pathways can be used to catabolize the simple sugar glucose, whereas ATP is formed during two processes, called substrate-level, and oxidative, phosphorylation. For example, glucose is degraded via glycolysis to yield two molecules of pyruvic acid. During this process, there is a net yield of two molecules of ATP formed by substrate-level phosphorylation per molecule of glucose entering the pathway (Fig. 2.7). Depending upon the parasite species, pyruvate may be: (1) transaminated to alanine, (2) reduced to lactic acid, or (3) further catabolized to acetate, or ethanol and CO_2. The specific end products will depend upon the enzymatic capabilities of the parasite. In addition, during glycolysis, two molecules of reduced nicotinamide adenine dinucleotide (NADH) are also produced as each glucose molecule is degraded; the NADH can then be oxidized in the mitochondria to yield additional ATP by oxidative phosphorylation. If the organism has a functional mitochondrion, pyruvate can be converted into acetyl-coenzyme A (acetyl-CoA) and enter the tricarboxylic acid cycle (TCA, also known as the citric acid cycle) (Fig. 2.8). In the formation of acetyl-CoA from pyruvate, an additional two molecules of NADH are generated. These can also

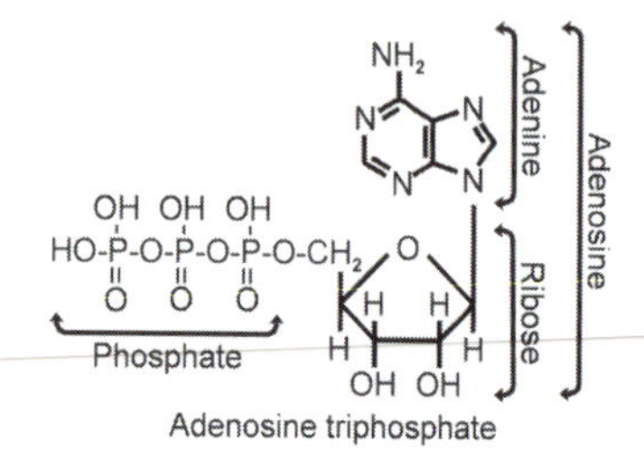

Box 2.1 ATP at work!

Energy transformations are the heart and soul of metabolism. Catabolic reactions are involved with molecular degradation and the concomitant release of energy. Anabolic reactions act in an opposite manner. During the catabolic transfer inside a cell, at least a portion of the energy will be captured in adenosine triphosphate, or ATP, which is the so-called 'universal energy currency' (see chemical structure left). In several metabolic pathways, living systems have evolved a mechanism for incorporating energy released at particular metabolic steps into a high-energy phosphate bond. This involves the combining of a phosphate group and ADP (adenosine diphosphate) to form ATP. Some heat will be lost in the process; however, a mole of ATP possesses much potential energy (approximately 7300 calories). This is the energy that will be used for, among other things, nucleic acid, protein, and polysaccharide synthesis. A simple example is the anabolic formation of the disaccharide sucrose. First, ATP must transfer its terminal phosphate group to glucose. This results in the formation of glucose 6-phosphate (G6P) and ADP. Then G6P reacts with fructose (a monosaccharide) to yield sucrose (glucose-fructose) and a free phosphate. The formation of sucrose from glucose and fructose in the presence of the appropriate enzyme catalyst requires over 5 kilocalories of energy in the form of ATP. This is only one of many energy-requiring reactions in a cell. The catabolic process of converting 1 mole of glucose to CO_2 and water will yield 38 moles of ATP or about 277 kilocalories of potential energy. On demand, the chemical energy in these high-energy phosphate bonds is released to do work. Some interesting trivia – did you know that ATP cannot be stockpiled? Interestingly, the average human body, at rest, i.e., your average 'couch potato', requires approximately 45 kg of ATP/day, yet, at any given time, the human body has < 1 g of ATP present!! That means that every second, in every cell, an estimated 10 million molecules of ATP must be made from ADP and phosphate and, since it cannot be stockpiled, an equivalent number must be hydrolyzed. Think about what must be happening in a really active person!

be oxidized in the mitochondrion, yielding a further three molecules of ATP per molecule of pyruvate catabolized to acetyl-CoA. The acetyl-CoA combines with the dicarboxylic acid, oxaloacetate, to form citrate, a tricarboxylic acid. In the TCA cycle, there is an additional net yield of two ATP molecules synthesized by substrate-level phosphorylation per glucose molecule that is catabolized. However, in addition to the ATP produced by this process, reducing equivalents are produced at several sites in the form of NADH and flavin adenine dinucleotide ($FADH_2$) that pass their electrons to the cytochromes and through the cytochrome system (also called the electron transport chain) in the mitochondrion (Fig. 2.9). In this pathway, there will be two or three ATP molecules formed per ½ O_2 molecule reduced by NADH or $FADH_2$ to H_2O. The exact yield of ATP will depend upon the initial site at which the electrons enter the cytochrome system (Fig. 2.9). If pyruvate is completely catabolized via the TCA cycle, three molecules of CO_2 and H_2O will be produced, and a yield of 15 additional molecules of ATP will be generated per molecule of pyruvate catabolized by the complete **oxidation** of NADH and $FADH_2$. This process is referred to as oxidative phosphorylation. In contrast to substrate-level phosphorylation, there is a much greater yield of ATP per molecule of carbohydrate consumed (Table 2.7). A summary of glycolysis and the TCA

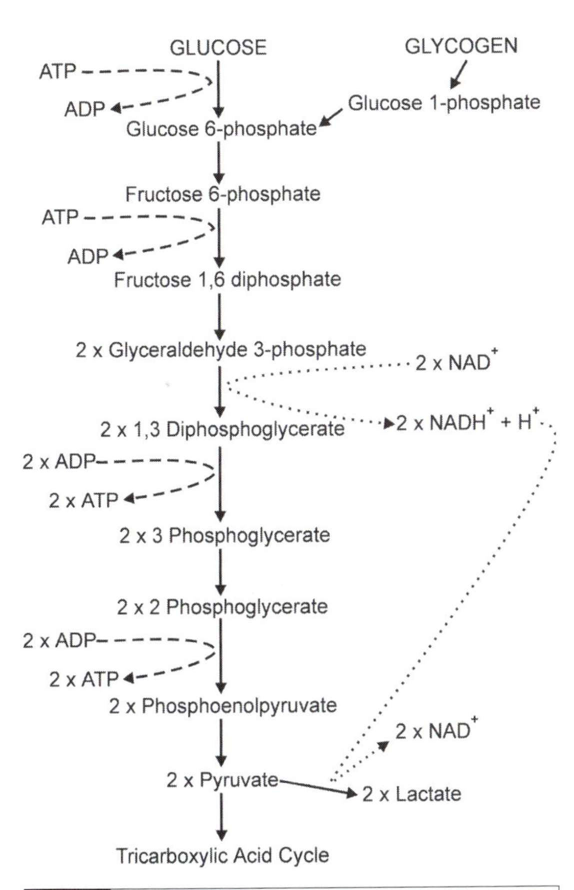

Fig. 2.7 Glycolysis. The arrows indicate the direction of the enzymatic pathway. ADP, adenosine diphosphate; ATP, adenosine triphosphate; NAD, nicotinamide adenine dinucleotide; NADH, reduced nicotinamide adenine dinucleotide. The notation '2 × . . .' means that there are two molecules.

cycle that can be involved in energy production is provided in Fig. 2.10. The yield of ATP in each part of the pathway is shown in Table 2.7.

2.4.3 Carbohydrate metabolism in parasites

In many parasites, such as *Leishmania*, *Ascaris*, and *Fasciola*, CO_2 can be fixed into phosphoenolpyruvate (PEP) by PEP carboxykinase to form oxaloacetate that is reduced rapidly to malate. Malate can then enter the mitochondria and be utilized for energy production. However, in other organisms lacking a mitochondrion, e.g., *Entamoeba* spp., there is incomplete catabolism of glucose (or carbohydrate) and the ATP yield is greatly reduced. In the Kinetoplastida, the enzymes of glycolysis are contained in a unique organelle called the glycosome rather than in the cytosol of the cell (Opperdoes, 1991). It is speculated that the glycosome evolved to allow for a more efficient transfer of glycolytic intermediates between enzymes in the pathway in order to meet the high energy demands of rapidly growing parasites solely by substrate-level phosphorylation.

In other parasites, the tricarboxylic acid cycle is incomplete and, in addition to the generation of electrons that are passed into the mitochondria, a variety of dicarboxylic acids (succinate, malate, and/or fumarate) are produced as catabolic end products. Therefore, there is incomplete oxidation of glucose and less ATP is produced per molecule of carbohydrate consumed (Bryant, 1989). Many of the helminths are known to produce organic acids or alcohol as catabolic end products (Prichard, 1989). For example, the major end products of anaerobic carbohydrate catabolism in *F. hepatica* are propionate, butyrate, and acetate, with traces of succinate and lactate (Prichard, 1989). The anaerobic end products of *Hymenolepis diminuta* are succinate, lactate, and acetate. With *Ascaris suum* adults, 2-methylvaleric acid, 2-methylbutyric acid, and succinate are produced, plus smaller amounts of acetate, propionate, and N-valeric acid. Electrons must enter the cytochrome system at three different sites in order to generate maximum yields of ATP (Fig. 2.9). Therefore, parasitic organisms that have an incomplete cytochrome system usually have only one or two electron transport sites and will also have reduced yields of ATP. Different parasite species, as well as different stages in a parasite's life cycle, will utilize different nutrients as an energy source, produce different catabolic end products, and have net ATP yields varying from two to 38 ATP molecules per metabolite consumed.

A comparison in the metabolites produced in two different stages in the life cycle of the African trypanosomes, as well as between the African trypanosomes and the American trypanosome, is shown in Table 2.8 (Gutteridge & Coombs, 1977; Opperdoes, 1991). The trypomastigote stage of *T. cruzi* has a functional mitochondrion with a TCA cycle and a cytochrome system. In contrast, the bloodstream stages of the African trypanosomes

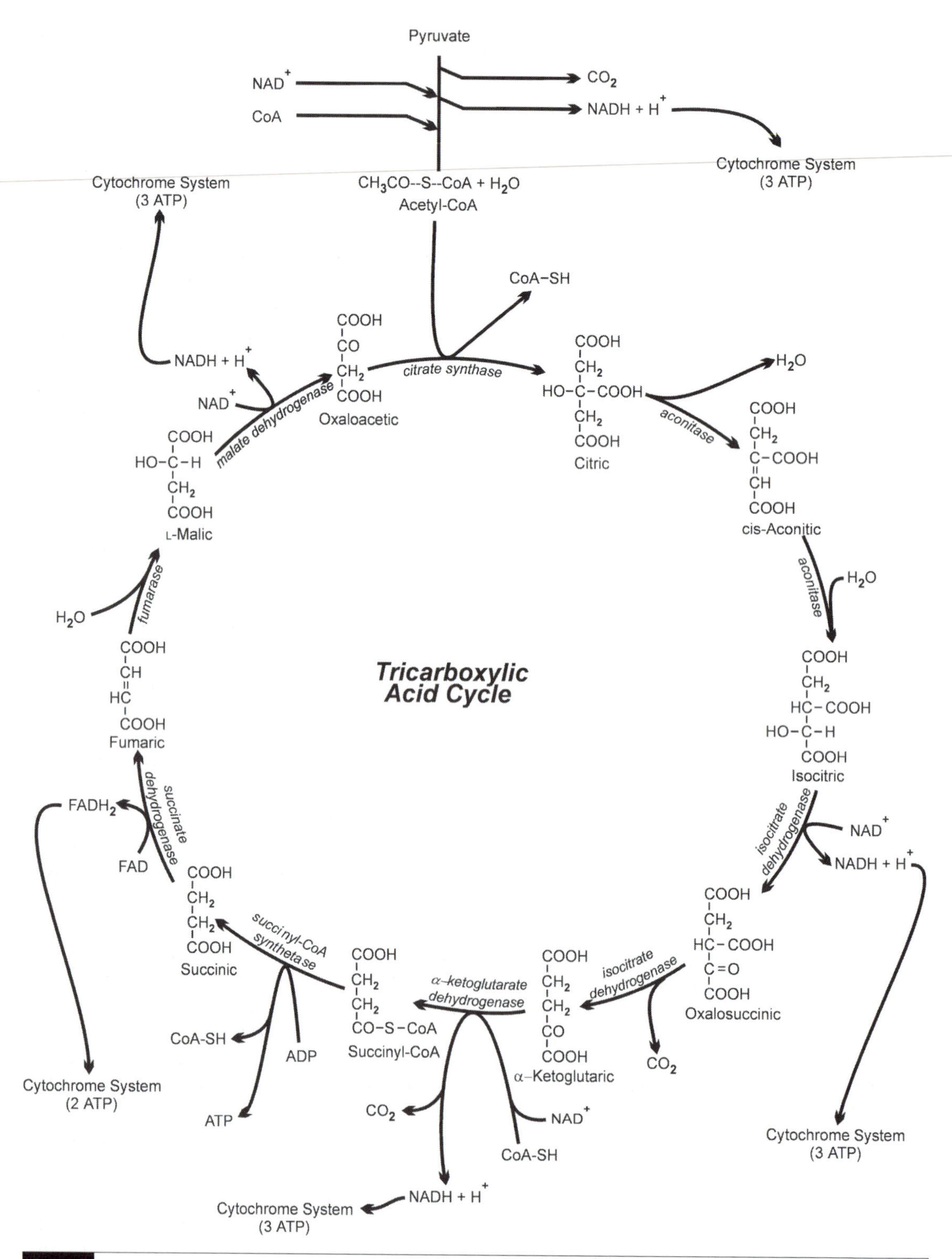

Fig. 2.8 The tricarboxylic acid cycle (citric acid cycle). The arrows indicate the direction of the pathway. The enzymes involved at each step in the pathway are shown in smaller print. For a description of the cytochrome system see Fig. 2.9 and the text. CoA, coenzyme A; CoA-SH, free coenzyme A; FAD, flavin adenine dinucleotide; $FADH_2$, reduced flavin adenine dinucleotide; NAD, nicotinamide adenine dinucleotide; NADH, reduced nicotinamide adenine dinucleotide.

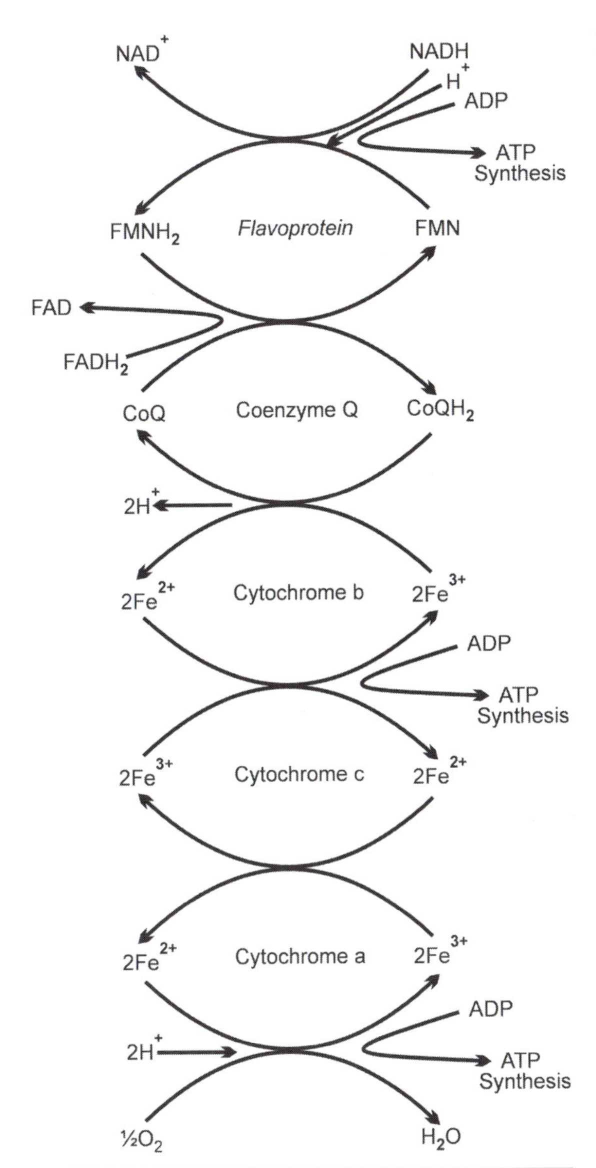

Fig. 2.9 The cytochrome system (electron transport chain) coupled to oxidative phosphorylation. The flavoprotein, coenzyme Q, and cytochromes b, c, and a are all proteins with active groups that are involved in the transport of electrons, ultimately to oxygen, with the subsequent formation of water. FAD, flavin adenine dinucleotide; $FADH_2$, reduced flavin adenine dinucleotide; ADP, adenosine diphosphate; FMN, flavin mononucleotide; $FMNH_2$, reduced flavin mononucleotide; NAD, nicotinamide adenine dinucleotide; NADH, reduced nicotinamide adenine dinucleotide.

lack a TCA cycle and cytochromes. These different metabolic profiles are characteristic of a particular parasite species and presumably reflect evolutionary adaptation to the parasite's *in vivo* habitat. In the bloodstream stages, the African trypanosome is in an environment rich in glucose and other necessary nutrients. Its mammalian host is, except when it has very high numbers of trypomastigotes, capable of metabolizing pyruvate for its own energy needs and, furthermore, is able to maintain homeostasis and its acid–base balance. Each individual African trypanosome consumes its own weight in glucose per hour and employs substrate-level phosphorylation to obtain a net yield of only two ATP molecules per molecule of glucose utilized. In contrast, in the vector hindgut, the **procyclic** stage of the African trypanosome is in an environment that is poor in glucose and has a low O_2 tension. In the hindgut, the procyclic stage utilizes proline, in addition to glucose, as its primary energy source (von Brand, 1966; Gutteridge & Coombs, 1977). Proline is an amino acid that is presumably derived from proteins in the tsetse's blood meal. The procyclic stage also has a functional mitochondrion and, therefore, can generate additional ATP through oxidative phosphorylation (Table 2.9). The procyclic stages have a reduced metabolic rate and live in an environment with a low O_2 concentration and at a lower environmental temperature. Nonetheless, there is obviously sufficient O_2 present to permit the mitochondrion to function at the lower catabolic rate.

In summary, both the glycolytic pathway and the TCA cycle can, in addition to producing ATP, supply the cell with intermediates for amino acid, fatty acid, and lipid biosynthesis. Acetyl-CoA is a key intermediate in fatty acid and lipid synthesis. Pyruvate, oxaloacetic acid, and α-ketoglutarate can be transaminated to the amino acids alanine, aspartate, and glutamate. Most parasitic organisms will use some combination of both substrate-level and oxidative phosphorylation to obtain needed energy. Only those parasitic organisms growing in an anaerobic environment, e.g., *Entamoeba histolytica* and *Trichomonas* spp. (see Chapter 3), lack mitochondria and rely solely on substrate-level phosphorylation for their growth needs. The trichomonads are interesting in that

Table 2.7 Potential energy yield per catabolic pathway[a]

Catabolic pathway	Energy (ATP) yield from: Substrate phosphorylation[b]	Oxidative phosphorylation[c]	Total ATP yield/pathway
Glycolysis:			
glucose + 2NAD → 2pyruvate + 2($NADH + H^+$)	2ATP	2($NADH + H^+$) = 6ATP	8
2pyruvate + 2NAD → 2acetyl-CoA + $2CO_2$ + 2($NADH + H^+$)	0	2($NADH + H^+$) = 6ATP	6
TCA cycle:			
2acetyl-CoA + 6NAD + 2FAD → $4CO_2$ + 6($NADH + H^+$) + $2FADH_2$	2ATP	6($NADH + H^+$) = 18ATP and $2FADH_2$ = 4ATP	24
Summary (Glycolysis + TCA cycles):			
Glucose + 10NAD + 2FAD → $6CO_2$ + 10($NADH + H^+$) + $2FADH_2$			38

Notes:

[a] Abbreviations: TCA, tricarboxylic acid cycle; ATP, adenosine triphosphate; $NADH + H^+$, reduced nicotinamide adenine dinucleotide; $FADH_2$, reduced flavin adenine dinucleotide.

[b] Substrate phosphorylation occurs in the absence of the transport of electrons through the cytochrome system. An example is the formation of ATP during the enzymatic conversion of 1,3-diphosphoglycerate and adenosine diphosphate (ADP) to 3-phosphoglycerate during glycolysis.

[c] Oxidative phosphorylation involves the transport of electrons through the cytochrome system coupled to the synthesis of ATP from ADP and inorganic phosphate (P_i). The summary reactions are: $NADH + H^+ + \frac{1}{2} O_2 + 3ADP + 3P_i \rightarrow NAD + H_2O + 3ATP$ and $FADH_2 + \frac{1}{2} O_2 + 2ADP + 2P_i \rightarrow FAD + H_2O + 2ATP$.

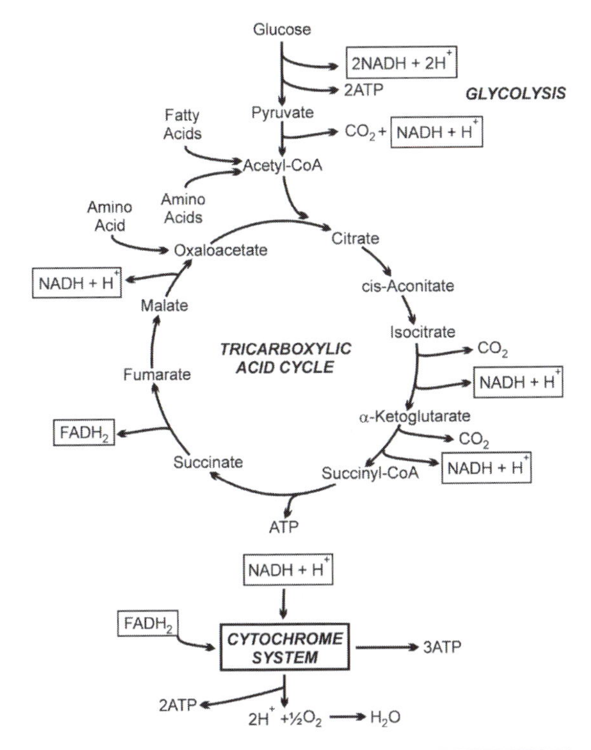

Fig. 2.10 Summary of the roles of the glycolytic pathway, the tricarboxylic acid cycle, and the cytochrome system in the complete oxidation of glucose to CO_2 and H_2O. In this cycle, glucose is first split, during glycolysis, into two molecules of pyruvate. The two pyruvate molecules are further metabolized to two acetyl-CoA and two CO_2 molecules. Note that it requires two complete metabolic cycles of the tricarboxylic acid cycle for both acetyl-CoA molecules to be catabolized. ATP can be generated by substrate-level phosphorylation at three steps in the entire pathway giving a net yield of four ATP molecules/molecule of glucose catabolized. In contrast, some 34 ATP molecules are synthesized/molecule of glucose utilized by the transfer of electrons from NADH and $FADH_2$ into the electron transport chain coupled to oxidative phosphorylation. The net maximum ATP yield per glucose molecule for substrate-level and oxidative phosphorylation is therefore 38 molecules. However, this maximum yield is not always achieved. For example, in some cells, the cytochrome system is not complete, or the cell does not completely oxidize glucose resulting in energy-rich end products such as succinate, acetate, etc. that are excreted into the environment. Note that both amino acids and fatty acids can also enter the tricarboxylic acid cycle and, therefore, be a source of energy.

they have a unique mechanism (and organelle, the **hydrogenosome**) for maintenance of their oxidation–**reduction** balance (Müller, 1991).

In the prior discussion, the classical glycolytic pathway of carbohydrate utilization has been described briefly. However, we must point out that other catabolic pathways also occur in some parasitic organisms. For example, the enzymes for a portion of, or the complete, pentose phosphate pathway (PPP) (Fig. 2.11) are present in many parasites. This pathway supplies intermediates for aromatic amino acids and phosphorylated carbohydrates used in the formation of glycoproteins. More important, it provides ribose 5-phosphate, a necessary precursor in nucleotide biosynthesis. As shown in Figs. 2.7 and 2.11, the first step in both glycolysis and the pentose phosphate pathway involve the phosphorylation of glucose to glucose 6-phosphate. Glucose 6-phosphate can be converted to fructose 6-phosphate which enters the glycolytic pathway or it can be oxidized to 6-phosphogluconate which enters the pentose phosphate pathway and is then enzymatically converted to a variety of phosphorylated intermediates, CO_2, and reducing power in the form of reduced nicotinamide dinucleotide phosphate (NADPH).

2.4.4 Non-carbohydrate metabolism

Although glycolysis, the PPP, and the TCA cycles can supply some amino acids and nucleotide precursors needed for protein and nucleic acid synthesis, parasites that reside in the rich environment of their host will primarily employ salvage mechanisms to supply their nutritional needs. They generally lack the genetic capability, and thus the enzymes, necessary to synthesize all of their amino acids, purines, pyrimidines, or lipids *de novo*. They therefore must obtain these essential nutrients directly from their host, and use preformed precursors in their anabolic pathways. The generalized **salvage pathway** for purines in parasitic organisms (Fig. 2.12) has recently been reviewed by Craig & Eakin (1997). Since phosphorylated intermediates generally cannot be transported across the cell membrane, the parasite must first take up the non-phosphorylated purine and pyrimidine precursors. They are then phosphorylated, converted to other

Table 2.8 A comparison of the end products formed during catabolism of glucose by *Trypanosoma cruzi* and the *Trypanosoma brucei* group

Species	Stage	End products
T. cruzi	trypomastigote	succinate, pyruvate, alanine, acetate, CO_2
T. brucei group	trypomastigote (long slender stage)	pyruvate, alanine
	promastigote	succinate, pyruvate, alanine, acetate, CO_2[a]

Notes:

[a] The promastigote form can also catabolize proline resulting in alanine, aspartate, and CO_2.

Table 2.9 A metabolic comparison between the mammalian long slender and short stumpy bloodstream stages and the procyclic (culture forms) insect gut stage of the African trypanosomes

		Mammalian bloodstream stages	
Metabolic function	Insect	Long slender	Short stumpy
Glycolysis	+	+	+
TCA cycle	+	–	partial
Glycerol oxidase	low	+	+
Proline oxidase	+	–	low
Cytochrome chain	+	–	–
Cyanide sensitivity	+	–	–
Mitochondrial structure	plate-like cristae	limited or no cristae	tubular cristae

needed purines and pyrimidines, and ultimately enzymatically converted into the appropriate nucleotides required for nucleic acid synthesis. Many parasitic organisms can synthesize pyrimidines *de novo*. However, even when the *de novo* pathways exist, it is estimated that a salvage pathway also meets much of their pyrimidine requirements. In contrast, energy-rich carbohydrates, e.g., glucose and essential amino acids such as tryptophan and phenylalanine, must be obtained directly from the host. Most parasitic organisms require a complex set of nutrients if they are cultured *in vitro*. Therefore, not only will the required energy sources and metabolic pathways vary between different parasites in different habitats, but so too will the parasite's sources of amino acids, lipids, etc. For example, *Plasmodium* spp., while residing within a RBC, will obtain many of their essential amino acids from the proteolytic digestion of the host's hemoglobin. In contrast, African trypanosomes in the blood of their hosts have active amino acid transport mechanisms that allow them to compete successfully with their hosts for essential amino acids. The African trypanosomes must therefore obtain almost all of the amino acids required for protein synthesis, plus both purines and pyrimidines and an assortment of enzyme co-factors, either directly from the host or by salvage pathways. The same is true for the parasitic helminths. Amino acids are also required by some parasites not only for protein synthesis but also as precursors to peptide hormones and neurotransmitters, e.g., adrenaline (epinephrine) and serotonin (5-hydroxytryptamine). It has been demonstrated that these neurohormones can be synthesized by both free-living and parasitic helminths. Dopamine and noradrenaline (**catecholamines**) are synthesized from phenylalanine and tyrosine. Similarly, serotonin (an **indolamine**) is synthesized from tryptophan. In contrast to the parasitic helminths, however, none of the parasitic protozoa are known to produce these neurohormones. The parasitic flagellates will, on the other hand, cata-

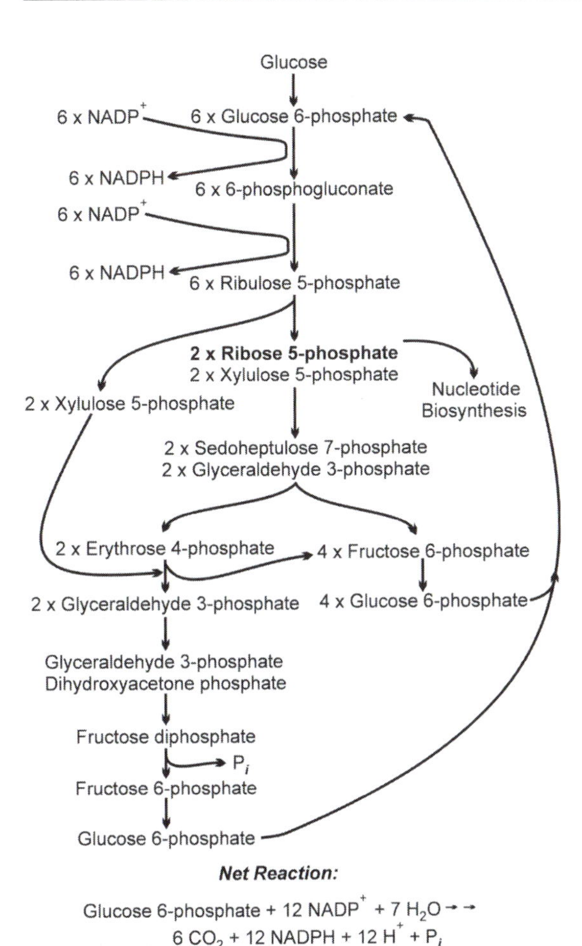

Fig. 2.11 The pentose phosphate pathway. A cycle in which glucose can be oxidized to CO_2. Two points to note are (1) the synthesis of ribose 5-phosphate which is involved in the *de novo* synthesis of nucleotides, and (2) generation of reducing power in the form of NADPH which is utilized in the cytoplasm in the reductive synthesis of long-chain fatty acids and cholesterol. Electrons are not transferred to the electron transport chain in the mitochondrion. Therefore, this pathway is not directly involved in the generation of ATP by oxidative phosphorylation. NADP, nicotinamide adenine dinucleotide phosphate; NADPH, reduced nicotinamide adenine dinucleotide phosphate. The notation '6 × . . ., 4 × . . .' etc. refers to the number of molecules.

bolize aromatic amino acids (phenylalanine, tyrosine, and tryptophan) to a number of different catabolites. In helminths, the catecholamines and indolamines have been shown to play a role in reproductive behavior, in feeding and locomotor activities, as well as in **ecdysis** and development in the nematodes. Finally, serotonin has been shown

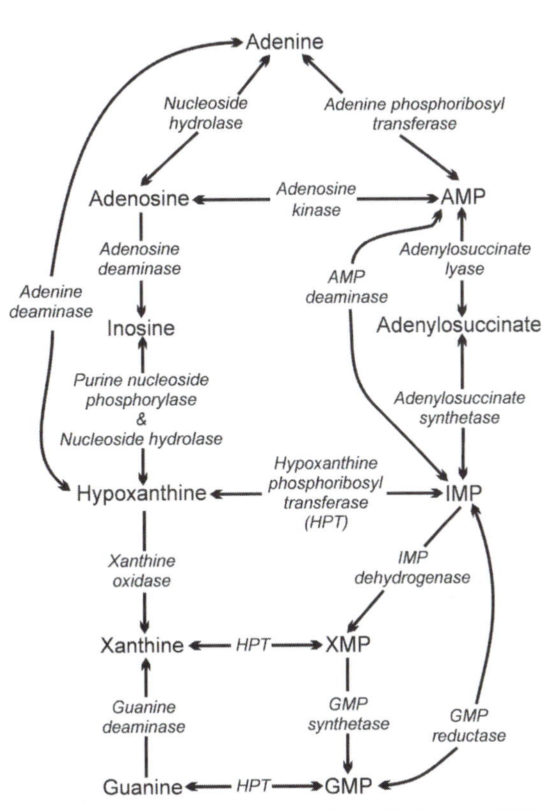

Fig. 2.12 Generalized diagram for the salvage pathways of purine in animals. The substrates involved are shown in large bold print. The enzymes involved at each step are in smaller, italicized print. AMP, adenosine monophosphate; GMP, guanosine monophosphate; IMP, inosine monophosphate; XMP, xanthine monophosphate.

to play a role in cyclic AMP mediated control of carbohydrate metabolism in several helminthic species.

2.5 Metabolism in an environmental context

Section 2.4 was intended to introduce some key biochemical pathways. It suggested that the biochemical pathways in any successful species of parasite have evolved to best fit the *in vivo* ecological habitat in which the parasite resides. It can, therefore, be concluded that although the general metabolic needs of parasites for their maintenance, growth, and reproduction are similar to those of all living forms, the specific details (pathways) will vary greatly. In fact, it is these metabolic differences between the host and parasite that have allowed biochemists to develop chemotherapeutic agents that are selective for the parasite. The differences in metabolic patterns appear to reflect the particular *in vivo* habitat in which the parasite resides. Table 2.10 compares the pathways involved in energy transformation in a number of parasitic organisms. It can be seen that *Entamoeba histolytica*, a protozoan which resides in a **microaerophilic**, or anaerobic, environment of the large intestine, does not employ a complete TCA cycle or possess cytochromes, and mitochondria. *Trichomonas vaginalis,* a protozoan that lives in the urogenital system of humans, similarly does not have a mitochondrion. In order for *T. vaginalis* to maintain its oxidation–reduction balance, electrons are transferred to hydrogen in the hydrogenosome, and H_2 gas is released (Müller, 1991). Both *E. histolytica* and *T. vaginalis* must obtain their ATP by substrate-level phosphorylation. In contrast, *T. cruzi* and the procyclic stage of the African trypanosomes (both protozoans) use a functional TCA cycle, and a cytochrome system, for the generation of ATP (Table 2.9). The parasitic helminths *Ascaris* spp. and *Fasciola hepatica* have partial TCA cycles and a cytochrome system. **Unsaturated** organic acids in these species, not O_2, act as the terminal electron acceptors. The differences in the energy-transforming pathways of some parasitic organisms are shown in Table 2.10.

The metabolic (energy) needs of short-lived and free-living larval stages of some digeneans (see Chapter 4) will differ from the metabolic requirements of their adult counterparts that might reside, for example, in the blood vessels of their definitive hosts. Similarly, the free-living stages of hookworm larvae in the soil will differ metabolically from their tissue-dwelling and migrating larval stages, as well as from the egg-laying adults attached to the gut of their hosts. A comparison of the enzymatic differences between the migrating 2nd and 3rd larval stages of *Ascaris suum* with the gut inhabiting adult stage shows that the larval stages, in an internal **aerobic** environment of the host, have a TCA cycle, a complete cytochrome system, and the terminal electron acceptor is O_2. In contrast, in the microaerophilic environment of the host gut, the adult has only a partial TCA cycle and cytochrome system and uses unsaturated organic acids as the terminal electron acceptor (Saz, 1981; Komuniecki & Komuniecki, 1989; Kita *et al.*, 1997).

The dividing bloodstream forms of *Trypanosoma brucei* lack cytochromes and a functional cytochrome system, and obtain their ATP solely by substrate-level phosphorylation. In contrast, the trypomastigote form of *T. cruzi*, residing in the same habitat, has a fully functional mitochondrion. It would appear at first glance, as if these two, closely related organisms, residing in identical habitats, should have very similar metabolic requirements. However, in contrast to the trypomastigote of *T. cruzi*, the long slender trypanosome form of *T. brucei* is rapidly dividing. Once the switch from the dividing, long slender stage to the non-dividing, short stumpy trypomastigote occurs, there are biochemical changes that take place in the African trypanosomes (Table 2.9). In the short stumpy form, enzymes required for mitochondrial function and oxidative phosphorylation are present. It would appear that the short stumpy stage of the African blood trypanosomes is more analogous to the trypomastigote stage of *T. cruzi* (Tables 2.8 and 2.9) than the long slender form. Both the short stumpy stage of the African trypanosomes and the trypomastigote stage of *T. cruzi* are non-dividing stages. Moreover, both are believed to be the stages required for growth and development in the insect vector. It seems that in the Kinetoplastida, mitochondrial function and oxidative phosphorylation are a prerequisite for growth in the gut of the vector. All members of the Kinetoplastida that have a stage in which the flagellates reside in the insect gut, have a functional mitochondrion and, presumably, obtain a significant portion of their energy requirements from oxidative phosphorylation. Laboratory strains of the African trypanosomes that are syringe-passaged through mice in the absence of a vector lose mitochondrial function, and eventually their ability to grow as procyclic forms. In addition, *T. evansi*, a member of the *brucei* group that is mechanically passaged by biting flies, has lost mitochondrial function and its ability to grow as a procyclic form. Therefore, it would appear that members of the Kinetoplastida that reside for all, or part, of their life cycle at lower temperatures

Table 2.10 Pathways of energy generation in some parasitic organisms

Species	Environment	Glycolytic enzymes (organelle involved)	TCA enzymes	Cytochromes	Terminal electron acceptor (organelle involved)
Entamoeba histolytica	human large intestine, anaerobic	yes	partial	no	small organic acids, H_2
Trichomonas vaginalis	human genitourinary tract, anaerobic	yes	partial	no	H_2 (hydrogenosomes)
African trypanosomes, dividing blood forms	blood, lymph, CSF, aerobic	yes (glycosomes)	no	no	O_2
African trypanosomes, vector forms	gut, microaerophilic	yes (glycosomes)	yes	yes	O_2 (mitochrondrion)
Malaria	blood, RBC stages, aerobic	yes	partial	yes	O_2 (mitochondria)
Fasciola hepatica	liver	yes	partial	yes (partial-complex II)	unsaturated organic acids (mitochrondria)
Ascaris suum	intestine	yes	partial	yes (partial-complex II)	unsaturated organic acids (mitochrondria)

and in microaerophilic environments of the insect gut, require a functional mitochondrion in order to obtain sufficient energy for growth. In contrast, trypanosomes that spend all, or part, of their life cycle in the carbohydrate-rich, aerobic, and high-temperature environment of mammalian blood, can obtain sufficient energy for growth by substrate-level phosphorylation.

Unfortunately, we know too little about the basic metabolic patterns and requirements of most parasitic organisms, and even less about the molecular mechanisms involved in the transition from one stage to the next in their life cycles. For most host–parasite associations, we have little or no information on the microhabitat requirements of the parasite, the metabolic changes that occur over time in specific habitats, or the metabolic drain upon the host resources. This information is critical to our understanding of host–parasite interactions and, ultimately, host pathology. As previously noted, parasites can either systemically, or in another localized tissue site, drastically alter their biochemical habitat. The physiological interactions of the host and parasite can lead to hypoxia, acidosis, and nutrient and co-factor depletion, as well as localized changes in temperature due to both parasite and elevated host metabolism. Many of these physiological interactions are similar to those observed at sites of inflammation. We are still a long way from having a good understanding of the biochemical and molecular interactions that take place between a host and its parasites.

References

Adams, S. H. & Boyce, W. T. (1995) Stress, personality, and infectious illness: an integrative approach. *Medicine, Exercise, Nutrition and Health*, **4**, 146–156.

Alafiatayo, R. A., Crawley, B., Oppenheim, B. A. & Pentreath, V. M. (1993) Endotoxins and the pathogenesis of *Trypanosoma brucei brucei* infection in mice. *Parasitology*, **107**, 49–53.

Bryant, C. (1989) Oxygen and the lower Metazoa. In *Comparative Biochemistry of Parasitic Helminths*, ed. E.-M. Bennet, C. Behm & C. Bryant, pp. 55–64. London: Chapman & Hall.

Clark, I. A., Al Yaman, F. M. & Jacobson, L. S. (1997) The biological basis of malaria disease. *International Journal for Parasitology*, **27**, 1237–1249.

Craig, III, S. P. & Eakin, A. E. (1997) Purine salvage enzymes of parasites as targets for structure-based inhibitor design. *Parasitology Today*, **13**, 238–241.

Crompton, D. W. T. (1991) Nutritional interactions between host and parasites. In *Parasite–Host Associations: Coexistence or Conflict*, ed. C. A. Toft, A. Aeschlimann & L. Bolis, pp. 228–257. Oxford: Oxford University Press.

De Jong-Brink, M. (1995) How schistosomes profit from the stress responses they elicit in their hosts. *Advances in Parasitology*, **35**, 178–254.

Donelson, J. E., Hill, K. L. & El-Sayed, N. M. A. (1998) Multiple mechanisms of immune evasion by African trypanosomes. *Molecular and Biochemical Parasitology*, **9**, 51–66.

Gomes, C. M. C., Goto, H., Corbett, C. E. P. & Gidlund, M. (1997) Insulin-like growth factor (IGF-1) is a growth promoting factor for *Leishmania* promastigotes. *Acta Tropica*, **64**, 225–228.

Gutteridge, W. E. & Coombs, G. H. (1977) *Biochemistry of Parasitic Protozoa*. Baltimore: University Park Press.

Hiramoto, R. N., Rogers, C. F., Demissie, S., Hsueh, C.-M., Hiramoto, N. S., Lorden, J. F. & Ghanta, V. K. (1997) Psychoneuro-endocrine immunology: site of recognition, learning and memory in the immune system and the brain. *International Journal of Neuroscience*, **92**, 259–286.

Hoffman, J. A., Kafatis, F. C., Janeway, Jr., C. A. & Ezekowitz, R. A. B. (1999) Phylogenetic perspectives in innate immunity. *Science*, **284**, 1313–1318.

Janeway, Jr., C. A. & Travers, P. (1994) *Immunobiology: The Immune System in Health and Disease*. New York: Current Biology. Garland.

Kaplan, D., Ferrari, I., Bergami, P. L., Mahler, E., Levitus, G., Chiale, P., Hoebeke, J., Van Regenmortel, M. H. V. & Levin, M. J. (1997) Antibodies to ribosomal P proteins of *Trypanosoma cruzi* in Chagas disease possess functional autoreactivity with heart tissue and differ from anti-autoantibodies in Lupus. *Proceeding of the National Academy of Science*, **94**, 10301–10306.

Khansari, D. N., Murgo, A. J. & Faith, R. E. (1990) Effect of stress on the immune system. *Immunology Today*, **11**, 170–175.

Kirby, G. M., Pelkonen, P., Vatanasapt, V., Camus, A. M.,

Wild, C. P. & Lang, M. A. (1994) Association of liver fluke (*Opisthorchis viverrini*) infestation with increased expression of cytochrome P450 and carcinogen metabolism in male hamster liver. *Molecular Carcinogenesis,* **11**, 81–89.

Kita, K., Hirawake, H. & Takamiya, S. (1997) Cytochromes in the respiratory chain of helminth mitochondria. *International Journal for Parasitology,* **27**, 617–630.

Kojima, S. (1998) Schistosomes. In *Topley & Wilson's Microbiology and Microbial Infections,* 9th edn, vol. 5, *Parasitology*, ed. F. E. G. Cox, J. Kreier & D. Wakelin, pp. 479–504. London: Edward Arnold.

Komuniecki, R. & Komuniecki, P. R. (1989) *Ascaris suum*: a useful model for anaerobic mitochondrial metabolism and the transition in aerobic–anaerobic developing parasitic helminths. In *Comparative Biochemistry of Parasitic Helminths,* ed. E.-M. Bennet, C. Behm & C. Bryant, pp. 1–12. London: Chapman & Hall.

Landsteiner, K. (1962) *The Specificity of Serological Reactions*. New York: Dover Publications.

Liew, F. Y. & Cox, F. E. G. (eds.) (1998) *Immunology of Intracellular Parasitism*, Chemical Immunology Series, vol. 70. Basel: Karger.

Müller, M. (1991) Energy metabolism of anaerobic parasitic protists. In *Biochemical Protozoology,* ed. G. Coombs & M. North, pp. 80–91. London: Taylor & Francis.

Müller, N. & Ackenheil, M. (1998) Psychoneuroimmunology and the cytokine action in the CNS: implications for psychiatric disorders. *Progress in Neuro-psychopharmacology and Biological Psychiatry,* **22**, 1–33.

Opperdoes, F. R. (1991) Glycosomes. In *Biochemical Protozoology,* ed. G. Coombs & M. North, pp. 134–144. London: Taylor & Francis.

Prichard, R. K. (1989) How do parasitic helminths use and survive oxygen and oxygen metabolites. In *Comparative Biochemistry of Parasitic Helminths,* ed. E.-M. Bennet, C. Behm & C. Bryant, pp. 67–68. London: Chapman & Hall.

Saz, H. J. (1981) Energy generation in parasitic helminths. In *The Biochemistry of Parasites,* ed. G.M. Slutzky, pp. 177–189. Oxford: Pergamon Press.

Seed, J. R. (1993) Immunoecology: origins of an idea. *Journal of Parasitology,* **79**, 470–471.

Von Brand, T. (1966) *Biochemistry of Parasites*. New York: Academic Press.

Wakelin, D. (1984) *Immunity to Parasites: How Animals Control Parasite Infections*. London: Edward Arnold.

Wakelin, D. (1993) *Trichinella spiralis*: immunity, ecology, and evolution. *Journal of Parasitology,* **79**, 488–494.

Wassom, D. L. (1993) Immunoecological succession in host–parasite communities. *Journal of Parasitology,* **79**, 483–487.

Willadsen, P. & Jongejan, F. (1999) Immunology of tick–host interactions and the control of ticks and tick-borne diseases. *Parasitology Today*, **15**, 258–262.

Zwilling, B. S. (1992) Stress affects disease outcomes. *ASM News,* **58**, 23–25.

Chapter 3

The Protozoa

3.1 Introduction

The Protozoa are unicellular, eukaryotic organisms possessing organelles such as a true nucleus, mitochondria and chloroplasts, Golgi, etc., found in higher plants and animals (Fig. 3.1). They can be found in almost all possible habitats and include both free-living and parasitic forms. The parasitic protozoans inhabit both plant and animal hosts at all phylogenetic levels, and the intracellular parasites can be found in almost all cell types. The parasitic protozoans are considered as microparasites because of their small size; they divide within their definitive host and generally cause acute, rather than chronic, diseases.

In the classification scheme of Corliss (1984), there are six phyla with parasitic forms. However, with the use of the newer molecular tools, it is now recognized that at least one of these phyla, the Myxozoa, is not a protozoan but a metazoan (see section 9.3). In addition, these new molecular tools have raised questions concerning the taxonomic position of the phylum Ascetospora, which has multicellular spores, and the phylum Microsporida (Box 3.1). The major protozoan phyla to be discussed in this text are the Sarcomastigophora, which includes both the flagellated and the amoeboid parasites, the Apicomplexa, all of which are intracellular parasites, and the Ciliophora in which occur the ciliated parasites (Box 3.2). A simplified evolutionary scheme for this kingdom, based on the work of Molyneux & Ashford (1983), Kreier & Baker (1987), and Cavalier-Smith (1993), is shown in Fig. 3.2. We emphasize that this scheme is under constant revision and, with the use of molecular procedures, it will continue to be revised for the foreseeable future.

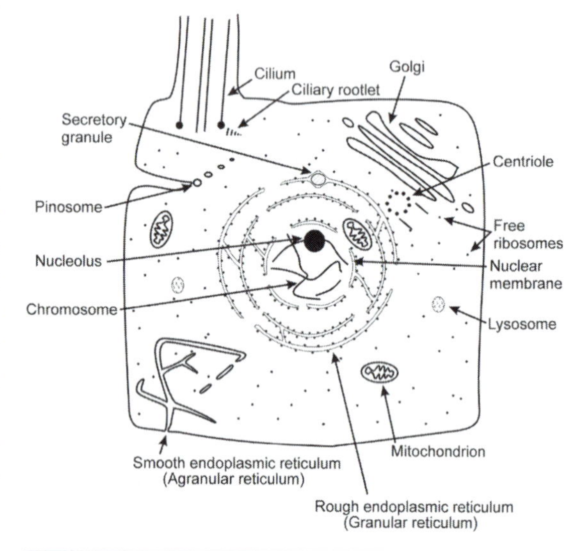

Fig. 3.1 The structure of a generalized eukaryotic cell.

The Kingdom Protozoa is a group of very old and diverse unicellular organisms that presumably originated from **prokaryotic** cells first through parasitic, and then mutualistic, associations between two or more prokaryotic cells (Margulis, 1981). There is considerable evidence to suggest that both the chloroplast and mitochondrion have a prokaryotic and a symbiotic origin. The most primitive of the protozoan groups are believed to be members of the Sarcomastigophora. From a sarcomastigophoran ancestor, it is believed that the other protozoan phyla evolved. It has been suggested that the flagellates and ciliates

Box 3.1 The Microsporida (Microspora)

The Microsporida are obligate, intracellular, spore-forming parasites that infect every major animal group from protozoans to humans. They can inhabit all major tissue types as well as a majority of the host's organ systems. Therefore, the pathology observed depends upon both the location and the extent of the infection in the host.

The Microsporida are considered to be primitive eukaryotic cells. They contain a membrane-bound nucleus but lack centrioles or mitochondria and have the prokaryotic 70S ribosome. Current taxonomy is based on the morphology of the spore, which contains a unique organelle, the polar apparatus. The taxonomic status of the prokaryotes and the primitive eukaryotes is currently uncertain but evidence suggests that the Microsporida belong in a separate Kingdom (Fig. 3.2) or at least in a totally separate phylum in the Kingdom Protozoa. The Microsporida contain dozens of genera and over 200 recognized species.

The life cycle of the Microsporida has three basic phases. There are no known vectors. First is a transmission or infective phase that includes the spore stage, and the oral uptake or inhalation of the spores by a new host. Transmission is followed by the proliferation (or merogony) phase with growth and division of the meront, and, finally, the sporogonic phase. Sporogony involves the development of a sporoblast, the sporont, and the mature spores.

Spores are swallowed or inhaled by a new host. The spore contains one or two nuclei in the sporoplasm, depending upon the species. Anterior to the sporoplasm, there is a polar body called a polaroplast which is believed to be involved in the extrusion of the polar tubule. The polar tubule (or filament) is coiled around the sporoplasm and the polaroplast. It is thought to be derived from a specialized Golgi apparatus. When the spore reaches the intestine, the polar tubule is extruded from the spore and penetrates a host cell. A pressure build-up in the spore, triggered by a calcium influx, is believed to cause the extrusion. The sporoplasm moves along the hollow polar tubule to emerge as a small amoeba in the cytoplasm of a host cell. This process is analogous to injection (of the sporoplasm) by a syringe (spore) and needle (polar tubule). In this site, the microsporidian replicates either in a parasitophorous vacuole or directly in the cytoplasm. The amoeba grows into a cell with two to eight nuclei. Each multinucleate cell then divides and forms a uninucleate meront. The process of merogony may be repeated, or the merozoites may migrate within the host ultimately reaching their final tissue site. Once intracellular, the parasite divides repeatedly and the enlarged host cell becomes filled with sporonts (a pseudocyst). Each sporont matures into one or more spores having a thick three-layered spore coat. Transmission depends upon the location of the pseudocyst in the host. If the pseudocyst is within the deep tissues of its host, the spores can only be released upon the death of the host. In contrast, if the pseudocysts (spores) are present in tissues such as the skin, or the intestinal epithelium, they can be released directly into the external environment. Some microsporidian species are also known to be transmitted via an egg to a larva.

Microsporidian infections have economic and public-health importance. It was Pasteur who first showed that the microsporidian *Nosema bombycis* caused significant pathology in the silkworm. This disease resulted in significant economic losses to silkworm farmers of France. In addition to demonstrating that *N. bombycis* produces significant disease, Pasteur also showed that transmission occurred via the egg, as well as by silkworm consumption of spore-contaminated leaves. Finally, he taught the silkworm farmers how to control the infection by identifying, and then removing, the infected silkworms from their farms. Other microsporidian species also cause disease in economically important animals. *Nosema apis* can infect and kill honeybees. Other genera and species can cause significant pathology and kill crustaceans, and marine and freshwater fishes. The Microsporida are also known to infect other parasites such as flatworms, leeches, and even parasitic protozoa.

Recently, the Microsporida have been recognized as human parasites. Some five different genera of Microsporida are known to infect humans. In one study, over 30% of homosexual men were serologically positive for the microsporidian *Encephalitozoon cuniculi*, suggesting that the infection might be fairly widespread. Microsporidiosis is now recognized as an important, emerging infectious disease. Although originally believed to be solely an opportunistic infection in immunocompromised individuals, particularly AIDS patients, Microsporida are suggested to have caused an acute, self-limiting diarrhea in an immunocompetent individual (reviewed in Fedorko & Hijazi, 1996). The recognition that microsporidiosis is an important human disease has led to increased interest in this fascinating group of parasites.

Box 3.2 | Kingdom Protozoa

The major taxonomic divisions of the parasitic protozoa within the Kingdom Protozoa are outlined and described below. This taxonomic scheme is similar to that proposed by Corliss (1984). However, several phyla (Archezoa or Microsporida, Ascetospora, Myxozoa) are not discussed because recent taxonomic research suggests that they do not belong within the Kingdom Protozoa. For example, the phylum Archezoa or Microsporida is now considered by some taxonomists to be a more primitive kingdom, and the phylum Myxozoa, because it has multicellular spores, is thought to be metazoan in nature.

Kingdom PROTOZOA: Single-celled, eukaryotic organisms, without plant-like chloroplasts, generally motile, and usually phagotrophic. The parasitic protozoa can most simply be divided into four groups based upon type of motility: (1) those that move by flagella, (2) those that move by amoeboid motion or pseudopodia, (3) those that move by cilia, and (4) protozoa that have no obvious organelle used in locomotion. These simple taxonomic groupings will be a main characteristic in the separation of the parasitic protozoa.

Phylum SARCOMASTIGOPHORA: Cells that move by flagella, pseudopodia (amoeboid movement), or both

Subphylum MASTIGOPHORA: Cells with one or more flagella

Class ZOOMASTIGOPHOREA: Chloroplast absent, one to many flagella, amoeboid forms with or without flagella, sexual reproduction in some groups

Order KINETOPLASTIDA: One or two flagella typically with a paraxial rod, a single large mitochondrion containing up to 20% of the cell's DNA in an aggregated network
Suborder BODONINA: (*Cryptobia*, *Bodo*)
Suborder TRYPANOSOMATINA: (*Trypanosoma*, *Leishmania*)
Order RETORTAMONADIDA: Two to four flagella (*Chilomastix*)
Order DIPLOMONADIDA: One to four flagella per nucleus (*Giardia*)
Order TRICHOMONADIDA: Four to six flagella (*Histomonas*, *Trichomonas*)
Order HYPERMASTIGIDA: Many flagella (insect flagellates)
Subphylum OPALINATA: Numerous cilia, flagellated gametes, cytostome absent
Class OPALINATEA
Order OPALINIDA: (*Opalina*)
Subphylum SARCODINA: Amoeboid motion by pseudopodia. Flagella when present generally restricted to developmental stages in the life cycle.
Superclass RHIZOPODEA
Class LOBOSEA
Order AMOEBIDA: No flagellated stage (*Entamoeba*)
Order SCHIZOPYRENIDA: A temporary flagellated stage usually present (*Naegleria*)
Phylum APICOMPLEXA: All intracellular, flagella on sexual stages only.
Class SPOROZOASIDA
Subclass GREGARINASINA: Parasites of invertebrates and lower vertebrates (*Gregarina*)
Subclass COCCIDIASINA: Most parasitic in vertebrates.
Order EUCOCCIDIORIDA
Suborder ADELEORINA: (*Haemogregarina*)
Suborder EIMERIORINA: (*Eimeria*, *Isospora*, *Toxoplasma*, *Cryptosporidium*)
Suborder HAEMOSPORORINA: Vector transmitted (*Plasmodium*)
Subclass PIROPLASMASINA: Ticks are the only known vectors
Order PIROPLASMORIDA: (*Babesia*, *Theileria*)
Phylum CILIOPHORA: Move by simple or compound cilia, two types of nuclei, cytostome usually present (*Balantidium*, *Nyctotherus*, rumen ciliates)

For further details on the various taxonomic groups within the Kingdom see the older texts by Kudo (1966), Manwell (1968), and the more recent *An Illustrated Guide to the Protozoa* edited by Lee, Hutner & Bovee (1985). To further review the parasitic protozoa one might begin by reading Kreier & Baker (1987) as well as Cox (1993). Finally, for encyclopedic descriptions of mammalian trypanosomes and malarial parasites examine the marvelous texts by Hoare (1972) and Garnham (1966).

separated early from the Sarcodina and the Apicomplexa (Fig 3.2). Despite their early evolutionary origin, it must be assumed that none of the present phyla closely resembles the original cell precursor of the modern protozoans. Indeed, based on a variety of taxonomic characteristics, including gene sequence homologies, the phyla are only distantly related. This is reflected in the wide range of habitats in which the symbiotic protozoans reside, their varied morphologies,

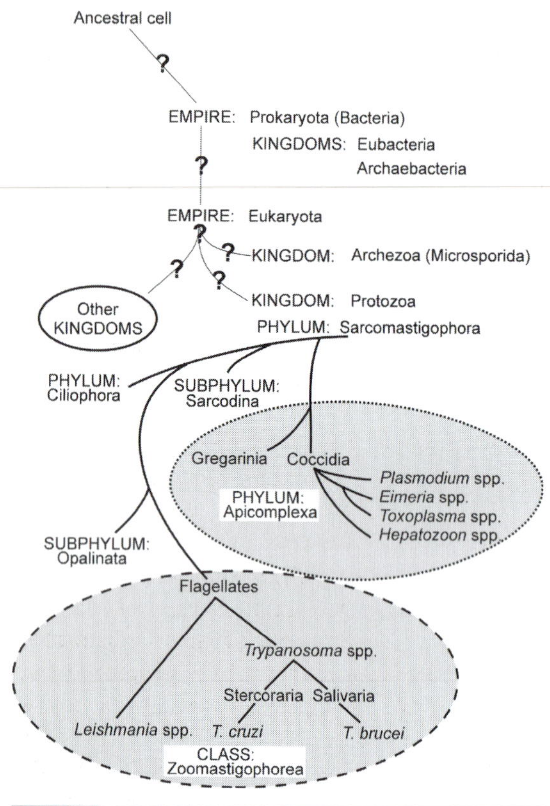

Fig. 3.2 Simplified evolutionary tree showing the suggested evolutionary relationship between the main parasitic Protozoan phyla. Note that the Protozoa are placed in a separate kingdom.

their divergent modes of motility and reproduction, and the great diversity in their nutritional requirements and metabolic patterns.

3.1.1 Form

Many protozoans use a gliding type motion, **pseudopodia**, or external **flagella** or **cilia** for movement. Their cell surfaces may be a naked cell membrane, may have numerous flagella, or may be covered with cilia. In addition, during their life cycles, they may form **cysts** or **spores** with thick cell walls, containing stages with reduced metabolic activity. These cyst stages are usually released into the environment, and must survive outside until acquired by a new host. Protozoans possess the typical eukaryotic cell membrane. There is a complex infrastructure of microtubules, microfilaments, and other organelles associated with the cell membrane in flagellated and ciliated forms. The cell membrane and associated infrastructure is often referred to as the pellicle.

Within the cytoplasm, there is a variety of eukaryotic organelles, including mitochondria, Golgi, etc. The number, type, and even position of these organelles vary between the different taxonomic groups, and many appear to be associated with the parasite's habitat. For example, the parasitic amoeba *Entamoeba histolytica*, which lives in the anaerobic environment of the intestine, does not possess mitochondria. In contrast, the flagellated African trypanosomes, in their insect vector, have a completely functional mitochondrion. However, when in the mammalian host, these trypanosomes possess a non-functional mitochondrion that lacks cristae and cytochromes. Only members of the subphylum Phytomastigophorea have chloroplasts and most of the species in this group are free-living. Although parasitic protozoans have a nucleus with a nuclear membrane, the presence or absence of a true mitotic apparatus and the presence of condensed chromosomes during division vary among the different phyla. Also, within some protozoan phyla are groups with unique cellular structures. For example, the ciliates have a subpellicle organization of microfilaments, microtubules, etc., that are associated with the cilia and are involved in the coordination of their movement; some flagellates possess hydrogenosomes, organelles that produce hydrogen as a molecular by-product.

3.1.2 Function

3.1.2.1 Nutrient uptake

Protozoans obtain their nutrients by a variety of mechanisms. The most direct method is by the active transport of small molecular nutrients across the cell membrane. Some employ phagocytosis of bacteria, red blood cells (RBCs), other host cells, or debris that can be digested internally as nutrients. Another mechanism used in some groups is **pinocytosis**. For example, the trypanosomes transport large-molecular-weight compounds, such as transferin, from the host into the flagellar pocket where the specific receptor for transferin is located. These molecules are then ingested by pinocytosis into the cell cytoplasm where they are digested. The flagellar pocket is thus a distinct organelle with a specific function. Finally, protozoans such as the ciliates have a **cytostome** (or mouth) into which food particles are swept by the action of cilia.

Some protozoans have a specialized organelle for eliminating wastes and removing excess water. The **contractile vacuole** is a rhythmically pulsating vesicle that opens to the outside through a small pore in the cell membrane. The infrastructure of the contractile vacuole insures that water and wastes flow out from the vacuole into the external environment.

3.1.2.2 Metabolism

The metabolism among different protozoan groups is as varied as the organelles used for feeding and locomotion. Because of the different habitats in which protozoans live, this should not be surprising. We have already noted that the parasitic amoebae, which inhabit the gut, lack mitochondria. They also have an anaerobic type of metabolism in which oxidative phosphorylation does not play a role. Similarly, African trypanosomes within the mammalian host do not have functional mitochondria and obtain their energy by substrate-level phosphorylation. Although they use the oxygen within their habitat to maintain the appropriate oxidation–reduction balance, there is no energy gained during this process. During the vector phase of their life cycle, however, these flagellated parasites have a functional mitochondrion and, therefore, possess an active oxidative phosphorylation pathway. Similarly, the pathways involved in carbohydrate catabolism, as well as amino acid and nucleic acid metabolism, are equally varied (Gutteridge & Coombs, 1977; Coombs & North, 1991; Bryant, 1993). However, these pathways are similar to those observed in other animal cells, e.g., glucose is catabolized to CO_2 and water through glycolysis, the pentose phosphate shunt, the tricarboxylic acid cycle, and the cytochrome system (see Chapter 2). The trypanosomes and most apicomplexans do not store carbohydrates, whereas *Entamoeba* stores glycogen. Other carbohydrates such as starch and amylopectin are found in other protozoan groups.

3.1.3 Habitat

The protozoans are usually classified according to their method of motility. As noted earlier, these include amoeboid, flagellated, and ciliated forms. Habitat (free-living or parasitic), other morphological features, and physiology are then used to differentiate these large, basic groups. More recently, the tools of molecular biology have allowed the protozoan systematists to draw further evolutionary and phylogenetic implications based on nucleotide-sequence homologies. Most often, these are based on the sequence homologies of small ribosomal RNA. A second procedure is to compare the sequence homology for genes coding for specific enzymes or structural proteins. However, for the purposes of this text, it is best to consider habitat as a key systematic character because it is the organism's habitat that helps to define the ecological relationship between the host and parasite. In considering the habitat of parasitic protozoans, it must be recognized that essentially all of the possible different sites within, or on, a host can be occupied (see also Box 3.3). These include external surfaces such as the skin and gills, as well as internal sites, e.g., the blood, lymph, brain, liver, intestine, etc. In addition, protozoans have successfully adapted to both extracellular and intracellular locations. Their habitat may also define the environmental factors that influence the parasite's life cycle. For example, the fact that a parasite resides within the blood of a vertebrate host will largely determine the method of transmission to the next host, as well as routes of entry and exit. Habitat also defines the nature of energy resources available to the parasite and, to some extent, the level of **competition** for these resources between the host and the parasite. It ultimately, therefore, defines the nature of those factors that lead to clinical illness and pathology.

3.1.4 Reproduction

The process of reproduction among the parasitic protozoans is highly varied. Some groups employ only asexual reproduction. In its simplest form, this involves binary, or transverse, fission in which one cell divides equally into two daughter cells. A variant of **binary fission**, known as **multiple fission**, is associated with some parasitic

Box 3.3 | Protozoans as hosts

In addition to discussing protozoans as parasites, is also necessary to recognize that these same organisms can act as hosts for a number of other microorganisms (Cheng, 1970; Trager, 1970; Miles, 1988). For example, *Mixotrichia paradoxa* is a flagellated protozoan that superficially resembles a hypermastigid, an organism with what seems to be thousands of hair-like flagella. This flagellated protozoan is, however, host to two different bacterial symbionts, both of which are bound to the protozoan's surface. One is a spirochaete and the second a bacillus. Both bacteria are arranged on the surface of *M. paradoxa* in a highly organized pattern, actually fitting into distinct and separate surface receptors. Even more amazing is that the spirochaetes move in a coordinated manner and are clearly involved in propelling the flagellate through its environment. *Crithidia oncopelti*, a flagellated parasite of insects, is host to a bacillus that supplies it with a variety of nutrients. Most *Crithidia* species require a number of essential amino acids, vitamins, and other nutrients. However, *C. oncopelti* requires only one amino acid and several vitamins. If the *Crithidia* is freed of its bacterial symbiont by antibiotics, it then has the nutritional requirements of other *Crithidia* species. An equally intimate relationship is illustrated by the ciliated protozoan *Paramecium bursaria* and its algal symbiont. Within the cytoplasm of this ciliate, there are numerous symbiotic *Chlorella* sp., a green alga. The ciliate both protects and transports the *Chlorella* cells, and the photosynthetic alga supplements the host's diet. What is clear from these examples is that not only have protozoans evolved parasitic modes of life, they also serve as habitats for other microorganisms.

protozoans and occurs when a second division takes place before the first division is completed. This leads to a common cytoplasm, more than one set of organelles, and long chains of connected cells; there may also be a ring, or rosette, of organisms, all joined through a common cytoplasmic bridge.

There are three types of asexual reproduction in the Apicomplexa, i.e., **merogony** (also called **schizogony**), **sporogony**, and **gametogony**. Following nuclear division in merogony (or schizogony) the individual nuclei move to the cell's periphery. When nuclear division is completed, the cytoplasmic membrane then surrounds each nucleus and the daughter cells bud from the parent. Each schizont, or **merozoite**, then contains a nucleus with the appropriate organelles and cytoplasm. In sporogony, a **zygote** undergoes multiple asexual divisions, with the formation of **sporozoites**. Finally, gametogony leads to the formation of gametes.

Sexual reproduction also occurs in many protozoan groups. It is a standard part of the life cycle of apicomplexans, the ciliophorans, and some flagellates. The kinds of sexual reproduction, however, vary widely. In some, for example, **micro-** and **macrogametes** are involved and, in other groups, the gametes are identical. In the ciliates, true gametes are not formed; rather, following the temporary fusion of the two cells (**conjugation**), there is an exchange of micronuclei. Some parasitic flagellates are believed to undergo sexual recombination in the insect vector. However, the sexual process has never been observed. Data on recombinant gene frequency suggest that sexual reproduction in these forms is a rare phenomenon and that reproduction usually takes place by asexual division. These differences in reproductive patterns between various groups (sexual recombination versus asexual clonal division) have considerable significance for our understanding of their evolutionary biology.

3.1.5 Success

As previously noted, the parasitic protozoans are able to survive successfully and reproduce in almost all animal species. They are a remarkably successful group of organisms that have evolved complex and closely integrated relationships with their hosts. Different protozoan species vary widely in their range of hosts. For example, *Plasmodium falciparum* is believed to parasitize only humans. In contrast, the African and American trypanosomes are able to infect an exceedingly wide range of mammals. The success of the African trypanosomes can be further emphasized by noting that even after large control programs to reduce the extent of human trypanosomiasis, it persists in the same geographical ranges in Africa. This is also true for the *Plasmodium* species infecting humans. Despite all the malaria control programs of the past 75 years, the prevalence of malaria is believed to be as high today as at any time in human history. The success of the protozoan parasites not only makes them fascinating organisms to study biologically, but also of enormous economic and public health importance.

3.2 Phylum Sarcomastigophora

3.2.1 General considerations

Sarcomastigophorans are a highly diverse group of single-celled, eukaryotic organisms that move by flagella (cilia), or pseudopodia, or both. This phylum was first proposed by Honigberg and Balamuth in 1963, and is based on the assumptions that all amoeboid organisms evolved from flagellated protozoa by the loss of the flagella. The close association between the flagellates and the amoebae can be seen in present-day organisms such as *Naegleria* whose species can be either rapidly swimming flagellated forms or amoeboid stages which move by pseudopodia.

There are three subphyla within the Sarcomastigophora (Fig. 3.3). Mastigophoran trophozoites possess one or more flagella. Asexual reproduction by binary fission is the general mode of reproduction, although sexual reproduction has been demonstrated to occur in some groups (Molyneux & Ashford, 1983; Seed & Hall, 1992). Sarcodinians have pseudopodia, and

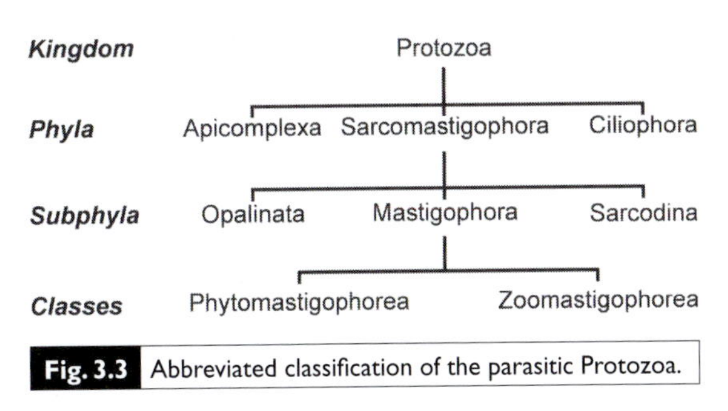

Fig. 3.3 Abbreviated classification of the parasitic Protozoa.

exhibit some form of amoeboid movement. Flagella, if present, are restricted to developmental, or sexual, stages. Reproduction is generally asexual by fission and sexual reproduction is less common. Most species in this subphylum are free-living, not parasitic. The Opalinata are characterized by the presence of numerous cilia, arranged in oblique rows, covering the entire surface of the cell. Organisms within this subphylum are distinguished from ciliates in the phylum Ciliophora by the absence of a cytostome and by their mode of nuclear division. The Opalinata are included within the Sarcomastigophora because of the structural similarity between the flagellum and cilia, as well as the presence of flagellated gametes within their life cycles. All species in this subphylum are parasitic.

Members of the Sarcomastigophora are distinguished not only in their locomotory organelles, but also in their mechanism for obtaining food. The Sarcodina phagocytize food particles, whereas the mastigophorans and the Opalinata obtain nutrients by pinocytosis and by the active transport of small-molecular-weight compounds across their cell membranes. Cytostomes or 'mouthparts' have been reported in some parasitic flagellates. In addition to their diversity in locomotion and methods for obtaining essential nutrients, the Sarcomastigophora also occupy many highly diverse habitats and can differ extensively in their mode of transmission. Because of the great diversity among Sarcomastigophorans, many investigators believe that the group has a **polyphyletic** origin. The evolutionary relationships between members of the three subphyla, however, will only become clear by additional comparative studies using the modern tools of molecular biology and biochemistry.

3.2.2 Subphylum Mastigophora

3.2.2.1 General considerations

Organisms within this group are characterized by having one or more flagella present in the **trophozoite** stage of their life cycle. There are two classes within the subphylum. The Phytomastigophorea are mostly free-living, contain chloroplasts, and will not be discussed further. The Zoomastigophorea possess one to many flagella, rarely reproduce sexually, do not contain chloroplasts, and are mostly parasitic. The Zoomastigophorea infect almost all of the animal phyla, occupying a great variety of different habitats within their many hosts. They are found in the intestine (*Giardia lamblia*), the reproductive tract (*Trichomonas vaginalis*), deep body tissues (*Leishmania donovani*), as well as both intracellularly (*Leishmania* spp.) and extracellularly (*Trypanosoma brucei*) in the blood–vascular system. Transmission can be direct, i.e., host-to-host, as occurs in the sexual transmission of *Trypanosoma equiperdum* or *Trichomonas vaginalis*. It can be indirect by the accidental ingestion of infective stages from contaminated feces (*Giardia, Crithidia*), by scratching infective stages into the lesion produced by the bite of an insect vector (*Trypanosoma cruzi*), or by the injection of the parasite during a blood meal of an insect vector (*T. brucei*). Reproduction occurs primarily by asexual binary fission. However, sexual reproduction has been described in a number of species, e.g., *T. brucei*. There is also considerable diversity in the cell structure and the organelles observed in the different orders. It is apparent that the Zoomastigophorea is a highly diverse group of organisms with only one common characteristic, i.e., they all have a flagellum at some stage in their life cycle.

3.2.2.2 Form and function

Members of the Zoomastigophorea are classified on the basis of the number of their flagella. There are five orders. Members of the Kinetoplastida possess one or two flagella, typically with a **paraxial rod**. They have, at some stage in their life cycle, a single mitochondrion which contains up to 20% of the cell's DNA. Representative genera in this order include *Trypanosoma* and *Leishmania*. The Retortamonadida have cells with two to four flagella/nucleus. One of the flagella is directed posteriorly. Species of *Chilomastix* and *Retortamonas* are included in this order. The third order, Diplomonadida, includes those protozoans with one to four flagella/nucleus. Included here are *Giardia* and *Hexamita*. Members of the Trichomonadida have cells with four to six flagella/nucleus, sometimes with an **undulating membrane**. Examples of genera in this order include *Trichomonas* and *Histomonas*. The Hypermastigida possess numerous flagella and are all symbiotic in insects such as termites.

The prominent external cell structures present on members of the Zoomastigophorea, therefore, include flagella and an **axostyle**, or an undulating membrane, or both. As noted, the number and arrangement of the flagella are key taxonomic features. Several different arrangements of these cell structures are shown in Fig. 3.4. Organisms such as *Giardia* have an adhesive disc that allows them to adhere to the epithelial lining of the gut. In some members of the Kinetoplastida, the flagellum, in addition to its use in locomotion, is also an organelle of attachment to the host substratum. Finally, in some species of Hypermastigida, there are specific surface receptors for the attachment of two bacterial symbionts; one of these is a bacillus, and the other is a **spirochaete**. The spirochaete appears to assist the protozoan in locomotion (Box 3.3).

The Zoomastigophorea have typical eukaryotic organelles such as nucleus, mitochondrion, and one or more flagella; there are also unique organelles associated with several of the orders. For example, in the Kinetoplastida, the single mitochondrion contains up to 20% of the cell's DNA that is localized in one area of the mitochondrion called the kinetoplast (Fig. 3.5). This DNA is arranged into densely packed strands of coiled and twisted DNA and codes for a number of the mitochondrial proteins. The kinetoplast is closely associated with the flagellar pocket and the **basal body** of the flagellum. The Kinetoplastida have a second unique organelle called a glycosome. This membrane-bound organelle contains a majority of the enzymes involved in glycolysis (Blum, 1991; Opperdoes, 1991; Clayton & Michels, 1996;

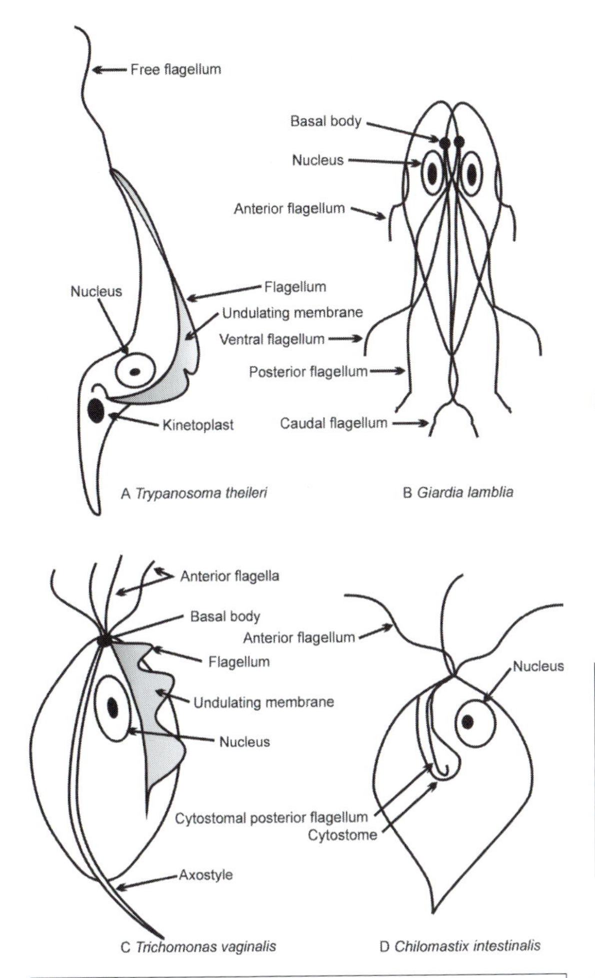

Fig. 3.4 Diagrams of the flagellar arrangements of four representative species of the four orders contained within the class Zoomastigophorea. (A) *Trypanosoma theileri*, a member of the order Kinetoplastida with one or two flagella; (B) *Giardia lamblia*, a member of the order Diplomonadida having one to four flagella; (C) *Trichomonas vaginalis*, a member of the order Trichomonadida having four to six flagella; and (D) *Chilomastix intestinalis*, a member of the order Retortamonadida having two to four flagella with one turned posteriorly.

Michels *et al.*, 1997). Finally, the Kinetoplastida possess a ring of microtubules located just under the cell membrane. These microtubules extend the length of the cell and are connected by a series of short fibrils, creating a ladder-type appearance (Fig. 3.5). This microtubule–microfibril complex is linked to the cell membrane and remains together as a discrete unit following cell lysis. It

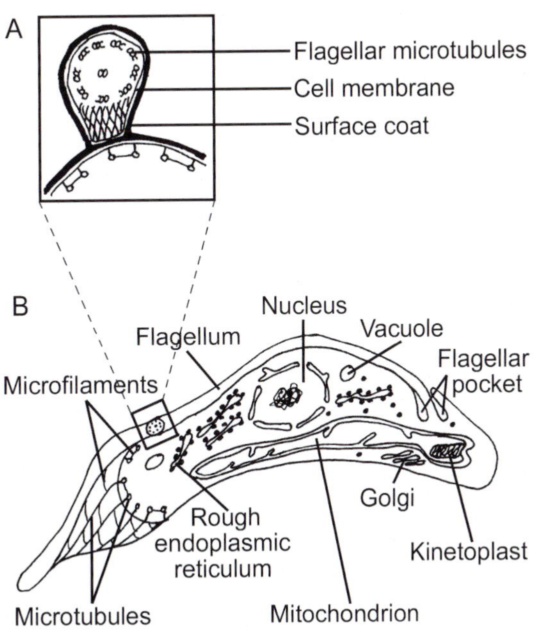

Fig. 3.5 Diagrammatic sketch of the ultrastructure of the African trypanosomes. A longitudinal section of an intermediate blood stage trypanosome is shown. The insert is a cross-sectional blow-up of the box on the main figure. It shows the relationship between the flagellum, the cell membrane, and the associated microtubule–microfilament complex.

presumably gives structural support to the trypanosome's cell membrane

The trichomonads, such as *Trichomonas vaginalis*, do not have a mitochondrion. They obtain all their energy by substrate-level phosphorylation and maintain their oxidation–reduction balance by reducing hydrogen in a special membrane-bound organelle called a hydrogenosome. In this organelle, protons are reduced to H_2. The hydrogenosome is thought to be unique to this group and appears adapted to the organism's life in an anaerobic environment (Müller, 1991).

The Zoomastigophorea obtain a majority of the energy sources and nutrients required for their anabolic pathways from their host by direct transport across the cell membrane. In the Kinetoplastida, however, cytostomes involved in food uptake have been observed in some species. We have previously described how pinocytosis of large-molecular-weight proteins occurs in the flagellar pocket of these organisms.

From this very brief description of the Zoomastigophorea, it must be obvious that just as the external features of the individual species differ extensively, so do the organelles found in the cytoplasm and the various physiological or biochemical pathways.

3.2.2.3 Life cycle

It should not surprise the student to learn that the life cycles of members of the Zoomastigophorea are also quite varied. Organisms such as *Giardia* spp. are transferred from host to host by a resistant stage (cyst) that is passed in feces (Box 3.4). New hosts are thus directly infected by the oral ingestion of the cyst in contaminated water or

Box 3.4 | Giardiasis

Fifty years ago, Saturday afternoons might find a gaggle of youngsters at the 'picture show' watching Tom Mix and his wonder horse chasing outlaws and saving 'damsels in distress'. In almost all of these movies, the Sons of the Pioneers would be featured singing a western ballad. One of the most popular of these tunes was a song called 'Cool, Clear Water'. We can all conjure up visions of rippling cold water, a rustling aspen grove, and fluffy clouds adorning an otherwise clear blue sky. The only problem in this scene is with the rippling water because you must not drink it, even though it may look irresistible. If you succumb to the urge, and do drink it, you may come down with an uncomfortable case of the 'trots' brought on by an unseen, and unfriendly, flagellated protozoan called *Giardia lamblia*.

This parasite has a long history, having been first described by Antony von Leeuwenhoek from his own feces in 1681! There is still some dispute, however, of the parasite's proper scientific name since it is referred to as *G. duodenalis* by some of the leading investigators in the field (see Thompson *et al.*, 1993).

The pear-shaped parasite is striking in appearance, almost like a cartoon character (Fig. 3.4B). It has two prominent nuclei, a pair of axonemes (axostyles), four pairs of flagella, and an adhesive disk set in a convex ventral surface. The life cycle is direct, with a cyst as the infective stage. The parasite is believed to have a wide range of reservoir hosts, including dogs, cats, sheep, and beaver. The cysts must be ingested in contaminated food or water, hence the injunction regarding cool, clear water, especially in North America, where the parasite is known to cause what is called hiker's, or picnicker's, disease or 'beaver fever'! *Giardia lamblia* has been found in 81% of raw surface water supplies entering 66 treatment plants in 14 states in the USA and in one province in Canada. In addition, *G. lamblia* was detected in 17% of the filtered water samples from these same plants. From 1984 through 1990, it has been estimated that there were 25 outbreaks of waterborne disease involving 3486 individuals due to *G. lamblia*; from 1991 through 1994, nine of 36 waterborne diarrhreal outbreaks, caused by a known etiological agent, were due to *G. lamblia* (Marshall *et al.*, 1997; Steiner *et al.*, 1997). According to the World Health Organization, giardiasis is now recognized as one of the 10 most common parasitic diseases of humans, infecting some 200 million people in both developed and developing countries around the world.

Although giardiasis is not a severe pathogen in immunocompetent individuals, waterborne disease outbreaks can have significant direct and indirect economic consequences to the community. There are the direct medical and pharmaceutical expenses, but also the indirect expenses due to the necessity for homes and businesses to obtain *Giardia*-free water, loss of work and leisure time while ill, loss of school time due to absenteeism, loss of restaurant and bar business in the area affected, etc. An excellent analysis of an outbreak in Pennsylvania (USA) has been described in the text by Harrington *et al.* (1991).

Following ingestion of cysts, excystation occurs in the small intestine, after being prompted by the acidic pH of the stomach. Following attachment via the ventral disk, sometimes in large numbers, the trophozoites multiply asexually by binary fission. Cyst production will ensue as early as 4 days following initial infection. A single stool may contain as many as 300 million cysts that can survive for up to 3 months in water!

Giardia lamblia is usually not a life-threatening pathogen. However, it can become severe in young children and immunocompromised individuals. A frequent clinical feature of giardiasis is a foul-smelling stool (steatorrhea), which is affected by faulty fat absorption in the small intestine. Other symptoms include malaise, abdominal cramps, weight loss, and flatulence. The disease usually runs its course in a few weeks, but the diarrhea may become chronic and last for months, with frequent relapses. Several drugs are efficacious in treating the disease, although, with re-infection and retreatment in children, there is a risk of toxicity problems.

food. Other species are passed directly from one host to another during sexual intercourse, e.g., *Trichomonas vaginalis* and *Trypanosoma equiperdum*. In contrast, many of the Kinetoplastida have indirect life cycles. Generally, vertebrates are considered the definitive host and the parasite is transmitted from one host to another by a blood-sucking vector. In some cases, the vector acts purely in a mechanical manner, i.e., the insect feeds on an infected animal, its mouthparts become contaminated by the parasite, and it transfers the parasite mechanically to a new host at the next blood meal (*T. evansi*). Members of the *brucei* group of African trypanosomes, however, undergo a complex morphological and biochemical transformation in their vector. These trypanosomes are acquired in a blood meal; they migrate into the hindgut and, after development, move into the salivary glands where they eventually mature into infective stages. The insect, at its next blood meal, injects the infective **metacyclic** forms into the new host. This cycle in the vector takes 3 to 4 weeks to complete, depending upon external humidity and temperature. The various morphological stages observed in the different life cycles of members of the Kinetoplastida are shown in Fig. 3.6.

In the remainder of this section, we will contrast the life cycles of two kinetoplastids, namely *Trypanosoma cruzi*, the American trypanosome (see also Box 3.5), and the African trypanosome, *T. brucei* (Molyneux & Ashford, 1983; Seed & Hall, 1992). These two species are classified as either **stercorarian** or **salivarian** depending on the location within which the parasite develops to an infective metacyclic stage (Fig. 3.7). We will begin the life cycle of *T. cruzi* as an intracellular amastigote parasite in the muscle (or other tissue) of its mammalian host (Fig. 3.8). In the vertebrate host, the amastigotes reproduce by binary fission and their numbers increase until the cell lyses and releases the daughter amastigotes that can then either infect other cells, or transform into trypomastigotes. The trypomastigote stage does not divide and is found in extracellular sites, predominantly the blood. It is this stage that is acquired by the blood-sucking reduviid vectors. In the vector, the trypomastigotes travel to the midgut and develop first into promastigotes and then into epimastigotes. The epimastigotes divide by binary fission. The larger epimastigote forms migrate to the rectum and adhere to the epithelial lining of the rectal gland (posterior station). Here, the epimastigotes transform into infective metacyclic

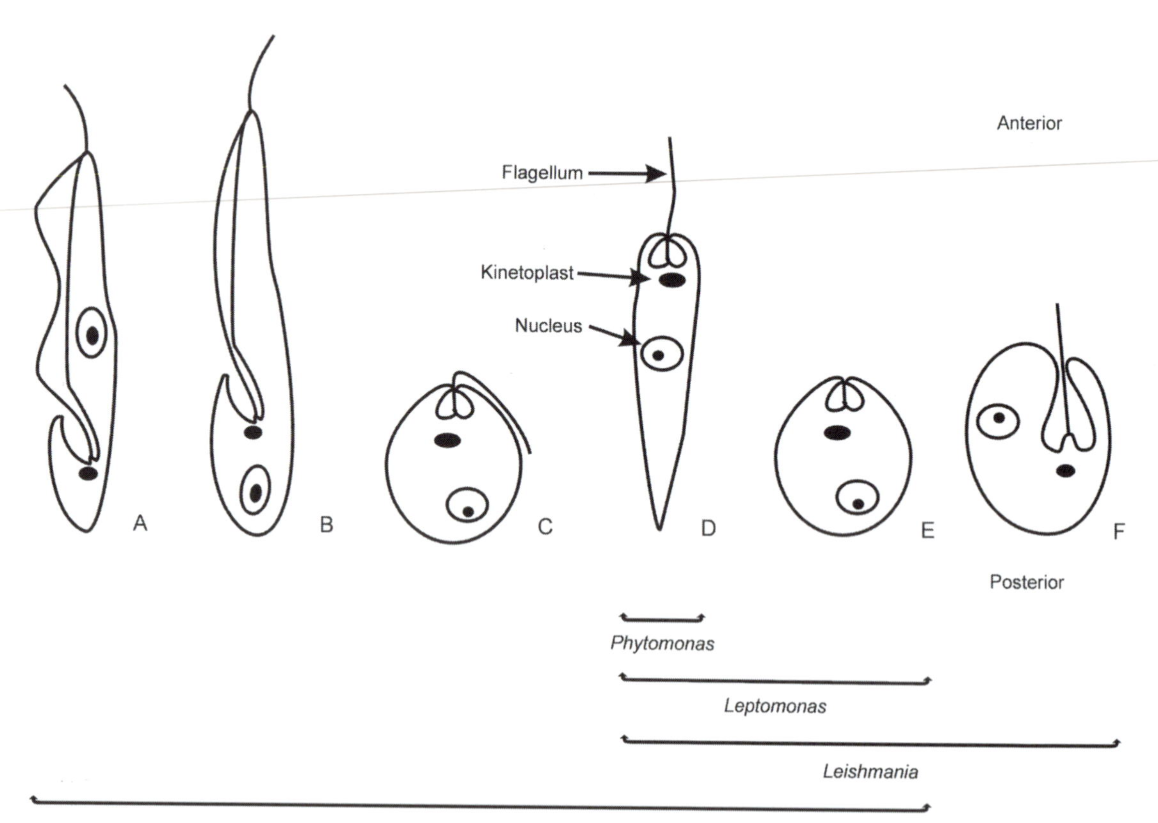

Fig. 3.6 Morphological stages present in some genera of the Trypanosomatidae. (A) Trypomastigote stage, (B) epimastigote, (C) sphaeromastigote, (D) promastigote, (E) amastigote, and (F) paramastigote.

stages that are usually found free in the lumen. The metacyclic forms do not divide. The development of the ingested trypanosomes into infective metacyclics takes 7 to 14 days depending on the vector and the external temperature. The metacyclics are released in the vector's feces, which are deposited on the skin near the bite. There is no evidence that metacyclic forms can enter through intact skin; rather, when the bite itches, and the host scratches, the metacyclic trypanosomes in the contaminated feces are forced into the lesion left by the insect's proboscis. Infection can also occur by the ingestion of infective feces or an infected vector since metacyclics can penetrate mucosal surfaces. Members of the Stercoraria, therefore, develop in the posterior station of the vector, the hindgut and rectum, and transmission is either by contamination of the bite with vector feces containing infective metacyclics or by the ingestion of the vector, or its feces, during grooming.

In contrast, in salivarian trypanosomes, the development of infective metacyclics occurs in the salivary glands or mouthparts (anterior station) of their vector. Trypanosomes of this type are transmitted during feeding by the insect vector. In the African trypanosomes, the **short stumpy**, non-dividing trypomastigote stage is acquired in the tsetse's (*Glossina* spp.) blood meal. It travels to the midgut and then to the hindgut as a procyclic form that maintains the trypomastigote morphology. After a period of development, the procyclics migrate anteriorly to a region of the midgut where they divide and multiply. Between 14 and 21 days later, the parasites enter the proventricular area of the gut. From there, they

Box 3.5 | *Trypanosoma cruzi*

Trypanosoma cruzi, the causative agent of American trypanosomiasis, is a zoonotic disease that can infect over 150 species of mammals, including humans (Molyneux & Ashford, 1983). It is estimated that approximately 13 to 14 million people in Central and South America are infected with this organism. Approximately 10% of those infected will develop a serious disease involving cardiovascular pathology, or nerve damage leading to malfunction of the colon, or both. In some Latin American countries, *T. cruzi* is the main cause of heart failure. Triatomid bugs, commonly referred to as the kissing bug because it usually bites on the face, and usually at night, transmit the disease. The bite is said to be painless, so the insect obtains its blood meal while the host peacefully sleeps. The vector takes up the trypanosome during the blood meal and, within a relatively short time, the parasite moves to the hindgut where it multiplies and matures into an infective metacyclic form. While the vector feeds, it defecates, depositing fecal material containing infective forms onto the skin near the bite. The insect bite itches and the individual literally scratches the infective metacyclics into the wound. Once within the host, the metacyclic forms invade a variety of cell types, including macrophages and muscle cells. Following entry into the host cells, the parasites transform into amastigotes and escape from the phagosome, but remain within the cytoplasm of the host cell. The infected cells eventually burst, releasing the amastigotes, which invade surrounding cells, and the cycle is repeated. Subsequently, however, trypomastigote forms are released from the infected cells into the blood. These trypomastigotes do not divide, but are infective to the vector; when they reach the vector's hindgut the process of division and eventual maturation to the infective metacyclic stage is repeated.

Pathology in the mammalian host is not fully understood, but is believed to induce both direct cell destruction as well as immunopathological responses. The infection is difficult to treat due to the lack of effective chemotherapeutic agents. A vaccine is not available. Therefore, current control measures involve methods to prevent humans from becoming infected. Primary emphasis is on insect control through the use of insecticides, and better housing (screens on windows, metal roofs, etc) in order to eliminate vector habitats.

This host–parasite association is an amazingly successful one. *Trypanosoma cruzi* is able to infect a large variety of hosts. In addition, various triatomid species, adapted to both domestic and field habitats, can transmit the disease. Moreover, the infection causes serious disease in about 10% of the human population and is chronic, lasting for a number of years, thereby insuring a long-term carrier state that increases the opportunity for transmission.

migrate anteriorly into the salivary glands where they transform into epimastigotes and again multiply. The epimastigotes attach themselves by their flagella to microvilli that line the epithelium of the salivary gland. In this site, the epimastigotes transform to non-dividing, infective, metacyclic forms that are released from their attachment to the microvilli. They can then be found in the lumen of the salivary gland. The entire cycle from the time of the first blood meal containing short stumpy trypanosomes to infective metacyclics takes approximately 3 to 4 weeks. The metacyclics are then injected in the saliva into a new host during a blood meal. Once in the

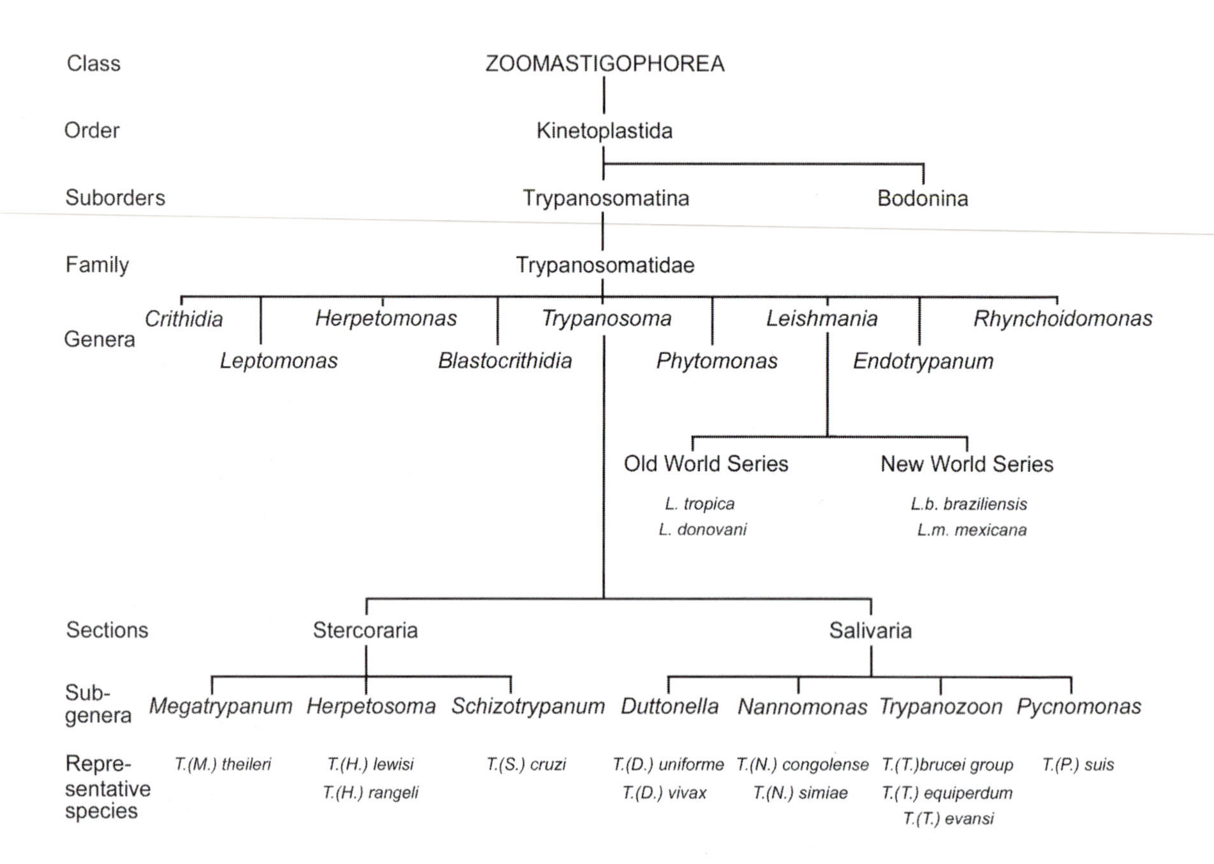

Fig. 3.7 Abbreviated classification of the family Trypanosomatidae.

mammalian host, the metacyclics transform to a **long slender stage** (trypomastigote), which divides rapidly. In the African trypanosomes, the life cycle is therefore totally extracellular. As the parasitemia reaches its peak, a percentage of the long slender forms transform into a short stumpy, non-dividing, trypomastigote stage that is infective for the vector. The survival time of the infective forms has not been absolutely established, but it is estimated to be at a minimum of 1 day.

In the life cycle of *Trypanosoma rangeli*, the parasites are able to develop to infective metacyclic forms in both salivary glands and in rectal sites of their South American vectors. In addition, fish trypanosomes, which are transmitted by leeches, develop in the anterior portion of the leech gut. Similarly, phytomonads also develop in the salivary glands of their insect vectors.

In all members of the Kinetoplastida, reproduction is primarily, if not totally, via asexual binary fission. 'Sex' has only been reported to occur in the African trypanosomes, and is assumed to occur in the vector. Even then, however, it would appear to be a rare event and has never been observed directly.

3.2.2.4 Diversity

In addition to the examples discussed above, there are also several trypanosome species that infect amphibians. In *Rana clamitans* from Louisiana, *Trypanosoma rotatorium* has two different cycles in its blood parasitemia. During cold winter months, when frogs are out of the water and inactive, it is difficult to find any trypanosomes in the peripheral circulation of the frog host (Seed *et al.*, 1968*a*). The frogs do not lose their infection but, rather, the trypanosomes are sequestered in the deep, vascular beds of the internal organs. When the weather warms, trypanosomes move to the peripheral circulation. During this time, the frogs spend significant time periods in the water. Since division of trypanosomes has never been observed in these frogs, this represents a true seasonal cycle

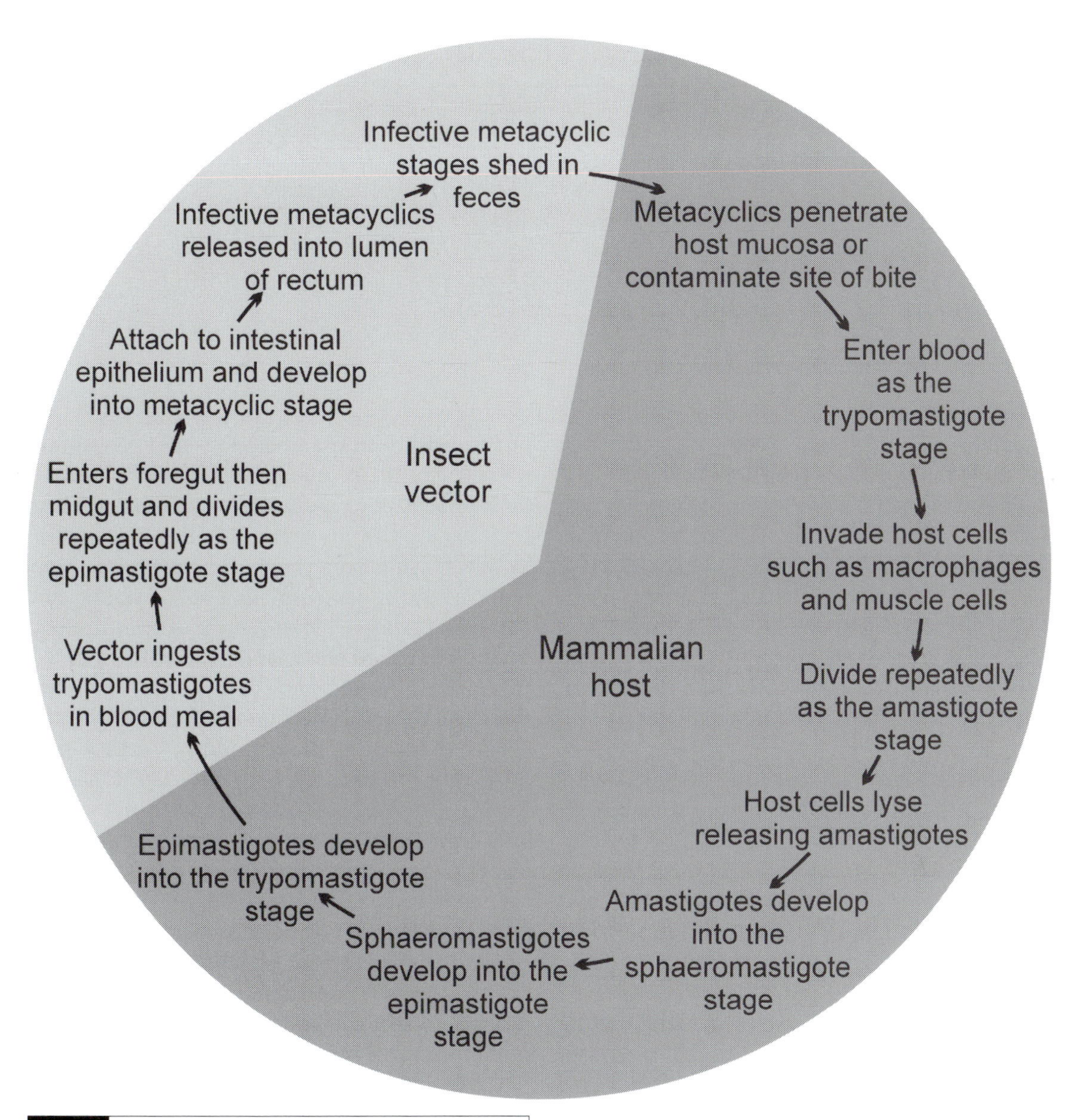

Fig. 3.8 Life cycle of *Trypanosoma cruzi*.

of migration of the trypanosomes through the frog's vascular bed. It is important to recognize that *T. rotatorium* is leech-transmitted; therefore, it is significant that peripheral parasitemia is seasonally high when the frogs are in the water and have the greatest probability for contact with the leech vector.

In addition to the seasonal cycle in parasitemia, *T. rotatorium* and a second, unidentified, trypanosome species, both have a 24-hour rhythm in their peripheral parasitemia in *R. clamitans.* *Trypanosoma rotatorium* reaches a peak in parasitemia during the daylight hours when the frogs are often in the water along the shoreline, maximizing potential contact between frogs with high parasitemias and the leech vector (Seed *et al.*, 1968*b*). In contrast, the second trypanosome species has its peak parasitemia during the night (Seed *et al.*, 1968*b*). At night, frogs are usually out of the water at the edge of the shoreline. It is at this time that many blood-sucking insects are foraging most actively and it is believed that this second species of trypanosome is transmitted by an insect vector.

If one considers these frog–parasite examples

Table 3.1 Examples of the diversity of host–parasite relationships among members of the Zoomastigophorea

Parasite	Host	Habitat	Host–parasite relationship
Trypanosoma theileri	cattle	blood, tissue	generally non-pathogenic, chronic infection
Trichomonas hominis	human	intestine	commensal
Trichomonas vaginalis	human	vagina	pathogen
Trichomonas tenax	human	mouth	commensal
Giardia lamblia	human	intestine	pathogen
Trichonympha corbula	termite, woodroach	intestine	mutualist

as integrated units, one recognizes that the success of the parasite is dependent upon the coupling of activity cycles of the frog community, the vector community, and the trypanosome population. These cycles demonstrate the very intimate interactions between the host's, the vector's, and the parasite's physiology. They are, therefore, excellent examples of a long evolutionary history between zoomastigophoreans and their hosts. (See Chapter 13 for additional examples of synchronization in life cycles.)

The flagellated animal parasites have managed to colonize successfully almost every available site within their animal hosts (Tables 3.1, 3.2; see also Box 3.6 and Fig. 3.9). Thus, there are flagellated parasites that occupy the gut, vagina, seminal fluid, blood, lymph, cerebrospinal fluid, muscle cells, macrophages, and even lymphocytes. Table 3.2 illustrates some of the habitats occupied by the Kinetoplastida (see also Hoare, 1972). The Zoomastigophorea have also successfully evolved an array of different strategies to avoid the defense mechanisms of their host. These range from sequestration in intracellular sites, to the phenomenon of antigenic variation.

3.2.3 Subphylum Sarcodina

3.2.3.1 General considerations

The Sarcodina are thought to have evolved from flagellated ancestors by the loss of their flagellum; they are probably polyphyletic in origin. Amoeboid movement now characterizes the group. Flagella, when present, are restricted to developmental or sexual stages. For example, *Naegleria* spp. have a free-living, flagellated stage, but transform into amoebae in their vertebrate hosts (Fig. 3.10). Reproduction is usually asexual, by simple fission; sexual reproduction, when it occurs, usually involves flagellated gametes. However, amoeboid gametes have also been observed. Most sarcodinians are free-living, not parasitic. Both free-living and parasitic forms feed by engulfing food particles (phagocytosis).

A majority of the parasitic amoebae lives in the intestine of their hosts. They may, however, invade deeper tissues, as can occur during infection of humans by *Entamoeba histolytica*. The amoebae have been found in abscesses primarily in the liver and brain, but have also occasionally been observed at other tissue sites. The amoebae located in these deep tissue sites are not involved in the amoeba's life cycle. They are at a dead-end since transmission always occurs by excretion of environmentally resistant cysts in the stool. Acquisition of amoebae is generally by a fecal–oral route in which cysts are ingested by a new host.

Table 3.3 lists some of the more common amoeboid parasites of humans and other hosts. It should be noted that *E. dispar*, a non-pathogenic parasite of humans, has only recently been separated taxonomically from the pathogenic *E. histolytica*. This separation is based on the association of distinct biochemical and molecular markers with pathogenic and non-pathogenic isolates (Martínez-Paloma & Espinosa-Cantellano, 1998). Also, the flagellated amoebae of *Histomonas* and *Dientamoeba* are now considered to be trichomonads, not members of the Sarcodina.

All symbiotic amoebae are members of the superclass Rhizopoda, class Lobosea, and the orders Amoebida or Schizopyrenida (Fig. 3.11).

Table 3.2 Examples of the diversity in hosts, habitat, and methods for transmission in members of the Trypanosomatidae

Parasite	Hosts	Habitat	Method of transmission
Herpetomonas muscarum	house flies	intestine, hemocoel	cyst-like stage; fecal/oral route
Crithidia fasciculata	*Culex* mosquitoes	intestine	cyst-like stage; fecal/oral route
Phytomonas	*Euphorbia* (milkweed)	plant latex	various species of insects
Leishmania tropica	humans, mammals	macrophages, intracellular	blood-feeding flies, *Phlebotomus* spp., sand fly
Trypanosoma brucei (Salivaria)	humans, many mammals	extracellular, all intra- and extravascular tissue fluids	blood-feeding flies, *Glossina* spp., tsetse
Trypanosoma cruzi (Stercoraria)	humans, many mammals	intracellular muscle, nervous tissue, macrophage, extracellular, blood	blood-feeding insects, reduviid bugs, kissing bug

Box 3.6 | Leishmaniasis

Leishmaniasis is a global disease of considerable proportion; it is apparently on the increase. Transmitted to humans by the bite of female phlebotomine sand flies, the causative agents are flagellated protozoans that become intracellular in a wide range of vertebrates, which, in turn, can act as reservoir hosts for human infections. There are at least six different species of *Leishmania*, probably seven different subspecies, and at least 30 different serotypes that have been identified, with most causing a unique form of clinical disease (Molyneux & Ashford, 1983).

The genus *Leishmania* includes both New and Old World species of human parasites. The Old World species include *Leishmania aethiopica, L. major*, and *L. tropica. Leishmania tropica* normally produces skin lesions (oriental sore) that can be self-limiting and capable of spontaneous cure (Fig. 3.9). Another Old World species is *L. donovani*, which causes visceral leishmaniasis (or kala azar), a potentially fatal disease. However, it should be noted that the etiological agent involved in several recent cases of visceral leishmaniasis also has been identified as *L. tropica*.

New World species includes *Leishmania braziliensis*. The subspecies in this group cause oriental sore, as well as a non-healing, mucocutaneous form of disease (espundia). A second New World species is *L. mexicana*. Again, the subspecies in this group produce diseases ranging from oriental sore to diffuse cutaneous leishmaniasis. *Leishmania donovani* has also been identified in the New World, where it produces an infantile form of visceral leishmaniasis.

According to recent accounts by the World Health Organization, the leishmaniases are now endemic to 88 countries on five continents, with 12 million cases, and an estimated 2 million new infections annually. In the latter group, fully 500000 are visceral leishmaniasis, with 90% of the victims living in Bangladesh, Brazil, India, Nepal, and Sudan. Most (90%) of the mucocutaneous infections occur in Bolivia, Brazil, and Peru. Approximately 1–1½ million new cases of cutaneous leishmaniasis occur annually, with 90% of these in Afghanistan, Brazil, Iran, Peru, Saudi Arabia, and Syria. Of increasing concern is the association between *L. donovani* and the HIV virus, especially in southern Europe, where 673 of the first 700 reported co-infections occurred. Of this number, most victims were males, young, and intravenous drug users.

A remarkable aspect of leishmaniasis is the propensity of the infective promastigote to invade cells of the macrophage lineage in their vertebrate hosts, for these cells are directly involved in the phagocytosis of microbes. Once inside the cell, the promastigotes transform into amastigotes that reside permanently within a phagolysosomal vacuole. The latter is formed by the fusion of the phagosomal vacuole containing the parasite with a lysosomal vacuole. These vacuoles are highly acidic (pH 4–5) and contain all the hydrolases (proteases, nucleases, etc.) released from the lysosomal vacuole. Clearly, the environment in which the amastigote resides is hostile. However, in the non-immune host, the parasite is able not only to survive, but to reproduce as well (see Alexander & Russell, 1992).

The clinical manifestations of the disease are associated with the particular species or subspecies of the parasite. In kala azar, the parasite is viscerotropic and there is an undulant fever, hepatosplenomegaly, and even dysentery. In contrast, *L. braziliensis braziliensis* is dermotropic and affects the mucocutaneous membranes of the mouth and nostrils, causing horribly disfiguring lesions. *Leishmania tropica* is also dermotropic, producing a self-limiting cutaneous infection. Although not normally considered serious, the lesions produced by *L. tropica* can result in scarring. Since multiple lesions can be found (up to 200) on exposed parts of the body, scarring can be significant, and can be of cosmetic concern, particularly for children. *Leishmania mexicana mexicana* also produces cutaneous lesions (oriental sore) predominantly on the ears, which are referred to as 'chiclero ulcers'. The name of the disease comes from the chicleros who harvest gum in the chicle tree plantations in Central America.

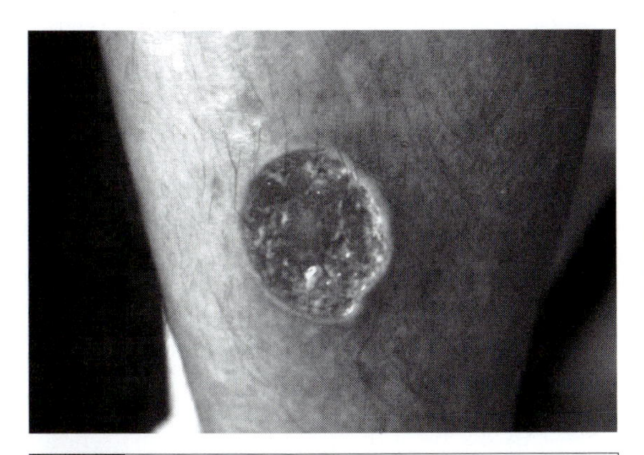

Fig. 3.9 Photograph showing a cutaneous lesion (oriental sore) on an individual from French Guyana. (Photograph courtesy of Jean-Phillipe Chippaux, Institut de Recherche pour le Développement, Paris, France.)

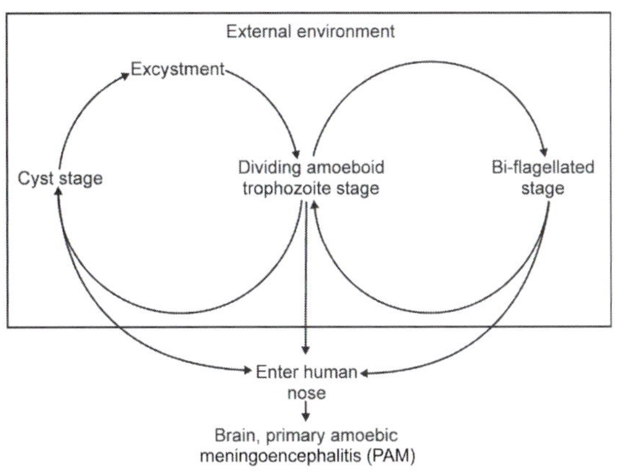

Fig. 3.10 Life cycle of *Naegleria fowleri*. This protozoan can exist in three different morphological stages in the external environment. The cyst and flagellated stages are infectious but do not divide. They are not found in the human host. The transition from the trophozoite to the flagellated or cyst stage is dependent upon external environmental conditions.

Members of the Lobosea reproduce solely by fission (usually binary), and move by lobose pseudopodia. All are phagotrophic and most produce resistant cyst stages for transmission. Species of Amoebida move only by pseudopodia and never have a flagellated stage. They are naked amoebae, never possessing any form of an external covering. Genera within this order include *Entamoeba*, *Endolimax*, and *Iodamoeba*. Species in these genera produce cysts that exit from their hosts in feces. They are directly transmitted by ingestion of fecal material containing cysts. An exception is *E. gingivalis*, an amoeba that inhabits the mouths of humans. This parasite does not produce cysts and is believed to be transmitted directly as a trophozoite. Species of Schizopyrenida are similar to the Amoebida, except that the former possess flagellated stages at some stage of their life cycles. The Schizopyrenida includes *Naegleria fowleri*, an important parasite of humans. Members of this genus form cysts, but also have flagellated stages in their life cycles. They contain a contractile vacuole in their free-living stages. Normally, free-living trophozoites of these species may enter the nose or mouth of humans while the potential hosts swim in contaminated water. Amoeboid stages migrate up the olfactory tract into the brain rapidly, causing an almost always fatal primary amoebic **meningoencephalitis**, or PAM (Martínez & Visvesvara, 1997).

3.2.3.2 Form and function

Parasitic amoebae within the Amoebida are naked, moving by lobose pseudopodia. In the

Table 3.3 Examples of host diversity and host–parasite relationships among species of *Entamoeba*

Species	Host(s)	Host–parasite relationship
E. blattae	termites, cockroaches	commensal
E. ranarum	frogs	pathogenic (?)
E. invadens	turtles, snakes	commensal, pathogenic
E. muris	rats, mice	commensal
E. bovis	cattle	commensal
E. coli	primates	commensal
E. dispar	humans	commensal
E. histolytica	humans	pathogenic

amoeboid stage, there are no observable external structures outside the cell membrane. Among the Schizopyrenida, typical eukaryotic flagella are present during part of their life cycle. In addition, as previously noted, the flagellated stage of *Naegleria* has a conspicuous contractile vacuole. At the ultrastructural level, a **glycocalyx** (surface or fuzzy coat) of varying thickness can be detected. The cell surface of *Entamoeba histolytica* also has both small (0.2 to 0.4 μm) and large (2 to 6 μm) circular openings that are involved in pinocytosis.

The cysts are smooth-walled, oval stages containing a variable number of nuclei and other internal structures. There are no obvious distinguishing external features and, except for size and internal structural details, can not be distinguished microscopically from each other at the species level.

The internal morphology of the amoebae is eukaryotic in design. Species of *Entamoeba* and *Naegleria* possess a nucleus that has an endosome, a nucleolus-like organelle that does not disappear during mitosis, and they have chromatin that is distributed around the inner surface of the nuclear membrane. The nuclear membrane remains intact during cell division. The cytoplasm of trophozoites contains food vacuoles, lysosomes, an endoplasmic reticulum, and ribosomes. Among species of *Entamoeba*, both the Golgi apparatus and mitochondria are absent. The absence of cytochromes and a mitochondrion limits the amoebae, metabolically, which means they must obtain energy by substrate-level phosphorylation. They are considered to be microaerophilic organisms in which glucose is catabolized to acetate, ethanol, and CO_2. They do not produce molecular hydrogen (Müller, 1991). The normally free-living amoebae, e.g., *Naegleria* spp., which are accidental parasites of humans, contain mitochondria. Chromatoid bodies (or bars), composed of ribonucleoprotein, can be observed by light microscopy in some trophozoites and in cysts when stained appropriately. The differences in the chromatoid bodies are used diagnostically at the species level. In the pre-cyst and young cyst stages, glycogen vacuoles can also be observed. Both glycogen vacuoles and the chromatoid bodies disappear as the cysts mature. The free-swimming stage of *Naegleria* spp. possesses two flagella.

Species of *Entamoeba* infecting humans can be differentiated diagnostically by their cyst morphology. There are differences in the number of nuclei per cyst, nuclear structure, and chromatoid bars. Size is also an important characteristic.

3.2.3.3 Life cycle

Entamoeba spp. spend most of their life cycles as trophozoites that reproduce asexually in the gut of their hosts. They are extracellular, rarely penetrating beyond the lumen of the gut. When they do occupy extra-intestinal sites, they invade through the epithelial lining of the intestine. Those amoebae that invade deeper tissue cause extensive pathology. Most species are not pathogenic but, rather, live commensally feeding on intestinal microorganisms. They reproduce by binary fission and do not reproduce sexually. Following a period of growth and reproduction, some trophozoites begin to encyst as they are carried down the intestinal tract. During this

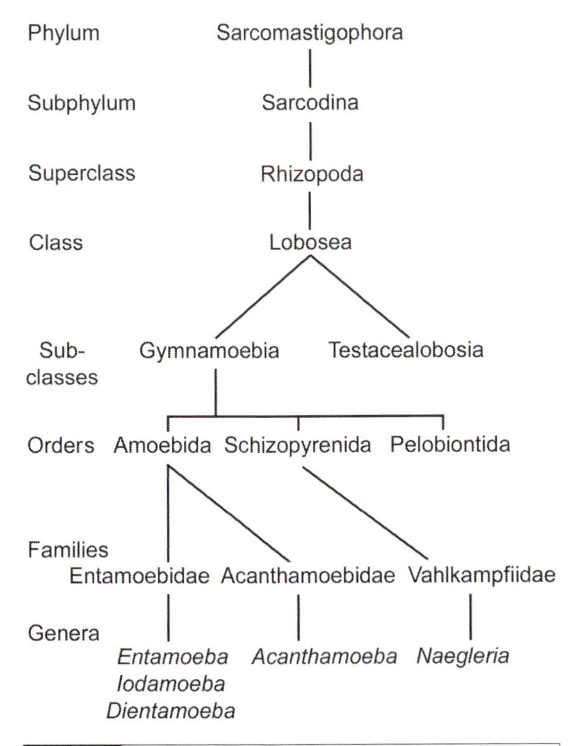

Fig. 3.11 Abbreviated taxonomic tree of some parasitic Sarcodina.

process, they lose any undigested food and condense into a sphere. This early stage of encystment is often referred to as the pre-cyst. Within the precyst stage is a large glycogen vacuole. Chromatoid bars also are formed during this period. The precyst secretes a thin, but tough, cyst wall. There is nuclear division during the cyst stages, and the glycogen vacuole and chromatoid bodies usually disappear. The mature cyst can remain viable outside the host in a moist, cool environment for up to 2 weeks, and in water for even a longer period.

When the cysts are ingested, they are capable of surviving the acid environment of the stomach. They only begin to excyst when they reach the small intestine and an alkaline environment. The multinucleate amoeba then begins to move in the cyst, the cyst wall weakens, and the amoeba emerges and quickly undergoes cytokinesis. The individual trophozoites are then passed down the intestinal tract and, upon reaching the appropriate habitat, they become established. They are site-specific, e.g., *E. histolytica* colonizes the colon of humans, primarily in the sigmoidal–rectal and appendix areas.

We have noted that *E. gingivalis* does not produce cysts, but is, instead, passed from host to host in the trophozoite stage. *Naegleria fowleri*, in contrast, will form cysts, but also has a flagellated, free-swimming stage. The flagellated stage is presumably important for **dispersal** of the parasite in contaminated water, thereby increasing the probability of contact with a potential host. The cyst stage, on the other hand, is assumed to increase the parasite's ability to survive under harsh, or extreme, environmental conditions.

3.2.3.4 Diversity

Many free-living amoebae are capable of surviving in a variety of habitats and under a wide range of environmental conditions. However, the commensal and parasitic amoebae are generally restricted to the gut of their invertebrate and vertebrate hosts. There are, of course, exceptions, such as *E. gingivalis* which is found in the mouth and the upper larynx, or *Naegleria* spp. that have the free-swimming flagellate stages in their life cycles (Table 3.4). There is even a species of amoeba that can invade opalinids, a protozoan that, in turn, is a commensal in frogs. These amoebae are sequestered within pockets in the cytoplasm of the opalinid cells.

Although generally restricted to the intestinal habitat, species of Sarcodina parasitize a very wide range of both invertebrate and vertebrate hosts. Examples of some of the different host–parasite combinations are provided in Table 3.3. The Sarcodina are thus a highly successful group of organisms that have evolved to grow and survive in a wide diversity of different habitats.

3.2.4 Subphylum Opalinata

3.2.4.1 General considerations

All members of this subphylum are commensals. They infect a variety of the lower vertebrates, being found primarily in the intestines of fishes, frogs, toads, and snakes. The opalinids have numerous longitudinal and oblique rows of cilia covering the entire body surface. Although superficially resembling ciliophorans, they do not have a cytostome, they possess micro- and macroflagellated gametes,

Table 3.4 Some examples of the diversity in habitats and transmission of species of the Sarcodina

Parasite	Host	Habitat	Stage of transmission
Entamoeba histolytica	humans	colon	cyst/feces
Entamoeba gingivalis	humans	mouth	trophozoite, no cyst
Naegleria fowleri[a]	free-living; soil, water (opportunistic pathogen)	polluted water (human brain)	cyst, trophozoite, flagellar stage
Acanthamoeba spp.	free-living; sewage (opportunistic pathogen)[b]	dust, polluted water (body)	cyst, trophozoite

Notes:
[a] For a review of these free-living opportunistic parasitic protozoa, see Martínez & Visvesvara (1997).
[b] Opportunistic pathogens in immunodeficient individuals.

and they have two to many nuclei of only a single type. The opalinids feed using a modified form of pinocytosis, and reproduction is by binary fission. Compared to other subphyla within the Sarcomastigophora, the Opalinata is small, with only four genera, each distinguished from the other on the basis of their size and shape, cilia arrangement, and the number and morphology of their nuclei.

3.2.4.2 Form and function

The distinguishing external feature of the opalinids is the characteristic morphology of the cilia that are arranged in longitudinal rows over the surface of the entire cell. The gametes are flagellated and differ in size (micro- and macrogametes). There are also two cyst stages that can occur, a gamontocyst from which the micro- and macrogametes will develop and a smaller, thick-walled **zygocyst**. In *Opalina ranarum*, the gamontocysts are released into the water by adult frogs. Tadpoles then ingest the gamontocysts. Mating occurs in the intestine of the tadpole and the zygote ultimately encysts. The so-called zygocysts are released in the tadpole's feces. Following ingestion by the adult frogs, excystation occurs in the intestine and the trophozoites divide asexually.

The cell surface structure is dominated by the cilia and associated kinetosomes, and they have typical eukaryotic type organelles. Unfortunately, there has been limited research on the ultrastructure, biochemistry, or physiology of this fascinating group of protozoans.

3.2.4.3 Life cycle

The life cycle of the opalinids includes both a sexual cycle with flagellated micro- and macrogametes, and an asexual cycle in which the cells divide by longitudinal binary fission. In species of *Opalina*, the entire life cycle is synchronized with the reproductive cycle of the host (Cheng, 1970; Wessenberg, 1978; Lee *et al.*, 1985). The life cycle of *Opalina ranarum* is shown in Fig. 3.12. During the non-breeding period, only trophozoites of *O. ranarum* are present in the intestine. During this period, the trophozoites reproduce solely by asexual division. However, when the frogs enter the water to mate, the opalinids increase their rate of fission. Shortly thereafter, small pre-cystic forms appear and then mature to encysted forms with an average of four nuclei/cyst. Meiosis is prezygotic and is believed to occur prior to cyst formation. After the frogs mate, the number of cysts (gamontocysts) increases and the cysts are passed out into the water with the feces. Thereafter, the number of the cysts in the intestine slowly decreases for several months until no cysts can be found. It has been shown experimentally that adrenalin and both sex and gonadotropic hormones can induce the encystment of *Opalina* in the adult frog.

Tadpoles ingest the cysts that are voided in the feces of the adult frog. Approximately 8 hours later, excystment occurs; small, multinucleate micro- or macrogametes emerge and then migrate to the cloaca where meiotic division occurs. The haploid micro- and macrogametes ultimately fuse to form a diploid zygote. The zygote then encysts;

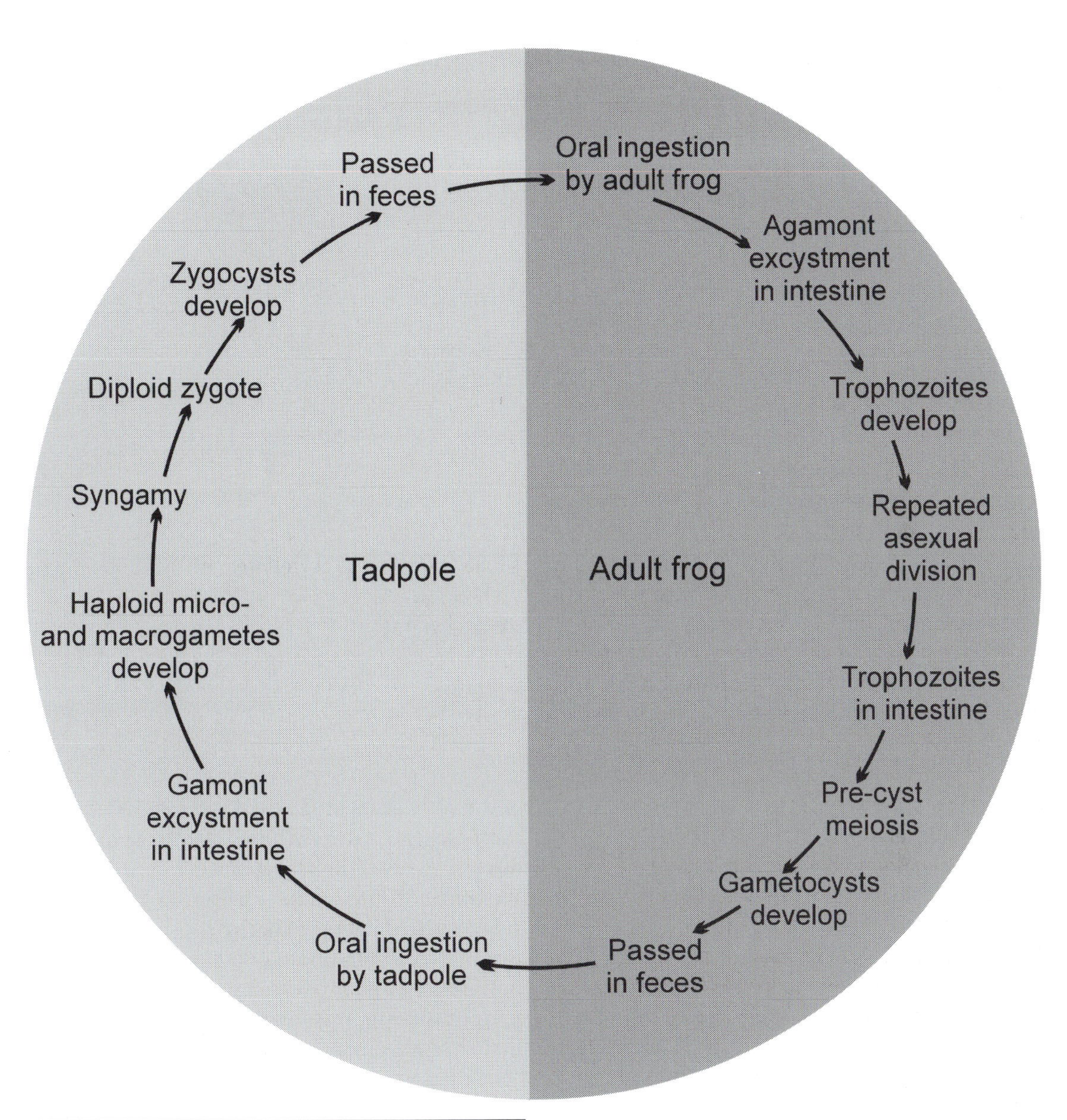

Fig. 3.12 Life cycle of *Opalina ranarum*. Meiosis occurs during the formation of the gamont in the adult frog. Gamete formation, syngamy, and zygocyst formation occur in the tadpole.

the zygocyst is passed in the feces and is later ingested by adult frogs. In the intestine of the adult frog, the zygote excysts, a uninucleate form escapes and divides repeatedly giving rise to trophozoites.

The entire life cycle of this opalinid is thus closely coordinated with the life cycle of its host. The timing of cyst formation, its release into the water which coincides with the time that the adult frog enters the water, and the time at which the new tadpoles emerge in the pond, are all highly synchronous and, presumably coevolved. Despite what is known about the opalinid's life cycle, this fascinating system still presents considerable opportunity for additional research.

3.2.4.4 Diversity

By far, this is the smallest subphylum within the Sarcomastigophora. The opalinids are a discrete group of protozoans believed to have evolved from a flagellated ancestor. They are quite distinct from the ciliates and presumably are monophyletic in origin. Moreover, compared with the other groups, they are relatively restricted in terms of their hosts, being found only in the lower vertebrates. Although there are few known genera, and the host range is limited, based upon the close reproductive coordination between the opalinids and their hosts, it would appear that they are an old, well-established group of protozoans, remarkably well-adapted to life in the intestine of their hosts.

3.3 Phylum Apicomplexa

3.3.1 General considerations

The Apicomplexa is accorded separate phylum status based on the usual absence of obvious external organelles involved in locomotion, although the sexual stages may be flagellated. The presence of flagellated microgametes in some of the Apicomplexa has suggested a possible phylogenetic link to the Zoomastigophorea. The presence of an **apical complex** (Fig. 3.13) also is diagnostic of the group, suggesting a monophyletic origin. Finally, most apicomplexans are intracellular for a majority of their life cycles.

Members of the Apicomplexa reproduce both sexually and asexually, and are thought to be haploid for most of their life cycle. Only the zygote is believed to be diploid. Many of the Apicomplexa form spores (or resistant stages) that are involved in transmission. These spores or **oocysts** have a thickened protective cell wall when released into the environment. In Apicomplexa that are vector-transmitted, the protective spore wall is reduced, or non-existent.

In this group, there are several extremely important parasites of humans and various domesticated animals. *Plasmodium* spp., the causative agents of malaria, infect over 300 million people worldwide and kill an estimated 1 million children annually. *Toxoplasma gondii* may infect even more humans. In some areas, the prevalence of *T. gondii* infection, based on serological assays, is in excess of 50% in the local population. The coccidia are important parasites of poultry, and can produce serious losses. *Cryptosporidium parvum*, an intestinal apicomplexan of humans and other animals, caused the largest waterborne disease

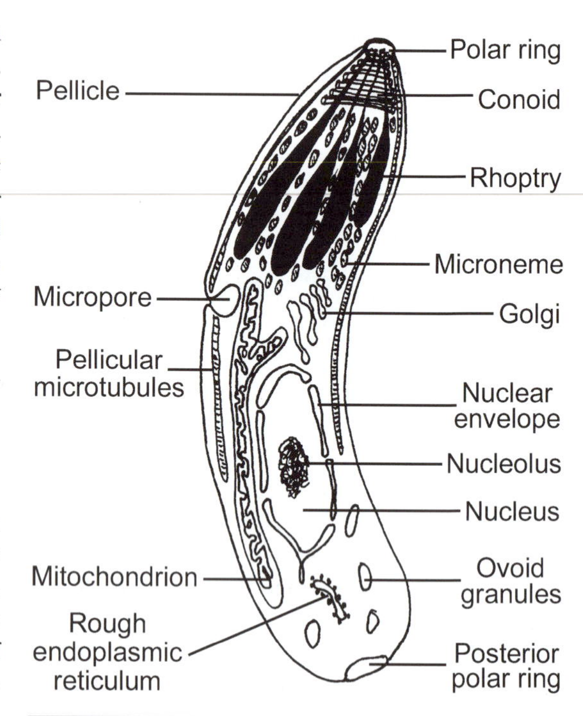

Fig. 3.13 Diagram of an apicomplexan merozoite showing key structures visible with the electron microscope.

outbreak ever recorded, with over 400 000 people estimated to have suffered from a diarrheal disease induced by this parasite in Milwaukee, Wisconsin (USA) (Solo-Gabriele & Neumeister, 1996; Smith & Rose, 1998). This localized epidemic resulted in approximately 5000 individuals being hospitalized and also caused a number of deaths in immunosuppressed individuals. It was estimated to have resulted in millions of dollars in economic losses. The piroplasms *Babesia* and *Theileria* produce serious disease in cattle and other farm animals, again resulting in great economic losses to the agricultural industry.

3.3.2 Form and function

The external structure of the vegetative stages of apicomplexans is unremarkable. The general form may vary from being amorphous (like an amoeba) to teardrop in shape. The sexual stages may be flagellated and the gametes (micro- and macrogametes) vary in size. Cysts, when produced, are usually thick-walled and round, but again, the gross external features, except for size, are of limited diagnostic value.

Internally, the most prominent feature of apicomplexans is the apical complex at the anterior end of the sporozoite and merozoite stages. This feature is characteristic of the phylum (Fig. 3.13). The complex has five distinct components: (1) polar rings, consisting of one or more electron-dense rings at the most anterior position of the cell; (2) the conoid, which is inside the polar ring and is composed of a number of coiled microtubules; (3) the micronemes, which are elongated tubular organelles arranged longitudinally in the anterior part of the cell; (4) rhoptries, which are tubular or saccular organelles extending from inside the conoid back longitudinally into the cell body (in some groups the rhoptries appear paired); and (5) subpellicular microtubules, which extend away from the polar ring into the posterior part of the cell. The function of the individual structures in the apical complex is not fully understood, but it has been suggested that the rhoptries are involved in secretion, and the entire complex may be involved in both attachment and penetration of the parasite into the host cell (Sam-Yellowe, 1996). It is assumed that the subpellicular microtubules give structural support and contribute to the general shape of the cell. They may also be involved in locomotion.

The class Sporozoasida can be divided into five principal groups. The structures of the apical complex, identified by electron microscopy, in each of the groups are listed in Table 3.5. In addition, Table 3.5 also differentiates between the various groups on the basis of their definitive hosts, vectors, etc. Note that sexual and asexual reproduction are believed to occur in most groups. The definitive hosts, the intracellular sites within the host, and the mode of transmission differ considerably between the groups. For example, the eimeriorinans are usually parasites of vertebrates and inhabit intestinal cells. They have no vector and transmission is direct, i.e., by the ingestion of oocysts in contaminated food or water. In contrast, vectors transmit the haemospororinans, e.g., *Plasmodium* spp. They are primarily parasites of vertebrates where they inhabit blood and hepatic cells. In the insect vector, the parasites occupy a number of sites, including the intestine, the body cavity, and the salivary glands.

In many species, a thick-walled oocyst is excreted into the external environment. Inside the mature oocyst, sporozoites are found within individual sporocysts. The sporozoite is the infective stage. The sporocyst wall is also thickened and the numbers of sporozoites within individual sporocysts vary. Similarly, the number of sporocysts per oocyst also varies between the different species and genera. For example, *Cyclospora* spp. have two sporocysts per oocyst, and each sporocyst contains two sporozoites. *Toxoplasma gondii* also has two sporocysts per oocyst, but each sporocyst contains four sporozoites. In contrast, *Eimeria* spp. possess four sporocysts per oocyst and each sporocyst contains only two sporozoites. There may be more than 16 sporozoites per sporocyst, as well as more than 16 sporocysts per oocyst. Both the number of sporozoites per sporocyst and the number of sporocysts per oocyst are diagnostic characters at the generic level. The sporozoites have a typical nucleus, mitochondria, Golgi, endoplasmic reticulum, microtubules beneath the pellicle, and various refractory globules (Fig. 3.13).

3.3.3 Life cycle

The basic life cycle of the apicomplexans has both sexual and asexual phases (Fig. 3.14). Sporozoites released from sporocysts, or from cells in the vector, invade a new host cell, and grow as trophozoites that, in turn, divide repeatedly by multiple fission (merogony) to form merozoites. Depending on the species, many generations of merozoites may be produced. Eventually, some of the merozoites will emerge, invade new cells, and differentiate into either micro- or macrogametes (via gametogenesis or gametogony). The merozoite may form a single microgamete, or it may divide repeatedly to form multiple microgametes. Microgamete formation follows two phases. There is an initial growth phase, with repeated nuclear divisions, and then a differentiation phase in which division occurs and microgametes develop. Following release of the micro- and macrogametes, the microgamete will fuse with a macrogamete forming a diploid zygote. Meiosis then occurs, followed by repeated divisions. The cycle is repeated with the formation of infective sporozoites (sporogony), and merozoites (merogony). There are several variations on the pattern.

The life cycle of *Plasmodium vivax* is shown in

Table 3.5 Some characteristics of the principal Apicomplexa groups in the class Sporozoasida

Groups[a] Representative genus	Eugregarinorida *Lankesteria*	Adeleorina *Hepatozoon*	Eimeriorina *Eimeria*	Haemospororina *Plasmodium*	Piroplasmorida *Babesia*
Structures:[b]					
Polar ring	unknown	unknown	present	present	present or absent
Conoid	present	present	present	absent	present or absent
Micronemes	unknown	present	present	present	present or absent
Rhoptries	present	present	present	present	present
Final hosts[c]	invertebrates	vertebrates	usually vertebrates	vertebrates	vertebrates
Habitat in vertebrate host	intestinal lumen or coelom	tissue and blood cells	mostly intestinal cells	blood and liver cells	blood, other tissue cells
Type vectors	usually none	ticks, mites, insects, leeches	usually none	dipterans	ticks
Method of transmission	ingestion of oocysts	ingestion of vector	ingestion of oocysts	infected vector bite	infected vector bite
Stages in life cycle[d]	G, S	M, G, S	M, G, S	M, G, S	M, G, S

Notes:

[a] See Fig. 3.18 for further details.

[b] See Fig. 3.13 for diagram of the cell structure of the Apicomplexa.

[c] In this Table, the vertebrate host is considered the final one. Also see Fig. 3.18 for further details.

[d] M, merogony; G, gametogony; S, sporogony.

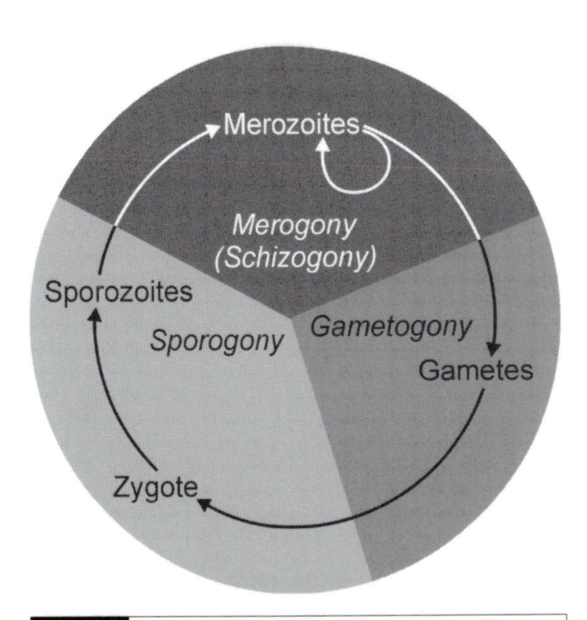

Fig. 3.14 Schematic representation of a generalized life cycle of members of the Apicomplexa.

Fig. 3.15. Note that sporogony occurs in the vector, whereas merogony occurs in the human host. Gametogenesis is initiated in the human host and is completed in the mosquito. The sporozoite injected into the host infects hepatic cells where there is repeated division and the formation of merozoites. The diagram shows that the liver merozoites can either re-infect new liver cells or initiate an infection in RBCs. The merozoite in the RBC divides repeatedly and new merozoites are released. These can then re-infect new RBCs. In this way, parasite numbers are greatly amplified. The number of merozoite generations is both enormous and, apparently, infinite. Note, however, that repeated cycles of division in the liver of the host do not occur in all species of *Plasmodium*. Finally, the parasite numbers are again amplified in the vector, with repeated divisions occurring during sporogony in the oocyst. The sporozoites emerge from the oocyst and migrate to the lumen of the salivary gland where, eventually, they can be injected into a new host (Vickerman & Cox, 1967).

In *Eimeria* spp., the life cycle (Fig. 3.16) differs in a number of ways from that of the *Plasmodium* spp. First, a thick-walled oocyst is released with the feces into the environment. Oocyst maturation and sporozoite formation occur in the external environment outside the host. Second, both merogony and gametogony are restricted to a single host. Finally, the number of merozoite generations is limited to only two or three merogonic cycles.

The final example of an apicomplexan life cycle is that of *Lankesteria culicis* (Fig. 3.17). In this case, mosquito larvae ingest oocysts, and the sporozoites are then released to invade gut cells. The sporozoite grows into a large trophozoite that eventually is released into the gut lumen. When the mosquito larva pupates, the mature trophozoites enter the Malpighian tubules where they pair and begin the process of encystment. Repeated divisions then occur and there is the formation of a large number of gametes that fuse. The zygotes then develop into oocysts with each oocyst containing eight sporozoites. The oocysts are released into the lumen of the tubules, enter the intestine, and are released into the environment with the feces. The major difference in this life cycle versus that of *Plasmodium* spp. and *Eimeria* spp. is the absence of merogony. Amplification of parasite numbers occurs during the processes of gametogony and sporogony.

3.3.4 Diversity

The Apicomplexa includes two classes, Perkinsasida and Sporozoasida, of which only the latter is of medical or veterinary importance. Perkinsasida is a small group with only one species that is parasitic in invertebrates. In contrast, the Sporozoasida is very large, with over 4000 known species. It has been estimated, however, that there may be in excess of 2 million species infecting all known animal groups.

The Sporozoasida contains almost all of the well-studied Apicomplexa. There are three subclasses in the Sporozoasida and each of these will be briefly discussed. For a more complete taxonomic classification, see the review by Levine (1985).

3.3.4.1 Subclass Gregarinasina

The subclass Gregarinasina (Fig. 3.18) contains numerous different species. One genus alone, *Gregarina*, contains approximately 260 species. Members are generally **monoxenous**, i.e., they infect only a single type of host. They are usually

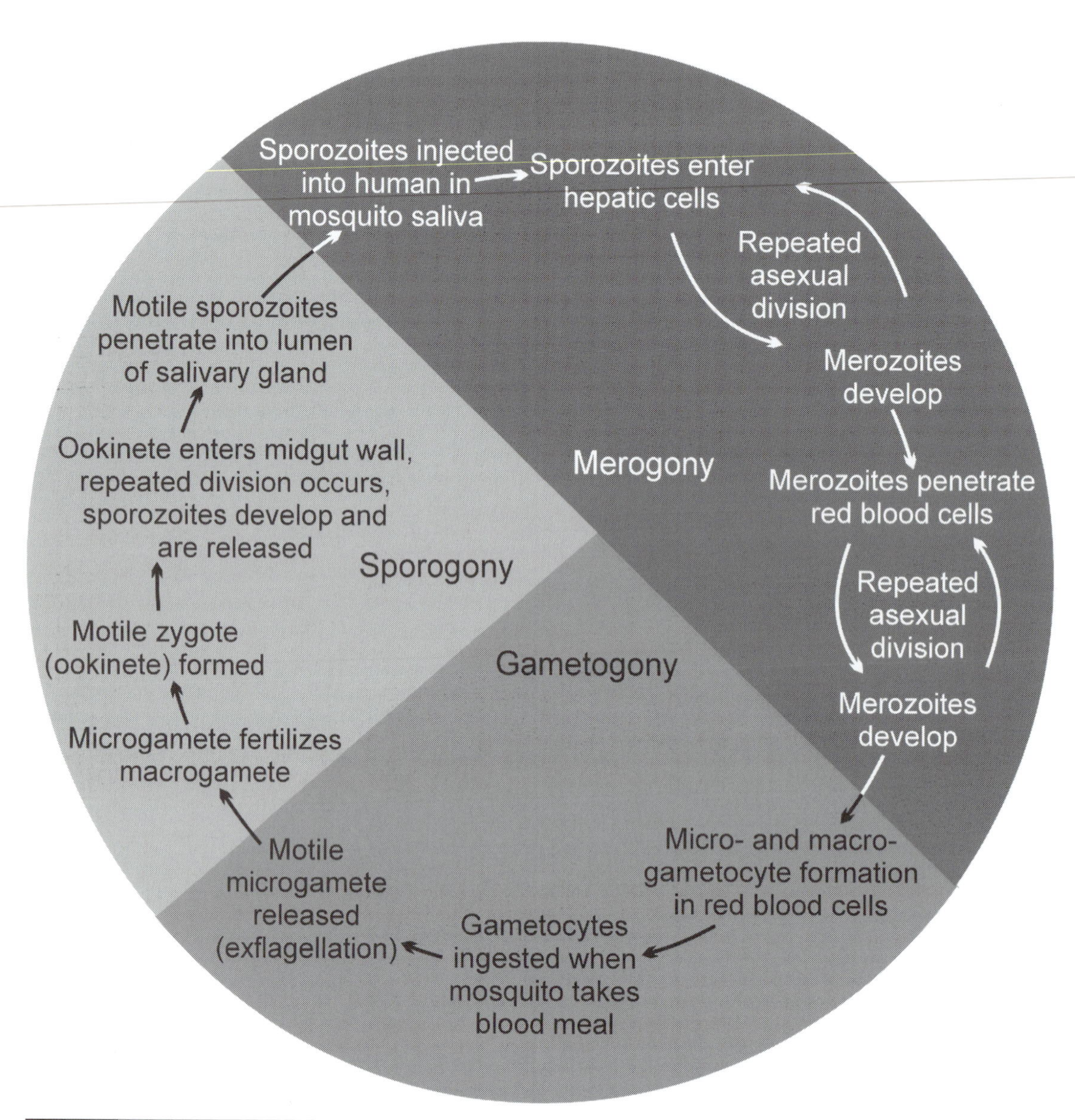

Fig. 3.15 Life cycle of *Plasmodium vivax* in man and the mosquito. Only the ookinete is diploid, all other stages are haploid.

parasites of the digestive tract or the hemocoel of invertebrates, particularly the insects. They are also parasites of echinoderms, molluscs, and annelids (Table 3.6).

Most gregarines do not have a merogony phase in their life cycle (Table 3.5). They increase in number solely through sporogony. The oocysts (spores) are released from the infected host and are transmitted to a new host by ingestion of the oocysts. Following ingestion, sporozoites emerge from the oocyst in the intestinal tract and enter an epithelial cell where they increase in size. For most gregarines, the trophozoites eventually leave the host cell but remain temporarily attached to it by their anterior end, called the mucron or epimerite.

The gregarines are divided into two suborders. In the Septatorina, the body of the parasite is segmented into an anterior and a posterior end. The anterior end contains a modified conoid that is involved in attachment of the trophozoite to the host cell. The anterior segment does not contain a

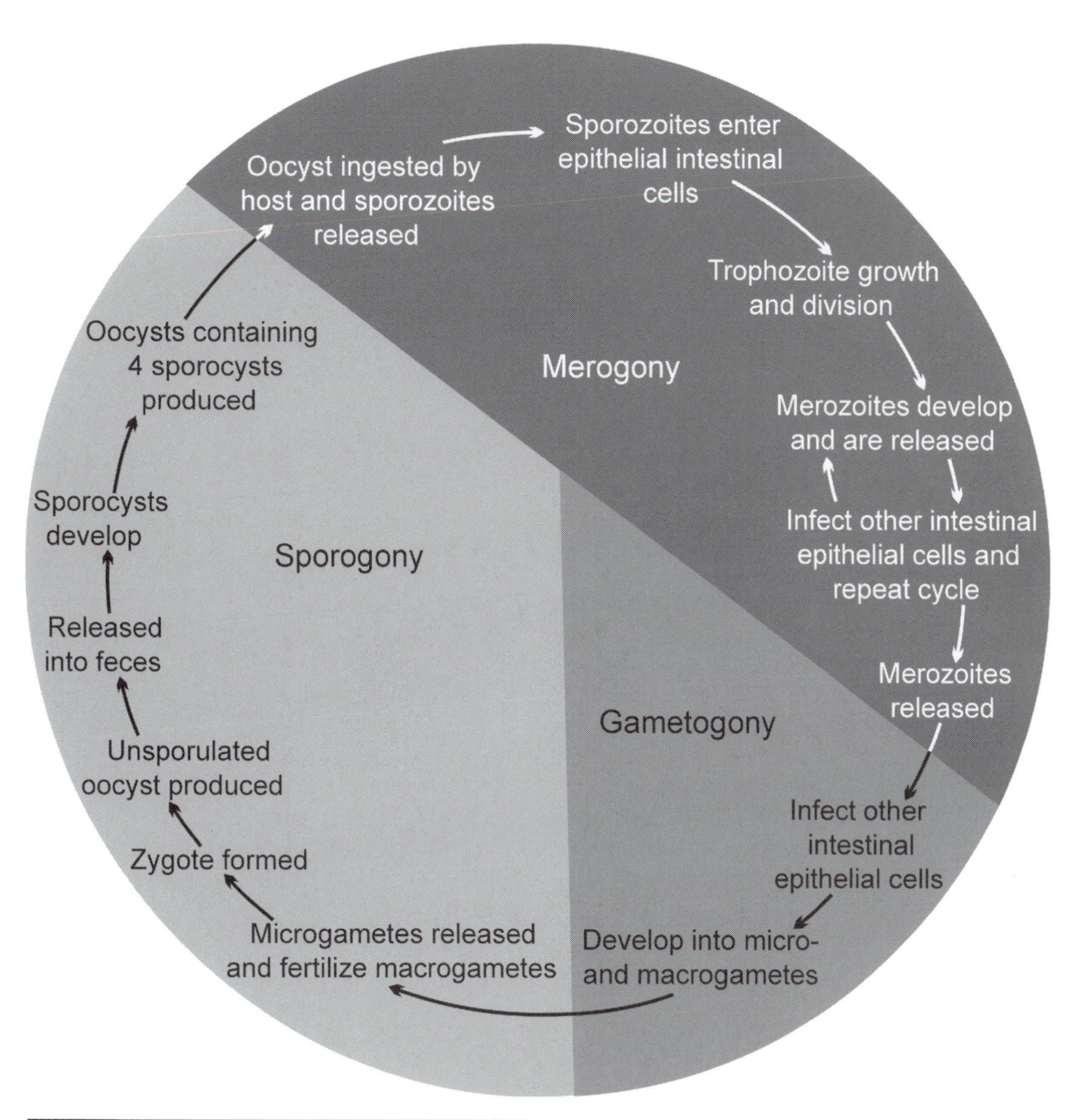

Fig. 3.16 Life cycle of *Eimeria* spp.

nucleus and is separated from the posterior end by a septum. The posterior segment contains only a single nucleus and therefore the anterior and posterior segments should not be considered two cells. In the Aseptatorina, the gregarines do not appear segmented and a septum is not visible. In the aseptate gregarines, the anterior attachment portion of the trophozoite is referred to as the mucron. In addition to the modified conoid apparatus, gregarines also contain mitochondria, a Golgi apparatus, and numerous granules presumably containing stored carbohydrates, lipids, etc.

The gregarines in both suborders eventually detach or break away from the host cell and wander in the gut or body cavity of the host. The wandering trophozoites will become gametocytes. Two gametocytes become attached and encyst. The gametocytes within the cyst (or gametocyst) undergo nuclear division and, eventually, individual nuclei bud off as gametes. The gametes within the cyst fuse in pairs to become zygotes. The zygotes encyst forming an oocyst that remains within the gametocyst. Division occurs within the oocyst (now a sporocyst) forming

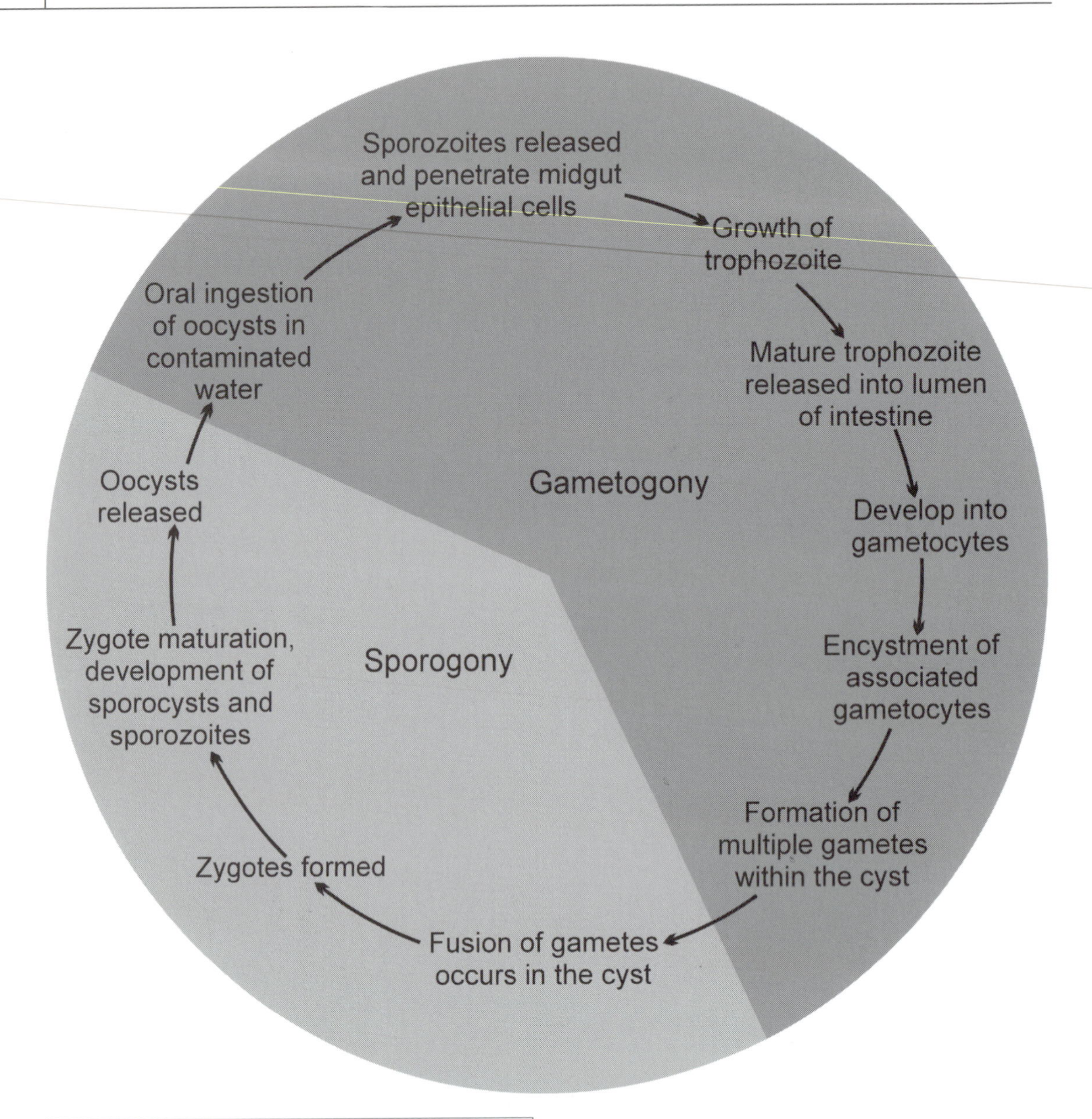

Fig. 3.17 Life cycle of *Lankesteria culicis* in *Aedes aegypti*. Note that the merogony phase of the life cycle is absent.

sporozoites and, ultimately, the individual sporocysts (or oocysts) are released from the gametocyst and the host into the external environment. The life cycle of the aseptate gregarine *Lankesteria culicis* is shown in Fig. 3.17.

The epithelial cells to which the gregarines attach are destroyed. However, unless the infection is massive or there is amplification by merogony, these parasites rarely seem to harm their host. The absence of significant pathology has suggested to some authors that the gregarine–host association is an old one.

3.3.4.2 Subclass Coccidiasina

There are three orders within this subclass (Fig. 3.18). Two are represented by only a few known species, mainly infecting marine annelids. The remaining order, Eucoccidiorida, has most of the known species in the subclass Coccidiasina. Not surprisingly, it also contains a majority of species that are of medical and veterinary importance.

Members of the suborder Adeleorina, or the haemogregarines, are distinguished by male and

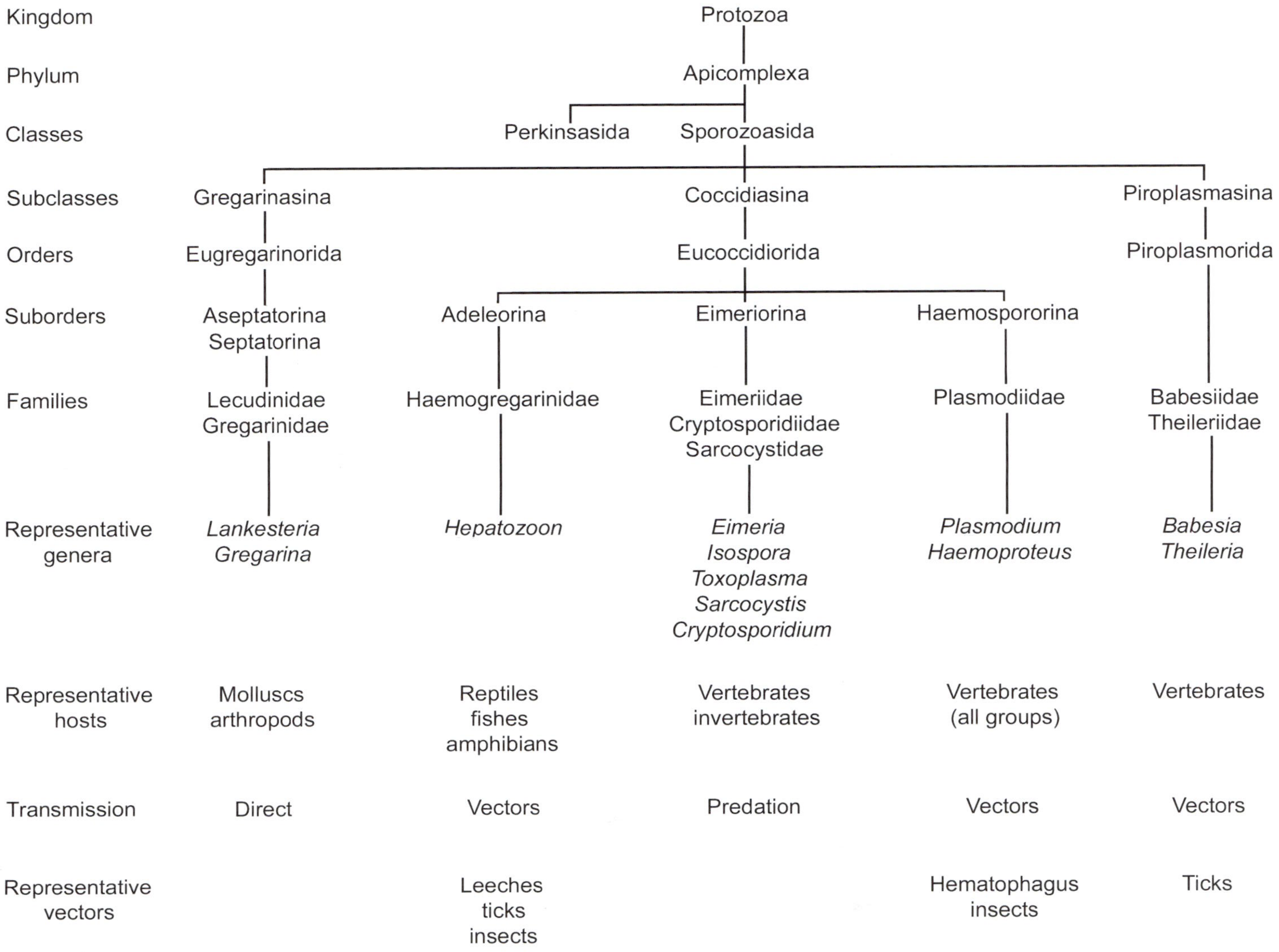

Kingdom			Protozoa		
Phylum			Apicomplexa		
Classes		Perkinsasida	Sporozoasida		
Subclasses	Gregarinasina		Coccidiasina		Piroplasmasina
Orders	Eugregarinorida		Eucoccidiorida		Piroplasmorida
Suborders	Aseptatorina Septatorina	Adeleorina	Eimeriorina	Haemospororina	
Families	Lecudinidae Gregarinidae	Haemogregarinidae	Eimeriidae Cryptosporidiidae Sarcocystidae	Plasmodiidae	Babesiidae Theileriidae
Representative genera	*Lankesteria* *Gregarina*	*Hepatozoon*	*Eimeria* *Isospora* *Toxoplasma* *Sarcocystis* *Cryptosporidium*	*Plasmodium* *Haemoproteus*	*Babesia* *Theileria*
Representative hosts	Molluscs arthropods	Reptiles fishes amphibians	Vertebrates invertebrates	Vertebrates (all groups)	Vertebrates
Transmission	Direct	Vectors	Predation	Vectors	Vectors
Representative vectors		Leeches ticks insects		Hematophagus insects	Ticks

Fig. 3.18 Abbreviated classification of some parasitic Apicomplexa.

Table 3.6 | Examples of species of gregarines

Gregarine	Body style[a]	Host	Habitat in host
Lecudina pellucida	AS	polychaetes	intestine
Monocystis ventrosa	AS	earthworms	intestine
Lankesteria ascidiae	AS	tunicates	intestine
Lankesteria culicis	AS	culicine mosquitoes	intestine
Apolocystis gigantea	AS	earthworms	seminal vesicles
Urospora chiridotae	AS	sea cucumbers	blood vessels
Nematopsis ostrearum	S	oysters, crabs	intestine
Gregarina blattarum	S	cockroaches	midgut
Cephaloidophora oliva	S	spiders, littoral crabs	intestine
Didymophyes gigantea	S	scarabaeid beetles	intestine
Actinocephalus parvus	S	flea larvae	intestine
Monoductus lunatus	S	millipedes	proventriculus and intestine

Notes:
[a] AS, aseptate, suborder Aseptatorina; S, septate, suborder Septatorina.

female gametes that develop in association with each other in a process called **syzygy**. They are primarily intracellular parasites of vertebrate RBCs, white blood cells (WBCs) and, on occasion, other cell types. They have an indirect life cycle, being transmitted by leeches, ticks, mites, and biting insects. Transmission may occur when a new host consumes an infected vector, or, in some cases, the vector's feces in which sporocysts are found. Infection is thus by various mechanisms, although in most cases the sporozoites actually enter the vertebrate through the mucus membrane. Haemogregarines follow the basic apicomplexan life cycle pattern, with sexual reproduction in the vector and asexual reproduction in the vertebrate host. Most species are not known to be pathogenic.

There are many important families and genera in the suborder Eimeriorina, more commonly known as the coccidia. We will restrict our discussion to groups of medical and veterinary importance since these are known best. Some species complete their life cycles in a single host. Sporocysts are passed in feces or urine, and are then ingested directly by a new host. Coccidians are usually distinguished on the basis of the number of sporocysts within the oocysts, as well as the number of sporozoites contained within each sporocyst (Table 3.7).

Cryptosporidium spp. are enteric coccidians. They are characterized by having a merogony phase, micro- and macrogametes, oocysts containing four naked sporozoites, and a direct or monoxenous life cycle (Fig. 3.19; see also O'Donoghue, 1995; Smith & Rose, 1998). *Cryptosporidium* spp. are considered as true coccidians and are placed in the same order as species of *Eimeria* and *Isospora*. However, recent phylogenetic studies suggest that *Cryptosporidium* spp. may be the most distantly related genus in the order. There have been over 20 different species of *Cryptosporidium* described, many based on differences in the host species from which they were isolated. At the present time, only six species are considered valid, based on oocyst morphology (size differences), host specificity, and site of infection (Table 3.8). Host-specificity studies suggest that species of *Cryptosporidium* are restricted to a single vertebrate class, but that cross-transmission can occur between hosts within a vertebrate class (O'Donoghue, 1995). Pathology is often associated with *Cryptosporidium* infections in both domesticated and wild animals. The pathology appears to be dependent on the route of inoculation, the age of the host, and the particular host species infected. *Cryptosporidium parvum* can cause a severe gastrointestinal disturbance in humans.

Species of *Eimeria* are generally thought to have a single host. Oocysts are produced and excreted into the environment. The oocysts are

Table 3.7 | Some subfamilies and genera of the Eimeriidae

Subfamily	Genus	Sporocysts/oocyst	Sporozoites/sporocyst
Cryptosporidiinae	*Cryptosporidium*	0	4
	Pfeifferinella	0	8
	Schellackia	0	8
Caryosporinae	*Mantonella*	1	4
	Caryospora	1	8
Cyclosporinae	*Cyclospora*	2	2
	Isospora	2	4
	Toxoplasma	2	4
	Sarcocystis	2	4
	Besnoitia	2	4
Eimeriinae	*Eimeria*	4	2
	Wenyonella	4	4
	Angeiocystis	4	8
Yakimovellinae	*Octosporella*	8	2
	Yakimovella	8	>16
Pythonellinae	*Hoarella*	16	2
	Pythonella	16	4
Barrouxinae	*Barrouxia*	>16	1
	Echinospora	>16	1
Aggregatinae	*Merocystis*	>16	2
	Ovivora	>16	>16

thick-walled and represent the environmentally resistant stage in the life cycle. There are four sporocysts per oocyst and two sporozoites per sporocyst (see Table 3.7). Sporozoites of *Eimeria* spp. infect cells of the intestinal mucosa of all classes of vertebrates and may cause serious disease in a number of domesticated animal species (Table 3.9). Human infections are not known. One, or a limited number of cycles of merogony characterize their life cycles.

Species of *Isospora* have two sporocysts per oocyst and four sporozoites in each sporocyst. Recent phylogenetic studies based on the use of molecular markers (18S ribosomal DNA sequences) show a close relationship between *Toxoplasma*, *Neospora*, and *Isospora*. A more distant relationship is seen between *Sarcocystis*, *Eimeria*, and *Cyclospora* (Carreno *et al.*, 1998). *Cryptosporidium* spp. appear to be the most taxonomically distant group based on the differences in their 18S ribosomal DNA sequences as well as in their life cycles. It has been stated that many, possibly all, of the Eimeriidae that have an oocyst containing two sporocysts, each of which contains four sporozoites, have two hosts in their life cycle. In many cases, the sexual stages occur in the definitive host, whereas the asexual stages are found in an alternate host. Carnivores often serve as the definitive host; alternate hosts are accidentally infected by ingesting oocysts excreted in the feces. Table 3.10 identifies a number of *Isospora*-type species, as well as several closely related species now accorded different generic status. Species of the *Isospora* type infect intestinal cells, as well as many other tissues in a variety of vertebrate species.

Based on molecular phylogeny, *Toxoplasma* spp. are related to, but are distinct from, the *Isospora*. The life cycle of *T. gondii* is shown in Fig. 3.20, and, as can be seen, this pathogen may infect many species of birds and mammals (Kreier & Baker, 1987; Cox, 1993). It has a worldwide distribution and a high prevalence in humans, estimated at approximately 23% of the world's population. Fortunately, however, the infection is usually benign. On the other hand, it may cause severe pathology in a developing fetus when infected *in*

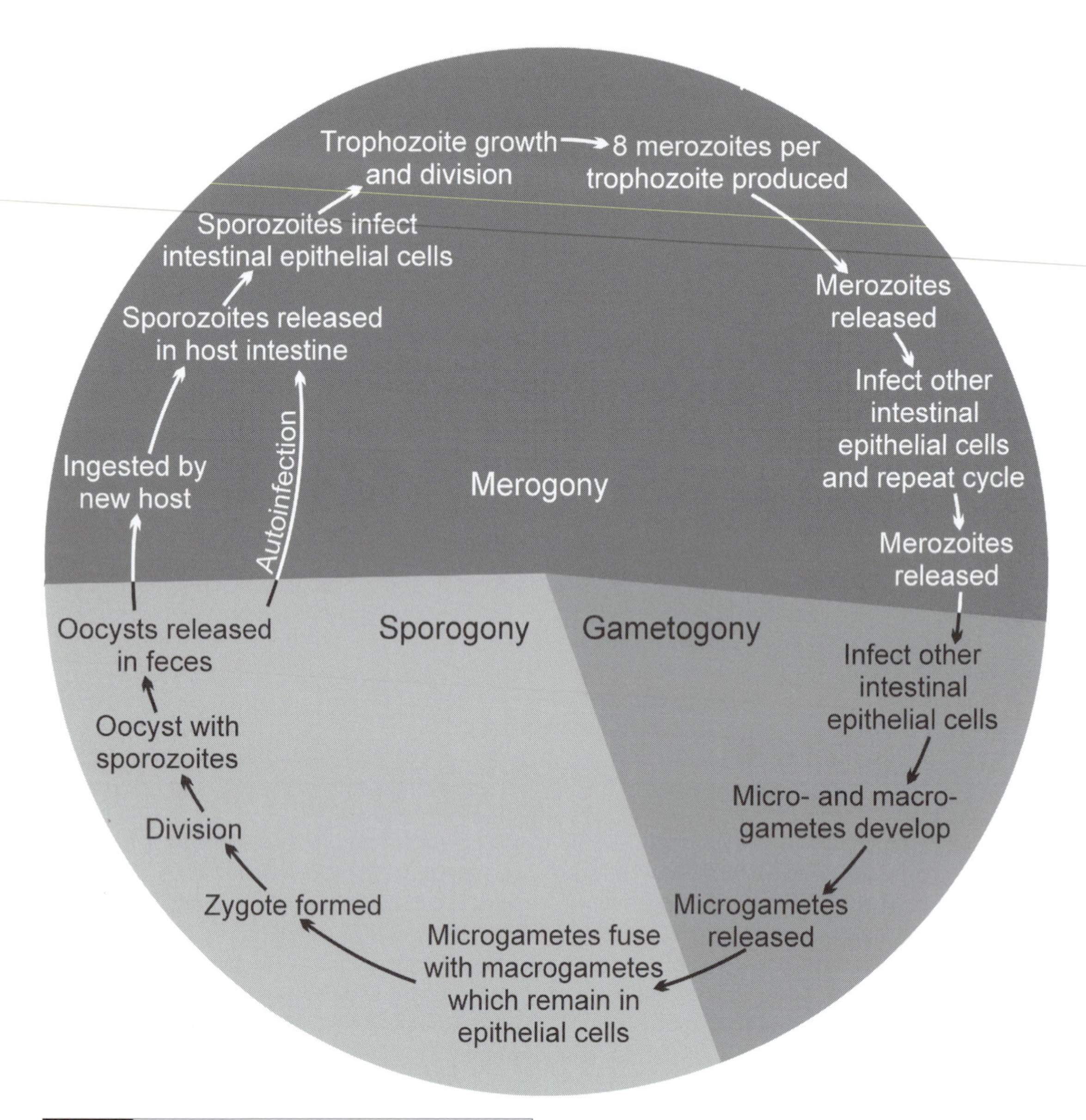

Fig. 3.19 Life cycle of *Cryptosporidium parvum*. Note that autoinfections can occur.

utero, as well as in severely immunosuppressed patients. Cats, the definitive hosts, become infected by either accidentally ingesting infective oocysts in contaminated soil, or by eating the tissues of infected prey in which are found **tachyzoites**. Although *T. gondii* reproduces asexually in a variety of different hosts, both merogony and gametogony are restricted to the intestine of the cat. Sporogony is completed in the feces. Alternative hosts, such as mice or humans, become infected by accidental ingestion of oocysts from cat feces or by the ingestion of **pseudocysts** in infected meat. Infective sporozoites released from ingested oocysts, or infective stages released from pseudocysts, will penetrate the gut wall and infect macrophages. The infected macrophages then migrate throughout the body. Macrophages are infected with a stage referred to as a trophozoite; it divides asexually by a process called **endodyogeny**, or internal budding, in which two daughter cells develop within a parent cell. The trophozoites divide repeatedly within the cell, eventually forming what is called a

Table 3.8 Somes species of *Cryptosporidium*

Species	Vertebrate host (example)[a]	Usual habitat	Clinical disease
C. muris	mammals (mice)	stomach	limited to none
C. parvum	mammals (cattle, humans)	intestine	highly pathogenic
C. meleagridis	turkeys	intestine	variable[b]
C. baileyi	chickens	trachea, bursa, cloaca	variable[b]
C. serpentis	reptiles	stomach	variable[b]
C. nasorum	fishes	stomach, intestine	limited to none

Notes:
[a] In many reports of infected host species, the species of *Cryptosporidium* was not identified.
[b] Depending upon host species, age of animal, route of inoculation, etc.

Table 3.9 Some important *Eimeria* species that infect domesticated animals

Species	Host	Habitat	Pathogenicity
E. necatrix	chickens	small intestine and ceca	high
E. tenella	chickens	ceca	high
E. meleagrimitis	turkeys	intestine	moderate
E. danailova	ducks	small intestine	moderate
E. ducephalae	ducks	small intestine	high
E. truncata	geese	kidney tubules	high
E. bovis	cattle	intestine	moderate
E. arloingi	sheep	small intestine	moderate
E. debliecki	pigs	intestine	moderate
E. stiedai	rabbits	bile duct	high
E. irresidua	rabbits	small intestine	moderate
E. canis	dogs	intestine	moderate
E. nieschulzi	rats	small intestine	moderate
E. falciformis	mice	intestine	moderate

Table 3.10 Some important species of the isosporan-type coccidia

Species	Final host	Habitat in final host	Intermediate hosts[a]	Habitat in intermediate host	Pathogenicity
Isospora suis	pigs	small intestine	—	—	moderate
Isospora bigemina	dogs	small intestine	—	—	moderate
Eimeria felis	cats	small intestine	—	—	moderate
Toxoplasma gondii	cats	small intestine	many vertebrates	many tissues	low[b]
Sarcocystis tenella	cats, dogs	intestine	sheep, goats	muscle	low
Sarcocystis lindermanni	cats, dogs	intestine	humans	muscle	low
Cryptosporidium parvum	humans	intestine	—	—	low[b]

Notes:
[a] A dash (–) denotes the absence of an intermediate host in these species.
[b] *Toxoplasma gondii* can be highly pathogenic to the human fetus, or in severely immunosuppressed adults. *Cryptosporidium parvum* normally produces a self-limiting, mild, diarrheal-type disease but can be highly pathogenic in the immunosuppressed host.

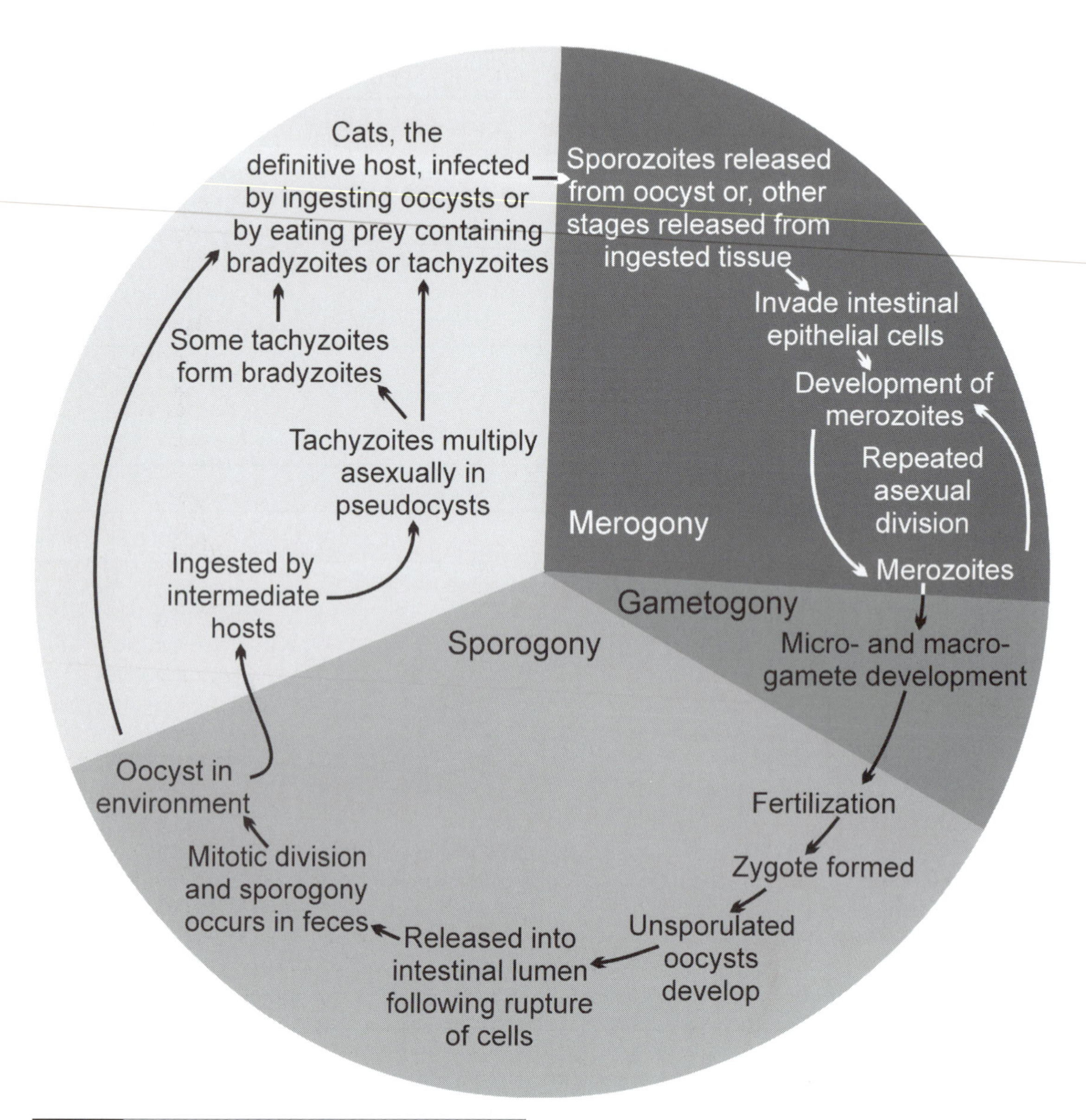

Fig. 3.20 Life cycle of *Toxoplasma gondii.* Note that there is a large number of possible intermediate hosts. They include humans, mice, other mammals, and birds. In the intermediate host, pseudocysts, containing the rapidly dividing tachyzoites, and true cysts, containing the slow-developing stage called bradyzoites, are both found in many host tissues.

pseudocyst. When the infected host cell dies and the trophozoites, now called tachyzoites, are released, they will infect cells in all tissues and organs of the body. When the proliferative stage ends, the tachyzoites enter new cells and develop slowly. The dividing stages occur in a true cyst with a thin, protective wall and are referred to as **bradyzoites**. The cysts can also be found throughout the body and, when fully mature, have no trace of the original host cell. The bradyzoites can remain viable in the cyst for several years. They may become active if the infected host is immunosuppressed, or if the tissue containing the cyst is consumed by a new host. Therefore, although the host immune response does not eliminate the infection, it would appear to control it and, as a result, the infected host may have few, or no, clinical symptoms (Alexander & Hunter, 1998).

Species of *Sarcocystis* are parasites in a wide

Table 3.11 Some important species of *Sarcocystis* of vertebrates

Species	Final host	Intermediate host
S. cuniculi	cats	rabbits
S. cymruensis	cats	rats
S. fusiformis	cats	water buffaloes
S. gigantea	cats	sheep
S. hirsuta	wild cats, cats	oxen
S. porcifelis	cats	pigs
S. leporum	cats, raccoons	cottontail rabbits
S. muris	cats, ferrets	mice
S. bertrami	dogs	horses, asses
S. capricanis	dogs	goats
S. equicanis	dogs	horses
S. fayeri	dogs	horses
S. gracilis	dogs	red deer
S. levinci	dogs	water buffaloes
S. tenella	dogs, wild canids	sheep
S. miesheriana	dogs, wild canids	pigs
S. cruzi	dogs, wild canids	oxen
S. hemionilatrantis	dogs, coyotes	mule deer
S. hominis	primates	oxen
S. suihominis	primates	pigs

variety of mammals and birds. They are most commonly observed in cattle, horses, sheep, pigs, monkeys, ducks, chickens, and humans (Table 3.11). Usually, the infection goes unnoticed and is only detected at necropsy. However, mild to fatal infections have been observed in mice. The stages of the life cycle in the definitive host are of the *Isospora* type in which merogony and gametogony occur in the intestinal epithelial cells of the host. The oocysts contain two sporocysts and four sporozoites per sporocyst. In the intermediate host, the stage observed is the **sarcocyst**, which can be quite large and contain numerous trophozoites. The cyst wall is complex, consisting of both host and parasite material. The final, or definitive, host must always be a carnivore or an omnivore since consuming muscle tissue from prey containing sarcocysts infects the host. Therefore, the oocyst characteristics are of the *Isospora* type, and the life cycle resembles that of *Toxoplasma*.

Members of the suborder Haemospororina are obligate intracellular parasites for most of their life cycles (Fig. 3.18). They are acquired from a vertebrate host by blood-feeding vectors such as mosquitoes (anophelines), sand flies (phlebotomines), or midges (*Culicoides*). Merogony and gametogony occur in the vertebrate host. Merogony takes place in tissues such as the liver and RBCs, whereas gametogony occurs only in RBCs. Fertilization and zygote formation are restricted to the insect vector.

The life cycles of all *Plasmodium* spp. are similar. There are many species infecting reptiles, birds, and mammals (Table 3.12). *Plasmodium* spp. infecting humans cause malaria, and, next to tuberculosis, it is the most significant public-health problem in the world today. There are four species that infect humans. They differ from each other in such characteristics as the morphology of the erythrocytic stages, the temporal duration of schizogony (merogony), etc. The different morphological stages of *Plasmodium* observed microscopically in stained human blood films are shown in Figs. 3.21 and 3.22. Several of the diagnostic differences used to distinguish microscopically between the human *Plasmodium* spp. in stained blood films are shown in Fig. 3.23 and are

Table 3.12 Examples of *Plasmodium* species and their hosts

Plasmodium species	Vertebrate host	Susceptible vectors
P. (Plasmodium) falciparum	humans	>60 anopheline species
P. (P.) hylobate	gibbons	unknown
P. (P.) cynomolgi[a]	Rhesus monkeys	>35 anopheline species
P. (P.) simium	howler monkeys	*Anopheles cruzi*
P. (Vinckeia) lemuris	black lemurs	unknown
P. (V.) chabaudi[a]	tree rats	*Anopheles stephensi*
P. (V.) berghei[a]	>30 different species of wild rodents	*Anopheles dureni* (plus several other anopheline species)
P. (V.) voltacium	West African bats	*Anopheles smithii*
P. (Haemamoeba) relictum	>100 different bird species	*Culex, Anopheles*
P. (H.) gallinaceum[a]	jungle fowl, domestic hens	*Aedes, Anopheles, Culex, Mansonia, Armigeres, Culiseta*
P. (Giovannolaia) durae	domestic turkeys	unknown
P. (G.) lophurae[a]	chickens, ducks, pheasants, turkeys, guinea fowl	*Aedes, Anopheles*, poorly in *Culex*
P. (Sauramoeba) diploglossi	American lizards	unknown
P. (S.) mexicanum	iguanid lizards	mites
P. (Ophidiella) wenyoni	snakes	unknown

Notes:
[a] Species of *Plasmodium* that infect vertebrates other than man and frequently used in experimental studies.

further outlined in Table 3.13. The current tools used to diagnose human malaria are more fully discussed in Box 3.7. All four species of *Plasmodium* that infect humans cause a serious, debilitating fever accompanied by violent shaking. However, usually only *P. falciparum* produces fatal infections. Malaria affects some 200–300 million people in the world and kills about 1% annually. It is estimated that over 40% of the world's population is at risk of infection. A current estimate of yearly deaths among children in Africa alone is about 1 million.

There are four species of *Plasmodium* that infect humans; *Plasmodium vivax* has the greatest prevalence. Although *P. falciparum* causes the greatest number of deaths, *Plasmodium vivax, P. malariae*, and *P. ovale* all cause significant morbidity. A colleague infected with vivax malaria mused that, during the fever period, you knew you probably would not die, but wished you would because you felt so bad. Moreover, malaria has a huge economic price for families and communities in areas where the disease is endemic. These include the cost of treatment, school absenteeism, and loss of workdays by family members caring for those sick with the disease.

Historically, it is well known that a combination of malaria and yellow fever significantly impeded progress on construction of the Panama Canal early in the twentieth century. It adversely affected the human settlement of a highly malarious area of southern Europe prior to vector control. Malaria caused significantly greater casualties among allied soldiers in the South Pacific during World War II than bullets, and some historians feel that the fall of the Roman Empire was in no small part due to the severe ramifications created by malaria.

Although various areas of the world have greatly reduced the prevalence of malaria, e.g., the USA and Europe, the worldwide prevalence has increased and, in certain areas, its geographical range has also increased. Malaria is, therefore, truly a re-emerging parasitic infection. The importation of malaria from endemic areas into areas with little or no current malaria is well documented. For example, over 7000 cases of malaria were diagnosed in Russian soldiers returning

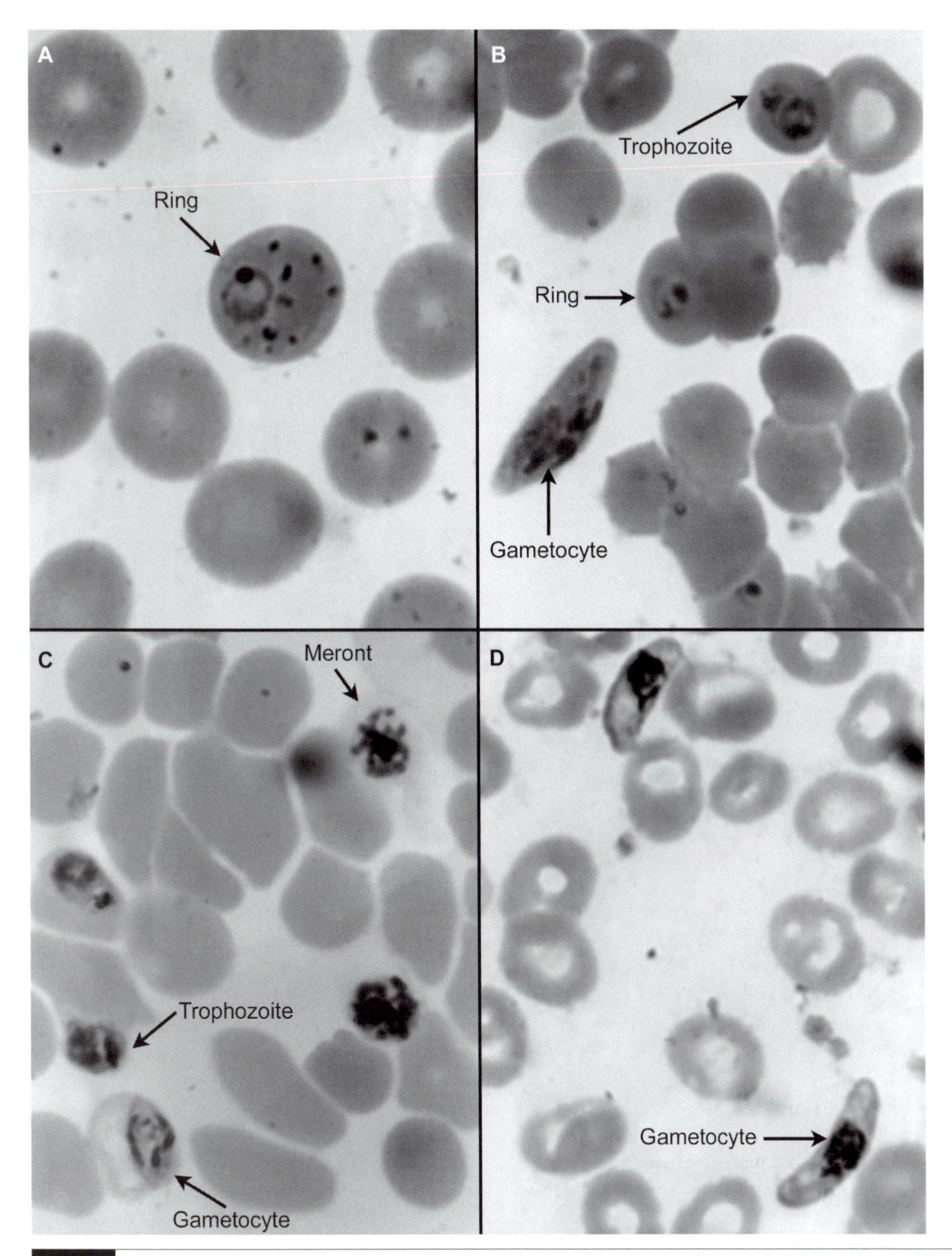

Fig. 3.21 Microscopic morphology of the various life cycle stages of *Plasmodium falciparum* in stained, thin blood smears. (A) The ring stage of *P. falciparum* in infected RBCs. Maurer's dots in the infected cells can also be seen. (B) Ring, trophozoite and the banana-shaped gametocyte stages are present. The gametocyte-infected RBC is greatly distorted in shape. (C) Trophozoite, a late-stage meront (or schizont), and an early gametocyte stage are seen. (D) Gametocyte-infected RBCs are seen. Again, note the banana shape of the infected cell. All of the different morphological stages observed in stained, thin blood smears of *P. falciparum*, except the free merozoites, can be compared. (Photographs courtesy of Mr. Purnomo Projodipuro and Dr. Michael J. Bangs, US Naval Medical Research Unit No. 2, Jakarta, Indonesia.)

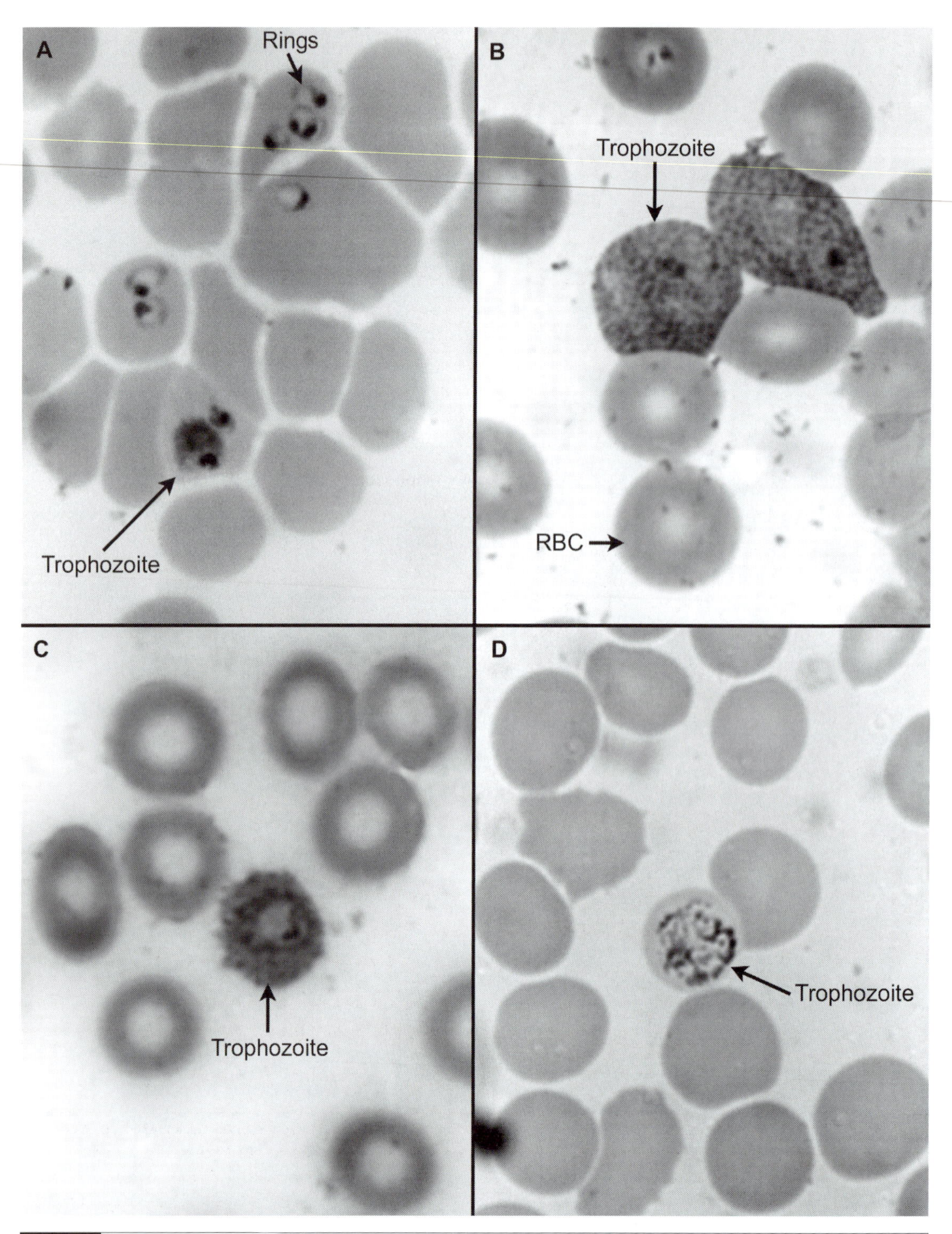

Fig. 3.22 Comparison of the microscopic morphology of the trophozoite stage in stained, thin blood smears of all four species of human malaria. (A) Ring and early trophozoite stages of *P. falciparum* are present. (B) Trophozoite stage of *P. vivax*. The trophozoite-infected RBC is irregular in shape, the RBC is enlarged, and there is abundant cytoplasmic pigment (Schüffner's dots). (C) Trophozoite stage of *P. ovale*. (D) The trophozoite stage of *P. malariae*. (Photographs courtesy of Mr. Purnomo Projodipuro and Dr. Michael J. Bangs, US Naval Medical Research Unit No. 2, Jakarta, Indonesia.)

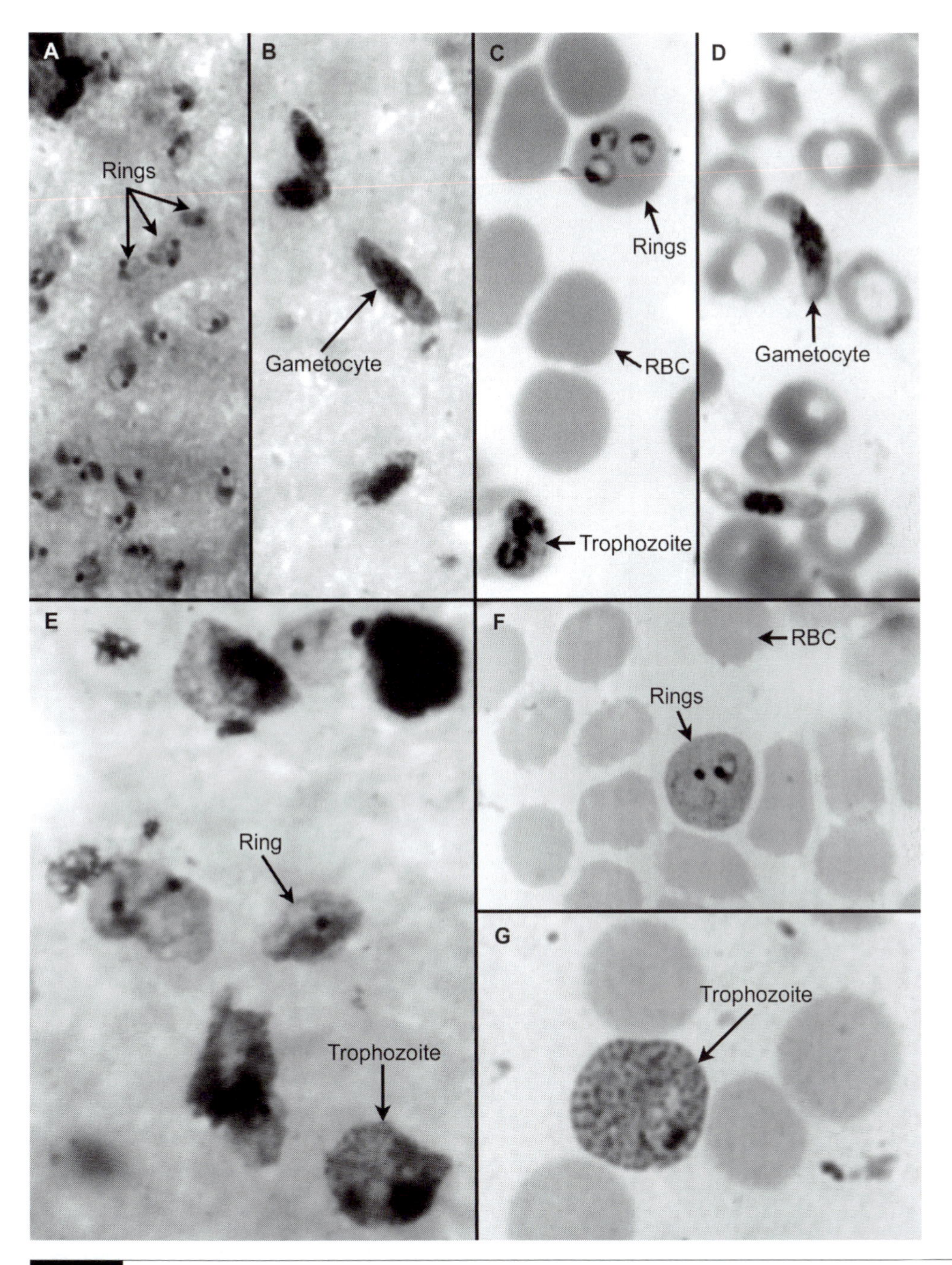

Fig. 3.23 Comparison of the microscopic morphology of *Plasmodium falciparum* and *P. vivax* in stained, thick and thin blood preparations. (A) Thick blood smear of *P. falciparum* in which numerous ring stages can be seen. (B) Thick blood smear of *P. falciparum* in which gametocytes are present. (C) Thin blood smear of *P. falciparum* in which the ring stages and a trophozoite can be seen. (D) Thin blood smear in which gametocytes are present. Note that in the thick blood smears (A, B), it is more difficult to see the morphological details of the various stages; however, many more infected RBCs can be detected. An experienced microscopist more easily detects an infection in a thick blood smear especially when the parasitemia is low. (E) Thick blood smear of *P. vivax* in which ring and trophozoite stages can be seen. (F) Thin blood smear of *P. vivax* showing rings and (G) a trophozoite stage in infected RBCs. Again, by comparing the ring stages in the thick and thin preparations, it can be observed that the morphological details are more difficult to see in thick preparations. (Photographs courtesy of Mr. Purnomo Projodipuro and Dr. Michael J. Bangs, US Naval Medical Research Unit No. 2, Jakarta, Indonesia.)

Table 3.13 Characteristics of human *Plasmodium* in stained peripheral blood erythrocytes

Species	Ring form trophozoite	Trophozoites	Merozoites	Gametocytes	Host erythrocyte
P. ovale	approx. 1/3 diameter of RBC, smaller than *P. vivax*, usually few in number	compact in appearance	4 to 12	similar to *P.vivax*, round or ovoid in shape	slightly enlarged, stippled
P. vivax	approx 1/3 diameter of RBC, multiple infections common	amoeboid in appearance, abundant cytoplasmic pigment in RBC	12 to 24	macro: fills enlarged cell, micro: (normal RBC size)	enlarged, stippled
P. malariae	approx. 1/3 diameter of RBC, usually only a single infection	compact sometimes band shape; abundant cyto-plasmic pigment in RBC	6 to 12	macro: fills cytoplasm, RBC size micro: fills cytoplasm, RBC size	heavy pigment
P. falciparum	small, often numerous infections	rarely observed	8 to 24	macro & micro: crescent shape, 1½ times normal RBC	[a]

Notes:
[a] The percentage of infected host RBCs is often higher in a falciparum infection than in the other three species of malaria.

Box 3.7 Diagnosis of human malaria

The diagnosis of human malaria is based on three different methods, i.e., microscopic, immunological, and molecular techniques. The immunological protocols can be divided into those that detect antibodies to *Plasmodium* antigens or procedures that directly assay for *Plasmodium* antigens in infected blood. Antibodies to *Plasmodium* can be detected by several different serological tests. The presence of anti-plasmodium antibody in infected serum indicates either a past or a present infection. It cannot distinguish between these two possibilities. It is, therefore, of limited value as a guide for treatment. In contrast, the direct detection of *Plasmodium* antigens in human serum indicates a current infection and can be used as a guide for disease management and chemotherapy. A commercial, simple-to-use antigen detection kit is now available. The test takes only 10 minutes to run, and can be easily used by healthcare workers.

A biochemical assay for the detection of *Plasmodium* lactate dehydrogenase (PLDH) is also being tested for use in the field. Similar to the use of the immunological assay that detects parasite antigens, this biochemical assay is also simple to use, can be performed rapidly, and indicates the presence of an active infection. At the present time, there is considerable interest in using antibody specific for the PLDH to capture the enzyme in infected blood, and then to combine it with the biochemical assay for detecting PLDH enzyme activity. This improves detection sensitivity many-fold.

The most sensitive assay for the detection of any parasitic organism is through the use of molecular tools. It is suggested that a single parasite could be detected in a 1.0 μl blood sample. This is at least a five-fold greater sensitivity than the microscopic examination of a thick blood smear by a competent microscopist. Molecular techniques, therefore, hold great promise. In these assays, specific parasite genes within a blood (or tissue) sample are amplified many times and then detected by hybridization of the amplified gene(s) with known parasite-specific gene probes. However, at the present time, molecular techniques cannot be performed in the field, they require expensive equipment and reagents, and they have not achieved their potential sensitivity.

The diagnostic 'gold standard' is still the microscopic detection of *Plasmodium* in infected blood using stained thick films. Several recent modifications permit easier detection as well as reducing the time required to examine a slide. One is by the use of acridine orange which stains the parasite. This makes it easier to detect in thick films. In addition, by centrifugation in a specifically designed and patented microcapillary tube, the parasites are concentrated in one area of the tube (the buffy coat area). The parasites and white blood cells are then stained with acridine orange. This protocol, the quantitative buffy coat (QBC) method, is fast, easy to perform, and easy to read. However, in comparison to the thick blood film assay, it requires extra equipment and the greater expense of the capillary tubes. In addition, it is apparently more difficult to distinguish between the different species of *Plasmodium* than by the use of the simple thick blood film. Unfortunately, the sensitivity of the serological assays (either the antibody, or the *Plasmodium* antigen detection protocols) is, to date, no better than the microscopic examination of stained thick blood films, which do not require the expensive reagents, and/or additional equipment. The most promising new diagnostic tool would appear to be combining the colorimetric PLDH assay with the capture of PLDH from infected blood by anti-PLDH. As noted, this quick, easy-to-use assay requires no special equipment and has the sensitivity of the microscopic examination of stained thick blood films. Further work is required to determine if this assay can distinguish between the four species of *Plasmodium* and to determine if the cost of the commercial kit is within the means of those countries where infections are endemic. At the present time, it appears that the microscopic examination of stained thick blood films will continue to be used in most endemic areas for some time into the future

from Afghanistan. Similar examples of US troops with malaria were reported following their return from the Korean and Vietnam wars. Infected individuals returning to formerly endemic areas that are still potentially receptive to malaria could lead to the re-establishment of the disease if the public-health services are not alert to the problem. Moreover, with the potential for global warming, there is also concern about an increase, as well as a change in the distribution, of the mosquito vectors for *Plasmodium*. It has been suggested that the global areas suitable for vector survival could easily expand and significantly increase the prevalence of malaria and other vector-borne diseases (Hyde, 1998).

Another note of concern is that there appears to be a possible synergy between **HIV** and malarial infections, and there is evidence to suggest that malaria exacerbates the severity of HIV. It has been shown experimentally that a *Plasmodium* antigen can increase the rate of HIV replication *in vitro*. If malaria can increase the rate of progression of an HIV-infected individual to an HIV infection, it could have a profound impact on morbidity and mortality rates in areas such as Africa, which have a high prevalence of both infections.

As we have noted, the morbidity associated with malaria is staggering. The debilitating nature of disease is due to the complicated nature of the pathophysiology (see Mehdis & Carter, 1995). Much is known about malaria, but a great deal remains to be learned. The pathophysiology considered here will include the **paroxysm**, anemia, pigmentation and organ enlargement, capillary **thromboses** and cerebral malaria, and renal failure.

The classic paroxysm begins with chills, convulsions, and shaking, collectively known as rigor. These symptoms are abruptly followed by a burning fever, violent headache, and nausea. The body temperature may reach 39.4–40.6 °C; there will be profuse sweating. After about 10 hours, the temperature returns to normal. The attacks typically begin between midnight and noon. The fevers are recurrent and peaks appear to be closely correlated with the release of the merozoites from infected RBCs. Since the duration of the erythrocytic merogony phase differs between three of the four *Plasmodium* spp., the length of time between paroxysms and fever peaks also differs (Table 3.14). *Plasmodium ovale* and *P. vivax* have a paroxysm and fever peak every 48 hours, whereas *P. malariae* has a longer erythrocytic merogony phase and its paroxysm occurs every 72 hours (Table 3.14). The differences in the time interval between paroxysms and their clinical severity has lead to different names for the disease produced by each *Plasmodium* spp. For example, the clinical disease produced by *P. falciparum* is referred to as malignant subtertian malaria and the disease produced by *P. vivax* as benign tertian malaria (Table 3.14). Evidence suggests the presence of a malaria toxin that, upon release, induces the production of a cytokine called the tumor necrosis factor (TNF). The most likely toxin candidate is a lipid moiety. TNF has been suggested to play a central role in malaria pathology in that it heightens the inflammatory response. High serum levels of TNF have been associated with cerebral malaria. It has also been shown that the 'malaria toxin' can increase the production of interferon (IFN) *in vitro*. It has been suggested that the pre-inflammatory cytokines induce an increase in nitric oxide (NO) that,

Table 3.14 Characteristics of malaria parasites (*Plasmodium* spp.) affecting humans

Species	Merogony, exoerythrocytic (days)	Disease	Merogony, erythrocytic (hours)	Distribution	Comments
P. ovale	9	ovale tertian	48	mainly tropical West Africa	rarest of the malarias
P. vivax	8	benign tertian	48	worldwide between latitudes 16° N and 20° S	most common of the malarias
P. malariae	14–15	quartan	72	worldwide but patchy, mainly tropical and subtropical	less frequent than *P. vivax*
P. falciparum	5.5	malignant subtertian	36–48	worldwide but mainly tropical and subtropical	greatest killer, causes cerebral malaria, also blackwater fever

in turn, ultimately induces the pathology observed in human malaria (Bordmann *et al.*, 1997; Clark *et al.*, 1997; Smith *et al.*, 1998).

The increase in merozoite numbers in host blood appears to occur synchronously, and the rate of development appears to be specific to different species. Thus, in *P. falciparum*, peak parasitemia is reached in 36 to 48 hours, with the synchronous release of the merozoites and periodic bouts of fever approximately every 2 days (malignant subtertian). In contrast, the merozoite increase in *P. malariae* takes 72 hours and there is a fever peak approximately every 3 days (quartan). Peak parasitemia (merozoite and also micro- and macrogamete numbers), a high fever with a high metabolic rate, and increased CO_2 release, have been hypothesized as important factors in parasite transmission. Peak parasitemia, including the number of gametocytes, has been suggested to occur during periods of maximum mosquito activity, and the increased exhalation of CO_2 may act as an attractant to vectors. It is possible, therefore, that parasite development and host pathology have evolved to maximize parasite transmission.

A serious consequence in destruction of the RBCs is anemia. There is also phagocytosis of both infected and uninfected RBCs, especially in the spleen, and this serves to exacerbate the anemia problem. Some evidence suggests that iron incorporation is slowed in **erythropoiesis** and that RBC production may be suppressed. Multiple vascular thromboses in the brain, especially in falciparum malaria, also add to the anemia problem. Moreover, the thromboses produce plasma loss from the blood vascular system. At necropsy, both the liver and spleen will be black and the white matter in the central nervous system may be slate gray in color. The pigment accumulation in the spleen and liver is due to the phagocytosis of **haemozoin** (a byproduct of hemoglobin metabolism by the trophozoite) by fixed macrophages. In the brain, it is the result of haemozoin accumulation produced when blood vessels burst because of the multiple vascular thromboses. **Splenomegaly** in malaria is due to hyperplasia or an increase in the number of cells, primarily macrophages.

During malaria, the surfaces of RBCs infected with trophozoites and meronts (schizonts; in the literature meronts are commonly referred to as schizonts or the schizont stage) change structurally and become sticky. The endothelial linings of capillaries and post-capillary venules also become sticky, primarily, in the brain. The sticky RBCs then become sequestered in the brain. This cytoadherence, as it is known, involves surface ligands on the infected RBCs as well as other specialized receptors on the endothelial cells lining the capillaries and post-capillary venules. Evidence suggests that cytoadherence is mediated in some way by the spleen because it does not occur in splenectomized monkeys. Another phenomenon, called rosetting, involves the binding of infected and uninfected RBCs. It occurs primarily in the brain of hosts in which sequestration is known to occur, e.g., in humans with *P. falciparum*. When cells become sticky, they cause the formation of multiple vascular thromboses, or clots. Following formation of the clots, pressure builds, and the vessels burst. There will then be localized hemorrhaging. When the blood supply is interrupted in these tissues, localized **anoxia** and cell death takes place, followed by convulsions, coma, and death. It must be emphasized that the explanation regarding mechanical blockage in cerebral malaria has a number of etiologies and that its pathophysiology is far from being clearly understood.

Another cause of death by falciparum malaria is renal failure. Here, there is acute tubular necrosis, resulting in a condition known as blackwater fever. Intravenous **hemolysis**, **hemoglobinemia**, and **hemoglobinuria** characterize the disease. The urine becomes black in color, and the disease is accompanied by chills, fever, jaundice, and vomiting.

As noted, with the exception of *P. falciparum*, death rarely occurs. Eventually the fever subsides and parasites become difficult to detect in the blood. The host immune response will not eliminate the infection but, with time, will keep it under control. Host immunity is age-related, possibly strain-specific, and, to be maintained, requires continuous exposure. Immunity involves T- and B-cells, as well as both Th-1 and Th-2 T-cells and their associated cytokine responses. The immune response is directed towards both the liver and the infected erythrocytes. Adaptive transfer and depletion experiments have demonstrated that CD-4 cells are required for host immunity (Smith *et al.*, 1998). Unfortunately, CD-4 cells are also suggested as being responsible for the pathology and the release of key cytokines. It has been known for a long time from work in both human and experimental animal models, that antibody can dramatically reduce parasitemia. It would appear that the IgG class of antibody molecules is the most important in immunity. In humans, there is a strong epidemiological association between the IgG-1 and IgG-3 subsets of antibody to the asexual blood stages and immunity. These antibodies are known to promote phagocytosis of RBCs containing the more mature stages of merogony. The opsonized infected RBCs are primarily cleared in the spleen of the host. The intact spleen appears to be critical for maintaining immunity to the *Plasmodium* spp. parasites. Unfortunately, immunity to human malaria is short-lived.

If an infected adult from an endemic area who shows few, or no, signs of malaria leaves for a relatively short time (a few months to a few years), and then returns to an endemic area, they appear to be as susceptible to re-infection as a previously unexposed individual. It has, therefore, been suggested that the development of a vaccine will be most difficult (but also see Engers & Godal, 1998). Experimental animals vaccinated with sporozoite or merozoite antigens may show good protection against a homologous challenge; however, a field vaccine for use in humans has not yet been developed. There are several possible reasons for this difficulty. One is that there appear to be numerous antigenic strains of *Plasmodium* spp. and some strains occur in only a very limited geographical area (Babiker & Walliker, 1997). Another reason is that *Plasmodium* spp. are capable of undergoing antigenic variation. Therefore, any vaccine for use in the field must include multiple antigenic epitopes. Moreover, the parasites, except for very short time intervals during their life cycle, reside within host cells. Although infected cells eventually possess surface antigens of parasite origin, they are expressed on the host cell surface primarily during the later stages of parasite development and the window of opportunity for the host to respond to these parasite antigens is, therefore, limited. Moreover, early in the infection, the *Plasmodium* spp. may have already invaded new host RBCs, before the host has adequately responded immunologically. Finally, the host is immunosuppressed during a malarial infection. It thus appears that the parasite may manipulate the host's immune response to its own advantage, thereby increasing the opportunity to successfully complete its life cycle. Whoever develops a vaccine will have to do better than 'Mother Nature' and thousands of years of evolution. Although complete protection and long-lived immunity does not occur naturally, the host has, in addition to the short-lived and age-related immunity, also evolved several RBC genotypes that confer greater resistance to *Plasmodium* spp. It appears that *Plasmodium* parasites grow poorly in RBCs from humans with a variety of RBC abnormalities. For example, the human gene for sickle-cell anemia confers resistance to *P. falciparum*.

3.3.4.3 Subclass Piroplasmasina

This subclass contains a single order, the Piroplasmorida, and four families, of which species from only two of the families will be discussed, the Babesiidae and Theileriidae (Fig. 3.18). The piroplasms are all intracellular parasites inhabiting erythrocytes and, sometimes, other vertebrate cells. They are known to infect birds and mammals. Humans are not normally infected, but can act as an accidental host. All species of these parasites are assumed to be transmitted by ticks.

Species of *Babesia* primarily infect mammals, but a few are known from birds and reptiles. Generally, members of this group inhabit only host RBCs. In the RBC, they divide by binary fission, or by budding, into two, possibly four, individuals. Recently, it has been reported that several species have a pre-erythrocytic stage producing a single generation of meronts in host lymphocytes. Some of the different species of *Babesia* are listed in Table 3.15.

The life cycle of *Babesia* spp. is shown in Fig. 3.24. Merogony occurs in the vertebrate host, primarily in RBCs, whereas gamete maturation and fusion, plus sporogony, occur in the tick vector. Gamete fusion and the formation of a motile zygote (**kinete**) take place in the gut. The kinete

Table 3.15 Some *Babesia* species causing pathology in domesticated animals

Species	Size	Main vertebrate host(s)	Host pathology	Common name	Main tick vector(s)	Geographical distribution
B. bigemina	2–5 μm	cattle, deer	moderate	redwater fever	*Boophilus, Haemaphysalis, Rhipicephalus*	worldwide
B. bovis	1–2 μm	cattle, deer	high	redwater fever	*Boophilus, Ixodes, Rhipicephalus*	worldwide
B. major	2–5 μm	cattle	low	—	*Boophilus*	Europe
B. divergens	1–2 μm	cattle	moderate	redwater fever	*Ixodes*	Europe
B. equi	1–2 μm	horses, zebras	high	biliary fever	*Dermacentor, Hyalomma, Rhipicephalus*	worldwide
B. ovis	1–2 μm	sheep, goats	low	—	*Rhipicephalus*	Europe, Asia, Africa
B. trautmanni	2–5 μm	pigs	low	—	*Rhipicephalus*	Europe, Asia, Africa
B. canis	2–5 μm	dogs	high	tick fever	*Rhipicephalus*	Europe, Asia, Africa, Americas
B. felis	1–2 μm	cats, lions, leopards	moderate	—	*Haemaphysalis*	Asia, Africa
B. microti	1–2 μm	rodents	low	—	*Ixodes*	worldwide

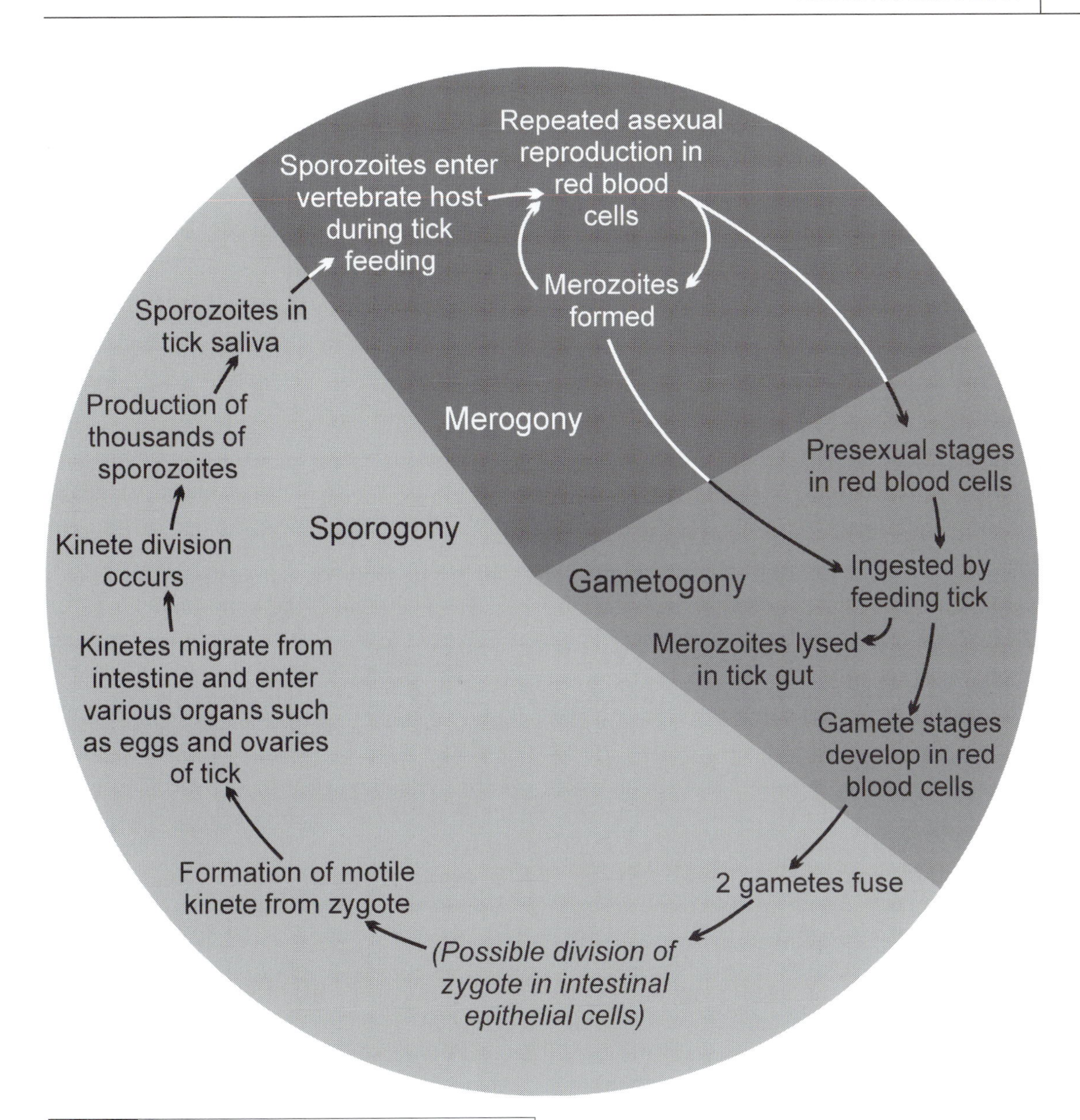

Fig. 3.24 Life cycle of *Babesia* spp. Note that by the kinetes entering the eggs and ovaries of the tick, the infection can be transferred from tick to tick without infecting a vertebrate host.

can migrate away from the gut and penetrate cells of the hemolymph, muscle, Malpighian tubules, and even into the eggs of the adult female tick. The intracellular kinetes then divide by multiple fission to produce additional kinetes. This process can be repeated until the kinetes migrate to the salivary glands where further divisions occur to form sporozoites. In the vector, *Babesia* spp. can be passed from stage to stage during its developmental cycle (Kreier & Baker, 1987; Cox, 1993). In addition, some species of *Babesia* can enter the eggs of the female tick and the infection can be maintained in the absence of the vertebrate host. It would appear that sporozoite formation in the salivary gland of the tick is only initiated after the vector begins to feed on a vertebrate host. When pathogenic, the disease usually produces fever, an enlarged spleen, and anemia. Infections can be very serious in splenectomized individuals.

Species of Theileriidae undergo merogony in RBCs and in other cells such as lymphocytes in the vertebrate host. The only two genera in the family are separated on the basis of the size and location of their meronts in exoerythrocytic cells. *Theileria* has small or medium-sized meronts in lymphocytes, whereas *Cytauxzoon* has larger meronts in phagocytic cells. Species of *Theileria* cause important diseases of domestic animals (Table 3.16).

A summary of the life cycle of *Theileria* is shown in Fig. 3.25. It appears similar to that of *Babesia* spp. with the following exception. Kinetes develop only in the salivary glands of the tick hosts, and transmission of the parasite cannot occur from an adult tick to the larval stage via the egg (Kreier & Baker, 1987; Cox, 1993). *Theileria* resides in both CD-4 and CD-8 T-cells as well as in B-cells, the very cells that are central to the host's immune response. In *T. parva*, division of the parasite is closely associated with division of the host cell. During mitosis, the parasite associates with the mitotic spindle and divides along with the host cell. Schizont nuclei are distributed randomly to both daughter cells. Therefore, the expansion of the parasite population can only occur during the clonal expansion of the infected host cells.

In highly susceptible cattle, *T. parva* produces a lymphoproliferative disease, causing death in 4 weeks or less. Experimental evidence has been presented to suggest that host T-cell responses actually potentiate the growth of the parasitized cells. The growth-promoting activity of the cytokine IL-2, and possibly IFN and IL-10, has been demonstrated *in vitro*. In addition, infected cells have been shown to contain messenger RNA transcripts for IL-10 and, in some cells, IL-2. Parasitized cells are reported to express cytokine receptors for these cytokines. It would, therefore, appear that the host's immune response significantly potentiates the growth of parasitized cells. Immunity to infection can occur in animals that survive an infection. It seems to be T-cell-mediated and involves the elimination of parasitized lymphocytes by cytotoxic T-cells or by inhibiting the penetration of sporozoites by antibody into susceptible cells. There is strong evidence for antigenic heterogeneity between strains of *T. parva* and, therefore, strain specificity in the host's immune response (Morrison & McKeever, 1998). The development of an effective vaccine will be difficult.

3.3.5 Host and habitat diversity

As previously noted, the Apicomplexa has approximately 4000 known species and it has been estimated that there may actually be millions of species. They infect almost all known animal groups. Although the Apicomplexa parasitize hosts throughout the animal kingdom, most species show some specificity, infecting only single genera or even species. A good example is *Plasmodium falciparum*, which infects only humans. On the other hand, *Toxoplasma gondii*, an isosporan, is not host-specific. It is worldwide in its distribution, and has been detected in cats, wild carnivores, dogs, horses, pigs, sheep, cattle, i.e., basically all domesticated animals, as well as rodents and birds.

The large number of known species of Apicomplexa, the large number of different host species and different habitats, as well as the variety of methods of transmission, indicate that this is an enormously successful group of organisms. Another way to measure their success is to examine their enormous public-health and economic importance. At the end of the nineteenth century, malaria occurred in more than 100 million people and was unquestionably one of the most serious public-health problems in the world. The economic and social consequences of this disease were many. By the late 1950s through the 1960s, endemic malaria had virtually been eliminated from the United States, and different groups, such as the World Health Organization, were openly speaking of worldwide malaria eradication. The reasons for this optimism were the development of cheap and effective insecticides, e.g., DDT, new inexpensive and relatively non-toxic chemotherapeutic agents, e.g., chloroquine, the belief that a vaccine was right around the corner, and finally, a better understanding of the ecology of the mosquito bringing hope of better vector control. However, that was then and this is now; by the 1990s, it was apparent that the prevalence of malaria worldwide had exceeded its pre-1900 level. Cases of malaria also continue to be imported into the southeastern and southwestern USA by migrant farm workers. There are many reasons for the re-emergence of malaria worldwide (Krogstad, 1996). One is the evolution of drug resistance by *Plasmodium* spp. Chloroquine resistance is now

Table 3.16 Some *Theileria* species causing pathology in domestic animals

Species	Main vertebrate host(s)	Host pathology	Common name	Main tick vector(s)	Geographical distribution
T. annulata	cattle, buffaloes	high	Mediterranean coast fever	*Hyalomma*	Europe, Africa, Asia
T. parva	cattle, buffaloes	high	East coast fever	*Rhipicephalus, Hyalomma*	Africa
T. mutans	cattle	low	—	*Amblyomma*	Europe, Africa, Asia, Australia
T. ovis	sheep, goats	low	—	*Rhipicephalus*	Africa
T. separata	sheep, goats	low	—	*Rhipicephalus*	Africa

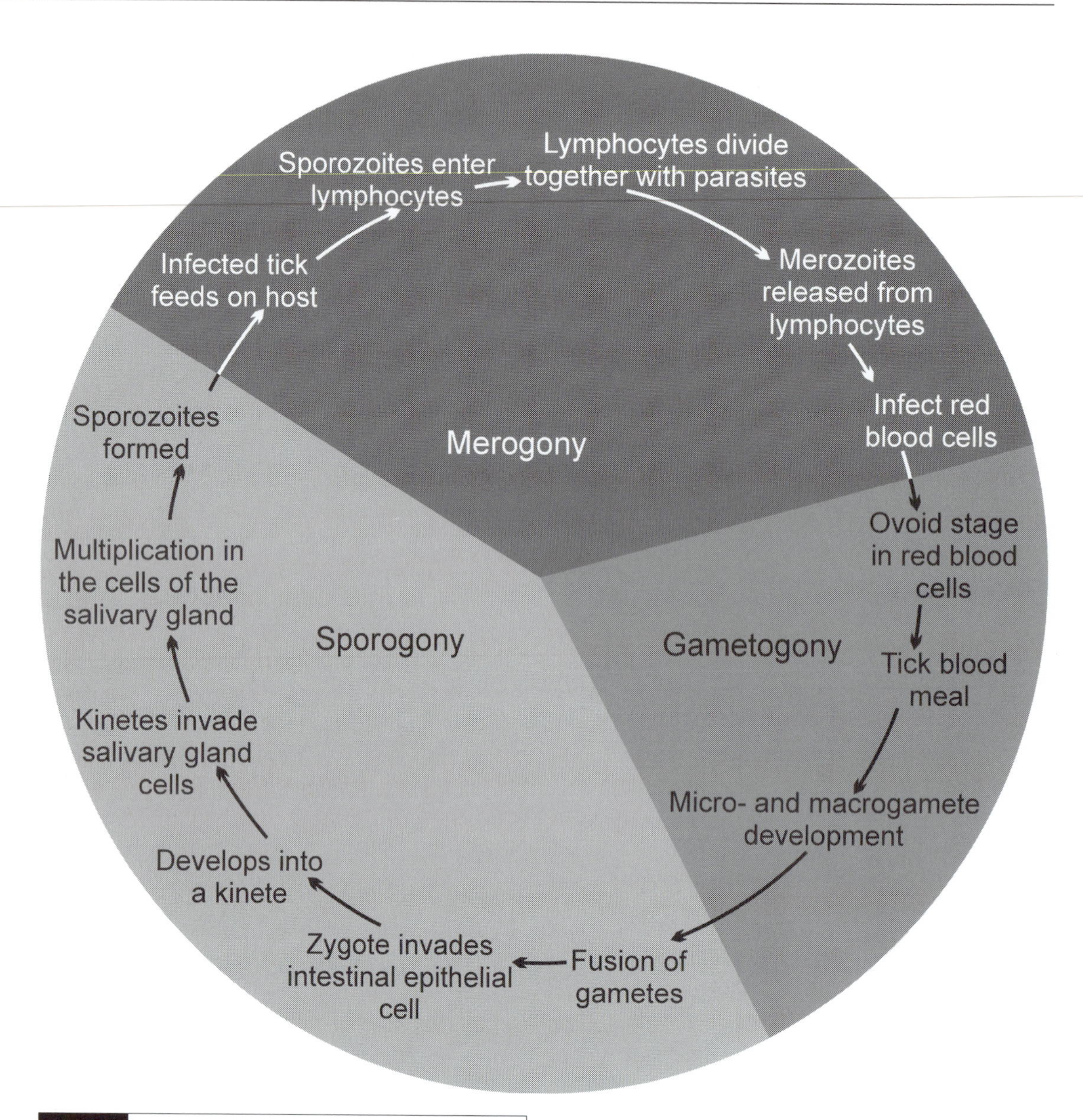

Fig. 3.25 Life cycle of a species of *Theileria*.

worldwide, and the scientific community is racing against rapidly evolving *Plasmodium* spp. to find new drugs (or drug cocktails) to which the parasite is susceptible. In addition, although millions of dollars of research money have gone into attempts to develop a vaccine, to date, a truly effective vaccine is not available. Part of the problem in vaccine development is the phenotypic variability of the *Plasmodium* spp. In addition, we still do not understand fully either the host's immune response to this parasite or the biology of vector transmission, as was previously believed. Finally, the development of insecticide resistance by the mosquito vector was not anticipated. This resistance, as well as the inability to use DDT due to its environmental toxicity, significantly decreased the success of vector-control efforts (Roberts *et al.*, 1997). The point here is that, even with large-scale attempts to eradicate the *Plasmodium* spp. and the scourge malaria causes, the parasite has adapted to new environmental conditions and continues to infect its human hosts. This is a remarkable success story, certainly from the perspective of the human *Plasmodium* spp.

Table 3.17 Some examples of ciliates that inhabit animal hosts

Ciliate	Host	Habitat
Sphaerophyra sol	ciliate	cytoplasm
Paenucleatrum sp.	cockroach	intestine
Tetrahymena limacis	garden slug	renal organ
Spirochona gemmipara	amphipod	gill plates
Ichthyophthirius multifiliis	fish	skin, gills, fins
Chilodonella hexastichia	grass carp	gills
Trichodina nobilis	grass carp	skin
Trichodina pediculus	hydra, fish	surface, gills
Trichodina californica	salmon	gills
Trichodina urinicola	amphibian	urinary bladder
Nyctotherus cordiformis	frog, toad	intestine
Balantidium spp.	frog, newt, iguana	intestine
Balantidium coli	human, pig	cecum, colon
Entodinium spp.	sheep, cattle	rumen

3.4 Phylum Ciliophora

3.4.1 General considerations

The ciliates are a large and diverse group of protozoans. They are distinguished from each other on the basis of their pattern of cilia and associated cortical structure, and on the basis of their nuclear structure and function. Ciliates have two morphologically different nuclei, a macronucleus that is involved in the day-to-day metabolism of the cell, and a micronucleus involved in sexual reproduction. The macronucleus is derived from the micronucleus and is polyploid. Generally, both nuclei can replicate when the cell divides by binary or multiple fission. During division, however, the nuclear membrane usually remains intact, and intranuclear microtubules are involved in the separation of the chromosomes. Sexual reproduction in the ciliates occurs by a unique process called conjugation in which ciliates pair and there is a temporary, partial fusion of the cells. This is then followed by an exchange of their micronuclei, after which the cells separate and eventually divide by binary fission.

Most free-living ciliates are predators and ingest nutrients through a cytostome or **cytopharynx** that is often surrounded by cilia. This structure and associated cilia are referred to as the oral apparatus. Following ingestion, food vacuoles are formed and the material within the vacuoles is digested. Waste products are excreted through the **cytoproct**, the cell's anus. In addition, ciliates have a contractile vacuole, an organelle believed to control intracellular water and ion concentrations.

There are three subphyla and eight different classes of ciliates. Most of these are free-living and can be found in many freshwater and marine environments. Within each of the subphyla, and in a majority of the classes, there are mutualistic, commensalistic, and parasitic species. These species appear to have evolved independently within the various taxonomic groups. They infect a variety of hosts (Table 3.17), including marine and terrestrial animals, and both invertebrates and vertebrates (Frank, 1984).

In contrast to the apicomplexans and sarcomastigophorans, which have a diversity of parasitic forms in humans, there is only a single ciliate in humans. *Balantidium coli* is a large, colon-dwelling species (Fig. 3.26) that is not usually pathogenic (Zaman, 1998).

3.4.2 Form and function

The most prominent external features of ciliates are the arrangement of their cilia and the associated cortex. The major component of the cortex

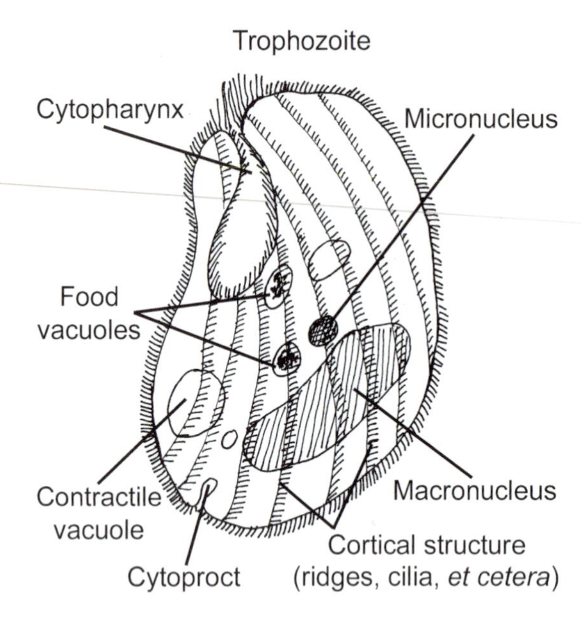

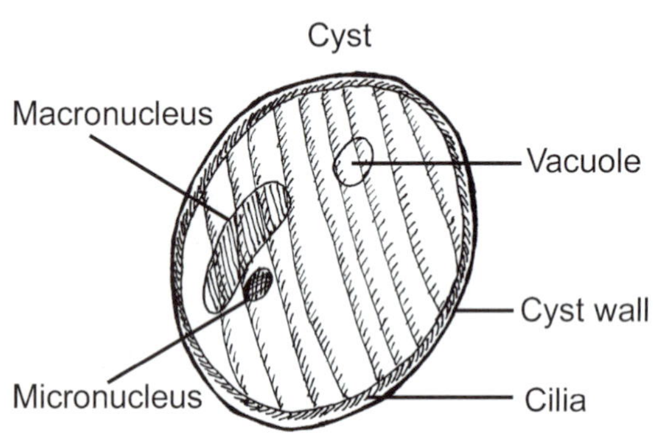

Fig. 3.26 Diagrammatic sketch of the trophozoite and cyst of the parasitic ciliate *Balantidium coli*.

is a **kinetid** that contains one or two **kinetosomes**, their associated fibrils, and cilia. The kinetids are linked by a highly organized and complex system of microtubules and fibrils (Fig. 3.27). The cortex can generally be divided into two basic regions. The oral apparatus is involved in food uptake and digestion, whereas the somatic region is involved in forming a protective coating, sensing the environment, locomotion, and attachment to the substrate. A large expenditure of energy is needed for these activities and mitochondria can be found located in grooves between the kinetids (Fig. 3.27). This close association presumably permits the most efficient transfer of energy from the mitochondria to the kinetids. The oral apparatus is composed of associated cilia and the cytostome. The kinetids in the oral apparatus can be simple, or structurally arranged into dikinetids or polykinetids. The associated cilia direct food particles towards the cytostome that is involved in the formation of food vacuoles. The somatic regions of the cell are characterized by the morphological arrangement of cilia on the surface of the cell. If one examines all the known ciliates, it seems that any pattern of shape and cilia arrangement that the human mind could design is present (see Small & Lynn, 1985). The microscopic morphology of the ciliate *Chilodonella hexastichia* is shown in Fig. 3.28A. Note the symmetry of the ciliary pattern. Considering the great diversity in these patterns, the complexity of ciliate surface structures, and their important and coordinated function in eating and movement, it is not difficult to imagine why biologist have long been fascinated by these protozoans.

Associated with the cortex are organelles called **extrusomes**. These structures release various materials to the exterior of the cell. For example, toxicysts extrude toxins used to capture prey, and microcysts release materials that help trap food, coat the cell membrane, or help coat the wall of the cyst. Other prominent external structures include the contractile vacuole, and the anus. Sessile or sedentary ciliates can attach either temporarily, or permanently, using a number of different devices. These structures may include a special set of cilia in the cortical region used in attachment, secreted products that form a stalk or an adhesive disk, or an elongated terminal body part that secretes a stalk. The elongated pedicel, or stalk, or both, can assume a variety of different shapes. The microscopic morphology of an adhesive basal disk on the aboral surface with its ring of radially arranged proteinaceous subpellicular elements (denticles) used by *Trichodina nobilis* for attachment to its host is shown in Fig. 3.28B. It provides an excellent example of the structural complexity of the ciliates.

Since the ciliates are eukaryotic cells, the cytoplasm also contains the normal eukaryotic organelles, i.e., mitochondria, nuclei, Golgi, etc. The most prominent ultrastructural features include those previously discussed, i.e., the cortex, the oral apparatus, and the micro- and macronucleus.

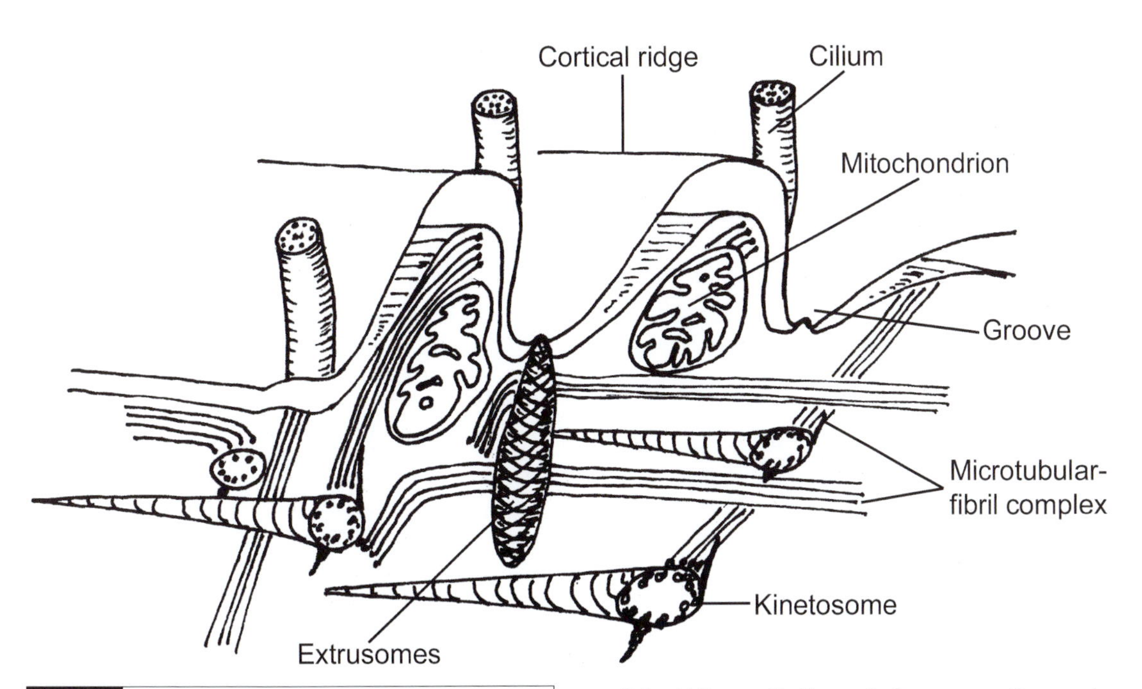

Fig. 3.27 Simplified diagram of the cortical ultrastructure of a ciliate. Note the overlap of microtubular and fibrous associates of kinetosomes to provide a structural latticework forming and supporting the cortex. Mitochondria underlie the cortical ridges. Rod-shaped extrusomes lie in grooves between kinetids.

There is very little information available on the biochemistry and physiology of the parasitic ciliates. It is known that *Balantidium coli* in humans can use O_2 even though it normally resides in the microaerophilic environment of the large intestine. It has also been shown that *B. coli* contains hydrolytic enzymes such as hyaluronidase which may be involved in the invasion of host tissue. Numerous starch granules and ingested bacterial cells, as well as RBCs and fat droplets, have been observed in the cytoplasm. It is assumed that the stored fats and starch are reserve energy sources and that ingested RBCs and bacteria, following their hydrolytic digestion, supply the cell with many of the required macromolecular building blocks.

3.4.3 Life cycle

Ciliates reproduce asexually by binary fission, or sexually in which two cells temporarily join and their micronuclei are exchanged. There are also environmentally resistant cyst stages in some groups.

Balantidium coli, the only human pathogen in the phylum, has a cosmopolitan distribution. The prevalence in most endemic locations is less than 1.0%. However, in some areas, prevalences as high as 5% to 6% have been recorded. *Balantidium coli* has also been isolated from monkeys, pigs, cats, and rodents. The infection is **anthropozoonotic**, with pigs suggested as the primary reservoir host (Esteban *et al.*, 1998). Normally, *B. coli* infections are asymptomatic and the organisms act as commensals. Humans appear to have a high degree of natural resistance to infection with *B. coli*. However, the ciliate can invade the mucosa and submucosa of the cecum and colon, producing ulcers. It usually does not penetrate beyond the intestinal wall into deeper body tissues and other organs, although several cases involving the liver, lung, or pleural cavity have been reported. The reason for the increased susceptibility of some individuals to clinical disease is not known, but may be related to a depressed immune status.

In the intestine, cysts are produced and then excreted in the feces. Division does not occur in the cysts, which can remain viable for weeks in moist feces. Cysts are ingested in contaminated food or water, and excyst in the intestine. The trophozoites (Fig. 3.26) are phagotrophic, feeding primarily on cell debris, starch grains, etc. Over time, the trophozoites may, as indicated previously, invade the intestinal mucosa (Fig. 3.29).

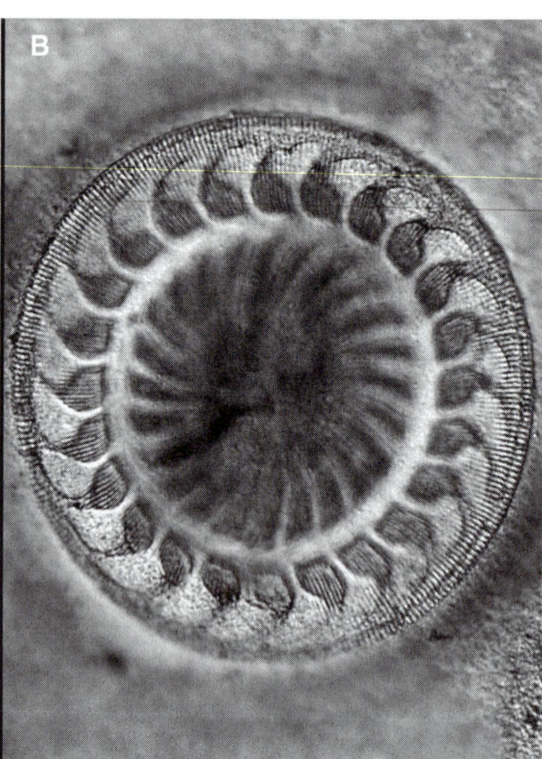

Fig. 3.28 (A) Microscopic morphology of the ciliary pattern of *Chilodonella hexastichia*. (B) Microscopic morphology of the adhesive basal disk on the aboral surface and its ring of radially arranged proteinaceous subpellicular elements (denticles) of *Trichodina nobilis*. This structure is used in the attachment of the ciliate to its host. (Photographs courtesy of Dr. Mohamed Meguid, National Water Resource Center, Egypt.)

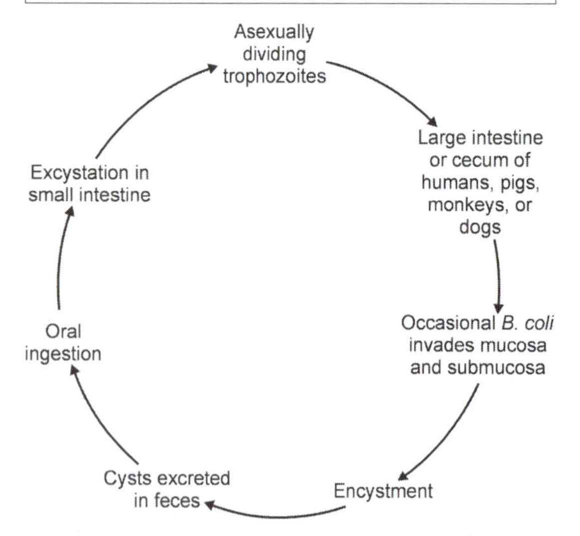

Fig. 3.29 Life cycle of *Balantidium coli*. It is a direct life cycle transmitted by the fecal–oral route. *B. coli* can also reproduce sexually by conjugation.

Sexual reproduction occurs by conjugation between trophozoites in the intestinal tract.

Nyctotherus cordiformis lives in the intestine and colon of the common tree frog, *Hyla versicolor*. The tadpole ingests cysts of *N. cordiformis* which excyst, and divide repeatedly until host **metamorphosis** begins, and then conjugation occurs. Sexual reproduction only occurs in tadpoles undergoing metamorphosis. Following metamorphosis, the ciliate matures into the adult trophozoite, complete with a micro- and macronucleus. Division again occurs in the adult frog and, eventually, precyst and cyst stages are excreted into the environment. There is obviously a very close association between host and parasite development.

The life cycle of *Ichthyophthirius multifiliis*, a parasite of freshwater fishes, differs from that of *B. coli* and *N. cordiformis* in at least two significant ways. First, free-swimming trophozoites of *I. multifiliis* invade the skin and gills of fishes, not the intestinal tract, forming small pustules. Second, reproduction does not occur on the host. When the pustules rupture, the freed forms eventually settle to the bottom of the lake or pond, secrete a thick external coat, and divide repeatedly within the cyst. After a short time (7–8 hours), division is

Table 3.18 | The ciliated epifauna on the amphipod *Gammarus locusta* in Danish waters

Ciliate	Ciliate habitat on amphipod
Zoothamnium sp.	antenna
Cothurnia gammari	plates of thorax
Zoothamnium hiketes[a]	spines, last abdominal segments
Lagenophrys sp.	maxillipeds
Heliochona sessilis	gills
Heliochona scheuteni	pleopods
Acineta foetida	host surface, feeds on other ciliates
Trochiloides sp.	host surface, bacterial feeder
Conidophrys pilisuctor	feeds on host tissue
Gymnodinioides inkystans	feeds on host tissue

Notes:

[a] *Zoothamnium hiketes* is also parasitized by the ciliate *Hypocoma parasitica*.

completed, and free-swimming, infective trophozoites (also called **tomites**) are released. Once in contact with a fish, the tomite bores into the skin and the pustule is formed. Within 3 days, the organism has developed into the trophozoite again and is ready for release back into the environment. For a microscopic study of the stages in the life cycle of *I. multifiliis*, see McCartny *et al.* (1985).

Ciliates demonstrate both site specificity and habitat diversity. A good example is found in the ciliate epifauna on the amphipod host *Gammarus locusta*, in Danish waters. Included are four species of ciliates that individually reside on the host's antenna, on the plates of the thorax, among the spines of the last abdominal segment, or on the maxillipeds. There are additional ciliates associated with the host's gills and the pleopods (Table 3.18). These ciliates generally are commensals feeding on bacteria that are attached to the surface of the host; however, at least one species feeds on host tissue in the exuvium. The free-swimming stages of the different ciliate species appear to settle only on *Gammarus*. This should insure the successful transmission of the ciliates from host to host. In addition, the free-swimming 'swarmers' are only released from *Gammarus* just prior to molting. This suggests that host hormones involved in molting control the release of swarmers. The timing of swarmer release presumably permits movement of the ciliates to the exoskeleton of a new host (Fenchel, 1987).

In summary, the transmission of most parasitic ciliates involves an environmentally protected component in the form of a cyst, or a cyst-like, stage. Asexual reproduction can take place as trophozoites in the host, or within the cyst. Sexual reproduction will occur in the host. Ciliates may be serious pathogens, non-pathogenic commensals, or mutuals (as in the case of the ruminant ciliates). The life cycle of many ciliates appears to be closely correlated with the development of their host, as in *N. cordiformis*, in which sexual reproduction occurs only during host metamorphosis.

3.4.4 Diversity

Most of the ciliates are free-living and can be found in both marine and freshwater environments. When symbiotic, ciliates may also inhabit a wide range of hosts. They are associated with most animal phyla, including invertebrates and all vertebrate classes (Table 3.17). They occupy a variety of different tissue sites, and most are ectocommensals. For example *Trichodina nobilis*, a peritrich, is found on the skin of the grass carp (Fig. 3.30A), while a second ciliate, *Chilodonella hexastichia*, is found on the gills of the same host (many *C. hexastichia* can be seen near, or on, the gill surface in Fig. 3.30B). Some species, such as *Trichodina pediculus*, are known to produce heavy infections on the gills of freshwater fishes causing serious injury. It appears that the ciliates have very successfully adapted to a variety of diverse

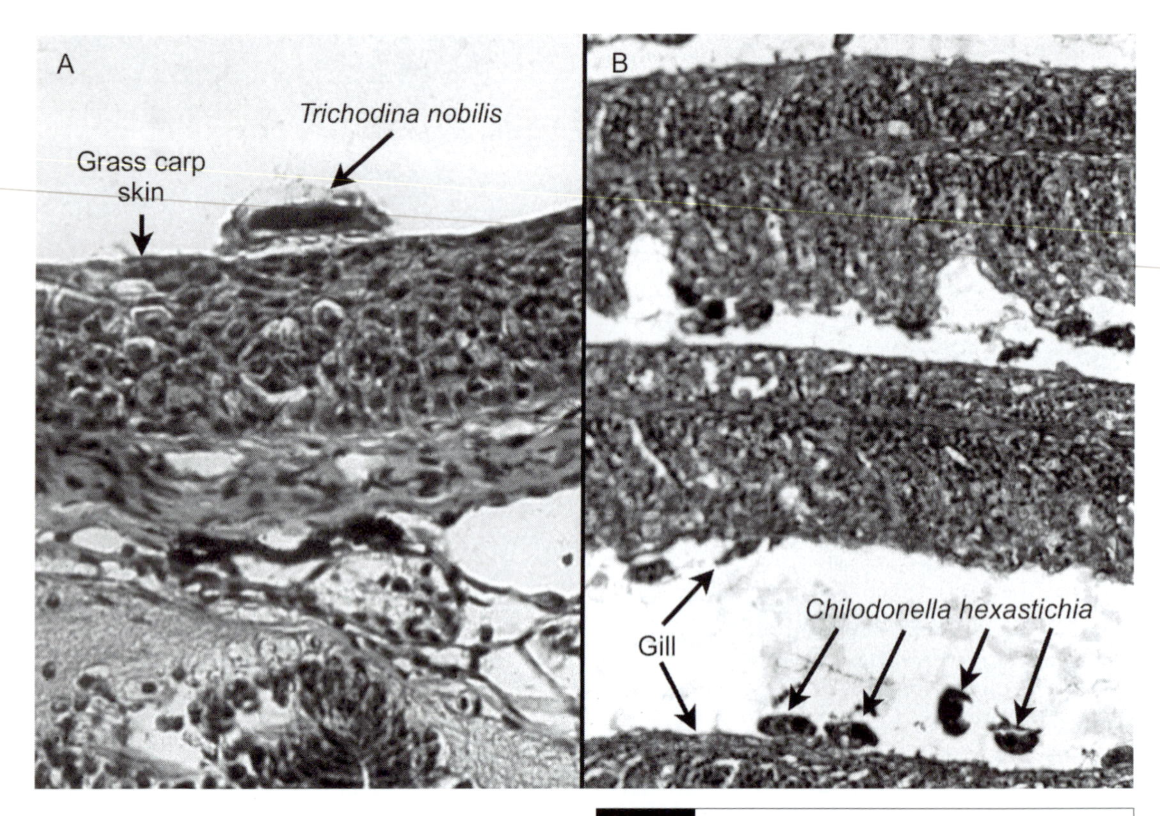

Fig. 3.30 (A) Microscopic section of the skin of the grass carp *Ctenopharyngodon idella* showing *Trichodina nobilis* on the surface; (B) microscopic section of the gills of the grass carp showing numerous *Chilodonella hexastichia* on or near the gill surface. (Photographs courtesy of Dr. Mohamed Meguid, National Water Resource Center, Egypt.)

habitats. The close relationship between some hosts and their ciliate symbionts would suggest that many of these associations are evolutionarily very old.

References

Alexander, J. & Hunter, C. A. (1998) Immunoregulation during toxoplasmosis. In *Immunology of Intracellular Parasitism,* Chemical Immunological Series, vol. 70, ed. F. Y. Liew & F. E. G. Cox, pp. 81–102. Basel: Karger.

Alexander, J. & Russell, D. J. (1992) The interaction of *Leishmania* species with macrophages. *Advances in Parasitology*, **31**, 175–254.

Babiker, H. A. & Walliker, D. (1997) Current views on the population structure of *Plasmodium falciparum*: implications for control. *Parasitology Today*, **13**, 262–267.

Blum, J. J. (1991) Intermediary metabolism of *Leishmania*. In *Biochemical Protozoology*, ed. G. Coombs & M. North, pp. 123–133. London: Taylor & Francis.

Bordmann, G., Favre, N. & Rudin, W. (1997) Malaria toxins: effects on murine spleen and bone marrow cell proliferation and cytokine production *in vitro*. *Parasitology*, **115**, 475–483.

Bryant, C. (1993) Biochemistry. In *Modern Parasitology: A Textbook of Parasitology*, ed. F. E. G. Cox, pp. 117–136. Oxford: Blackwell Scientific Publications.

Carreno, R. A., Schnitzler, B. E., Jeffries, A. C., Tenter, A. M., Johnson, A. M. & Barta, J. R. (1998) Phylogenetic analysis of coccidia based on 18S rDNA sequence comparison indicates that *Isospora* is most closely related to *Toxoplasma* and *Neospora*. *Journal of Eukaryotic Microbiology*, **45**, 184–188.

Cavalier-Smith, T. (1993) Protozoa and its 18 Phyla. *Microbiological Reviews*, **57**, 953–994.

Cheng, T. C. (1970) *Symbiosis: Organisms Living Together*. New York: Western Publishing.

Clark, I. A., Al Yaman, F. M. & Jacobson, L. S. (1997) The

biological basis of malarial disease. *International Journal for Parasitology*, **27**, 1237–1249.

Clayton, C. E. & Michels, P. (1996) Metabolic compartmentation in African trypanosomes. *Parasitology Today*, **12**, 465–471.

Coombs, G. & North, M. (eds.) (1991) *Biochemical Protozoology*. London: Taylor & Francis.

Corliss, J. O. (1984) The Kingdom Protista and its 45 Phyla. *Biosystems*, **17**, 87–126.

Cox, F. E. G. (1993) Parasitic Protozoa. In *Modern Parasitology: A Textbook of Parasitology*, ed. F. E. G. Cox, pp. 1–23. Oxford: Blackwell Scientific Publications.

Engers, H. D. & Godal, T. (1998) Malaria vaccine development: current status. *Parasitology Today*, **14**, 56–64.

Esteban, J.-G., Aguirre, C., Angles, R., Ash, L. & Mas-Coma, S. (1998) Balantidiasis in Aymara children from the northern Bolivian Altiplano. *American Journal of Tropical Medicine and Hygiene*, **59**, 922–927.

Fedorko, D. P. & Hijazi, Y. M. (1996) Application of molecular techniques to the diagnosis of microsporidial infection. *Emerging Infectious Diseases*, **2**, 183–191.

Fenchel, T. (1987) Symbiotic protozoa. In *Ecology of Protozoa*, pp. 161–166. Madison: Science Tech Publishers.

Frank, W. (1984) Non-hemoparasitic protozoans. In *Diseases of Amphibians and Reptiles*, ed. G. L. Hoff, F. L. Frye & E. R. Jacobson, pp. 259–384. New York: Plenum Press.

Garnham, P. C. C. (1966) *Malaria Parasites and other Haemosporida*. Oxford: Blackwell Scientific Publications.

Gutteridge, W. E. & Coombs, G. (1977) *Biochemistry of Parasitic Protozoa*. Baltimore: University Park Press.

Harrington, W., Krupnick, A. J. & Spofford, Jr., W. O. (1991) *Economics and Episodic Disease: The Benefits of Preventing a Giardiasis Outbreak*. Washington, DC: Resources for the Future.

Hoare, C. A. (1972) *The Trypanosomes of Mammals: A Zoological Monograph*. Oxford: Blackwell Scientific Publications.

Hyde, B. (1998) Infectious diseases: weather or not? *ASM News*, **64**, 316–317.

Kreier, J. P. & Baker, J .R. (1987) *Parasitic Protozoa*. Winchester: Allen & Unwin.

Krogstad, D. (1996) Malaria as a reemerging disease. *Epidemiological Reviews*, **18**, 77–89.

Kudo, R. (1966) *Protozoology*, 5th edn. Springfield: Charles C. Thomas.

Lee, J. J., Hutner, S. H. & Bovee, E. C. (eds.) (1985) *An Illustrated Guide to the Protozoa*. Lawrence: Society of Protozoologists.

Levine, N. D. (1985) Phylum II. Apicomplexa Levine, 1970. In *An Illustrated Guide to the Protozoa*, ed. J. J. Lee, S. H. Hutner & E. C. Bovee, pp. 322–374. Lawrence: Society of Protozoologists.

Manwell, R. G. (1968) *Introduction to Protozoology*. New York: Dover Publications.

Margulis, L. (1981) *Symbiosis in Cell Evolution: Life and its Environment on the Early Earth*. San Francisco: W. H. Freeman.

Marshall, M., Naumovitz, D., Ortega, Y. & Sterling, C. (1997) Waterborne protozoan pathogens. *Clinical Microbiological Reviews*, **10**, 67–85.

Martínez, A. J. & Visvesvara, G. S. (1997) Free-living amphizoic and opportunistic amebas. *Brain Pathology*, **7**, 583–598.

Martínez-Paloma, A. & Espinosa-Cantellano, M. (1998) Amoebiasis: new understanding and new goals. *Parasitology Today*, **14**, 1–2.

McCartny, J. B., Fortner, G. W. & Hansen, M. F. (1985) Scanning electron microscope studies on the life cycle of *Ichthyophthirius multifiliis*. *Journal of Parasitology*, **71**, 218–226.

McDonald, V. & Bancroft, G. J. (1998) Immunological control of *Cryptosporidium* infections. In *Immunology of Intracellular Parasitism*, Chemical Immunology Series, vol. 70, ed. F. Y. Liew & F. E. G. Cox, pp. 103–123. Basel: Karger.

Mehdis, K. N. & Carter, R. (1995) Clinical disease and pathogenesis in malaria. *Parasitology Today*, **11**, 1–16.

Michels, P. A. M., Hannaert, V. & Bakker, B. M. (1997) Glycolysis of kinetoplastida. In *Trypanosomiasis and Leishmaniasis*, ed. G. Hide, J. C. Mottram, G. H. Coombs & P. H. Holmes, pp. 133–148. Wallingford, UK: CAB International.

Miles, M. A. (1988) Viruses of parasitic protozoa. *Parasitology Today*, **4**, 289–290.

Molyneux, D. H. & Ashford, R. W. (1983) *The Biology of Trypanosoma and Leishmania, Parasites of Man and Domestic Animals*. London: Taylor & Francis.

Morrison, W. I. & McKeever, D. J. (1998) Immunobiology of infections with *Theileria parva* in cattle. In *Immunology of Intracellular Parasitism*, Chemical Immunology Series, vol. 70, ed. F. Y. Liew & F. E. G. Cox, pp. 163–185. Basel: Karger.

Müller, M. (1991) Energy metabolism of anaerobic parasitic protists. In *Biochemical Protozoology*, ed. G. Coombs & M. North, pp. 80–91. London: Taylor & Francis.

O'Donoghue, P. J. (1995) *Cryptosporidium* and cryptosporidiosis in man and animals. *International Journal for Parasitology*, **25**, 139–195.

Opperdoes, F. R. (1991) Glycosomes. In *Biochemical*

Protozoology, ed. G. Coombs & M. North, pp. 134–144. London: Taylor & Francis.

Roberts, D. R., Laughlin, L. L., Hsheih, P. & Legters, L. J. (1997) DDT, global strategies and malaria control crisis in South America. *Emerging Infectious Diseases*, **3**, 295–302.

Sam-Yellowe, T. Y. (1996) Rhoptry organelles of the Apicomplexa: their role in host cell invasion and intracellular survival. *Parasitology Today*, **12**, 308–316.

Seed, J. R. & Hall, J. E. (1992) Trypanosomes causing disease in man in Africa. In *Parasitic Protozoa*, 2nd edn, ed. J. P. Kreier & J. R. Baker, pp. 85–155. New York: Academic Press.

Seed, J. R., Bollinger, R. & Gam, A. A. (1968*a*) Studies of frog trypanosomiasis. II. Seasonal variations in the parasitemia levels of *Trypanosoma rotatorium* in *Rana clamitans* from Louisiana. *Tulane Studies in Zoology and Botany*, **15**, 54–69.

Seed, J. R., Southworth, C. & Mason, G. (1968*b*) Studies of frog trypanosomiasis. I. A 24 hour cycle in the parasitemia level of *Trypanosoma rotatorium* in *Rana clamitans*. *Journal of Parasitology*, **54**, 946–960.

Small, E. B. & Lynn, D. H. (1985) Phylum Ciliophora, Doffein, 1901. In *An Illustrated Guide to the Protozoa*, ed. J. J. Lee, S. H. Hutner & E .C. Bovee, pp. 393–575. Lawrence: Society of Protozoologists.

Smith, H. V. & Rose, J. B. (1998) Waterborne cryptosporidiosis: current status. *Parasitology Today*, **14**, 14–22.

Smith, N. C., Fell, A. & Good, M. F. (1998) The immune response to asexual blood stages of malaria parasites. In *Immunology of Intracellular Parasitism*, Chemical Immunology Series, vol. 70, ed. F. Y. Liew & F. E. G. Cox, pp. 144–162. Basel: Karger.

Solo-Gabriele, H. & Neumeister, S. (1996). US outbreaks of cryptosporidiosis. *Journal of the American Water Works Association*, **88**, 76–86.

Steiner, T. S., Thielman, N. M. & Guerrant, R. L. (1997) Protozoal agents: what are the dangers for the public water supply? *Annual Review of Medicine*, **48**, 329–340.

Thompson, R. C. A., Reynoldson, J. A. & Mendis, A. H. W. (1993) *Giardia* and giardiasis. *Advances in Parasitology*, **32**, 71–160.

Trager, W. (1970) *Symbiosis*. New York: Van Nostrand Reinhold.

Vickerman, K. & Cox, F. E. G. (1967) *The Protozoa*. Boston: Houghton Mifflin.

Wessenberg, H. S. (1978) Opalinata. In *Parasitic Protozoa*, vol. 2, ed. J. P. Kreier, pp. 551–581. New York: Academic Press.

Zaman, V. (1998) *Balantidium coli*. In *Topley & Wilson's Microbiology and Microbial Infections*, 9th edn, vol. 5, *Parasitology*, ed. F. E. G. Cox, J. P. Kreier & D. Wakelin, pp. 445–450. London: Edward Arnold.

Chapter 4

Platyhelminthes: the flatworms

The Platyhelminthes is a large and diverse group of organisms, some of which are free-living, but most of which are parasitic, living on, or in, most species of vertebrate animals and invertebrates as well. As the phylum name suggests, they are flattened dorsoventrally. They are without segmentation although cestodes, or tapeworms, superficially appear otherwise. Cestodes in reality are **modular iterations** (for a discussion of this concept, see Hughes, 1989), with each segment or **proglottid** being more like an individual within a colony since each is a complete sexual unit. Moreover, there is no coelom or peritoneum as there are in truly segmented animals such as the annelids. Platyhelminths may or may not possess an incomplete gut. They are without circulatory, skeletal, and respiratory systems. The functional and structural unit of their excretory/osmoregulatory system is a **protonephridium**, or flame cell (Fig. 4.1), so named for a tuft of cilia extending away from the cell body that resembles the flame of a burning candle. Most species are monoecious, but a few are dioecious.

The systematics of platyhelminths has changed significantly in the last decade or two because of new morphological information made available by electron microscopy, the application of cladistic techniques, and molecular systematics (Boerger & Kritsky, 1993; Brooks & McLennan, 1993; Rohde *et al.*, 1993; Hoberg *et al.*, 1997; Justine, 1998; Mariaux, 1998). According to the traditional literature, and prior to the cladistic revolution, the phylum Platyhelminthes was divided into four classes, namely Turbellaria, Monogenea, Trematoda, and Cestoda. Although most of these groups remain as units in the newly proposed systematic system, several that were included within these classes now form distinct groups based on their phylogenetic and evolutionary affinities. The classification system used here and depicted in Box 4.3 follows (with some modifications) the one proposed by Brooks (1989), Brooks & McLennan (1993), Boerger & Kritsky (1993), and Hoberg *et al.* (1997). In the present text, however, for practical reasons and for the sake of clarity (we hope), the groups of parasites will be discussed with very little emphasis on their taxonomic status.

4.1 Temnocephalidea

Temnocephalideans are a small group of ectocommensal organisms living on the surface or in the branchial chamber of crayfish, prawns, isopods, and other crustaceans, although they also have been found in turtles and molluscs. Most of their geographic distribution follows a Gondwanic pattern, as they are relatively common in South America, Australia, New Zealand, Madagascar, and India, although a few species have been reported in Europe. Temnocephalids are small and flattened, with a sucker at the posterior end and five to 12 finger-like tentacles at the anterior end (Fig. 4.2). They have two eyespots and are able to move about with a leech-like motion. The majority of species feed on small organisms such as bacteria, diatoms, protozoans, rotifers, nematodes, etc., found in or around the host and, although they are mobile, they do not normally leave their hosts. If experimentally removed from

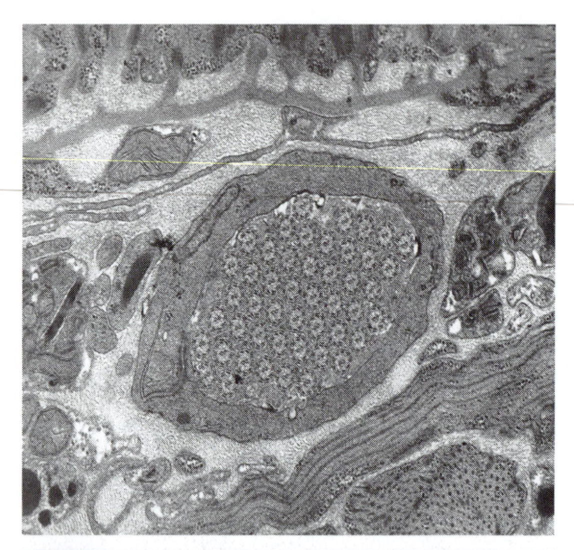

Fig. 4.1 Electron micrograph of a protonephridium showing a cross-section of the tuft of cilia extending away from the cell body. (Photograph courtesy of Darwin Wittrock.)

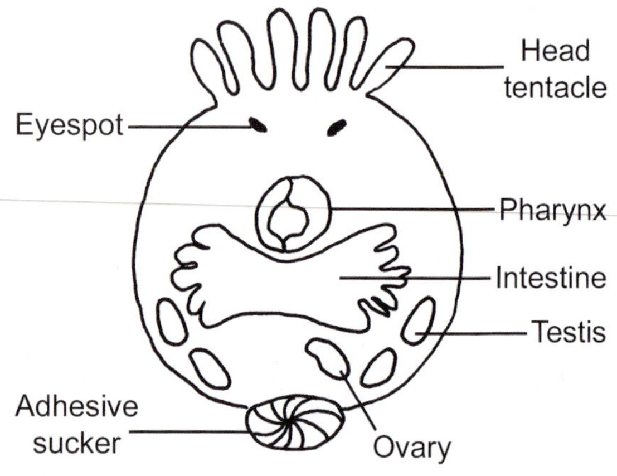

Fig. 4.2 A temnocephalidean illustrating several of the most prominent internal and external features. (Drawing courtesy of Lisa Esch-McCall.)

their host, some species die promptly, whereas others can survive for weeks (Avenant-Oldewage, 1993). Temnocephalids are monoecious. The eggs are situated within capsules attached to the exoskeleton of the host and development is direct. A miniature of the adult hatches from the eggs and matures to adulthood.

4.2 Udonellidea

The Udonellidea includes a very small number of parasites that previously were included with the Monogenea. Udonellideans are small, nearly cylindrical, with a posterior muscular sucker without hooks or anchors, and two small anterior suckers. They have a modified excretory system that consists of many pores and canals; this, together with the lack of hooks and the lack of a ciliated larva, defines the group and separates udonellideans from monogeneans. Very few species are known; one of them, *Udonella caligorum*, has a cosmopolitan distribution. Udonellideans are normally described as parasites of copepods (Fig. 4.3) that parasitize marine fishes. However, it seems that the worms use copepods only for transport, feeding on the surface of the fish. This indicates that they are thus parasites of fishes and not of copepods (Kabata, 1973; Byrnes, 1986; Aken'Ova & Lester, 1996). Regardless of the nature of their true host, udonellideans are ubiquitous on many caligid copepods in temperate marine waters.

4.3 Aspidobothrea

Aspidobothreans are always endoparasitic in molluscs and ectothermic aquatic vertebrates. There are three distinctive body forms. In some, a single large sucker divided into shallow depressions by muscular septa covers the ventral surface (Fig. 4.4). In others, there is a series of suckers distributed longitudinally, whereas in still others there is a ventral holdfast consisting of transverse ridges known as rugae. The digestive system is incomplete, like that of digeneans and monogeneans, except that it is in the form of a simple, blind sac. Digestion is extracellular. The excretory/osmoregulatory system includes numerous protonephridia and a bladder located at the posterior end. The nervous system is more complex than usual for a parasite, with a well-developed set of anterior nerves and a ladder-like peripheral system. Aspidobothreans also have a great diversity and number of receptors in their surface, which is a counter-trend to the generalized reduction of the

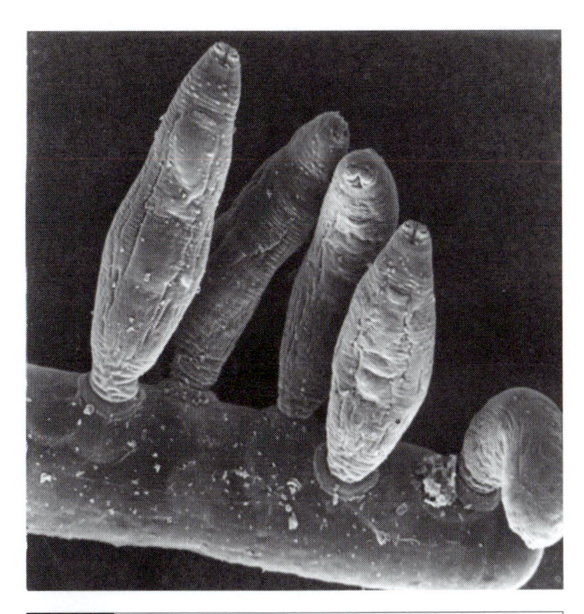

Fig. 4.3 *Udonella caligorum* attached to the egg sacs of the copepod *Lepeophtheirus mugiloidis* from Chile.

Fig. 4.4 The aspidobothrean *Trigonostoma callorhynchi* from the Chilean elephant fish *Callorhynchus callorhynchus*, showing the ventral surface of the parasite modified into an adhesive structure. (From Fernández *et al.*, 1986, with permission, *Biología Pesquera*, **15**, 63–73.)

nervous system of many parasitic organisms. Rohde (1989) suggested that in large aspidobothreans, such as *Lobatostoma*, there are between 20 000 and 40 000 receptors whose role is to sense the host's environment and avoid damaging the host. These parasites are all monoecious and their reproductive systems resemble those of digeneans (see below). Development is direct. From the egg hatches a **cotylocidium** larva that develops directly into an adult. Most aspidobothreans that parasitize molluscs have simple life cycles, without intermediate hosts, but species parasitic on fishes and turtles, e.g., *Lobatostoma* spp., require a snail intermediate host. In *L. manteri*, a snail ingests the egg; a larva hatches in the stomach, and migrates into the hepatopancreas where it develops into a pre-adult. When a fish ingests an infected snail, the parasite matures.

Aspidobothreans do not exhibit a high degree of specificity for either their molluscan or vertebrate hosts. For example, *Aspidogaster conchicola* is a common parasite in the pericardium of freshwater clams in Africa, Europe, and North America. In North America alone, it has been reported from over 70 different species of freshwater hosts, mostly bivalves (Hendrix *et al.*, 1985).

4.4 Digenea

4.4.1 General considerations

Most adult digeneans are endoparasitic and are found in all classes of vertebrates. Sometimes digeneans are also referred to as trematodes or digenetic trematodes which, whereas somewhat redundant, is acceptable vernacular. They usually, though not always, possess a pair of suckers, including a ventral acetabulum and an anterior oral sucker in the center of which is the mouth. Some have accessory suckers anteriorly. Digeneans have an incomplete digestive system. In terms of size and general morphology, there is no such thing as a typical digenean. They occur in virtually all shapes and sizes, from those that are microscopic to those that may range up to 10–12 m in total length. Some are thin and round, and others are ribbon-like. The genital openings are usually located between the two suckers, but this is variable as well. These flukes always have indirect life cycles, with as many as four hosts in a few species.

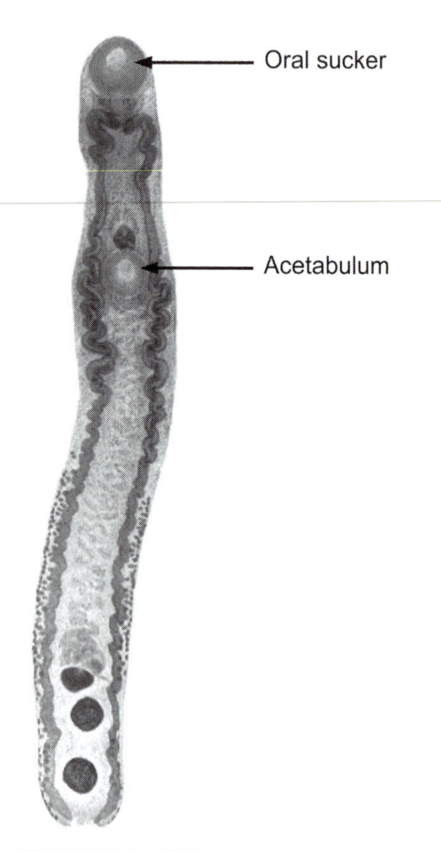

Fig. 4.5 External morphology of a distome digenean. (Photograph courtesy of Harvey Blankespoor, Hope College.)

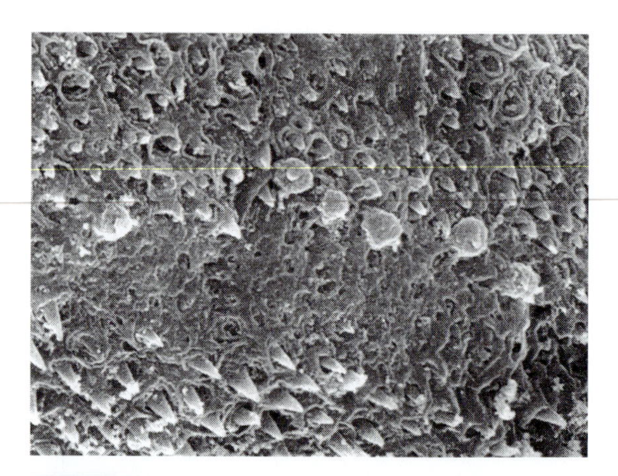

Fig. 4.6 Surface of *Austrobilharzia variglandis*, a schistosome fluke, illustrating raised tubercles (or bosses), papillae, and spines. (From Barber & Caira, 1995, with permission, *Journal of Parasitology*, **81**, 584–592.)

4.4.2 The outside

With a few exceptions, there are two suckers (**distome** digenean), including an oral sucker and a ventrally positioned **acetabulum**, usually located about halfway along the length of the body (Fig. 4.5). The acetabulum also may be at the posterior end (**amphistome** digenean). In some, there is a lone sucker at the anterior end (**monostome** digenean). In one family (Bucephalidae), there is a single sucker located in the middle of the body, into which the mouth opens (**gasterostome** digenean). In a number of taxa, the body is split into two distinguishable portions (**holostome** digenean).

The tegumental surface of all digeneans is **syncytial** (multinucleated). The covering is thus alive, and contains numerous mitochondria and many vesicles. The nuclei are below the surface of the tegument in **cytons**, along with Golgi apparatus, ribosomes, endoplasmic reticulum, and the other usual cytoplasmic organelles. The tegumental surface of some digeneans may be covered with spines, tubercles, or other characteristic morphological features (Fig. 4.6). Two layers of muscle lie below the tegument.

4.4.3 The inside

Digestive system. The tissue in which the internal organs are located is called the **parenchyma** and forms a matrix within which all the internal organs and organ systems occur. The incomplete digestive system opens into a mouth within the oral sucker. In many, the gut begins with a prepharynx, followed by a muscular pharynx, and then an esophagus that usually splits into two ceca that may or may not extend the length of the body (Fig. 4.7). The pharynx is missing in some. The ceca are generally simple canals lined with columnar epithelium; in some species they are highly branched, whereas in others they may or may not fuse posteriorly. The epithelium of some species is syncytial and of others is cellular. Functionally, these cells are both absorptive and secretory, releasing both proteoglycans and proteases (Fujino, 1993).

Excretory/osmoregulatory system. The removal of metabolic wastes in digeneans is effected by diffusion through the surface to the outside, via the intestinal ceca to the lumen, and by protoneph-

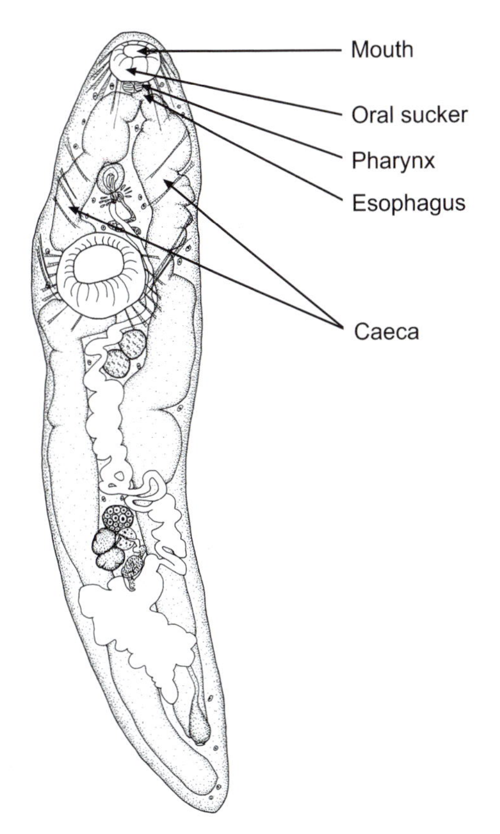

Fig. 4.7 The hemiurid digenean *Machidatrema akeh*, showing the typical intestinal morphology of a fluke. (Modified from León-Règagnon, 1998, with permission, *Journal of Parasitology*, **84**, 140–146.)

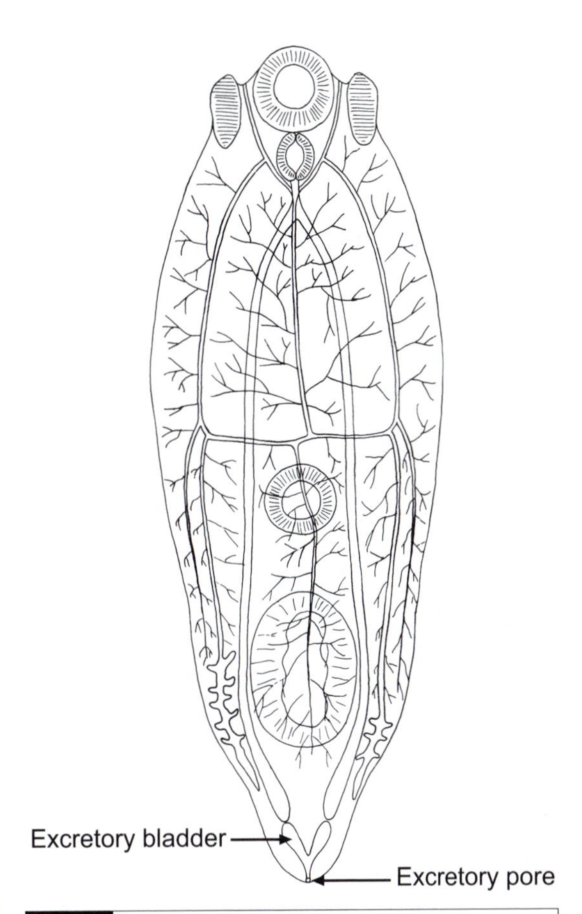

Fig. 4.8 The excretory system in the metacercaria of *Tylodelphys xenopi*. Notice the terminal Y-shaped, posterior bladder and the pore at the very posterior end. The terminus of each line on the drawing represents the location of an individual protonephridium. (From King & Van As, 1997, with permission, *Journal of Parasitology*, **83**, 287–295.)

ridia, or flame cells. Nitrogenous wastes primarily are in the form of ammonia. The pattern of organization of the flame cells is species-specific and bilateral; the complex arrangement of flame cells and their collecting ducts is illustrated in the metacercaria shown in Fig. 4.8. The collecting ducts from the flame cells fuse and empty into a posterior excretory bladder, which may (epitheliocystidians) or may not (anepitheliocystidians) be lined with epithelium. The bladder usually empties through a pore at the very posterior end of the body. There is evidence suggesting that the protonephridial system may have an osmoregulatory function as well.

Reproductive system. In monoecious flukes, the body of the worm contains a complete set of male and female reproductive organs (Fig. 4.9). Sperm are produced in paired testes located in the posterior portion of the body (the number and location of the testes will vary). Generally, the testes are of similar size. Sperm leave the testes through the vas efferens, which fuse to form the vas deferens, or sperm duct. The terminal portion of the male system usually ends in a cirrus sac and protrusible cirrus. Associated with the cirrus sac are the prostate gland and **seminal vesicle**. The final opening to the outside may lie in a genital atrium or depression on the ventral surface. During copulation, the cirrus is inserted into the gonopore of the female and sperm pass through the female's reproductive tract to the **seminal receptacle** where they are stored. Ova are produced in a

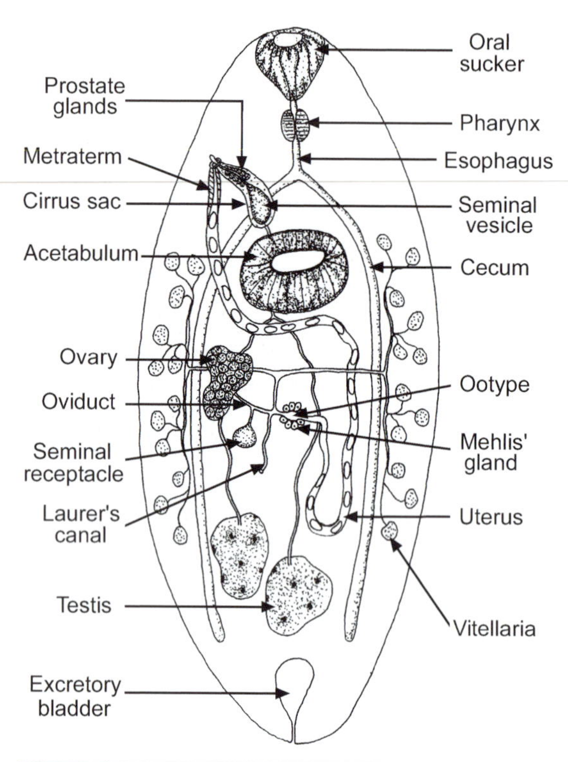

Fig. 4.9 Internal anatomy of a digenean. (Drawing courtesy of Derek Zelmer & Michael Barger, Wake Forest University.)

single ovary. They move into an oviduct and, as they pass by the opening of the seminal receptacle, they are fertilized. Close by, there may be a vestigial **Laurer's canal** connecting the oviduct to the surface of the digenean body. The diploid zygote continues into the **ootype** which is surrounded by the **Mehlis' gland**. Also emptying into the ootype is the vitelline duct which is connected to **vitellaria** that are usually scattered along both lateral borders of the body. The vitellaria have a dual function in that they supply yolk as well as produce substances that will become the eggshell. Enzymes released by the Mehlis' gland harden the eggshell as the egg passes through the ootype into the uterus. Usually, enormous numbers of eggs will be produced and packed into the uterus as they make their way to the outside (note that parasitologists refer inaccurately to the stage freed from the adult as an egg when in fact it is an embryo within a shell). So many eggs may be present inside the uterus that they make it difficult to see other structures within the body of stained specimens. The uterus of some digeneans possesses a muscular terminus called a **metraterm**.

Nervous system and sensory input. The central nervous system includes a primitive brain anteriorly from which extend at least one pair of longitudinal nerve cords connected at intervals by lateral commissures. There is also a highly organized peripheral system connected to the tegument and muscle layers, the gut, and the reproductive system. The neurotransmitters include most of those commonly associated with vertebrate animals, e.g., acetylcholine, noradrenaline, dopamine, serotonin, etc. The sensory physiology of flukes and other parasitic flatworms has been difficult to assess because of the generally small size of the organisms. Putative functions of external sensory organs include mechanoreception, chemoreception, and osmoreception. The existence of eyespots and experimental evidence of phototactic behavior strongly suggest photoreceptive capabilities. For thorough reviews of the neurophysiology of parasitic flatworms, including digeneans, students should consult Sukhdeo & Sukhdeo (1994) and Halton & Gustafsson (1996).

4.4.4 Nutrient uptake

Certain nutrients, including glucose and some amino acids, are absorbed through the tegument. Most evidence indicates that some combination of passive diffusion and active transport is involved in moving these nutrients through the parasites' surfaces. Feeding and digestion by means of the digestive system also is highly variable among the Digenea. Thus, the fluke's food within the definitive host may consist of blood, host tissue, mucus, host intestinal contents, or some combination of these. In most species studied, digestion in the gut ceca appears to be extracellular. In some species, such as the liver fluke *Fasciola hepatica*, digestion is both intra- and extracellular, whereas *Haplometra cylindracea*, a lung parasite of frogs, secretes an enzyme that predigests its food outside before ingestion, similar to some free-living flatworms. Enzymes secreted by the gut of digeneans include, among others, proteases, aminopeptidases, esterases, and phosphatases.

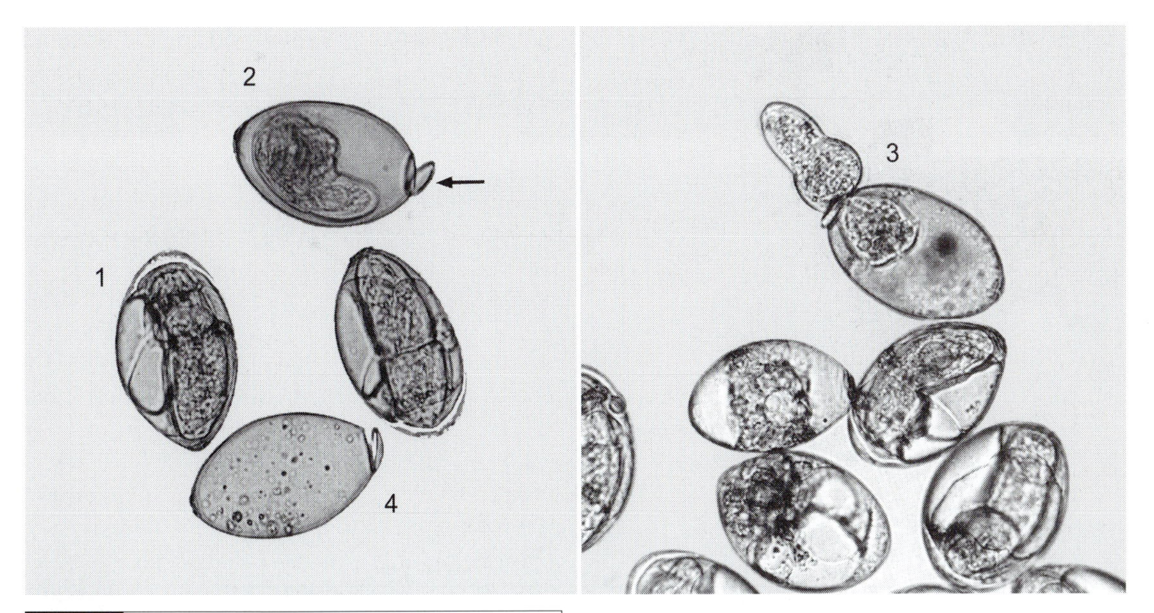

Fig. 4.10 Hatching sequence of a miracidium from an egg of a digenean. (1) Miracidium within an egg. (2) Operculum in the process of opening. (3) Miracidium emerging from the eggshell. (4) Empty eggshell with an open operculum. (Modified from Ataev *et al.*, 1998, with permission, *Journal of Parasitology*, **84**, 227–235.)

4.4.5 Metabolism

Whereas the energy metabolism of digeneans, cestodes, and nematodes is basically similar (see also Chapter 2), there are several species-specific variations that are reflected in both the efficiency and nature of waste products produced. In a well-studied species such as *F. hepatica*, there is also strong evidence suggesting that intermediary carbohydrate metabolism shows some ontogenetic change as well. The primary energy resource for digeneans is glucose and the storage form of the carbohydrate is glycogen. Indeed, the glycogen content in some species is known to be as high as 30% of the dry weight.

In *F. hepatica*, the primary metabolic waste products are lactate, acetate, and proprionate (Tielens *et al.*, 1984). The former is produced through glycolysis. The latter two are produced in mitochondria. *Fasciola hepatica* is able to fix CO_2 to phosphoenolpyruvate to form oxaloacetate which is then converted to malate; in the mitochondria, malate is metabolized to both acetate and proprionate, with the concomitant production of additional ATP in the process. There is still some controversy regarding the intermediary carbohydrate metabolism of the schistosomes. Some workers claim that at least some species are homolactic fermenters while others claim that oxidative phosphorylation occurs. Whatever the case, the carbohydrate metabolism in the schistosome species studied is nearly 20 times more efficient than in *F. hepatica* (Bueding & Fisher, 1982).

4.4.6 Development

Once fertilization has occurred and eggshell formation is under way (see section 4.4.3) the zygote proceeds with embryonic development. The first larval stage produced in this process is called a **miracidium**. Development of the miracidium in some species may be complete by the time the egg is shed, in which case the egg will hatch immediately. In other flukes, development inside the eggshell will not proceed until the egg reaches the outside. Access to the molluscan host is gained by the free-swimming, ciliated miracidium, or when an appropriate mollusc accidentally ingests an egg with an unciliated miracidium. In eggs that hatch, there is an operculated shell through which the miracidium emerges (Fig. 4.10). A fully developed miracidium will be covered by cilia, except for an apical papilla (Fig. 4.11A). A chemical process known as saponification can remove the

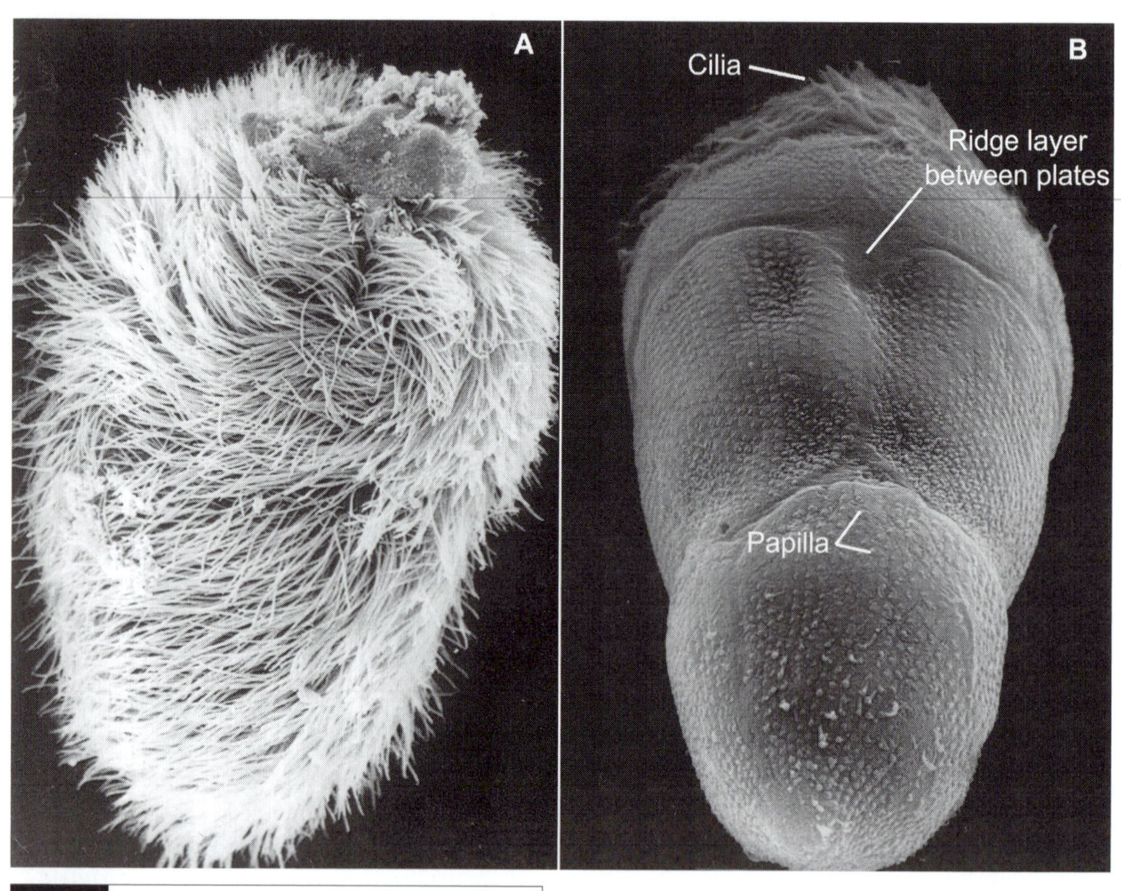

Fig. 4.11 (A) Fully developed miracidium covered with cilia; (B) following saponification, the cilia are removed revealing the ectodermal plates on the right. (From Ataev *et al.*, 1998, with permission, *Journal of Parasitology*, **84**, 227–235.)

cilia and reveal the ectodermal plates to which the cilia were attached (Fig. 4.11B).

The development of digeneans inside the molluscan host is by a specialized asexual process called **polyembryony**. In polyembryony, more than one embryo is ultimately derived from a single zygote, a kind of cloning. On shedding their cilia, miracidia migrate to species-specific sites within their hosts where they transform into **sporocysts**. (See Ataev *et al.* [1997] for a detailed description of miracidial migration and internal development by *Echinostoma caproni*.) In reality, sporocysts are little more than germinal sacs containing a mixed population of stem and somatic cells (Fig. 4.12). Stem cells continue to give rise to populations of both cell types, whereas somatic cells give rise to daughter sporocysts or to another developmental stage called a **redia** (Fig. 4.13). The redia resembles the sporocyst in containing populations of stem and somatic cells, but rediae are more complex morphologically than sporocysts in that they also possess a sucker, mouth, pharynx, and primitive gut. The somatic cell population may give rise to daughter rediae or to new developmental forms called **cercariae**. Cercariae, for the most part, emerge from their molluscan hosts and swim freely (usually) in the water column in search of another host. Morphologically, cercariae resemble the adults into which they will develop, i.e., they have a mouth, suckers, etc., except that the genitalia are generally absent. However, most also possess tails which propel the larvae in the water column. Cercariae form is highly variable (Fig. 4.14); anatomical features of importance include the position and number of the suckers, accessory structures such as eyespots, circumoral spines, etc., and tail morphology. In many species, cercariae penetrate another intermediate host where they become **metacercariae** that are

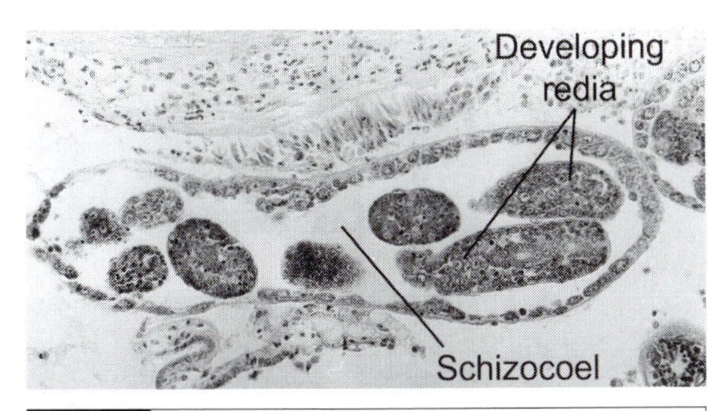

Fig. 4.12 A sporocyst showing the schizocoel, developing rediae, and germinal masses inside. (From Ataev *et al.*, 1998, with permission, *Journal of Parasitology*, **84**, 227–235.)

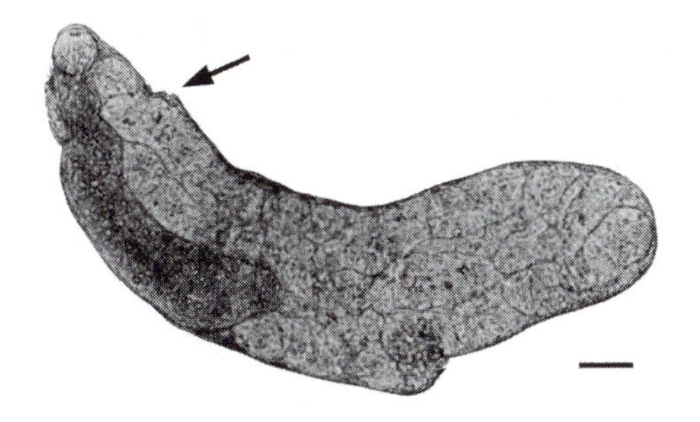

Fig. 4.13 Redia of the digenean *Echinostoma macrorchis*. The arrow indicates the birth pore. Scale = 50 μm. (From Lo, 1995, with permission, *Journal of Parasitology*, **81**, 569–576.)

usually sequestered in a cyst. Developmentally, a metacercaria is little more than an immature adult. When a metacercaria is ingested by a definitive host, it excysts and completes its development becoming a sexually mature adult digenean.

4.4.7 Life cycle

The life cycles of most digeneans are among the most complex in nature, and are usually linked, inextricably, to the feeding strategies of their definitive hosts. The life cycle patterns in Fig. 4.15 are stylized representations, suggesting an enormous number of variations. There are, however, two features of overall consistency. First, almost all life cycles include both free-living and parasitic stages. It is with the free-living forms, e.g., miracidia and cercariae, that one sees a number of morphological modifications (cilia and tails) for locating an appropriate host(s). Then, in some of the parasitic stages, host behavior may be modified in order to exploit trophic relationships (see also section 13.2.4). Second, substantial complexity also occurs within the molluscan intermediate host. Thus, as described above, the miracidium stage is always followed by a sporocyst, but then, depending on the species, the pattern is highly variable. Moreover, there is always an increase in parasite numbers within the molluscan host. From an ecological/evolutionary perspective, this increase in numbers reflects an amplification of the successful genome of the parasite. Many of these life cycle traits will be detailed in the discussion of the more common or economically important families, genera, or species of digeneans.

4.4.8 Diversity

Digeneans are a diverse group of parasites with respect to both their hosts and their habitat within the host. Consequently, a discussion of their diversity can be better accomplished by using a taxonomic approach. However, only some representative families of several of the orders will be considered. (A systematic outline for the group is provided in Box 4.3.)

PARAMPHISTOMIFORMES

Many paramphistomiform digeneans are robust, and sometimes are referred to as being 'fleshy'. Their cercariae encyst on aquatic vegetation, animals, or inanimate objects, after emerging from their snail hosts. In the Paramphistomidae, the acetabulum is located posteriorly (amphistomes) (Fig. 4.16), whereas in other families of the order an acetabulum is lacking (monostomes) (Fig. 4.17). Paramphistomids are parasitic in all groups of vertebrates. One of the most common species in cold-blooded vertebrates in North America is *Megalodiscus temperatus*, a parasite in the rectum of ranid frogs. Pulmonate snails are intermediate hosts. The cercariae released from the snails encyst on the heavily pigmented areas of a tadpole's skin, probably due to some sort of chemotactic response. When the tadpole undergoes metamorphosis, it consumes the shed skin and infects itself with the encysted metacercariae. On occasion, a tadpole may accidentally eat cercariae before they have the chance to encyst. When this happens, the parasite encysts immediately, becoming a metacercaria which then passes through the intestine. On reaching the rectum, it

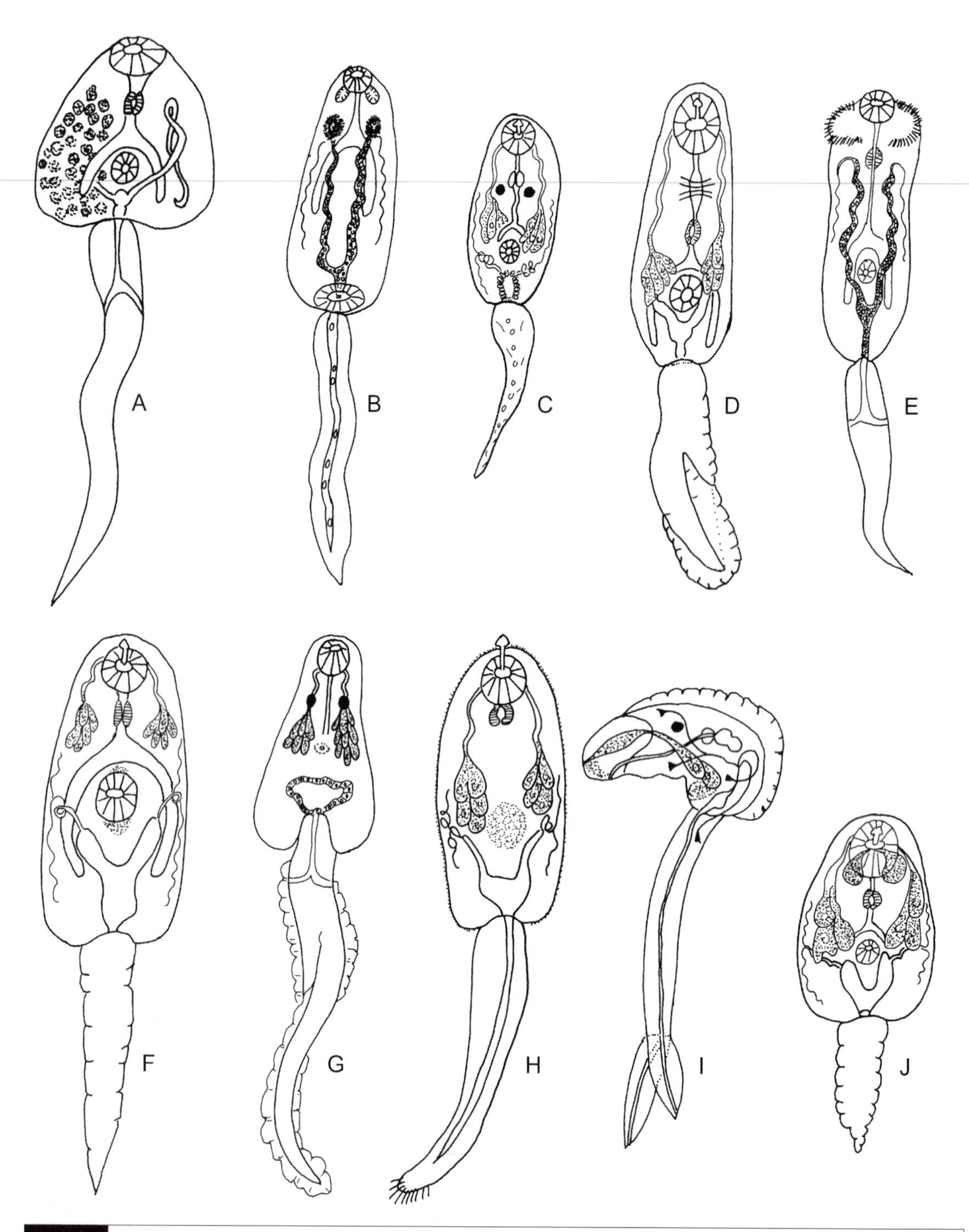

Fig. 4.14 Cercariae morphology is highly variable. Identification and naming of cercariae types are based on sucker placement, tail morphology, and a number of other topological features such as the presence of a stylet, eyespots, pharynx, or circumoral spines. (A) Gymnophalus; (B) Amphistome; (C) Ophthalmoxiphidio; (D) Ornatae; (E) Echinostome; (F) Armatae; (G) Parapleurolophocercous; (H) Ubiquita; (I) Brevifurcate-pharyngate; (J) Virgulate. (Drawings courtesy of Lisa Esch-McCall.)

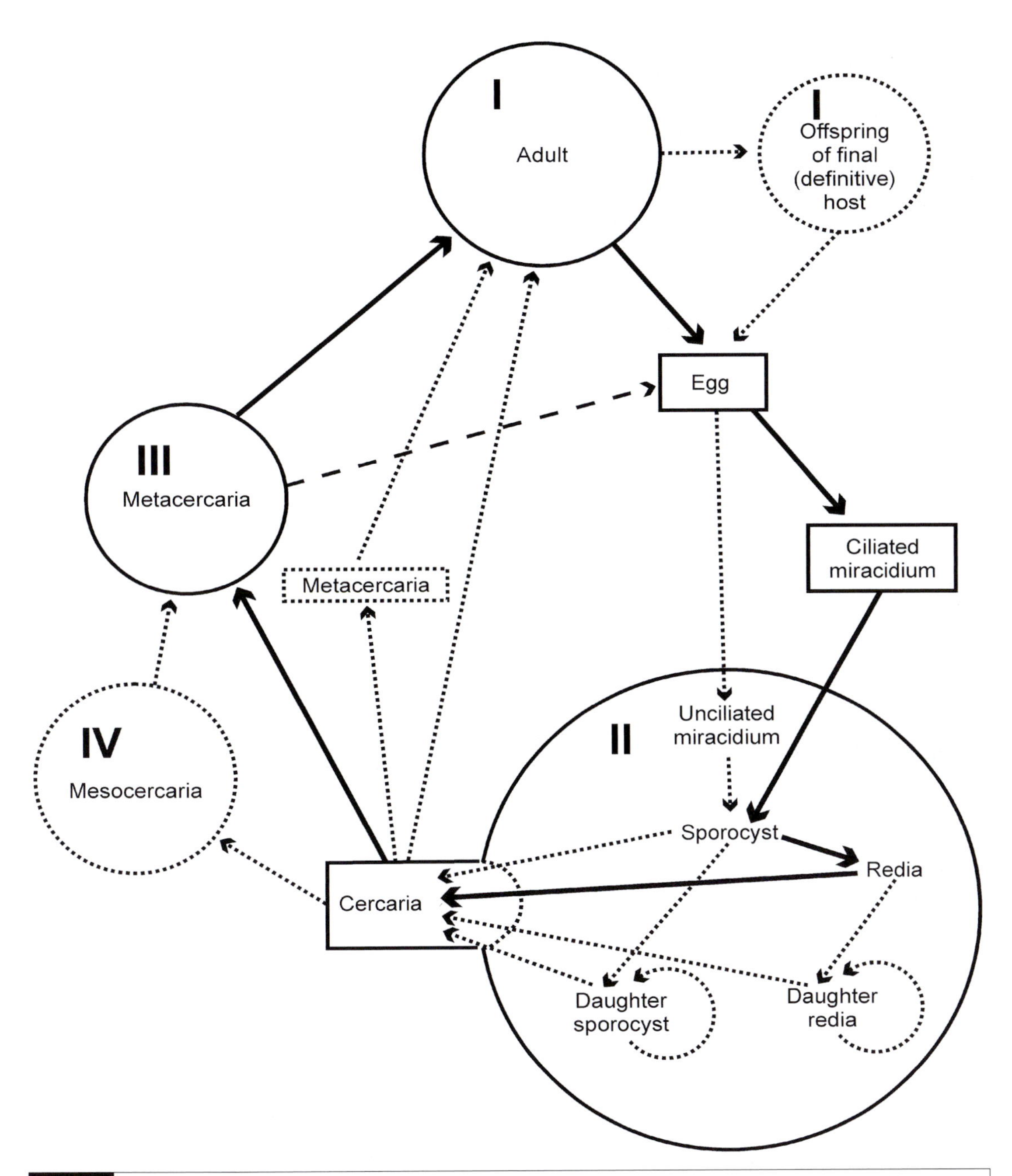

Fig. 4.15 Generalized life cycle of digeneans, with some variations on the basic pattern. Host types are represented by roman numerals (I = definitive host; II = molluscan first intermediate host; III = second intermediate host; and IV = paratenic or third intermediate host). Stages within hosts are enclosed in circles, whereas stages in the environment are enclosed within rectangles. Solid lines represent a 'typical life cycle', whereas dotted lines represent less common variations. Note that some metacercariae are found encysted in the environment (and hence are free-living) and that some cercariae do not leave their first intermediate hosts. The dashed line represents progenesis (see text for details). Some rare, alternative pathways are not included in this diagram.

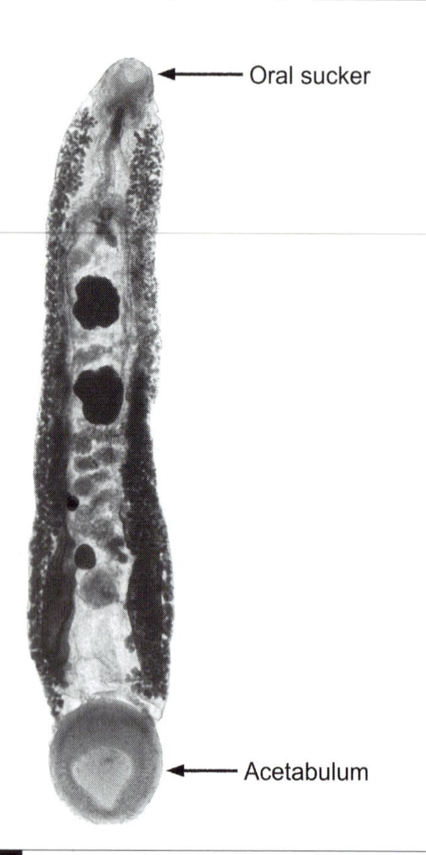

Fig. 4.16 *Wardius* sp., a paramphistomid digenean from a muskrat *Ondatra zibethicus*. (Photograph courtesy of Harvey Blankespoor, Hope College.)

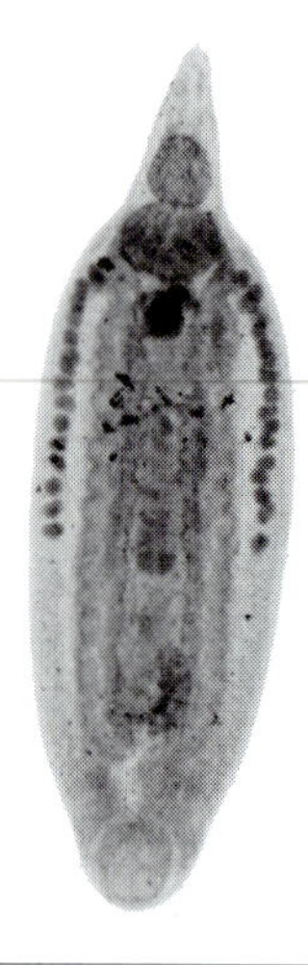

Fig. 4.17 *Teloporia* sp., a paramphistomid digenean from the spiny softshell turtle *Trionyx spinifera*. (Photograph courtesy of Harvey Blankespoor, Hope College.)

excysts, but remains undifferentiated. When the tadpole undergoes metamorphosis, a few may be lost, but others migrate back up into the stomach where they remain until the metamorphosed frog takes a meal. Then, they migrate back down into the rectum again and mature sexually.

A number of species within the Paramphistomidae parasitize the stomach or liver of domesticated and wild ruminants worldwide, causing an important disease known as **paramphistomiasis**. When the ruminant ingests metacercariae encysted on vegetation, the worms excyst in the duodenum, penetrate the mucosa, and migrate anteriorly through the tissues. On reaching the true stomach or abomasum, they cross the mucosa into the lumen and move to the rumen where they finally attach. The migration of the juveniles causes enteritis, diarrhea, and hemorrhage, physically affecting the host. In severe cases, when large quantities of metacercariae are ingested in a short time, death of the host is likely.

ECHINOSTOMIFORMES

The Echinostomiformes includes a number of families that often do not resemble each other much in the adult stage. Some of the most common families are the Fasciolidae, parasites of the liver, bile duct, and intestine of herbivorous mammals, including humans, the Cyclocoelidae which parasitize the air sacs and respiratory system of birds, and the Echinostomidae, intestinal parasites of reptiles, birds, and mammals.

With a cosmopolitan distribution, members of the family Fasciolidae are almost always of enormous size (up to 5 cm), and somewhat leaf-like in appearance. A few species occur in the intestine of their definitive host, but the majority are associated with the liver, gall bladder, and bile ducts. Most species are pathogenic, especially in cattle and sheep, where they may cause serious problems, not only through the mortality they may inflict, but also by the weight losses they almost inevitably induce. It must be remembered that weight loss to a cattle rancher means money lost when the herd is sold for slaughter.

Fasciola hepatica (Fig. 4.18) is typically found in the major sheep-raising countries of the world (also see Box 1.1). Adults occur in the bile ducts

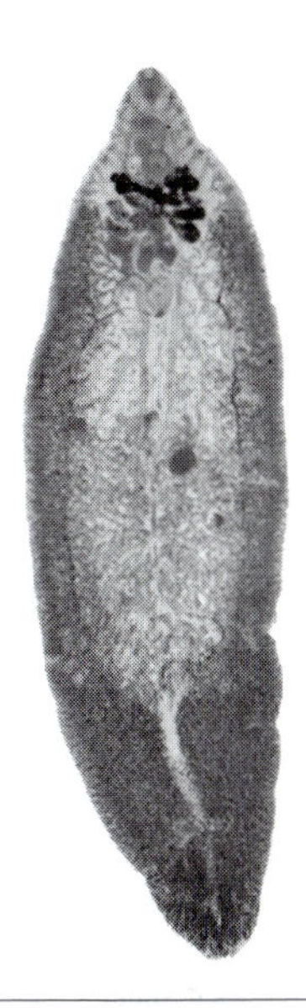

Fig. 4.18 The digenean *Fasciola hepatica* from the bile duct of a sheep. (Photograph courtesy of J. Teague Self.)

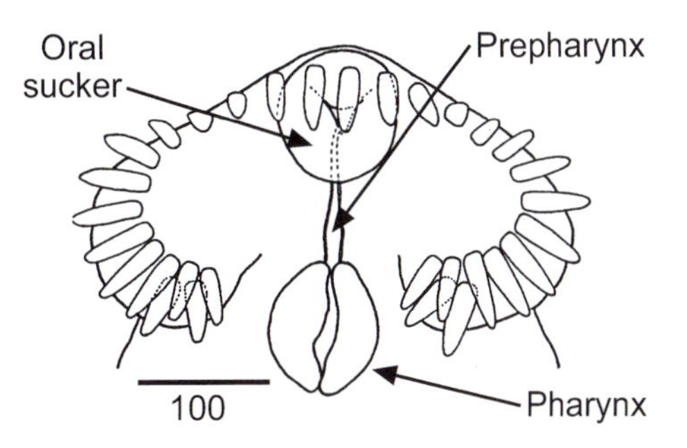

Fig. 4.19 Anterior end of the echinostomatid digenean *Himasthla limnodromi* showing a collar of spines surrounding the oral sucker. Units are μm. (Modified from Didyk & Burt, 1997, with permission, *Journal of Parasitology*, **83**, 1124–1127.)

where they release eggs that are shed in the feces. Aquatic snails are intermediate hosts, and cercariae released from the snail encyst on vegetation or inanimate objects, becoming metacercariae. When metacercariae are ingested by sheep, they excyst in the lumen of the small intestine under the influence of host's digestive enzymes. The parasites migrate through the gut wall into the body cavity, then into the liver and finally into the bile duct and gall bladder where they mature sexually. It is the liver migration that produces the severe pathology and weight loss. Fortunately, the liver is a resilient organ and has the capacity to repair itself once the parasite leaves. A point of interest in infections by *F. hepatica* concerns the migratory route taken once a metacercaria has excysted; this includes passage through the gut wall, the abdominal cavity, and obligatory migration in the liver. *Dicrocoelium dendriticum*, an unrelated fluke, also resides in the bile ducts and gall bladders of sheep. However, when metacercariae of this species excyst in the sheep's small intestine, they migrate to the bile duct and enter it directly. Metacercariae of each species are subjected to exactly the same set of potential environmental cues, yet they each respond to a special subset of these cues and take a completely different route to the final site of infection, a rather telling evolutionary story.

There are several other fasciolid flukes infecting cattle, sheep, and wild ungulates in various areas of the world. In North America, the huge *Fascioloides magna* occurs in the livers of deer, where it reaches up to 7 cm in length. Because of their size, they may cause extensive damage; parasitized white-tailed deer usually are smaller and have fewer antler points, which may affect their reproductive success. In the southeastern United States, local residents refer to *F. magna* as 'swamp butterflies' and are said to deep-fry them like they would hush puppies, a widely consumed cornbread cake in that part of the country. *Fasciola gigantica* and *Fasciolopsis buski* are widely distributed in cattle. The former is common in Asia, Africa, and Hawaii, whereas *F. buski* is abundant in the Far East where it infects up to 10 million humans. The usual source of infection for both cattle and humans is watercress, water caltrop, or other edible aquatic plants, on which metacercariae have encysted.

The Echinostomidae is a diverse group of flukes infecting the intestine and bile ducts of a wide range of reptiles, birds, and mammals, including humans on occasion. Both adults and cercariae possess either a single or a double row of spines (collar) that almost completely encircle the oral sucker and mouth (Fig. 4.19). There are at least 17 genera of echinostomatids and in excess of 100 species in the genus *Echinostoma*. Many species show very little specificity for their definitive host, but are more specific for their first intermediate host, normally a freshwater gastropod (Huffman & Fried, 1990). The life cycles of *Echinostoma* spp. involve two intermediate hosts. Cercariae that emerge from the molluscan host penetrate a variety of second intermediate hosts, both invertebrate and vertebrate, and encyst as metacercariae.

At least 15 species of echinostomes infect humans, mainly in the Far East and Southeast Asia. Humans become infected by eating raw snails, tadpoles, or freshwater fishes harboring metacercariae. Several species, including *E. caproni* and *E. paraensi*, have been adopted for experimental use in the laboratory. An excellent review of the former species is presented by Fried & Huffman (1996).

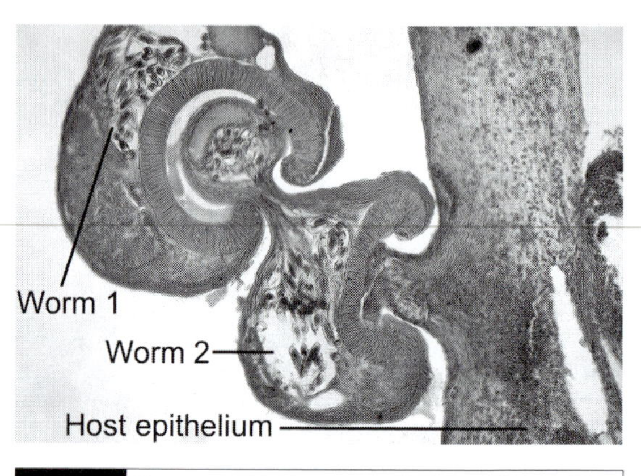

Fig. 4.20 Two adult individuals of *Halipegus occidualis* (Hemiuridae), one attached to the other, in the buccal cavity of the green frog *Rana clamitans*. (Photograph courtesy of Derek Zelmer, Wake Forest University.)

HEMIURIFORMES

Most of the species in this order are parasites in the stomachs, occasionally the intestines, of fishes. They are found mainly in marine teleosts, but some occur in elasmobranchs, freshwater fishes, and a few in amphibians and reptiles. Although none of the families, or even species in this group, has any epidemiological or economic relevance for humans, many are of biological interest.

An interesting aspect of many species in the order Hemiuriformes is their location in the definitive host. Although digeneans are considered as endoparasitic, several species of hemiuriforms are found on the gills or in the mouth of their definitive host. Most species in the family Syncoeliidae are found on the skin, or in buccal and branchial cavities of sharks and rays. Most digeneans in the family Accacoeliidae parasitize the gills and gut of *Mola*, a large, marine sunfish that lives in open, warm waters and has a cosmopolitan distribution. The diet of these fish is peculiar, consisting almost exclusively of cnidarians (jellyfishes), ctenophores (comb jellies), and salps (also a jelly-like organism). The few species found in hosts other than the sunfish parasitize fishes that also prey on jellyfishes and comb jellies. Not surprisingly, accacoeliid metacercariae are common in jellyfishes and comb jellies.

The Hemiuridae is the largest in number of species and most conspicuous family in the order. Most species are found in the gut of marine and freshwater fishes, although a few occur in the lungs or the intestines of sea snakes, and in the mouth, esophagus, and eustachian tubes of ranid frogs. Many species in this family, mainly those parasitic in marine fishes, exhibit very low host specificity and have worldwide distributions. A peculiarity of many hemiurids present in the stomach of fishes is their capacity to retract the posterior end of their body. The retractible section of the body is called the tail or **ecsoma**, and it appears to function as a feeding organ that is extruded when the pH or osmolarity of the stomach content is at a tolerable level (Gibson & Bray, 1979).

Hemiurids of the genus *Halipegus* are rather common in ranid frogs of South America, North America, and Europe. *Halipegus occidualis* and *H. eccentricus* occur in the buccal cavity and eustachian tubes, respectively, of ranid frogs in North America and have been studied extensively from an ecological perspective (see also Chapter 10) since both can be enumerated as adults without the necessity of killing their host (see Esch *et al.*, 1997, and references within). *Halipegus occidualis* has such a high affinity for the buccal cavity that the worms may actually attach to each other in an effort to locate at the preferred site of infection (Fig. 4.20). The stimulus is probably a specific chemical released in the mucus of the frog's mouth.

The Hemiuriformes can be large digeneans.

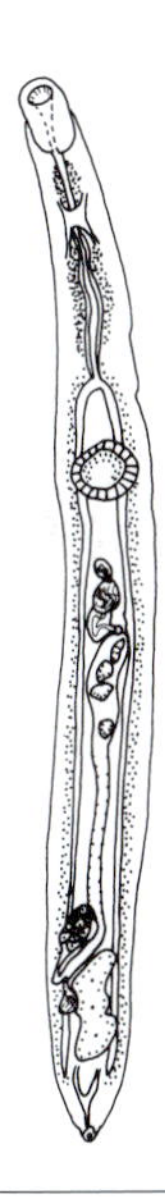

Fig. 4.21 The digenean *Urotrema burnsi* (Spirorchiidae) from the blood vessels of the heart and liver in the Australian turtle *Emydura krefftii*. (From Platt & Blair, 1996, with permission, *Journal of Parasitology*, **82**, 307–311.)

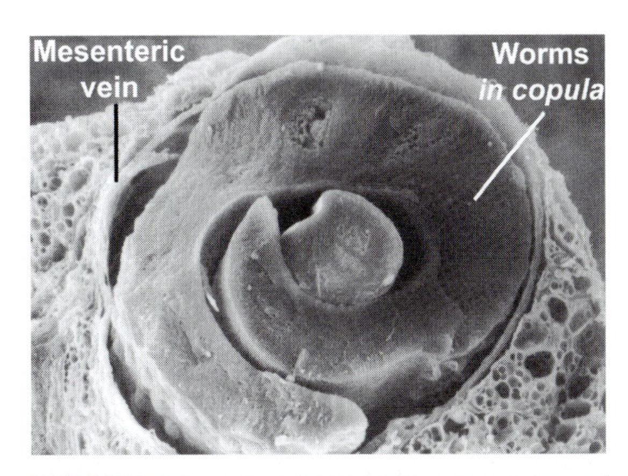

Fig. 4.22 Cross-section of male and female *Schistosoma mansoni* in copula within a mesenteric vein of an experimentally infected laboratory mouse. (From File, 1995, with permission, *Journal of Parasitology*, **81**, 234–238.)

Those who fish for marine billfish, wahoo, little tunny, or other similar pelagic fishes are familiar with *Hirudinella*, a large digenean found in the stomach. These flukes can change their shape considerably, ranging from 17 cm long and 3 cm wide when extended, to just 4 cm long when contracted.

STRIGEIFORMES

The Strigeiformes is another diverse group of digeneans. The adults parasitize all classes of vertebrates and some invertebrates as well. Location inside the definitive host also is quite heterogeneous.

Most of the species of Spirorchiidae inhabit the heart, large arteries, and other blood vessels of reptiles, primarily turtles. Their bodies are flat, lanceolate (Fig. 4.21), and the ventral sucker or acetabulum may be absent. Their life cycle includes a snail intermediate host, and cercariae released from the snail penetrate the mucus membranes of the eyes, nose, mouth, and cloaca of turtles. The Sanguinicolidae are small, flat digeneans without suckers, that inhabit the blood vessels and heart of freshwater and marine fishes. *Cardicola davisi* is a common sanguinicolid of trout in North America and may cause severe mortality in hatcheries. One of the problems faced by parasites in the circulatory system of vertebrates is how to get the eggs produced by the adult out into the environment. In many cases, the eggs escape from the host's blood vessels into the intestine and then into the environment with the host feces. In others, the eggs are carried by the blood into the capillaries of the gill filaments where the miracidium hatches and escapes the fish host through the very thin epithelium overlying the gill capillaries. The pathology produced by sanguinicolid flukes is caused primarily by eggs which occlude vessels in many areas of a host's circulatory system or by miracidia exiting the gill epithelium, and not by the adult flukes themselves (Smith, 1997).

Probably the best-known of all the digeneans inhabiting the circulatory systems of vertebrates are the schistosomes. These are dioecious parasites (an unusual feature among digeneans) that occupy the hepatic portal and pelvic veins of birds and mammals. The size of adult schistosomes is generally in the range of 10–15 mm, although *Gigantobilharzia acotyles*, a parasite of the black-headed gull, reaches 140–165 mm. Male schistosomes are considerably larger than females and possess a **gynecophoric canal** in which the female lies in permanent copula (Fig. 4.22). This canal is a ventral groove that begins immediately behind a pedunculated ventral sucker. Most schistosomes

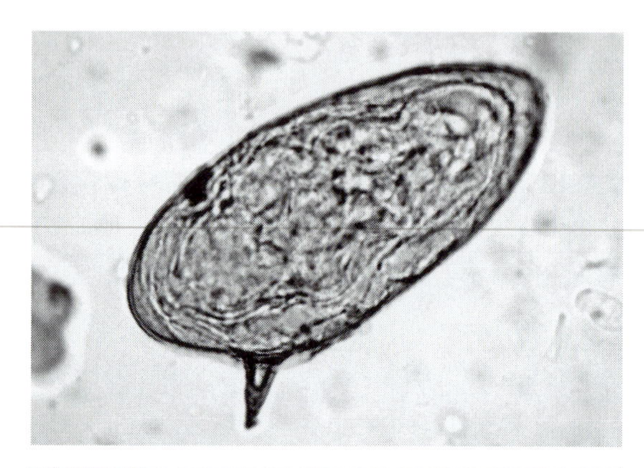

Fig. 4.23 Egg of *Schistosoma mansoni* showing large lateral spine. (Photograph courtesy of J. Teague Self.)

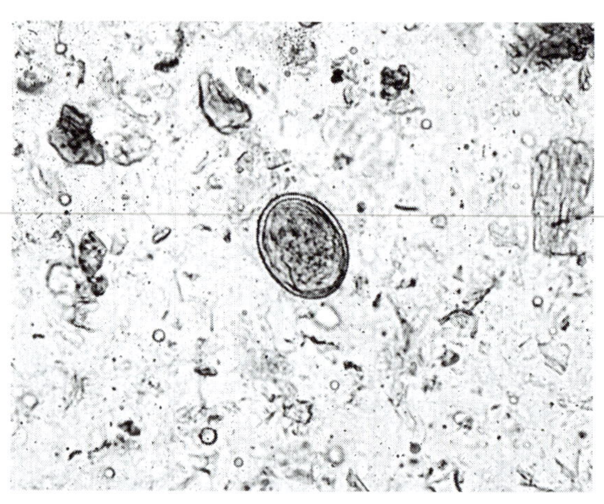

Fig. 4.24 Egg of *Schistosoma japonicum* that has no apparent spine. (Photograph courtesy of J. Teague Self.)

have a life span of between 5 and 30 years, but egg production is usually relatively low, being on the order of 300 eggs per day in some species. In single worm infections, males will develop normally, but females remain sexually immature. An elegant commentary on the origin of dioecy in the schistosomes is provided by Platt & Brooks (1997).

The life cycles of the various schistosome species, including those that infect humans, are essentially the same. Eggs produced by females are released to the outside in feces or urine, depending on the site of infection in the vertebrate host. The eggs of *Schistosoma mansoni* have a large lateral spine (Fig. 4.23). A terminal spine is characteristic of *S. haematobium*, whereas eggs of *S. japonicum* are much smaller and without a spine (Fig. 4.24). The mechanism involved with egg escape through the tissues is linked directly to the pathology associated with each parasite species and is considered in Box 4.1. Cercariae are usually released in 'puffs' from snail intermediate hosts and penetrate the skin of their definitive hosts directly. On penetration, the cercaria drops its tail, enters the circulation, and becomes a **schistosomule**. After 2–3 days in the circulatory system, the schistosomules stop in the blood sinuses of the liver where they remain for up to 15 days before moving into small venules at their species-specific, final sites of infection. Approximately 4 to 5 weeks after infection, they begin producing eggs and the life cycle is complete. The three primary species of schistosomes infecting humans can be characterized by their site preference inside the circulatory system, their geographical distribution, and the snails used as first intermediate hosts (see Box 4.1).

Box 4.1 | Schistosomiasis

Schistosomiasis, or bilharziasis, is a disease produced by dioecious digeneans that occupy the hepatic portal and pelvic venous systems of birds and mammals. At least three species, and probably a couple more, produce devastating problems for some 200 million humans in some 75 countries in the world. According to recent (1998) estimates by the World Health Organization, the disease is on the increase because of increasing population densities and changing water-resources development.

The epidemiology of the disease is inextricably linked with certain agricultural and religious practices and poor socioeconomic conditions where the parasite is endemic. For all three of the primary species infecting humans, *Schistosoma mansoni, S. haematobium*, and *S. japonicum*, there is, however, a single thread in the transmission of the parasites which ties them to humans, and that is the requirement by both the host and the parasite for water. Whether to drink, bathe, or irrigate crops, water is an absolute requisite in the lives of humans; likewise, an aquatic medium is required by the parasites' cercariae in order to gain access to humans and other hosts, as well as for the survival of the snail intermediate host. Assuming the presence of obligatory snail hosts for completing a key step in the parasites' life cycles, the mutual need for water by the parasite and the human host insures contact, whether in an Egyptian irrigation canal (for both *S. mansoni* and *S. haematobium*) (Fig. 1) or in a Taiwanese rice paddy (*S. japonicum*) (Fig. 2).

Schistosoma mansoni is widely distributed in Africa. It also occurs around the rim of the Caribbean, including Puerto Rico and the northern coast of South

Fig. 1 Women doing the daily family washing in an Egyptian irrigation canal, a common site of infection for *Schistosoma mansoni*. (Photograph courtesy of Dr. Mohamed Meguid, National Water Resource Center, Egypt.)

Fig. 2 A worker in a rice paddy in modern Taiwan, a common source of infection for *Schistosoma japonicum*. (Photograph courtesy of the Revd John McCall.)

America. Although it is not endemic to continental North America, many Puerto Ricans who live in several of the large cities on the continent suffer from the disease because of the constant movement of people in this community between the island and the continental USA. Monkeys and certain rodents are reservoir hosts for the parasite in nature. The parasites are typically associated with the inferior mesenteric veins in the definitive host, and only occasionally in the superior mesenteric or pelvic veins. The result is that 99% of the eggs produced by the female worms are released via feces, with the remaining 1% in the urine. The species can be identified by the relatively large size of the egg and a conspicuous lateral spine (Fig. 4.23). The snail intermediate host for *S. mansoni* is *Biomphalaria* spp. Because *Biomphalaria* is a quiet surface feeder, *S. mansoni* is restricted to the upper and lower reaches of the Nile Valley in Africa. This distribution was greatly extended in Egypt, however, with construction of the Aswan Dam on the Nile River in the 1950s (Fig. 3).

Fig. 3 The Aswan Dam on the Nile River in Egypt. (Photograph courtesy of Dr. Mohamed Meguid, National Water Resource Center, Egypt.)

The pathology and disease caused by *S. mansoni* are directly related to the passage of eggs through the tissues of the definitive host from the mesenteric veins to the lumen of the large intestine. There are several problems with which the host must contend during the course of the disease. Foremost is the antigenic nature of the eggshell, as well as the antigens (soluble egg antigen) produced and released through the shell by the miracidium developing inside. Amazingly, adults cause no direct problems for their hosts; the adults produce no fatty acids or cholesterol, but bind lipoproteins of host origin to their surface, becoming hidden from the host's immune system.

The disease begins with an explosive onset of rising temperature known as **katayama fever**. This is accompanied by chills, gastrointestinal pain, and dysentery. Hepatic fibrosis and cirrhosis will develop if eggs are carried via the hepatic portal veins to the liver (Fig. 4). Tumor necrosis factor produced by activated macrophages plays a significant role in the etiology of the disease in humans. Severe granulomatous reactions in the liver produce pseudotubercles not unlike those seen in tuberculosis. The main damage comes with severe

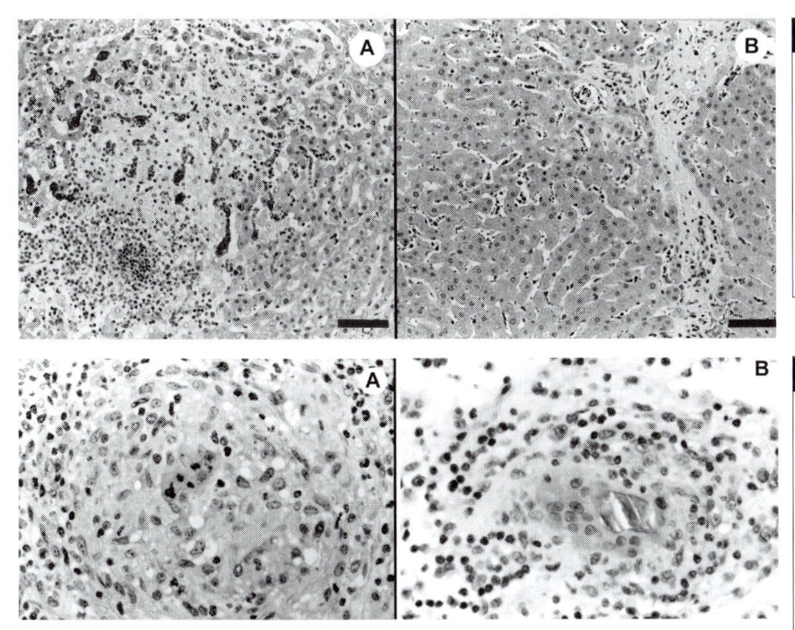

Fig. 4 (A) Hepatic necrosis and inflammatory infiltrate, with abundant eosinophils in a goat infected experimentally with *Schistosoma bovis*; (B) portal fibrosis in a goat liver infected with *S. bovis*. Scales = 50 μm. (From Lindberg *et al.*, 1997, with permission, *Journal of Parasitology*, **83**, 454–459.)

Fig. 5 Different appearances of anti-egg granulomas in the small intestine of goats experimentally infected with *Schistosoma bovis*. (A) Strong epithelioid–cellular granuloma around a degenerated egg; (B) microgranuloma in which the calcified remnant of an *S. bovis* egg can be seen. (From Lindberg *et al.*, 1997, with permission, *Journal of Parasitology*, **83**, 454–459.)

granulomatous inflammation (Fig. 5) and necrosis in the gut wall, followed by scarring and loss of functional integrity. Eggs may become lodged in the gut wall and produce pseudoabscesses. The eggs also cause extensive mechanical damage to the intestine affecting peristalsis. It should be remembered that peristalsis in the intestine involves constant contraction and relaxation of smooth muscle. This, in combination with the antigenicity of the miracidial metabolites and the eggshell, and the presence of a prominent spine, can lead to many serious problems for a host. It is interesting, however, that the pathology caused by the parasite actually is essential for the passage of schistosome eggs from the venous system to the lumen of the intestine (see Chapter 2). Whereas tumor necrosis factor has been related to the stimulation of egg production by females, it is apparently directly toxic to schistosomules. Some have linked the parasite eggs and rectal cancer, although a direct cause and effect has not been established.

Schistosoma haematobium has been known since 4000 BC and was referred to indirectly (pink or red urine) in the Ebers' Papyrus from 1500 BC (see Box 1.1). It is endemic to Africa and parts of Asia Minor. In the Nile Valley of Egypt, 10 million humans are infected, with approximately 40 million in Africa. It is distributed throughout the entire length of the Nile Valley since its snail host, *Bulinus* spp., is able to establish in both slow and relatively fast-moving water. Eggs of this species have a terminal spine; 99% of the eggs passed by humans appear in the urine because the adults are found primarily in the vesicular, prostatic, and uterine circulatory plexi. A few eggs may be shed in the feces if adults find their way into the inferior mesenteric veins. There are no reservoir hosts.

Because of the site of infection by the adults, an inevitable outcome of *S. haematobium* infection will be hematuria, or bloody urine. So prevalent are the parasites in certain endemic areas that younger children believe urine is supposed to be pink or red in color. Because of the parasites' association with the excretory system, there may be catastrophic nephritis and renal failure. There are reports of *S. haematobium* eggs found in cancers of the bladder, with this form of the disease being referred to as Egyptian infiltration cancer. The same kind of mechanical damage and inflammation provoked by *S. mansoni* are seen in association with *S. haematobium*. Again, both are related to the antigenicity of the eggshell and the metabolites produced by the developing miracidium as the parasite's eggs exit the venous system and traverse tissues of the urinary bladder.

Distributed in the Far East, primarily Japan, Taiwan, China, the Philippines, and Indonesia, *Schistosoma japonicum* has a prevalence of 10%–25% in some endemic areas. There are probably 1½ million people infected in China alone. Agricultural practices, such as the use of night soil (human feces) as fertilizer in rice fields, contribute to the widespread occurrence of this parasite. Reservoir hosts include a very wide range of mammals, from rats, dogs, and cats, to water buffalo. In one study in the Philippines (Hairston, 1965), it was estimated that if the schistosome was completely eliminated from a local village, it would return within a year to the same level of infection as before expulsion from humans, simply because it is so common in rice rats.

Schistosoma japonicum is found in the superior mesenteric venous system which drains the small intestine, so the pathology induced is not unlike that seen with *S. mansoni*. In the case of *S. japonicum*, however, eggs are also likely to end up in the liver, as well as the lungs and even the brain. Eggs of *S. japonicum* are much smaller and more round than those of the other two primary species; the lateral spine is quite tiny and may not even be discernible (Fig. 4.24). Fever, coughing, diarrhea, anemia, hepatomegaly, and splenomegaly are frequent consequences. Ascites and weight loss may be seen as well.

At least two other schistosomes have been reported from humans. *Schistosoma mekongi* has been observed in humans, as well as several species of carnivores, including dogs, in Kampuchea (Cambodia) and Laos. *Schistosoma mattheei* (= *S. intercalatum*) occurs in ruminants and primates, including humans, in Africa. There is some evidence that hybridization can occur between some cattle species and at least two of those causing schistosomiasis in humans, and that the resulting hybrids can infect humans (De Bont & Vercruysse, 1998).

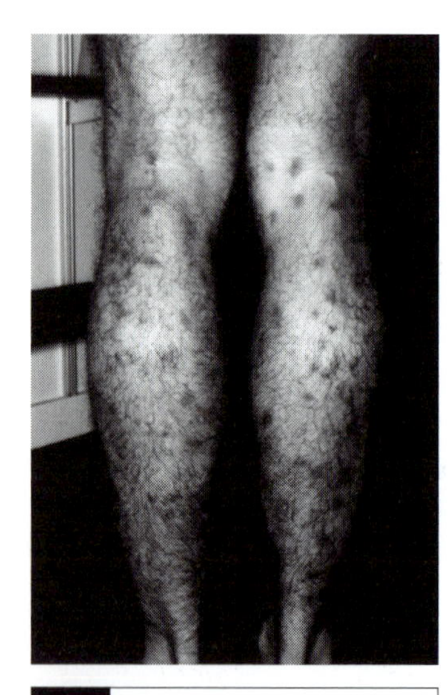

Fig. 6 A severe case of swimmer's itch. (Photograph [and legs!] courtesy of Russ Hobbs, Murdoch University.)

A number of avian schistosomes produce cercariae which penetrate the skin of humans, provoking localized hypersensitivity reactions known variously as swimmer's itch or clam digger's itch on a worldwide basis (Fig. 6). These conditions are most frequently seen along the east coast and in the central provinces of Canada, or in the northern states of the USA. In freshwater lakes of North America, swimmer's itch is frequently attributed to species of *Trichobilharzia* and *Gigantobilharzia*, normally common parasites of birds, ranging from ducks to cardinals and goldfinches. Along the coast, the causative agent is *Microbilharzia*, a parasite of gulls. Whereas these schistosomes are not serious or dangerous pathogens in humans, swimmer's itch can cause severe economic problems for owners of resorts on the many lakes where the parasites occur. Swimmer's itch can become so severe that areas in which the

Fig. 7 Sign on a beach area at Riding Mountain National Park, Canada, posted to warn bathers that they enter at their own risk.

disease is rampant signs may actually be posted, warning bathers to be aware of the potential risk (Fig. 7). The signs and symptoms of the infections quickly disappear, however, as the parasites are rapidly killed by a strong host immune response.

Some schistosomes also cause a much less serious, but extremely annoying, condition in humans, known as '**swimmer's itch**' or **cercarial dermatitis**. Although miracidia are usually specific for their snail host, the cercariae of many species do not discriminate much and often penetrate the skin of unsuitable hosts, including humans. Swimmer's itch is caused when cercariae of blood flukes that normally infect birds and mammals penetrate the skin of humans, sensitizing the area and causing the formation of small pustules and an itchy rash. These cercariae, however, do not make it into the circulatory system because they are destroyed by the host's immune system within a few days.

In a number of families within the order Strigeiformes, i.e., Strigeidae, Diplostomidae, Bolbophoridae, Neodiplostomidae, Cyathocotylidae, Liolopidae, and Proterodiplostomidae, the body of the adult may be divided into an anterior flattened, spoon-shaped portion and a posterior section shaped like a cylinder. The anterior portion has a pair of accessory suckers on each side of the oral sucker to aid in attachment, and an adhesive, or **tribocytic organ**, behind the ventral sucker. Besides helping with attachment, the tribocytic organ secretes proteolytic enzymes that digest host tissues on which the parasite then may graze. The posterior end of the body contains most of the reproductive structures.

The strigeids are cosmopolitan parasites in a number of avian hosts, primarily ducks and geese, and several mammals. *Cotylurus flabelliformis* is a common intestinal parasite of ducks in North America. Cercariae penetrate lymnaeid snails where they develop into a special kind of metacercaria called a tetracotyle. However, if the cercariae enter planorbid or physid snails, they actually may penetrate the sporocysts or rediae of other digenean species and then develop into tetracotyle metacercariae within them, a clear case of hyperparasitism. Ducks become infected when they eat snails harboring tetracotyle metacercariae.

Species in the Diplostomidae are intestinal parasites of birds and mammals. The group includes species of *Alaria* and *Uvulifer*, several of which have received substantial attention because of a distinctive characteristic associated with their life cycles or ecology. For example, *Alaria americanae* has a complicated four-host life cycle. Adults of this parasite are found in the intestines of canines in North America. Cercariae from snails penetrate the skin of tadpoles, developing into **mesocercaria**, an intermediate stage between a cercaria and a metacercaria. The mesocercaria remains within the body even after the tadpole undergoes metamorphosis to the adult stage. If a canine eats an infected tadpole or an adult frog, then the life cycle can be completed. This last step is generally unlikely, however, as canines do not normally prey on tadpoles and frogs. So, how is the life cycle completed? If the tadpole or adult frog is eaten by a snake or a rodent, a more ecologically sound alternative, the parasite (mesocercaria) is freed from the tissues by digestion, penetrates the gut wall, and takes up residence in the tissues as an unencysted mesocercaria. The snake or rodent thus becomes a paratenic host and, when preyed upon by canines, the parasite's life cycle is completed. This is yet another example of a parasite using a particular host to bridge an ecological, or trophic, gap. However, this complicated life cycle is still incomplete. Once into the canine, the mesocercaria penetrates the gut wall and moves into the coelom, passes through the diaphragm, and becomes established in the lung. There, it develops into a special metacercaria called a diplostomulum, which remains unencysted in the lungs for about 5 more weeks. It then migrates out of the lungs into the trachea and throat, where it is coughed up and swallowed, finally passing into the intestine and maturing sexually. If the definitions for intermediate and definitive hosts developed earlier are strictly applied, then the canine serves as both an intermediate and a definitive host. In *A. marcianae*, a related species, mesocercariae may be acquired by young pups in milk from a lactating bitch, an example of **transmammary transmission** (Shoop & Corkum, 1987; Shoop *et al.*, 1990).

The metacercariae of a number of species of Diplostomidae may induce either black spot disease or impaired vision, two potentially serious and even lethal conditions in fishes. Black spot in North American fishes is caused by the metacercariae of several species, including *Uvulifer ambloplitis* (the life cycle and ecology of this species are discussed in greater detail in Chapter 12). Metacercariae in the flesh of fishes (the second intermediate hosts) secrete their own cyst and provoke a strong tissue reaction by the host. The reaction includes mobilization of histiocytes and melanocytes that produce a darkly

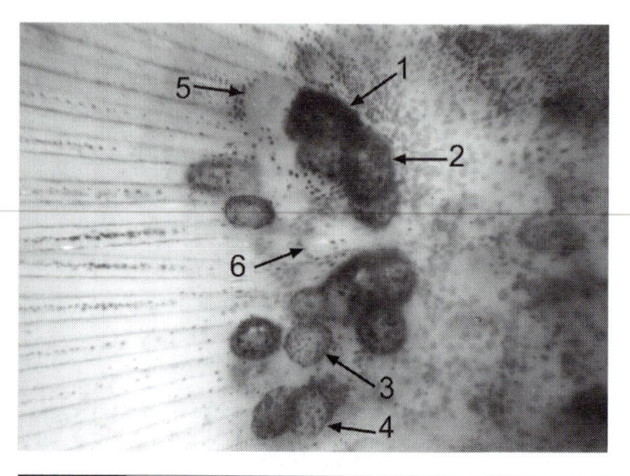

Fig. 4.25 A concentration of *Uvulifer ambloplitis* metacercariae at the base of the caudal spine of the bluegill sunfish *Lepomis macrochirus*, showing deeply pigmented (1, 2), partially pigmented (3, 4), and unpigmented (5, 6) cysts. Pigmentation reflects the time elapsed since the day of infection. (Photograph courtesy of A. Dennis Lemly.)

pigmented and thick cyst wall around the parasite, sequestering it even further (Fig. 4.25). Experimental evidence strongly suggests that the host reaction may be quite debilitating, to the extent that heavily infected young-of-the-year sunfishes, which do not feed in the winter, are unable to overwinter successfully (Lemly & Esch, 1984). Cercariae of several species of *Diplostomum* penetrate the eyes of fishes and develop into metacercariae, inducing cataractous lenses and impairing vision. With both black spot and impaired vision, successful completion of the life cycle depends on predation by fish-eating birds. Both conditions handicap the fish host, and facilitate transmission of the parasite to the avian definitive host.

Species in the Fellodistomidae are restricted to marine habitats, but some occasionally occur in freshwater fishes, and a handful reaches adulthood in marine molluscs. Species of *Fellodistomum* are common parasites of the gall bladder, whereas species of *Proctoeces* can be found both in the intestine of fishes and in molluscs.

Leucochloridium paradoxum is a brachylaimid fluke of birds, including the robin, *Turdus migratorius*, in North America. In its snail host, a sporocyst develops first in the hepatopancreas and then secondarily in the snail's tentacles. The two portions remain connected via a tube through which

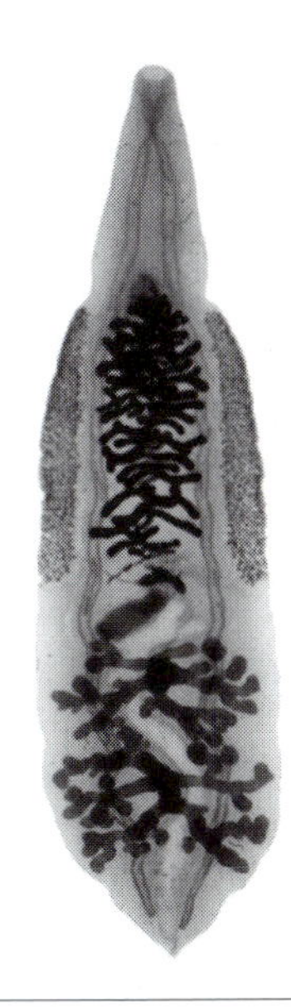

Fig. 4.26 *Clonorchis* (= *Opisthorchis*) *sinensis*, the Chinese liver fluke. (Photograph courtesy of J. Teague Self.)

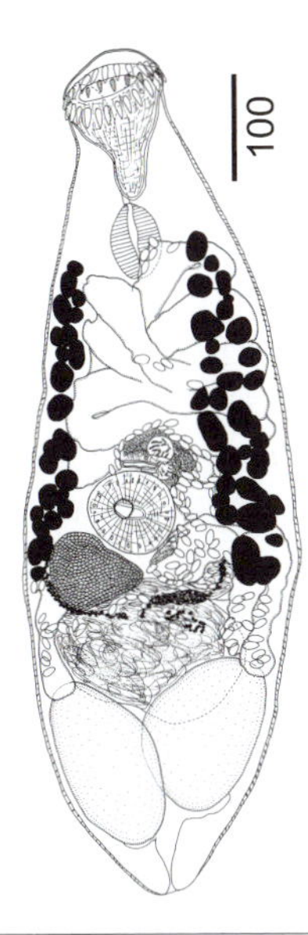

Fig. 4.27 The digenean *Ascocotyle nunezae* (Heterophyidae) from the great egret *Casmerodius albus*. Units are μm. (From Scholz *et al.*, 1997, with permission, *Journal of Parasitology*, **83**, 141–147.)

embryos are passed from the sporocyst in the hepatopancreas into that part of the sporocyst located in the tentacles. Cercariae develop within the sporocyst, in the tentacles, which become swollen, turn a bright color, and begin to pulsate. The bright coloration and the pulsating tentacles (caused by the sporocysts) create an irresistible spectacle for the unsuspecting predator and, 'voila', the parasite is passed to the unsuspecting robin when the infected snail falls prey to the avian host!

OPISTHORCHIIFORMES

Species of Opisthorchiidae infect the bile ducts, gall bladders, and livers, of reptiles, birds, and mammals. Perhaps the best known is *Clonorchis sinensis*, also called the Chinese liver fluke (Fig. 4.26). This parasite (sometimes referred to as *Opisthorchis sinensis*) infects nearly 30 million humans, mostly in the Far East, but also in many other parts of the world where fish are consumed raw, as sushi (see also Box 5.1). These flukes occur in the liver, bile duct, and gall bladder where a single adult may produce up to 4000 eggs per day. Several species of the freshwater snail *Parafossarulus* serve as first intermediate hosts. Cercariae emerge from the snail, penetrate the surface of fishes, and encyst in the muscles as metacercariae. More than 80 species of cyprinid fishes are known to harbor the parasite. There are several reservoir hosts for the fluke, including canines, felines, badgers, and mink. With the increase in aquaculture in many areas of the world, the parasite appears to be spreading as well. Infected humans may suffer minor to serious problems, including liver damage, vomiting, and diarrhea. Two other related parasites with similar life cycles and hosts are *Opisthorchis felineus* and *O. viverrini*. The former is common in southern, central, and eastern Europe, parts of the former Soviet Union, Japan, and India, whereas the latter is prevalent in Southeast Asia.

In terms of their life cycles and mammalian hosts, species in the Heterophyidae (Fig. 4.27) closely resemble opisthorchiids. These are intestinal parasites, however, and their numbers may reach into the thousands within a single host. Fortunately, they are quite small, but, if in sufficient number, they may cause abdominal discomfort, nausea, vomiting, and diarrhea. Adult worms may on occasion erode the intestinal mucosa and deposit eggs, which then may reach the circulatory system and spread to various parts of the body, especially the brain and heart. They may actually cause death in human hosts due to cardiac arrest if a sufficient number of eggs become entrapped in cardiac muscle. Common species in the family include *Heterophyes heterophyes* and *Metagonimus yokawagai*. The former is prevalent in Egypt, eastern Asia, and Hawaii, whereas the latter is common in the Balkan countries and the Far East. Most of these parasites are present in countries where fish is consumed raw and, like their opisthorchiid relatives, are increasing in prevalence because of the increasing consumption of raw fish on a worldwide basis (see also Box 5.1). A species common along the Gulf Coast of North America is *Phagicola longa*. Metacercariae of this species occur in the viscera and flesh of mullets, whereas adults are present in a number of birds, including the brown pelican. A sample of pelicans in an endemic area produced an average of 12 000 *P. longa* per host. Interestingly, mullets along the Pacific Coast of South America also are infected by *P. longa* and it is thought that brown pelicans migrating between the Gulf Coast and South America carry and spread the parasite.

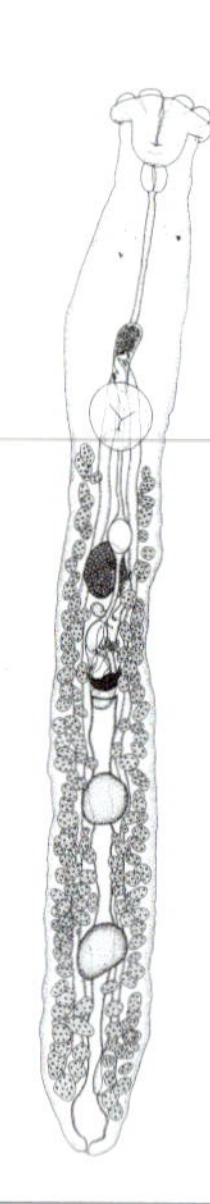

Fig. 4.28 The digenean *Crepidostomum percopsis* (Allocreadidae) from the trout perch *Percopsis omiscomaycus*. (From Nelson *et al.*, 1997, with permission, *Journal of Parasitology*, **83**, 1157–1160.)

PLAGIORCHIFORMES

The Plagiorchiformes includes the largest number of families among the digeneans and is probably the most diverse in terms of morphology, with very little superficial resemblance among those in the order.

Species of the Allocreadidae are referred to as the papillose flukes because some possess a series of prominent muscular papillae, or head lappets, surrounding the oral sucker (Fig. 4.28). These parasites are found in fishes and amphibians throughout the world. One of the better-studied and discussed in an ecological context elsewhere (see Chapter 11) is *Crepidostomum cooperi*; it is widely distributed in North America and commonly occurs in the pyloric ceca of freshwater centrarchid fishes. Sphaeriid clams and arthropods (mayflies, amphipods, crayfish) are intermediate hosts. Several allocreadids are neotenic, and their larva develops to sexual maturity in what were presumably their arthropod second intermediate hosts, e.g., *Allocreadium neotenicum* in dytiscid diving beetles.

Adult troglotrematids are generally found in the lungs, nasal passages, cranial cavities, and intestines of birds and mammals. Perhaps the best-known member of the family is *Paragonimus westermani*, which occurs in the lungs of a number of crab-eating mammals, including humans, in eastern Asia, parts of Africa, and Peru and Ecuador. Cercariae emerge from snail intermediate hosts usually in the late evening, which coincides with the maximum behavioral activity of crabs and crayfish, their next intermediate hosts. The cercariae crawl on the substratum much like inchworms; when they make contact with a crab or crayfish, they attach and penetrate through the arthrodial membranes in the joints, and encyst in its muscles. When a crab-eating mammal consumes the infected second intermediate host, the parasites excyst in the small intestine, penetrate the gut wall, then the diaphragm, and enter the lung where they mature sexually. Remarkably, they are usually able to locate a partner and generally encyst in pairs within the lung.

Paragonimiasis in humans is a zoonotic disease because the normal hosts in nature are felines and canines, and humans are infected secondarily. Nonetheless, there are several million human infections in the Far East where crabmeat is eaten raw and unsalted as a delicacy. Sometimes, live crabs may be marinated overnight in wine and then consumed raw. Women eat raw crabmeat, believing it will increase fertility, and Buddhist monks consume it because, to them, crabmeat is neither fish nor fowl. The pathology associated with *P. westermani* may be severe, producing abscesses similar to those seen in tuberculosis, and may even be misdiagnosed as that disease. The parasite may also become ectopic, wandering into such organs as the spleen, urinary bladder, eyes, and even the brain.

Nanophyetus salmincola is a troglotrematid that occurs in the northwestern areas of North America and in Siberia where it causes salmon-poisoning disease in canines (Millemann & Knapp, 1970). Whereas the adult fluke is virtually innocuous for the canine, it is the vector for *Neorickettsia helminthoeca*, a bacterial organism that causes the disease in the definitive host. The adult parasite is embedded in the intestinal wall of canines and other mammals, including humans, although raccoons and spotted skunks are apparently the primary reservoir hosts in nature. The stream snail *Oxytrema silicula*, and a wide range of fish species, especially salmonids, are the required intermediate hosts. The rickettsial infection in dogs is 90% lethal within 7 to 10 days, if untreated. The symptoms include a high fever, vomiting, and diarrhea. If the dog recovers, it is totally immune to another exposure to the rickettsial organism.

Another troglotrematid of interest is *Collyriclum faba*. Adults encyst in pairs subcutaneously around the cloaca of several species of birds (Blankespoor *et al.*, 1985), including English sparrows, grackles, swifts, etc. The cysts are slightly larger than an ordinary garden pea and appear in clusters of up to 15 in a single bird. Each cyst has a small hole that is plugged with a mucus secretion which dissolves when the bird dips its rump in water when taking a drink. Eggs are released at that point. The life cycle is completely unknown, although winged insects are probably the second intermediate hosts since swifts are infected with the parasite and their feeding is strictly on the wing.

Species in the Dicrocoelidae parasitize terrestrial or semi-terrestrial vertebrates and use terrestrial snails and arthropods as first and second intermediate hosts. Several dicrocoelids are known to produce some rather bizarre behavioral modifications in their arthropod hosts. The best known is *Dicrocoelium dendriticum* which has an apparently cosmopolitan distribution in sheep, cattle, pigs, goats, and cervids. Occurring in the bile ducts, adults produce eggs that are shed in the feces. Snails ingest the eggs and cercariae accumulate in the snail's modified ctenidia or mantle. Their accumulation in the 'lung' is an irritant and causes the snail to secrete mucus in which the cercariae become entangled. The infected snail releases masses of mucus, containing up to 500 cercariae, as slime balls. Apparently, these slime balls are attractive to ants (*Formica* spp.) which consume them. When this occurs, the cercariae penetrate the gut wall where they become encysted in the hemocoel as metacercariae, or at least most of the cercariae follow this route. One or two, however, migrate to the subesophageal ganglion and encyst there. These metacercariae provoke a striking behavioral change in the ant. Ants infected with *D. dendriticum* are unable to open their mouths when temperatures fall at dusk. If they happen to be grazing on a blade of grass, they become attached for the night. They must remain there, fixed in position by their closed jaws, until temperatures rise the next morning. Since sheep and other hosts graze extensively during the early evening and early morning hours, an infected ant caught on a blade of grass is thus more vulnerable to accidental ingestion. On excystment in the small intestine, the metacercariae migrate directly to the bile ducts where they take up residence and mature sexually. It is interesting that those metacercariae in the subesophageal ganglion are not infective for sheep; is this a case of altruistic behavior on their part? If all the cercariae in the slime ball come from sporocysts produced by a single egg, then the answer is no because they are genetically identical, but if the cercariae in a single slime ball come from sporocysts produced by several different eggs, then the answer could be yes because they are genetically different.

Brachylecithum mosquensis is a rather common dicrocoelid fluke in North American robins, *Turdus migratorius;* metacercariae of this parasite also alter the behavior of its second intermediate host, thereby increasing the probability of transmission. Metacercariae encyst in the subesophageal ganglia of carpenter ants, *Camponotus* spp.

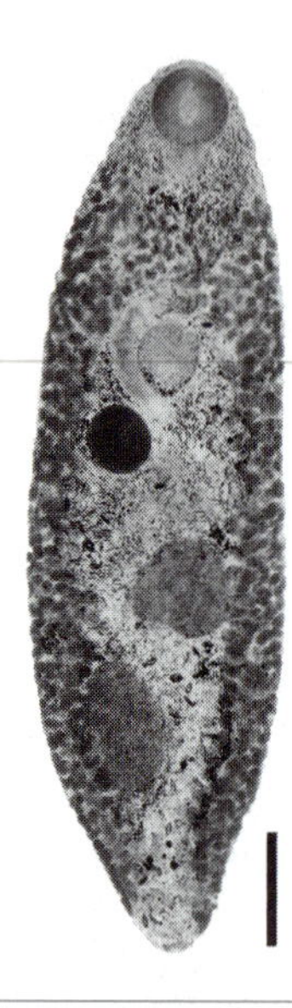

Fig. 4.29 The digenean *Plagiorchis muris* (Plagiorchiidae) from an infected human in Korea. Scale= 400 μm. (From Hong *et al.*, 1996, with permission, *Journal of Parasitology*, **82**, 647–649.)

Normally, carpenter ants are relatively secretive, avoiding bright sunlight. When infected with this parasite, however, they prefer bright light and become vulnerable to predation by robins.

The Plagiorchiidae is a large family of digeneans infecting vertebrates of all classes. Plagiorchiids typically use snails as a first intermediate host and an insect second intermediate host, although there are many other invertebrates that serve in this capacity as well. *Plagiorchis noblei* is a common parasite of yellowheaded and redwinged blackbirds in North America; lymnaeid snails and several larval insects (larval mosquitoes, and naiads of caddis flies and damselflies) are first and second intermediate hosts. *Plagiorchis muris* (Fig. 4.29) occurs in the small intestine of a very diverse group of hosts, including dogs and rats (and humans), as well as terrestrial and marine birds. *Haematoloechus* is a widely distributed genus of plagiorchiid found in the lungs of frogs. A number of different snails and dragonfly nymphs serve as intermediate hosts. Several species have been described from North American frogs. Most exhibit great morphological plasticity, depending on the host, making their specific identification difficult.

Opecoelid flukes typically occur in the intestines of marine fishes, and a few are found in freshwater piscine hosts. They usually have a three-host life cycle, one that includes a snail, an arthropod, and the piscine definitive host. A recent report by Barger & Esch (2000) has revealed the highly unique and abbreviated cycle of *Plagioporus sinitsini* in the freshwater stream snail *Elimia* sp. These investigators have found snails shedding sporocysts that contained daughter sporocysts, cercariae, and sexually mature adults with eggs and viable miracidia. It seems that this abbreviated life cycle is neither neoteny nor progenesis because the adult parasites inside the sporocysts are equivalent in size to adults occurring in stream fishes in the same habitats.

4.5 Monogenea

4.5.1 General considerations

Most monogeneans are ectoparasites of both marine and freshwater fishes. A few parasitize amphibians and one species, *Oculotrema hippopotami*, is found on the eyes of hippos. Monogeneans typically occur on the external surfaces of their hosts and exhibit remarkable host and site specificity. They possess an incomplete gut, are monoecious, and have direct life cycles. Most are usually small in size and equipped with a large, modified attachment organ at the posterior end. The morphology of this structure is truly spectacular, and provides insight as to the relationship between the attachment device of the parasite and the morphology of the site of infection on, or in, the host. Their life spans range from several days to several years.

4.5.2 The outside

Most monogeneans are thin, flattened, and range in size from about 0.3 mm to 20 mm. The anterior end of the body, also called the **prohaptor**, has various adhesive and feeding structures (Fig. 4.30). In some species, the prohaptor may have a number of cephalic or head glands that secrete a sticky adhesive substance, and shallow muscular suckers, all used for attachment. In other species, there is an oral sucker, with various degrees of muscularization that surrounds the mouth.

At the posterior end of the body is a distinctive structure called the **opisthaptor**, primarily

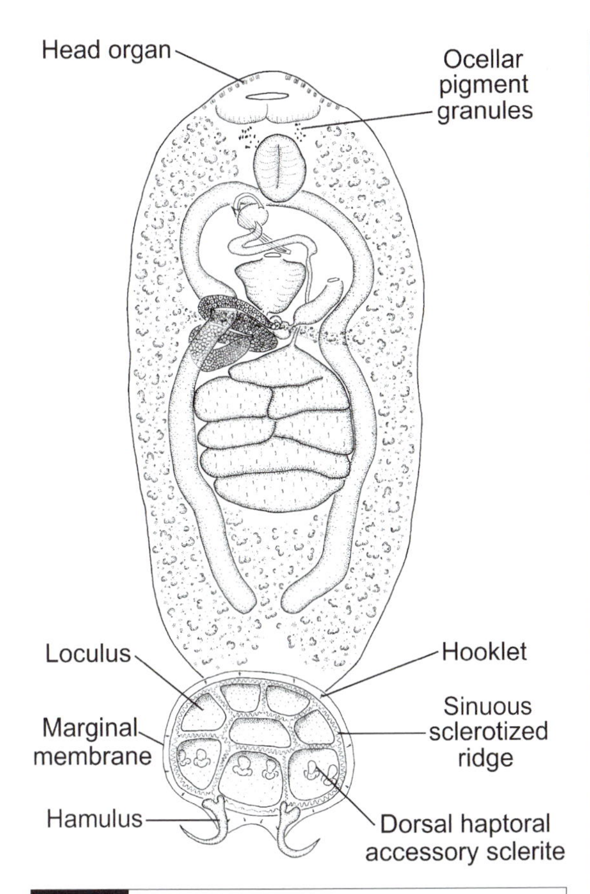

Fig. 4.30 The monogenean *Neoheterocotyle inpristi* from the gills of *Pristis pectinata* on the Florida coast. (From Chisolm, 1994, with permission, *Journal of Parasitology*, **80**, 960–965.).

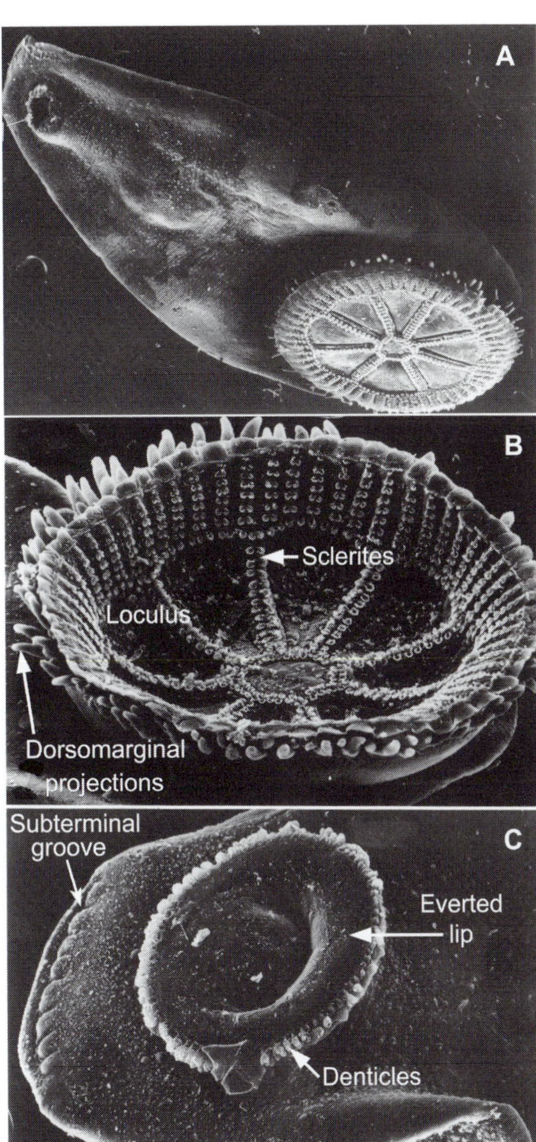

Fig. 4.31 The monogenean *Clemacotyle australis*. (A) Whole animal, ventral view showing haptor and mouth; (B) haptor with one of several loculi and other structures labeled; (C) ventral surface of anterior extremity showing subterminal groove equipped with flat, tegumental tooth-shaped markings. Note the everted lip of the mouth bearing tegumental denticles at the edge. (From Beverly-Burton & Whittington, 1995, with permission, *Journal of Parasitology*, **81**, 616–625.)

responsible for the attachment of the monogeneans to the host (Figs. 4.30, 4.31). The morphology of the opisthaptor is highly variable. It may have suckers in various degrees of development, large hooks called anchors (or hamuli), small hooks that are remnants from the larval stage, or complex clamps that may be either muscular or sclerotized. There are normally one to three pairs of anchors that often have a connecting bar or accessory sclerites supporting them. The hooks are generally very small and, like the anchors, are made of keratin. These hooks and those of eucestodes and cestodarians are similar chemically, whereas hooks of digeneans are composed of crystallized actin. Clamps work by pinching at their attachment site, something like clothespins, and their numbers vary from eight to several hundred, depending on the species.

Because monogeneans are mostly ectoparasitic, their survival depends on the efficiency of their attachment organs. Consequently, the structure of their attachment organs, mainly the opisthaptor, has been largely determined by the morphology of the attachment site, to the extent that the two

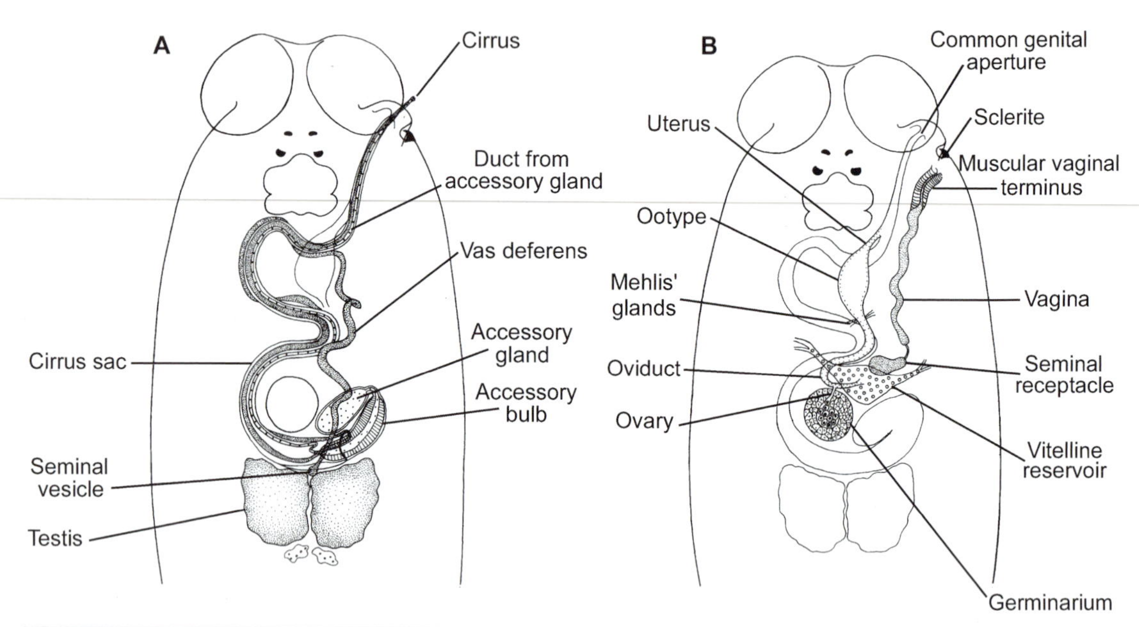

Fig. 4.32 Reproductive systems of the monogenean *Metabenedeniella parva* from the dorsal fin of *Diagramma pictum* on the Great Barrier Reef, Australia. (A) Male; (B) female. (From Horton & Whittington, 1994, with permission, *Journal of Parasitology*, **80**, 998–1007.)

almost work as a lock-and-key mechanism. An indication of this specialization is the high degree of host and site specificity exhibited by most monogeneans, in particular, those that parasitize the gills of fishes.

On the anterior surface of the body there may be up to four pigmented eyespots, or photoreceptors. There also may be an elaborately organized set of external papillae and underlying nerve endings which probably function as mechanoreceptors, especially in association with the opisthaptor of some species, e.g., *Entobdella* spp. The surface of monogeneans, as in digeneans and cestodes, is syncytial, with the nuclei and many other organelles located in cytons below the tegumental surface.

4.5.3 The inside

Digestive system. The incomplete digestive system is simple and includes a mouth, buccal funnel, prepharynx, a muscular and glandular pharynx, esophagus, and paired ceca which may possess diverticula, or fenestrations. The pharynx is a powerful sucking structure that brings food into the gut. The alimentary tract is in the shape of a wishbone, although sometimes the two branches may unite near the posterior end of the body. There is no anus. Once food is internalized, digestion is both extra- and intracellular.

Excretory/osmoregulatory system. The excretory/osmoregulatory system is generally quite simple. The flame cells of the protonephridia are connected to two lateral main collecting ducts that swing down the length of the body and then back toward the anterior end, where paired contractile bladders help empty them to the surface via excretory pores. These pores are lateral and located near the anterior end of the organism. This is a major distinguishing character between monogeneans and digeneans, because in digeneans a single bladder and excretory pore are located posteriorly.

Reproductive system. Monogeneans are monoecious, but the reproductive systems are entirely separate (Fig. 4.32) and cross-fertilization is the rule in most cases. The number of testes is variable, depending on the species, with as few as one and as many as two hundred. Most species, however, have a single testis. Arising from each testis is a vas efferens that follows into an ejaculatory duct and then continues into a genital atrium. In many species, the tissues around the ejaculatory duct are thickened and form a penis-like structure armed with hooks distally. In some, there is a complex, sclerotized copulatory organ that joins with the ejaculatory duct. The morphology of the hooks of the opisthaptor and of the sclerotized copulatory organ are characteristic of each species.

The female reproductive system is complex and highly variable. It contains a single ovary of variable shape. Haploid ova move from the ovary into a short oviduct, which, in turn, empties into an ootype that is surrounded by the Mehlis' gland. Before the oviduct enters the ootype, the vitelline duct, the vaginal canal, and a genitointestinal canal may join it. Sperm stored in the seminal receptacle fertilize the ova as they pass through the oviduct on their way to the ootype. Once inside the ootype, the zygote begins development to the egg stage. Secretions from the Mehlis' gland apparently function in lubricating the ootype, facilitating movement of the developing egg. Secretions from the vitelline glands, or vitellaria, which are scattered laterally up and down the length of the body on both sides, contain a tanning protein that contributes to sclerotizing the eggshell. Egg production can be quite rapid, with several being produced within just a few seconds. Eggs leaving the ootype move into the uterus that opens into the genital atrium. The genitointestinal canal, mentioned earlier, connects the female reproductive system to the gut. Although its function is not known, there is speculation that it may be a vestigial structure through which eggs were passed into the intestine to be expelled through the mouth.

Nervous system and sensory input. The nervous system of monogeneans is similar to that of most other platyhelminths, with paired cephalic ganglia located anteriorly and usually a pair of nerve cords extending posteriorly. At intervals, the nerves are connected laterally. A variety of sensory cells and structures are located in the tegument. The most prominent of these sensory structures are eyespots on the surfaces of oncomiracidia (the monogeneans' free-swimming larval stage), which usually disappear in the adult stage, although some species may possess eyes as adults. As might be expected, the opisthaptor is well innervated internally and probably possesses external papillae that function in mechanoreception as well. Monogeneans are known to produce a number of neurotransmitters, with serotonin being primary among them (Halton *et al.*, 1993).

4.5.4 Nutrient uptake

Monogeneans feed on various host tissues, including blood, mucus, and epidermal cells. Feeding is achieved by the sucking action of the pharynx, and in some species, e.g., *Entobdella soleae*, the pharynx can be everted. The pharyngeal glands of *E. soleae* secrete proteases that erode and lyse the epidermis which is then sucked inside (Kearn, 1971). In species that feed on blood, such as *Diclidophora* spp., digestion of hemoglobin is mostly intracellular. By-products of digestion, such as hematin, or undigestible compounds, are evacuated through the mouth. Nutrient uptake through the tegument probably occurs. The blood-feeding *Diclidophora merlangi*, for example, can absorb neutral amino acids through the tegument and these may supplement its diet (Halton, 1978).

4.5.5 Metabolism

The metabolism of monogeneans is basically unknown (Smyth, 1994). Living on the surface of fishes would suggest aerobic metabolic pathways, but this is speculative.

4.5.6 Development and life cycle

The life cycle of most monogeneans is direct, without intermediate hosts, and includes an egg, a larval stage called an **oncomiracidium** (Fig. 4.33), and the adult. The oncomiracidium hatches from the egg. These larvae are covered with cilia, have one or two pairs of eyespots, and possess numerous hooks at the posterior end, which are retained by the adult, becoming the smaller hooks on the opisthaptor. In many respects, these larvae resemble the hexacanth embryos of eucestodes. Oncomiracidia are excellent swimmers, but their life span is brief. Mucus secretions apparently influence host location. Once attached to an appropriate host, they lose their cilia and develop directly to the adult stage.

The exceptions to this pattern are the gyrodactylids, all of which are viviparous. Larvae are retained in the uterus until they develop into functional juveniles. Even more interesting, within a developing juvenile there is a second juvenile, and

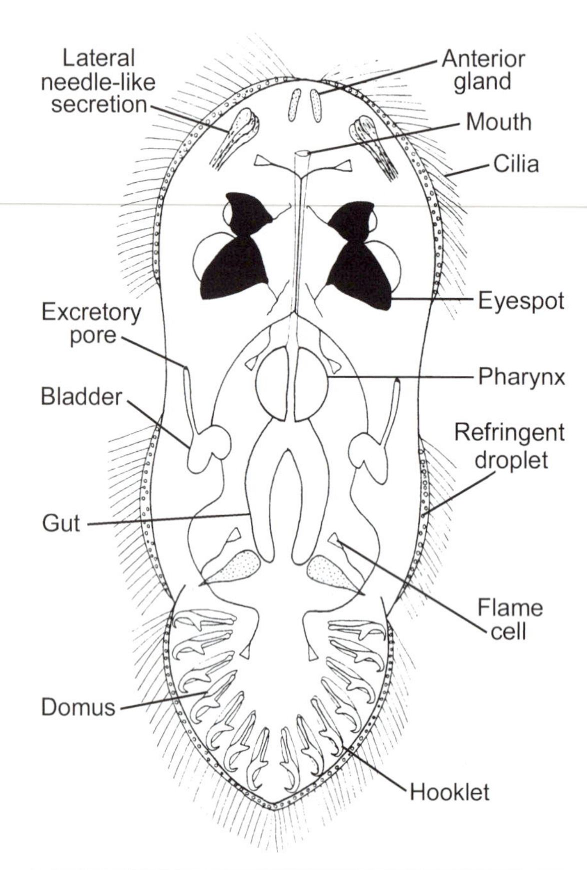

Fig. 4.33 Oncomiracidium of the monogenean *Clemacotyle australis*, from the branchial cavity of the white-spotted eagle ray *Aetobatis narinari*. (From Beverly-Burton & Whittington, 1995, with permission, *Journal of Parasitology*, **81**, 616–625.)

a third within the second, and a fourth within the third, much like nested boxes. As soon as the first juvenile is born, it begins to feed and gives birth to the juveniles remaining inside. This process can be completed very rapidly, within 24 hours. Only after the first juvenile has given birth to all the others can one of its own ova become fertilized, then to repeat the reproductive sequence.

4.5.7 Diversity

Monogeneans are a diverse group of parasites that exhibit a relatively high degree of host specificity when compared to other groups of parasites. Unfortunately, because monogeneans do not pose a zoonotic problem for humans, and because most of them are not pathogenic, very little is known about the biology of most species. However, interest in the group is increasing because of the realization that monogeneans may

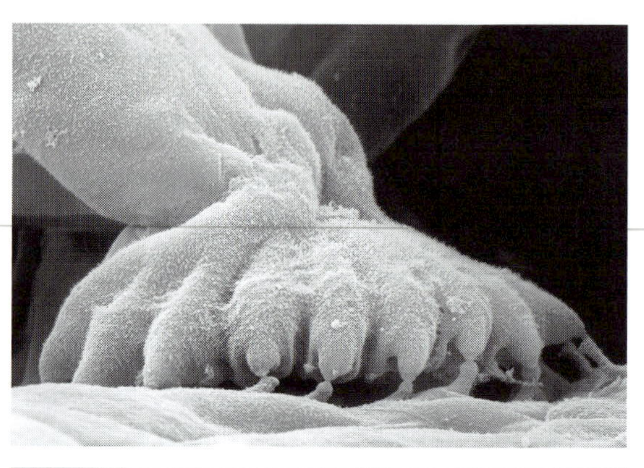

Fig. 4.34 The monogenean *Gyrodactylus colemanensis* attached to the surface of a rainbow trout *Oncorhyncus mykiss*. (From Cone & Cusack, 1988, with permission, *Canadian Journal of Zoology*, **66**, 409–415.)

be excellent models for some ecological studies, e.g., see Rohde (1993) and references therein.

Gyrodactylids are very small monogeneans, measuring 1 mm or less in length, that are parasitic on the gills and skin of marine and freshwater fishes (Fig. 4.34). Their opisthaptor is relatively simple, with two median anchors and 16 small hooks. They are unique among the monogeneans because they are viviparous. The lack of a swimming oncomiracidium does not appear to hinder the worm's dispersal, which occurs directly by physical contact between hosts and via water that is circulated through a host's gills, to which gyrodactylids are attached. The highest rate of reproduction in gyrodactylids occurs during the breeding period of the host because the proximity of a potential host clearly increases the chances of successful transfer.

Many gyrodactylids are known to cause fish mortalities in hatcheries and rearing ponds, mainly in Europe and the former Soviet Union, where high densities of the host facilitate parasite dispersal. Malmberg (1993), in an extensive review of gyrodactylosis in Scandinavia, suggested that **genetic drift** under conditions of aquaculture might produce new pathogenic forms. Gyrodactylids damage the gill epithelium by feeding on it and by the mechanical action of the anchors and hooks of the opisthaptor. This causes epithelial proliferation and increases the production of mucus in the gills that may eventually lead to host death by functional failure of the respiratory epithelium, causing asphyxia (Schaperclaus, 1991).

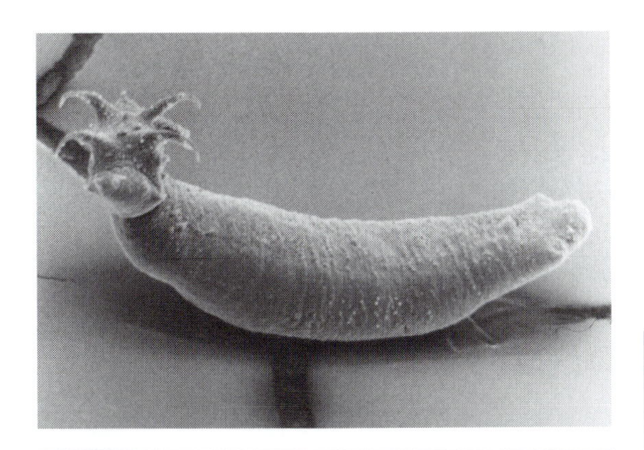

Fig. 4.35 The dactylogyrid monogenean *Urocleidus adspectus*, from the gills of a fish. (With permission, from Cone, D. [1995] Monogenea [Phylum Platyhelminthes]. In *Fish Diseases and Disorders*, Vol. 1, *Protozoan and Metazoan Infections*, ed. P. T. K. Woo. Wallingford: CAB International.)

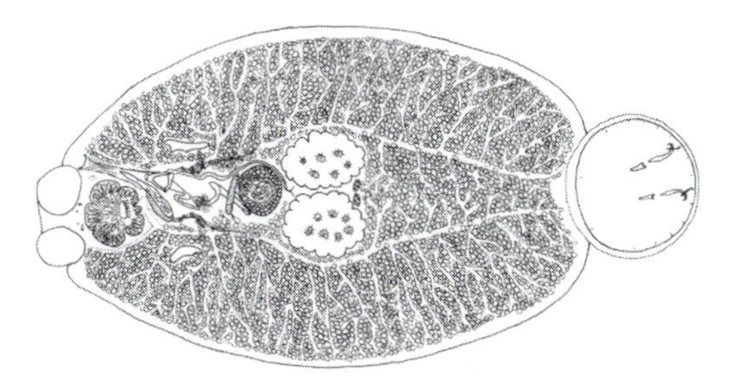

Fig. 4.36 The monogenean *Neobenedenia girellae* from the body surface, fins, and eyes of fishes from Japan. (From Ogawa *et al.*, 1995, with permission, *Journal of Parasitology*, **81**, 223–227.)

Members of the Dactylogyridae (not to be confused with the Gyrodactylidae), and, in particular, species of *Dactylogyrus*, are of great economic significance as pathogens of fishes in hatcheries and rearing ponds throughout the world. They parasitize the gills of fishes, and the damage inflicted is very similar to that caused by gyrodactylids. Dactylogyrids are small monogeneans measuring up to 2 mm in length and, superficially, they resemble gyrodactylids (Fig. 4.35). Their life cycle, however, follows the typical monogenean pattern, with an egg and a free-swimming oncomiracidium. It is fairly common to find more than one species on a given host, in which case they exhibit marked site specifity. In Europe, for example, the common carp may simultaneously harbor three different species, each with a specific location on a single gill filament. For example, *D. vastator* prefers the outer tips of the gill filaments, *D. extensus* is found half the way along the length of the gill filament, and *D. anchoratus* prefers the basal region. Incidentally, all three species are highly pathogenic.

The Capsalidae, including species of *Neobenedenia* (Fig. 4.36), are relatively large monogeneans parasitic on the skin, mouth, and nostrils of marine fishes. Massive attacks by some species can produce the same kind of damage as by the previously described gyrodactylids.

As previously emphasized, one of the most distinctive ecological features of monogeneans is their relatively high host and site specificity. Many monogenean families are very narrow in their range of hosts, and are normally associated with related hosts. For example, several families are exclusively parasitic on elasmobranchs. The Loimoidae and Hexabothriidae parasitize the gills, whereas the Microbothriidae attach to the skin and nostrils of elasmobranchs. The opisthaptor of microbothriids has no anchors or hooks of any kind and attachment is provided by the presence of a muscular cup or flaps. It makes one wonder if this relatively harmless attachment apparatus is related to the rather delicate nature of the host's nostrils. The Dionchidae are exclusive parasites of the gills of remoras, and the Mazocraeidae are gill parasites of herring and mackerels. The opisthaptor of this latter group is rather peculiar because the clamps used for attachment are not located in a distinctive opisthaptor as in most groups, rather they are mounted directly on the sides of the body creating a very strange appearance.

Members of the Diplozoidae are among the most bizarre of all monogeneans. During their development, juveniles, called **diporpa**, grow independently. Unless a diporpa larva encounters another, it will die. If they come into contact, they attach to each other and fuse permanently (Fig. 4.37) which also stimulates sexual maturation.

Although most monogeneans are parasites of marine and freshwater fishes, a few parasitize amphibians and reptiles. Species of *Sphyranura* are parasitic on the gills and skin of caudate amphibians such as waterdogs, whereas species in the Polystomatidae parasitize the urinary bladder, nasal cavities, mouth, pharynx, or esophagus of amphibians and reptiles.

In those monogeneans parasitic on amphibians,

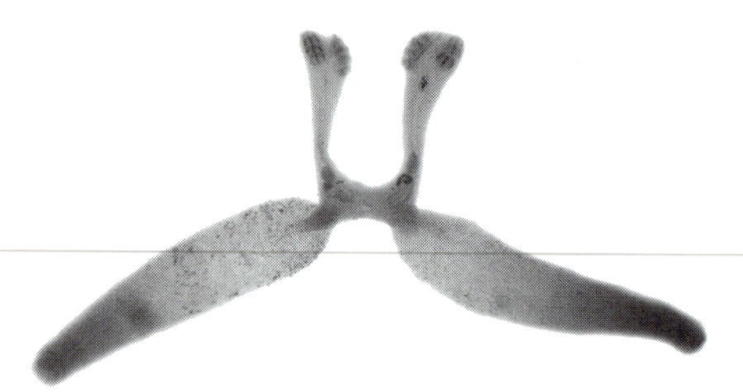

Fig. 4.37 The monogenean *Diplozoon algipteuris* from the gills of *Labeo forskalii*. (Photograph courtesy of Walter Boeger, Universidade Federal do Parana.)

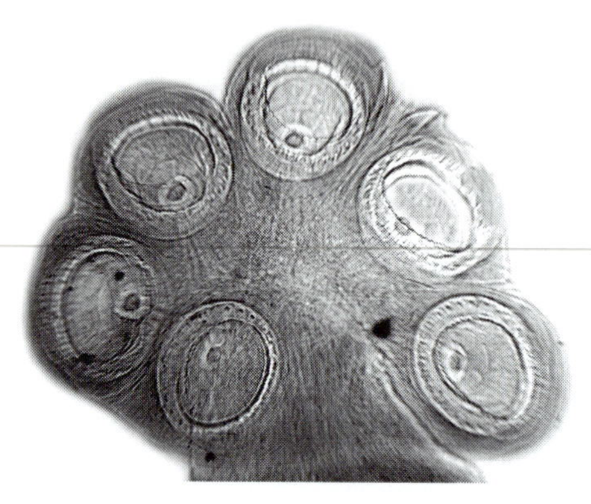

Fig. 4.38 Opisthaptor of *Polystoma* sp. from the urinary bladder of the mud turtle *Kinosternon subrubrum*. (Photograph courtesy of Walter Boeger, Universidade Federal do Parana.)

the amphibious nature of the hosts (in and out of the water) imposes some restrictions on the nature of the parasite's habitat, and the skin is not a desirable attachment site. The degree of terrestriality of the host also drives the reproductive strategy of the parasite, although they still rely on an aquatic oncomiracidium for dispersal. Because many amphibians only visit bodies of water during their breeding period, parasite reproduction is closely coordinated with that of the host, thus ensuring successful **colonization** of new hosts by the oncomiracidia. One of the most fascinating polystomatids in this regard is *Polystoma integerrimum*, a common species in the urinary bladders of frogs and toads in Europe (Fig. 4.38). This parasite reproduces in synchrony with its host and is apparently stimulated to do so by gonadotrophins produced by the host during its reproductive cycle. Other polystomatids of interest include *Pseudodiplorchis americanus*, a parasite of the spadefoot toad (*Scaphiopus couchii)* and *Neodiplorchis scaphiopodis*, a parasite of two other species of spadefoot toads (*S. mutiplicatus* and *S. bombifrons)*, all of which are sympatric in the driest parts of southwestern North America (for an elegant review, see Tinsley, 1995). All of these species lead very precarious, but highly successful existences. The toads spend most of the year underground, in a state of torpor. They emerge and breed only during a few nights in ephemeral pools created by brief desert rainstorms. While in torpor, oncomiracidia develop completely inside eggs within the uterus of the adult parasite. When the toad emerges from torpor and enters the ephemeral pools of water, the monogenean releases its eggs, they pass in the toad's urine, and hatch in the water in a matter of seconds. When the free-swimming oncomiracidia contact a toad, they attach to the skin, crawl towards the snout, and make their way into the nares, then into the lung. After a brief stay, most migrate back up the trachea and into the mouth. They then pass through the intestinal tract and finally make their way into the urinary bladder. The pattern of reproduction of the parasites and the toads is thus highly synchronized. Evidence suggests that both reproductive cycles are directly influenced by the toad's gonadotrophic hormones, a truly elegant example of the highly integrated physiological relationship between a host and its parasites (Tinsley & Jackson, 1988)!

4.6 Gyrocotylidea

The gyrocotyleans are oblong, hermaphroditic platyhelminths, without an intestine. The anterior end has a small, cup-shaped holdfast organ; the posterior end bears the attachment organ, which may be in the form of a frilled, rosette-like structure (*Gyrocotyle;* Fig. 4.39), or a simple, funnel-like structure without frills (*Gyrocotyloides*). Nutrients are absorbed through the tegument and the protonephridium is the functional unit of the excretory system. The male reproductive system is relatively simple, with testes and a vas deferens. The female reproductive system includes an ovary located in the posterior end of the body, with an oviduct that becomes the ootype before enlarging to form the uterus. Vitelline glands, Mehlis' gland, and seminal receptacle are also present. The genital pores are near

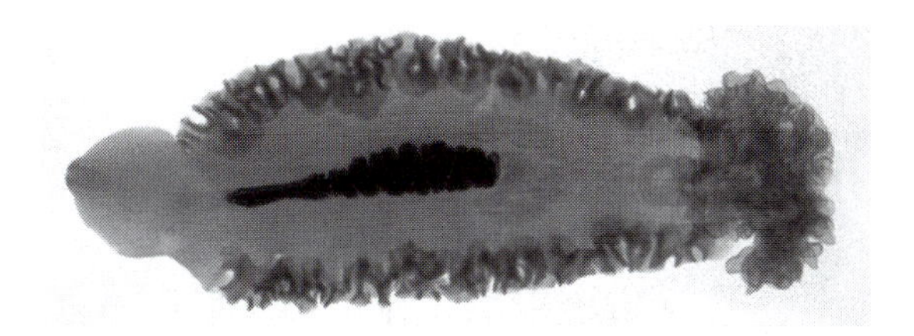

Fig. 4.39 The cestodarian *Gyrocotyle* sp. from a holocephalan fish. (Photograph courtesy of Todd Huspeni, University of California, Santa Barbara.)

the anterior end of the body. Gyrocotyleans have a free-swimming decacanth larva, the **lycophore**, with 10 hooks in the posterior end. The life cycle is probably direct, but has not been completed experimentally. Adult gyrocotylids have at least 10 different sensory receptors and their larvae have eight. Interestingly, the receptors in the adult and the larva are different, suggesting extreme specialization for their respective habitats.

Gyrocotylids are parasites in the spiral intestine of primitive cartilaginous fishes, the Holocephali, with strict specificity for their hosts. The population structure of gyrocotylids in the spiral intestine of the host seems to be controlled by density-dependent factors, with most of the infected hosts having only two worms, a few with one, and some others with three. It is unusual to find more than three individuals in a host and, when this occurs, the parasites are smaller than usual (Fernández *et al.*, 1987; Williams *et al.*, 1987).

4.7 Amphilinidea

Amphilinids are flattened and unsegmented platyhelminths, with a proboscis-like holdfast at the anterior end (Fig. 4.40). They live in the coelomic cavity of their hosts and some can become relatively large, measuring up to 380 mm in length. They lack a digestive system. The male reproductive system is simple, consisting of paired testes and a vas deferens. The female reproductive system consists of an ovary, oviduct, ootype, Mehlis' gland, and uterus. A vagina is present and the vitelline follicles are poorly developed. The life cycle of amphilinids is complex, with ciliated larvae possessing 10 hooks, and a crustacean intermediate host.

Amphilinideans are parasites of the coelomic

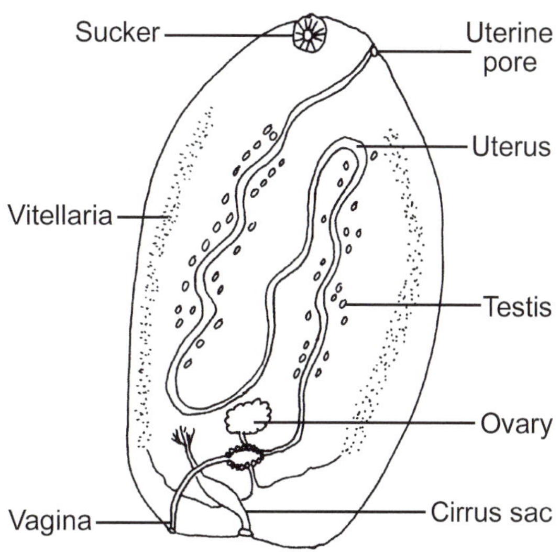

Fig. 4.40 Diagram of an amphilinidean. (Drawing courtesy of Lisa Esch-McCall.)

cavity of primitive fishes and turtles on all continents, excluding Antarctica. Species of *Amphilina* parasitize the body cavity of sturgeons (*Acipenser* spp.) and use amphipods as intermediate hosts. The parasite seems to have adverse and debilitating effects on the host, affecting the levels of hemoglobin, minerals such as zinc, and body fat. *Austramphilina elongata* parasitizes the body cavity of Australian freshwater turtles, and crayfish and freshwater shrimp are the intermediate hosts. The amphilideans and gyrocotylideans represent an interesting evolutionary link between the monogeneans and eucestodes.

4.8 Eucestoda

4.8.1 General considerations

All adults in the Eucestoda, except one species (a pseudophyllidean in tubificid annelids), are parasitic in the intestine (or adjacent structures, e.g., under the koilin lining or in the cloaca) or its accessory ducts of vertebrate animals. Except for a couple of species, all tapeworms are monoecious. All species are without an intestine and nutrients are absorbed through a syncytial, **microtrich**-covered, tegumental surface. The unit of structure and function in the excretory (osmoregulatory) system is the protonephridium. Their life cycles are complicated, involving two, or sometimes three, hosts.

4.8.2 The outside

The body plan of eucestodes (usually referred to simply as cestodes, a practice that will be followed here) includes a **scolex**, neck, and **strobila** (Fig. 4.41). The scolex is a specialized attachment device located at the anterior end. The morphology is highly variable and may have suction organs, hooks, spines, tentacles, glands, or any combination of these structures. There are three main types of suction devices, i.e., **bothria**, **phyllidea** (or bothridia), and **acetabula**. Bothria are slit-like grooves with weak attachment power. The common number is two per scolex, as in the Diphyllobothriidae. Phyllidea are highly muscular, may assume several shapes (although ear-like is common), and their margins are thin and flexible, usually matching the structure of their attachment site. The scolices of Tetraphyllidea, for example, have four phyllidea. Acetabula are basically suckers or suction cups distributed around the scolex; there are normally four and they are characteristic of the Cyclophyllidea. Some of the Cyclophyllidea may also have a dome-shaped structure, the **rostellum**, at the tip of the scolex, which may or may not be armed with hooks. If present, hooks are arranged in one or more circles around the rostellum.

Immediately posterior to the scolex is an unsegmented neck region that internally contains a large population of undifferentiated stem cells that give rise to the proglottids. The entire worm behind the scolex is collectively called the strobila. Sometimes the proglottids also are referred to as segments. Although tapeworms with differentiated proglottids appear to be segmented, tissues such as the tegument and muscles are continuous between the proglottids, without division and, therefore, are not true segments. Immediately behind the neck are the youngest proglottids, followed by mature and then gravid proglottids. As new proglottids are formed in the neck, the previous ones move posteriorly in a continuous process. In some species, eggs are shed from the strobila while the proglottid remains attached (**anapolysis**). Then, when these gravid proglottids are spent, they detach and rapidly disintegrate. In other species, the proglottid may detach before releasing the eggs and leave the host intact with the feces (**apolysis**). Most cestodes have many ploglottids and these

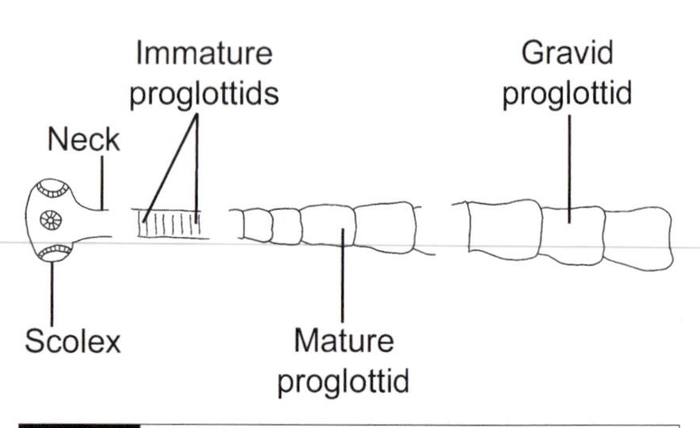

Fig. 4.41 Adult cestode showing the typical body plan with scolex and strobila of a polyzoic tapeworm. (Drawing courtesy of Lisa Esch-McCall.)

species are sometimes called **polyzoic** cestodes. In a few tapeworms, however, all of the reproductive organs and the scolex are within a single body, without a strobila, lacking all signs of segmentation; these tapeworms are called **monozoic** cestodes and most of them belong to the Caryophyllidea.

The tegument of cestodes is responsible for the absorption of all nutrients. In many respects, the tapeworm's surface resembles that of a digenean, except that the cestode tegument is completely covered by minute microvilli called microtrichs (Fig. 4.42); scatttered within the microtrich border are ciliated sensory structures which function as tactile receptors, chemoreceptors, etc. The microtrichs are quite similar to the microvilli found in the intestinal mucosa of animals and, similarly, function to increase the absorptive area of the tapeworm. The tegument is covered by a glycocalyx which is a carbohydrate-rich surface that seems to be involved with a number of biochemical functions, e.g., inhibition of some of the host's digestive enzymes and absorption of cations and bile salts. The tegument contains numerous vesicles whose origin has long been debated. Some suggest that they are pinocytotic, whereas others argue that they are produced within the subtegumental cytons whose cytoplasm is continuous with that of the tegument. The tegument also contains nuclei and other typical cell organelles, e.g., ribosomes, rough endoplasmic reticulum, Golgi apparatus, etc. Mitochondria are also abundant within the cytoplasm of the distal tegument, but these typically possess few cristae, suggestive of the largely

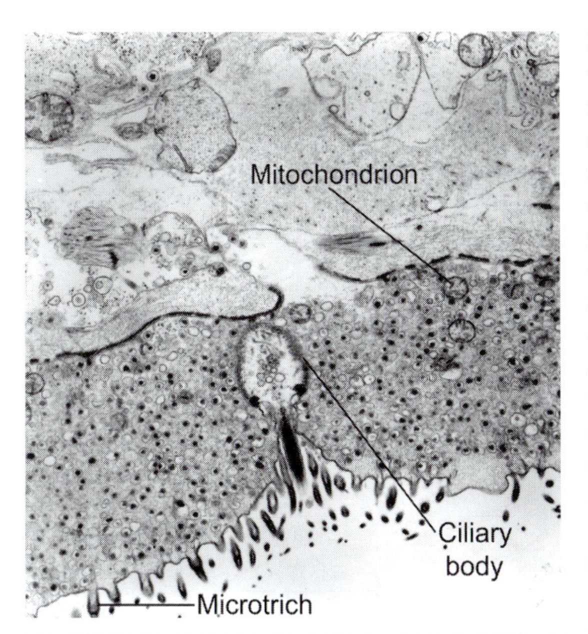

Fig. 4.42 Transmission electron micrograph of the surface of the pseudophyllidean cestode *Bothriocephalus acheilognathi*, the so-called Asian tapeworm. Externally on the tegument are microtrichs and a sensory structure showing the terminal cilium; as can be seen, the tegument is packed with various vesicles and mitochondria. (Modified from Granath *et al.*, 1983, with permission, *Transactions of the American Microscopical Society*, **102**, 240–250.)

anaerobic metabolism that is characteristic of cestodes.

4.8.3 The inside

Immediately below the syncytial tegument there is a layer of circular and one or more longitudinal muscle layers. A number of transverse, longitudinal, dorsoventral, and sometimes radial, muscle fibers may be present; their pattern and distribution is constant within a species, but highly variable among different groups. The scolex possesses a well-developed muscular system in association with its attachment structures.

Within the parenchyma of cestodes are calcareous corpuscles. These are concentrations of various mineral salts within an organic matrix. Their function is not known, although some propose that their contents may act to buffer metabolic acids produced during metabolism of lipids and carbohydrates. Others speculate they serve as reserves for phosphates used in phosphorylation or for CO_2 in carbon dioxide fixation.

Excretory/osmoregulatory system. The excretory/osmoregulatory system consists of protonephridia that connect with longitudinal collecting canals, one pair dorsal and one pair ventral, on each side of the strobila. Normally, a transverse duct at the posterior end of each proglottid connects the ventral canals. The dorsal and ventral osmoregulatory canals merge in the scolex, which means that they are actually continuous. The excretory fluid present in the canals of *Hymenolepis diminuta* contains glucose, lactic acid, soluble proteins, urea, and ammonia (Webster & Wilson, 1970). It appears that short-chain organic acids, the main end products of metabolism, may be excreted through the tegument.

Reproductive system. Most cestodes are monoecious, with each proglottid having one set of male organs and one set of female organs. In some species, e.g., *Moniezia*, however, there is a duplicate set of male and female organs in each proglottid and, in a few, there are two sets of male and one set of female organs, e.g., *Diplophallus*. Species of *Shipleya* and *Dioecocestus* are dioecious and the mechanism of sex determination in these species may depend on the interaction between two or more strobilas. For example, if only one strobila is present in a host, it is typically female; but if two strobilas are present, one is female and the other is male.

Both self- and cross-fertilization occur. However, self-fertilization is normally avoided because either the male reproductive system matures before the female counterpart (**protandry**), or because the female reproductive organs mature before the male organs (**protogyny**). Transfer of sperm normally occurs between different proglottids of the same strobila, or between proglottids in different strobilas.

The reproductive organs of cestodes are highly variable in their structure, arrangement, and distribution among the different taxonomic groups. Figure 4.43 depicts the mature proglottid morphology of four species of cestodes, a proteocephalan, a pseudophyllidean, and two cyclophyllideans. There are some basic differences in the plan as to ventral versus lateral openings of the reproductive systems, lateral versus posterior vitellaria, compact versus scattered vitellaria, numbers of testes, and several other more specific

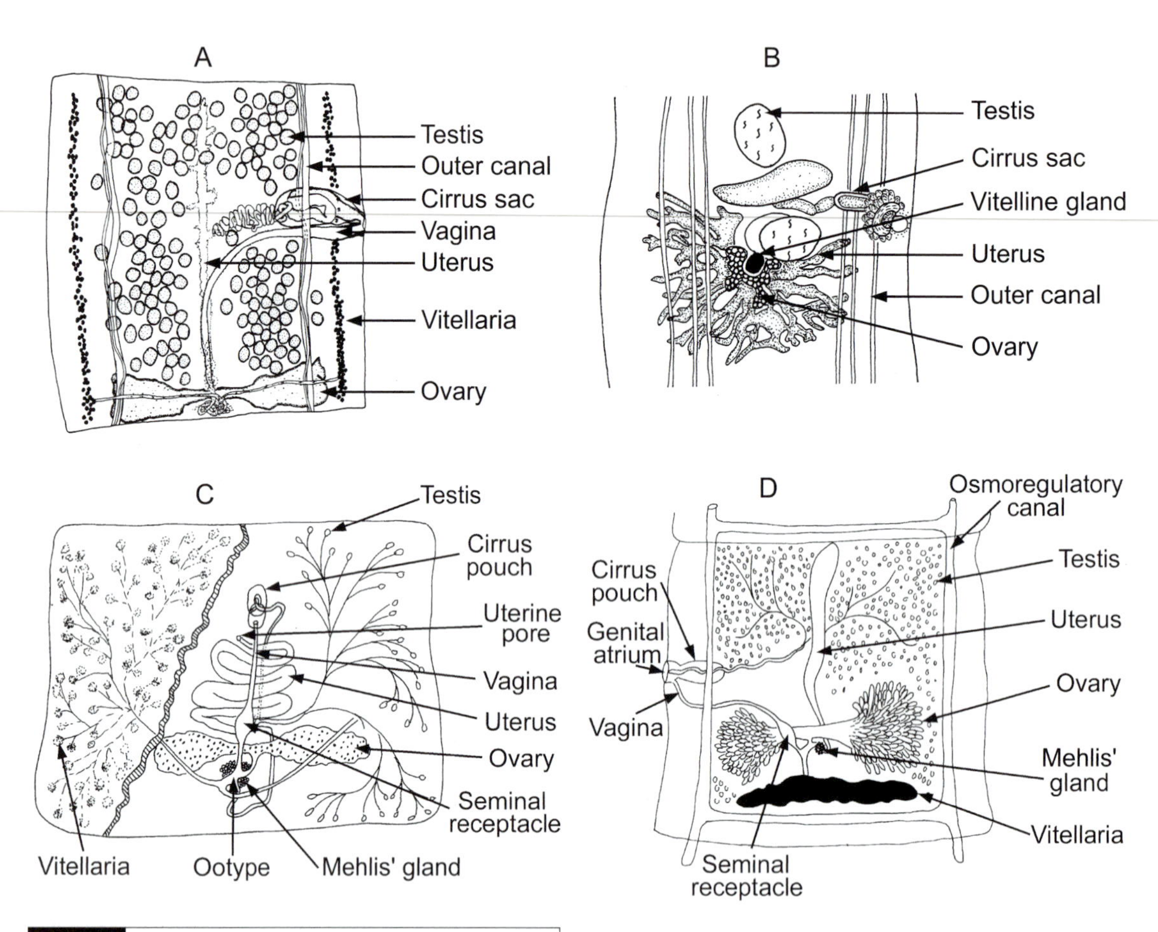

Fig. 4.43 Proglottid morphology of different cestodes. (A) The proteocephalan *Proteocephalus brooksi*; (B) the cyclophyllidean *Fimbriasacculus africanensis*; (C) generalized pseudophyllidean proglottid; (D) generalized taeniid proglottid. ((A) Modified from García-Prieto *et al.*, 1996, with permission, *Journal of Parasitology*, **82**, 992–997; (B) modified from Alexander & McLaughlin, 1996, with permission, *Journal of Parasitology*, **82**, 907–909; (C, D) drawings courtesy of Lisa Esch-McCall.)

characters. Despite this variability, some basic structures remain in common and provide a basis for which the variations can be considered herein. The male reproductive system includes one or more testes, each with a vas efferens which unite to form a common vas deferens that directs the sperm towards a cirrus located in the cirrus sac. The cirrus is a protrusible copulatory organ that evaginates through the genital atrium. The vas deferens may be a simple tube or may be modified in several ways to store sperm.

The female reproductive system is more complex than that of the male. A single ovary is continuous with an oviduct that is, in turn, joined by the vagina carrying implanted sperm, and by the vitelline duct from the vitelline glands. The oviduct then merges into an ootype surrounded by the Mehlis' gland. The ootype is connected to the uterus, which may be a single sac, or a simple or a convoluted tube. In some groups, it may be replaced by other structures such as the paruterine organ of the Mesocestoididae. The vagina may be simple or may include a seminal receptacle. The vitelline glands, which contribute yolk and shell material to the embryo, may be grouped into a single mass, or may be dispersed, forming a variety of structural patterns within the proglottid.

Sperm transfer normally occurs from the cirrus into the vagina. Some species, however, lack a vagina. In these cases, the cirrus is forced through the body wall and sperm are deposited into the parenchyma. It is unclear how the sperm then find their way into the female's reproductive system.

One important difference among the various groups is the structure of the egg and the process of shell formation. Both the vitelline follicles and the Mehlis' gland contribute to shell formation. In groups like the Pseudophyllidea, a thick capsule similar to the eggs of digeneans covers the egg. In others, the egg capsule may be very thin, or even absent.

Nervous system and sensory input. The nervous system in all cestodes except *Diphyllobothrium* consists of a pair of cerebral ganglia located in the scolex and two main nerve cords that extend the length of the strobila (Halton & Gustafsson, 1996). Species of *Diphyllobothrium* possess a single ganglion and two nerve cords. The longitudinal nerves are connected by transversal comissures present in each proglottid. A number of smaller nerves emanating from this ladder-like system innervate the muscles, sensory structures, and various reproductive structures such as the cirrus and vagina. Cestodes are without a gut and, probably as a consequence, the tegument is well innervated. The nervous system also is highly developed in the scolex, with nerves associated with the various structures used for attachment, e.g., acetabula, bothria, bothridia, and rostellum. A number of sensory receptors for both chemical and physical stimuli are present in cestodes, and probably function in the same manner as those in digeneans, e.g., mechanoreception, chemoreception, and osmoreception. Eyespots or other photoreceptors apparently are lacking in cestodes.

4.8.4 Nutrient uptake

Because cestodes do not have an intestine, all nutrients must be taken in through the surface. The uptake process is facilitated by the presence of microtrichs on the outside which increase the absorptive area. Nutrient uptake is accomplished by active transport, mediated diffusion, and simple diffusion (Pappas & Read, 1975). Most cestodes do not produce digestive enzymes and rely on the host to provide all their nutrient needs. Glucose, galactose, and amino acids are actively transported across the tegument. Purines and pyrimidines, necessary for the synthesis of DNA and RNA, are absorbed by facilitated diffusion. Although it is not clear how lipids are absorbed, a form of diffusion may be involved.

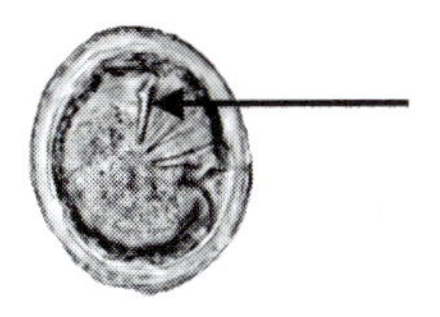

Fig. 4.44 An oncosphere or hexacanth embryo from the cestode *Choanotaenia atopa*. The arrow points to one of the six embryonic hooks. (From Rausch & McKown, 1994, with permission, *Journal of Parasitology*, **80**, 317–320.)

4.8.5 Metabolism

Respiration is primarily anaerobic. In general, intermediary carbohydrate metabolism occurs in two ways among the cestodes, making them similar to digeneans. One is of the homolactate type, wherein glucose is converted to lactate via the glycolytic pathway. The second is via malate dismutation, whereby glucose is converted to phosphoenolpyruvate (PEP), then CO_2 is fixed to PEP and malate is formed. Malate is then metabolized to either proprionate or acetate with the generation of additional ATP in the process. Some species excrete succinate. Whereas oxygen uptake can be demonstrated in cestodes, there is no evidence to indicate that oxygen is an electron acceptor as in the classical electron transfer system that operates in the mitochondria of most organisms. Since the major energy resource for tapeworms is glucose, many species store considerable quantities of glycogen, up to 50% of their dry weight. Protein and lipid metabolism are not important energy sources for tapeworms.

4.8.6 Development and life cycles

Most cestodes have life cycles that follow one of two major patterns, but with many variations in each. These variations will be considered in connection with specific species and will be detailed subsequently.

In the first group are included the Pseudophyllidea and Proteocephalidea. Life cycles of species in these two orders are inextricably connected with first intermediate hosts that are totally aquatic, namely zooplanktonic microcrustaceans, mostly copepods. Within the eggs are **oncospheres**, or **hexacanth** (six-hooked) embryos (Fig. 4.44). In some species (Fig. 4.45), free-floating eggs in the water column must be consumed by the first intermediate hosts. In other species, the

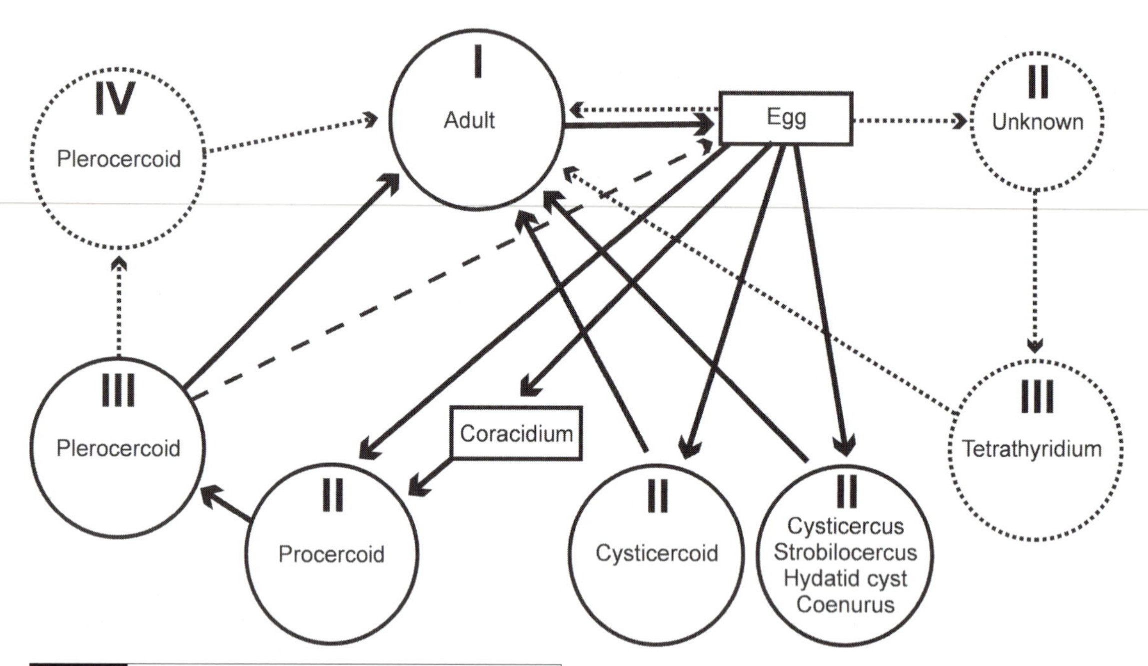

Fig. 4.45 Generalized life cycle of cestodes. Host types are represented by roman numerals (I = final/definitive host; II = first intermediate host; III = second intermediate host; IV = paratenic or third intermediate host). Solid lines represent a 'typical life cycle', whereas dotted lines represent less common variations. Dashed line represents a form of progenesis in which the plerocercoid is thought to be neotenic, i.e., *Glaridacris* sp. Circles indicate stages found within hosts and rectangles are free-living stages.

eggshell has an operculum and a larval stage hatches. This larva is called a **coracidium** and is covered by cilia that enable it to actively swim in the water column. As for miracidia and oncomiracidia, it is a short-lived stage; it must be eaten by a microcrustacean within 24 to 36 hours, or it will die. When the first intermediate host eats either an egg or a coracidium, the parasite penetrates the gut wall and enters the hemocoel. There, it transforms into a **procercoid**, a nondescript larval stage that resembles a cigar, possessing a knob-like structure called a **cercomer** on one end. Retained on the cercomer are the six hooks of the hexacanth embryo; this is similar to monogenean development where the juvenile adult retains embryonic hooks of the oncomiracidium. Next, the infected microcrustacean must be eaten by an appropriate planktivorous host, usually, but not always, a fish. The procercoid penetrates the gut wall of the fish and then migrates to species-specific locations within the new host. Once the final site of infection is reached, the procercoid grows rapidly and transforms into the next stage, known as a **plerocercoid**. In general, the scolex that develops on the plerocercoid closely resembles that of the adult. The overall size to which a plerocercoid will grow is also species-specific. In some, they become quite large relative to the size of the host and in others they remain relatively small. The biomass of *Schistocephalus solidus* plerocercoids, for example, may equal that of the entire intermediate host (Fig. 4.46). The life cycle is completed when this host is eaten by a final host.

The second life cycle pattern (see Fig. 4.45) is generally associated with a first intermediate host which is mostly terrestrial in character, though certainly not always. Among the cyclophyllideans, there are two basic body plans associated with the larval stage in the intermediate host, but with many variants on the two as well. In some species, eggs are shed via apolytic proglottids. Within the eggs are the hexacanth embryos, or oncospheres. When these eggs are ingested, always by an invertebrate and many times an insect, the oncosphere emerges and penetrates the gut wall. On entering the hemocoel, the parasite transforms into a **cysticercoid** which has a fully withdrawn scolex

Fig. 4.46 Plerocercoids of the cestode *Schistocephalus solidus* from the body cavity of a fish. (Photograph courtesy of Tim Goater, Malaspina College.)

resembling that of the adult into which the parasite will eventually develop. Membranes that were present when the oncosphere was still inside the eggshell surround the scolex. To complete the life cycle, the infected host must be eaten by the definitive host. The second type of larval body plan among cyclophyllideans is always associated with a mammalian second intermediate host (see Box 4.2). In these species, the eggshell is characteristically striated. In reality, the shell consists of hundreds of tiny blocks held together with cement that will be dissolved through action of the intermediate host's digestive enzymes in the small intestine. Once freed, the oncosphere then uses its six embryonic hooks to assist in penetrating the gut wall. There, the larva is picked up in the circulatory system and transported to the appropriate site of infection; although some assert that the embryo is transported first in the lymphatic system and then in the circulatory system. The larval stage that develops is one of several different types (see Box 4.2), but all were probably derived evolutionarily from one, the **cysticercus**, or bladder worm. The others are the **strobilocercus**, **coenurus**, **unilocular hydatid cyst**, and **multilocular hydatid cyst**. Cysticerci can be found in a number of organs and tissues in mammals, including the heart, skeletal muscles, mesenteries, and brain, depending on the species of parasite. The cysticercus may or may not be encysted by a hyaline capsule of host origin, but this is also related to the species of cestode as well as the site of infection. The cysticercus has an invaginated scolex, complete with four suckers. There is always a rostellum, generally with a crown of hooks that are in the shape of rose-thorns. Large and small hooks usually alternate in two rows. Associated with the cysticercus, there is always a bladder filled with a transudate of host origin. The life cycle of the parasite is completed when flesh containing the cysticercus is consumed by a definitive host. Once freed from intermediate host tissue by digestion in the small intestine of the definitive host, the scolex of the cysticercus evaginates, attaches to the gut wall using the suckers and rostellar hooks, and undergoes **strobilization**.

Box 4.2 When humans get in the way

The larval stages of at least four species of taeniid cestodes can wreak havoc in humans if eggs of these tapeworms are ingested accidentally. The larval stage of a fifth species, the pseudophyllidean *Diphyllobothrium mansonoides*, can also infect humans, but the manner of infection may be different. In any event, if humans get in the way, the outcome can range from simply very disconcerting, to disastrous.

Taenia solium is one of the rare species of parasites in which a single human may serve as both the definitive and intermediate host. The adults of *T. solium* are restricted to the small intestine of humans where they can grow to a length of 4–6 m. Eggs released from gravid proglottids are usually ingested by swine where cysticerci develop in the skeletal and cardiac muscles. Humans will then become infected with an adult tapeworm when poorly cooked pork is consumed. However, in many places in the world where hogs are raised for human

Fig. 1 Pigs wandering and feeding in the yard of a home near Huatlatlauca, Mexico. (Photograph courtesy of Mirna Huerta.)

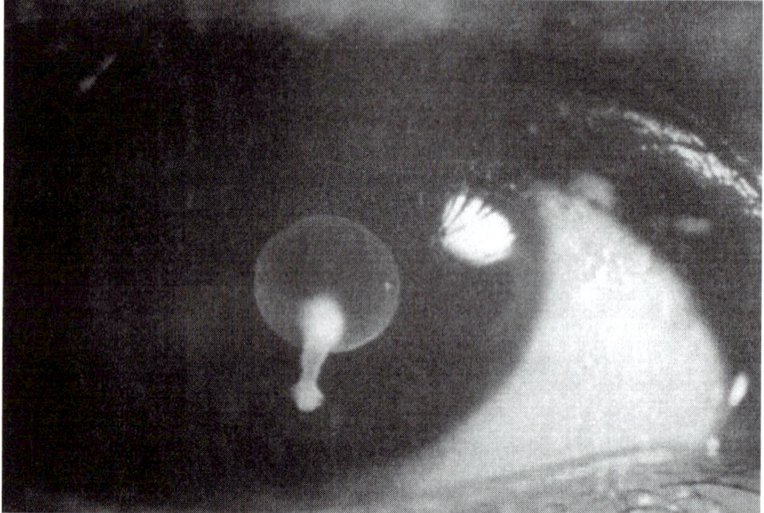

Fig. 2 A cysticercus of *Taenia solium* sequestered in the eye of a human. (Photograph courtesy of Mirna Huerta.)

consumption, sanitary conditions are primitive. Food, water, or soil, may become contaminated with eggs from human feces and humans might ingest these accidentally. In many of these infection foci, pigs and humans live, literally, side by side (Fig. 1). Indoor plumbing facilities in these rural communities are usually absent, and latrines, even when available, may not be used on a regular basis. Human feces containing eggs of *T. solium* are, therefore, readily accessible to pigs. Indeed, human fecal material may become a prime food source for swine living in such primitive conditions. In this manner, *T. solium* eggs are readily accessible to humans in these typically rural settings. If ingested by humans (or pigs), the eggs will hatch and oncospheres will penetrate the gut wall where they then migrate to a number of sites, including subcutaneous tissues, heart muscle, and the brain. Cysticerci are capable of developing in any of these sites in humans, whereas in the natural intermediate host (pigs), intramuscular sites are the most common. Figure 2 in this Box, for example, shows a fully developed cysticercus in the eye of a human. However, if the brain is involved (Fig.

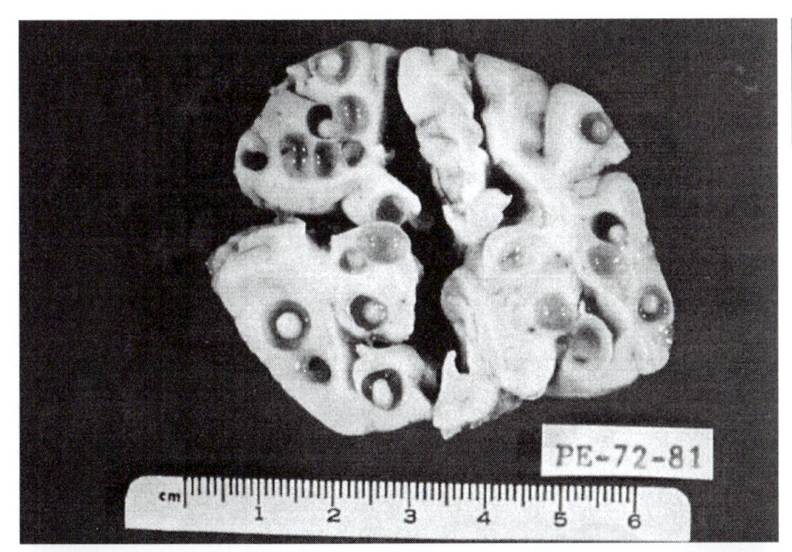

Fig. 3 Cysticerci of *Taenia solium* in the brain of a human at necropsy. (Photograph courtesy of Mirna Huerta.)

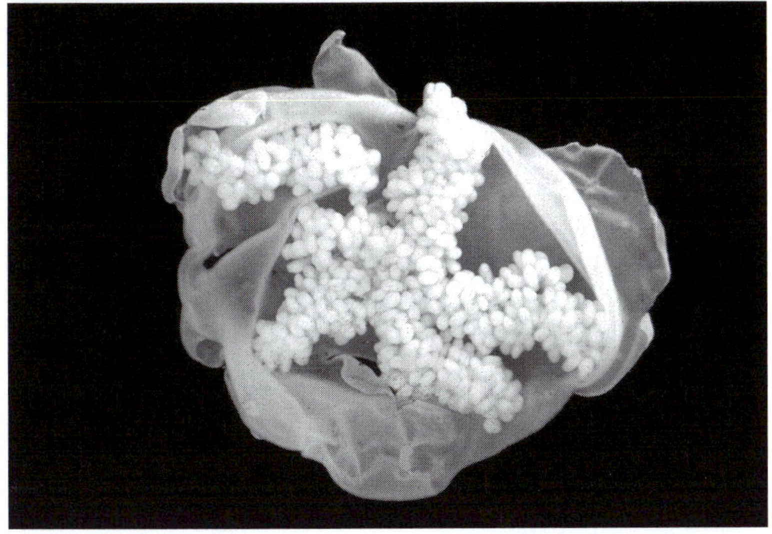

Fig. 4 The ruptured coenurus of *Taenia multiceps* removed from an experimentally infected laboratory mouse.

3), then a condition known as neurocysticercosis will develop. This disease is the leading cause of epilepsy in many developing countries, certainly those in the Western Hemisphere. Most troublesome, as a prominent Latin American parasitologist, Dr. Carlos Larralde, reminded one of us, is that these epileptic seizures can be induced by a single cysticercus in the brain! Dr. Larralde was asked about egg dispersal while we were traveling through a local barrio in rural Mexico. He glanced outside our air-conditioned van as a gust of wind kicked up a cloud of dust and rhetorically remarked, 'Do you think *T. solium* eggs might be carried with such a cloud of dust?' The answer was obvious!

Taenia multiceps occurs in canines as an adult and employs jackrabbits *Lepus californicus* as the intermediate host. The larval stage is a coenurus that occurs subcutaneously and intramuscularly in the rabbit. This larva, which may grow to the size of a grapefruit or a cantaloupe, possesses many scolices (Fig. 4) that remain attached to the internal surface of the bladder (multiceps: *multi* = many, *ceps* = head, referring to the scolices). The bladder is filled with a fluid transudate originating in the tissues of the host. In tissues other than the brain, this bladder is surrounded by a hyaline, adventitious capsule of host origin. The bladder appears to cause little harm to the jackrabbit, but a strain of *T. multiceps* in sheep will produce gid, or staggers, during which the animal walks with an unsteady gait or turns helplessly in circles because of coenuri that become localized in the brain. *Taenia multiceps* also has been recorded in the brains of humans, mostly children; fortunately, it is rare.

Three other taeniid cestodes that cause considerable harm to their intermediate hosts, including man, are *Echinococcus granulosus*, *E. mutilocularis*, and *E. vogeli*. Adults of all three species are found in canines, with *E. granulosus* primarily in wolves, *E. multilocularis* in foxes, and *E. vogeli* in bush dogs. It appears that most canine species, including the domesticated dog, can serve as the definitive host for *E. granulosus* and *E. multilocularis*. Adults of all *Echinococcus* species are tiny, averaging 2–3 mm in total length; the entire worm consists of a scolex, and three proglottids, an immature, a mature, and a gravid one.

Adults of *E. granulosus* pass gravid proglottids in the canine host feces and, in the sylvatic, or natural, cycle, the eggs are accidentally consumed by deer, moose, and caribou. The oncospheres emerge from the eggs, penetrate the gut wall, and are picked up in the circulation. They exit in the capillary beds of various organs of the body, primarily the liver and lung, where they develop into hydatid cysts. These unilocular larvae can reach enormous size and contain extraordinary quantities of fluid. A single hydatid cyst may reach 50 cm in diameter, and contain several liters of fluid, along with up to a million protoscolices! If the organization of the hydatid cyst wall is examined microscopically, an adventitious capsule of host origin will be seen on the outside. Immediately inside is a so-called laminated layer produced by the parasite and inside that is a very thin, syncytial germinal membrane. From the germinal membrane, the parasite produces buds that will drop off as brood capsules and float free as 'hydatid sand' in the hydatid fluid. Within the brood capsules will develop the infective protoscolices, which are little more than invaginated scolices. Note that in this species, budding can occur both exogenously and endogenously. The life cycle is completed when the larvae are consumed by the canine definitive host.

The parasite becomes dangerous for humans when the normal sylvatic cycle is interrupted by the domesticated dog in which the adult parasite can develop, or by sheep, cattle, hogs, or horses, where the hydatid cyst can develop, usually in the liver or lungs. If eggs of the parasite are accidentally ingested by humans, then hydatid cysts will develop in the liver, lungs, or even the brain, reaching the same size as those in any natural intermediate host. In addition, the parasites are quite long-lived in their intermediate hosts. The long life of the parasite means that the host is exposed for extended periods to small, but persistent, quantities of parasite antigens which begin to accumulate in the hydatid fluid. If the hydatid cyst ruptures, the human host is suddenly exposed to large quantities of antigen to which it has developed a strong hypersensitivity over a long period of time. The result can be classic anaphylactic shock and rapid death.

Also dangerous is *E. multilocularis*. The sylvatic cycle of this parasite typically involves species of rodents in the genus *Microtus* as intermediate hosts and foxes as the final hosts. The sylvatic cycle is broken when humans accidentally

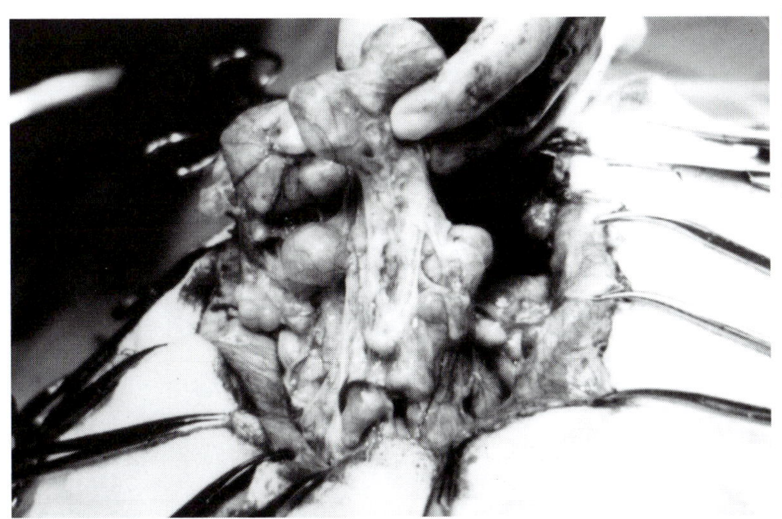

Fig. 5 A polycystic hydatid infection from the abdominal cavity of a patient in Colombia. (From Rausch & D'Alessandro, 1999, with permission, *Journal of Parasitology*, **85**, 410–418.)

consume eggs passed by their domesticated dogs that have become infected by eating parasitized rodents. The hydatid cysts typically develop in the liver, lungs, and abdominal cavity. These hydatid cysts lack a laminated layer. Apparently for this reason, the cysts are able to bud exogenously, or 'metastasize', and are referred to as multilocular or alveolar cysts. Via the exogenous budding, they spread from the initial site of infection into other locations. They may, for example, occupy the liver as a primary site of infection, but metastasize into the lungs. Figure 5 in this box shows a polycystic hydatid infection in the abdominal cavity of a patient in Colombia.

Rausch & D'Alessandro (1999) have elegantly described the histogenesis associated with the hydatid cysts of *E. vogeli*. These authors have modified diagrams of Blanchard (1889) portending to illustrate the exogenous budding of *Echinococcus granulosus* (Fig. 6), a process which, if it occurs, is limited to inappropriate primary hosts.

The control of echinococcosis under normal conditions should not be difficult since praziquantel is a most efficacious and easily administered drug. There are two problems that make control perplexing, however. One is that in many rural areas, diagnoses are not routinely attempted and the drug may not be readily available due to depressed socioeconomic conditions. Another problem is with the sylvatic character of the parasite. It is the normal pathway in nature, but the parasite also is easily accessible to the domestic canines commonly kept in rural parts of the world where intermediate hosts, such as reindeer, are butchered. As for *E. granulosus*, praziquantel is effective against *E. multilocularis*, but diagnosis, drug delivery, and cost in rural areas, are frequent, and significant, obstacles to successful treatment.

A pseudophyllidean cestode of concern to humans in certain areas of the world is *Diphyllobothrium mansonoides*. This parasite is normally found in felines, but may occur in humans where it produces a condition known as sparganosis. The plerocercoid is also known as a sparganum and acts as a form of larval migrans as it moves under the skin of its second intermediate host. Man may acquire the parasite by ingesting poorly cooked frog or snake meat containing the plerocercoid, by accidentally ingesting a copepod (the first intermediate

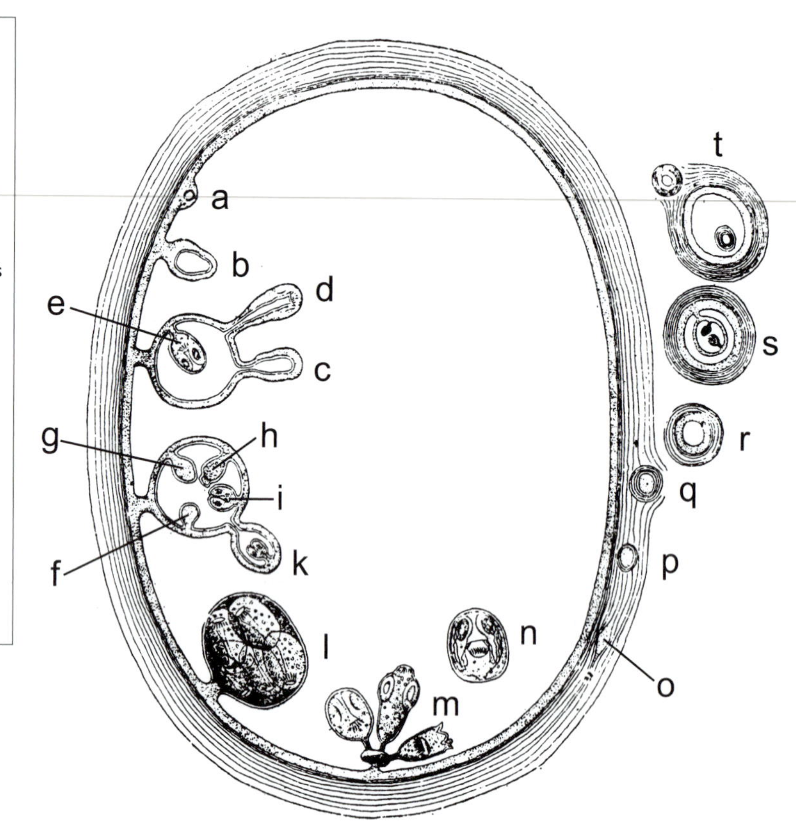

Fig. 6 Diagram illustrating the processes of endogenous and exogenous budding in hydatid cysts. The sequence begins with endogenous budding from the germinal lining at *(a)*. Brood capsule formation occurs *(e–l)*, but protoscolices can bud from the lining of the brood capsule as well as the germinal lining (not depicted). Exogenous budding *(o–t)* is shown from the laminated layer, but it is now known that this is an acellular layer outside the germinal lining from which exogenous budding takes place. (Diagram, originally published by Blanchard [1889]. Modified from Rausch & D'Alessandro [1999], with permission, *Journal of Parasitology*, **85**, 410–418.)

host) infected with a procercoid, or by having a poultice of frog flesh containing the plerocercoid applied to an open wound. Once the worm is subcutaneous, it has the capacity to migrate extensively and may grow up to 30 cm in length. When laboratory rats are experimentally infected with *D. mansonoides*, they grow to enormous sizes under the influence of a compound produced by the spargana. This molecule, produced by the parasite and known as **plerocercoid growth factor**, mimics the effects of human growth hormone (somatotrophic hormone). The plerocercoid growth factor works mainly as a cysteine proteinase, facilitating the tissue migration of the sparganum. Apparently, it also acts on the same cell-surface receptor sites affected by the human growth hormone, inducing growth of the host. There are suggestions that the plerocercoid growth factor may shorten the life span of the host and increase the likelihood of the parasite's transmission to the definitive host. So similar in action are human growth hormone and plerocercoid growth factor, that some have suggested the possibility of horizontal gene transfer from the human or vertebrate host to the parasite.

In all of these cestode species, humans are accidental hosts, but the parasites can nonetheless develop without apparent difficulty. In all of these cases, the adult parasite in the definitive host is harmless but, as is the case of many larval helminths in the tissues of their natural or accidental hosts, they can be quite harmful, and potentially lethal. With any of these parasites, when humans get in the way, they certainly will suffer the consequences.

4.8.7 Diversity

Cestodes are a morphologically diverse group, parasitic in the digestive system of vertebrates, although in one of the most primitive groups, the Caryophyllidea, adults are also found in aquatic annelids. Because of their diversity and numbers, only some representative families and orders will be discussed. (A systematic outline of the group is given in Box 4.3.)

CARYOPHYLLIDEA

Caryophyllideans are small cestodes, parasitic in the intestine of freshwater teleost fishes, mainly cyprinids, catfishes, and catostomids, although a few parasitize the coelom of freshwater oligochaetes. The scolex is very simple, with shallow depressions or bothria (Fig. 4.47). The body lacks proglottids or any sign of segmentation (monozoic); internally, these cestodes resemble pseudophyllideans. The life cycle involves two hosts, the fish definitive host and an oligochaete intermediate host infected with the procercoid stage. *Glaridacris catostomi* is a common parasite in the stomach and small intestine of suckers (*Catostomus* spp.) in North America. Species of *Caryophyllaeus*, also known as cloverworms because the scolex resembles a clover leaf, are common parasites of carp, bream, and other cultured cyprinids in Europe. Intermediate hosts are tubificids and other oligochaetes that are castrated when infected. Several species of *Archigetes*, however, mature sexually while in the oligochaete host. *Archigetes sieboldi* is a common species found in the coelom of aquatic oligochaetes in both Europe and North America. In order to release their eggs, gravid worms escape the oligochaete through the body wall and die, liberating the eggs as their body disintegrates on the substratum of a pond or lake. Species of *Archigetes*, then, appear to be neotenic procercoids.

SPATHEBOTHRIIDEA

Spathebothriideans are a small group of parasites of marine and freshwater fishes. Their scolex is extremely simple, without hooks of any kind. Their most distinctive trait, however, is the lack of apparent segmentation, even though a complete, linear series of reproductive structures is present.

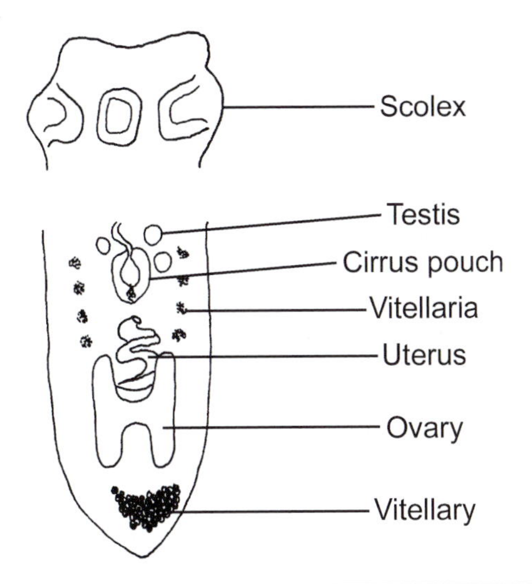

Fig. 4.47 An adult caryophyllidean cestode. (Drawing courtesy of Lisa Esch-McCall.)

PSEUDOPHYLLIDEA

Adult pseudophyllideans are found in all classes of vertebrates. Most are polyzoic, showing distinct segmentation. Their scolices are spatulate, possessing paired bothria. On occasion, there may be two sets of reproductive organs within a single proglottid. Their size is variable, ranging from a few mm to more than 30 m in length in the case of *Hexagonoporus*, a tapeworm of sperm whales.

Diphyllobothriids infect both birds and mammals, including humans. The broad fish tapeworm, *Diphyllobothrium latum*, is a prominent member of the family, which extends in distribution from the northern United States and Canada, to parts of Scandinavia, Siberia, south into the Balkans, Japan, China, and Korea, and even South America (Ravenga, 1993). Its host specificity is wide, variously infecting felids, canids, urcids, mustelids, pinnipeds, and humans. It is estimated that there are up to nine million humans infected worldwide. The parasite is quite common in a few localities, with 100% of the populations in some villages being infected. Humans acquire the parasite by eating raw or poorly-cooked fish, including salmon, northern pike, walleye, and other fishes, all infected with the plerocercoid larval stage. *Diphyllobothrium latum* has a typical pseudophyllidean cycle that includes a free-swimming

Fig. 4.48 Plerocercoid of the cestode *Ligula intestinalis* next to the fathead minnow *Pimephales promelas* from which it was removed. (Photograph courtesy of Michael Riggs, Wake Forest University.)

coracidium, procercoids in copepods, plerocercoids in fishes, and adults in mammals. *Diphyllobothrium latum* is a large tapeworm, reaching 8 to 10 m in length and possessing up to 3000 proglottids. In humans, diphyllobothriasis results in abdominal discomfort, diarrhea, and nausea. It may also cause pernicious anemia since the tapeworm is quite literally a sink for vitamin B12, which in humans is converted into a coenzyme necessary for nucleic acid synthesis and the maturation of red blood cells.

Ligula intestinalis is a fairly cosmopolitan parasite in fish-eating birds. Copepods are first intermediate hosts and harbor the procercoid stage, whereas fishes are second intermediate hosts carrying the plerocercoid stage in the coelom. These parasites produce a chemical agent which disrupts the gonadal/pituitary axis of the host, causing it to cease growing (Arme & Owen, 1967). However, the plerocercoids continue increasing in size and soon their biomass is sufficient to cause the fish to swim upside down (Fig. 4.48). In this position, the fishes lose any semblance of protective coloration and become vulnerable to predation by their fish-eating, avian definitive hosts.

All species of Bothriocephalidae are parasitic in fishes, amphibians, and reptiles. *Bothriocephalus acheilognathi*, the Asian tapeworm, was first described in Japan in the 1920s. Since that time, it has spread throughout the world, first into eastern Europe, then the United Kingdom and, by the mid-1970s, into North America, including Mexico. This inexorable migration was made possible by a number of factors, but in no small part because of parasite's lack of specificity at both the intermediate and definitive host levels (Marcogliese & Esch, 1989). As an adult, it has been reported from 40 different host species and, as a procercoid, in five different species of copepods. Another reason for its ability to colonize so successfully is the rapid increase in aquaculture throughout the world, coupled with the rather indiscriminant shipment of cultured fishes between countries, an objectionable practice which is finally receiving the appropriate attention of various political entities. An interesting characteristic of *B. acheilognathi* is that it exhibits indeterminate growth, i.e., the growth of the parasite is related to host size. For example, in the mosquito fish *Gambusia affinis*, the adult may reach 2 cm in total length, whereas in a large carp, they may reach 10 times the size they attain in the mosquito fish. (For some interesting aspects of the population biology of *B. acheilognathi*, see section 11.1.3.)

The plerocercoid larvae of several pseudophyllidean tapeworms are capable of infecting humans and other vertebrates that are not their normal intermediate hosts, causing a condition generally known as **sparganosis**, a form of **larval migrans** (see Box 4.2). Wild animals normally become infected by ingesting copepods infected with procercoids. In humans, sparganosis is normally caused by the ingestion of plerocercoids of several species of *Diphyllobothrium*. In eastern Asia, the most common species is probably *D. erinacei*, a parasite of carnivores, whereas in North America it is probably *D. mansonoides*, a parasite of felines. Most of the pathology associated with sparganosis derives from the secretions and wandering of the plerocercoids throughout the body.

HAPLOBOTHRIIDEA

The Haplobothriidea is a very small group with only one genus and species, but with some unique developmental characteristics. *Haplobothrium globuliformae* parasitizes the bowfin *Amia calva*, a primitive holostean fish in North America. The life cycle of *H. globuliformae* is similar in some respects to that of pseudophyllideans. It includes a free-swimming coracidium, a procercoid within a cyclopoid copepod, a plerocercoid within bullheads (*Ameiurus* spp.), and an adult in the bowfin. The plerocercoid within the liver of the bullhead is bulbose, or globular, and appears to have four suckers. These are not suckers, however, but are the surfaces of spinose, retractable/protractile tentacles. In many ways, the scolex resembles those present in trypanorhynchs, which are parasitic in the spiral valves of sharks. Once the plerocercoid reaches the bowfin intestine, it strobilates. However, no sexual development occurs within proglottids of what is termed the primary strobila. Instead, these proglottids drop off from the primary strobila, form a typical pseudophyllidean scolex at one end and begin to develop a secondary strobila. Within each new proglottid in the secondary strobila, a complete set of male and female reproductive organs then develops. This parasite thus possesses two scolex types, one with four tentacles resembling those of trypanorhynchs, and another that is spatulate, with paired bothria, resembling pseudophyllidean scolices. The reproductive system in proglottids of the secondary strobila morphologically resembles that of a pseudophyllidean cestode.

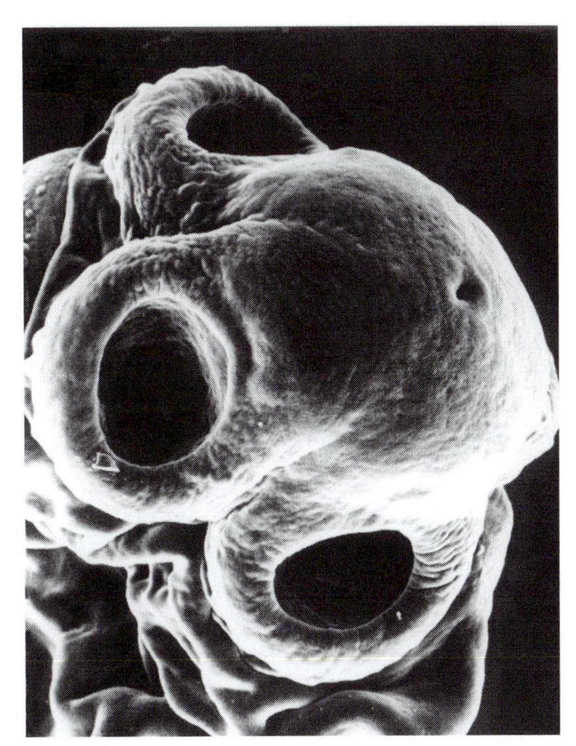

Fig. 4.49 Scolex of the cestode *Proteocephalus ambloplitis*. The pore on the tip is the opening of the apical end organ that secretes the tissue-dissolving enzyme(s) allowing the plerocercoid to migrate through parenteric host tissues and organs. (Photograph courtesy of James Coggins, University of Wisconsin–Milwaukee.)

PROTEOCEPHALIDEA

Proteocephalidean cestodes are commonly found in fishes, amphibians, and reptiles throughout the world. They resemble cyclophyllideans in having a bulbous scolex with four acetabula. Perhaps the best-studied member of the group in North America is *Proteocephalus ambloplitis*, the so-called bass tapeworm. Its life cycle will be detailed elsewhere (see section 10.3.2). Parenteric plerocercoids of *P. ambloplitis* cause host castration when they penetrate the host's gonads and wander extensively within other organs and tissues of the visceral mass, causing substantial damage. These plerocercoids possess a terminal pore on the tip of the scolex through which they secrete proteolytic enzymes that enable them to move through host tissues (Fig. 4.49). The bass tapeworm has apparently spread from North America, where it was originally endemic, to other parts of the world where the largemouth bass *Micropterus salmoides* has been introduced in recent years through ill-conceived stocking programs.

Another well-studied cestode in this order is *Proteocephalus parallacticus*, which is a cold-water stenotherm infecting lake trout in North America. Copepods are the intermediate hosts and, if consumed by a lake trout, procercoids develop into plerocercoids enterically. If the water temperature is $>12\,^{\circ}C$, development will continue to the adult stage. But, if water temperatures are $<12\,^{\circ}C$, the then the larva remains as an enteric plerocercoid until temperatures rise above the $12\,^{\circ}C$ threshold, then development to the adult stage proceeds. If a fish other than a lake trout eats the infected copepod, a plerocercoid will develop enterically. However, it will remain as such unless that fish is

Fig. 4.50 Scolex of the cestode *Otobothrium* sp., a trypanorhynch from the spiral intestine of sharks. (Photograph courtesy of Janine Caira, University of Connecticut.)

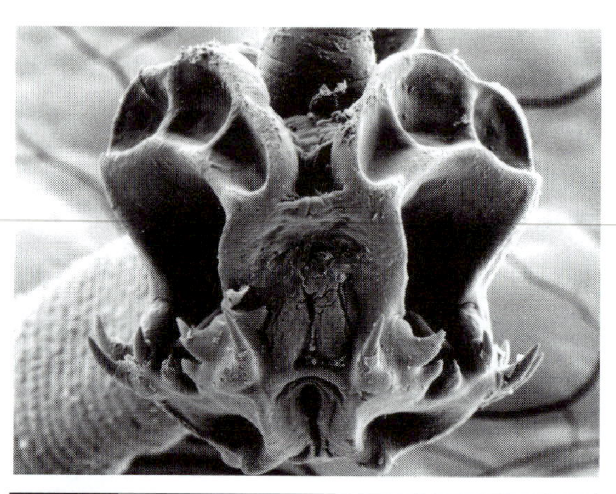

Fig. 4.51 Scolex of the cestode *Phoreiobothrium* sp., a tetraphyllidean from the spiral intestine of sharks. (Photograph courtesy of Janine Caira, University of Connecticut.)

eaten by a lake trout, in which case development to the adult stage then will be either constrained or stimulated by the water temperature (Freeman, 1964).

TRYPANORHYNCHA

The Trypanorhyncha is a large order of cestodes, the adults of which occur in the spiral valves of elasmobranchs worldwide (Fig. 4.50). The scolices of these cestodes are remarkable for their four tentacles, each armed with large numbers of hooks, in addition to two or four bothridia. Each tentacle is capable of being retracted into an internal tentacle sheath. Two life-cycle patterns are characteristic. In some species, procercoids develop in copepods, plerocercoids in planktivorous fishes, and adults in elasmobranchs. In other species, the sequence of hosts involves a filter-feeding mollusc and an elasmobranch definitive host, or two teleost fishes and an elasmobranch. Trypanorhynch plerocercoids typically are surrounded by a fleshy capsule or vesicle called the **blastocyst**. A number of commercially important fishes serve as second intermediate hosts. In many cases, the plerocercoids are located in the flesh and infected fishes must be discarded because of their 'wormy' condition, even though trypanorhynchs cannot infect humans. In temperate oceans of the Southern Hemisphere, species of *Hepatoxilon* and *Grillotia* are ubiquitous in teleost fishes. *Poecilancistrium caryophyllum* is common along the Gulf Coast of North America where the bull shark and related elasmobranchs are the definitive hosts, copepods are first intermediate hosts and sea trout are the second intermediate hosts. Indeed, fishermen along the Gulf Coast are quite familiar with these 'wormy trout'.

TETRAPHYLLIDEA

Tetraphyllideans are also exclusively parasitic in elasmobranchs and have a cosmopolitan distribution. The shape of the scolex is remarkable, possessing highly modified bothridia that may or may not include hooks, spines, and suckers (Fig. 4.51). In tetraphyllideans, host specificity for the definitive host is relatively high. It appears that scolex morphology is responsible to some extent for this phenomenon, with the shape of the bothridia fitting perfectly among the folds of the mucosa in the spiral valve of the host. *Echeneibothrium maculatum*, for example, occurs in the intestine of the ray *Raja montagui* but not in the closely related *R. naevus*. Similar 'lock-and-key' patterns between the morphology of the cestode's scolex and the host's mucosa have been described for other tetraphyllidean–elasmobranch associations (Williams, 1966; Williams *et al*., 1970).

CYCLOPHYLLIDEA

Species of Cyclophyllidea parasitize amphibians, reptiles, birds, and mammals throughout the world. Indeed, most of the tapeworm species found in birds and mammals belong to this order. Similarly, most of the species common in human

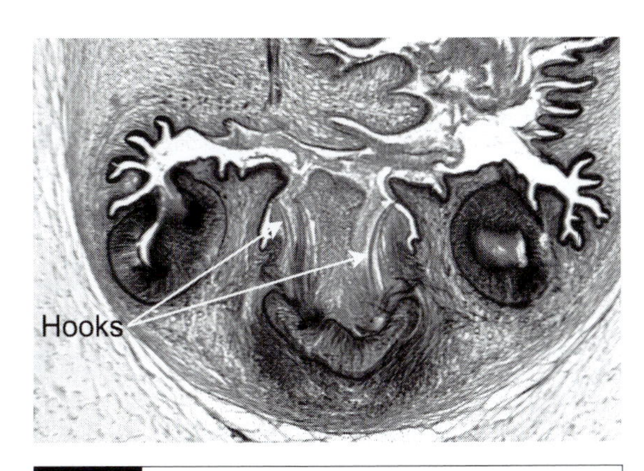

Fig. 4.52 Scolex of *Taenia crassiceps* showing two of the four suckers, the rostellum, and some rose-thorn-shaped hooks. (Photograph courtesy of Philip Mount.)

and domesticated animals are cyclophyllideans. The scolices of cyclophyllidean cestodes are quite variable morphologically, but most are somewhat bulbous, possessing four acetabula. A rostellum may or may not be present and, if present, may or may not be armed with hooks (Fig. 4.52). In some species, the acetabula may be spinose around the rim. All species are monoecious, except for *Dioecocestus* spp., a parasite of grebes.

Taeniid cestodes are usually large in size, with the beef tapeworm of humans reaching 10 m. Most species have a rostellum and, when present, it is not retractable. *Taenia saginata*, the beef tapeworm of humans, lacks a rostellum and hooks, even though these structures are present in all other taeniids. Some authors refer to *Taenia saginata* as *Taeniarhynchus saginatus*, but *Taenia saginata* is the correct designation, having been given this dispensation because of its long-standing priority. Some taeniids that infect humans have the strong potential for doing serious harm (see Box 4.2). The adults of *Taenia saginata* have about 10 000 eggs packed in each gravid proglottid when the intact proglottid is shed in the feces. On occasion, however, the free proglottid may actively crawl out through the anus. Eggs can be dispersed widely by flies and other insects once the proglottid is shed and disintegrates (Lawson & Gemmell, 1990). On ingestion of an egg by an appropriate intermediate host, an active oncosphere emerges, penetrates the gut wall, gains access to the circulatory system, and is carried to skeletal muscle sites where it moves into the tissues and develops into a cysticercus. This bladder worm measures about 10 mm and is also called *Cysticercus bovis*, having been given both generic and species names long before the parasite's life cycle was worked out by Kuchenmeister in 1851. When the larva is eaten by humans in rare or poorly cooked beef, the cycle is complete. The adult tapeworm is innocuous in humans from the standpoint of harm and little damage is inflicted by the cysticercus in infected cattle.

In contrast, *Taenia solium*, the pork tapeworm, is a highly dangerous parasite for humans. As an adult, it is about half the length of its bovine counterpart. Eggs are shed in gravid proglottids. On ingestion by hogs, the oncospheres emerge and penetrate the gut wall where they move into both cardiac and skeletal muscles before developing into cysticerci, which are also known as *Cysticercus cellulosae*. The adult tapeworm develops in humans when they consume poorly cooked pork infected with cysticerci. The adult tapeworm in humans is benign. However, in many parts of the world where pork is a prime source of protein, and socioeconomic conditions are poor, humans also are likely to ingest the eggs of *T. solium*. When this happens, the eggs hatch in the intestine and the oncospheres migrate via the circulatory system to the skeletal and cardiac muscle where they develop into cysticerci. However, they are also known to migrate into the brain tissue, causing neurocysticercosis (see Box 4.2; Fig. 4.53). In many Third World countries, the presence of cysticerci of *T. solium* in the brain is the leading cause of epilepsy (Flisser, 1988, 1998).

Taenia taeniaeformis, *Taenia pisiformis*, and *Taenia crassiceps* are all cosmopolitan taeniids. *Taenia taeniaeformis* uses felines as definitive hosts, whereas the latter two occur in canines. The larval stage of *T. taeniaeformis* is a strobilocercus (Fig. 4.54), also known as *Cysticercus fasciolaris*. This robust larva has a typical taeniid bladder at the posterior end and what appears to be a segmented neck, although there are no true proglottids. The intermediate hosts are rats where the larvae are confined to the livers. *Taenia pisiformis* is a common cestode in canines, with cysticerci occurring unencysted in the abdominal cavities of rabbits. *Taenia crassiceps* is unusual because the cysticerci bud exogenously in their microtine intermediate hosts. Typically, these larvae occur

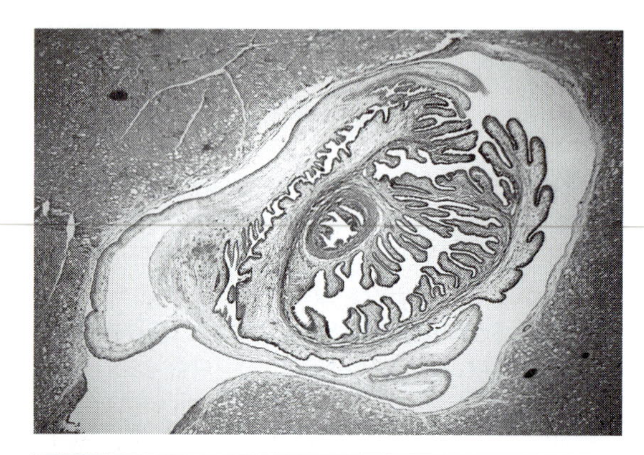

Fig. 4.53 Cysticercus stage of the cestode *Taenia solium*, also known as *Cysticercus cellulosae*. (Photograph courtesy of Aline Aluja, Universidad Nacional Autónoma de México.)

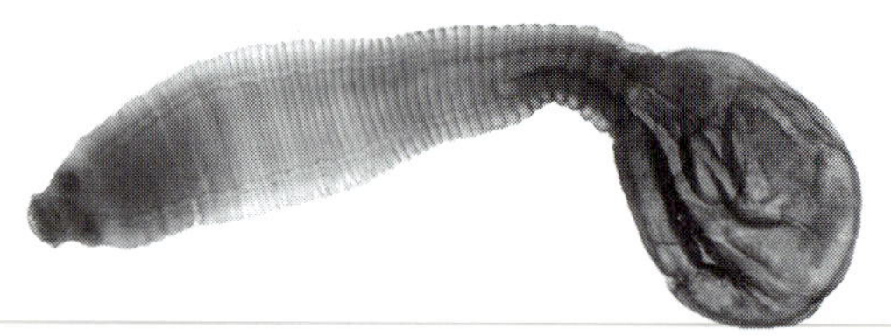

Fig. 4.54 Strobilocercus stage of the cestode *Taenia taeniaeformis*. (Photograph courtesy of Harvey Blankespoor, Hope College.)

encysted in subcutaneous locations, but may also occur in the abdominal cavity. When inoculated by a syringe and needle into the abdominal cavity of a laboratory mouse, a single small bud will develop into an infective cysticercus which then begins budding exogenously from the region of the bladder (Freeman, 1962). Budding is possible because the bladder wall possesses large populations of undifferentiated stem cells. Each daughter bud will develop into an infective cysticercus that, in turn, also produces new daughter buds. From the single bud initially inoculated, the infrapopulation may reach into the thousands before the host dies of thirst or starvation because its belly has grown so distended that the animal cannot reach a water-bottle drain or the food box. Because of its prodigious reproductive capacity, this parasite has become a favorite for the experimentalist who requires large quantities of parasite tissue for biochemical or immunological studies.

Echinococcus granulosus is one of the smallest cestodes, measuring between 3 and 6 mm in total length. Adults parasitize canines; herbivorous mammals, including humans, are intermediate hosts for the larval hydatid cyst (see Box 4.2). A sylvatic life cycle (the cycle which occurs in nature) may involve several canine–herbivore combinations, depending on the geographic location in the world, e.g., wolf–moose, wolf–reindeer, dingo –wallaby. Dogs and domestic ungulates, such as sheep, horses, and swine, provide a potential life-cycle exposure to humans. If humans ingest eggs, they will develop unilocular hydatid cysts, producing a potentially lethal condition known as **hydatidosis**. The cysts normally establish in the liver and lungs, where they grow at a rate of about 1 cm per year. Adults of *E. multilocularis* are also diminutive, occurring in canines, primarily foxes. The multilocular, or alveolar, hydatid cysts occur in microtine rodents (see Box 4.2), but humans may become infected as well.

The mesocestoidid cestodes are an unusual group of cyclophyllideans. They are typically found as adults in mammals such as opossums and raccoons, and birds. The scolex has four acetabula and lacks a rostellum. The larval stage is called a **tetrathyridium** and is found in the body cavities of reptiles and rodents. It is assumed that these larvae are in the second intermediate host because the hosts cannot be infected with eggs and the larva can be used to experimentally produce adults in laboratory animals. The first intermediate host has not been identified yet. However, adults of some *Mesocestoides* species can undergo asexual reproduction in the definitive host when the scolex divides longitudinally (Smyth, 1987).

As adults, cestodes of the cosmopolitan Anoplocephalidae occur in herbivorous mammals. They always possess two complete sets of reproductive organs in each proglottid. In some species, the adults measure 6 m in length. Commonly occurring in sheep-raising countries in the world are *Moniezia expansa* and *M. benedeni*. Eggs of both of these species infect oribatid (soil) mites where cysticercoids develop in the hemocoel. Sheep are infected when they accidentally ingest infected mites while grazing. *Cittotaenia* spp. are common anoplocephalids in rabbits and *Bertiella* occurs in monkeys, apes, and, occasionally, humans.

Hymenolepidid cestodes have many different

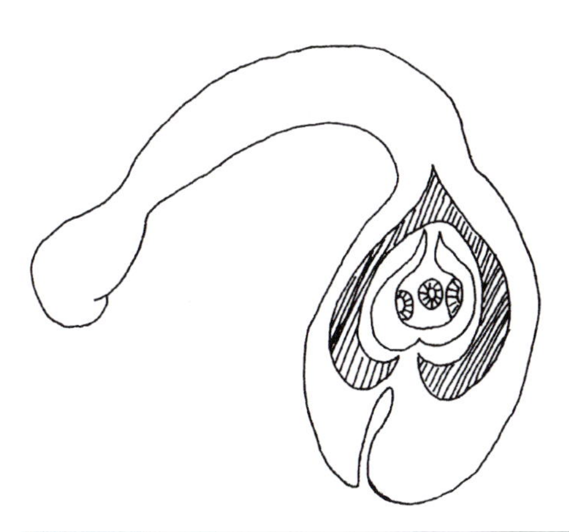

Fig. 4.55 A hymenolepidid cysticercoid. (Drawing courtesy of Lisa Esch-McCall.)

species, infecting a wide range of birds and mammals throughout the world. Arthropods are obligate intermediate hosts for all species except *Vampirolepis nana* (the tapeworm formerly known as *Hymenolepis nana*). *Hymenolepis diminuta* is probably the most widely studied of all tapeworm species, primarily because of the ease with which it can be maintained in the laboratory and its relatively large size (up to 90 cm). Rodents, primarily rats, serve as the definitive hosts and beetles, mostly in the genera *Tenebrio* or *Tribolium*, are the intermediate hosts where cysticercoids (Fig. 4.55) develop within the hemocoel. In several areas of the world, *H. diminuta* occurs with some frequency in humans. *Vampirolepis nana* is the so-called dwarf tapeworm because of the diminutive size of its strobila, on average 30–32 mm in length. It has a cosmopolitan distribution, using rodents and sometimes humans as the definitive host and grain beetles as intermediate hosts. An unusual feature of this parasite is its ability to bypass the arthropod intermediate host and complete development from an egg to the adult stage in the definitive host. If eggs are ingested by a definitive host, they hatch within the intestine, develop into cysticercoids between villi of the mucosa, emerge into the lumen again, and develop rapidly into adults. Humans may become infected, with prevalences as high as 25% in endemic areas.

The dilepidid cestodes infect an enormous range of birds and mammals. One of the most widespread members of the family is *Dipylidium caninum*, a parasite of dogs, cats, and occasionally humans. It possesses a retractable rostellum, with several rows of hooks. There are two sets of reproductive organs in each proglottid. Early in its development, the uterus disappears and the eggs become enclosed within packets. The gravid proglottids are very active and can frequently be seen exiting the anus, then migrating in the fur in the perianal region of an infected canine. When the proglottids dry, the freed egg packets, which resemble tiny grains of rice, will be consumed by fleas, where cysticercoids develop in the hemocoel. When a flea bites, so does the dog, and, in the process, becomes infected when it ingests the infected flea. Humans, in particular children, may also become infected by ingesting fleas.

4.9 Phylogenetic relationships

The Platyhelminthes is a diverse group of organisms showing great diversity in anatomy, life history, size, and habitat. The group includes free-living, commensal, mutualistic, and parasitic species. Platyhelminthes occupy an important position in the animal kingdom due to their basal position among the higher Metazoa phyla. This means that Platyhelminthes are likely ancestors of all the metazoan phyla.

Box 4.3 Classification of the Platyhelminthes

The classification of the Platyhelminthes presented here focuses on the parasitic groups discussed in this chapter. This classification is based (with some modifications) on the work of Brooks & McLennan (1993) on Platyhelminthes and Digenea, of Hoberg *et al.* (1997, 1999) on Eucestoda, and of Boeger & Kritsky (1993) on Monogenea. The diagnoses are adapted from the synapomorphies (derived or novel characters) and characteristics listed by these authors. We should caution the reader that the hierarchical system is rather complex. Do not be discouraged by this. The system comports to reflect the natural relationships and true evolutionary pathways of the different groups. Try to focus on the groups themselves and keep in mind that their position in the hierarchy may change over time as new characters (morphological and molecular) are examined and more species are included in the cladistic analyses.

Due to the large size of the group, only the higher taxonomic levels are diagnosed. For detailed diagnoses of the orders, please consult the original reports.

Phylum Platyhelminthes

Subphylum Rhabdocoela

Superclass Cercomeria

With a a doliiform pharynx, saccate gut, and posterior adhesive organ; adults without locomotory cilia.

Sub-superclass Temnocephalidea

With cephalic tentacles. Representative genus: *Temnocephala*.

Sub-superclass Neodermata

Larval ciliated epidermis shed at the end of larval stage; presence of a post-larval syncytial neodermis; genital pores in anterior half of the body; Mehlis' gland present.

Class Udonellidea

With a secondary protonephridian system formed by pores and canals; phoretic on ectoparasitic copepods. (The older literature places this group within the Monogenea.)

Representative genus: *Udonella*.

Class Cercomeridea

With an oral sucker; uterus with lateral coiling; adult intestine bifurcate.

Subclass Trematoda

With a posterior adhesive organ in the form of a sucker; male genital pore opens into an atrium; pharynx of the adults near the oral sucker; presence of a Laurer's canal.

Infraclass Aspidobothrea

Neodermis with specialized microvilli and microtubules; posterior sucker is hypertrophied and divided into compartments.

Representative genera: *Aspidogaster*, *Cotylogaster*, *Lobatostoma*, *Multicotyle*.

Infraclass Digenea

First larval stage is a miracidium; life cycle with one or more sporocyst generations and cercarial stage; gut development does not occur until redial or cercarial stage (paedomorphic development).

Orders Echinostomatiformes, Haploporiformes, Hemiuriformes, Heronimiformes, Lepocreadiiformes, Opisthorchiformes, Paramphistomiformes, Plagiorchiformes, Strigeiformes, Transversotrematiformes.

Subclass Cercomeromorphae

Posterior adhesive organ, called a cercomer, armed with hooks; cercomer of the larvae with 16 hooks.

Infraclass Monogenea

Larval stage is an oncomiracidium; oncomiracidium with three ciliary epidermal bands; oncomiracidium with 16 marginal hooks; adults with a single testis. The Monogenea are divided into three major taxa: Polyonchoinea, Polystomatoinea, and Oligonchoinea, each one of which comprises several orders.

Oligonchoinea: Orders Chimaericolidea, Diclybothriidea, Mazocraeidea.

Polyonchoinea: Orders Capsalidea, Dactylogyridea, Gyrodactylidea, Monocotylidea, Montchadskyellidea.

Polystomatoinea: Order Polystomatoidea.

Infraclass Cestodaria

No intestine; cercomer paedomorphic, reduced in size, and partially invaginated; oral sucker and pharynx vestigial; adult with multiple testes in two lateral bands; larval cercomer with 10 hooks.

Cohort Gyrocotylidea

Rosette with a funnel at the posterior end of the body; margins of the body crenulated. Representative genera: *Gyrocotyle, Gyrocotyloides*.

Cohort Cestoidea

Male and female genital pores close to each other; cercomer totally invaginated during development; hooks on larval cercomer of two sizes.

Subcohort Amphilinidea

Genital pores at posterior end; uterus N-shaped; adults parasitic in body cavity. Representative genera: *Amphilina, Austramphilina*.

Subcohort Eucestoda

Body of adults polyzoic; six-hooked larval cercomer lost during ontogeny; first larval stage is an hexacanth embryo; tegument covered with microtrichs; life cycles with more than one host.

Orders Caryophyllidea, Cyclophyllidea, Diphyllidea, Lecanicephalidea, Nippotaenidea, Onchobothriidea, Phylobothriidea, Proteocephalidea, Pseudophylidea, Spathebothriidea, Tetrabothriidea, Trypanorhyncha.

The evolutionary history of Platyhelminthes has not been completely deciphered. Evolutionary studies within the group, using morphological and molecular data, combined with cladistic methods, show significant concordance. These studies agree, for the most part, on a phylogenetic tree (Fig. 4.56) in which the free-living, commensal, and mutualistic groups of platyhelminths, e.g., Catenulida, Acoela, Macrostomatida, Polycladida, etc., occupy a basal position, whereas the parasitic groups, i.e., Aspidobothrea, Digenea, Monogenea, Gyrocotylidea, Amphilinidea, and Eucestoda, occupy a terminal position.

All the parasitic groups, collectively called Neodermata, resemble each other in structural and developmental aspects of their epidermal cilia, sensory receptors, flame cells, and sperm. Moreover, the epidermis of the larva is replaced in the adult by a syncytial 'neodermis', a characteristic feature that led to the collective name of Neodermata.

The phylogenetic relationships within the groups of parasitic Platyhelminthes have received much attention in the last decade, both from the morphological and molecular point of view (Ehlers, 1985, 1986; Boeger & Kritsky, 1993; Brooks

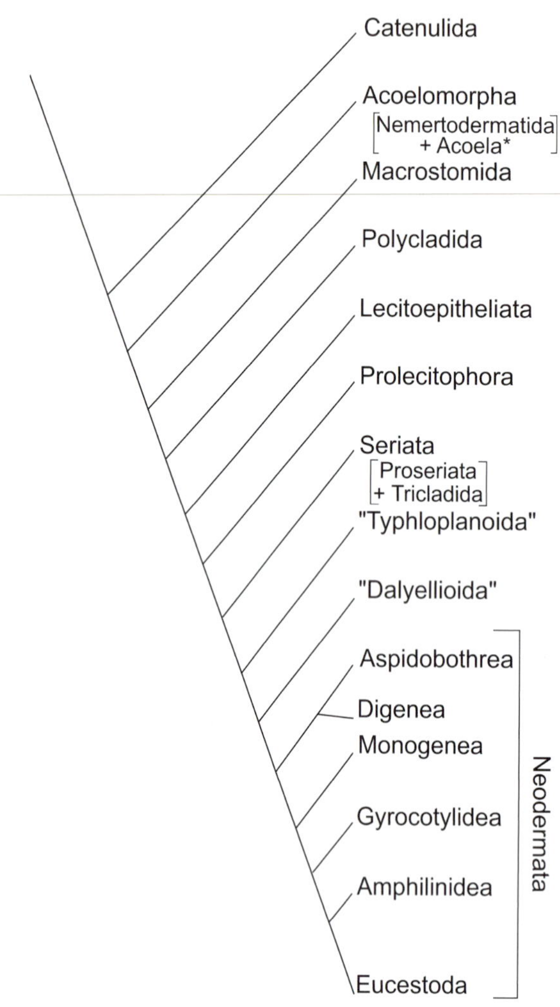

Fig. 4.56 Cladogram depicting relationships of the Platyhelminthes. (Based on several sources.) * Recent studies (Ruiz-Trillo *et al.*, 1999) indicate that the Acoela do not belong within the Platyhelminthes; they are likely to be placed in their own phylum.

& McLennan, 1993; Rohde *et al.*, 1993; Hoberg *et al.*, 1997, Justine, 1998; Mariaux, 1998; Hoberg *et al.*, 1999). The abundance of characters generated by combining morphological and molecular data has allowed researchers to propose a more evolutionarily sound classification of the Platyhelminthes, especially of the parasitic forms (see Box 4.3). The phylogenetic relationships of the Temnocephalidea and Udonellidea are not yet resolved. The Aspidobothrea occupy an important position at the base of the platyhelminth parasitic groups, and it is likely that they are the most primitive members of the Neodermata. The Digenea are closely related to the aspidobothreans and probably evolved from them. Although the Monogenea have been historically treated as closely related to the Digenea and Aspidobothrea, most researchers now consider the Monogenea to be the sister group of the Cestodaria, the clade that includes the Gyrocotylidea, Amphilinidea, and Eucestoda.

Current phylogenetic and evolutionary studies of Platyhelminthes are not restricted to the major groups discussed above. A plethora of studies has been devoted to phylogenetic analyses of families and genera, mainly within the Digenea, Monogenea, and Eucestoda. Their detail is beyond the scope of this book, but anyone interested should consult the primary literature, most of which has been summarized up to 1993 by Brooks & McLennan (1993).

The basic principle of phylogenetic studies is that their hypotheses are open to continuous testing, modification, and refinement. Recent phylogenetic studies of Platyhelminthes and other metazoans using 18S ribosomal DNA genes indicate that the Acoela, one of the most basal and primitive groups included in the Platyhelminthes (Fig. 4.56), are not really platyhelminths (Ruiz-Trillo *et al.*, 1999). They seem to be positioned even earlier in the evolutionary tree, representing one of the earliest branches within the bilateral animals with living descendants. The Acoela, then, are a living relic of the transition between radially symmetrical animals such as jellyfish, and the more complex bilateral organisms. These findings indicate that the Acoela should be placed in their own phylum, probably evolving just prior to the Platyhelminthes.

Even with this radical change at the base of the evolutionary tree of the Platyhelminthes, the evolutionary sequence of the parasitic groups remains basically unchanged, having evolved from the more basal Platyhelminthes.

References

Aken'Ova, T. O. & Lester, R. J. G. (1996) *Udonella myliobati* n. comb. (Platyhelminthes: Udonellidae) and its occurrence in Australia. *Journal of Parasitology*, **82**, 1017–1023.

Arme, C. & Owen, R. W. (1967) Infections of three-spined stickleback, *Gasterosteus aculeatus* L., with the plerocercoid larvae of *Schistocephalus solidus* (Muller, 1776), with special reference to pathological effects. *Parasitology*, **57**, 301–314.

Ataev, G. L., Dobrovolskij, A. A., Fournier, A. & Jourdane, J. (1997) Migration and development of mother sporocysts of *Echinostoma caproni* (Digenea: Echinostomatidae). *Journal of Parasitology*, **83**, 444–453.

Avenant-Oldewage, A. (1993) Occurrence of *Temnocephala chaeropsis* on *Cherax tenuimanus* imported into South Africa and notes on its infestation of an indigenous crab. *South African Journal of Science*, **89**, 427–428.

Barger, M. & Esch, G. W. (2000) *Plagioporus sinitsini* (Digenea: Opecoelidae): A one-host life cycle. *Journal of Parasitology*, **86**, 150–153.

Blanchard, R. (1889) *Traité de Zoologie Médicale*, vol. 1. Paris: Baillière.

Blankespoor, H. D., Esch, G. W. & Johnson, W. C. (1985) Some observations on the biology of *Collyriclum faba* (Bremser in Schmalz, 1831). *Journal of Parasitology*, **71**, 469–471.

Boeger, W. A. & Kritsky, D. C. (1993) Phylogeny and a revised classification of the Monogenoidea Bychowsky, 1937 (Platyhelminthes). *Systematic Parasitology*, **26**, 1–32.

Brooks, D. R. (1989) The phylogeny of the Cercomeria (Platyhelminthes: Rhabdocoela) and general evolutionary principles. *Journal of Parasitology*, **75**, 606–616.

Brooks, D. R. & McLennan, D. A. (1993) *Parascript, Parasites and the Language of Evolution*. Washington, DC: Smithsonian Institution Press.

Bueding, E. & Fisher, J. (1982) Metabolic requirements of schistosomes. *Journal of Parasitology*, **33**, 208–212.

Byrnes, T. (1986) Five species of Monogenea from Australian bream, *Acanthopagrus* spp. *Australian Journal of Zoology*, **34**, 65–86.

De Bont, J. & Vercruysse, J. (1998) Schistosomiasis in cattle. *Advances in Parasitology*, **41**, 285–364.

Ehlers, U. (1985) Phylogenetic relationships among the Platyhelminthes. In *The Origins and Relationships of Lower Invertebrates*, ed. C. Morris, J. D. George, R. Gibson & H. M. Platt, pp. 143–158. Oxford: Oxford University Press.

Ehlers, U. (1986) Comments on a phylogenetic system of Platyhelminthes. *Hydrobiologia*, **132**, 1–12.

Esch, G. W., Wetzel, E. J., Zelmer, D. A. & Schotthoefer, A. M. (1997) Long-term changes in parasite population and community structure: a case history. *American Midland Naturalist*, **137**, 369–387.

Fernández, J., Villalba, C. & Albiña, A. (1987) Parásitos del pejegallo, *Callorhynchus callorhynchus* (L.), en Chile: Aspectos biológicos y sistemáticos. *Biología Pesquera*, **15**, 63–73.

Flisser, A. (1988) Neurocysticercosis in Mexico. *Parasitology Today*, **4**, 131–137.

Flisser, A. (1998) Larval cestodes. In *Microbiology and Microbiology Infections*, ed. F. E. G. Cox, J. P. Kreier & D. Wakelin, pp. 539–560. London: Edward Arnold.

Freeman, R. S. (1962) Studies on the biology of *Taenia crassiceps* (Zeder, 1800) Rudolphi, 1810 (Cestoda). *Canadian Journal of Zoology*, **40**, 969–990.

Freeman, R. S. (1964) On the biology of *Proteocephalus parallacticus* Maclulich (Cestoda) in Algonquin Park, Canada. *Canadian Journal of Zoology*, **42**, 387–408.

Fried, B. & Huffman, J. E. (1996) The biology of the intestinal trematode *Echinostoma caproni*. *Advances in Parasitology*, **38**, 312–368.

Fujino, T. (1993) Ultrastructure and function of alimentary systems in parasitic helminths. *Japanese Journal of Parasitology*, **42**, 277–294.

Gibson, D. I. & Bray, R. A. (1979) The Hemiuroidea: terminology, systematics and evolution. *Bulletin of the British Museum (Natural History), Zoology series*, **36**, 35–146.

Hairston, N. (1965) On the mathematical analysis of schistosome populations. *Bulletin of the World Health Organization*, **33**, 45–62.

Halton, D. W. (1978) Trans-tegumental absorption of L-alanine and L-leucine by a monogenean, *Diclidophora merlangi*. *Parasitology*, **76**, 29–37.

Halton, D. W. & Gustafsson, M. K. S. (1996) Functional morphology of the platyhelminth nervous system. *Parasitology* (Supplement), **113**, S47–S72.

Halton, D. W., Maule, A. G. & Shaw, C. (1993) Neuronal mediators in monogenean parasites. *Bulletin Français de la Pêche et Pisciculture*, **328**, 82–104.

Hendrix, S. S., Vidrine, M. F. & Hantenstine, R. H. (1985) A list of records of freshwater aspidogastrids (Trematoda) and their hosts in North America. *Proceedings of the Helminthological Society of Washington*, **52**, 289–296.

Hoberg, E. P., Mariaux, J., Justine, J.-L., Brooks, D. R. & Weekes, P. J. (1997) Phylogeny of the orders of the

Eucestoda (Cercomeromorphae) based on comparative morphology: historical perspectives and a new working hypothesis. *Journal of Parasitology*, **83**, 1128–1147.

Hoberg, E. P., Gardner, S. L. & Campbell, R. A. (1999) Systematics of the Eucestoda: advances towards a new phylogenetic paradigm, and observations on the early diversification of tapeworms and vertebrates. *Systematic Parasitology*, **42**, 1–12.

Huffman, J. & Fried, B. (1990) *Echinostoma* and echinostomiasis. *Advances in Parasitolology*, **29**, 215–269.

Hughes, R. N. (1989) *A Functional Biology of Clonal Animals*. London: Chapman & Hall.

Justine, J.-L. (1998) Spermatozoa as phylogenetic characters for the Eucestoda. *Journal of Parasitology*, **84**, 385–408.

Kabata, Z. (1973) Distribution of *Udonella caligorum* Johnston, 1835 (Monogenea: Udonellidae) on *Caligus elongatus* Nordman, 1932 (Copepoda: Caligidae). *Journal of the Fisheries Research Board of Canada*, **30**, 1793–1798.

Kearn, G. C. (1971) The physiology and behavior of the monogenean skin parasite *Entobdella soleae* in relation to its host (*Solea solea*). In *Ecology and Physiology of Parasites*, ed. A. M. Fallis, pp. 161–187. Toronto: University of Toronto Press.

Lawson, J. R. & Gemmell, M. A. (1990) Transmission of taeniid eggs via blowflies to intermediate hosts. *Parasitology*, **100**, 143–146.

Lemly, A. D. & Esch, G. W. (1984) Effects of the trematode *Uvulifer ambloplitis* on juvenile bluegill sunfish, *Lepomis macrochirus*: ecological implications. *Journal of Parasitology*, **70**, 475–492.

Malmberg, G. (1993) Gyrodactylidae and gyrodactylosis of Salmonidae. *Bulletin Français de la Pêche et Pisciculture*, **328**, 5–46.

Marcogliese, D. J. & Esch, G. W. (1989) Experimental and natural infection of planktonic and benthic copepods by the Asian tapeworm, *Bothriocephalus acheilognathi*. *Proceedings of the Helminthological Society of Washington*, **56**, 151–155.

Mariaux, J. (1998) A molecular phylogeny of the Eucestoda. *Journal of Parasitology*, **84**, 114–124.

Millemann, R. E. & Knapp, S. E. (1970) Biology of *Nanophyetus salmincola* and salmon poisoning disease. *Advances in Parasitology*, **8**, 1–41.

Pappas, P. W. & Read, C. P. (1975) Membrane transport in helminth parasites. *Experimental Parasitology*, **37**, 469–530.

Platt, T. R. & Brooks, D. R. (1997) Evolution of the schistosomes (Digenea: Schistosomatoidea): the origin of dioecy and colonization of the venous system. *Journal of Parasitology*, **83**, 1035–1044.

Rausch, R. L. & D'Alessandro, A. (1999) Histogenesis in the metacestode of *Echinococcus vogeli* and mechanism of pathogenesis in polycystic hydatid disease. *Journal of Parasitology*, **85**, 410–418.

Ravenga, J. E. (1993) *Diphyllobothrium dendritcum* and *Diphyllobothrium latum* in fishes from southern Argentina: association, abundance, distribution, pathological effects, and risk of human infection. *Journal of Parasitology*, **79**, 379–383.

Rohde, K. (1989) At least eight types of of sense receptors in an endoparasitic flatworm: a counter-trend to sacculinization. *Naturwissenschaften*, **76**, 383–385.

Rohde, K. (1993) *Ecology of Marine Parasites*, 2nd edn. Wallingford: CAB International.

Rohde, K., Hefford, K., Ellis, J. T., Baverstock, P. R., Johnson, A. M., Watson, N. A. & Dittman, S. (1993) Contributions to the phylogeny of Platyhelminthes based on partial sequencing of 18S ribosomal DNA. *International Journal for Parasitology*, **23**, 705–724.

Ruiz-Trillo, I., Riutort, M., Littlewood, D. T. J., Herniou, E. A. & Baguña, J. (1999) Acoel flatworms: earliest extant bilaterian metazoans, not members of Platyhelminthes. *Science*, **283**, 1919–1923.

Schaperclaus, W. (1991) *Fish Diseases*, 5th edn. New Delhi: Oxonian Press.

Shoop, W. L. & Corkum, K. C. (1987) Maternal transmission by *Alaria marcianae* (Trematoda) and the concept of amphiparatenesis. *Journal of Parasitology*, **73**, 110–113.

Shoop, W. L., Font, W. F. & Malatesta, P. F. (1990) Transmammary transmission of mesocercariae of *Alaria marcianae* (Trematoda) in experimentally infected primates. *Journal of Parasitology*, **76**, 869–873.

Smith, J. W. (1997) The blood flukes (Digenea: Sanguinicolidae and Spirorchidae) of cold-blooded vertebrates: Part 1. A review of the literature published since 1971, and bibliography. *Helminthological Abstracts*, **66**, 255–294.

Smyth, J. D. (1987) Asexual and sexual differentiation in cestodes: especially *Mesocestoides* and *Echinococcus*. In *Molecular Paradigms for Eradicating Parasites*, ed. A.J. MacInnis, pp. 19–34. New York: Alan R. Liss.

Smyth, J. D. (1994) *Introduction to Animal Parasitology*. Cambridge: Cambridge University Press.

Sukhdeo, S. C. & Sukhdeo, M. V. K. (1994) Mesenchyme cells in *Fasciola hepatica* (Platyhelminthes): primitive glia? *Tissue and Cell*, **26**, 123–131.

Tielens, A. G. M, Van den Heuvel, J. M. & Van den Bergh, S. G. (1984) The energy metabolism of *Fasciola hepatica* during its development in the final host. *Molecular Biochemistry and Parasitology*, **13**, 301–307.

Tinsley, R. C. (1995) Parasitic disease in amphibians: control by the regulation of worm burdens. *Parasitology* (Supplement), **111**, S153–S178.

Tinsley, R. C. & Jackson, H. C. (1988) Pulsed transmission of *Pseudodiplorchis americanus* between desert hosts (*Scaphiopus couchii*). *Parasitology*, **97**, 437–452.

Webster, L. A. & Wilson, R. A. (1970) The chemical composition of protonephridial canal fluid from the cestode *Hymenolepis diminuta*. *Comparative Biochemistry and Physiology*, **35**, 201–209.

Williams, H. H. (1966) The ecology, functional morphology and taxonomy of *Echeneibothrium* Benaden, 1849 (Cestoda: Tetraphyllidea), a revision of the genus and comments on *Discobothrium* Beneden, 1870, *Pseudanthobothrium* Baer, 1956, and *Phormobothrium* Alexander, 1963. *Parasitology*, **56**, 227–285.

Williams, H. H., Colin, J. A. & Halvorsen, O. (1987) Biology of gyrocotylideans with emphasis on reproduction, population ecology and phylogeny. *Parasitology*, **95**, 173–207.

Williams H. H., McVicar, A. H. & Ralph, R. (1970) The alimentary canal of fish as an environment for helminth parasites. In *Aspects of Fish Parasitology*, Symposium of the British Society of Parasitology, vol. 8, ed. A. E. R. Taylor & R. Muller, pp. 43–77. Oxford: Blackwell Scientific Publications.

ADDITIONAL REFERENCES

Arme, C. & Pappas, P. W. (1983) *Biology of the Eucestoda*, vols. 1 and 2. New York: Academic Press.

Barrett, J. (1981) *Biochemistry of Parasitic Helminths*. London: Macmillan.

Bryant, C. & Behm, C. A. (1989) *Biochemical Adaptation in Parasites*. London: Chapman & Hall.

Bychowsky, B. E. (1957) *Monogenetic Trematodes, Their Systematics and Phylogeny*. Washington, DC: American Institute of Biological Sciences.

Clayton, D. H. & Moore, J. (1997) *Host–Parasite Evolution: General Principles and Avian Models*. Oxford: Oxford University Press.

Dawes, B. (1946) *The Trematoda*. Cambridge: Cambridge University Press.

Dogiel, V. A. (1964) *General Parasitology*. Edinburgh: Oliver & Boyd.

Hoffman, G. L. (1967) *Parasites of North American Freshwater Fishes*. Berkeley: University of Californa Press.

Schell, S. C. (1985) *Handbook of Trematodes of North America North of Mexico*. Moscow: University of Idaho Press.

Schmidt, G. D. (1986) *Handbook of Tapeworm Identification*. Boca Raton: CRC Press.

Smyth, J. D. (1990) *In Vitro Cultivation of Parasitic Helminths*. Boca Raton: CRC Press.

Smyth, J. D. & Halton, D. W. (1983) *The Physiology of Trematodes*. Cambridge: Cambridge University Press.

Smyth, J. D. & McManus, D. P. (1989) *Physiology and Biochemistry of Cestodes*. Cambridge: Cambridge University Press.

Thompson, R. C. A. (1986) *The Biology of* Echinococcus *and Hydatid Disease*. London: Allen & Unwin.

Wardle, R. A. & McLeod, J. A. (1952) *The Zoology of Tapeworms*. Minneapolis: University of Minnesota Press.

Yamaguti, S. (1963) *Systema Helminthum*. New York: Interscience Publishers.

Chapter 5

Nematoda: the roundworms

5.1 General considerations

The Nematoda (Greek: *nema* = thread) are commonly called the roundworms. The names 'Nemathelminthes' or 'Nemata' are sometimes used in place of 'Nematoda' though neither seems to have a strong following. Current estimates of between 16000 and 20000 described species place the nematodes third (behind the phyla Arthropoda and Mollusca) as the most species-rich phylum in the Kingdom Animalia. Most scientists recognize that the vast majority of nematodes remain undescribed and suggest that, when all is said and done, the phylum will be the most speciose in the Kingdom. Nematodes do not fossilize well and the oldest reports are from the Tertiary, about 30–50 million **ybp**. Nematodes have a long relationship with people as records of 'roundworm' symptoms date back to approximately 2700 BC (Bird & Bird, 1991) (see also Box 1.1).

Nematodes are now recognized as perhaps the most significant **metazoan** parasite associated with humans. However, most nematodes are free-living, not parasitic. These free-living forms are found in a wide variety of aquatic (both marine and freshwater) and terrestrial habitats and, perhaps, exploit a greater array of habitats than any other metazoan. Many free-living nematodes are **detritivores** or **decomposers** and play a disproportionately large role in recycling chemicals and organic nutrients. Others feed on bacteria and microorganisms and are important in food web relationships. Some, called entomopathogenic nematodes, may be important in regulating insect populations and thus, although parasitic, may have a beneficial impact to people. Many are not beneficial, however, and parasitic forms are found in many of the eukaryotes. As plant-parasites, nematodes have had, and continue to have, a significant negative impact on people. Entire crops can be virtually destroyed by nematodes in a single season. Finally, of course, nematodes can, and do, infect almost all phyla of animals. It is those parasitic in vertebrates (approximately 5800 species), on which we will focus in this chapter. When necessary, however, we will include information derived from other types of nematodes when such information on animal-parasitic forms is scanty or lacking.

Nematodes are typically dioecious, often exhibit **sexual dimorphism**, and vary from <1 mm to (rarely) >1 m in length. (The largest nematode reported to date is *Placentonema gigantissima* from the placenta of sperm whales *Physeter catodon*; female worms are approximately 8 m long and about 2.5 cm in diameter.) They are **pseudocoelomates** and are bilaterally symmetrical. A complex cuticle that must be molted four times prior to reaching sexual maturity covers them externally. They are characterized by **eutely** meaning that growth is due to increased cell size rather than an increase in cell number. They have a complete, but simple, digestive system, a nervous system, a putative excretory system, and a rather large reproductive system. They lack respiratory and circulatory systems. The overall body form is a cylinder tapered at both ends and, as such, tends to be rather uniform. In regard to their 'supposed' uniformity, a colleague relates the following story. As a postdoctoral student examining the parasites in bobwhite quail (*Colinus virginianus*), she happened to find a specimen of *Dispharynx nasuta* (a nematode with distinctive structures on the head known as cordons). After mounting the specimen on a microscope slide, an eminent ecologist happened by the lab. She asked him to look at the specimen – his comment: 'My God, a nematode with a feature!'. Because of their apparent uniformity, some consider the nematodes as conservative and evolutionary dead-ends. Others, noting the extraordinary array of habitats invaded and the diversity of life-history strategies, consider them as evolutionary radicals.

5.2 Form and function

5.2.1 The outside

Body shape. The body of most nematodes is a naked cylinder with tapered ends. Typically there are no distinct divisions of the body and arbitrary reference must be made to a head and a tail (Fig. 5.1). Amongst the nematode parasites of vertebrates, there exist exceptions to the general uniform cylindrical body shape (Fig. 5.2) but these are few. Much more common are variations in the external morphology of the anterior or posterior (or both) ends (Figs. 5.3, 5.4).

Anteriorly, the primitive number of lips is thought to be six but in extant nematodes, the number may range from zero to six. In addition, lateral lips, called pseudolabia, may be present, particularly on those with a reduced number of lips. The lips open internally to a buccal capsule lined by cuticle; this cuticle may be very hard (**sclerotized**). Some, such as the hookworms, have a highly modified buccal capsule containing tooth-like structures. Often, it is some combination of lips or the buccal capsule that nematodes employ to attach to the host (Fig. 5.3A, D). Rather prominent anteriad structures on some nematodes are cordons or cuticular projections found in many nematodes that live under the **koilin lining** of the gizzards of birds (Fig. 5.3B). No specific function is known for these structures but, since they mostly characterize worms associated with the gizzard lining, their presence may be related to the potentialities or constraints associated with that microhabitat. Finally, some nematodes have lateral projections, called cephalic alae, associated with the anterior end (Fig. 5.3C).

Also found anteriorly are sensory structures that may include some combination of glands, papillae, or bristles. **Amphids** occur in pairs and open internally into the amphidial glands, the largest of the anterior sense organs. Though often reduced in nematodes parasitic in animals, amphids nonetheless are thought to be important as chemosensory structures. Amphids are also secretory though the function of the secretory proteins seems variable (Riga *et al.*, 1995). Associated with the lips are labial papillae but these are often fused accompanying a reduction in the number of lips. At about the same level as the amphids are cephalic papillae or bristles. Slightly posterior to these is a pair of cervical papillae known as **deirids.** Papillae or bristles are thought to serve a tactile function and their number and location is an important taxonomic character.

Following the cephalic structures, some nematodes may have some combination of spines, lateral alae, additional sensory papillae, or other modifications of the body. This region may also be important for taxonomic purposes, at least in the trichostrongyles, as this is where the cuticular ridges (the **synlophe**) are most prominent. For most nematodes, however, this is a region largely devoid of any cuticular elaborations.

Posteriorly may be found pore-like **phasmids**, caudal papillae, and, on some males, caudal alae or an elaborate copulatory **bursa**. Phasmids are important for taxonomic purposes. They characterize the Rhabditea which includes most nematode parasites of animals. Structurally, phasmids are similar to amphids though the internal glands to which they connect are often smaller and, exceptionally, absent. Caudal papillae are also important taxonomically and, though more prominent in males, are found on both sexes. Caudal alae, smaller but otherwise similar to cephalic alae, are lateral extensions of the cuticle (Fig. 5.4A). Copulatory bursae are found on the males of two unrelated groups of nematodes. A fleshy bursa with no supporting rays is characteristic of the Dioctophymatida, while males in the Strongylida have well-developed bursae supported

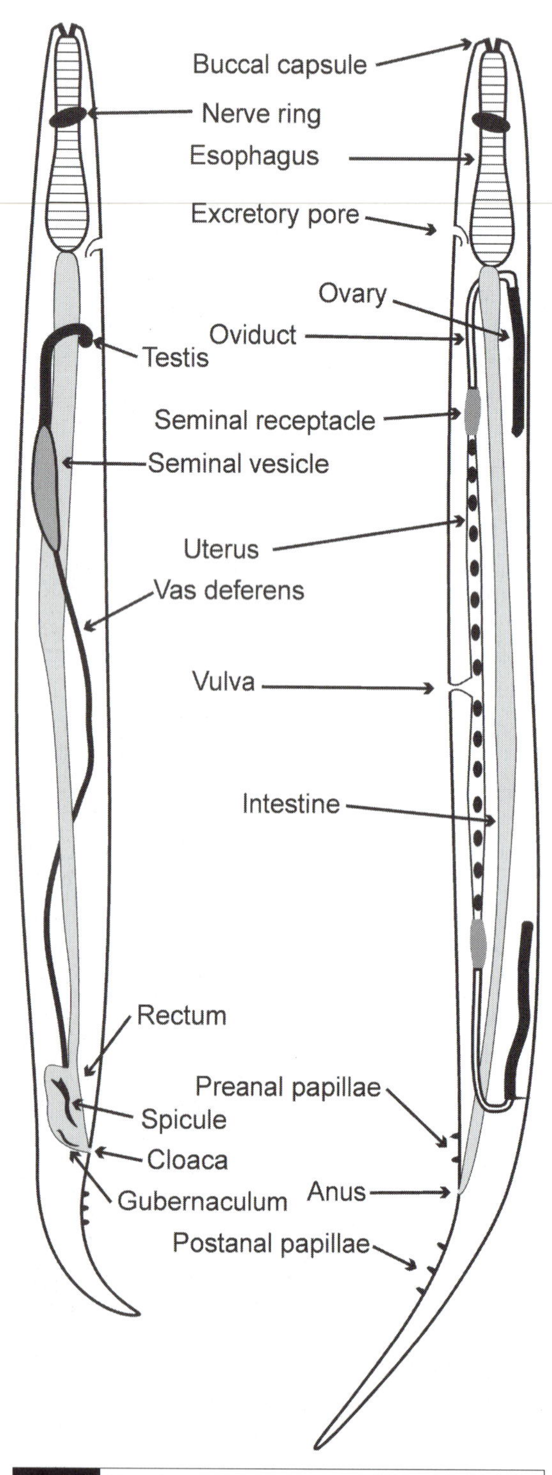

Fig. 5.1 Hypothetical male and female nematodes. The reproductive systems are much reduced in length for the sake of clarity; typically, structures such as the ovary and testis are many times longer than the body and are wrapped around the intestine.

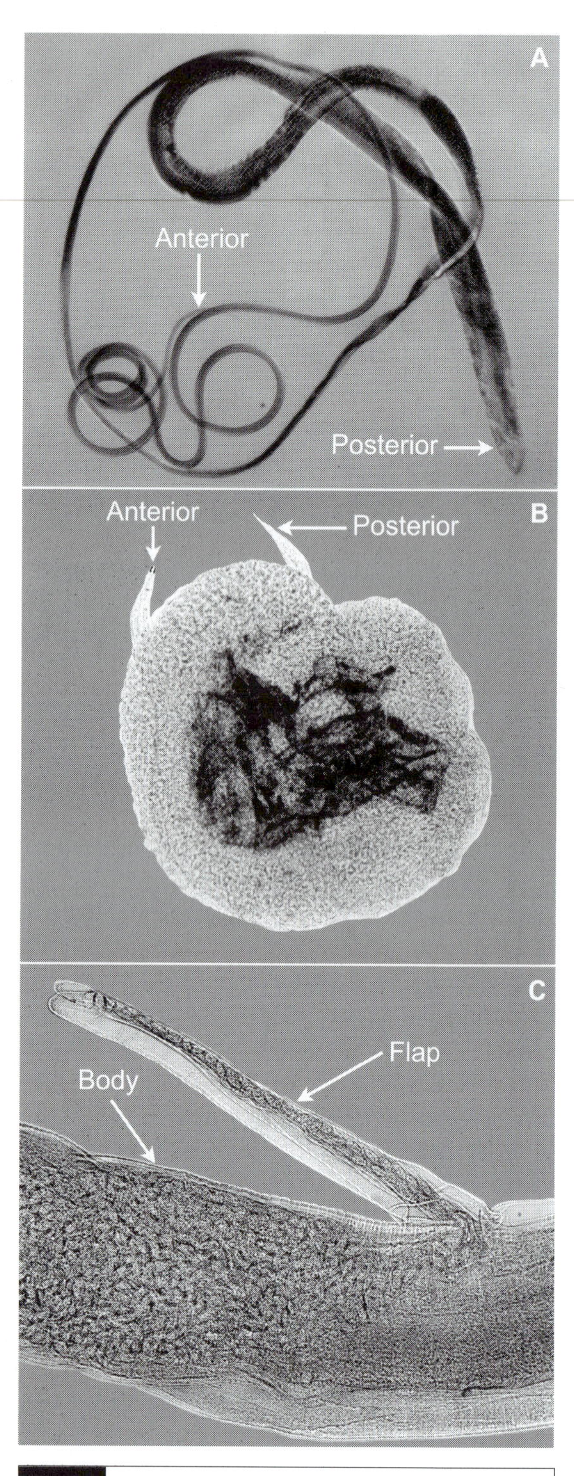

Fig. 5.2 Different body shapes or structures found in some nematodes. (A) *Trichuris*; (B) female *Tetrameres*; (C) vulvar flap on *Evaginurius*.

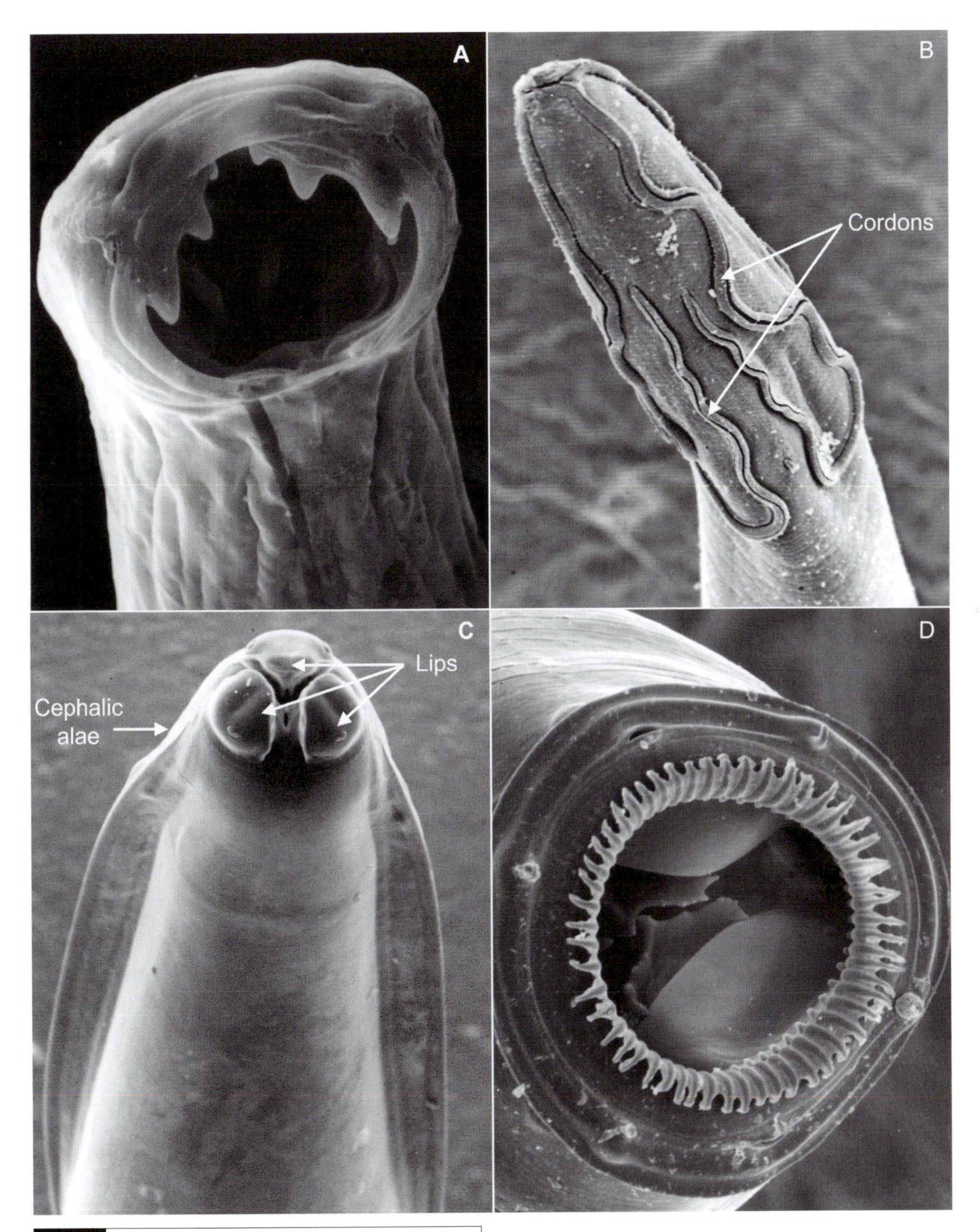

Fig. 5.3 Cephalic structures found in some nematodes. (A) Buccal capsule on *Ancylostoma*; (B) cordons on *Dispharynx*; (C) cephalic alae on *Toxocara*; (D) buccal capsule on *Triodontophorus*. (Photographs courtesy of Russ Hobbs, © Murdoch University.)

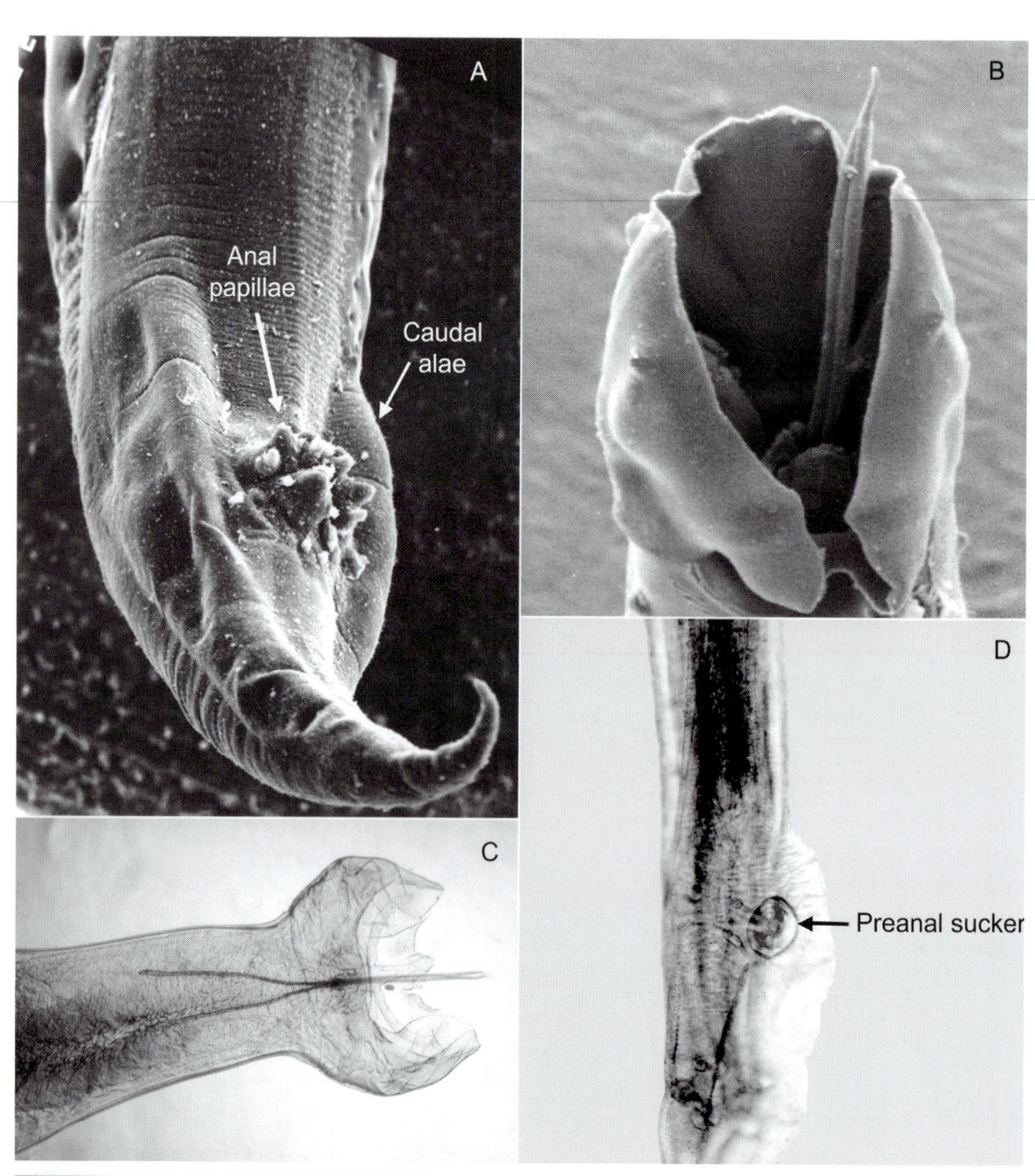

Fig. 5.4 Caudal structures found in some nematodes. (A) Caudal alae and anal papillae on *Eugenurus*; (B) scanning electron micrograph showing bursa on male *Chabertia* with extended spicule; (C) light microscopy photograph showing bursa on *Uncinaria* with extended spicule; (D) preanal sucker on *Heterakis*. ((A,B) Photographs courtesy of Russ Hobbs, ©Murdoch University; (C) photograph courtesy of Toni Raga, Universidad de Valencia.)

by finger-like rays (Fig. 5.4B, C). The males use the copulatory bursae to grasp the female during insemination.

Hanging on. Whereas some of the groups discussed previously (e.g., the Platyhelminthes) and later (e.g., the Arthropoda) have specialized structures known as holdfasts to maintain their position in or on a host, such structures are uncommon in the nematodes parasitic in animals. Most nematodes in vertebrates live either in the tissues or in the gut. Tissue-dwelling worms probably have no need for discrete holdfasts. Their habitat, with respect to motility, is benign. On the other hand, nematodes parasitic along the gut tube must face peristalsis, or some other form of muscular contractions, at some point in their lives.

There are certainly exceptions to this, for example the oral plates and teeth of the hookworms make pretty impressive holdfast structures (Fig. 5.3A). Others, such as *Trichuris*, thread the narrow anterior portion of the body through the mucosa to maintain their position (Fig. 5.2A). Still others, such as female *Tetrameres* (Fig. 5.2B), which live in the glands of the proventriculus of birds, may rely simply on their massive girth (attained after entering the glands) to hold them in place. Finally, secretions from the worm have been speculated to act as a 'biochemical holdfast' by reducing intestinal peristalsis. Initially, acetylcholinesterase secretions from the amphidial or secretory–excretory glands were thought to mediate this action (e.g., Ogilvie & Jones, 1971). More recent studies dispute the importance of acetylcholinesterase in the role of modulation, but not the concept of a biochemical holdfast. For example, Lee (1996) suggests that a 'vasoactive intestinal polypeptide-like protein' might cause a reduction in gastrointestinal motility.

Otherwise, nematodes pretty much lack adaptations for hanging on to the host. (Cuticular structures such as alae are thought to be used for hydrodynamic purposes or for added stability when crawling.)

Body wall. The outermost layer of the nematode body wall is the epicuticle. This is followed internally by the cuticle itself which is a non-cellular structure secreted by syncytial epidermal cells. Chemically, it is composed of proteins, notably collagen, with some other glycoproteins thrown in on occasion. Stability of the cuticle is provided mainly through disulfide bonds. In cross-section, there are three main layers: an external cortical zone, an intervening median zone, and a basal fibrous zone. However, the number, and substructure, of layers varies between species and between different developmental stages within a species (Neuhaus *et al.*, 1996; Martínez & de Souza, 1997).

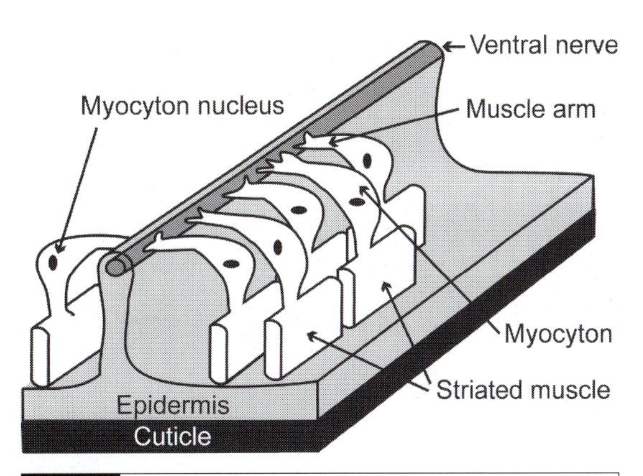

Fig. 5.5 Schematic diagram of somatic musculature in the nematodes showing the arrangement of the contractile and non-contractile elements of individual muscle cells. All muscle cells in the lower hemisphere attach to the ventral nerve while all muscle cells in the upper hemisphere attach to the dorsal nerve.

The cuticle is an important structure in the Nematoda – it is the first line of defense against pathogens and anthelmintics. It also acts as a hydroskeleton antagonizing the longitudinal muscles, it protects against desiccation and digestion, and it may play a role in nutrition and excretion. Blaxter *et al.* (1992) even suggested a secretory role for the cuticle that may be involved in antigenic activity.

A basement membrane marks the division between the cuticle and the underlying epidermis, internal to which are longitudinal muscles. The muscle cells have two distinct parts: a contractile U-shaped projection (composed of striated fibers) inserting on the epidermis and the other a non-contractile **myocyton** (or cell body) containing the nucleus. Arising from the myocyton are muscle cell arms, terminating in non-contractile, digit-like projections that insert directly on the nerve cords (Fig. 5.5). While not unique to the Nematoda, this relationship between muscle and nerve is unusual.

5.2.2 The inside

The pseudocoel. Between the somatic musculature and the digestive system is the fluid-filled pseudocoel. It contains hemolymph, an almost cell-free fluid. Some cells, called **coelomocytes**, do occur (usually a total ranging from two to six), and these range from microscopic to approximately 5

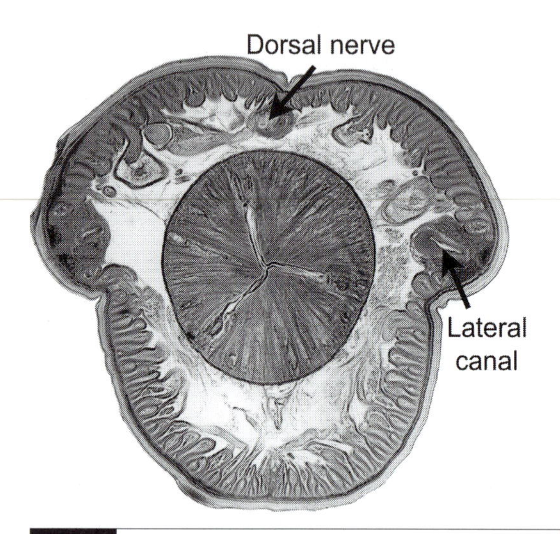

Fig. 5.6 Cross-section through the esophageal region of *Ascaris suum*. Note the powerful musculature associated with the esophagus and its triradiate lumen. The lateral canals, embedded in epidermis, are clearly visible as is the dorsal nerve, also embedded in epidermis. Several individual muscle cells can be seen distinctly.

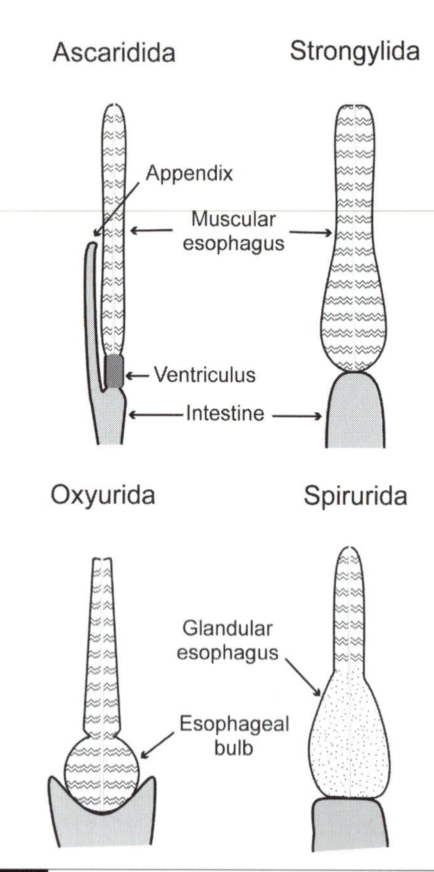

Fig. 5.7 A variety of esophageal patterns seen in some nematodes. Although the glands are most prominent in the spirurids, all nematodes have glands associated with the muscular portion of the esophagus.

mm in length. Their function is not clear but the cells may be multifunctional having a role in synthesis, storage, and secretion. The fluid-filled pseudocoel has two primary functions: (1) it is a primary component of the hydrostatic skeleton and thus is crucial for movement, and (2) in the absence of a circulatory system, it serves to transfer solutes, e.g., trehalose (Behm, 1997), from tissue to tissue.

Digestive system. Unlike any of the parasites discussed thus far, the Nematoda have a complete digestive system meaning that there are two openings. The nematode digestive system can be thought of as a tube within a tube, beginning anteriorly with a mouth (**stomodaeum**) and ending posteriorly with a subterminal **proctodaeum** (anus in females, **cloaca** in males). The mouth leads directly into a buccal capsule that may be simple (most nematodes) or complex (hookworms) and this, in turn, connects with the pharynx (sometimes called an esophagus). These structures are lined by cuticle that is continuous with the cuticle covering their body. Therefore, when they molt their body covering, they also must molt the lining of their foregut and hindgut. The pharynx is triradiate in cross-section (Fig. 5.6) and is often very muscular. It sucks food into the worm and propels it into, and along, the intestine. Variable numbers of glands are interspersed with the muscles of the pharynx and, in nematodes such as the spirurids, the pharynx is often divided into a muscular anterior portion and a glandular posterior portion. Pharyngeal glands are associated with the digestive system in that their secretions may be involved in such functions as anticoagulation (Pritchard, 1995). The morphology of the buccal capsule, the pharynx, and structures associated with the pharynx can be important taxonomically (Fig. 5.7).

The foregut terminates with a valve and is followed by a relatively unspecialized, uncuticular-

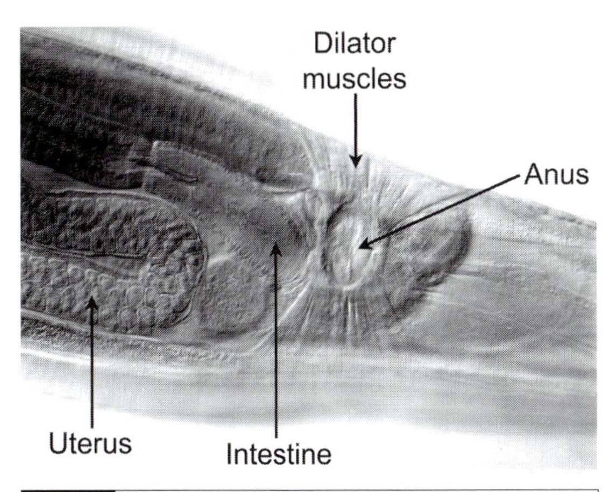

Fig. 5.8 Rectal dilator muscles in a female *Oesophagostomum*.

ized midgut composed of a single layer of columnar epithelial cells bearing microvilli. In some forms such as *Ancylostoma*, parts of the midgut are syncytial and in others, such as *Haemonchus*, the entire intestinal epithelium is syncytial. Being pseudocoelomates, nematodes lack any muscle associated with the midgut and propulsion of food is due to the muscular activity of their pharynx coupled with their high internal hydrostatic pressure and possibly movement of their body. The high internal pressure seems to obviate regional specialization along the gut; for example, under experimental conditions, *Ascaris lumbricoides* empties its gut about every 3 minutes (Crofton, 1966), the free-living nematode *Caenorhabditis elegans* empties its gut about every 45 seconds (Bird & Bird, 1991). That same high internal pressure requires some rather impressive muscles to open the anus/cloaca (Fig 5.8). The hindgut is essentially a cuticularized tube connecting the posterior intestine with the anus in females or the cloaca in males. Rectal glands, if present, open into the hindgut. The vas deferens opens into the male rectum hence the designation of the opening as a cloaca. Also in the posterior hindgut of males are pouches that contain, or house, the **spicules** and, if present, the **gubernaculum**.

Osmoregulatory/excretory system. Osmoregulatory function in nematodes appears correlated with their environment. Free-living forms (and this would include the larvae of parasitic forms) appear capable of moving significant amounts of water across their cuticle. Adult nematodes are often held to be **osmoconformers**, but Davey (1995) suggests that short-term osmoregulation may occur in some species.

Nematodes possess a secretory–excretory system composed of one or two **renette cells**, unique to the phylum. The secretory–excretory system is typically some variation on an H-shape with two longitudinal canals lying in the epidermis, connected transversely near the anterior end, and extending to the 'excretory pore' via an 'excretory canal' (Fig. 5.9). Variations include, for example, additional glandular cells or the extent of the distribution of the lateral canals. Once considered simply an excretory system, no clear demonstration of an excretory function has yet been made. Instead, available evidence suggests a primarily secretory function that may have a role in molting, pheromone secretion, digestion, or secretion of epicuticular glycoproteins.

Several nematodes have been shown to excrete organic acids across their cuticles and Sims *et al.* (1996) suggest that this may be common in nematodes. The intestine, however, probably serves as the primary means for removal of nitrogenous waste products, principally ammonia.

Reproductive system. Animal-parasitic nematodes are typically dioecious and often show sexual dimorphism. Generally, the female is larger than the male whereas the male tends to have a curved tail in contrast to the straight posterior end found in females. Sex determination is chromosomally based in some species. For example, Špakulová *et al.* (1995) note that sex determination in *Heterakis gallinarum* is based on an XX (female)/XO (male) system. Recently, Gemmill *et al.* (1997) suggested that the sexuality of some nematodes, i.e., the production of sexually reproducing adults, might be determined by the immune status of the host.

Males (Fig. 5.1A) typically have a single testis that is solid and thread-like. This continues as a seminal vesicle that becomes a vas deferens before terminating in the cloaca as an ejaculatory duct. Accessory structures often include spicules (single or paired), a gubernaculum, and, in the

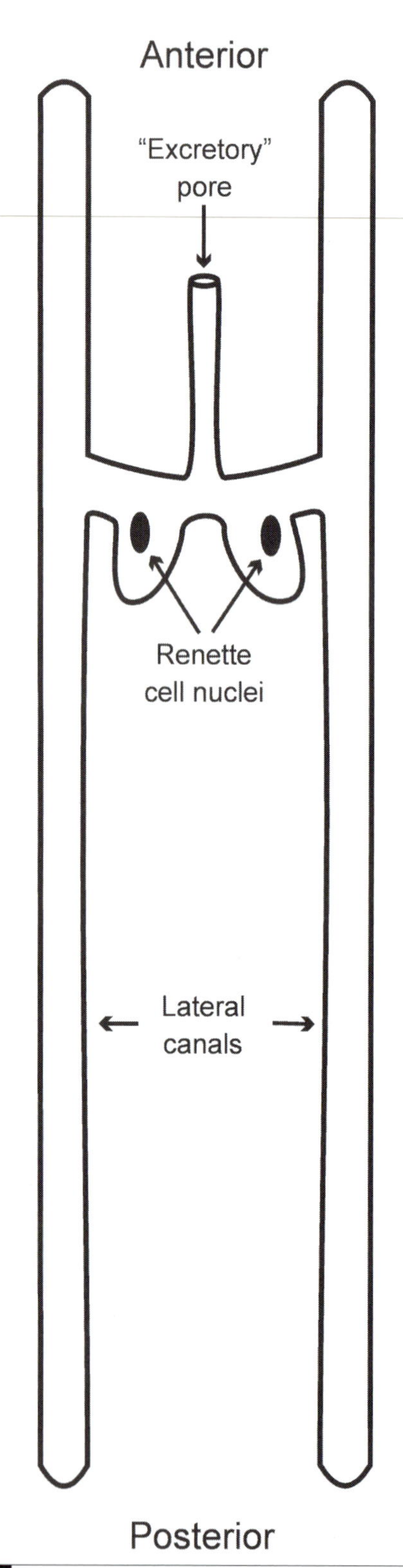

Fig. 5.9 A hypothetical renette showing an 'H' pattern. The 'excretory pore' is always found in the anterior of the worm. Lateral canals can be seen in cross-section in Fig. 5.6.

Dioctophymatida and Strongylida, copulatory bursae. Spicules are chitinized structures often surrounded by a sheath. They vary in number (one or two), length, and shape and are used to hold the female's vulva open (against the high hydrostatic pressure) while the non-flagellated sperm pass from male to female. The gubernaculum, if present, is a sclerotized portion of the rectum and it appears to act as a guide for the spicules as they pass through the cloaca. Bursae are, in a way, like holdfast organs except that they function solely to bind the male tightly to the female during copulation. The size and shape of the spicules and gubernaculum are extremely important taxonomic characters in distinguishing between various taxa of nematodes.

An interesting feature of male nematodes is that their sperm lack flagella or cilia. Movement, when it has been observed, is likened to amoeboid locomotion using temporary pseudopodia. Some crustaceans similarly lack flagellated sperm.

Unlike males, the female reproductive system (Fig. 5.1B) is often paired, a condition known as **didelphic**. Proximally, a thread-like ovary merges with a slightly larger oviduct and this grades into the uterus. The uterus is muscular, particularly at the distal ends where it may be so muscular that it forms a distinct structure known as the **ovijector**. Ovijectors from the two uteri unite to form a short vagina that opens to the exterior through a ventral (usually) vulva. An important, but structurally obscure, region is the junction of the oviduct and uterus. This area serves as a seminal receptacle and is the site where fertilization occurs. After fertilization, eggshell formation begins. The eggshell of most nematodes seems to be composed of three layers although a fourth has been recognized in some ascarids (Bruňanská, 1997). Eggs of some groups, e.g., the oxyurids and trichurids, have one or two opercula, thought to facilitate hatching. In the filarids, the females are **ovoviviparous** and give birth to live larvae known as microfilariae. Eggshells in these worms are often delicate, composed of a vitelline membrane and a thin, chitinous shell. Often, these microfilariae retain an egg membrane as a sheath and, in a sense, they represent highly developed embryos.

Nervous system and sensory input. A dominant structure in the nervous system of nematodes is the circumesophageal nerve ring. It is the best that the nematodes can put forth as a 'brain'. As its name implies, it is always located in the anterior portion of the body in conjunction with the pharynx. It is composed of few cells (four neural and four glial cells in *Ascaris*) and, arising anteriorly and posteriorly, is a series of nerves. Anteriorly directed nerves innervate the amphids and the cephalic papillae. Posteriorly directed are the prominent dorsal and ventral nerve cords along with less-prominent lateral nerves. There may be variable numbers of ganglia associated with these cords as well as commissures connecting the dorsal and ventral cords. The most posteriad ganglion in the ventral cord is called the preanal ganglion. It gives rise to additional anteriorly directed nerves that innervate the rectum and form what is called a posterior nerve ring. The posterior nerves also service the caudal papillae and phasmids (if present). In the free-living, and much-studied, *Caenorhabditis elegans*, a number of behavioral mutants have been identified and these appear to be associated with specific nerve cells.

The nervous system is tightly integrated with sensory input, i.e., in the analysis of the nematode's environment. As noted previously, sensory structures are mainly papillae and glands. Papillae have sensory endings that are modified cilia and whose function is probably tactile. Glands are either amphids, or phasmids, or both; their sensory endings are also cilia and their function is largely chemoreception and secretion. Where muscles are innervated, acetylcholine is thought to be the primary neurotransmitter. Various other neurotransmitters have been found in the nematodes, but their function remains obscure.

5.2.3 Nutrient uptake

Although we are restricting our discussion to nematodes parasitic in vertebrates, there exists a large variety in the way they acquire nutrients. As free-living larvae, they feed on bacteria and possibly other microorganisms. As 3rd stage larvae, enclosed within the 2nd stage molt, they are seemingly incapable of feeding although Ogbogu & Storey (1996) suggest that the 3rd stage larvae of *Litosomoides carinii* may feed by a transcuticular route. Adults parasitize almost all organs and tissues and there is a concomitant variety of foods available to them ranging from various tissues to food ingested by the host.

Food is taken in through the mouth, which may be relatively simple, as might be expected in those that graze, or quite complex as in those that attach and feed on body fluids. The nematode pharynx is composed of a syncytium of radial muscle. The strong musculature associated with the pharynx (Figs. 5.6, 5.7) is necessary to 'pump' food against the high internal pressure of the worm. It is also the pumping of the pharynx that helps propel the food through the remainder of the worm's gut. Like so many other structures in the nematodes, the shape of the pharynx is often of diagnostic value.

5.2.4 Metabolism

Given the diversity of nematodes and the habitats they occupy, metabolic challenges are great. For example, many adult nematodes live in oxygen-poor environments yet have eggs or larvae that occur in comparatively oxygen-rich environments. The eggs of many animal-parasitic nematodes that are shed in feces have been examined and all require oxygen for continued development. When the pO_2 is reduced, development slows; under anaerobic conditions, development ceases. Interestingly however, although development may be curtailed, the eggs (and even the infective larvae) of some species that infect the host by an oral route are resistant to oxygen deprivation, a possible hedge against poor environmental conditions. On the other hand, the larvae of some skin-penetrating species such as *Strongyloides* are very sensitive to a lack of oxygen. Little is known about the intermediary metabolism of nematode larvae. There are reports that glycolysis, the TCA cycle, and cytochrome systems may be present, and functional (Kita *et al.*, 1997). That some, such as *Strongyloides*, are so dependent on oxygen suggests that the glycolytic pathway is poorly developed in oxygen-sensitive larvae.

Adult nematodes basically have two options with regard to intermediary carbohydrate metabolism. The first is to process glucose solely through glycolysis. This is a catabolic process, independent of oxygen, and occurs in the cytosol. In energetic terms, this is extremely wasteful since glycolysis alone yields a net of only two ATP from a molecule of glucose. (Aerobically, when glucose is metabolized through glycolysis, the TCA cycle, and the electron transport system resulting in CO_2 and H_2O, 36 to 38 ATP are produced. For this complete degradation to occur pyruvate must enter the mitochondria and oxygen must be present.) However, many nematodes live in glucose-rich environments and we have already noted that they pass food through their guts rapidly. Thus, the 'wasteful' (this process extracts only about 2% of the energy present in glucose) use of glucose may pose no problems for them. The end product of glycolysis is pyruvate, this is reduced (and NAD^+ reoxidized) typically to lactate. Nematodes employing this form of catabolism are called homolactic fermenters.

An alternative is to develop different metabolic pathways, to become what are called CO_2 fixers. Following this route is similar to mammalian catabolism to the point of phosphoenolpyruvate. Instead of being converted to pyruvate, the phosphoenolpyruvate formed during glycolysis is carboxylated to oxaloacetate. The oxaloacetate is reduced to malate which then enters the mitochondria. In the mitochondria, malate is metabolized to succinate and pyruvate. This yields additional ATP beyond simple glycolysis. The end products are succinate, which is excreted as an organic acid, and pyruvate, which is rarely excreted. In addition, other organic acids such as lactic, proprionic and succinic may be excreted. Some nematodes have some of the enzymes characteristic of the TCA cycle and still others have elements of the cytochrome system. For more complete details of these processes, the student should consult Chapter 2.

Protein catabolism is not well known in the nematodes but ammonotelic, ureotelic, and even uricotelic excretion are known. Since ammonia is toxic, it would only be expected where suitable quantities of water were available. For example, *Ascaris lumbricoides* is predominantly ammonotelic; interestingly however, when water becomes limiting, it switches to ureotelic excretion.

Anabolism, or synthesis, is the other side of the story to catabolism. Anabolic pathways have received little study in the nematodes. Glycogen is the major energy reserve and it is synthesized from carbohydrates ingested during feeding. Typically, higher glycogen reserves are found in those nematodes that do not have direct access to the host's carbohydrate reserves, e.g., *Ascaris* sp., compared to those with more direct access, e.g., *Ancylostoma* sp. When provided with suitable substrates, e.g., acetate, glycine, and glucose, some nematodes can synthesize a wide variety of amino acids. Others have been shown to synthesize fatty acids.

5.2.5 Development

The study of nematode development has a long history. Indeed, the first elucidation of fertilization and meiosis in animals was provided by studies on ascarid nematodes. Nematode zygotes undergo deterministic cleavage. Many, but not all, nematode eggs are laid in the single or two-cell stage. Exceptions include eggs being laid with fully developed larvae or even ovoviviparity (whereby the female produces live offspring). The first cell division (resulting in two blastomeres) is important because one of the daughters will go on to produce only somatic tissue, the other will produce some somatic tissue and all germinal tissue. The blastomere whose fate is to produce only somatic tissue undergoes the unusual process of chromosome diminution. In this process, parts of the chromosomes fragment resulting in the loss of much genetic information. Chromosome diminution occurs several times so that, by the 64-cell stage, only two cells retain the complete genetic code. These two cells will produce the gametes for the next generation. All of the other cells, which may have lost up to 80% of their genetic coding through chromosome diminution, will produce all of the somatic tissue.

Embryogenesis progresses through a typical morula and blastula stage prior to gastrulation (accomplished through epiboly and invagination). Once organogenesis is complete, mitosis ceases and, with the rare exception of some intestinal, or epidermal cells, or both, there is nuclear constancy throughout life (eutely).

The eggshell of nematodes is important because it often acts as a barrier both within the host and also in the free-living environment. Typically, the nematode eggshell is divided into three layers produced by the egg itself: an inner lipid layer, a middle chitinous layer, and an outer vitelline layer. Added to these egg secretions may be one or two uterine layers (secreted by the uterus). The uterine layers often consist of pores, spaces, proteins, or some combination of these structures. Although the layers of nematode eggs are permeable to gases and lipid solvents, they are nonetheless quite resistant to many substances. For example, the eggs of *Ascaris* spp. can survive in solutions such as 9% sulfuric acid, 14% hydrochloric acid, and even 12% formalin (Bird & Bird, 1991). They are also resistant to environmental extremes. For example, a high concentration of trehalose (Behm, 1997) is thought to provide nematode eggs with the ability to survive environmental stresses such as cold and desiccation.

Excepting those that are ovoviviparous, and a very few others, most nematodes produce an egg that will be deposited in the host's feces. Once in the environment, and after embryonation (if necessary) the egg may hatch into a 1st stage larva (such as occurs in the strongyles) or it may remain as a resistant stage (such as in some ascarids). Often, the fecundity of egg-laying nematodes can be rather impressive – female *Ascaris* may produce 200 000 eggs per day.

Hatching in the environment may be spontaneous, but it seems hard to deny some environmental role. In any event, the actual process of hatching involves the breakdown of the lipid layer (possibly enzymatic) and chitinous layer (due to the secretion of chitinases). The secretion of other enzymes also helps to degrade the shell to the point where the larvae can eclose. A 1st stage larva hatches from the egg and is fully free-living, feeding on microorganisms. It will molt to become a 2nd stage larva that is also free-living and feeds on microorganisms. At some point, this larval form will molt but the cuticle remains as a sheath surrounding the 3rd stage larva. Although free-living, this stage is incapable of feeding and must survive on stored energy reserves, primarily lipids (Medica & Sukhdeo, 1997). This is usually the infective stage and, depending on the parasite's life-history strategy, this stage must penetrate or be eaten by the host. (Rarely, e.g., in some metastrongyles, the 1st stage larva will penetrate an intermediate host and develop to the infective stage within that host.)

Alternatively, an egg may remain as a resistant stage in the environment until it is ingested. Once ingested by an appropriate host, it will require some host-specific stimuli to induce hatching. For example, *Ascaris* eggs require a fairly specific temperature, redox potential, and CO_2 concentration to induce hatching. Hatching under such conditions may result in an infective 3rd stage larva for those parasites having a direct life cycle or it may result in a 1st stage larva, which will develop to the infective 3rd larval stage, in those forms that use an intermediate host.

Some infective 3rd stages penetrate the final host directly and lose their sheaths in the process; others, ingested directly or via an intermediate or paratenic host, must exsheath. Like hatching of ingested eggs, exsheathment often requires specific stimuli and, in fact, the stimuli to exsheath are often the same as those required for egg hatching.

Hypobiosis, sometimes called **arrest** or **developmental arrest**, is common to many nematodes and may involve either the egg or the sheathed 3rd stage infective larva. Functionally, it acts as a hedge against extinction since the appropriate stage is resistant to environmental vagaries or physiological degradation (Fetterer & Rhoads, 1996).

5.2.6 Life cycle

Perhaps no single group of related organisms on earth possess greater life history variability than do members of the Nematoda. In Figure 5.10, we attempt to show a broad range of possible nematode life cycles (but only those parasitic in vertebrates). Admittedly, the figure is rather complex but, even then, it still lacks some variants. For example, omitted are the metastrongyles which have 1st stage larvae that penetrate an intermediate host. Nonetheless, the illustration serves, we hope, to emphasize our point regarding diversity and complexity. We think it would be foolish to try and memorize the diagram, instead, we urge you to follow the routes of nematodes emphasized in lecture or lab, and contrast them against other possibilities.

Some important general considerations follow. Life cycles involving eggs only, whether ingested by an intermediate or final host, are unlikely to have any evolved adaptations to facilitate reaching the appropriate host. In contrast, many life cycles involve free-living larval stages and, often, these stages may show adaptations to facilitate

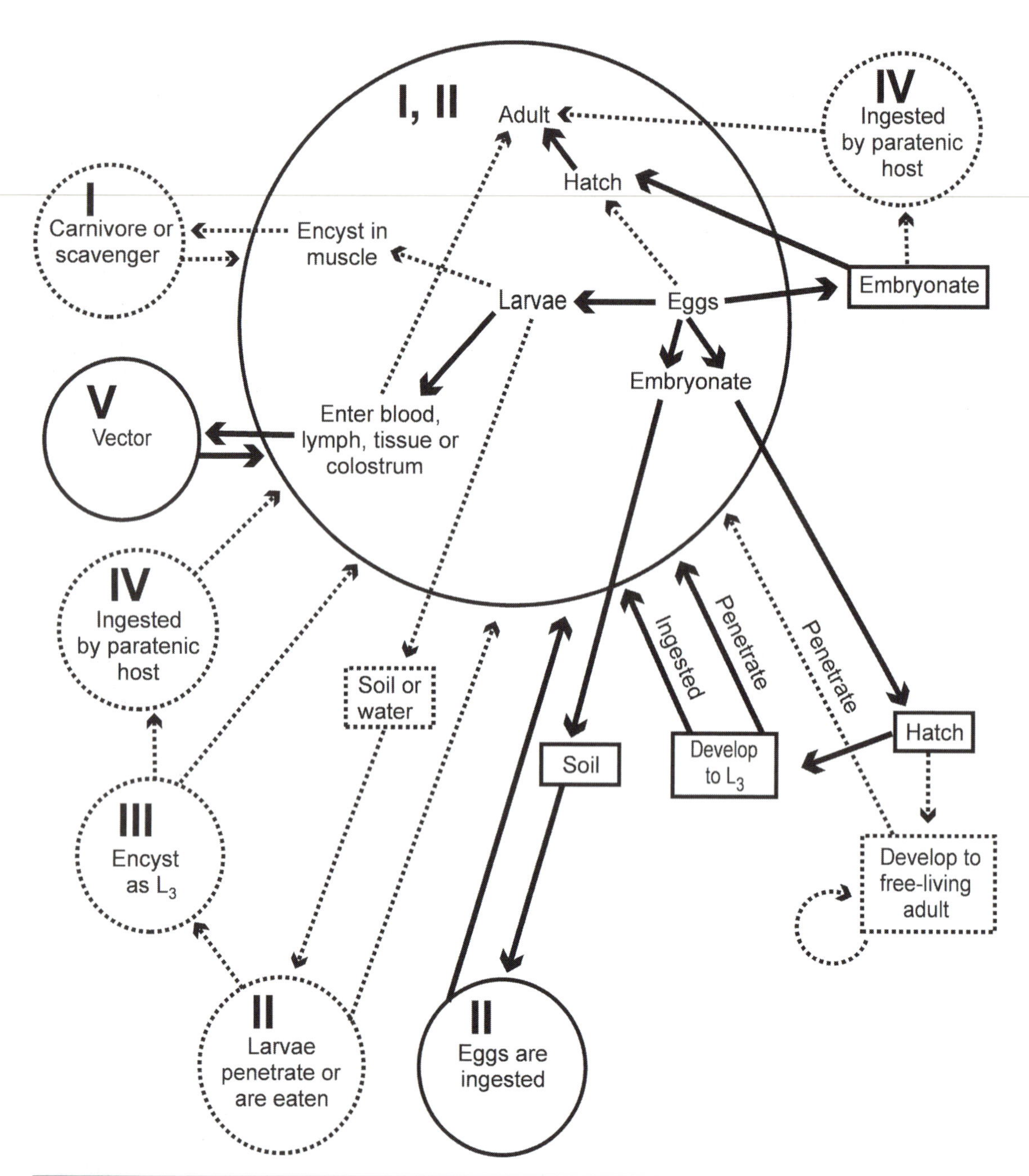

Fig. 5.10 Alternative pathways found in the life cycles of the parasitic Nematoda. Host types are represented by roman numerals (I = definitive host, II = first intermediate host, III = second intermediate host, IV = paratenic host, V = hematophagous or tissue-eating micropredator or 'vector'). Stages within circles are found within hosts and stages in rectangles are found in the environment. Despite the obvious complexity, there are really four basic patterns of interest (solid lines) and all begin with the fate of the egg. If the egg leaves the body unembryonated, it is ingested by the host and the larva hatches and passes through all larval stages within the final host. If the egg embryonates (at least partially) prior to leaving the definitive host, there are two possible fates. First, it can hatch in the external environment and develop to the infective 3rd stage larva. This may then penetrate the definitive host directly or be eaten by the definitive host. Second, an intermediate host may ingest the embryonated egg. In this case, the larvae develop to the infective stage in the intermediate host. Finally, the eggs may hatch inside the definitive host and larvae will be ingested by a micropredator; after a period of development, the larvae will be infective to a definitive host when the micropredator feeds again.

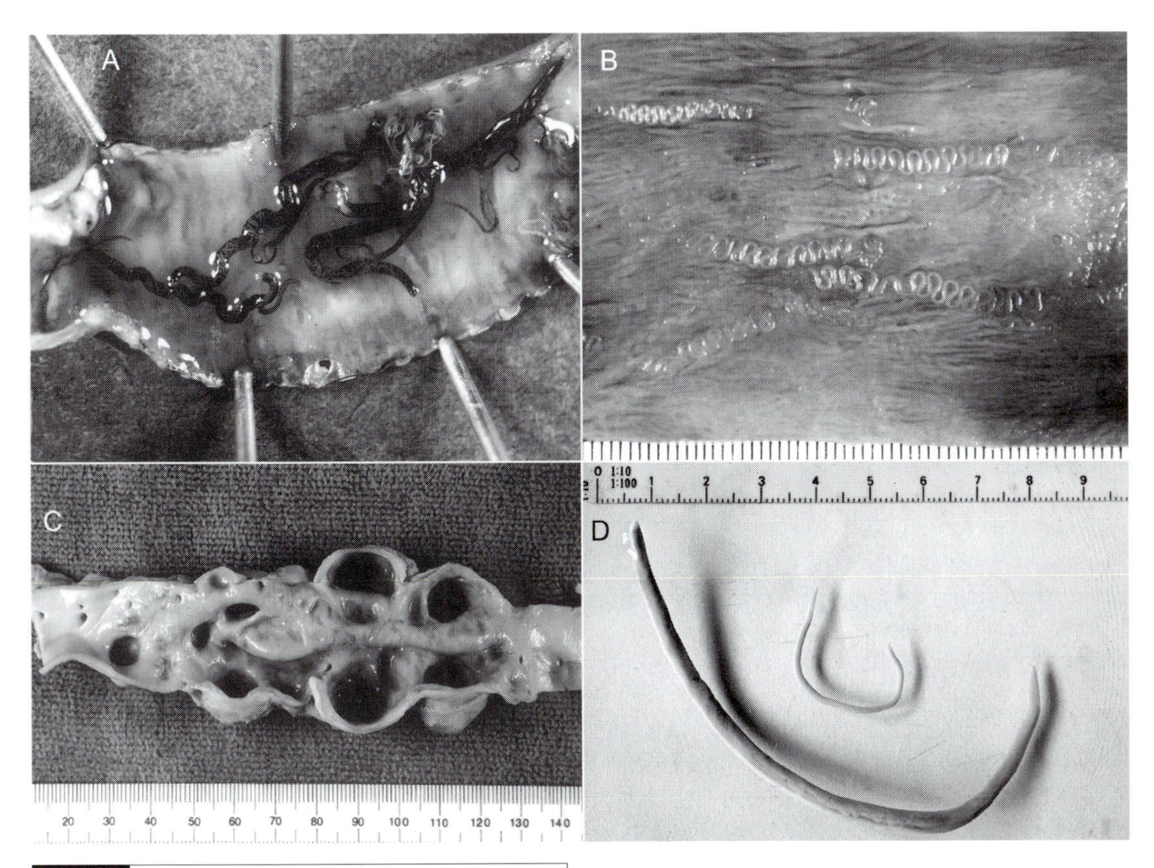

Fig. 5.11 Some non-intestinal habitats used by members of the Nematoda. (A) *Syngamus trachea* in the trachea of a pheasant *Phasianus colchicus*; (B) *Gongylonema pulchrum* in the esophagus of a white-tailed deer *Odocoileus virginianus*; (C) lesions in the dorsal aorta of a coyote *Canis latrans* produced by *Spirocerca lupi*; (D) female and larva of *Anisakis physeteris* from the stomach of a sperm whale *Physeter catodon*. ((A) Photograph courtesy of Bill Samuel, University of Alberta; (B, C) photographs courtesy of Danny B. Pence, Texas Tech. University; (D) Photograph courtesy of Toni Raga, Universidad de Valencia.)

transmission. For example, the infective stages of many strongyles exhibit various behavioral taxes that have them migrate to the tips of the blades of grass ensuring they are more likely to be ingested by their herbivorous definitive host. Even when the infective stages are not free-living, as in many filarids, the infective stages may show migrations between the visceral and peripheral blood in response to the feeding habits of the appropriate vector.

Some nematodes, such as *Trichinella*, use the same individual as both intermediate and definitive host. As such, they must involve carnivores or scavengers and essentially go from final host to final host (the two circles with 'I' in Fig. 5.10). Others, ostensibly, can persist in the same host e.g., pinworms, depicted completely within the large circle in Fig. 5.10.

5.3 Diversity

Considering all of the nematodes, e.g., those free-living, those parasitic in plants, those parasitic in invertebrates, and those parasitic in vertebrates, they must be regarded as the most diverse creatures on earth. We would even argue that the focus of this chapter, the nematodes parasitic in vertebrates, might deserve similar consideration. All vertebrates studied, regardless of their habitat, seem to have nematodes. Furthermore, unlike some groups studied thus far, e.g., the Eucestoda, or next, e.g., the Acanthocephala, where the adults are pretty much restricted to the small intestine, nematodes exploit just about any tissue in the vertebrate body (Fig. 5.11). Also unlike many other parasites, intermediate hosts (when they occur) can range from molluscs, to arthropods, to vertebrates. Later, we provide a 'working' classification of the nematodes parasitic in animals. Below, we discuss in more detail some examples of those parasitic in vertebrates. Because of their remarkable diversity, it seems logical to discuss them systematically.

5.3.1 The enoplean nematodes

The Trichurida and Dioctophymatida are the main parasitic forms found as adults in vertebrates. The trichurids, which have several representatives found commonly in humans, will be considered first.

TRICHURIDA

The major characteristic of the trichurids is the possession of a unique esophageal structure known as a **stichosome**. It is composed of large, individual glandular cells surrounding the thin-walled esophagus (Fig. 5.12). Other characteristics are bi-operculate eggs, a single gonad in both genders, and males having one, or no, spicule or spicule sheath.

Trichuris is worldwide in distribution but with a tendency to occur in the moist tropics (perhaps because of its direct life cycle?). They are found in the large intestine and cecum of many terrestrial mammals. Their common name, whipworms, derives from their rather distinctive body shape in which the anterior end is very much thread-like compared to a more robust posterior (Fig. 5.2A). The thread-like anterior end induces the formation of an epithelial syncytium by the host and this effectively locks the parasite in place. *Trichuris trichiura* is the species found in humans and other primates. In heavy infections, it can cause a prolapse of the rectum and can ultimately cause death.

Capillaria is also worldwide in distribution and can parasitize all classes of vertebrates. Most systematists feel that there are probably several genera contained within '*Capillaria*' but, for convenience, we will use the single genus. Unlike *Trichuris*, they are found in various parts of the gut, in urinary tracts, and in such organs as the liver and spleen. This is a large group with about 300 described species. A variety of life cycles is found. Some appear to require an intermediate host, e.g., *C. caudinflata*, a potentially serious

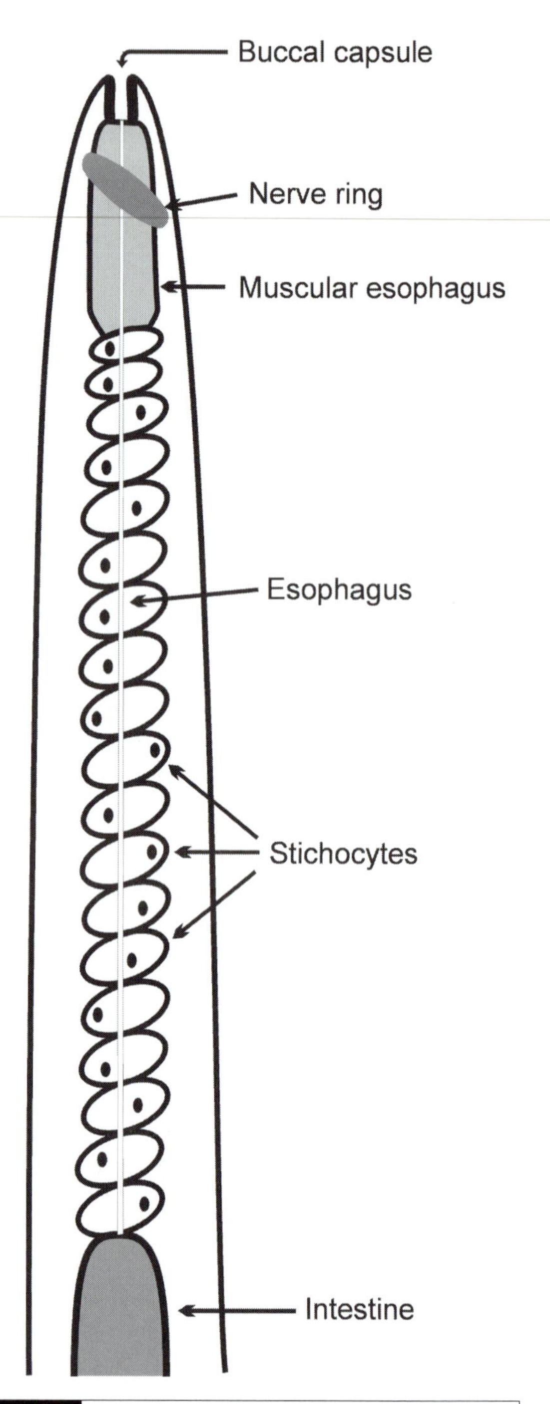

Fig. 5.12 A stichosome, composed of individual stichocyte cells, is characteristic of the Trichurida. The stichosome can often be quite long; the thread-like portion of *Trichuris* (Fig. 5.2A) is mostly stichosome.

Box 5.1 | Do you like sushi?

In the Far East, it is called 'sushi' or 'shashimi'; in South America it is called 'ceviche' ('cebiche'). We don't care what you call it, it's raw fish! *Raw!* (And, don't laugh people in the Balkans, Scandinavia, and residents of other northern climates – 'delicacies' such as 'pickled herring' and 'smoked fish' are almost as bad!) Depending on your ethnic circumstances, on your palate's fancy, or perhaps on a challenge from a friend, you may find these sorts of foods on the delectable side. The price one might pay for partaking in such a gastronomic repast is, however, at the very least, abdominal discomfort. It can be much worse, severe abdominal pain, vomiting and diarrhea. Why?

The answer to this question resides with the fact that a large number of digeneans, at least one cestode, and several nematodes use fish as their intermediate hosts and these particular worms can develop, partially or completely, in humans. Among the digeneans, e.g., *Heterophyes heterophyes, Metagonimus yokogawai, Watsonius watsoni,* etc., the metacercariae occur encysted in the flesh of the fish and the adults are very small, enteric forms that, in small numbers, are relatively innocuous. But, in large doses, they can be highly pathogenic, producing some of the more extreme clinical manifestations mentioned above.

At least one cestode, *Diphyllobothrium* sp., can also be acquired by humans through the consumption of raw, or undercooked, fish. Dr. Robert Desowitz, in his delightful book titled *New Guinea Tapeworms and Jewish Grandmothers*, describes how unsuspecting Jewish grandmothers prepare gefilte fish, which is essentially minced fish, pressed into balls, then cooked until it is 'just right'. According to Desowitz, cooking it until it is 'just right' is the problem because, to determine when it is ready to be served, it must be sampled. This, of course, opens the possibility of ingesting, accidentally, an infective plerocercoid of *Diphyllobothrium* and acquiring a 10-m adult tapeworm in the process. So much for gefilte fish!

Amongst the nematodes are other reasons for questioning the consumption of raw, or undercooked, fish. Within the Anisakidae are several species known to cause anisakiasis, a disease in humans that is especially difficult to diagnose since the symptoms it produces resemble closely those associated with the common ulcer. Anisakiasis is caused by *Anisakis simplex, Pseudoterranova decipiens,* and, probably, *Contracaecum osculatum*. Humans are not normal hosts for any of these parasites but, by eating raw, or undercooked, fish, they intervene in the normal food web involving fishes and marine mammals. If humans consumes the flesh of a fish with infective larvae, the larvae excyst and attach, most frequently, to the stomach. Within 24 hours of acquiring the parasite, the infected person will suffer from intense gastric pain. Vomiting and diarrhea may accompany this. If the parasites attach to the stomach wall, they can be removed by endoscopy but, if they are in the small intestine, intrusive surgery may be needed to remove the worms.

As fads come and go, so too will sushi bars, but history suggests that raw, salted, smoked, or pickled fish will continue to be consumed in ethnic cuisine. Sadly, the consumption of most of these parasites by humans is a dead-end for the parasites. Sushi anyone?

pathogen of gallinaceous birds. Others, e.g., *C. hepatica*, produce two types of eggs; one form passes into the environment and develops into infective stages, the other form is retained in the liver and must be ingested by a carnivore or scavenger. *Capillaria philippinensis*, described originally from humans, uses fishes as intermediate hosts and the definitive hosts are probably piscivorous birds (Anderson, 1992). This species, which is known to cause mortality in humans, is thus probably acquired accidentally from consuming raw, or undercooked, fish (see Box 5.1).

Trichinella is composed of two morphological species and, probably, at least four biochemical species. In total, the genus is cosmopolitan infecting a variety of homeotherms. These are very small worms with an interesting life cycle. Adults live in the small intestine, mate (following which the males usually die) and the female produces 1st stage larvae that penetrate veins or lymphatics and are subsequently carried to striated muscle. There, the larvae penetrate muscle cells and survive as intracellular parasites (Fig. 5.13; see also Fig. 2.5). Infected muscles cells become altered and develop into nurse cells that nourish and protect the larvae. Ultimately, when an appropriate carnivore or scavenger ingests the infected muscle, the larvae are digested out of the nurse cell in the small intestine. There, they invade the intestinal mucosa, undergo four molts, and emerge into the intestine as adults. The entire process takes as little as 36 hours. *Trichinella* is a serious pathogen of humans, particularly in northern climates. However, the most serious outbreaks, prior to the publication of this book, were in France and Italy where at least three species of *Trichinella* were involved and the source of the infection was horsemeat (de Carneri *et al.*, 1989). (Given what we think we know of the diet of horses, this kind of makes you wonder how the horses became infected!)

DIOCTOPHYMATIDA

A major characteristic of dioctophymatids is the production of a very thick-shelled egg with one or two polar opercula. They are also among the largest of the nematodes. All use oligochaetes as intermediate hosts and some use paratenic hosts. None are of significant concern to humans. *Dioctophyme renale* normally infects the right kidney (the reason for the predilection for the right kidney is unknown) of mink (*Mustela vison*) and effectively turns the kidney into an empty capsule. Worms in hosts other than *Mustela* spp. frequently seem to get lost and are often found in the body cavity. Eggs are shed in the urine. This species can cause pathology in mustelids and canids, both wild and domestic (Fig. 5.14).

5.3.2 The rhabditean nematodes

Most nematodes parasitic in vertebrates are found among the orders comprising the subclass Rhabditea (formerly called the Phasmidea or Secernentea), although some free-living, plant-parasitic and invertebrate-parasitic forms are also included. The nematodes in this group are ubiquitous in distribution and include most of the nematodes of human concern.

RHABDITIDA

Among the Rhabditida are the lung parasites of amphibians and some reptiles belonging to the genus *Rhabdias*. Worms in this genus differ from most nematodes in being **protandrous hermaphrodites**, i.e., an individual is a functional male first and then becomes a functional female. In some members of the genus, eggs are coughed up from the respiratory system, are swallowed and pass in the host feces. These develop through the normal sequence of parasitic nematode larvae resulting in infective 3rd stage larvae. These will orally infect (reptiles), or penetrate the skin (amphibians) of a suitable definitive host. This developmental pattern is called **homogonic**, e.g., there is no alternation in the life cycle with a free-living adult generation. An alternative, **heterogonic** development results in a free-living, dioecious adult. Subsequent to mating, the free-living female produces a few eggs that develop into larvae and which hatch and consume her internal organs (**matricidal endotoky**). The larvae, which break through her cuticle, are infective to the vertebrate. Parasitic individuals are haploid whereas free-living individuals are diploid.

Members of *Strongyloides* are worldwide in distribution and can be found as parasites in the gut mucosa of most vertebrates. Like *Rhabdias*, *Strongyloides* spp. have interesting life cycles involv-

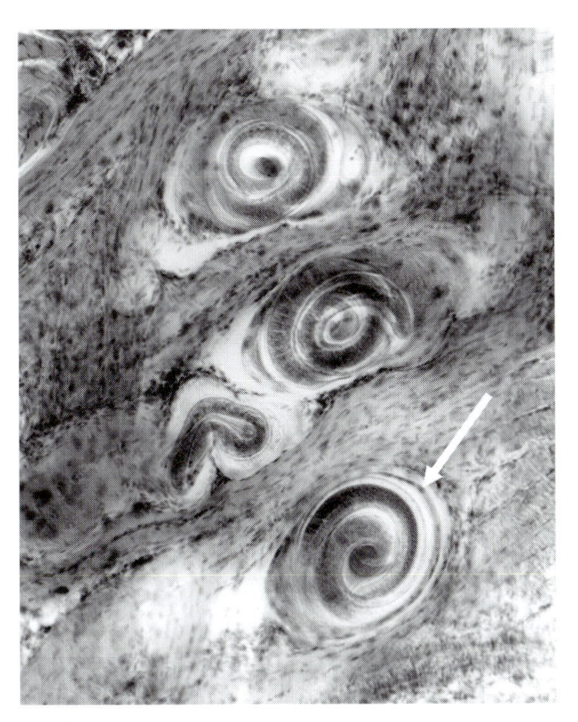

Fig. 5.13 Larvae of the nematode *Trichinella spiralis* encysted in the muscles of an experimentally infected mouse. The arrow points to a portion of the cyst wall. (See also Fig. 2.5.)

ing both homogonic and heterogonic development. The life cycles differ in some important respects however. Parasitic females produce genotypically female eggs by mitotic parthenogenesis (but both male and female larvae result). Eggs (or 1st stage larvae) passed in the feces develop through a sequence of molts to yield infective larvae that penetrate a suitable host. Alternatively, the larvae will develop into free-living larvae of both genders. These will then molt several times to become free-living adults. To produce eggs, the free-living female must copulate with a free-living male; the pronuclei do not fuse however and eggs are produced by meiotic parthenogenesis. These eggs result in infective larvae that will penetrate the host and mature as females. At least in *S. stercoralis*, eggs can hatch in the intestine, develop to the infective stage, and reinfect the same host (autoinfection). Further, both transplacental and transmammary infections have been proposed for some members of this genus. *Strongyloides stercoralis* can be a serious parasite in humans; most of the pathology is associated with the migration of

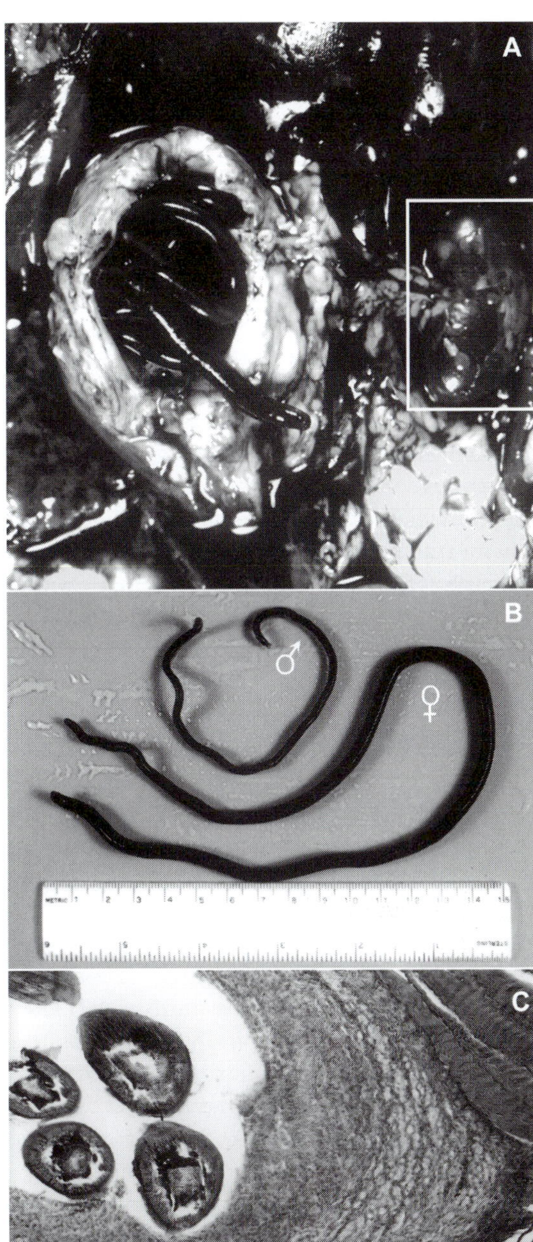

Fig. 5.14 The nematode *Dioctophyme renale*. (A) *D. renale* in the right kidney of an experimentally infected ferret; the box surrounds the normal-sized, uninfected left kidney; (B) these are large and robust worms with females (bottom) being about 30 cm in length and males about half that size; (C) encysted larvae in pumpkinseed (*Lepomis gibbosus*), a fish paratenic host. (Photographs courtesy of Lena Measures, Institute Maurice-Lamontagne de Pêches et Océans.)

larvae through the lungs or the destruction of intestinal tissue by the females.

Homogonic/heterogonic development may be an important adaptation to changing ecological conditions. In the absence of appropriate hosts, the parasite can persist in the environment awaiting the arrival of suitable hosts.

STRONGYLIDA

These are commonly called the bursate nematodes referring to the conspicuous copulatory bursa found on the posterior end of the males (Figs. 5.4B, C, 5.15). Most have a direct life cycle although some species may use paratenic hosts. Exceptions are the metastrongyles, or lungworms, which often use an intermediate host. When nematodes of medical or veterinary importance are considered, the Strongylida figure prominently. Most of the serious nematodes of domestic animals belong to this group and, collectively, enormous amounts of money and other resources are spent attempting to find effective, long-lasting cures.

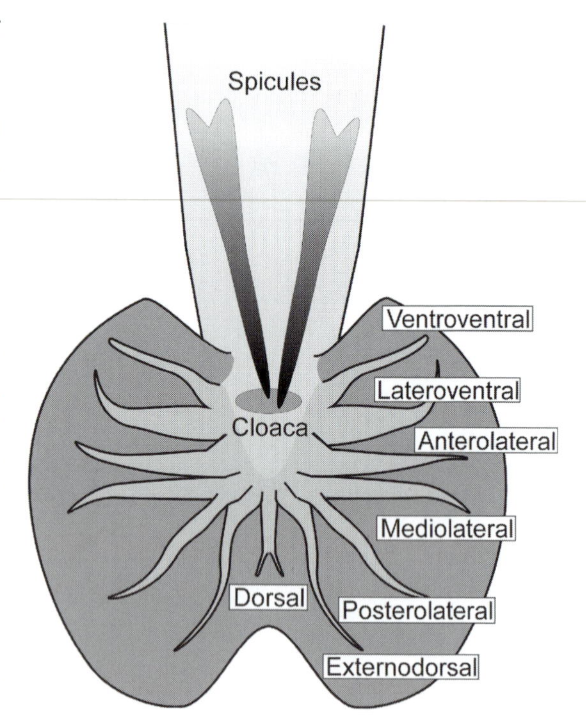

Fig. 5.15 Ventral view of an hypothetical bursa on a male strongyle nematode showing a possible arrangement of the fleshy rays. The position and relationship of the rays is of diagnostic importance.

A number of strongyle nematodes are capable of arrest where infective, unsheathed larvae survive in the tissues for long periods of time. In female hosts, these larvae may be passed to the offspring via transmammary or transplacental routes.

This is a very large group of nematodes and we will consider only three representatives. Members of the Ancylostomatidae typically infect their host by direct penetration across the host's skin and exhibit a rather extensive migration within the host's body. Members of the Trichostrongylidae are typically ingested as infective 3rd stage larvae and do not show significant migration through tissues. Members of the Protostrongylidae typically use a gastropod as an intermediate host.

Ancylostomatidae. As adults, hookworms are found in the small intestine of a wide variety of mammals. The life cycle is direct with thin-shelled eggs being passed in the host's feces. Females produce up to about 25 000 eggs per day. Under suitable environmental conditions, the eggs hatch and pass through a sequence of free-living larval stages resulting in 3rd stage infective larvae. The larvae lose their sheaths when penetrating the mammal and make their way to the lymphatic system. Ultimately they arrive in the lungs where they penetrate into the alveoli and develop to the 4th stage. They then migrate up the trachea, are coughed up and swallowed to the intestine. A final molt results in adults in the small intestine where they attach to the mucosa with their rather impressive buccal capsule (Fig. 5.3A). Unless the infection is massive, little pathology seems to be associated with the migration of the larvae through the body. Instead, the significant pathology is attributed to the blood feeding of the adults; each worm can ingest 0.2 to 0.3 ml every day. Blood loss may result in iron-deficiency anemia. Obviously, the extent of blood loss is a function of the severity of infection.

The main hookworms of humans are *Ancylostoma duodenale* and *Necator americanus*. Approximately 1.2 billion people, living primarily in warm, humid climates, are infected with one, or both, of these two species. The World Health Organization (1994) estimates that 44 million pregnant women are infected and, because there is no consensus on teratogenic effects that drugs may have on the fetus, they are mostly left untreated. Since women in developing countries spend approximately half their lives either pregnant or lactating, there can be highly significant anemia in these, as yet untreatable, individuals. In 1995, an estimated 65000 people died from hookworm infections.

Ecological conditions may be very important for the transmission of some hookworms. Lilley *et al.* (1997) attribute a significant increase in human hookworm infection in Haiti to deforestation along a riverbank. They postulate that the deforestation resulted in flooding and silting providing an ideal microhabitat for the development of hookworm larvae.

Canine (*Ancylostoma caninum, A. braziliense*) and feline (*A. braziliense, A. tubaeformae*) hookworms can also infect humans. This results in a condition called cutaneous larval migrans. What appears to happen is that the infective larvae penetrate the epidermal layer of the human but cannot proceed further. They migrate through the tissue causing extreme itching until they are finally killed by the human's immune response. Although not pathogenic, the severe itching may provoke scratching of the infected area thus allowing for possible bacterial invasion.

Although the common route of infection by hookworms is through the skin, some show the ability to infect via paratenic hosts, oral routes, via transmammary routes and, possibly, via transplacental routes. *Uncinaria lucasi*, parasitic in northern fur seals (*Callorhinus ursinus*) and northern sea lions (*Eumetopias jubata*), is an excellent example of a rather unique mode of transmission. The hosts spend most of their lives at sea, coming on shore at rookeries on islands in the Bering Sea only during a short breeding season (June to August). When adults arrive at the rookeries they may have 3rd stage larvae in their blubber and other tissues, but they never have adults in their intestines. Previously infected or not, 3rd stage larvae that have survived the winter in the soil of the rookery will penetrate the host's flippers. These too are distributed to various parts of the body accumulating in the blubber and, in females, the mammary glands as well. Newly born pups are infected when nursing from their infected mothers (see also Box 5.2). After migrating through the lungs and being coughed up and swallowed, these larvae will develop to adults in the small intestine of the pups and will produce eggs that are passed in the feces. The adult worms are short-lived, however, and all will die in several months. The 3rd stage larvae that have survived in the rookeries over winter will also infect the pups (as they did in the adults). These larvae, also as in the adult hosts, will remain in host tissues. When the hosts return to the sea, none (newly born or adult) have intestinal infections, but most have tissue phases. In short, the source of perpetuation of the nematode is from the short-lived intestinal phases found only in the pups and derived from their mother's colostrum.

Trichostrongylidae. This is an extremely diverse family of nematodes being found in all vertebrates except fishes; they reach their peak diversity in mammals. Although mainly parasitic in the gastrointestinal tract, they may occupy other diverse habitats ranging from lungs to bile ducts. The normal mode of infection is by the ingestion of infective 3rd stage larvae. Like other strongyles, however, other routes are known, e.g., cutaneous, transplacental, transmammary. The most typical life-cycle pattern found in those that parasitize herbivorous mammals is for the adult female to deposit eggs that pass out with the host's feces. Under appropriate environmental conditions, the egg hatches to a 1st stage larva. This and the subsequent 2nd stage are free-living and feed primarily on bacteria. The infective 3rd stage larva is also free-living but cannot feed due to the sheath covering its body. These infective stages often migrate to the tips of ground-dwelling plants where they are more likely to be ingested by grazing herbivores. Once ingested, cues from the host, e.g., CO_2 tension and pH, cause the larvae to secrete an exsheathing fluid, e.g., leucine aminopeptidase, which acts on the sheath very close to the anterior end. The result is the popping-off of the anterior cap, liberating the worm. Typically, these exsheathed larvae

Box 5.2 I bequeath to you . . .

A much-heralded feature of most parasites is the complexity of their life cycles. Concomitant with complexity is the notion that most parasite life cycles are risky and thus the probability of transmission or colonization is low. Some authors even suggest that the high fecundity of most parasites can be attributed to the enormous risk involved in reaching maturity. The question arises: 'Can parasites do anything to enhance the probability of reaching the correct host and subsequently producing offspring?'.

The answer is an emphatic 'Yes'. In Chapter 13 of this book, you will learn that parasites can do much to ensure perpetuation. For example, they can alter the behavior of their hosts (both intermediate and definitive) facilitating transfer, some exploit food webs, some larval forms move into areas where the appropriate hosts are most likely to occur, some become progenic, and some are autoinfective. These are but a few ways in which parasites enhance their reproductive chances. Perhaps the surest way to ensure perpetuation is to never enter the free-living environment.

Common to some nematodes (mostly hookworms, ascarids, and metastrongyles), but found also in alariid digeneans, is the process of transmammary or transplacental colonization. Obviously, given the routes of infection, these processes are restricted to parasites of mammals. These are the most direct of all possible parasitic life cycles. In transplacental transfer, infective stages pass directly into the offspring while those offspring are still *in utero*. The result is that the offspring are actually born infected with the parasite. In transmammary transfer, the offspring are born uninfected, but acquire infective stages while nursing. Certainly neither risk of environmental vagaries nor risk of finding an intermediate (or several intermediate) host(s) here! From a parasite's perspective, one can't but think that this is the 'way to go' but, anthropomorphically, and from a host's perspective, one can't but think: 'Gee, thanks mom!'.

Alternatives in the life cycle of these worms lead to some interesting evolutionary questions. Consider hookworms for example. They have what are traditionally considered direct life cycles meaning that no intermediate host is involved. Basically, infective stages in the environment penetrate the skin or are ingested. In either case however, they must come in contact with the appropriate host. Why retain this risky pattern when direct transfer from mother to offspring is so certain and positive? Perhaps it is a hedge ensuring genetic mixing of populations. Perhaps it makes male hosts meaningful. This latter may need some explanation. Assuming a 50:50 sex ratio, if the parasite can only be transferred from mother to offspring, then all infected male offspring are biological sinks meaning that one-half of the potential habitats for the worm suprapopulation can be colonized but cannot participate in transmission. If nature is conservative, and this scenario is correct, then, through evolutionary time, male hosts should not become infected.

It is easy to see why transplacental or transmammary transmission might occur in some types of hosts, particularly those in aquatic environments. In this chapter, we explain the life cycle of *Uncinaria lucasi* parasitic in fur seals. With such a limited time on land, transmammary transfer is probably of paramount importance in ensuring colonization of new hosts.

invade the mucosa, develop to the 4th stage, and return to the lumen where they undergo a final molt to become adults.

Haemonchus contortus is regarded as perhaps the most pathogenic parasite found in a domesticated animal. It is primarily a parasite of sheep, but has been reported to survive and reproduce in other ruminants. It is a parasite with a worldwide distribution, being found wherever sheep occur. Throughout its distribution, it is a source of major economic loss. As an adult, it inhabits the abomasum (the 'true stomach') and, like the hookworms, the major pathology is attributed to the voracious blood feeding habits of the adults.

Free-living stages of *H. contortus* seem to have fairly specific environmental requirements for survival. Perhaps for this reason, arrested development (early in the 4th stage) is very common (up to 100%) in this species. The arrested larvae may act as a hedge against poor environmental conditions (and thus equate to an 'overwintering' population) in temperate climates. It is, at present, unclear what stimulates the arrest. Both immunological and environmental stimuli have been suggested; perhaps both are involved.

Heavily infected hosts often die as a direct consequence of *H. contortus*. Those that survive often develop a resistance known as a 'self-cure'. This phenomenon, in which the extant population of worms is spontaneously expelled, is thought to be elicited by a newly ingested population of infective larvae. Sometimes, all worms (resident adults and newly ingested larvae) are expelled, sometimes only the adults are lost and the larvae re-establish the population, and sometimes the adults are lost but the larvae are retained in the arrested state. The ability to elicit the 'self-cure' phenomenon may be related to the strain of sheep, the strain of the worms, or both.

Due to the significant economic impact of *H. contortus*, much veterinary research is afforded them. At the time of writing this book, one of the major goals facing researchers was that of trying to develop genetically resistant strains of sheep. The impetus for such research seems to be the development, in some strains of worms, of resistance to many drugs, principally, Ivermectin®, the drug of choice.

Protostrongylidae. Unlike the other strongyles discussed, the protostrongyles are one of the groups having an indirect life cycle. They are a small group, found only in mammals (and then primarily in ruminants) and most live associated with the lungs of their hosts. In a typical life cycle, unembryonated eggs are shed into the lungs where they develop to 1st stage larvae. These larvae make their way up the respiratory system, are swallowed and pass out of the host in the feces. Once in the environment, the 1st stage larvae penetrate the head-foot of a terrestrial mollusc in which they will develop to the infective 3rd stage larvae. The life cycle is completed when a ruminant accidentally ingests the infected mollusc as it grazes.

Protostrongyles are important to wildlife the world over although some of the lungworms of domestic animals, e.g., *Metastrongylus apri* and *M. pudendotectus* in swine, may have an economic impact on humans. *Protostrongylus rushi* and *P. stilesi* commonly co-occur in Rocky Mountain bighorn sheep (*Ovis canadensis*); more rarely they are reported from mountain goats (*Oreamnos americanus*). As adults, *P. rushi* is found primarily in the bronchi, while *P. stilesi* is found in the parenchyma of the lungs. It is the latter species that is most often associated with morbidity and mortality. In heavy infections, a worm-induced pneumonia may develop, causing significant morbidity, sometimes mortality.

Protostrongylus spp. seem to initiate an immune response in infected adult hosts resulting in newly ingested larvae being sequestered in the lungs. In a pregnant ewe, these larvae can make their way across the placenta and lodge in the liver of the fetus. At lamb birth, the larvae migrate to the lungs where they develop to adults. Eggs deposited by the females, and the larvae that hatch from them, produce a substantial inflammatory response ultimately resulting in granulomas. Various species of bacteria and viruses invade this diseased tissue resulting in suppurative bronchopneumonia. Often, this results in death of the young lamb (Fig. 5.16A, B).

Bighorn sheep are gregarious animals and the herd has a strong tendency to reuse the same habitat when they alternate between winter and summer ranges. This results in a continued

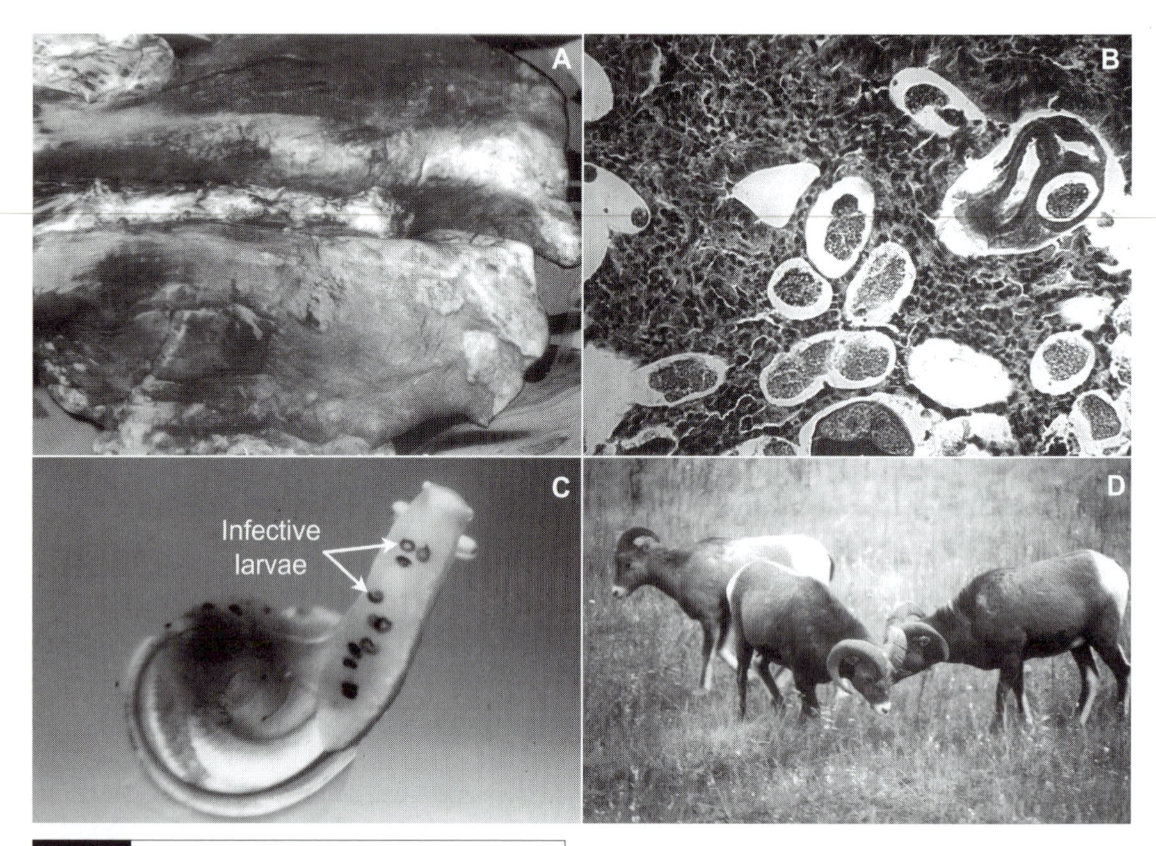

Fig. 5.16 (A) Lungs of bighorn sheep infected with the nematode *Protostrongylus stilesi*. Although the worms are not visible, inflammation of the lungs is evident. (B) Peribronchiole inflammation associated with adult and eggs of *Protostrongylus stilesi* in the lung of a bighorn sheep. (C) Infective larvae of *Protostrongylus* in the head-foot of a terrestrial snail. (D) Bighorn sheep (*Ovis canadensis*) grazing near a natural salt lick near Jasper National Park, Alberta. ((A, B) Photographs courtesy of Bill Samuel, University of Alberta; (C) photograph courtesy of Rob Kralka.)

seeding of those habitats with 1st stage larvae and therefore it is expected that the prevalence of infected snails would be high, increasing significantly the probability of transmission. Further, there may be foci of very high prevalence associated with natural salt licks, areas frequented by bighorns (Fig. 5.16C, D).

OXYURIDA

All oxyurids have a direct life cycle and live in the posterior intestine of their hosts. This is the only group of nematodes having species that mature in vertebrates and invertebrates (arthropods). Oxyurids infect all vertebrates although they are not particularly common in fishes. A prominent feature is a large esophageal bulb (Figs. 5.7, 5.17). The oxyurids are of little importance as agents of morbidity or mortality but, because of their life cycles, they are very interesting biologically. A typical life cycle would have a female crawling out of the anus (usually nocturnally) and depositing eggs in the perianal region of the host. These eggs develop (two molts within the egg) rapidly (time measured in hours). During the course of grooming, the eggs are ingested, hatch in the small intestine, undergo two additional molts and migrate to the large intestine, appendix, or colon where they reside as adults. It has been suggested that retroinfection may occur whereby 3rd stage larvae hatch from the eggs in the perianal region and then crawl back through the anus to continue development. This has not been demonstrated conclusively.

A reproductive adaptation seen in some oxyurids is **poecilogyny**, a process where two different types of females are produced. One type of female produces thin-shelled eggs containing fully devel-

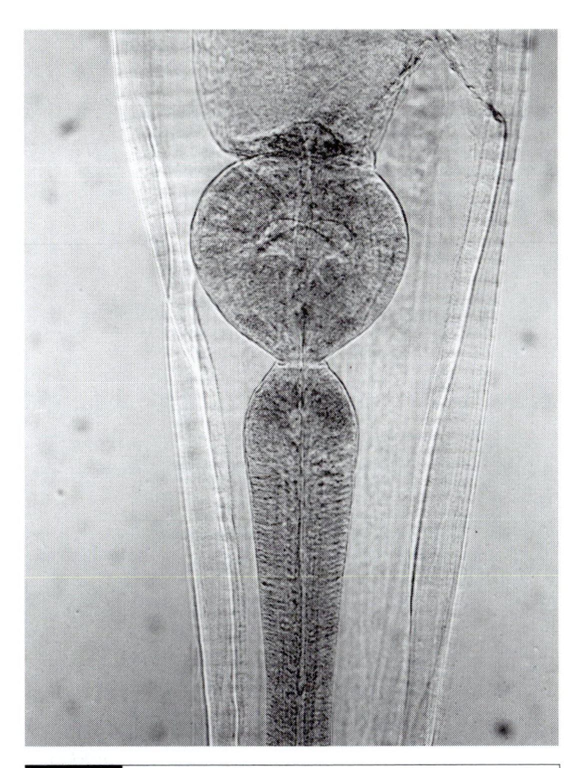

Fig. 5.17 Esophageal bulb of *Wellcomia*, a structure characteristic of oxyurid nematodes.

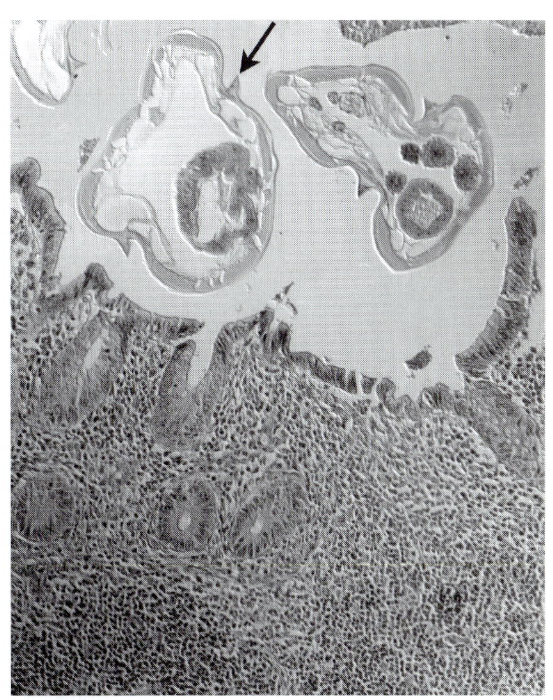

Fig. 5.18 Female *Enterobius vermicularis* in a human appendix. What appear to be spines (arrow) is nothing more than a cross-section through the lateral alae of the nematode.

oped larvae. These may play a role in autoinfection. The other type of female produces thick-shelled, unembryonated eggs that pass out with the feces and embryonate in the environment. These play a role in colonization of new hosts. An elaboration on this theme is found in *Gyrinicola batrachiensis*. In this species, female worms are didelphic, whereby one branch of the reproductive system produces the thin-shelled eggs (autoinfective) while the other branch produces thick-shelled, unembryonated eggs that pass into the environment, embryonate and play a role in colonization.

Another adaptation in the life cycle of oxyurids is **haplodiploidy**. Unfertilized eggs give rise to haploid males and fertilized eggs give rise to diploid females. This process, rare in nematodes, is rather common in many other invertebrate animals.

Oxyuridae. *Enterobius vermicularis* is perhaps the best known of the pinworms as, in temperate climates, it is cosmopolitan in humans. Whereas many parasites of humans are indicative of poor socioeconomic conditions, this cannot be said for pinworms. When females crawl out of the anus to deposit eggs, they cause intense itching. Frequently, especially in young children, this leads to scratching and the ultimate transference of the eggs to the mouth. This maintains the parasite in the children. Other members of the family, or schoolmates of the infected child, can become infected by inhalation. Soiled clothes from children may contain many eggs which, being very small and light, readily become airborne. Since a female can produce up to 16000 eggs, it is not difficult to see how this parasite is adapted for group transmission. It is a rare family indeed that has school-age children and never had an experience with pinworm! Other than the discomfort of itching, and perhaps embarrassment of being infected with 'worms', pinworms are of little significance to human health. Sometimes, large numbers of worms may infect the appendix causing a verminous appendicitis (Fig. 5.18). Interestingly, in this 'high tech' world of ours, diagnosis of pinworms in humans remains 'low tech'. Basically, a piece of transparent tape is applied in the perianal region and, when removed, it is then placed on a microscope slide. Diagnosis is made when either eggs, or the worms themselves, are observed microscopically to be stuck to the tape.

ASCARIDIDA

This is a very large and diverse order of nematodes found in animals the world over. Adults are parasitic in the digestive tract (mostly the small intestine) in all vertebrate groups. As nematodes go, the majority of ascarids are large, stout, and they generally have three prominent lips. Some have direct life cycles; however, using an intermediate or paratenic (or both) host is pretty common. In fact, some noted authorities (e.g., Anderson, 1992) suggest that the use of an intermediate host in the ascarids is the norm. Intermediate hosts may be either invertebrates or vertebrates. Some ascarids exhibit transplacental or transmammary transmission and, although progenesis (see Chapter 4) has not been reported, some nematodes, particularly among the ascarids, show precocious development of larvae in the intermediate host. Below, we discuss some of the more well known families of the Ascaridida and, when warranted, a couple of species within some families.

Heterakidae. The members of this family infect reptiles, amphibians and birds and are characterized by having a large preanal sucker possessing a cuticularized rim (Fig. 5.4D). *Heterakis gallinarum* is a ubiquitous parasite living in the ceca of poultry (primarily chickens and turkeys). Eggs are passed in the feces and may remain viable for very long periods of time. When ingested by the definitive host, the eggs hatch in the small intestine and the larvae migrate to the ceca where they become mature adults. It is also common for paratenic hosts (annelids and arthropods) to be included in the life cycle. This is an important parasite in agriculture because it transmits the flagellated protozoan *Histomonas meleagridis*, the etiological agent of a disease called 'blackhead'. Blackhead is a very serious pathogen to poultry, particularly in turkey poults. The histomonads are transmitted from infected bird to uninfected bird in the larva found within the egg that is shed in the feces.

Ascarididae. Members of this family are mainly parasites in the small intestine of terrestrial vertebrates and their transmission may be direct or, more commonly, indirect. Other than being comparatively large and robust, they have few distinguishing features. *Ascaris lumbricoides*, a serious parasite in humans, has been studied extensively. This species is one of the major parasites reported in humans, with an estimated worldwide prevalence >1 billion. It is most common in (but is certainly not restricted to) developing countries, where 100% of the inhabitants in some villages may be infected. It frequently co-occurs with *Trichuris trichiura*, perhaps because of similar environmental requirements for development outside the host. The life cycle is direct, but certainly not simple. Embryonated eggs are ingested and hatch as sheathed, 3rd stage larvae. These larvae then penetrate the intestine and enter the hepatic portal system prior to passing on to the liver. In the liver, they shed the sheath. These larvae migrate to the lungs where they molt to the 4th stage. Larvae then migrate up the trachea, are coughed up and swallowed and the larvae molt again in the small intestine to become adults. This rather bizarre intra-host migration (ending up where they began) is used to suggest that *Ascaris* originally had an intermediate host. The migratory phase in the human (or the pig in the case of the sibling species *Ascaris suum*) replaces that which had originally occurred in the intermediate host. *Toxascaris leonina*, another member of the family, bears further evidence for the loss of a required intermediate host. Dogs and cats can become infected directly by ingesting embryonated eggs. In such cases, there is a tissue phase before the worms enter the small intestine and mature. If 3rd stage larvae are ingested from eating an infected rodent, the tissue phase is eliminated.

The migratory phase is one source of pathology in ascariasis. Sometimes, larvae seem to get lost when leaving the lungs. They may migrate into the right side of the heart and be distributed by the arterial system to other tissues and organs causing local host reactions. Even when the larvae follow the traditional route, pathology associated with their molt can occur in the lungs. When large numbers of larvae penetrate the lungs, secondary infection and clogging of air passages can

Fig. 5.19 A single infrapopulation of the nematode *Ascaris suum*. (Photograph courtesy of Russ Hobbs, © Murdoch University.)

occur. When the larvae molt in the lungs, the shed cuticle and residual molting hormone can induce local hypersensitivity resulting in pneumonia. Large numbers of adults in the intestine can also be pathogenic by causing intestinal blockage (Figs. 5.19, 5.20A, B). In addition, when crowded, or sometimes in response to anthelminthic treatment, the adult worms may wander to the bile or pancreatic ducts causing blockage. In more extreme, but less pathogenic wanderings, adults may leave the host by mouth, nose, or anus. (Imagine watching a 15 to 40 cm long worm crawling out of a colleague's nose!) Even environmental changes may induce migration. Figure 5.21 shows an Atlantic cod (*Gadus morhua*) with *Hysterothylacium aduncum* (an anisakid) crawling out of virtually every body orifice. This appears to happen commonly when fishermen do not place fish on ice soon after capture.

Human ascariasis responds well to treatment but it is difficult to eradicate. Females shed up to 200 000 eggs per day (some authors, e.g., Olsen [1974], say >2 million eggs per day but this seems a bit too much!) and can live for up to a year. The result is >70 million eggs per female during her life. Added to this is the remarkable resistance that ascarid eggs exhibit. Embryonated eggs can survive up to 10 years provided they are not in direct sunlight and do not get too warm. Evidence suggests that in China alone, ascarids consume annually the equivalent of >127 tonnes of rice.

According to estimates from the World Health Organization (WHO), 60 000 people died from ascariasis in 1995. Recently, *Ascaris* was used in an attempted murder. Four students at McGill University in Quebec, Canada were each given between 300 000 and 400 000 eggs by a disgruntled room-mate. All four became critically ill but ultimately recovered.

Toxocara canis and *Toxocara cati* (found mainly in canids and felids, respectively) also have direct life cycles with pronounced tissue migrations. Both are capable of using paratenic hosts and *T. canis*, at least, is capable of transplacental and transmammary transmission. Because their life cycles are direct and they undergo extensive tissue migrations, they can be a concern to humans. When humans ingest eggs, the larvae begin their typical migration across the small intestine. They do not proceed through normal development however, and wander through the body causing visceral larval migrans. Most commonly, these wandering larvae infect the liver, but any organ, including the brain, can become infected. The degree of pathology in humans is related to the number of larvae present and their ultimate location. Ultimately, the larvae will be killed by a strong host reaction.

Anisakidae. Two common anisakids of marine mammals are *Pseudoterranova decipiens* and *Anisakis simplex*. The life cycles are indirect, involving marine crustaceans as intermediate hosts and, very frequently, a variety of fishes and cephalopods as paratenic hosts. In *P. decipiens*, at least, for which the definitive hosts are pinnipeds, paratenesis can be extreme. Microcrustaceans, e.g., copepods, ingest 2nd stage larvae. These infected intermediate hosts may then be ingested by fish or by macrocrustaceans. In either case, the infected fish or macrocrustaceans may be ingested by an additional fish prior to that fish being eaten by a seal or sea lion. Interestingly, *P. decipiens* is a precocious nematode. Most of the growth takes place in the 3rd larval stage and, when the mammal becomes infected, maturation of the worms is extremely rapid.

Anisakids are important to humans on a number of counts: economically, politically, and pathologically. First, paratenic hosts are often fishes important for human consumption. Because the larvae in *P. decipiens* are so large and visible, it is unlikely that humans would consume them. However, it is also unlikely that people

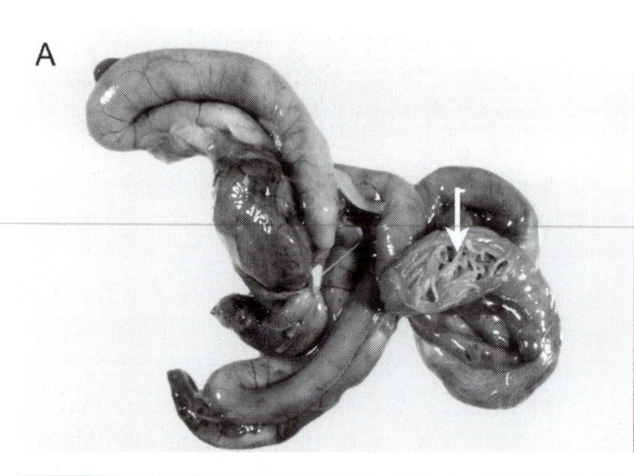

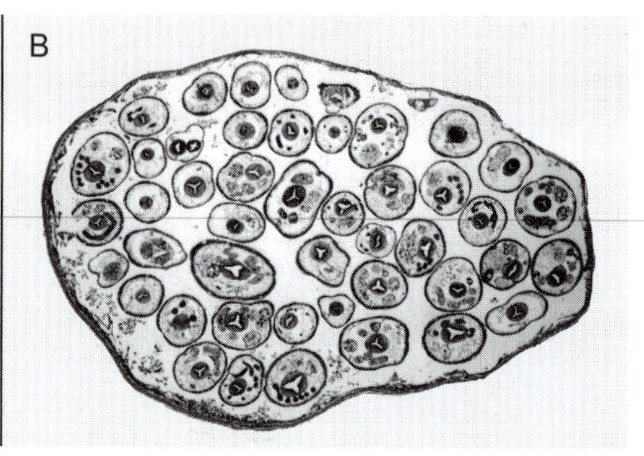

Fig. 5.20 (A) The nematode *Ascaridia platyceri* (arrow) in the small intestine of a pet regent parrot *Polytelis anthopeplus*; (B) cross-section of the same parrot's intestine showing a large number of nematodes *in situ*. (Photographs courtesy of Russ Hobbs, © Murdoch University.)

would buy a fillet of cod (or other desirable fish) if it were riddled with large (20–50 mm), conspicuous larvae (Fig. 5.22). Because so many fishes of commercial importance are infected, it is estimated that the increased processing cost attributable to *P. decipiens* infections (due to manual removal of the larva) in Atlantic Canada in 1984 was approximately $30 million (Malouf, 1986). A second consideration is political. Many fishermen in the North Atlantic know that seals and sea lions carry the worms that ultimately can make their catch of fish less profitable. Clearly, the fishermen would like to 'do away' with the seals. Also, since fishing is primarily a summertime activity, many fishermen support their families historically by hunting seals during winter. To the fishermen, this was an ideal situation since they had a source of income from the seal pelts during the winter and they were simultaneously reducing (or so they thought) the source of parasites that would ultimately end up in lost profits from fishing. (That they ever really reduced *P. decipiens* infections in fish has never been demonstrated!) With the public becoming ever more conscientious about conservation and cruelty to animals, clearly the stage is set for political mayhem. Finally, with an increase in the consumption of raw fish (see Box 5.1), the likelihood of humans becoming infected is on the rise. Because of its large size, this is unlikely in *P. decipiens*. However, *A. simplex* larvae are much smaller, much less conspicuous and often are found in herring (*Alosa*). Eating raw, undercooked, or pickled herring, infected with these larvae, will result in anisakiasis in humans. Annually, >1000 human cases of anisakiasis are reported. The worms do not mature in humans and the pathology is mostly associated with the larvae penetrating the small intestine or stomach. The degree of discomfort is likely a consequence of the number of larvae penetrating. In rare cases, humans have died from anisakiasis due to severe peritonitis.

SPIRURIDA

In some ways, the spirurids are pretty conservative nematodes. For example, with respect to their life cycles, virtually all but the gnathostomes produce eggs containing fully developed 1st stage larvae requiring an arthropod as an intermediate host. (Gnathostomes produce undeveloped eggs.) In other ways, the spirurids are the most diverse of the nematodes. For example, it is among the spirurids that one finds the greatest disparity in size, the greatest diversity of body ornamentation and the greatest diversity of microhabitats inhabited by the adult parasites. It is among the spirurids that species identified for eradication by the WHO occur. In fact, one of them, *Dracunculus medinensis*, may be extinct shortly after this book is published. We will discuss several of the more prominent families.

Dracunculidae. Commonly called the 'guinea worm', *Dracunculus medinensis* was earmarked by the WHO for eradication by the year 2000. (There goes 'Biodiversity'!) It is the ease of breaking this parasite's life cycle that leads to such a lofty goal. The parasite, in which the female may be almost a meter in length and contain >1 million larvae,

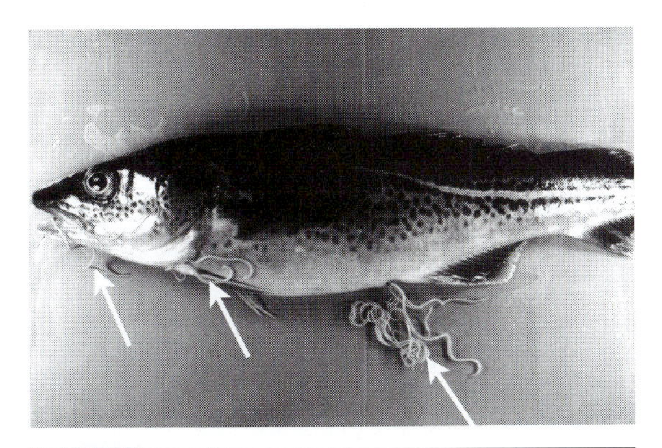

Fig. 5.21 *Hysterothylacium aduncum* adults crawling out of body orifices from an Atlantic cod *Gadus morhua* that had not been placed on ice after capture. (Photograph courtesy of Lena Measures, Institute Maurice-Lamontagne de Pêches et Océans.)

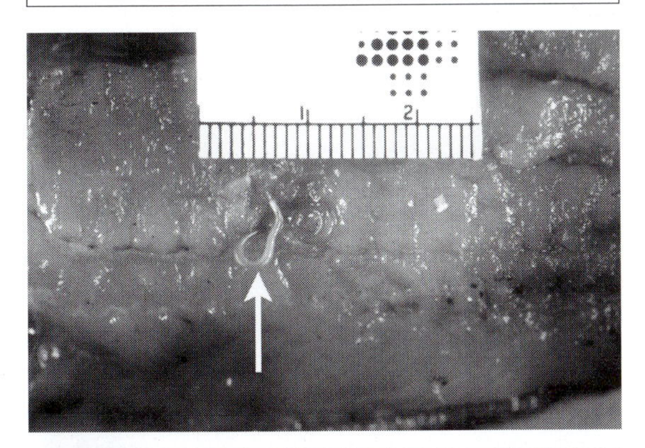

Fig. 5.22 Infective larvae of the nematode *Pseudoterranova decipiens* in a fillet of Atlantic cod *Gadus morhua*. (Photograph courtesy of Lena Measures, Institute Maurice-Lamontagne de Pêches et Océans.)

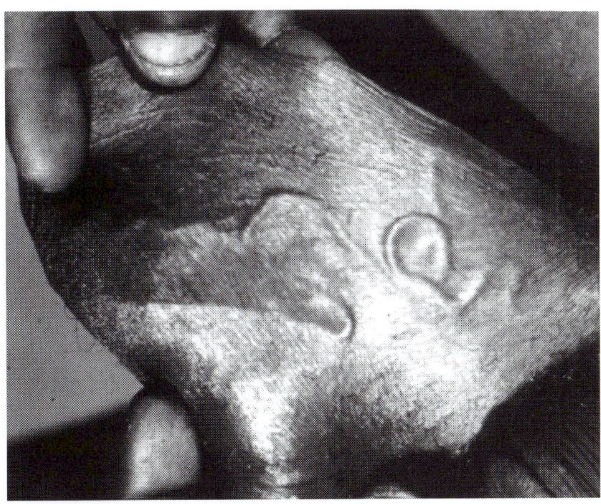

Fig. 5.23 A female *Dracunculus medinensis* in the subcutaneous tissue of a scrotum. (Photograph courtesy of the Armed Forces Institute of Pathology.)

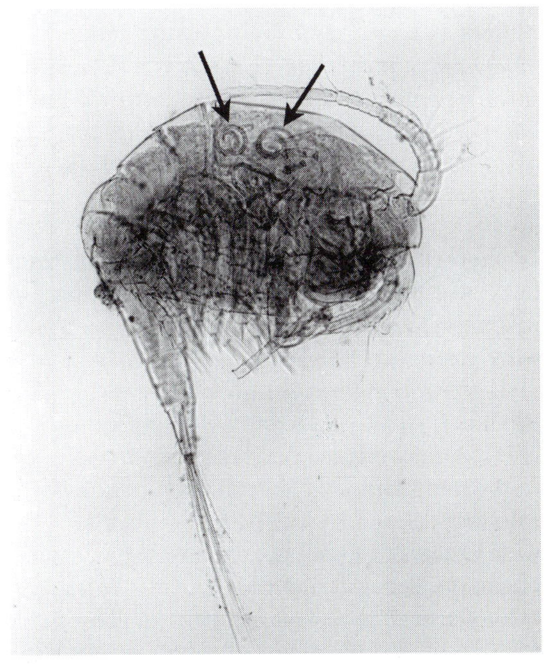

Fig. 5.24 Infective 3rd stage larvae (arrows) of the nematode *Dracunculus medinensis* in the hemocoel of *Cyclops* sp. (Photograph courtesy of the Armed Forces Institute of Pathology.)

lives under the skin. It can infect humans, other primates, and dogs and cats. It has been reported from humans primarily in sub-Saharan Africa, the Egyptian subcontinent and much of Southeast Asia. A predominant environmental feature in the areas where it is endemic is that clean drinking water tends to be very scarce. Gravid females, most frequently found under the skin of appendages or even a scrotum (Fig. 5.23), provoke the formation of a blister. When the blister is immersed in water, it bursts releasing large quantities of larvae. The larvae move vigorously in the water column and their movement tends to attract the intermediate host, a copepod. The copepod ingests larvae that subsequently penetrate into the hemocoel of the arthropod and develop to the infective 3rd stage (Fig. 5.24). The human becomes infected by drinking water containing infected copepods.

Although the migration of the worm is known to cause much pain and discomfort, it is the penetration of the skin, by the gravid female, that causes unbearable pain. The result, depending on the location of penetration, is partial or complete disability for up to several months. Once the worm sheds all her larvae, she will die and must be removed carefully (Fig. 5.25) or the ulcer will often become the source of a bacterial infection. A further complication is when the female's body is ruptured during removal as this too often leads to a bacterial infection.

Prevention is both inexpensive and easy. Filtering drinking water with fine-meshed cloth (about 0.15 mm mesh) will trap the copepods. The only way to become infected is through oral ingestion of the infected copepod. Obviously, if they have been removed by filtration, they cannot be ingested. At the beginning of the 1980s, there were an estimated 10 to 15 million cases of dracunculiasis; in 1986, there were an estimated 3.6 million cases. In 1997, only slightly more than 58 000 cases were reported to the WHO, of which >99% were in Africa. At the time of writing of this book, only Yemen is known to be endemic outside of Africa. Clearly, this looks like a 'win' for the WHO.

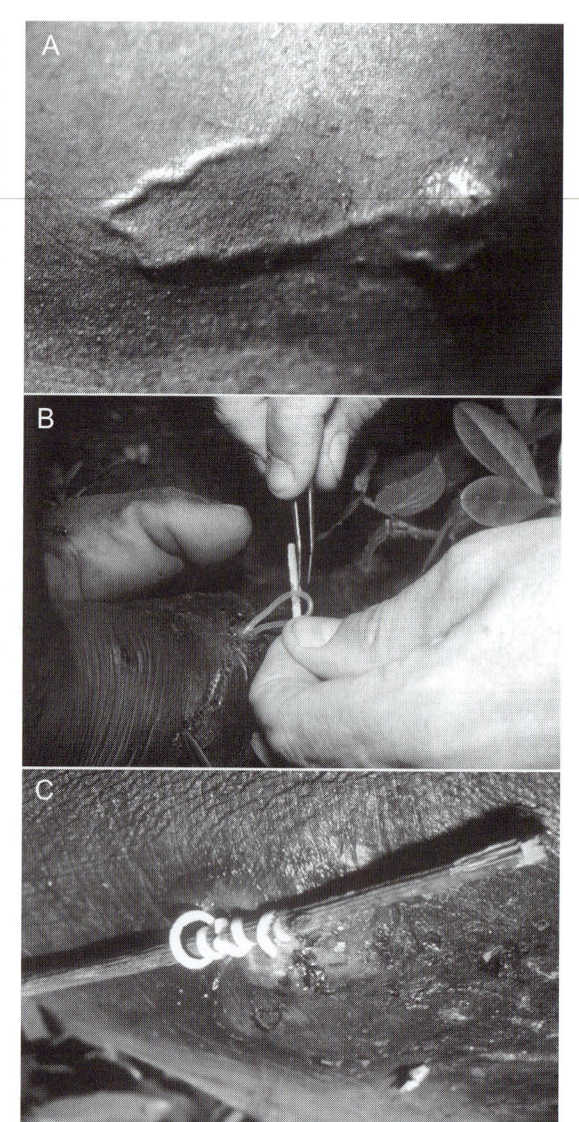

Fig. 5.25 Old-fashioned but effective – removing an adult female *Dracunculus medinensis*. Once a blister is formed (A) the posterior of the worm can be grasped (tweezers help!) (B) and wound onto a stick (C). Remembering that she can be up to a meter in length, and that if her body is ruptured there will be excruciating pain, this procedure is tedious and slow! It can actually take several weeks to remove an entire worm. ((A, C) Photographs courtesy of Jean-Phillipe Chippaux, Institut de Recherche pour le Développement, Paris, France; (B) photograph courtesy of Daniel Heuclin, Vaux en Couhé, Couhé Verac, France.)

Onchocercidae. *Dirofilaria immitis*, or heartworm, has been reported from a large number of mammals and is a serious pathogen in canids the world over (Fig. 5.26A, B). The adults are found mainly in the right side of the heart, sometimes in the vena cava or pulmonary artery. Microfilariae released from the female are found in the peripheral and visceral circulatory system (Fig 5.27). These 1st stage larvae are ingested by a wide variety of mosquitoes. Within the mosquito, the larvae undergo two molts and develop to the infective stage in the Malpighian tubules. These migrate to the salivary glands and are available for inoculation when the mosquito again feeds. The precise route taken by the infective larvae in the mammalian host is unclear. Infected dogs have an unmistakable 'cough' and typically suffer from pulmonary insufficiency. The disease is often fatal and prevention of infection by systemic filaricides (which kill microfilariae) is the best option.

Lymphatic filariasis is caused by two of the most disfiguring nematodes parasitic in humans: *Brugia malayi* (Brugian filariasis) and *Wuchereria*

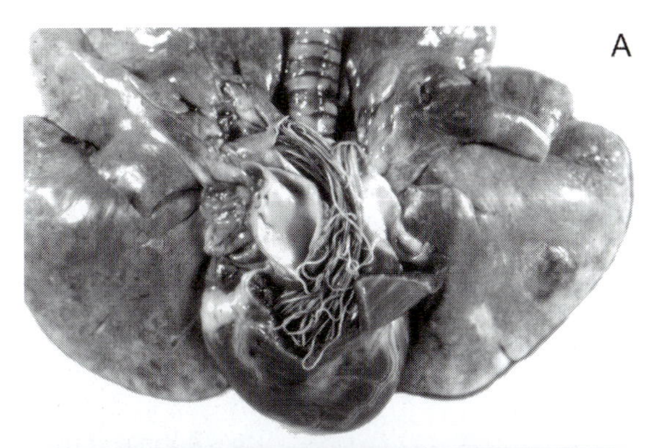

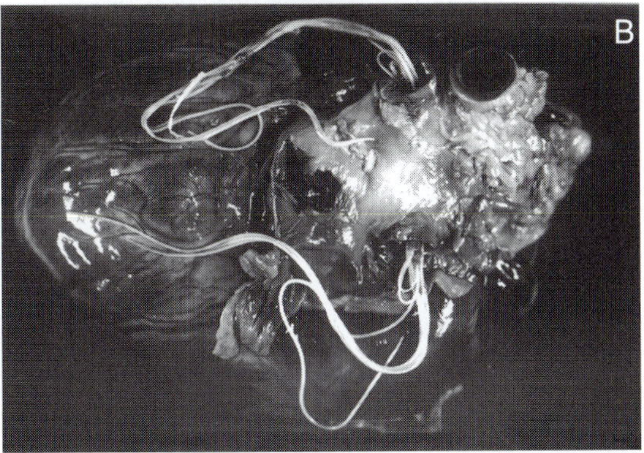

Fig. 5.26 (A) The heartworm nematode *Dirofilaria immitis* in the heart of a dog *Canis familiaris* from Australia; (B) *Dirofilaria immitis* in the heart of a red wolf *Canis rufus* from the southeastern USA. ((A) Photograph courtesy of Russ Hobbs, ©Murdoch University; (B) photograph courtesy of Danny B. Pence, Texas Tech. University.)

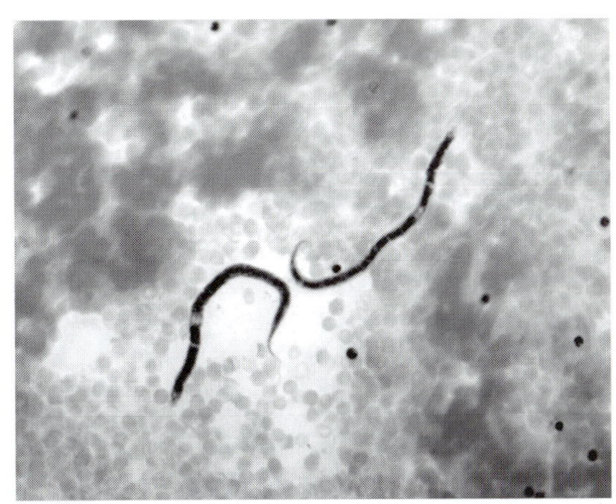

Fig. 5.27 Microfilariae of *Dirofilaria immitis* from the peripheral circulation of a dog.

bancrofti (Bancroftian filariasis). It is estimated that >120 million people are infected and that 20% of the world's population is at risk. As adults they occur in the lymphatic system causing blockage that often leads to a condition known as elephantiasis. Although generally similar, there are differences (other than morphological) between the two species. Brugian filariasis is most common in Southeast Asia, mainly causes elephantiasis in the legs (Fig. 5.28), and is found in a very diverse array of wild and domestic animals. Bancroftian filariasis is tropical and global in distribution, often causes elephantiasis of the genitalia (Fig. 5.29), and is pretty much restricted to humans (although one species of monkey has been infected experimentally). The life cycles of the species are similar. Microfilariae, circulating in the blood, are ingested by mosquitoes of the genera *Culex, Mansonia, Aedes,* and *Anopheles* along with a blood meal. The larvae develop to the infective stage in the thoracic muscles, migrate to the mouthparts, and are injected when the vector takes a blood meal. Once in the final host, the larvae enter the lymphatic vessels or lymph nodes and mature. An interesting feature about the circulating microfilariae of *B. malayi* and *W. bancrofti* is that several strains show a marked diurnal periodicity. In one strain, microfilariae are found in the peripheral circulation during the day and day-biting mosquitoes transmit that strain. At night, microfilariae are absent or rare in the peripheral circulation. In the other strain, microfilariae are most common in the peripheral circulation at night and are transmitted by night-biting mosquitoes (see also Chapter 13). Although grossly disfiguring, the parasites are not particularly pathogenic. Further, 'elephantiasis' results from repeated infections over long periods of time and is the product of a complex immunogenic reaction. There is evidence that, in the absence of new infections, the symptoms will disappear after the adults die.

Filariasis is another parasitic disease targeted for elimination by the WHO. Not quite as easy to solve as *D. medinensis*, eradication of lymphatic filariasis requires drug therapy. In 1998, SmithKline Beecham agreed to make their drug Albendazole®, shown to be 99% effective against the parasites, available free of charge. This is a rather impressive commitment since the treat-

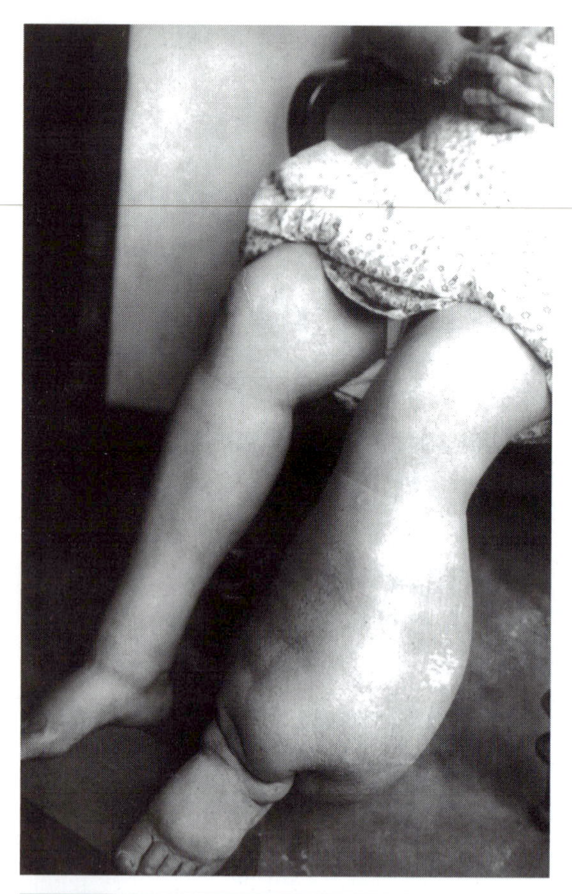

Fig. 5.28 Unilateral elephantiasis of the left leg. (Photograph courtesy of the Armed Forces Institute of Pathology.)

Fig. 5.29 Severe scrotal elephantiasis. (Photograph courtesy of Charles Allen, Wake Forest University.)

ment regime calls for all people in infected areas getting a dose each year for 4 to 5 years. The WHO estimates the program will run for at least 20 years and we might expect elimination of lymphatic filariasis about 2020. An added benefit is that the drug is also effective against many intestinal parasites such as hookworm and, therefore, treated individuals will reap multiple rewards.

Adults of *Onchocerca volvulus*, unlike *B. malayi* and *W. bancrofti*, occur in subcutaneous nodules known as onchocercomas (Figs. 5.30, 5.31). They survive in these nodules for approximately 10 years. This parasite is widely distributed in tropical Africa. It is also reported from the Americas and Yemen. Approximately 18 million people are infected with *O. volvulus*. The most important source of pathology is the microfilariae, which, unlike *B. malayi* and *W. bancrofti*, are found mainly in the host's tissues. When the microfilariae infect the cornea, iris, and optic nerve, the result is often irreversible blindness, the most severe form of pathology (Fig. 5.32). The vectors are various species of black flies (*Simulium*) whose larvae and pupae require fast-moving streams for development. Combining the major pathology to humans with the breeding requirements of the vectors results in the common name 'river blindness'. The microfilariae cause extreme itching which results in scratching, often to the point of bleeding. These host-produced lesions are a potential source for bacterial infiltration; they also invoke severe psychological trauma in infected humans. Additional pathology, again due to microfilaremia, includes pruritis, discoloration, and loss of elasticity of the skin. The latter can produce elephantiasis visibly similar to, though typically less dramatic, than that caused by *B. malayi* and *W. bancrofti* (Fig. 5.33).

Fig. 5.30 Adult *Onchocerca volvulus* in an onchocercoma. (Photograph courtesy of WHO and OCP.)

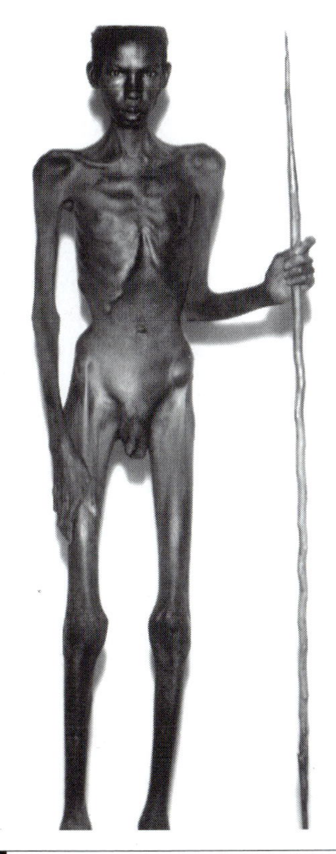

Fig. 5.31 This individual has a heavy onchocercal infection. (Photograph courtesy of the Armed Forces Institute of Pathology.)

Fig. 5.32 Blindness as a result of *Onchocerca volvulus* microfilariae. (Photograph courtesy of WHO and OCP.)

In 1974, recognizing the very significant impact of onchocerciasis in the West African savanna zone, the WHO, along with other participating organizations, launched the Onchocerciasis Control Programme (OCP). In 1986, an Extension area, consisting of Guinea, Guinea-Bissau, Senegal, and Sierra Leone, was added to the original countries (Côte d'Ivoire, Benin, Ghana, Togo, and Burkina Faso). The initial program involved aerial application of a variety of insecticides (larvacides) to reduce the black fly population to a point where transmission of microfilariae was unlikely. This plan would continue until microfilariae in humans in the region died out. When Ivermectin® became available for human use in 1987, large-scale distribution of the drug was added to the regime. At the time of this writing, onchocerciasis has been virtually eliminated in the original OCP countries and, in the Extension area, new infections are almost nonexistent. The socioeconomic impact of the program has been profound. An estimated 25 million hectares will now be available for resettlement and cultivation resulting in the production

Fig. 5.33 Scrotal elephantiasis caused by inguinal onchocercal lymphadenitis. (Photograph courtesy of the Armed Forces Institute of Pathology.)

of enough food to feed an estimated 17 million people (Fig. 5.34). The success of the OCP, in countries where onchocerciasis was most severe, has led to the African Programme for Onchocerciasis Control (APOC). Its goal is to remove onchocerciasis as a disease of public health concern throughout the remainder of Africa where some 15 million people are infected.

The last of the more common filarial worms in humans to be considered in this chapter is *Loa loa*. This parasite is restricted to certain areas in sub-Saharan Africa. The adults typically are found in subcutaneous nodules in the shoulder and neck regions. There is a localized hypersensitivity reaction causing the formation of what are known as fugitive, or calabar, swellings. As 'fugitive' implies, these nodules are ephemeral. Adult worms can often be seen wandering under the cornea of the eye causing transient blindness. The vectors for these worms are species of *Chrysops*, commonly known as horse flies.

Fig. 5.34 Once a productive village, now deserted because of onchocerciasis. With the eradication of the nematode, such fertile areas will once again contribute to feeding the populace. (Photograph courtesy of WHO and OCP.)

5.4 Phylogenetic relationships

Because nematodes have a poor fossil record and developmental/embryological details are scant, their relationship to other phyla is based largely on structural characteristics of the adults. A convenient classification used by some authors proposes a relationship whereby the Nematoda are considered a class of the Aschelminthes (cavity worms). The phylum Aschelminthes, then, would include such forms as Rotifera, Gastrotricha, and Acanthocephala. Other authors propose that the nematodes deserve phylum status and suggest affinities with phyla such as the Acanthocephala and Nematomorpha. Recent advances in molecular systematics may clarify the phylogenetic relationships. Recognizing that 18S rDNA sequences evolve rapidly in the rhabditan nematodes, Aguinaldo *et al.* (1997) used only the slowest evolving sequences from representative taxa (for nematodes, *Trichinella spiralis*, an enoplean). Their conclusion is a clade called the Ecdysozoa, comprising animals that molt. Therefore, nematodes would be allied to the Arthropoda, Kinorhyncha, Priapulida, Nematomorpha, Onychophora, and Tardigrada. The widespread acceptance of such a proposal awaits further verification and the test of time.

Until recently, the Nematoda have commonly been divided into two classes, the Phasmidia and the Aphasmidia. (To avoid confusion with the order Phasmida in the insects, the earlier names Secernentea [=Phasmidia] and Adenophorea [=Aphasmidia] are now used.) Andrassy (1976), questioning whether this dichotomy reflected a true evolutionary relationship, proposed that the Adenophorea were polyphyletic and proposed a third class, the Torquentia. Inglis (1983) concurred and recognized the Enoplida (=Adenophorea), Chromadorida (=Torquentia), and Secernentea as independently derived groups. Adamson (1987) performed a cladistic analysis on the three classes and concluded that the Enoplea (=Adenophorea) should be regarded as a sister group to the Chromadorida–Rhabditea (=Secernentea) line and that the Chromadorida were paraphyletic.

Kampfer *et al.* (1998), using 18S rDNA data from 28 species of nematodes, support division of the nematodes into two monophyletic classes, the Secernentea and the Adenophorea. They do, however, allow for the possibility of deep divisions within the Adenophorea. In direct contrast, Blaxter *et al.* (1998) also examined small subunit rDNA from 53 different nematodes. They conclude that the Adenophorea are clearly paraphyletic and that the Secernentea are a natural grouping within the Adenophorea. Their analyses indicate five clades in the Nematoda and, importantly, each clade includes a diversity of trophic adaptations. This suggests that parasitism has evolved many times in the Nematoda.

Even at lower levels of resolution, the phylogeny of the nematodes is controversial. Nadler (1995), for example, used full-length sequences of 18S rDNA and 300 nucleotides of cytochrome oxidase II to resolve 13 ascaridoid species. His analyses were mostly consistent with current taxonomic assignments at lower ranks, but 'inconsistent with most proposed arrangements at higher taxonomic levels'. In a more recent analysis of a larger data set, Nadler & Hudspeth (1998) argue 'some key features emphasized by previous workers represent ancestral states or highly homoplastic characters'. It is highly likely that the same argument applies to other groups of nematodes.

In summary, it is clear that we have much to learn about the evolutionary relationships amongst nematodes. To that end, the classificatory scheme we use in Box 5.3 is one of 'convenience', based on morphological data which are more readily available than nucleotide sequences. It certainly will change and is not intended to reflect true evolutionary relationships.

Box 5.3 Classification of the Nematoda

This classification is one of convenience and is unlikely to reflect true evolutionary relationships. We consider it simply as a way to list the major groups of nematodes parasitic in animals, 'a place to hang your hat, as it were'. It is based on gross, visible morphology and it follows, loosely, the proposals of Adamson (1987) and Anderson (1992). The Subclass Tylenchia and the Orders Drilonematida, Mermithida, and Rhigonematida (identified with asterisks) are parasitic exclusively in invertebrates and lower-level systematics are not included.

Class Rhabditea

Amphids distinct, three esophageal glands present, phasmids present.

*Subclass Tylenchia

Stomatostyle (buccal stylet) present. Mostly parasitic in or on plants and fungi, some free-living, few parasitic in insects, even fewer in other arthropods and annelids.

Subclass Rhabditia

One or two lateral canals. Many free-living, others parasitic in plants and animals.

Order Rhabditida

Most members have six lips leading to a muscular esophagus. Both sexes have conical tails and males have two spicules. A gubernaculum is often present. Most are free-living, those parasitic in animals are found in amphibians, reptiles, birds and mammals.

Families: Rhabdiasidae ,Rhabditidae, Strongyloididae

*Order Drilonematida

Often with cephalic hooks, phasmids appear like suckers. Parasitic in annelids.

*Order Rhigonematida

Buccal cavity with cuticular modifications. Parasites of diplopods.

Order Strongylida

Typically slender worms the males of which usually have a copulatory bursa supported by muscular rays. Renette cell H-shaped. Although rare in fishes, the bursate nematodes are found in all classes of vertebrates.

Families: Amidostomatidae, Ancylostomatidae, Angiostrongylidae, Chabertiidae, Crenosomatidae, Diaphanocephalidae, Dictyocaulidae, Filaroididae, Heligmonellidae, Heligmosomidae, Metastrongylidae, Molieidae, Ornithostrongylidae, Protostrongylidae, Pseudaliidae, Skrjabingylidae, Strongylacanthidae, Strongylidae, Syngamidae, Trichostrongylidae

Order Oxyurida

Prominent posterior bulb on esophagus with chitinized valve. Renette cell X-shaped. Males have one or no spicule, life cycle is direct. The only major group of nematodes with adults in both vertebrates and invertebrates.

Families: Heteroxynematidae, Oxyuridae, Pharyngodonidae

Order Ascaridida

Large worms with three prominent lips. Numerous labial and caudal papillae. Life cycles direct or indirect. Parasitic in all groups of vertebrates.

Families: Anisakidae, Ascarididae, Ascaridiidae, Atractidae, Cosmocercidae, Cucullanidae, Heterakidae, Kathlaniidae, Maupasinidae, Quimperiidae, Seuratidae, Subuluroidae

Order Spirurida

Esophagus distinctly divided into anterior muscular region and posterior glandular region. Parasites in all vertebrate classes with arthropods as intermediate hosts.

Families: Acuariidae, Anguillicolidae, Aproctidae, Camallanidae, Cystidicolidae, Desmidocercidae, Diplotriaenoidae, Dracunculidae, Filariidae, Gnathostomatidae, Gongylonematidae, Habronematidae, Hartertiidae, Hedruridae, Micropleuridae, Onchocercidae, Philometridae, Physalopteridae, Rictulariidae, Rhabdochonidae, Spirocercidae, Spiruridae, Tetrameridae, Thelaziidae

Class Enoplea

Amphids may be indistinct in parasitic species. Phasmids absent. Excretory system usually lacks canals.

Order Dioctophymatida

Large worms lacking lips. Esophageal glands well-developed.

Families: Dioctophymatidae, Soboliphymatidae

*Order Mermithida
Free-living as adults but parasitic in invertebrates, particularly insects, as larvae.
Order Muspiceida
Uncommon parasites in which males are unknown. Females have a much-reduced digestive tract.
Families: Muspiceidae, Robertdollfusidae
Order Trichurida
Two-part esophagus with a short, muscular anterior portion followed by a long glandular portion consisting of a narrow tube into which open naked gland cells.
Families: Cystoopsidae, Trichinellidae, Trichuridae

References

Adamson, M. L. (1987) Phylogenetic analysis of the higher classification of the Nematoda. *Canadian Journal of Zoology*, **65**, 1478–1482.

Aguinaldo, A. M. A., Turbeville, J. M., Linford, L. S., Rivera, M. C., Garey, J. R., Raff, R. A. & Lake, J. A. (1997) Evidence for a clade of nematodes, arthropods and other moulting animals. *Nature*, **387**, 489–493.

Anderson, R. C. (1992) *Nematode Parasites of Vertebrates Their Development and Transmission*. Cambridge: CAB International.

Andrassy, I. (1976) *Evolution as a Basis for the Sytematization of Nematodes*. Budapest: Akademiai Kiado.

Behm, C. A. (1997) The role of trehalose in the physiology of nematodes. *International Journal for Parasitology*, **27**, 215–229.

Bird, A. F. & Bird, J. (1991) *The Structure of Nematodes*, 2nd edn. San Diego: Academic Press.

Blaxter, M. L., Page, A. P., Rudin, W. & Maizels, R. M. (1992) Nematode surface coats: actively evading immunity. *Parasitology Today*, **8**, 243–247.

Blaxter, M. L., De Ley, P., Garey, J. R., Liu, L. X., Scheideman, P., Vierstraete, A., Vanfleteren, J. R., Mackey, L. Y., Dorris, M., Frisse, L. M., Vida, J. T. & Thomas, W. K. (1998) A molecular evolutionary framework for the phylum Nematoda. *Nature*, **392**, 71–75.

Bruňanská, M. (1997) *Toxocara canis* (Nematoda: Ascaridae): the fine structure of the oviduct, oviduct–uterine junction and uterus. *Folia Parasitologica*, **44**, 55–61.

de Carneri, I., Angelle, T., Dupouy-Camet, J. & Pozio, E. (1989) Different aetiological agents cause the European outbreaks of horsemeat induced human trichinellosis. In *Trichinellosis. Proceedings of the 7th International Conference on Trichinellosis*, ed. C. E. Tanner, A. R. Martínez-Fernández & F. Bolas-Fernández, pp. 387–391. Madrid: Consejo Superior de Investigaciones Científicas Press.

Crofton, H. D. (1966) *Nematodes*. London: Hutchinson.

Davey, K. G. (1995) Water, water compartments and water regulation in some nematodes parasitic in vertebrates. *Journal of Nematology*, **27**, 433–440.

Fetterer, R. H. & Rhoads, M. L. (1996) The role of the sheath in resistance of *Haemonchus contortus* infective-stage larvae to proteolytic digestion. *Veterinary Parasitology*, **64**, 267–276.

Gemmill, A.W., Viney, M. E. & Read, A. F. (1997) Host immune status determines sexuality in a parasitic nematode. *Evolution*, **51**, 393–401.

Inglis, W. G. (1983) An outline classification of the phylum Nematoda. *Australian Journal of Zoology*, **31**, 243–255.

Kampfer, S., Sturmbauer, C. & Ott, J. (1998) Phylogenetic analysis of rDNA sequences from adenophorean nematodes and implications for the Adenophorea–Secernentea controversy. *Invertebrate Biology*, **117**, 29–36.

Kita, K., Hirawake, H. & Takamiya, S. (1997) Cytochromes in the respiratory chain of helminth mitochondria. *International Journal for Parasitology*, **27**, 617–630.

Lee, D. L. (1996) Why do some nematode parasites of the alimentary tract secrete acetylcholinesterase? *International Journal for Parasitology*, **26**, 499–505.

Lilley, B., Lammie, P., Dickerson, J. & Eberhard, M. (1997) An increase in hookworm infection temporally associated with ecologic change. *Emerging Infectious Disease*, **3**, 391–393.

Malouf, A. H. (1986) Seals and sealing in Canada. *Report of the Royal Commission*, vol. 3. Ottawa: Canadian Government Publishing Centre.

Martínez, A. M. B. & de Souza, W. (1997) A freeze-fracture and deep-etch study of the cuticle and hypodermis of infective larvae of *Strongyloides venezuelensis* (Nematoda). *International Journal for Parasitology*, **27**, 289–297.

Medica, D. L. & Sukhdeo, M. V. K. (1997) Role of lipids in the transmission of the infective stage (L3) of *Strongylus vulgaris* (Nematoda: Strongylida). *Journal of Parasitology*, **83**, 775–779.

Nadler, S. A. (1995) Advantages and disadvantages of molecular phylogenetics: a case study of ascaridoid nematodes. *Journal of Nematology*, **27**, 423–432.

Nadler, S. A. & Hudspeth, D. S. S. (1998) Ribosomal DNA and phylogeny of the Ascaridoidea (Nemata: Secernentea): implications for morphological evolution and classification. *Molecular Phylogenetics and Evolution*, **10**, 221–236.

Neuhaus, B., Bresciani, J., Christensen, C. & Frandsen, F. (1996) Ultrastructure and development of the body cuticle of *Oesophagostomum dentatum* (Strongylida, Nematoda). *Journal of Parasitology*, **82**, 820–828.

Ogbogu, V. C. & Storey, D. M. (1996) Ultrastructure of the alimentary tract of third-stage larvae of *Litomosoides carinii*. *Journal of Helminthology*, **70**, 223–229.

Ogilvie, B. M. & Jones, V. E. (1971) *Nippostrongylus brasiliensis*: a review of immunity and the host/parasite relationship in the rat. *Experimental Parasitology*, **29**, 138–177.

Olsen, O. W. (1974) *Animal Parasites: Their Biology and Life Cycles*, 2nd edn, Baltimore: University Park Press.

Pritchard, D. (1995) The survival strategies of hookworms. *Parasitology Today*, **11**, 255–259.

Riga, E., Perry, R. N., Barrett, J. & Johnston, M. R. L. (1995) Biochemical analyses on single amphidial glands, excretory–secretory gland cells, pharyngeal glands and their secretions from the avian nematode *Syngamus trachea*. *International Journal for Parasitology*, **25**, 1151–1158.

Sims, S. M., Ho, N. F. H., Geary, T. G., Thomas, E. M., Day, J. S., Barsuhn, C. L. & Thompson, D. P. (1996) Influence of organic acid excretion on cuticle pH and drug absorption by *Haemonchus contortus*. *International Journal for Parasitology*, **26**, 25–35.

Špakulová, M., Mutafova, T. & Král, J. (1995) Cytogenetic study of *Heterakis gallinarum* (Nematoda, Heterakidae). *Biologia Bratislava*, **50**, 605–609.

World Health Organization (1994) WHO calls experts to define a strategy to control hookworms in women. Press Release WHO/94, 5 December 1994.

ADDITIONAL REFERENCES

Chitwood, B. G. & Chitwood, M. G. (1950) (Reprinted 1974.) *An Introduction to Nematology*. Baltimore: University Park Press.

Anderson, R. C., Chabaud, A. G. & Willmont, S. (eds.) (1974–1983) *CIH Keys to the Nematode Parasites of Vertebrates*, nos. 1–10. Wallingford: CAB International.

Faust, E. C., Beaver, P. C. & Jung, R. C. (1975) *Animal Agents and Vectors of Human Disease*. Philadelphia: Lea and Febiger.

Harrison, F. W. & Ruppert, E. E. (eds.) (1991) *Microscopic Anatomy of Invertebrates*, vol. 4, *Aschelminthes*. New York: Wiley–Liss.

Poiner, G. O. Jr. (1983) *The Natural History of Nematodes*. Englewood Cliffs: Prentice-Hall.

Rogers, W. P. (1962) *The Nature of Parasitism*. New York: Academic Press.

Skrjabin, K. I. (ed.) (various years) *Essentials of Nematodology*. Washington, DC, USA: Israel Program for Scientific Translations.

Yamaguti, S. (1961) *Systema Helminthum III: The Nematodes of Vertebrates*, Parts 1 and 2. New York: Interscience Publishers.

Chapter 6

Acanthocephala: the thorny-headed worms

6.1 General considerations

The acanthocephalans (Greek, *acantho* = thorn, *cephala* = head), or thorny-headed worms, are a relatively small group of parasites comprising about 1100 species. Acanthocephalans were first described in 1684 by Francesco Redi, an Italian physician (see Box 1.1), who reported finding white worms (probably *Acanthocephalus anguillae*) with hooked proboscides in the intestines of European eels. Since then, adult acanthocephalans have been found in the intestines of fishes, amphibians, reptiles, birds, and mammals, worldwide.

Adult acanthocephalans are usually white or cream-colored, although on occasion they may be stained by the intestinal content of the host. They vary in length from around 1 mm to more than 60 cm, have an eversible, hooked proboscis and a characteristic body wall with a lacunar system; they are dioecious and generally sexually dimorphic. There is no mouth, intestine, or conventional circulatory system, and an excretory system is present only in one family, the oligacanthorhynchids. The life cycle includes an arthropod intermediate host, a vertebrate definitive host and, in some species, a paratenic host.

6.2 Form and function

6.2.1 The outside

Body shape and holdfast. The body of most acanthocephalans resembles an elongated tube, tapering at both ends, with the thorny proboscis at one end. Externally, the body can be divided into the **proboscis**, the anterior retractile organ bearing hooks; the neck, a smooth region immediately posterior to the proboscis; and the trunk, which comprises most of the body (Fig. 6.1).

The shape of the proboscis is highly variable, from spherical to cylindrical depending on the species (Figs. 6.2, 6.3). The roots of the sclerotized hooks are embedded in a thin, muscular wall under the tegument, and their size, shape, number, and distribution are important characters used in classification. The proboscis is a hollow structure filled with fluid, which can be totally invaginated by the proboscis inverter muscles into a muscular sac called the proboscis receptacle. Evagination of the proboscis is achieved by hydraulic pressure created from within the lacunar system when the proboscis receptacle contracts. A nerve ganglion called the cerebral ganglion (or brain) is located inside the proboscis receptacle and an apical sensory organ may be present at the tip of the proboscis.

The neck is the smooth area located between the most posterior hooks of the proboscis and an infolding of the body wall that marks the beginning of the trunk (Fig. 6.1). The neck may contain lateral sense organs, and can be retracted by a pair of neck retractor muscles that attach to an infolding of the body wall.

The trunk, like the proboscis and neck, is covered by a tegument with internal circular and longitudinal muscular layers. In some species, foldings of the tegument may give the appearance of segmentation. The surface can be smooth or covered with sclerotized spines that help in the attachment to the host's intestine (Fig. 6.1). The

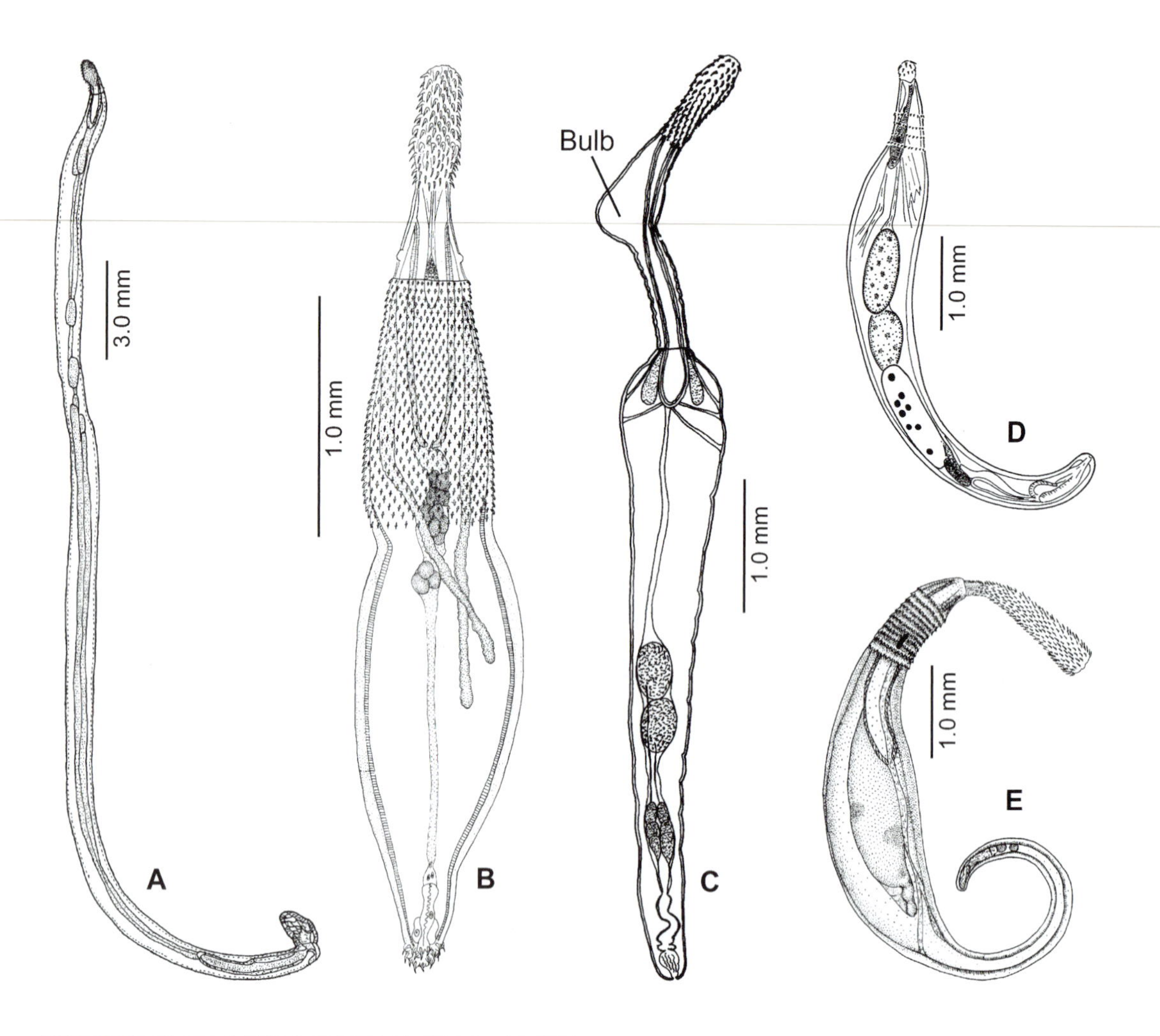

Fig. 6.1 Different types of acanthocephalan bodies (all males). (A) *Centrorhynchus conspectus* from owls in eastern USA; (B) *Corynosoma bipapillum* from Bonaparte's gull in Alaska; (C) *Pomphorhynchus patagonicus* from a fish in Argentina; (D) *Quadrigyrus nickoli* from a fish in Colombia; (E) *Polyacanthorhynchus kenyensis* from a paratenic host in Kenya. ((A) From Richardson & Nickol, 1995, with permission, *Journal of Parasitology*, **81**, 767–772; (B) from Schmidt, 1965, with permission, *Journal of Parasitology*, **51**, 814–816; (C) from Ortubay *et al.*, 1991, with permission, *Journal of Parasitology*, **77**, 353–356; (D) from Schmidt & Hugghins, 1973, with permission, *Journal of Parasitology*, **59**, 829–835; (E) from Schmidt & Canaris, 1967, with permission, *Journal of Parasitology*, **53**, 634–637.)

body wall of the trunk encloses the body cavity that contains the reproductive and excretory organs, the **ligament sacs**, and the genital ganglia in the male. Because acanthocephalans lack a digestive system, the trunk participates in the absorption and distribution of nutrients obtained from the host's intestine.

In most acanthocephalans, attachment is accomplished by the insertion of the proboscis into the intestinal wall of the host. In some species such as *Bolbosoma* spp., *Polymorphus* spp., and *Corynosoma* spp. spines on the trunk also aid in attaching to the host intestine (Fig. 6.1B). Attachment is not permanent in most species; repositioning is possible by introversion of the proboscis, followed by reinflation using the hydrostatic pressure produced by the lacunar system. Acanthocephalans such as *Pomphorhynchus* spp. cannot move inside the host's intestine because a region of their neck inflates or swells during the initial stages of attachment forming a bulb of significant size that becomes completely embedded in the intestinal wall (Fig. 6.1C). Although this morphological adaptation provides a stronger or more efficient attachment

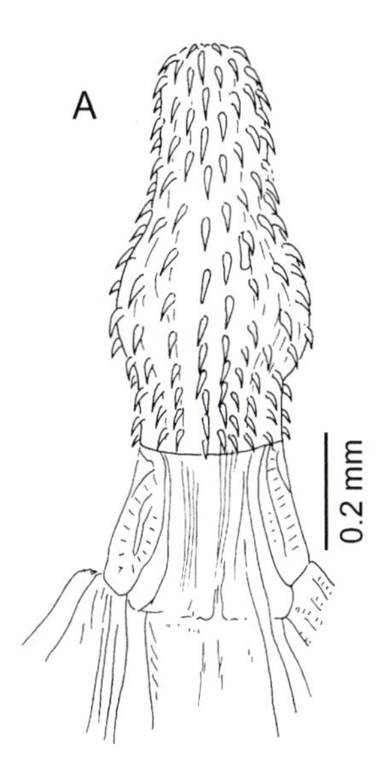

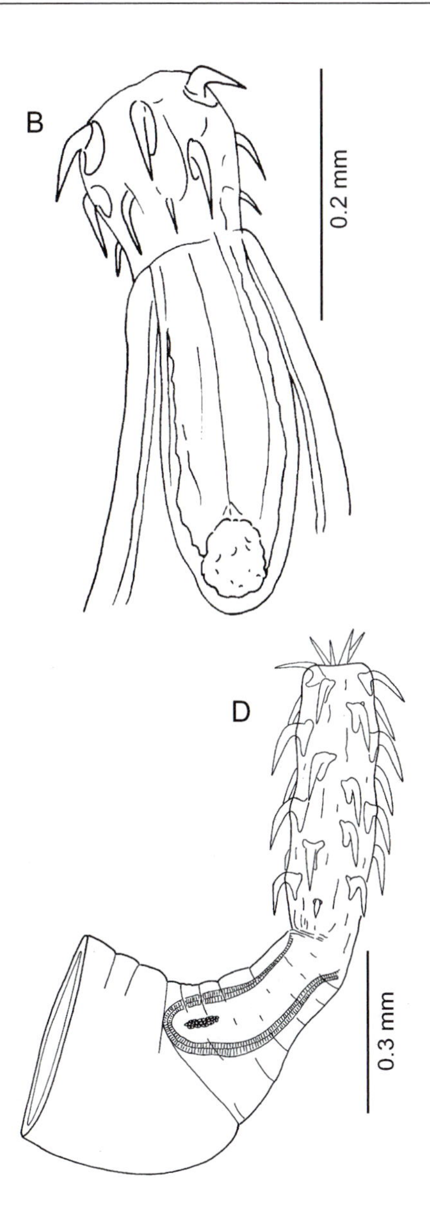

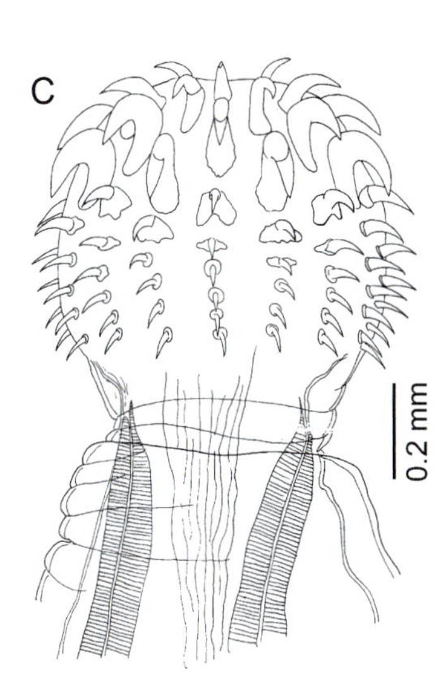

Fig. 6.2 Different types of acanthocephalan proboscides. (A) *Polymorphus spindlatus* from a heron in Peru; (B) *Neoechinorhynchus venustus* from a fish in northwestern USA; (C) *Sphaerechinorhynchus serpenticola* from a cobra in Borneo; (D) *Paracanthocephalus rauschi* from an otter in Alaska. ((A) From Amin & Heckmann, 1991, with permission, *Journal of Parasitology*, **77**, 201–205; (B) from Amin & Heckmann, 1992, with permission, *Journal of Parasitology*, **78**, 34–39; (C) from Schmidt & Kuntz, 1966, with permission, *Journal of Parasitology*, **52**, 913–916; (D) from Schmidt, 1969, with permission, *Canadian Journal of Zoology*, **47**, 383–385.)

mechanism, it prevents the worm from moving or migrating in the intestine.

Body wall. The body wall of acanthocephalans is a syncytium organized into three distinct sections, e.g., an outer tegument, a middle group of circular muscles, and an inner group of longitudinal muscles (Fig. 6.4A). A complex system of interconnecting canals called the lacunar system extends throughout these three sections (Fig. 6.4).

The **tegument** of acanthocephalans fulfills many functions, including protection, inactivation of the host's digestive enzymes, osmoregulation, and acquisition of nutrients. The most

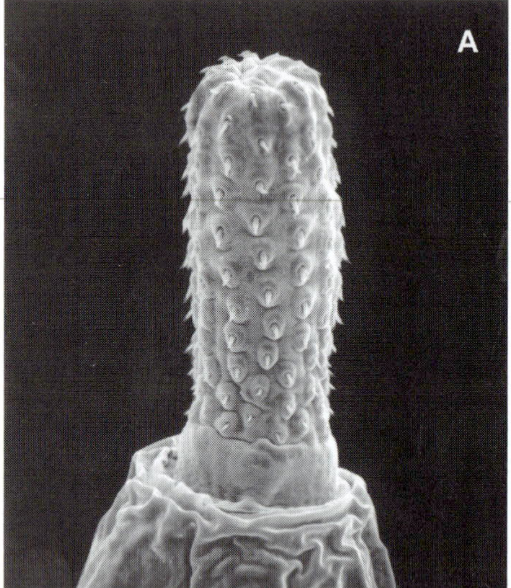

Fig. 6.3 Scanning electron micrographs of acanthocephalan proboscides. (A) *Moniliformis moniliformis* from rats; (B) *Acanthocephalus caspanensis* from a toad in Chile; (C) unidentified acanthocephalan from a flounder in Chile. ((A) Photograph courtesy of Tom Dunagan, Southern Illinois University at Carbondale.)

external layer of the tegument is a plasma membrane covered by a carbohydrate-rich glycocalyx and with numerous infoldings or crypts that open to the surface by pores. The glycocalyx and pores act as a sieve that allows particles of less than 8.5 nm to reach the crypts, where they undergo pinocytosis. Underlying the plasma membrane there is a fibrous stratum without cellular boundaries and pierced by numerous canals. This layer is followed by a highly invaginated basal plasma membrane that may be involved in water and ion transport, and by a fine basal lamina. Mitochondria, Golgi complexes, lysosomes, ribosomes, glycogen deposits, lipid droplets, and various vesicles are found in the tegument; rough endoplasmic reticulum is also present, but it is closely associated with the nuclei. There are few nuclei in the syncytial tegument; they are more or less fixed in position, and their numbers are approximately constant for each species, at least in their early developmental stages.

The **lacunar system** is a network of fluid-filled canals distributed throughout the tegument and muscle layers (Fig. 6.4). The lacunar fluid is moved by muscle contraction. In the proboscis and neck, the lacunar system has canals that connect with two sac-like structures called **lemnisci** (Fig. 6.5). The lemnisci extend from the base of the neck into the body cavity. The precise function of the lemnisci is unknown, but they may be involved in the hydraulic mechanism of proboscis eversion and may serve as reservoirs for lacunar fluid. Another hypothesis, suggesting that they may be reservoirs for nutrients and a metabolic site, has not been confirmed. The lacunar system in the trunk of acanthocephalans is totally separate from that of the proboscis and neck, and the extensive network of interconnecting canals includes several longitudinal and transverse canals.

The structure of the longitudinal and circular muscles of the body wall is unusual in acantho-

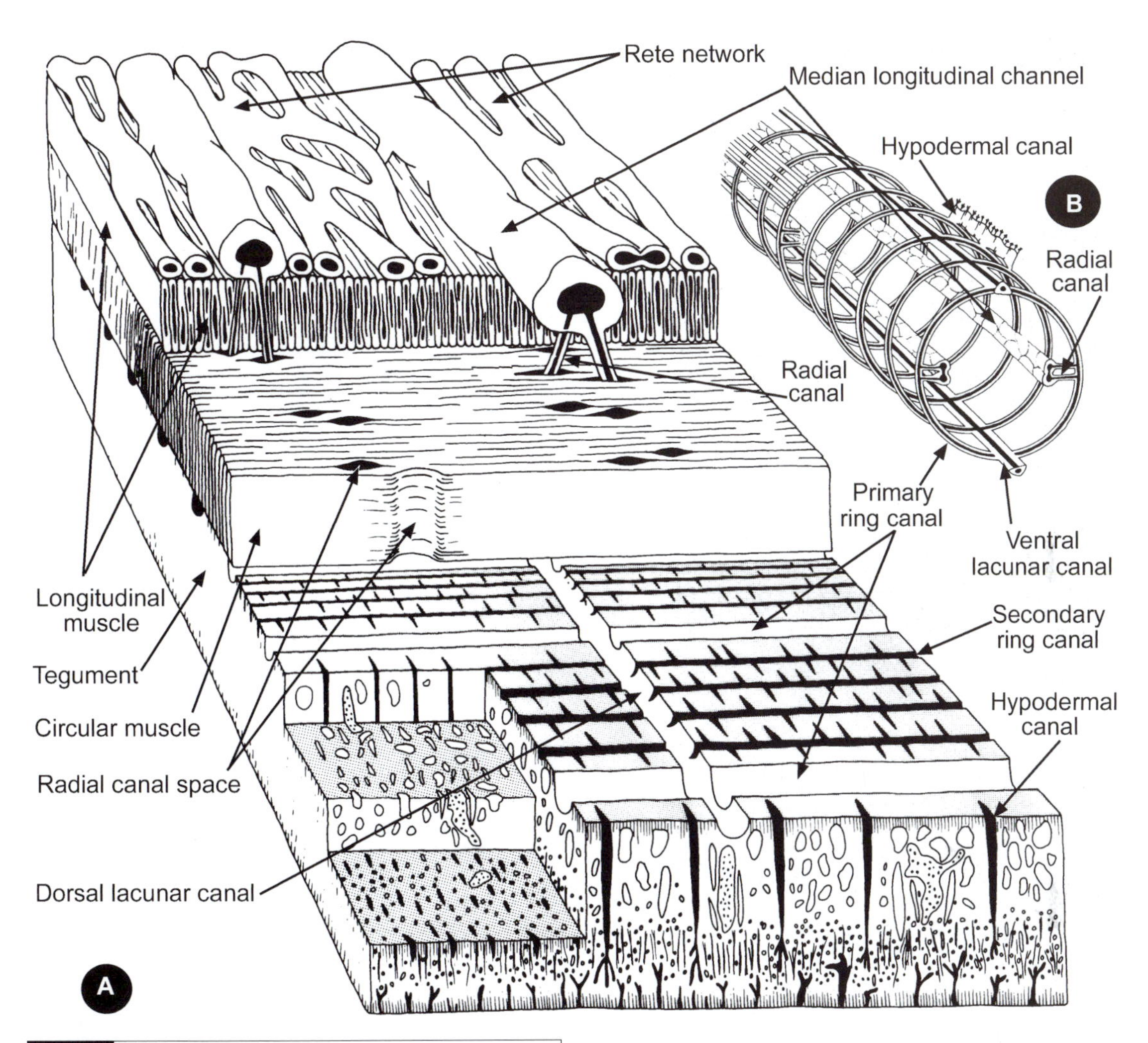

Fig. 6.4 The body wall and lacunar system of the acanthocephalan *Macracanthorhynchus hirudinaceus.* (A) Diagrammatic model of the body wall, sectioned in various ways, to show the relationship between the lacunar system, rete system, muscles, and tegument; (B) model of the lacunar system. (Modified from Dunagan, T.T. & Miller, D.M. [1991] Acanthocephala. In *Microscopic Anatomy of Invertebrates*, Vol. 4, *Aschelminthes*, ed. F. W. Harrison & E. E. Ruppert, pp. 299–332. ©1991 John Wiley & Sons. Reprinted by permission of Wiley-Liss, Inc., a division of John Wiley & Sons, Inc.)

cephalans. The muscles are hollow and tubular, with a number of anastomosing connections, and the lumen of each muscle fiber is continuous with the lacunar system. The interplay of the lacunar system and hollow muscles probably encourages the efficient transport of nutrients and waste products throughout the organism, while simultaneously serving as a hydrostatic skeleton. The muscles of the body wall also are closely associated with the **rete system** (Fig. 6.4A). The rete system is a network of branching and anastomosing tubules lying on the medial surface of the longitudinal muscles or between the longitudinal and circular muscle layers that seems to be involved in the initiation of muscle contraction.

6.2.2 The inside

Osmoregulatory and excretory system. Most acanthocephalans lack an excretory system and waste materials are eliminated by diffusion through the body wall, probably through the pores in the tegument. The only exception includes members of the Oligacanthorhynchidae, in which two protonephridia are present. Each protonephridium is formed by flame bulbs with

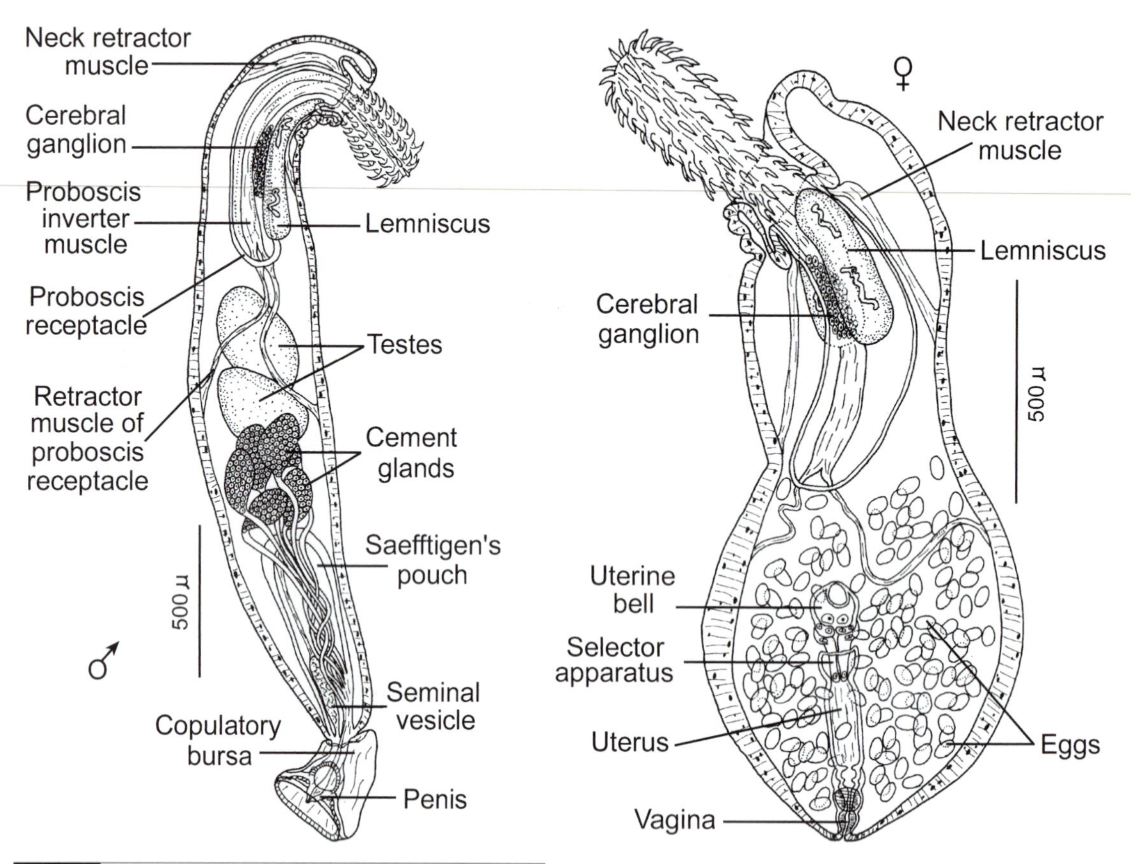

Fig. 6.5 Male and female *Echinorhynchus lageniformis*, from a flounder, illustrating the basic morphology of acanthocephalans. (Modified from Olson & Pratt, 1971, with permission, *Journal of Parasitology*, **57**, 143–149.)

tufts of cilia that may, or may not, be encapsulated, depending on the species (Dunagan & Miller, 1986). In males, the protonephridia are connected to the vas deferens and empty to the outside through it; in females, the protonephridia are attached to the uterine bell and empty to the outside through the uterus and vagina. Acanthocephalans are poor osmoregulators and the osmotic pressure of their pseudocoelomic fluid is close to that of the host's intestine.

Reproductive system. Acanthocephalans are dioecious and exhibit sexual dimorphism in size, with females being usually larger than males (Fig. 6.6). Both sexes have one or two ligament sacs that extend from the proboscis receptacle to near the genital pore, forming an envelope that surrounds the gonads and accessory organs. The ligament sacs are permanent in some species, whereas in others they disappear as the individuals mature sexually.

Males have two testes, each with a vas efferens through which spermatozoa travel to a common vas deferens, or to a penis, or both. There are two important accessory organs in males, the **cement glands** and the **copulatory bursa** (Fig. 6.5). The cement glands may or may not be syncytial; if they are not syncytial, they range from one to eight in number. The cells produce copulatory cement, which, in some species, may be stored in a cement reservoir. The number and shape of the cement glands are important characters used in the classification of acanthocephalans. The copulatory bursa is a bell-shaped structure that is everted only during copulation and is used to hold the female in place during insemination (Fig. 6.7).

The female reproductive system (Fig. 6.5) consists of an ovary which, early in development and usually before the acanthocephalan reaches the definitive host, fragments into ovarian balls that

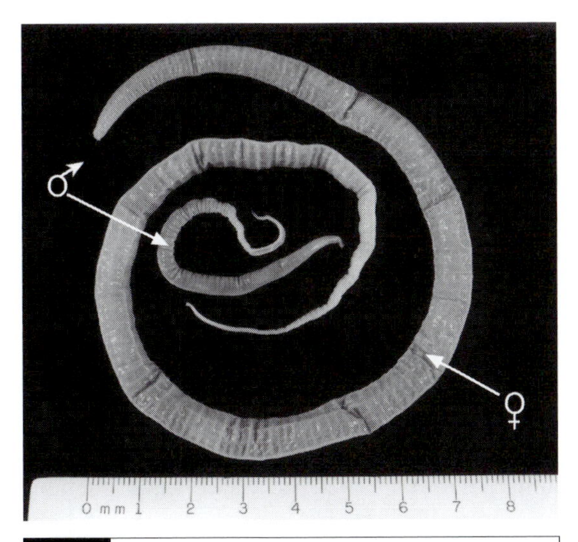

Fig. 6.6 Male and female individuals of the acanthocephalan *Oligacanthorhynchus tortuosa* showing sexual dimorphism in size. The flat appearance is the normal condition in the host. Also, the lacunar canals can be seen clearly in the female individual. (Photograph courtesy of Tom Dunagan, Southern Illinois University at Carbondale.)

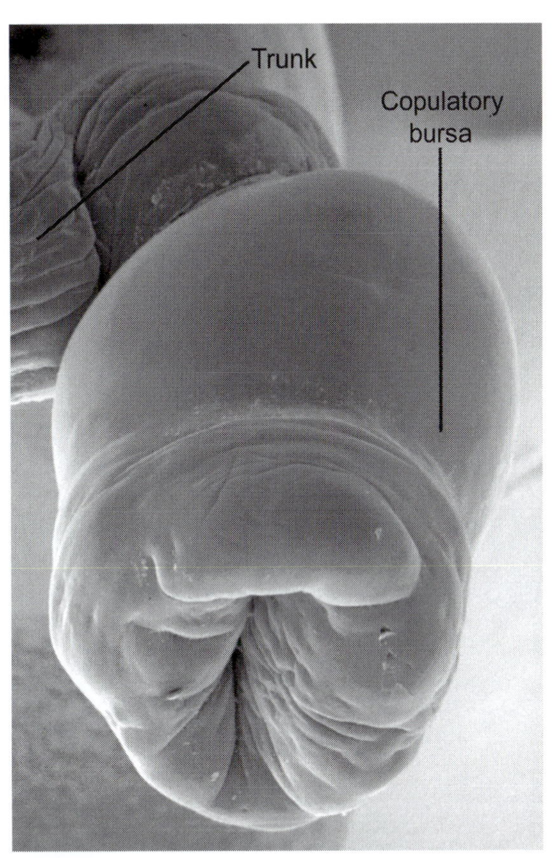

Fig. 6.7 Male of *Serrasentis sagittifer* with the copulatory bursa everted. (Photograph courtesy of Tom Dunagan, Southern Illinois University at Carbondale.)

float freely within the ligament sac. The posterior end of the ligament sac is attached to the **uterine bell**, a muscular, funnel-shaped organ that allows mature fertilized eggs to pass through into the uterus and vagina and to the outside via the gonopore. Immature eggs are returned to the ligament sac. The selection process appears to be size-dependent, with only the mature, larger eggs passing into the uterus (Whitfield, 1970).

During copulation in the host's intestine, the male's copulatory bursa (Fig. 6.7) is everted by the hydrostatic pressure of fluid forced into its lacunar system by a muscular sac called the **Saefftigen's pouch** (Fig. 6.5). The bursa is then wrapped around the posterior end of the female and the penis is introduced into the female gonopore. The released spermatozoa migrate through the vagina, the uterus, and the uterine bell into the ligament sac, where fertilization of the oocytes in the ovarian balls occurs. After the transfer of sperm is completed, the male plugs the posterior end of the female with cement from the cement glands, forming a **copulatory cap** (Fig. 6.8). The copulatory cap appears to last only for a few days, but may act as a temporary occlusion to prevent loss of sperm. The cap may also be used as a kind of 'chastity belt', guaranteeing that sperm from a given male will not encounter competitors, i.e., sperm from other males, at least for some time (see Box 6.1).

Nervous system and sensory input. The nervous system of acanthocephalans is relatively simple. Male worms have three different types of ganglia, i.e., cerebral, genital, and bursal ganglia. Females have only a cerebral ganglion located in the body wall. Nerves from the cerebral ganglion reach the two lateral sense organs of the neck and the apical sense organ of the proboscis, if present. In males, the two genital ganglia are located along the bursal muscle and appear to control the musculature associated with the reproductive system. Several nerve tracts connect the cerebral ganglion with the genital ganglia. The bursal ganglion, also

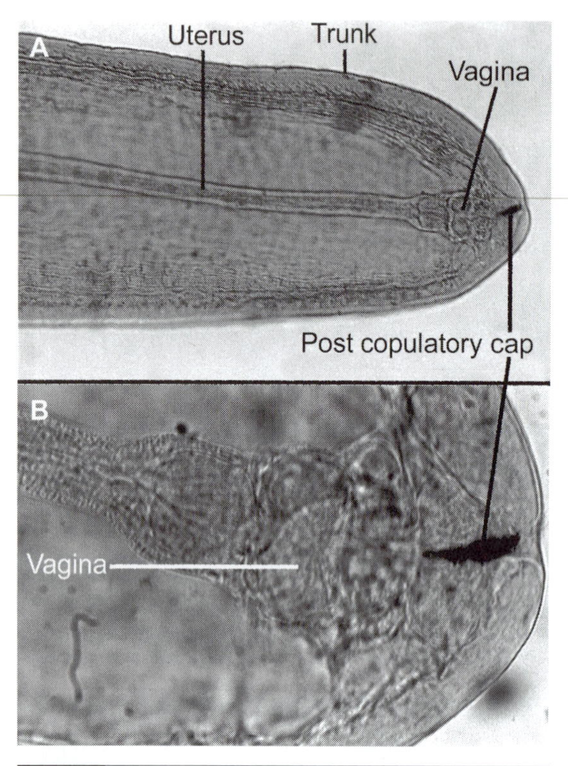

Fig. 6.8 (A) Female of *Pomphorhynchus patagonicus*, from a fish in Argentina, showing the posterior end of the reproductive system and a copulatory cap; (B) detail of the posterior end with the copulatory cap. (Modified from Semenas *et al.*, 1992, with permission, *Research and Reviews in Parasitology*, **52**, 89–93.)

found only in males, innervates the muscles of the bursa and the adjacent body wall.

Near the cerebral ganglion there is a large multinucleated cell known as a sensory support cell with processes that extend to the lateral and apical sensory organs (Miller & Dunagan, 1983). Although its function is not known, it may be part of the sensory network.

6.2.3 Nutrient uptake

Most of the knowledge about feeding, metabolism, and physiology of acanthocephalans comes from research on three species, *Moniliformis moniliformis*, a rat parasite and an excellent laboratory model, *Polymorphus minutus*, a parasite of ducks and chickens widely used as a laboratory model, and *Macracanthorhynchus hirudinaceus*, a rather large parasite of pigs.

Because acanthocephalans lack a digestive system, nutrients are acquired through the body surface from the contents of the host intestine. Some sugars, nucleotides, amino acids, and some triglycerides are regularly absorbed. *Moniliformis moniliformis*, for example, absorbs glucose, mannose, fructose, and galactose. The transport of sugars is accomplished by facilitated diffusion. Immediately after absorption, sugars are phosphorylated into hexose phosphate, and although some of it is used for general metabolism, e.g., glycogenesis and other synthetic processes, much is converted into the disaccharide trehalose.

The tegument of *M. moniliformis* contains aminopeptidases that can cleave small peptides into amino acids, which are then absorbed (Uglem *et al.*, 1973). The mechanism of lipid absorption is poorly known, although absorption of triglycerides begins at the anterior half of the proboscis (Taraschewski & Mackenstedt, 1991). Adult worms accumulate large quantities of neutral lipids, but they are not used as energy sources. It appears that acanthocephalans can control the absorption of lipids but do not have selective mechanisms to pick and choose specific ones. Nucleotides such as thymine are also absorbed and incorporated into mitochondrial DNA in the body wall, and nuclear DNA in ovarian balls and testes.

6.2.4 Metabolism

Most of the metabolic work done on acanthocephalans has focused on *Moniliformis moniliformis* and *Macracanthorhynchus hirudinaceus*. Both of these parasites are large, providing ample tissue for biochemical and physiological studies. *Moniliformis moniliformis* can be maintained with relative ease for laboratory purposes in rats and cockroaches.

Most of the research has been directed at the acanthocephalan's intermediary carbohydrate metabolism and related activities. It appears that their glycogen content ranges between 22% and 24% of the dry weight in adult worms removed from unfasted hosts. Moreover, glycogen is apparently the primary energy resource, with up to 75% being lost during a single 24-hour fasting period. Trehalose is a non-reducing disaccharide that comprises 2%–4% of the solid tissues of *M. moniliformis*. The function of trehalose in acanthocephalans is problematic. It has been suggested that

Box 6.1 | **Homosexual rape and the chastity belt**

During copulation in acanthocephalans, after the transfer of sperm is completed, cement from the cement glands of the male plugs the vagina and posterior end of the female forming a copulatory cap. The external cap appears to last only a few days but the cement packed into the genital tract of the female may last longer.

In the context of sexual selection and parental investment, acanthocephalans conform to the model where one sex (female) invests considerably more than the other. According to this model, males should compete among themselves to mate with females. It is possible then, that cement glands and capping behavior evolved in response to sexual selection (Abele & Gilchrist, 1977). The capping of females would prevent subsequent inseminations by other males at least for a few days, a sort of chastity belt, ensuring that sperm from the capping male will encounter and fertilize mature oocytes without competition from sperm from another male.

Copulatory caps are not unique to acanthocephalans. Copulatory plugs have also been found in female insects and snakes, where they prevent rival males from copulating with the same female (Devine, 1975). An interesting twist in this story, however, is that copulatory caps have also been found over the posterior end of male acanthocephalans (for references see Crompton, 1985) and they may represent another manifestation of the competition among males for the limited female resource. The capping of males, or homosexual rape, effectively, albeit temporarily, eliminates other males from the pool competing for the female resource. Although it is possible that homosexual rape may be the result of poor sex recognition, no sperm from the capping males has been found in the capped males, indicating that capping occurred without previous insemination.

In evolutionary terms, Abele & Gilchrist (1977) postulate that 'sperm competition may have led to the evolution of cement glands and capping behavior and that this may represent a preadaptation that under sexual selection may have assumed the additional function of removing male competitors from the reproductive pool'.

trehalose may be involved in maintaining internal osmolality, but there is also evidence that it may function as a 'shuttle' in the transfer of glucose to glycogen.

A functional TCA cycle is absent in acanthocephalans as is the case in a number of other intestinal helminths, and the worms probably operate as facultative anaerobes. In some acanthocephalans, the main wastes of carbohydrate metabolism are lactate and succinate, the latter being produced by carbon dioxide fixation via phosphoenolpyruvate carboxykinase, and the former as a product of glycolysis. In other species, various combinations and quantities of ethanol, lactate, butyrate, and acetate are excreted. Lipid metabolism as an energy resource is apparently unimportant. For additional information in this area, see Starling (1985).

6.2.5 Development

Following copulation, the fertilized eggs undergo early development in the ligament sac or body cavity of the female worm. Cleavage of the egg is spiral, although this pattern sometimes is distorted, giving the impression of determinate cleavage. When cell division starts, the polar

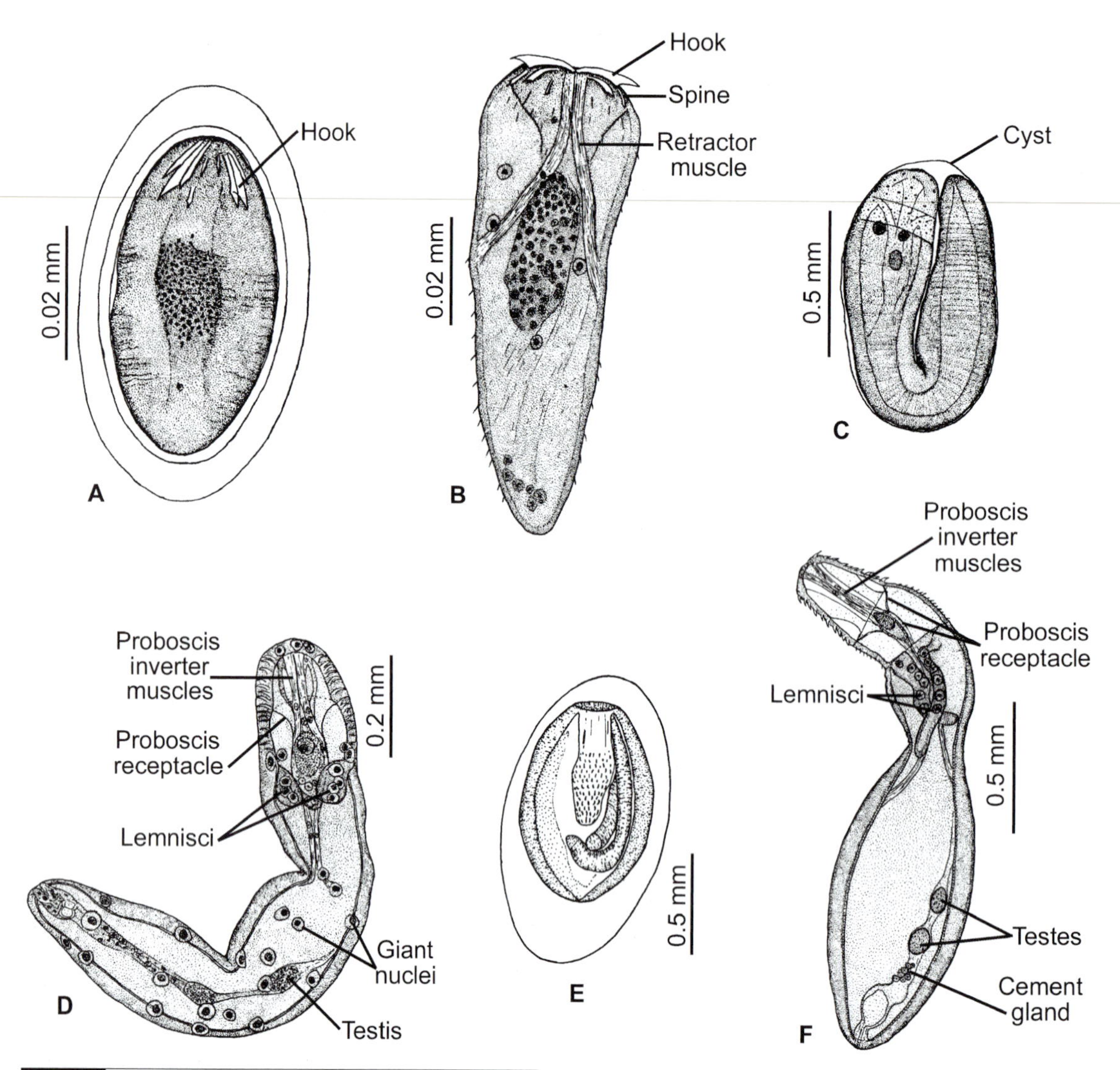

Fig. 6.9 Developmental stages of the acanthocephalan *Mediorhynchus grandis*, a parasite of grackles, crows, robins, and other birds. (A) Mature egg containing acanthor; (B) mature acanthor removed from the egg; (C) acanthella, age 24 days; (D) acanthella removed from the cyst, age 25 days; (E) cystacanth, age 27–30 days; (F) cystacanth removed from the cyst, age 27–30 days. (Modified from Moore, 1962, with permission, *Journal of Parasitology*, **48**, 76–86.)

bodies at one end of the embryo mark the future anterior end of the animal. In all other phyla, however, they mark the posterior end. Early in cell division, the cell membranes disappear and the embryo becomes syncytial.

When egg development is completed, the embryo, or **acanthor**, is surrounded by three or four membranes. The mature eggs leave the female's body through the gonopore and reach the external environment with the host feces (Fig. 6.9A). The acanthor inside the eggs is the infective stage for the next host in the life cycle and will not develop any further until the appropriate first intermediate host ingests it. The acanthor is also a resting, resistant stage. Under normal conditions, and depending on the species, an acanthor inside an egg can remain viable for several months. The acanthor is generally elongated and armed at the anterior end with six to eight hooks or spines; this spinose structure, or **aclid organ**, is used to penetrate the intestine of the first intermediate host (Fig. 6.9B).

When a suitable arthropod intermediate host

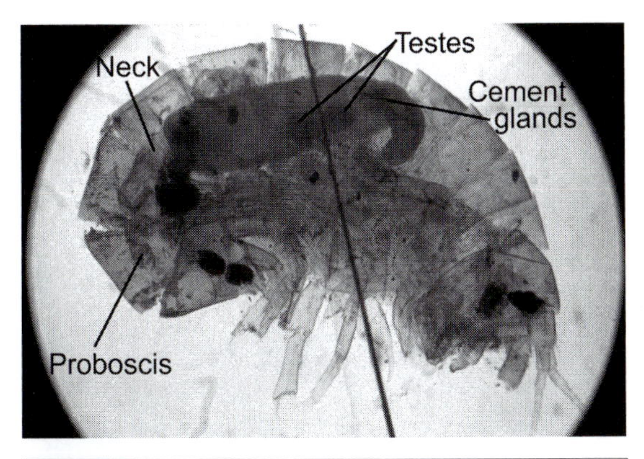

Fig. 6.10 Cystacanth of the acanthocephalan *Pomphorhynchus patagonicus* in the hemocoel of the amphipod *Hyalella azteca* from Argentina. (Photograph courtesy of Liliana Semenas, Universidad Nacional del Comahue.)

ingests an egg, the acanthor is released and bores through the intestinal wall with the aclid organ until it reaches the host hemocoel. In most cases, however, the acanthor stops under the serosa of the intestine. In either place, the acanthor begins to absorb nutrients from the host, grows, and develops primordia of all organs present in the adult. This growing stage is known as the **acanthella** (Fig. 6.9C, D). At the end of the acanthella stage, when development is completed, the worm is known as a **cystacanth**, which is the infective stage for the next host in the life cycle (Figs. 6.9E, F; 6.10). When the intermediate host infected with the cystacanth is eaten by a suitable vertebrate definitive host, the worm excysts, attaches to the wall of the intestine, and matures sexually.

6.2.6 Life cycle

The life cycle of acanthocephalans is indirect, with an arthropod intermediate host, normally an insect, a crustacean, or a myriapod, and a vertebrate definitive host (Fig. 6.11). Infection of both the intermediate and the definitive hosts is passive, which means that the potential host must eat the infective stages to become infected. In the life cycle of some species, a paratenic host intercedes between the intermediate host and the definitive host. As for other paratenic hosts, although they are not required to complete the life cycle, they usually bridge an important ecological gap at the trophic level between the intermediate and definitive hosts.

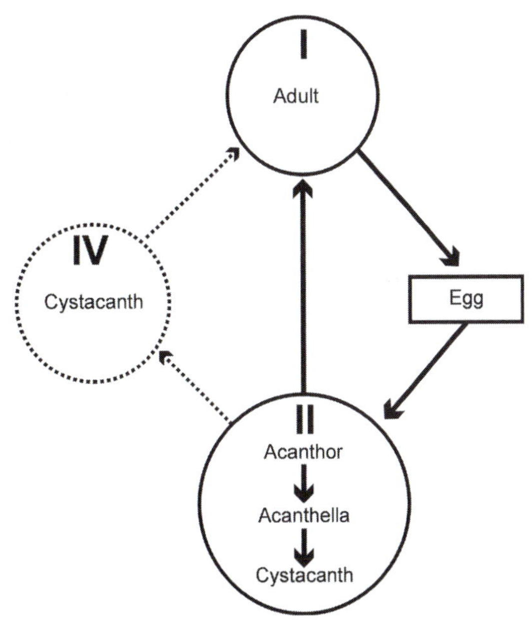

Fig. 6.11 Generalized life cycle of acanthocephalans. Host types are represented by roman numerals (I = definitive host, II = intermediate host, IV = paratenic host). Solid lines represent a 'typical life cycle', whereas dotted lines represent a less common variation. Stages within the host are enclosed in circles while stages in the environment are enclosed within rectangles.

The above pattern, in which there is an arthropod intermediate host, a vertebrate definitive host, and occasionally a vertebrate paratenic host, seems to be the norm for acanthocephalans. There are a few exceptions, one of them being *Neoechinorhynchus emydis*, a parasite of freshwater turtles that uses ostracods as intermediate hosts. Snails such as *Campeloma rufum* serve as paratenic hosts. However, in experimental infections, Hopp (1954) showed that cystacanths from snails were infective to turtles, but not cystacanths from ostracods. This indicates that the snail is required to complete the development of the cystacanth, making the snail a true second intermediate host and not a paratenic host.

6.3 Diversity

Acanthocephalans exhibit great variability in terms of definitive hosts. In contrast, they are very uniform in their site location within the

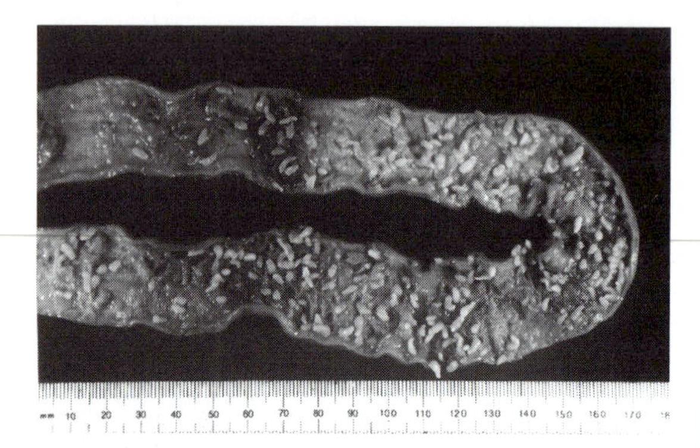

Fig. 6.12 The acanthocephalan *Oncicola canis* in the intestine of a coyote *Canis latrans* from Texas. (Photograph courtesy of Danny B. Pence, Texas Tech. University.)

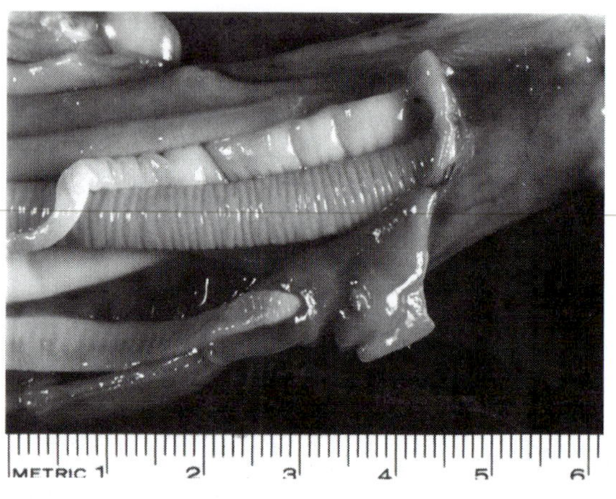

Fig. 6.13 The acanthocephalan *Macracanthorhynchus hirudinaceus* in a pig intestine. (Photograph courtesy of Tom Dunagan, Southern Illinois University at Carbondale.)

definitive host. Thus, whereas adults of these worms are found in representatives of all vertebrate classes, they almost always are constrained within rather narrow, lengthwise limits in the host's intestine. It should be noted, however, that several species undergo regular movement within the intestine presumably in response to host feeding behavior.

ARCHIACANTHOCEPHALA

Adult archiacanthocephalans parasitize birds and mammals, while their intermediate hosts are insects and myriapods. Species of *Oncicola, Macracanthorhynchus,* and *Moniliformis* are common in terrestrial mammals worldwide. *Oncicola* spp. occur in carnivorous mammals, mainly canines and felines. For example, *O. canis* parasitizes both feral and domesticated dogs, as well as coyotes in North America (Fig. 6.12), and, although the intermediate hosts are not known, armadillos, wild turkeys, and bobwhite quail reportedly serve as paratenic hosts. In Australia, *O. pomastomi* parasitizes dingoes (*Canis familiaris dingo*) and feral cats, with leopard cats serving as definitive hosts in Borneo, Malaysia, and probably the Philippines. Schmidt (1983) speculated that leopard cats from Malaysia, Borneo, and the Philippines were the original hosts and that migrating birds, which serve as paratenic hosts for *O. pomastomi,* were responsible for introducing the parasite into Australia.

A distinctive and cosmopolitan archiacanthocephalan widely used in laboratory research is *M. moniliformis;* rats are definitive hosts and cockroaches intermediate hosts. Interestingly, the gross morphology of these acanthocephalans gives them an almost tapeworm-like appearance, with the proboscis resembling a scolex and the trunk similar to a strobila. Although this evolutionary 'misrepresentation' is little more than anecdotal, it is nonetheless of interest since it raises some questions regarding the vagaries of selection pressure in the host gut.

One of the most robust of all acanthocephalans is *Macracanthorhynchus hirudinaceus*, a cosmopolitan parasite of pigs (Fig. 6.13) that uses beetles (including dung beetles) as intermediate hosts. *Macracanthorhynchus ingens* has been found in North American raccoons, mink, skunks, etc., and employs woodroaches as intermediate hosts. In Asia, *M. catulinus* is found in carnivorous mammals as adults. Both *M. ingens* and *M. catulinus* use a number of poikilothermic vertebrates, primarily snakes, as paratenic hosts.

Both *Macracanthorhynchus hirudinaceus* and *Moniliformis moniliformis* have been reported in humans, although rarely, and without causing appreciative harm (see also section 6.4). Eggs of a species of *Moniliformis*, presumably *M. clarki*, have been found in 2000-year-old human coprolites at an archeological location in Danger Cave, Utah (Moore *et al.*, 1969). Normally, adults of *M. clarki* occur in rodents and use camel crickets (*Ceutophilus utahensis*) as intermediate hosts.

PALEACANTHOCEPHALA

Paleacanthocephalans are the most diverse of the acanthocephalans in terms of their definitive hosts, parasitizing all five major groups of vertebrates. Another common feature among these parasites is that most of them infect hosts such as fishes or migratory waterfowl that are linked somehow to aquatic habitats, and employ aquatic arthropods as intermediate hosts. Four of the most common species found in freshwater fishes include *Acanthocephalus dirus* and *Pomphorhynchus bulbocolli* in North America and *A. lusci* and *P. laevis* in Europe. The two *Acanthocephalus* congeners use freshwater isopods as intermediate hosts, whereas cystacanths of *P. bulbocolli* and *P. laevis* occur in freshwater amphipods. Morphologically, the two species of *Acanthocephalus* possess a bulbous proboscis on a short, stubby neck, whereas *Pomphorhynchus* spp. have an elongated neck and a prominent proboscis. *Acanthocephalus anguillae*, which occurs in European but not North American eels, was probably the first species of acanthocephalan ever described, being initially recognized by Redi in 1684. It uses isopods as intermediate hosts.

Flesh-eating birds worldwide are commonly parasitized by species of *Centrorhynchus* (Fig. 6.1A) and *Sphaerirostris* that use insects and terrestrial isopods as intermediate hosts. However, frogs, lizards, and snakes, acting as paratenic hosts, are responsible for the successful completion of the life cycle in most cases. *Plagiorhynchus cylindratus* is a well-known species found in passerine birds, usually robins and starlings. Species of *Polymorphus*, whose intermediate hosts are freshwater amphipods, are abundant in ducks and other waterfowl around the world (Fig. 6.2A).

Species of *Corynosoma* and *Bolbosoma* infect a range of marine mammals and birds (Figs. 6.1B, 6.14). While *Bolbosoma* spp. use copepods and euphausids (krill) as intermediate hosts, and whales as definitive hosts, *Corynosoma* spp. typically use amphipods as intermediate hosts and a range of fishes as paratenic hosts in order to reach their definitive hosts: seals and piscivorous birds.

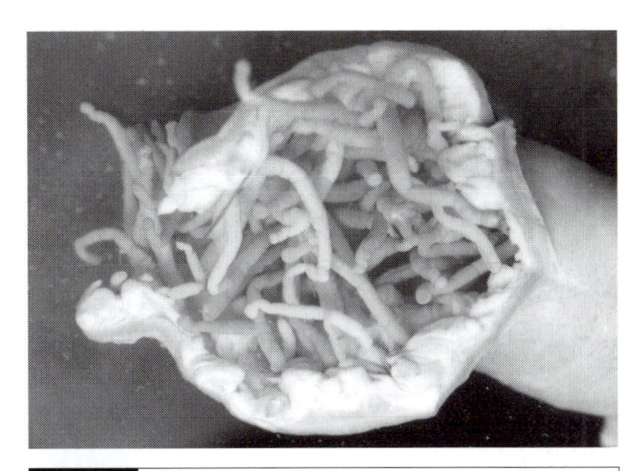

Fig. 6.14 The acanthocephalan *Bolbosoma* sp. in the intestine of a minke whale *Balaenoptera acurostrata* found stranded off the coast of Alaska. Each centimeter of intestine contained ≈144 acanthocephalans. (Photograph courtesy of Robert L. Rausch, University of Washington.)

One interesting paleacanthocephalan is *Megapriapus ungriai*, a parasite of South American potamotrygonid freshwater stingrays, and the only acanthocephalan known from elasmobranchs. Potamotrygonid stingrays originally inhabited the Pacific Ocean, but after the rising of the Andes mountain chain, they apparently became isolated in a freshwater ecosystem (see section 15.4.2) (Brooks *et al.*, 1981; Brooks, 1992). The phylogenetic position of the genus *Megapriapus* and its single species remains unclear, as well as how, when, and where it became a parasite of freshwater stingrays.

EOACANTHOCEPHALA

The eoacanthocephalans use amphibians, fishes, and reptiles as definitive hosts, but most commonly members of the latter two classes of vertebrates. In fishes, *Neoechinorhynchus cylindratus* is distributed throughout North America. It is a rather common species that occurs primarily in centrarchid fishes, with ostracods as the intermediate hosts. It exhibits an unusual life history in which the definitive host can also act as a paratenic host. For example, if a fish consumes an infected ostracod in which the cystacanth is not fully developed (infective), the parasite will migrate to a parenteric site in the fish, usually the liver or mesenteries, where it will complete its development to the infective cystacanth stage. This particular fish then, becomes a paratenic host carrying the infective cystacanth for a piscivorous fish. *Neoechinoryhnchus rutili* is the European counterpart of *N. cylindratus*, although it also has been found in cyprinid fishes in both Nebraska and Wyoming in North America. *Neoechinoryhnchus rutili* differs from *N. cylindratus* in that it does not use paratenic hosts in its life cycle, relying instead on ostracods as intermediate hosts.

An interesting complex of *Neoechinorhynchus* species has been described from several species of North American turtles. Among these eoacanthocephalans are *N. emydis, N. emyditoides, N. pseudemydis, N. chrysemydis*, and *N. chelonos*. In a study of the community structure of these acanthocephalans in the yellow-bellied slider turtle *Trachemys s. scripta* in South Carolina (USA), Aho *et al.* (1992) identified four of these species based solely on the morphological characteristics of the female worms. A starch-gel electrophoretic analysis of several enzyme loci within the same group of worms revealed the same four species. However, the individuals in these four species did not consistently match morphologically and electrophoretically. For example, one individual could be classified as one species morphologically but as a different one according to its enzymes, suggesting a high degree of species plasticity within this group of congeners, or the existence of sibling species, or both.

POLYACANTHOCEPHALA

The Polyacanthocephala is a small group that includes just a handful of species in the genus *Polyacanthorhynchus*. Three of these species parasitize species of South American caimans (Alligatoridae). The definitive host of a fourth species, *Polyacanthorhynchus kenyensis* (Fig. 6.1E), found in Africa, is not known. This species is only known from cystacanths found in fishes (paratenic hosts) in two lakes in Kenya. Because there are no alligatorids in these lakes, the definitive host might be a fish (Schmidt & Canaris, 1967; Amin & Dezfuli, 1995).

6.4 Acanthocephalans and humans

Although acanthocephalans do not normally parasitize humans, a few records exist both in recent and not so recent times (Moore *et al.*, 1969; Schmidt, 1971). The first report of a human infection by an acanthocephalan in recent times dates back to 1859, when *Macracanthorhynchus hirudinaceus* (a pig parasite) was found in a child in Prague. Apparently, this type of infection was also common in Russia where scarabeid larvae (the intermediate hosts) were eaten raw. Several other occasional reports of this parasite emanate from such places as Thailand, Brazil, and the USA. *Moniliformis moniliformis*, the rat parasite that uses cockroaches as intermediate hosts, has also been found repeatedly infecting humans throughout the world (guess how people get infected!). An interesting case that shows the extent of scientific curiosity was a paper published by B. Grassi and S. Calandruccio in 1888 reporting on the symptoms shown by the second author after voluntarily infecting himself with this parasite. Other parasites not so commonly found in humans include *Acanthocephalus bufonis*, a parasite of frogs in Indonesia that was found in the small intestine of a man during a routine autopsy. Two additional reports of acanthocephalans in humans, *Acanthocephalus rauschi* and *Corynosoma strumosum*, come from Alaskan Inuit. Infection in these cases probably occurred through the consumption of raw fish infected with cystacanths, a common practice among the Inuit (and sushi eaters too!).

6.5 Phylogenetic relationships

The phylogenetic relationships of the Acanthocephala with other animal groups are not absolutely clear, although recent phylogenetic analyses support the hypothesis that Acanthocephala and Rotifera (a small group of sessile or planktonic freshwater organisms) are sister taxa. Conway-Morris & Crompton (1982) originally suggested that Priapulida (a small group of free-living, marine, benthic worms) was the most likely sister group of the Acanthocephala. These authors indicated that the immediate ancestor of acanthocephalans might have been a marine interstitial meiofaunal worm, similar to some of the priapulid worms found in the Burgess Shale, a mid-Cambrian fossil locality in southern British Columbia, Canada. Because the first vertebrates, i.e., jawless fish, did not appear until the end of the Cambrian, it is thought that the well-represented arthropods may have been the first hosts used by acanthocephalans in that period. The incorporation of a vertebrate host to produce the present two-host life cycle was probably a later development, accomplished by the mid Paleozoic, when fishes and amphibians evolved.

Phylogenetic analyses of morphological data (Lorenzen, 1985) and 18S ribosomal RNA (Winnepenninckx *et al.*, 1995; Garey *et al.*, 1996) indicate that acanthocephalans and rotifers in the class Bdelloidea are sister groups and are more closely related to each other than to any other group of organisms. This area of research is in rapid flux and new findings will probably help clarify the position of acanthocephalans among the invertebrates and their probable time of divergence from the ancestral form, including their pattern and mechanism of evolution.

Phylogenetic analysis within the acanthocephalans based on 18S ribosomal DNA sequences indicates that the Acanthocephala include three monophyletic clades representing the three major recognized classes, i.e., Archiacanthocephala, Paleacanthocephala, and Eoacanthocephala, suggesting that the current classification of acanthocephalans is indeed natural (Near *et al.*, 1998). This latter study also suggests that acanthocephalans appear to have a strict pattern of historical association with their arthropod intermediate hosts but not with their vertebrate definitive hosts, meaning that the arthropod–acanthocephalan parasitic association is an old one in geological time.

Box 6.2 Classification of the Acanthocephala

The classification of the Acanthocephala presented here follows the one proposed by Amin (1987). This scheme is widely accepted by most researchers and it is based on the work of A. Meyer, H. J. Van Cleave, Y. J. Golvan, W. L. Bullock and O. Amin, well-known parasitologists who have made significant contributions to the study of this group.

Class Archiacanthocephala

Main longitudinal lacunar canals dorsal and ventral or just dorsal; few hypodermal nuclei; giant nuclei present in lemnisci and cement glands; two ligament sacs in the female; protonephridia present in one family; cement glands separate, pyriform. Parasites of birds and mammals; intermediate hosts are insects or myriapods.

Order Moniliformida

Proboscis cylindrical with long rows of hooks; proboscis receptacle double-walled; proboscis retractor muscles pierce posterior or postero-ventral end of the receptacle; protonephridia absent. Parasites of mammals and occasionally birds.

Family Moniliformidae

Order Gigantorhynchida

Proboscis truncate, cone-shaped, with rows of rooted hooks on the anterior portion and rootless spines on the basal portion; proboscis receptacle single-walled; proboscis retractor muscles pierce ventral wall of the receptacle; protonephridia absent. Parasites of birds and mammals.

Family Gigantorhynchidae

Order Oligacanthorhynchida

Proboscis subspherical, with nearly spiral rows of few hooks each; proboscis receptacle single-walled; proboscis retractor muscle pierces dorsal wall of receptacle; protonephridia present. Parasites of mammals and rarely birds.

Family Oligacanthorhynchidae

Order Apororhynchida

Proboscis large, globular, with diminute spine like hooks (which may not pierce the surface of the proboscis) arranged in several spiral rows; proboscis not retractable; neck absent or reduced; protonephridia absent. Parasites of birds.

Family Apororhynchidae

Class Paleacanthocephala

Main longitudinal lacunar canals lateral; nuclei of lemnisci, cement glands, and body wall fragmented; proboscis receptacle double-walled; spines present on trunk of some species; single ligament sac of female not persistent throughout life; protonephridia absent; cement glands separate, tubular to spheroid. Parasites of fishes, amphibians, reptiles, birds, and mammals.

Order Echinorhynchida

Proboscis cylindrical to spheroid; sensory papillae present or absent; proboscis receptacle double-walled; brain near middle or posterior end of receptacle. Parasites of fishes and amphibians.

Families Echinorhynchidae, Fessisentidae, Heteracanthocephalidae, Heterosentidae, Hypoechinorhynchidae, Illiosentidae, Pomphorhynchidae, Rhadinorhynchidae, Cavisomidae, Arythmacanthidae

Order Polymorphida

Proboscis spheroid to cylindrical, armed with numerous hooks in alternating longitudinal rows; proboscis receptacle double-walled, with brain near center. Parasites of reptiles (rare), birds, and mammals.

Families Centrorhynchidae, Plagiorhynchidae, Polymorphidae

Class Eoacanthocephala

Main longitudinal lacunar canal dorsal and ventral; nuclei of lemnisci and body wall few, giant; proboscis receptacle single-walled; two persistent ligament sacs in female; protonephridia absent; cement gland single, syncytial, with several nuclei; cement reservoir appended. Parasites of fish, and occasionally amphibians and reptiles.

Order Gyracanthocephalida

Trunk small or medium size, spined; proboscis small, spheroid, with a few spiral rows of hooks. Parasites of freshwater and marine fishes.

Family Quadrigyridae

Order Neoechinorhynchida

Trunk small to large, unarmed; proboscis spheroid to elongated, with hooks arranged variously. Parasites of fishes and other aquatic vertebrates.

Families Neoechinorhynchidae, Tenuisentidae, Dendronucleatidae

Class Polyacanthocephala

Trunk spinose; hypodermic nuclei many and small; main longitudinal lacunar canals dorsal and ventral; many hooks in longitudinal rows; proboscis receptacle single-walled; cement glands elongated, with giant nuclei; two ligament sacs in the female; protonephridia absent. Parasites of fishes and crocodilians.

Order Polyacanthorhynchida

With characters of the class.

Family Polyacanthorhynchidae

References

Abele, L. G. & Gilchrist, S. (1977) Homosexual rape and sexual selection in acanthocephalan worms. *Science*, **197**, 81–83.

Aho, J. M., Mulvey, M., Jacobson, K. C. & Esch, G. W. (1992) Genetic differentiation among congeneric acanthocephalans in the yellow-bellied slider turtle. *Journal of Parasitology*, **78**, 974–981.

Amin, O. M. (1987) Key to the families and subfamilies of Acanthocephala, with the erection of a new class (Polyacanthocephala) and a new order (Polyacanthorhynchida). *Journal of Parasitology*, **73**, 1216–1219.

Amin, O. M. & Dezfuli, B. S. (1995) Taxonomic notes on *Polyacanthorhynchus kenyensis* (Acanthocephala: Polyacanthorhynchidae) from Lake Naivasha, Kenya. *Journal of Parasitology*, **81**, 76–79.

Brooks, D. R. (1992) Origins, diversification, and historical structure of the helminth fauna inhabiting neotropical freshwater stingrays (Potamotrygonidae). *Journal of Parasitology*, **78**, 588–595.

Brooks, D. R., Thorson, T. B. & Mayes, M. A. (1981) Freshwater stingrays (Potamotrygonidae) and their helminth parasites: testing hypothesis of evolution and coevolution. In *Advances in Cladistics: Proceedings of the First Meeting of the Willi Henig Society*, ed. V. A. Funk & D. R . Brooks, pp. 147–175. New York: New York Botanical Garden.

Conway-Morris, S. & Crompton, D. W. T. (1982) The origins and evolution of the Acanthocephala. *Biological Reviews of the Cambridge Philosophical Society*, **57**, 85–115.

Crompton, D. W. T. (1985) Reproduction. In *Biology of the Acanthocephala*, ed. D. W. T. Crompton & B. B. Nickol, pp. 213–271. Cambridge: Cambridge University Press.

Devine, M. C. (1975) Copulatory plugs in snakes: enforced chastiy. *Science*, **187**, 844–845.

Dunagan, T. T. & Miller, D. M. (1986) A review of protonephridial excretory systems in Acanthocephala. *Journal of Parasitology*, **72**, 621–632.

Garey, R. J., Nonnemacher, M. R., Near, T. J. & Nadler, S. A. (1996) Molecular evidence for Acanthocephala as a subtaxon of Rotifera. *Journal of Molecular Evolution*, **43**, 287–292.

Grassi, B. & Calandruccio, S. (1888) Über einen *Echinorhynchus*, welcher auch in Menschen parasitirt und dessen Zwischenwirt ein Blaps ist. *Zentralblatt für Bakteriologie, Parasitenkunde, Infektionskrankheiten und Hygiene (abt. Originale)*, **3**, 521–525.

Hopp, W. B. (1954) Studies on the morphology and life cycle of *Neoechinorhynchus emydis* (Leidy), an acanthocephalan parasite of the map turtle, *Graptemys geographica* (LeSueur). *Journal of Parasitology*, **40**, 284–299.

Lorenzen, S. (1985) Phylogenetic aspects of pseudocoelomate evolution. In *The Origins and Relationships of Lower Invertebrates*, ed. S. Conway-Morris, J. D. George, R. Gibson & H. M. Platt, pp. 210–223. Oxford: Clarendon Press.

Miller, D. M. & Dunagan, T. T. (1983) A support cell to the apical and lateral sensory organs in *Macracanthorhynchus hirudinaceus* (Acanthocephala). *Journal of Parasitology*, **69**, 534–538.

Moore, J. G., Fry, G. F. & Englert, E. Jr. (1969) Thorny-headed worm infection in North American prehistoric man. *Science*, **163**, 1324–1325.

Near, T. J., Garey, J. R. & Nadler, S. A. (1998) Phylogenetic relationships of the Acanthocephala inferred from 18S ribosomal DNA sequences. *Molecular Phylogenetics and Evolution*, **10**, 287–298.

Schmidt, G. D. (1971) Acanthocephalan infections of man, with two new records. *Journal of Parasitology*, **57**, 582–584.

Schmidt, G. D. (1983) What is *Echinorhynchus pomatostomi* Johnston and Cleland, 1912? *Journal of Parasitology*, **69**, 397–399.

Schmidt, G. D. & Canaris, A. G. (1967) Acanthocephala from Kenya with description of two new species. *Journal of Parasitology*, **53**, 634–637.

Starling, J. A. 1985. Feeding, nutrition, and metabolism. In *Biology of Acanthocephala*, ed. D. W. T. Crompton & B. B. Nickol, pp. 125–212. Cambridge: Cambridge University Press.

Taraschewski, H. & Mackenstedt, U. (1991) Autoradiographic and morphological studies on the uptake of the triglyceride [3H]-glyceroltriolate by acanthocephalans. *Parasitological Research*, **77**, 247–254.

Uglem, G. L., Pappas, P. W. & Read, C. P. (1973) Surface aminopeptidase in *Moniliformis dubius* and its relation to amino acid uptake. *Parasitology*, **67**, 185–195.

Whitfield, P. J. (1970) The egg sorting function of the uterine bell of *Polymorphus minutus* (Acanthocephala). *Parasitology*, **61**, 111–126.

Winnepenninckx, B., Backeljau, T., Mackey, L. Y., Brooks, J. M., De Watcher, R., Kumar, S. & Garey, J. R. (1995) 18S rRNA data indicate that aschelminthes are polyphyletic in origin and consist of at least three distinct clades. *Molecular Biology and Evolution*, **12**, 1132–1137.

ADDITIONAL REFERENCES

Crompton, D. W. T. (1970) *An Ecological Approach to Acanthocephalan Physiology.* Cambridge: Cambridge University Press.

Crompton, D. W. T. & Nickol, B. B. (eds.) (1985) *Biology of the Acanthocephala*. Cambridge: Cambridge University Press.

Dunagan, T. T. & Miller, D. M. (1991) Acanthocephala. In *Microscopic Anatomy of Invertebrates*, vol. 4, *Aschelminthes*, ed. F. W. Harrison & E. E. Ruppert, pp. 299–332. New York: Wiley–Liss.

Petrochenko, V. I. (1956, 1958) *Acanthocephala of Domestic and Wild Animals*, vols. 1 and 2. Moscow: Akademii Nauk SSSR. [English translations: Israel Program for Scientific Translations, 1971.]

Schmidt, G. D. (1972) Revision of the class Archiacanthocephala Meyer, 1931 (Phylum Acanthocephala), with emphasis on Oligacanthorhynchidae Southwell and MacFie, 1925. *Journal of Parasitology*, **58**, 290–297.

Yamaguti, S. (1963) *Systema Helminthum*, vol. 5, *Acanthocephala*. New York: Wiley Interscience.

Chapter 7

Pentastomida: the tongue worms

7.1 General considerations

The pentastomids, linguatulids, or tongue worms, are a small group of parasites that includes about 100 species. Most adult pentastomids are found in the lungs of reptiles, mostly snakes, although a few species are found in the air sacs of marine birds, and in the nasopharynx and sinuses of canines and felines. *Raillietiella bufonis* is the only species known from an amphibian host. Larval stages or nymphs are found in a few insects and a wide variety of vertebrate hosts, including humans. Adult pentastomids are generally elongated, measure 1–16 cm in length, and their cuticle contains chitin. They have a complete digestive system with a mouth and an anus, but there is no respiratory, excretory, or circulatory system. The life cycle is indirect, with a definitive and an intermediate host.

Studies on the phylogenetic relationships of the Pentastomida, a group traditionally treated as a separate phylum, suggest that they are truly arthropods, with possible affinities to the Crustacea (Wingstrand, 1972; Riley *et al.*, 1978; Abele *et al.*, 1989; Storch & Jamieson, 1992). However, because of the lack of consensus about their definitive position within the Arthropoda, the Pentastomida will be discussed as a self-contained group without any specific taxonomic rank.

7.2 Form and function

7.2.1 The outside

Body shape and holdfast. Most pentastomids are elongated and cylindrical, but occasionally they may be flattened dorsoventrally. The body seems to be segmented, but this segmentation is only superficial, forming distinct **annuli**, or false segments (Figs. 7.1, 7.2, 7.3). The anterior end of the body, also called a head or cephalothorax, bears frontal and dorsal sensory papillae, frontal glands, hook glands, the mouth, and four openings with curved, retractile hooks (Figs. 7.4, 7.5). In the more primitive group (the Cephalobaenida), the retractile hooks are located at the tip of four protuberances (or legs) and the mouth is located at the tip of a fifth projection, also called a snout. (The name pentastomid, meaning five mouths, was erroneously coined in the belief that each of the five small protuberances had a mouth.) The mouth is constantly held open because its lining is hardened or sclerotized. The shape of the sclerotized lining, or **cadre**, is an important character used in classification. The retractile hooks are controlled by muscles and are used both to tear the host tissues and to attach to the host.

Body wall. The body surface is covered by a cuticle that contains chitin. In pentastomids the cuticle is soft, transparent, and flexible, similar to the cuticle of the larvae of holometabolous insects. These characteristics of the cuticle allow the parasites to move through very restricted

Fig. 7.1 The pentastomid *Porocephalus crotali* in the lung of a rattlesnake *Crotalus atrox*. The uterine coils of the females are clearly visible through the transparent body wall. (Photograph courtesy of John Riley, University of Dundee.)

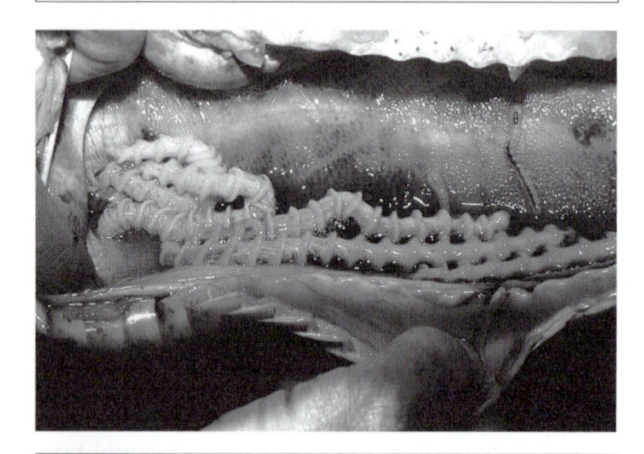

Fig. 7.2 The pentastomid *Armillifer armillatus* in the lung of a Gaboon viper. (Photograph courtesy of John Riley, University of Dundee.)

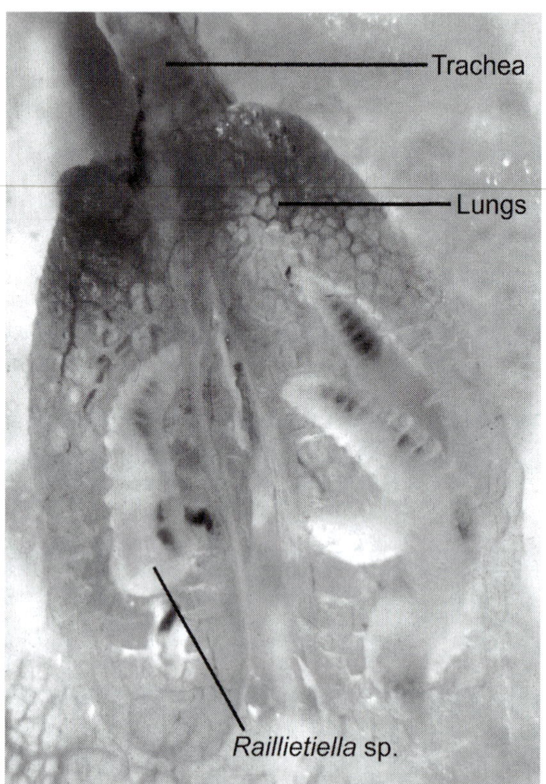

Fig. 7.3 The pentastomid *Raillietiella* sp. in the lungs of a gecko *Hemidactylus brooki* from Dominican Republic. (Photograph courtesy of John Riley, University of Dundee.)

spaces during their migration inside the host. The body surface may be annulated, appear serrated, or have spines (Figs. 7.2, 7.3, 7.4, 7.5); it also has a large number of subparietal glands. These glands, together with the frontal and hook glands of the head, produce a secretion that appears to protect the pentastomid body from the host immune system (see Box 7.1) (Riley *et al.*, 1979; Jones *et al.*, 1992; Riley, 1992).

Pentastomids have a well-developed muscular system, similar to other arthropods. However, because the cuticle is soft and flexible, it does not provide a suitable skeleton upon which muscles can act. The coelomic fluid within the body, however, provides a hydrostatic skeleton for support and locomotion. Movement, then, is accomplished by changes in the shape of the body, using the fluid skeleton as the basis for muscle antagonism. Because there is no circulatory or respiratory system, peristaltic movements of the body wall agitate the coelomic fluid, which may be used in the transport of gases. Gas exchange probably occurs through the cuticle.

7.2.2 The inside

Digestive system. The digestive system is divided into a foregut, midgut, and hindgut. As in arthropods, the fore- and hindgut are lined with cuticle. The foregut includes the buccal cavity, the pharynx, and the esophagus and is separated from the midgut by a valve that prevents the backflow of ingested food. The midgut is a straight tube without compartments and is where extracellular digestion occurs. The hindgut usually ends with an anus, except in the avian parasite *Reighardia sternae* that lacks one.

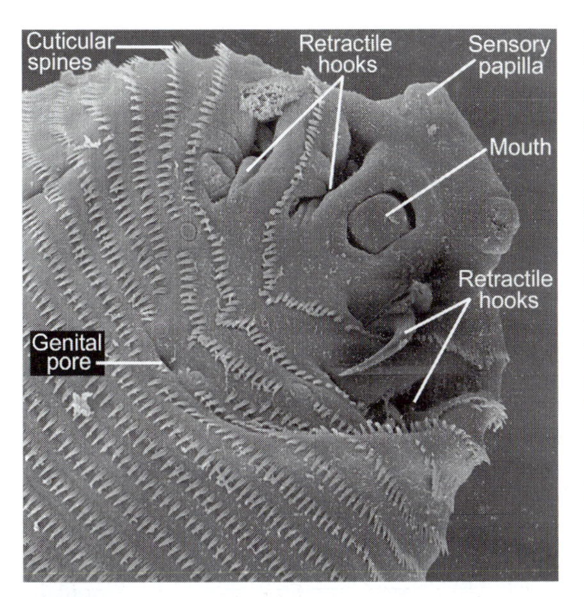

Fig. 7.4 Nymph of the pentastomid *Linguatula serrata* from the lungs and liver of an intermediate host, the Chilean deer *Pudu pudu*. (From Fernández & Villalba, 1986, with permission, *Parasitología al Día*, **10**, 29–30.)

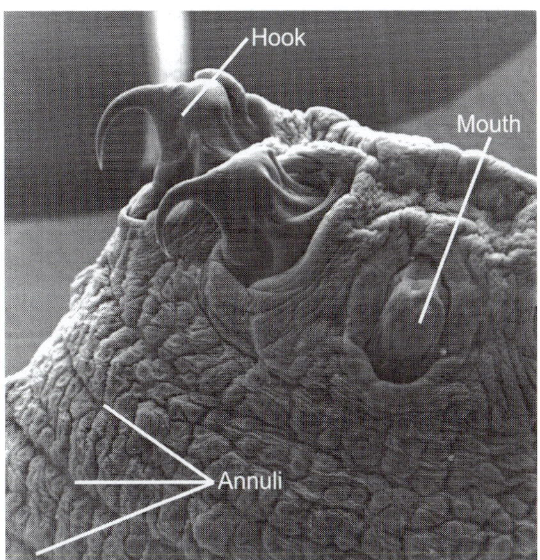

Fig. 7.5 Nymph of the pentastomid *Porocephalus crotali* from a rat intermediate host. (Photograph courtesy of John Riley, University of Dundee.)

Box 7.1 Pentastomids: masters of deception

If a parasite is to survive inside a host, it has to be able to avoid, suppress, or modulate the host's immune response. Adult pentastomids living in the lungs of vertebrates seem to have evolved a very effective mechanism to protect themselves from the attacks of their host's immune system.

One of the most striking characteristics of pentastomids is the profusion of glands covering their entire body, i.e., dorsal, hook, and subparietal glands. These glands constantly discharge a lamellate secretion into the cuticle that seems to protect the parasite surface from the host immune response. The mechanism of 'evasion' is rather interesting, although not new among parasites.

The lungs of tetrapod hosts are continuously covered by a membranous surfactant secreted by alveolar cells that seems to have a double purpose. First, this membranous surfactant lowers the alveolar surface tension at the air–water interface and facilitates the diffusion of gases, and, second, the surfactant functions as an immunosupressor, protecting the delicate gas-exchanging regions of the lungs from detrimental immune reactions to inhaled particles and microorganisms. This membranous surfactant consists of approximately 8% protein, 2% carbohydrates, and 90% lipids, of which 23% are neutral lipids and 74% are phospholipids.

Now, amazingly, but not surprisingly, it appears that the overall lipid composition of the lamellate secretions produced by the pentastomids' glands (see section 7.2.1) is very similar to that of the vertebrate pulmonary surfactant, although the proportion of the various components is different. Moreover, ultrastructural and behavioral studies cannot distinguish between membranes produced by the lamellate secretions of the parasites and the pulmonary membranous surfactant produced by the lungs of the vertebrate host.

Because most adult pentastomids are long-lived and do not seem to elicit an inflammatory response in the lungs, it is very likely that the host immune system cannot distinguish between its own membranous surfactant and the lamellate secretions of the pentastomids. It appears then, that lipids in the lamellate secretions of pentastomids mimic the pulmonary surfactant of the host and suppress the potential host immune response against the parasite (Riley, 1992).

Osmoregulatory and excretory system. Pentastomids lack an excretory system but they have specialized, epidermal ion-transporting cells called **ionocytes**, or **chloride cells**, that are responsible for osmotic and ionic regulation. These cells are presumably involved in the secretion of excess electrolytes (mainly sodium and chloride) contained in the blood and lymph they consume. These ion-transporting cells are protected from the host immune system by the secretions of the subparietal, frontal, and hook glands. Pentastomids are likely to regulate hypoosmotically, maintaining their ion concentration lower than that of the host. Other waste products, e.g., nitrogen, diffuse across the cuticle of the parasite probably in the form of ammonium ions.

Reproductive system. Pentastomids are dioecious and exhibit some degree of sexual dimorphism, with males being smaller than females (Fig. 7.6). The male has a single tubular testis (except *Linguatula* that has two) that occupies ⅓ to ½ of the body cavity and connects to the seminal vesicle, which stores the sperm until copulation. The seminal vesicle is continuous with a pair of muscular ejaculatory organs, each of which has a duct or vas deferens that reaches into a terminal penis (Fig. 7.7). The penis fits into a sclerotized dilator organ. The dilator is a complicated and often misunderstood organ in pentastomes that seems to guide the penis during copulation (Storch, 1993).

Females have a single ovary that occupies most of the body cavity. The ovary leads to one or two oviducts that continue into the uterus. There is a short muscular vagina that opens to the exterior through the gonopore. Attached to the uterus is a pair of seminal receptacles where the male deposits its sperm during copulation (Fig. 7.7).

Females apparently copulate only once in their lifetime. Males probably copulate more than once and with different females. Copulation occurs when both sexes are approximately the same size, which means that females are not yet fully mature, and their uterus is still undeveloped. After copulation, the sperm are stored in the seminal receptacle for several months until her reproductive structures are fully developed, at which time the seminal receptacle provides sperm for the fertilization of oocytes. Since females only copulate once in their lifetime, the total number of sperm held in the seminal receptacle is the limiting factor for egg production. In the Porocephalida, the uterus is tubular, elongated, extensively folded, and occupies most of the hemocoel. In this group, egg production is massive and continuous, with the uterus containing eggs in different stages of development. For example, *Linguatula serrata* (Fig. 7.4) has about 500 000 eggs at a time, and produces several million eggs in a lifetime. In the more primitive Cephalobaenidae, however, the uterus is saccate and eggs are accumulated and stored before being released. Females of *Reighardia sternae* (Fig. 7.6), for example, have about 2900 eggs in the uterus, the total number of eggs produced in their lifetime. After the eggs are released, the female dies (Banaja *et al.*, 1975).

Nervous system and sensory input. The organization of the nervous system differs between the two groups of pentastomids. The more derived Porocephalida have a pair of ventral nerve cords, and all the ganglia are fused into a subesophageal or cerebral ganglion. In the more primitive Cephalobaenida, three separate ganglia form the cerebral ganglion, with two more located ante-

Fig. 7.6 Adult male and female *Reighardia sternae* from the air sacs of a herring gull *Larus argentatus*. Note the marked sexual dimorphism in size between the sexes. (Photograph courtesy of John Riley, University of Dundee.)

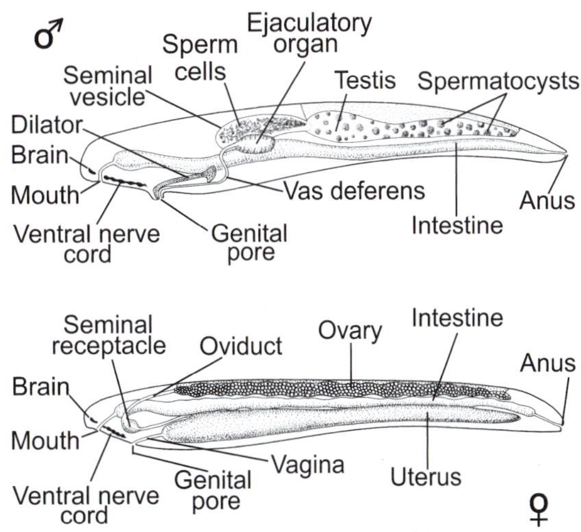

Fig. 7.7 Body plan of a generalized cephalobaenid pentastome, male and female. (Modified from Riley, J. [1983] Recent advances in our understanding of pentastomid reproductive biology. *Parasitology*, **86**, 59–83. Reprinted with the permission of Cambridge University Press.)

riorly along the ventral nerve cord, a pattern similar to the primitive arthropod nervous system. The only apparent sensory structures in pentastomids are the frontal and dorsal papillae of the cephalothorax, and the chains of lateral papillae located in between the abdominal annuli.

7.2.3 Nutrient uptake

Adult pentastomids living in the lungs and air sacs of reptiles and birds feed on blood sucked in from ruptured capillaries and lymph, whereas those living in the nasopharyngeal cavities of mammals feed on mucus and sloughed cells. Encysted nymphs in the intermediate hosts feed on blood, lymph, and lymphoid cells. Early larval stages feed predominantly on eosinophils recruited to the infection site by the inflammatory response of the host. Digestion in adult pentastomids seems to be mostly extracellular because hematin, the end product of hemoglobin digestion, accumulates in the intestinal lumen, causing it to become dark. Some intracellular digestion also must occur, however, because iron also accumulates in certain gastrodermal cells that are shed periodically into the intestinal lumen.

7.2.4 Metabolism

Very little is known about the biochemistry and metabolism of pentastomids. It appears that glycogen is the main storage product, being sequestered mainly within the striated muscles and the gastrodermis. Although details on the catabolism of *Kiricephalus pattoni* are unknown, it seems that this species, at least, is capable of some form of oxidative metabolism. It is likely that other species have similar capabilities.

7.2.5 Development

Within the uterus of a female, the fertilized oocyte develops into a **primary larva**, which is enclosed by three layers in the Cephalobaenida and four in the Porocephalida. The innermost layer is chitinous, covered by mucus, and has a characteristic opening called the **facette**. The mucus of the innermost layers is produced by the embryo's dorsal organ (discussed in the next paragraph) and flows out through the facette, surrounding the chitinous stratum. The middle layer is the chorion followed by a third layer that becomes the eggshell. The fourth layer present in Porocephalida is secreted by the reproductive tract of the female and forms a hyaline capsule when exposed to water (Storch, 1993).

By the time the egg is shed by the female pentastomid into the host's respiratory tract, it contains a fully developed primary larva that is infective to the next host in the life cycle (Fig.

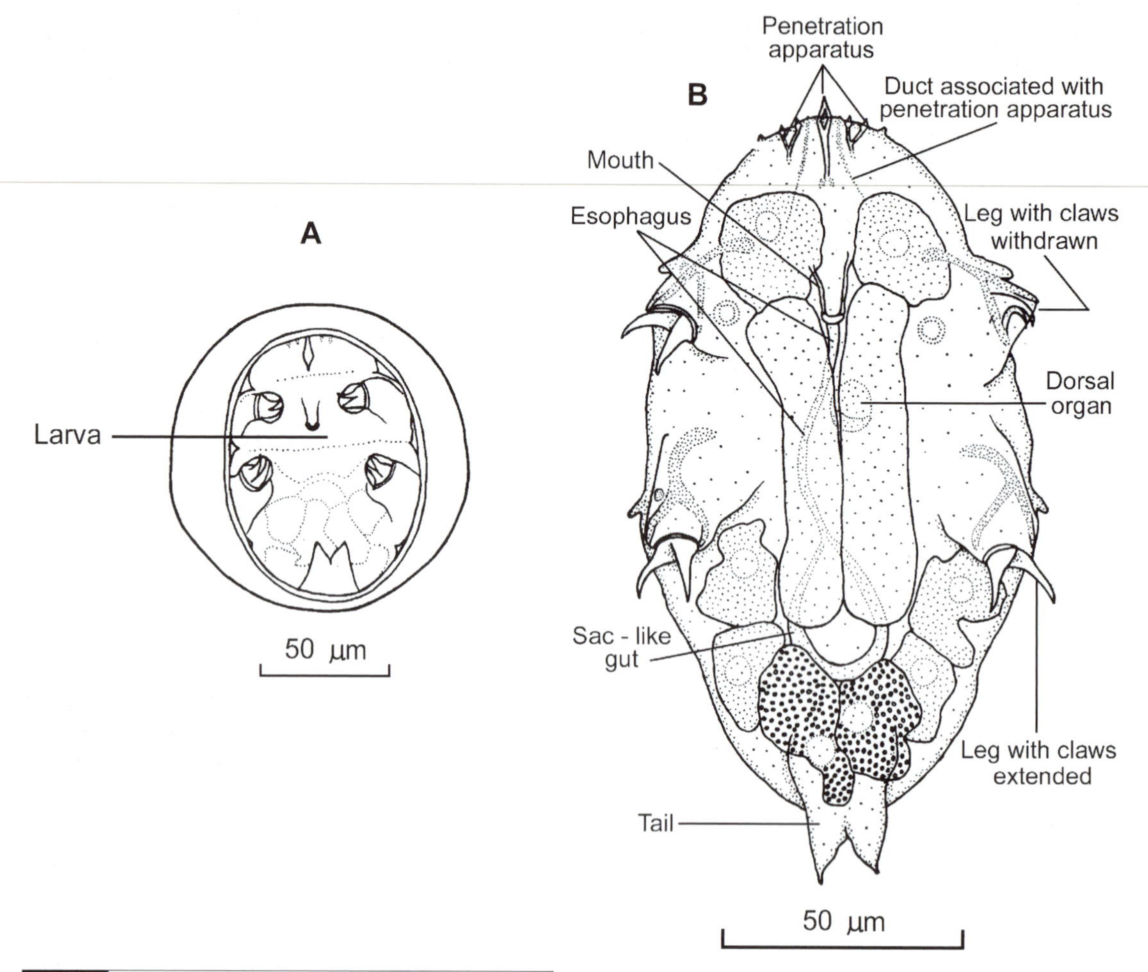

Fig. 7.8 Egg and primary larva of the pentastomid *Porocephalus crotali*. (A) Mature egg containing a fully developed primary larva; (B) ventral view of the primary larva. (Modified from Esslinger, 1962, with permission, *Journal of Parasitology*, **48**, 457–462.)

7.8A). This larva is ovoid, has a short bifurcate tail, four stumpy legs each with a pair of claws, and a penetration apparatus at the anterior end (Fig. 7.8B). The penetration apparatus, together with the clawed legs, is used to tear through the tissues of the intermediate host during the migration of the primary larva from the host intestine to the abdomen. Two ducts open within the penetration apparatus and are thought to secrete histolytic enzymes that aid in the migration process. Between the anterior pair of legs, there is a simple mouth that continues into an esophagus ending in a blind sac. In some species, a thin hindgut is present. An important characteristic of all the pentastomid larvae is the presence of a **dorsal organ** (Fig. 7.8B), which secretes a mucoid substance that becomes part of the first protective layer surrounding the larva. The dorsal organ is formed by a number of gland cells surrounding a central, hollow vesicle that opens to the surface by a dorsal pore.

When a suitable intermediate host ingests an infective egg, the primary larva hatches in the host's intestine, penetrates the intestinal wall using the penetration apparatus, and migrates into the abdominal cavity. After a number of molts in the host's abdomen, the primary larva becomes quiescent and metamorphoses into a **nymph**, which is a miniature version of the adult (Fig. 7.9). The nymph is the infective stage for the definitive host. When the intermediate host is eaten by a suitable definitive host, the nymph is freed in the

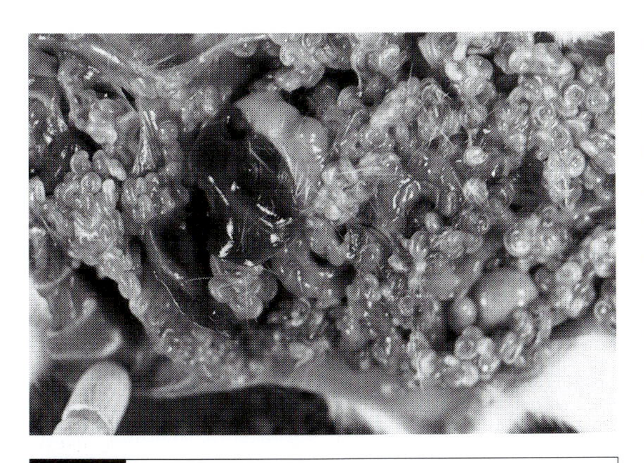

Fig. 7.9 Heavy infection by *Porocephalus crotali* nymphs in a rat intermediate host. (Photograph courtesy of John Riley, University of Dundee.)

digestive tract, penetrates the intestinal wall, and bores into the lung where it matures. In the case of *Linguatula*, the freed nymph migrates to the nasopharyngeal cavity directly up the esophagus from the stomach, without boring through tissues.

7.2.6 Life cycle

Although just a few life cycles are known for pentastomids, most species require an intermediate and a definitive host. The intermediate host in most cases is a vertebrate (fish, amphibian, reptile, or mammal), or an insect in some species of *Raillietiella*. Definitive hosts are reptiles, mainly snakes, mammals in the case of species of *Linguatula*, and sea birds for species of *Reighardia*. In those species found in the lungs of the host, eggs pass up the trachea of the host, are swallowed, and eliminated with the host feces. It is also likely that some eggs may reach the outside with the host sputum. In those species that inhabit the nasal passages and sinuses of the host, eggs are discharged with nasal secretions and while sneezing.

Intermediate hosts become infected by consuming the eggs, whereas preying upon infected intermediate hosts infects definitive hosts. Exceptions to this pattern include *Reighardia sternae, R. lomviae,* and *Subtriquetra subtriquetra*. *Reighardia sternae* (Fig. 7.6), a parasite of the air sacs of gulls and terns, has a direct life cycle. It appears that females of these pentastomids induce coughing and vomiting in the infected host and infection of other hosts occurs when they ingest the egg-infected vomit of a parasitized bird. *Reighardia lomviae*, a parasite of guillemots, has a similar life cycle, but the pattern of prevalence (very high in young birds with a sharp decline in adults by the end of the breeding season) suggests that transmission probably occurs between infected parents and the young. Infection is likely to occur when the parents feed their young with fish that were held in their mouths and thereby contaminated with pentastomid eggs (Banaja *et al.*, 1975, 1976). *Subtriquetra subtriquetra*, a parasite in the nasopharynx of South American crocodiles, is also unusual. It is the only species known to have a free-living larva that actively searches for its fish intermediate host (Winch & Riley, 1986).

7.3 Diversity

Most adult pentastomids parasitize the lungs of their reptilian hosts. The few species that use non-reptilian definitive hosts, however, occur in other structures. *Reighardia sternae* (Fig. 7.6), a parasite of seagulls and terns, and *R. lomviae,* a parasite of guillemots and puffins, are found in the air sacs of their hosts, and are the only pentastomids known from birds. Similarly, species of *Linguatula*, the only pentastomids to parasitize mammals, occur in the sinuses and nasal passages of their hosts.

Pentastomid larvae, unlike the adults, are more diverse in terms of site selection. For example, the larva of *Linguatula* (Fig. 7.4) prefers mesenteric and lymph nodes; *Porocephalus crotali* in mice and rat intermediate hosts prefers the fatty tissue around the intestine and reproductive organs (Fig. 7.9). The larva of *Sebekia* is found free in the body cavity among the viscera and *Subtriquetra* prefers the swim bladder of the fish intermediate host. The species of *Raillietiella* that use insects as intermediate hosts are found on the surface of the viscera or in the fat bodies (for more details, see Riley, 1986).

Approximately 90% of adult pentastomids parasitize reptiles and 70% of these species are found in snakes. Fishes harbor only larval stages and so do amphibians, with the exception of *Raillietiella bufonis,* a parasite of frogs. Reptiles and mammals are the most common intermediate hosts.

Armillifer is a common parasite in the lungs of snakes (Fig. 7.2). *Armillifer armillatus* and *A. grandis* infect pythons and vipers in Africa, whereas *A. moniliformis* infects pythons in Southeast Asia. A wide range of mammals, including rats, cats, monkeys, giraffes, antelopes, and sometimes

Box 7.2 | Pentastomids and humans

Although pentastomids are not common parasites of humans, they can become a zoonotic problem in some parts of the world. Humans can develop two types of pentastomiasis, visceral and nasopharyngeal.

Visceral pentastomiasis occurs when humans ingest infective eggs. Nymphs then develop and become established in the internal organs. Several species are responsible for this condition but the most common are *Linguatula serrata, Armillifer armillatus,* and *A. moniliformis. Linguatula serrata* is common in areas of the Middle East. In most instances, infections go unnoticed and are discovered during routine autopsies. Occasionally, nymphs may develop in unusual places such as the eye, but the patients recover completely after surgical removal of the misplaced parasite. Most cases of visceral pentastomiasis, however, are caused by *A. armillatus* in Central and West Africa and *A. moniliformis* in Southeast Asia. Nymphs of *Armillifer* develop and encapsulate in a thin-walled cyst, mainly in the liver, intestinal wall, and mesenteries. In older infections, these cysts calcify and can be detected using X-rays. Infections are probably acquired by the consumption of food or water contaminated with snake feces containing eggs, or by eating inadequately cleaned and undercooked snake meat. The first scenario (contaminated water or food) is common in desert areas where the few water holes are shared by humans and other animals, increasing the chances of transmission. The second mode of transmission (snake meat) is common among ethnic groups and tribes where snake meat is a routine source of animal protein. Among Malaysian aborigines who consume snake meat, for example, the prevalence of visceral pentastomiasis by *A. moniliformis* can reach up to 45%. Although the pathology is restricted to inflammatory and granulomatous responses, an infection may sensitize a person in such a way that a subsequent infection may elicit a strong allergic response.

Nasopharyngeal pentastomiasis results from the ingestion of nymphs of *L. serrata* that, upon ingestion, migrate and establish in the nasal passages of humans. This condition, also known as halzoun, marrara syndrome, or nasopharyngeal linguatulosis, can have dramatic effects. Infections are acquired by eating raw or undercooked meat, or viscera, of domestic herbivores harboring encapsulated nymphs. In rural areas of Lebanon, for example, consumption of raw liver from goats and sheep is relatively common; in the Sudan, a similar dish consists of raw stomach, liver, lung, and trachea of sheep, goats, cattle, and camels. The symptoms of infection appear soon after the ingestion of infected meat, beginning as an itching sensation in the throat and ears, followed by congestion of the buccopharyngeal mucosa, larynx, Eustachian tubes, nasal mucosa, and even lips. Nasal and lachrymal discharges, together with violent coughing and sneezing, may help dislodge the nymphs and produce immediate relief of the symptoms. If complications arise, they may include abscesses of the auditory canal, facial swelling and paralysis, and, sometimes, asphyxiation and death.

humans, serve as intermediate hosts, although humans and big game animals represent dead-end hosts because they are rarely prey for snakes. Several species within the well-known genus *Porocephalus* parasitize snakes in North, Central, and South America; all of these species use mammals as intermediate hosts. *Porocephalus crotali* is probably the best known of these species (Fig. 7.1, 7.5, 7.9). It is commonly found in the lungs of crotalid snakes, especially in rattlesnakes in western North America, where rodents and small mammals such as raccoons, opossums, muskrats, and armadillos, are intermediate hosts.

Species of *Sebekia* parasitize North, Central and South American crocodilians and use a variety of fishes as intermediate hosts. *Sebekia mississippiensis* is found in the bronchi of the American alligator (*Alligator mississippiensis*) and nymphs have been found in freshwater fishes, snakes, turtles, and mammals (Overstreet *et al.*, 1985).

Linguatula is a relatively common parasite (Fig. 7.4) found in the nasal sinuses and nasal passages of carnivorous mammals such as canids, felids, and hyaenids. Large grazing herbivores, and sometimes humans, are intermediate hosts. *Linguatula serrata* is a cosmopolitan parasite of dogs, wolves, foxes, and, on occasion, humans, with cattle, sheep, and goats serving as intermediate hosts.

Although most pentastomids occupy the lumen of the host lungs and a few occupy the sinuses and nasal passages, they are normally able to move freely, changing their feeding site. However, certain porocephalid females, i.e., *Leiperia, Waddycephalus, Kiricephalus, Cubirea, Parasambonia,* and *Elenia*, are permanently buried in the epithelium of the lungs and are unable to move. A host reaction at the attachment site of these pentastomids further encloses the anterior end with a fibrotic capsule, sequestering the parasite.

7.4 Phylogenetic relationships

The position of the Pentastomida has long been discussed. Historically, the pentastomids have been regarded as a distinct and isolated phylum with obscure relationships. Among the proposed relationships, they have been considered as related to the Tardigrada and Onycophora, or derived from annelids. Most modern taxonomists, however, recognize a strong relationship with the arthropods. Work by Wingstrand (1972), Riley *et al.* (1978), Abele *et al.* (1989), and Storch & Jamieson (1992) on the phylogenetic relationships of pentastomids, using both morphological and molecular data, supports the view that pentastomids are arthropods, with strong affinities to the Crustacea, in particular the Branchiura.

From an evolutionary point of view, and assuming that pentastomids arose from crustacean-type ancestors, Riley *et al.* (1978) propose that pentastomids became highly adapted to reptilian hosts early in their evolutionary history because about 80% of the genera and 90% of the living species parasitize reptiles. The association could have originated either in the Paleozoic among the freshwater reptiles of the Carboniferous and Permian, or in the Mesozoic, when a considerable proportion of the then-existing reptiles became secondarily adapted to a marine existence.

Box 7.3 Classification of Pentastomida

Although current evidence suggests that the pentastomids are probably arthropods, we treat them as an independent group, without assigning them any specific taxonomic ranks, since a consensus about their exact position has not been reached yet. The classification scheme follows Riley (1986). Recently, however, Almeida & Christoffersen (1999) analyzed the relationships among pentastomid groups using phylogenetic methods. Their study indicates that the current classification is not completely accurate in terms of their evolutionary relationships and that some changes are needed.

Pentastomida *incertae sedis*

Order Cephalobaenida

Mouth almost terminal, anterior to hooks; female genital pore at the anterior end of the abdomen.

Family Cephalobaenida. Parasites of snakes, lizards and amphibians. *Cephalobaena, Raillietella*

Family Reighardiidae. Parasites of marine birds. *Reighardia*

Order Porocephalida

Mouth subterminal, between or below the level of the anterior hooks; female genital pore near the posterior end of the body.

Family Sebekidae. Parasites of crocodilians and chelonians. *Sebekia, Alofia, Leiperia*

Family Subtriquetridae. Parasites of crocodilians. *Subtriquetra*

Family Sambonidae. Parasites of monitor lizards and snakes. *Sambonia, Elenia, Waddycephalus, Parasambonia*

Family Diesingidae. Parasites of chelonians. *Diesingia*

Family Porocephalidae. Parasites of snakes. *Porocephalus, Kiricephalus*

Family Armilliferidae. Parasites of snakes. *Armillifer, Cubirea, Gigliolella*

Family Linguatulidae. Parasites of mammals. *Linguatula*

References

Abele, L. G., Kim, W. & Felgenhauer B. E. (1989) Molecular evidence for inclusion of the phylum Pentastomida in the Crustacea. *Molecular Biology and Evolution*, **6**, 685–691.

Almeida, W. O & Christoffersen, M. L. (1999) Evolutionary pathways in Pentastomida. *Journal of Parasitology*, **85**, 695–704.

Banaja, A. A., James, J. L. & Riley J. (1975) An experimental investigation of a direct life-cycle in *Reighardia sternae* (Diesing, 1864) a pentastomid parasite of the herring gull (*Larus argentatus*). *Parasitology*, **71**, 493–503.

Banaja, A. A., James, J. L. & Riley J. (1976) Some observations on egg production and autoreinfection of *Reighardia sternae* (Diesing, 1864) a pentastomid parasite of the herring gull (*Larus argentatus*). *Parasitology*, **72**, 81–91.

Jones, D. A. C., Henderson, R. J. & Riley, J. (1992) Preliminary characterization of the lipid and protein components of the protective surface membranes of a pentastomid *Porocephalus crotali*. *Parasitology*, **104**, 469–478.

Overstreet, R. M., Self, J. T. & Vliet, K. A. (1985) The pentastomid *Sebekia mississippiensis* sp. n. in the American Alligator and other hosts. *Proceedings of the Helminthological Society of Washington*, **52**, 266–277.

Riley, J. (1986) The biology of pentastomids. *Advances in Parasitology*, **25**, 48–128.

Riley, J. (1992) Pentastomids and the immune response. *Parasitology Today*, **8**, 133–137.

Riley, J., Banaja, A. A. & James, J. L. (1978) The phylogenetic relationships of the Pentastomida: the case for their inclusion within the Crustacea. *International Journal for Parasitology*, **8**, 245–254.

Riley, J., James, J. L. & Banaja, A. A. (1979) The possible role of the frontal and sub-parietal gland systems of the pentastomid *Reighardia sternae* (Diesing, 1864) in the evasion of the host immune response. *Parasitology*, **78**, 53–66.

Storch, V. (1993) Pentastomida. In *Microscopic Anatomy of Invertebrates*, vol. 12, *Onycophora, Chilopoda, and Lesser Protostomata*, ed. F. W. Harrison & M. E. Rice, pp. 115–142. New York: Wiley–Liss.

Storch, V. & Jamieson, B. G. M. (1992) Further spermatological evidence for including the Pentastomida (tongue worms) in the Crustacea. *International Journal for Parasitology*, **22**, 95–108.

Winch, J. M. & Riley, J. (1986) Studies on the behavior and development in fish of *Subtriquetra subtriquetra*: a uniquely free-living pentastomid larva from a crocodilian. *Parasitology*, **93**, 81–98.

Wingstrand, K. G. (1972) Comparative spermatology of a pentastomid, *Raillietella hemidactyli*, and a branchiuran crustacean, *Argulus foliaceus*, with a discussion of pentastomid relationships. *Kongelige Danske Videnskab Selskab Biologiske Skrifter*, **19**, 1–72.

Chapter 8

The Arthropoda

The Arthropoda is the largest and most diverse of all the animal phyla. About 1 million species of arthropods have been described, comprising between 50% and 75% of all living species. Most are free-living and are found in almost any conceivable habitat. A substantial number of insects, crustaceans, and arachnids are parasitic, and many others are micropredators serving as vectors, transmitting infective stages of parasites or microorganisms to vertebrates.

Arthropods are coelomates, bilaterally symmetrical, and segmented, with jointed appendages. Segmentation of the body, or **metamerism**, is an important feature of all arthropods. The body is divided into similar parts, arranged in a linear series along the antero-posterior axis. In most adult arthropods, the original segmentation is masked because the body segments are combined into functional groups forming the different regions of the body. This process is known as **tagmatism**. The different regions of the body formed by the fusion of segments are called **tagmata** (singular = **tagma**). The body of arthropods usually is formed of between 10 and 25 segments, which are grouped into two or three tagmata. These tagmata are called the prosoma and opisthosoma in chelicerates; head, thorax, and abdomen in insects and most crustaceans; and, cephalothorax and abdomen in some other crustaceans. Tagmatism has often been lauded as being instrumental in the overall success of arthropods because it enables the appendages of all its segments to perform more efficiently and to operate as a unit. Another important trend in arthropods, often important in their success, is **cephalization**, the concentration of several vital functions into the anterior end, or head. Cephalization generally involves the location of the feeding structures, the brain, and most sensory receptors.

A continuous chitinous exoskeleton, or cuticle, covers the surface of the body and appendages, protecting and supporting the organism. The exoskeleton consists of hardened regions, or plates, connected by thinner, and flexible, articular membranes that permit the movement of the jointed appendages. The presence of an exoskeleton plays a significant role in the evolution and success of arthropods but, at the same time, severely constrains growth, which is accomplished by a number of molts.

Besides the most characteristic arthropodan features already mentioned (metamerism, tagmatism, cephalization, and exoskeleton), arthropods have a well-developed nervous system with sensory structures, a complete digestive system, an open circulatory system with a hemocoel, and varied respiratory and osmoregulatory structures. The sexes are separate (dioecious), fertilization is internal, females produce eggs rich in yolk, and an embryo usually hatches into some sort of feeding larva that molts repeatedly to form the adult.

Our understanding of the phylogeny of arthropods is in a state of constant flux, and there is little agreement among some authors regarding their origin and evolution; perhaps not too surprising considering the enormous diversity within the group (see section 8.4). One of the widely accepted classification systems recognizes four main divisions as most likely to represent the main evolutionary lines. These include the Trilobitomorpha (extinct trilobites), Chelicerata, Uniramia, and Crustacea. However, studies on the phylogenetic relationships of the Pentastomida (see Chapter 7), a group of parasites traditionally treated as a separate phylum, suggest that they are truly arthropods, and probably a sister group of the Branchiura within the Crustacea. Because of the lack of consensus about their definitive position within the Arthropoda, the Pentastomida were discussed as a self-contained group (previous chapter) with the belief that further studies will clarify their phylogenetic affinities and evolutionary history within the Arthropoda.

8.1 Crustacea

8.1.1 General considerations

The Crustacea are morphologically diverse and include familiar forms such as crabs, barnacles, shrimps, crayfish, and pill bugs (woodlice). Most crustaceans are aquatic and mainly marine. Of the 50 000 or so species of Crustacea currently known, about 3000 are parasites, infecting a wide array of hosts, from cnidarians to vertebrates. Some of these parasites are so highly modified that, superficially, they do not resemble arthropods at all.

In the classification system adopted in this book (Bowman & Abele, 1982) the Crustacea are divided into six classes, namely Cephalocarida, Remipedia, Branchiopoda (daphnia, brine shrimp), Ostracoda (ostracods), Maxillopoda (copepods, barnacles, fish lice), and Malacostraca (isopods, amphipods, shrimps, crabs, crayfishes) (see Box 8.4). Only the latter two classes include parasitic species and, accordingly, most of what follows will focus on them. Most of the parasitic crustaceans are copepods. Some are ectoparasitic on marine and freshwater fishes, whereas others are ecto- or endoparasites of marine invertebrates. Branchiurans, or fish lice, are parasites of marine and freshwater fishes. Cirripedians (barnacles), amphipods, and isopods are all parasites of marine organisms.

Many of the morphological adaptations to parasitism include reduction or loss of segmentation and appendages and, in many, development of a worm-like body. Generally, females have more morphological adaptations to parasitism than males; in many cases, males are significantly smaller than females and some are dwarfs, living permanently attached to their female partner.

8.1.2 The outside

Body shape. A typical crustacean has a head, a thorax, and an abdomen, although often one or several thoracic segments are fused to the head, forming a cephalothorax, which is normally covered by a **carapace**. The head has five segments with five pairs of appendages. The **antennules** (1st antennae of some authors) and **antennae** (2nd antennae) are usually sensory structures, but in some parasitic species they are used for attachment to the host. Next, the **mandibles**, **maxillules** (1st maxillae), and **maxillae** (2nd maxillae) are all used as feeding appendages, except in some copepods where they are also used for attachment. Behind the head is the thorax and the thoracic appendages, or **pereiopods**. Most thoracic appendages are used in locomotion, although one or more of them may be incorporated into the mouthparts and used for feeding. When this happens, they are called **maxillipeds**. The abdomen bearing abdominal appendages, or **pleopods**, follows the thorax, and it ends in a **telson**. The telson is flanked by the most posterior abdominal appendages, known as **uropods**. Pereiopods and pleopods can be modified for walking, swimming, or copulation.

Parasitic copepods are the most morphologically diverse of all crustaceans, ranging from the characteristic free-living body plan to highly modified forms that show no resemblance to typical copepods or even crustaceans (Fig. 8.1). Several key morphological changes or trends can be seen among copepods that have adopted a parasitic life style. First, there is a reduction in locomotory appendages. Locomotion in free-living copepods is achieved mainly by the antennules, antennae, and thoracic legs, which are moved in a rapid, oar-like motion. In parasitic species, however, the antennules are reduced and the antennae of most are equipped with claws used for attachment. With some exceptions, the tho-

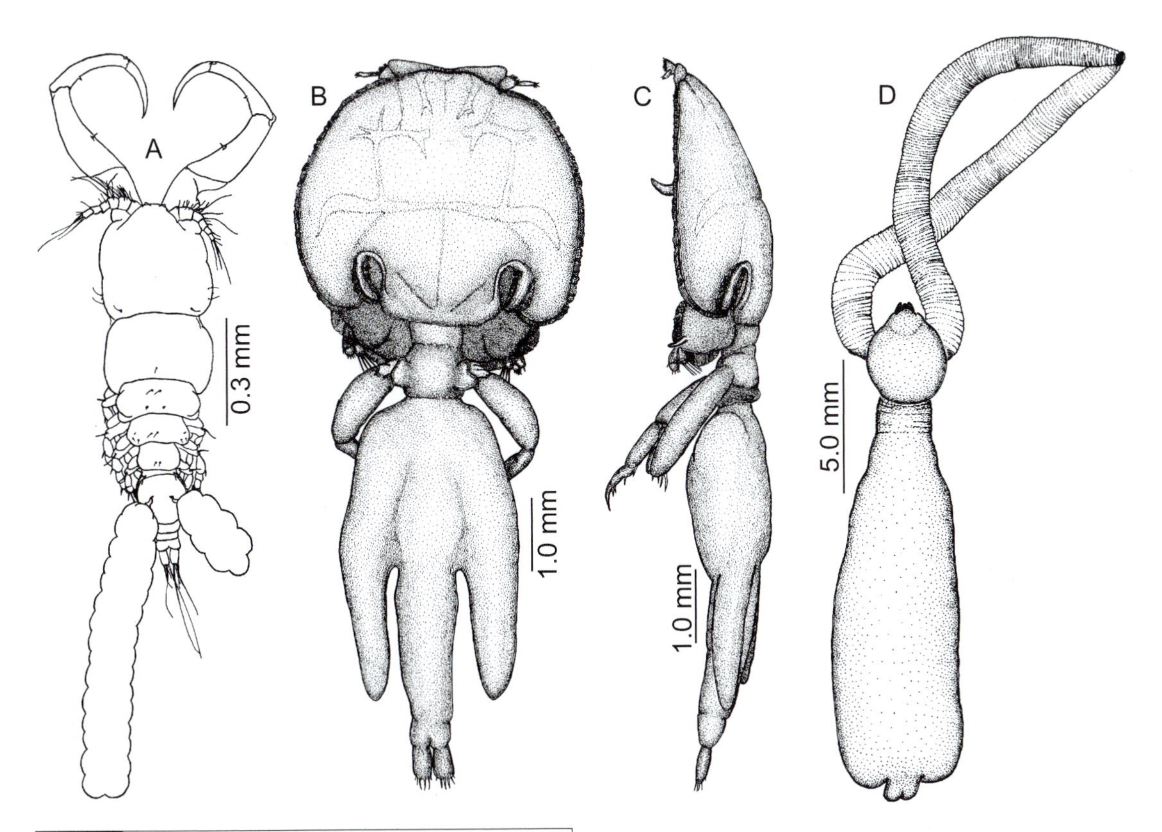

Fig. 8.1 Morphological variability in copepods. (A) Female of *Ergasilus versicolor* from the gills of a catfish, dorsal view; (B) female of *Paralebion elongatus* from the mouth of whitetip sharks, dorsal view; (C) female of *P. elongatus*, lateral view showing compressed body; (D) female of *Ommatokoita elongata* from the eye of a sleeper shark, dorsal view. ((A) From Roberts, 1969, with permission, *Journal of the Fisheries Research Board of Canada*, **26**, 997–1011; (B, C) from Benz et *al.*, 1992, with permission, *Journal of Parasitology*, **78**, 1027–1035; (D) from Benz et *al.*, 1998, with permission, *Journal of Parasitology*, **84**, 1271–1274.)

racic legs also are reduced in most copepod species. Second, there is an increase in the size of parasitic species accompanied by changes in their body proportions. Free-living copepods normally range from 0.5 to 2 mm in length. Although some parasitic groups still remain relatively small, most are larger than their free-living counterparts. One species, parasitic in whales, exceeds 30 cm. Third, there is fusion of body segments and loss of evidence of segmentation. In less modified copepods such as the Ergasilidae (Fig. 8.1A), Bomolochidae, and Lichomolgidae, segmentation is still apparent and the body maintains many of its free-living characteristics; the Lernaeidae, Pennellidae, and Lernaeopodidae (Fig. 8.1D), on the other hand, have lost almost all evidence of segmentation. Finally, because of the need to grasp a host, most parasitic copepods had to either modify existing appendages or develop new structures for this purpose. These modifications are discussed in the next section on holdfast mechanisms.

Similar morphological trends can be observed in other crustacean groups with free-living, as well as parasitic, species such as isopods, amphipods, and cirripedes. Isopods parasitic on fishes, such as the Cymothoidae and Gnathiidae, have relatively few deviations from the free-living body plan (Fig. 8.2). In groups such as the epicaridean isopods, however, some of the adult females have lost most of their external segmentation and appendages (see also Fig. 8.17).

One of the few amphipod groups that is truly parasitic is the Cyamidae. Cyamids are ectoparasites of whales in which the abdomen is vestigial,

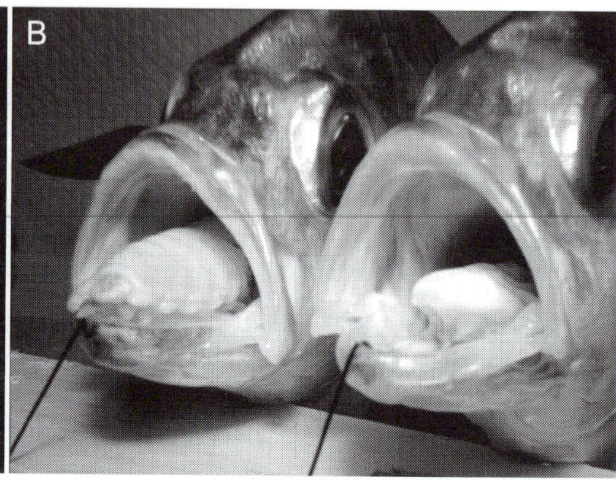

Fig. 8.2 (A) Male and female individuals of the isopod *Cymothoa oestrum* from the buccal cavity of the horse eye jack *Caranx latus*. (B) The oral cavity of the fish on the left is greatly obliterated by the presence of the isopod attached to the tongue. The fish on the right is not infected. (From McDermott, 1974, with permission, *Newsletter Bermuda Biological Station for Research*, **4**, 2.)

most legs are modified for attachment, and the body is flattened dorso-ventrally (like isopods) instead of laterally as in the typical amphipod body plan (Fig. 8.3). This is clearly an important adaptation for their ectoparasitic habitat.

The parasitic Cirripedia are probably the least variable group in terms of their morphological characteristics. Rhizocephalan cirripedes, for example, do not have any appendages or exhibit segmentation, not even vestigial, and their bodies form root-like processes that ramify through the tissues of the host (Fig. 8.4). Only their first larval stage retains crustacean-like characteristics.

The Branchiura are exclusively parasitic and do not have any free-living counterparts. Morphologically, they resemble caligid-like copepods. Although the abdomen of the branchiurans has lost all traces of segmentation, the thorax still bears four pairs of legs used for swimming (Fig. 8.5).

Hanging on. The development of a holdfast organ is an important adaptation to parasitism in ectoparasites. Again, copepods exhibit the highest diversity of attachment structures within the Crustacea. Caligid- and bomolochid-like copepods are ectoparasites of fishes, relying primarily on their modified cephalothorax and, to some extent, their antennae for attachment. The cephalothorax is concave on the ventral side and has a membrane on its perimeter that, when applied to the surface of the host, seals the ventral cavity (Fig. 8.1B, C). This cavity is then partially evacuated of water creating a vacuum that presses the copepod against the surface of the host. This kind of attachment allows the copepods to move freely on the surface of the fish.

Chondracanthid and ergasilid copepods rely exclusively on their antennae for attachment (Fig. 8.1A). The Lernaeopodidae, a highly modified group of ectoparasitic copepods of fishes, developed an efficient structure for attachment called the **bulla**. The bulla extrudes through the cephalothorax of the copepod and becomes embedded in a specially excavated cavity in the tissues of the fish. The maxillae are also used for attachment and are normally fused to the bulla (Fig. 8.1D). Some other copepods parasitic on fishes, e.g., Pennellidae, Lernaeidae, and Sphyriidae, have also developed unique holdfast structures to anchor the copepod permanently to the host. The holdfasts of many species reach deep into some of the host's organs causing extensive destruction of the tissues they penetrate (Fig. 8.6; see also Figs. 8.9, 8.14). In some species, the holdfast may have additional functions and may be used for feeding (see section 8.1.4).

Copepods such as the Philichthyidae are very specialized endoparasites of the sinuses and lateral lines of teleosts and elasmobranchs (Fig. 8.7). Because the copepods live in a cavity where there is little chance of being removed, all

Fig. 8.3 The amphipod *Cyamus* sp. from a grey whale. (Photograph courtesy of Tim Goater, Malaspina College.)

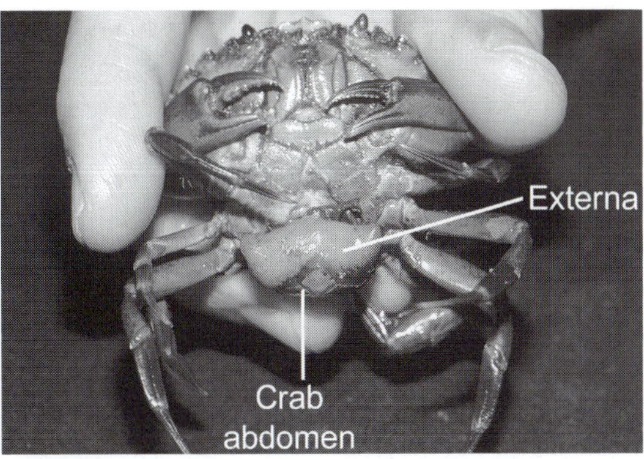

Fig. 8.4 Green crab *Carcinus maenas* infected with the rhizocephalan *Sacculina carcini*. Note the externa of *S. carcini* under the crab's abdomen. (Photograph courtesy of Todd Huspeni, University of California, Santa Barbara.)

appendages, including those that could be used for attachment, are reduced.

The structures used for attachment in the other crustaceans groups are not as varied as those used by copepods. In some Branchiura, for example, the maxillulae can be modified into large sucking discs (Fig. 8.5), similar to the suctional capability of the cephalothorax in caligid-like copepods. In ectoparasitic isopods and amphipods, modified legs with claws are used for attachment to the host. On the other hand, many of the endoparasitic species of amphipods, isopods, and cirripedes are found in tissues or closed cavities where there is no risk of removal and, consequently, their bodies have lost most, or all, of their appendages, including those traditionally used for attachment.

Body wall. The chitinous exoskeleton of crustaceans consists of several layers containing protein, lipid, and polysaccharides, all secreted by the underlying epidermis. The outermost layer of the exoskeleton is the epicuticle, which includes mostly chemically inert proteins. Beneath the epicuticle lies the endocuticle, which is formed by three layers. The outer layer is calcified and pigmented with tanned proteins; the middle layer is relatively thin, calcified, untanned, and unpigmented; the inner layer is thin, uncalcified, and untanned. Beneath the endocuticle is the epidermis that secretes all the overlying cuticular layers. Under the epidermis are clusters of tegumental glands that open via ducts into the surface of the exoskeleton. The secretions from these glands are involved in the production of the epicuticle. The middle and inner layers of the endocuticle contain chitin and protein, but the protein is not sclerotized, i.e., hardened. These layers are membranous and flexible, and become the articular membranes of the junctions.

Despite their abundance, little is known about the nature of the cuticle in parasitic crustaceans. In the least-modified forms, the cuticle probably conforms to the basic pattern found in most free-living crustaceans but, in the highly modified parasitic crustaceans, it is not always clear what kind of changes occur to adapt to their particular life styles. During the earlier life-cycle stages, parasitic copepods grow like all other crustaceans, by stepwise increase in size through a series of molts. However, once they reach sexual maturity, cope-

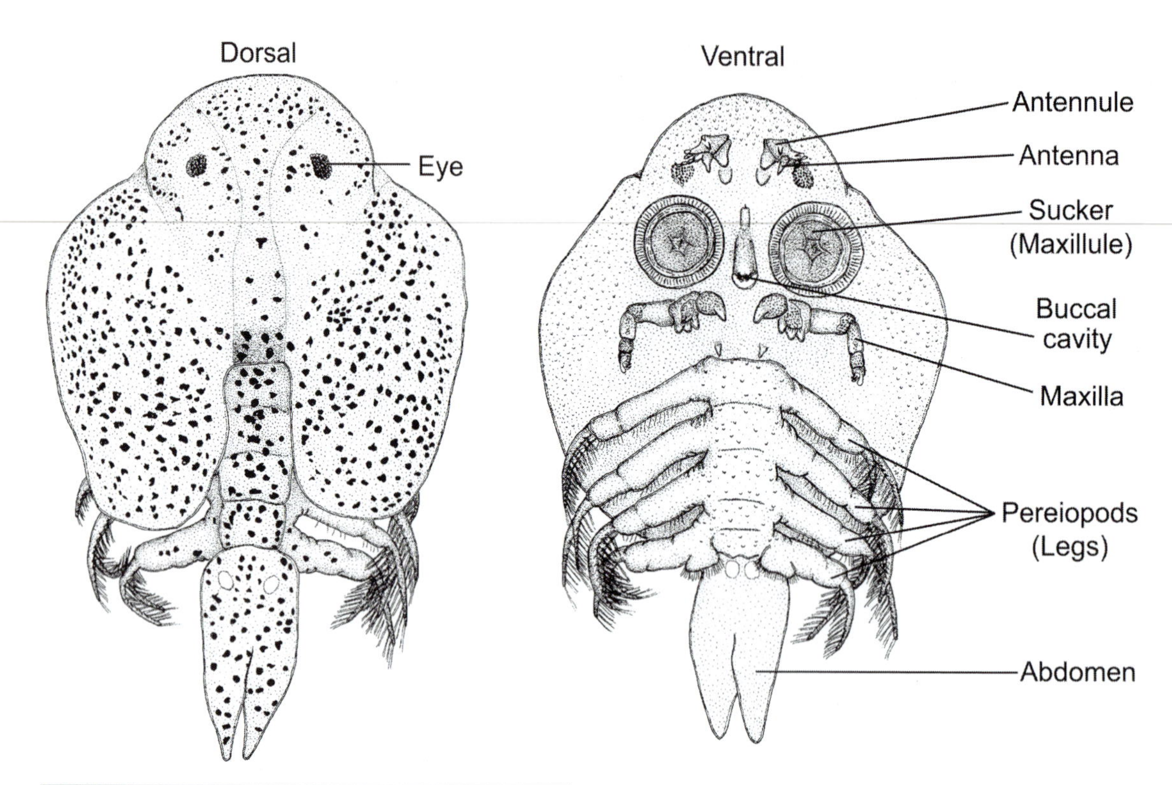

Fig. 8.5 Female individual of the branchiuran *Argulus melanosticus* from the California grunion *Leuresthes tenuis*. (Modified from Benz *et al.*, 1995, with permission, *Journal of Parasitology*, **81**, 754–761.)

pods stop molting, but many continue to grow gradually. It is not quite clear how this is accomplished, but the epicuticle of at least one species, *Pennella elegans*, does not contain amino acids with sulfur or aromatic rings, which are normally involved in hardening. The proteins that make up the cuticle of *P. elegans* also contain many di- and trityrosine links. It appears that these links help stabilize resilin, a rubber-like compound present in the cuticle of arthropods, that helps keep the cuticle flexible (Kannupandi, 1976*a*, 1976*b*). A similar study on the cuticle of *Caligus savala* did not show any special features, but somehow the hardening of the cuticle does not occur until the copepod reaches its definitive size.

In the more modified copepods, the cuticle has also been modified in response to the particular habitat they occupy. Some endoparasitic copepods have a thin, modified exoskeleton, in which only the epicuticle is present, and in some mesoparasitic copepods, there are numerous microvillosities that seem to be involved in the absorption of nutrients (see section 8.1.4).

8.1.3 The inside

Digestive system. In most Crustacea, the digestive tract is a fairly straight tube without many modifications. As in all arthropods, it is divided into foregut, midgut, and hindgut. The foregut includes the esophagus and stomach. The stomach may be equipped with chitinous ridges, teeth, or calcareous ossicles that help grind food. The midgut varies in size and contains a variable number of ceca or diverticula where enzymes are secreted and food is absorbed. Sometimes the ceca are modified to form a large solid digestive gland, or hepatopancreas. The hindgut or intestine is a single narrow tube that terminates in an anus. The gut is equipped with muscles used in defecation and, in some cases, there are dilator muscles used for anal pumping. These muscles also seem to aid in intestinal peristalsis and in the circulation of fluids through the body.

This generalized digestive system describes the

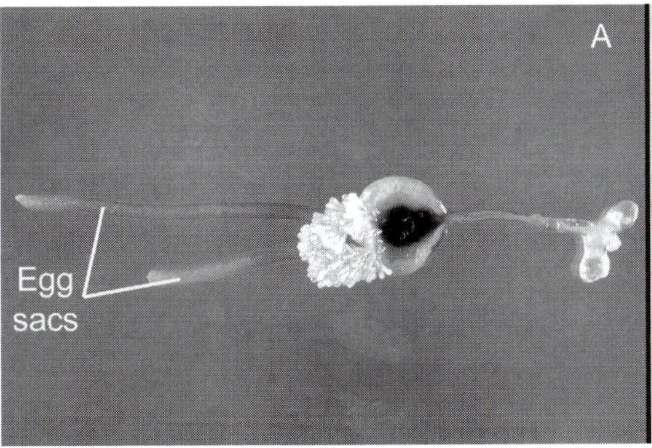

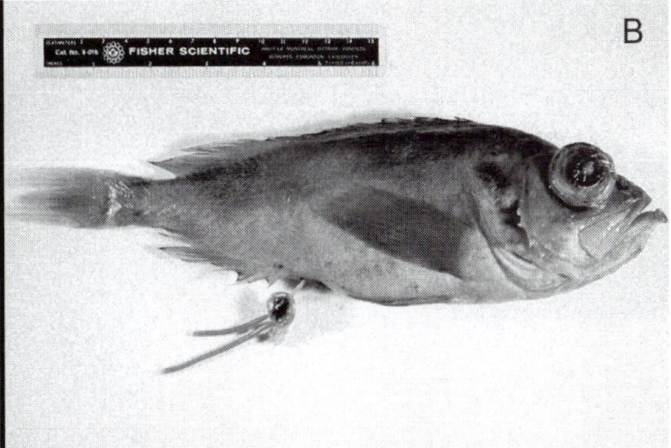

Fig. 8.6 (A) Female individual of the copepod *Sphyrion lumpi*; (B) female of *S. lumpi* parasitizing the red fish *Sebastes mentella*. (Photographs courtesy of Jonathan D. W. Moran, Fisheries and Oceans Pacific Biological Station, Nanaimo.)

basic pattern found in many parasitic crustaceans. In the most modified groups, such as the rhizocephalan cirripedes, a digestive system is absent and nutrients are absorbed through the body wall.

Osmoregulatory and excretory system. The organs responsible for excretion in crustaceans are the antennal glands located near the antennae, and the maxillary glands located near the maxillae. Both types of glands are found in larval crustaceans, but only one type persists in the adults. Nitrogenous wastes from the blood are filtered into the excretory gland and secreted. Ammonia is the main product of excretion, although amines, urea, and uric acid are also excreted. The urine produced is isosmotic with the blood of the crustacean and it seems that these glands do not function in osmoregulation. (In large, free-living crustaceans, the gills usually function as osmoregulatory organs, secreting salts in marine species and absorbing salts in freshwater ones.) Crustaceans also have **nephrocytes**, specialized cells capable of picking up and accumulating waste materials. Nephrocytes are particularly common in the gills and in the bases of the legs.

Circulatory system. The circulatory system in crustaceans is open. A simple dorsal heart is surrounded by a pericardial sinus. Blood enters the dorsal heart through lateral openings or **ostia**, and leaves through one or more arteries. The blood flows from the arteries into a system of open spaces called sinuses where it bathes the tissues directly. The blood then returns to the pericardial sinus and heart aided by the movements of the body. Parasitic copepods do not have a heart, and body movements and the gyrations of the gut move their blood. Likewise, highly modified crustaceans like the rhizocephalan cirripedes do not have a circulatory system.

The blood often has one or more respiratory pigments such as hemoglobin or hemocyanin, and one or more types of motile cells involved in clotting and phagocytosis.

Respiratory system. In small crustaceans and most parasitic species, gas exchange occurs through the body surface or by way of foliacean appendages that increase the body surface area. Branchiurans, for example, have cuticular extensions on the edge of the carapace that facilitate gas exchange. Cyamids and other much larger crustaceans have gills and various appendages that create water currents to facilitate gas exchange on their surface.

Reproductive system. Most crustaceans are dioecious and many are also sexually dimorphic. Sexual dimorphism normally involves differences in body size and the degree of development of certain appendages. Males often have various appendages modified as clasping organs and, in some, the males are dwarfs that live permanently attached to the female's genital segment (Fig. 8.8).

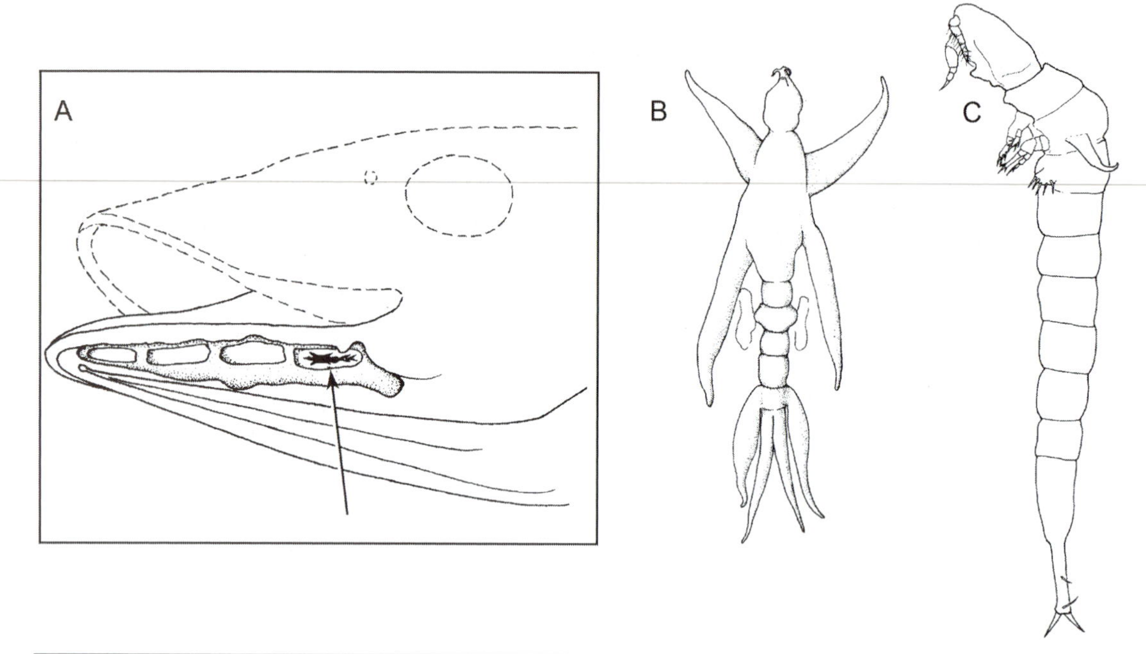

Fig. 8.7 The copepod *Colobomatus quadrifarius*. (A) Ventrolateral view of the mandibular canal of a haemulid fish showing the position of the copepod; (B) ventral view of a female copepod; (C) lateral view of a male copepod, showing a less modified body. (From Cressey & Schotte, 1983, with permission, *Proceedings of the Biological Society of Washington*, **96**, 189–201.)

In males, paired testes empty into sperm ducts that may have special seminal vesicles to store sperm. Females have paired ovaries that continue into the uterus and empty into oviducts and may or may not have seminal receptacles to store sperm. During copulation, the male clasps the female using modified appendages such as antennae, mouthparts, or thoracic appendages. A penis, or other modified abdominal appendage, is used to deposit the sperm within the seminal receptacle of the female or on the surface near the female gonopore. The sperm of most crustaceans lack flagella and, therefore, cannot swim. In some groups, the sperm are packed into a **spermatophore** that is delivered to the female.

Nervous system and sensory input. The nervous system of crustaceans displays a tendency toward concentration and fusion of elements. In the primitive condition, there is a supraesophageal ganglion or brain, nerves that supply the cephalic sense organs, a pair of nerve trunks that connect the brain with a pair of subesophageal ganglia, and a pair of ventral cords that lie beneath the digestive system connecting the segmental ganglia. In most advanced crustaceans, including many of the parasitic species, there is a tendency toward fusion of the segmental ganglia, and of the two ventral cords.

The sense organs of most Crustacea include eyes, sensory hairs, proprioceptors, and statocysts. The eyes can be naupliar or compound. A naupliar eye is a characteristic feature of the nauplius larva and it may degenerate or persist in the adult. It contains a few photoreceptors and is probably used for orientation, enabling the animal to determine the direction of light. Compound eyes are common in adult crustaceans and possess a variable number of ommatidia, depending on the species. Adult copepods have only a naupliar eye and some of the parasitic species lack eyes completely. Ectoparasitic isopods and amphipods retain their compound eyes, and branchiurans such as *Argulus* have a pair of compound eyes and a median naupliar eye.

Sensory hairs are located over the body surface, especially the appendages. **Aesthetascs** for example, are chemoreceptors usually present in the antennules and are important in food and

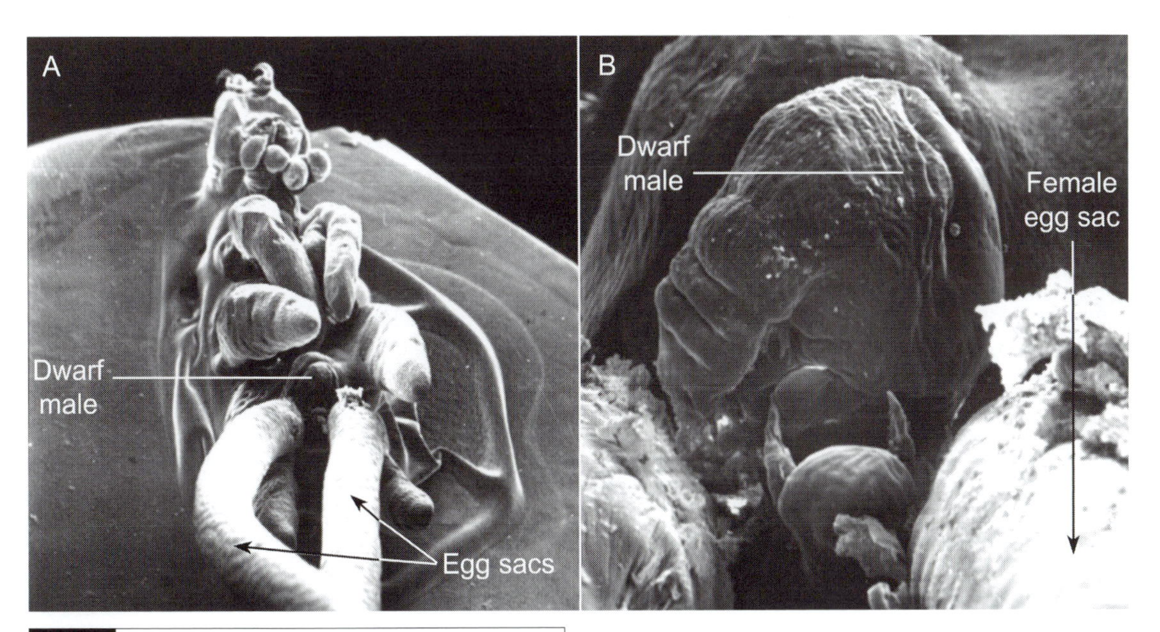

Fig. 8.8 The copepod *Chondracanthus palpifer* from the Chilean hake *Merluccius australis*. (A) Ventral view of a female with a dwarf male attached to her genital segment (arrow); (B) detail of the dwarf male attached to the female.

mate recognition. In parasitic copepods, they may be important in host recognition. Caligid copepods have a little-known sensory organ located in the middle of the anterior margin of the carapace. Its structure suggests a chemosensory function, but it also may be involved in host recognition (Kabata, 1974). Proprioceptors and stretch receptors are found in muscles, tendons, or articular membranes, and seem to be important in reflexes associated with body control during movement.

8.1.4 Nutrient uptake

What is ingested by the different parasitic crustaceans, and how, depends mostly on the type of oral appendages present and the location of the parasite on the host. Caligid-like copepods have a tubular mouth with a pair of mandibles that bear a sharp blade on one side and teeth on the other. The mandibles pierce and tear off pieces of the host tissues, which are then sucked up by the tubular mouth. *Ergasilus sieboldi*, a copepod parasite on the gills of fishes, ingests gill epithelium, mucus, erythrocytes, and white blood cells, in this order of preference.

In pennellid copepods, the holdfast of adult females is embedded in major blood vessels, trunk musculature, or the visceral cavity of the host and serves to anchor the copepods while they feed on blood or tissue fluids. The holdfast of *Cardiodectes medusaeus*, a parasite of lanternfishes, reaches into the bulbus arteriosus of the host's heart (Fig. 8.9). The holdfast not only functions as an anchoring device, but also acts in the digestion of host erythrocytes as well as in the detoxification and storage of iron freed from the catabolism of hemoglobin (Perkins, 1985). Species of *Phrixocephalus*, on the other hand, attach to the eyes of fishes. In *P. cincinnatus,* a parasite of flatfishes, the holdfast penetrates the cornea until it reaches the back of the eye and becomes embedded in the choroid, establishing a network of branched, intertwined rootlets. Development of the holdfast causes the formation of a large hematoma on which the parasite feeds. However, unlike *C. medusaeus*, the holdfast of *P. cincinnatus* does not function in digestion, and the blood is digested primarily in the intestine (Perkins, 1994). Interestingly, the epicuticle of highly modified copepods, including the holdfast of these two pennellid copepods, has abundant microvillosities and it has been suggested that they may function in the absorption of nutrients from the tissues of the host (Gotto, 1979; Bresciani, 1986).

Cyamid amphipods (Fig. 8.3) parasitic on the

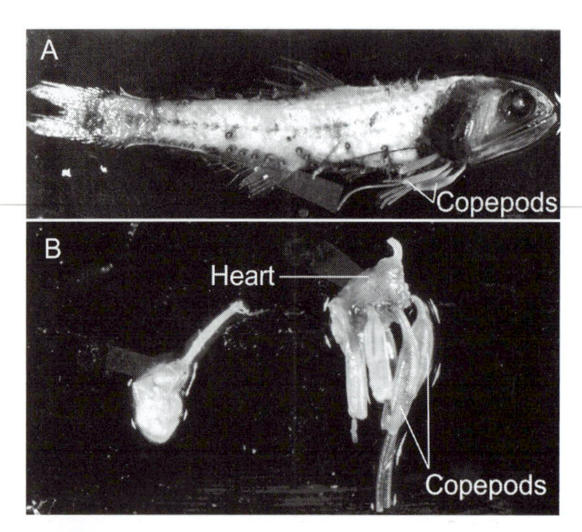

Fig. 8.9 (A) The lanternfish *Stenobrachius leucopsarus* parasitized by four individuals of the copepod *Cardiodectes medusaeus*. The head of the copepod is embedded in the heart of the host. (B) Bulbus arteriosus of the heart of the infected fish. Note the greater size of the infected bulbus arteriosus (right) when compared to an uninfected one (left). (Photographs courtesy of Mike Moser, University of California, Berkeley.)

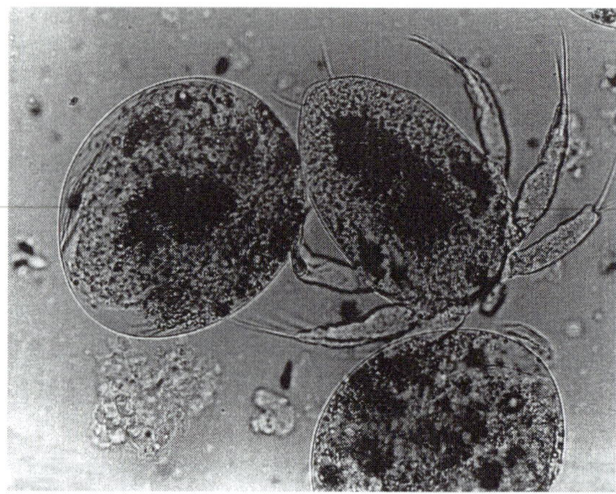

Fig. 8.10 Embryonated eggs and nauplius larvae of the copepod *Sarcotaces verrucosus*. (Photograph courtesy of Raúl González, Instituto de Biología Marina y Pesquera 'Alte. Storni', Argentina.)

surface and external cavities of whales and dolphins feed mainly on host skin. There is evidence that even juvenile cyamids still present in the brood chamber of the female might leave the chamber, feed on the host's skin, and then return to the brood chamber (Rowntree, 1996). Most parasitic isopods feed by sucking blood and other body fluids. Rhizocephalan cirripedes (Fig. 8.4), as well as other endoparasitic crustaceans that have lost most of their appendages and internal organs, probably absorb nutrients through their body wall. The cuticle of rhizocephalans shows acid phosphatase activity, suggesting that nutrients are probably absorbed through it (Bresciani & Dantzer, 1980). Branchiurans (Fig. 8.5) generally feed by rasping at the host integument with the serrated mandibles and burying the proboscis into a blood vessel to suck blood. Other species, however, feed primarily on mucus, epithelial cells, and extracellular fluid obtained from the surface of the fish host.

8.1.5 Development

Most crustaceans retain their fertilized eggs during embryonation within a specialized brood chamber or within an egg sac. Isopods and amphipods brood their eggs in a ventral brood chamber, or **marsupium**. Copepods retain and brood their eggs in an egg sac produced by secretions from the oviduct as the eggs pass by. The egg sacs remain attached to the female genital segment (Figs. 8.1A, 8.8A). Female branchiurans, on the other hand, do not retain their eggs. They leave the host temporarily and deposit their eggs on objects found on the substratum. In other highly modified crustaceans such as the sarcotacid copepods, the body of the female becomes a brooding chamber.

Development in crustaceans is indirect. Usually, a planktonic, free-living larva hatches from the eggs. This larva, known as a **nauplius**, is the most basic and primitive larval stage among crustaceans (Fig. 8.10). It has three pairs of appendages (antennules, antennae, and mandibles), a naupliar eye, and no apparent segmentation. As the nauplius molts, segmentation becomes apparent and more appendages are developed. When a full complement of functional appendages has been acquired, the nauplius becomes a postlarva. The postlarva is usually quite similar to the adult in its general appearance and increases in size through successive molts until it reaches sexual maturity. In most parasitic crustaceans, the postlarva is also parasitic.

Although many crustaceans conform to the basic nauplius, postlarva, and adult pattern described above, there are differences among and within the groups regarding the stage of development at which the larva hatches from the egg and the number of molts in each stage. Branchiurans, for example, do not have any naupliar stages and the egg hatches into a juvenile stage, also parasitic, that grows by molting until it becomes an adult. In most copepods, a free-living nauplius is followed by a postlarval stage known as a **copepodid** if it is free-living, or a **chalimus** if it is parasitic. At the end of this stage, the copepod normally settles definitively on a host, becoming an adult. Isopods do not have a naupliar stage. Instead, a well-developed postlarva called a **manca** hatches from the egg. Cyamid amphipods do not have a naupliar stage either, and the eggs are retained in the brood chamber. Even after hatching, the juvenile cyamids remain in the brood chamber.

The eggs of rhizocephalan cirripedes hatch into a nauplius larva that metamorphoses into a **cypris** larva. In one group of rhizocephalans, the cypris larva is the infective stage for the host, whereas in other rhizocephalans the cypris develops further into a **kentrogon**, which is the infective stage. Details of the life cycle of rhizocephalans will be discussed in the next section. In the Ascothoracica, a group of cirripedes parasitic in echinoderms and corals, development is highly variable, and what hatches from the egg can be a nauplius, a cypris larva, or a juvenile, depending on the species.

8.1.6 Life cycles

The life cycles of free-living crustaceans are diverse and include a wealth of larval types. Parasitic crustaceans are no exception to this pattern. Most branchiurans have a rather direct life cycle, without a nauplius larva. Unlike most crustaceans, the females do not retain their fertilized eggs. Instead, they place them on objects on the substratum. The egg hatches into a juvenile stage that is parasitic and must search for a host.

Most parasitic copepods require only one host to complete their life cycles, but the Pennellidae require two. The basic life cycle of most free-living copepods includes five or six naupliar stages and six copepodid (or chalimus) stages before reaching adulthood (Kabata, 1981). Parasitic copepods follow the basic plan, but most species have reduced the number of larval stages. Perhaps the most unusual pattern among copepod parasites of fishes is the two-host life cycle of the Pennellidae. Pennellids have one or two naupliar stages, followed by an infective copepodid that seeks and attaches to the intermediate host. On this host, the copepodid (a chalimus, technically) molts several times; with the last molt it becomes a free-living stage that seeks out the definitive host. The intermediate and definitive hosts may belong to the same taxonomic species or they may be different species. *Pennella*, for example, uses cephalopods and fishes, whereas the hosts for *Cardiodectes* are pelagic gastropods and fishes.

Isopods normally brood their eggs until late in development when a postlarva known as a manca hatches. In parasitic isopods, the manca is given different names. The eggs of gnathid isopods, for example, hatch into a parasitic larva known as a **praniza** that attaches to a fish host, sucks blood, and molts until it becomes an adult. The adults become free-living, leave the host, and settle in the benthos, where they mate and start over. Cymothoid isopods also have a relatively simple life cycle. A well-developed larva (or juvenile) leaves the brood pouch and swims until it finds a suitable host. Unlike most other crustaceans, however, the cymothoids are protandrous hermaphrodites, i.e., the individuals first develop into males and later into females. Thus, as soon as the larva finds a suitable host, it loses its swimming setae and begins to develop into a male. Epicaridean isopods have life cycles that include two hosts, with three different larval stages. They exhibit **epigametic sex determination**, where the biotic environment determines the sex of the individuals. The first isopod to infect a definitive host becomes a female, and any further isopods that settle on this particular host become males, often much smaller than the female. In some species, the male becomes a parasite of the female and lives in her brood sac. Interestingly, cymothoid and epicaridean isopods can also undergo **sex reversal**. If the female dies or is removed from the host, the male that remains in direct contact with the host can transform into a female.

In the highly modified rhizocephalan cirripedes such as *Sacculina*, the life cycle is also unique (Fig. 8.11). The egg hatches into a nauplius larva that metamorphoses into a male or female cypris larva. The female cypris attaches to a suitable crab host and sheds most of its appendages. Whatever

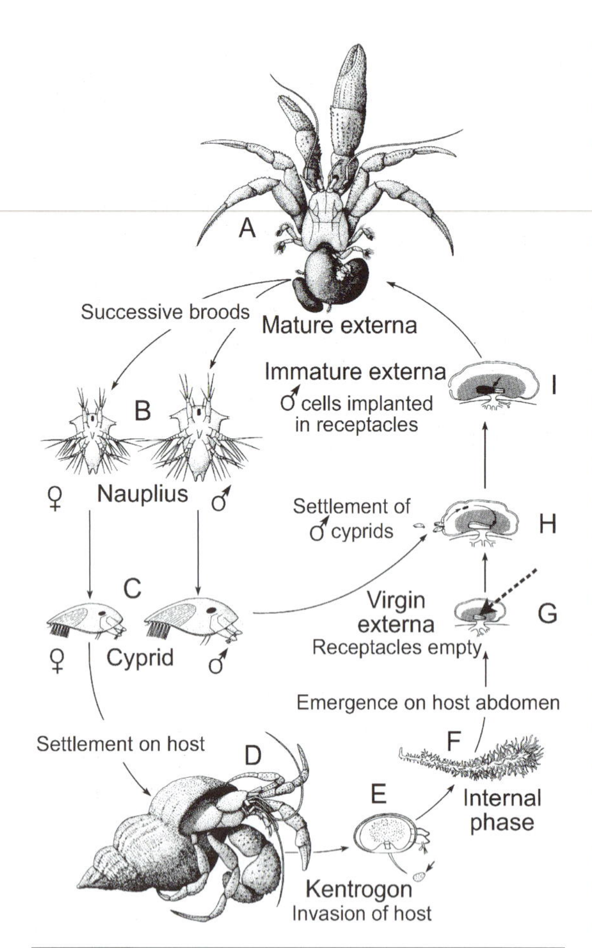

Fig. 8.11 The life cycle of a generalized kentrogonid rhizocephalan. (A) Externa releases nauplii; (B) nauplii molt into cyprids; (C) female cyprids settle on potential hosts, while male cyprids settle only on juvenile virgin externa; (D) the settled female cyprid metamorphoses into a kentrogon; (E) the kentrogon injects the parasitic cells; (F) the parasitic cells multiply and form a branched trophic root system; (G) the parasite emerges on the abdomen of the host as a virgin externa with empty male receptacles (dashed line); (H) male cyprids settle in the externa, metamorphose into trichogons and migrate into the male receptacles; (I) the implanted trichogons develop into hyperparasitic dwarf males. (Modified from Høeg, J.T. [1992] Rhizocephala. In *Microscopic Anatomy of Invertebrates*, Vol. 9: *Crustacea*, ed. F. W. Harrison & A. G. Humes, pp. 313–345. Copyright © 1992 Wiley-Liss, Inc. Reprinted by permission of Wiley-Liss, Inc., a division of John Wiley & Sons, Inc.)

is left, normally a mass of undifferentiated cells, becomes an infective larval stage known as the kentrogon. The kentrogon injects the undifferentiated mass of cells into the hemocoel of the crab. This mass grows, forms root-like processes that extend throughout the crab forming an **interna**, which eventually breaks to the outside through the cuticle on the ventral side of the crab. This external mass is called an **externa** and contains the gonads of the female rhizocephalan. The male cypris larva will eventually attach to this female externa, metamorphose into a **trichogon** larva, penetrate the externa, and then implant itself in a male receptacle. One or more trichogon larvae normally remain in the male receptacle throughout the life of the female parasite, producing spermatozoa to fertilize the eggs produced by the female externa (Raibaut & Trilles, 1993). In most kentrogonids, the externa is attached to the ventral surface of the host abdomen and liberates nauplii that develop into the morphologically distinct male and female cypris. Interestingly, for a long time, the male cells present in the rhizocephalan females were thought to be testes and the rhizocephalans were thought to be hermaphrodites. The discovery of these 'male parts' inside the female makes the rhizocephalans a classic example of **cryptogonochoristic sexuality**, where sexes are separate, but hidden.

8.1.7 Diversity

Because of the high diversity shown by parasitic crustaceans and the uniqueness of many of their adaptations to a parasitic life style, this section illustrating their success and adaptive radiation will be approached from a taxonomic perspective.

COPEPODA

The parasitic life style probably has evolved independently many times among copepods. As a result, an overwhelming diversity of body forms, habitats, development patterns, and life cycles is evident, making generalizations on aspects of copepod biology almost impossible. Copepods parasitize a wide variety of hosts including cnidarians, annelids, molluscs, arthropods, echinoderms, ascidians, fishes, amphibians, and even marine mammals. Most of the species are marine, but some are parasitic on freshwater fishes. Fishes are probably the most conspicuous hosts and, consequently, their copepod parasites are relatively well known.

Caligid and trebiid copepods, for example, parasitize the surface of fishes (Figs. 8.1B, C; 8.12). They are able to move freely, feeding on mucus and sloughed tissues. Their bodies are flattened

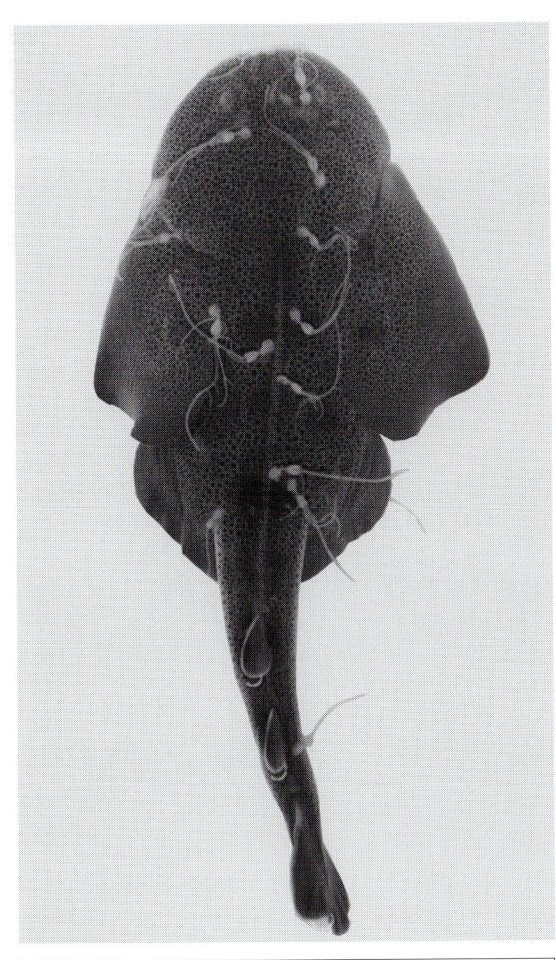

Fig. 8.12 The copepod *Trebius shiinoi* on the surface of an embryo of the Japanese angel shark *Squatina japonica*. Trebiid copepods are normally found on the surface of the body or branchial chambers of the host. This species is unusual in that it parasitizes the surface of shark embryos while still in the uterus. (From Nagasawa *et al.*, 1998, with permission, *Journal of Parasitology*, **84**, 1218–1230.)

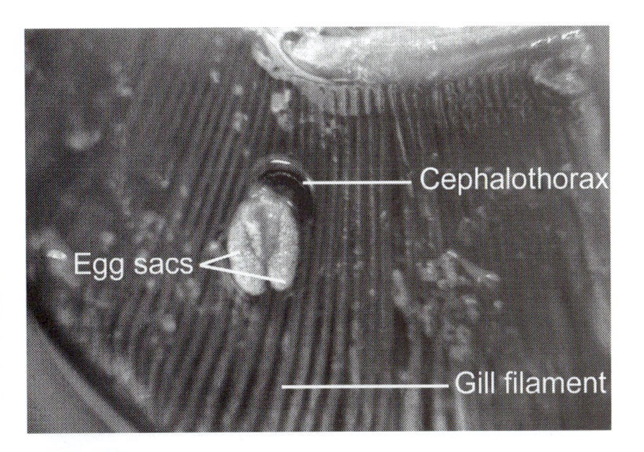

Fig. 8.13 The copepod *Naobranchia lizae* from the gill filaments of the grey mullet *Mugil cephalus* in Chile. The copepod attaches to a gill filament using the maxillae, while the flexible cephalothorax 'sweeps' the surface of the filaments, grazing.

dorso-ventrally, using their carapace as a sort of suction cup for attachment to the host. They still resemble their free-living counterparts and, unlike most other parasitic copepods, many caligids are capable of swimming and can switch hosts.

The gills of fishes are also a good site for copepod parasites (Fig. 8.13). Gill filaments and gill arches provide suitable anchoring sites and an excellent source of nutrients that includes mucus, epithelial cells, and blood. Species of *Ergasilus* (Fig. 8.1A) attach to the gill filaments using their modified antennae and feed mainly on mucus and tissue from the gills; they can also live on other areas of the fish, but only if those areas are free of scales. Ergasilids are among the least-modified copepods and, like caligids, many are able to swim and, therefore, change hosts. Some species in the Ergasilidae are completely free-living. Several species of Chondracanthidae (Fig. 8.8) are found not only on the gill arches, but also in the mouth cavity of fishes, including the tongue. In this group of copepods, the females attach to the host using their antennae. They eventually become permanently attached because a host tissue reaction surrounds the cephalothorax of the parasite completely. Apparently, the extrabuccal digestion carried out by the female copepod at the point of attachment irritates the tissues around the attachment area, leading to encapsulation of the anterior end of the copepod. The females are highly modified, with no external segmentation and have a white, fleshy appearance. The males, on the other hand, are dwarfs, retain many non-parasitic characteristics, and remain attached to the genital segment of the female for life (Fig. 8.8).

Perhaps the most striking of all the parasitic copepods are the so-called **mesoparasites**. The body of these copepods remains largely outside the host, but the cephalothorax transforms into an anchoring device that penetrates the body surface and may reach well into the body cavity, frequently invading the circulatory system.

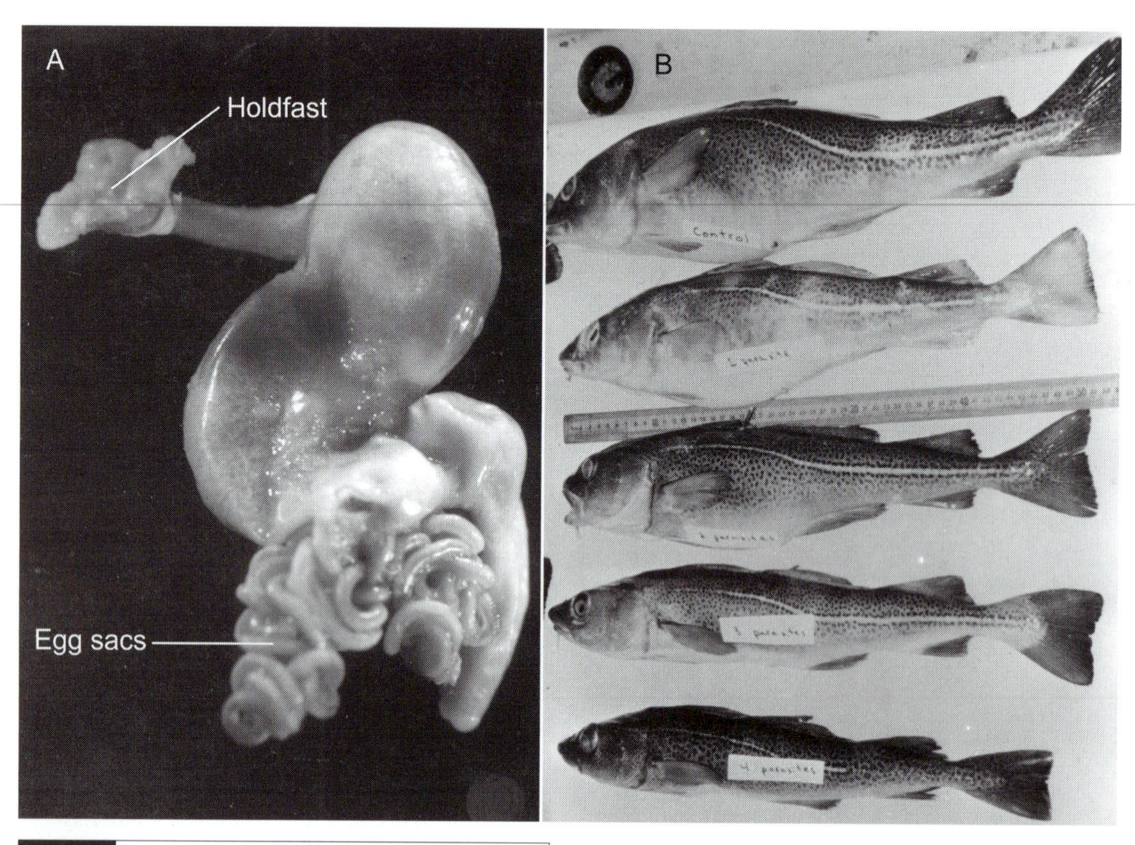

Fig. 8.14 (A) Female *Lernaeocera branchialis* from the Atlantic cod *Gadus morhua*; (B) effect of *L. branchialis* on the growth of parasitized cod 9 months post infection. All fish were approximately the same size when the infection started. The fish on top was not infected (control), the fishes below had 1, 2, 3, and 4 copepods, respectively. (From Khan, 1988, with permission, *Journal of Parasitology*, **74**, 586–599.)

Species of *Lernaeocera* develop anchoring systems that reach the aorta, branchial vessels, and even the heart. *Lernaeocera branchialis*, for example, a parasite of the Atlantic cod *Gadus morhua*, attaches to the branchial area, but the anchoring structure grows and penetrates the fish body until it reaches the bulbus arteriosus of the heart. The large trunk, bearing the reproductive organs and the egg sacs, remains on the surface of the fish. The parasite severely affects the fitness and survival of the host and is likely to have a significant impact on commercial fisheries (Fig. 8.14). The presence of the copepod is normally associated with anemia, weight loss, reduction in fat content and liver weight, and a decrease in reproductive potential (Khan, 1988). Khan *et al.* (1990) found that up to 33% of cods infected with the copepod died over a period of 4 years.

The attachment structure of *Cardiodectes medusaeus*, a parasite of lanternfishes, also reaches the bulbus arteriosus of the host and feeds on blood. In multiple infections, the heart becomes enlarged, and the flow of blood through it may be affected (see Fig. 8.9B) (Moser & Taylor, 1978). The anchoring process of *Phrixocephalus* is highly branched and complex reaching into the retinal area of eyes to feed on blood. In the Lernaeidae, a small group that parasitizes freshwater fishes, the attachment processes are not as complex as the previous ones forming a true anchor that is embedded in the flesh of the host. *Lernaea cyprinacea*, for example, is a ubiquitous species where females can reach up to 12 cm.

Small body cavities that communicate to the outside, such as the pores of the lateral line, the frontal mucus passages, and sinuses of teleosts and elasmobranchs, are the preferred sites of philichthyid copepods (Fig. 8.7). Because of this protected habitat, these copepods do not require

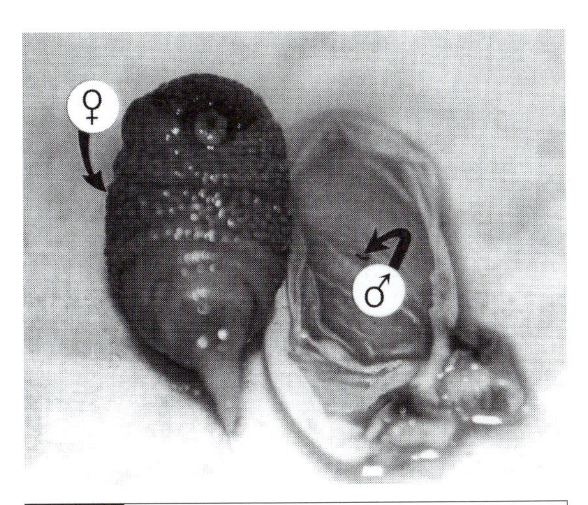

Fig. 8.15 Individual female of the copepod *Sarcotaces verrucosus* from the body cavity of the sandperch *Pseudopercis semifasciata* in Argentina. The female measures more than 5 cm whereas the male measures less than 3 mm. (Photograph courtesy of Raúl González, Instituto de Biología Marina y Pesquera 'Alte. Storni', Argentina.)

many specialized attachment structures and most of their appendages have been lost. They have developed, however, body processes that seem to be useful in maintaining their position inside the cavities they occupy (Fig. 8.7B). Males are less modified than females and retain most of their appendages and segmentation (Fig. 8.7C).

One of the copepod groups that can be considered as truly endoparasitic is the Sarcotacidae. Sarcotacids live in cysts in the muscles or abdominal cavity of fishes where they feed on blood from the vascular wall of the cyst. They are highly modified, most appendages are vestigial, and a variable number of dwarf males are found with a female in each cyst (Fig. 8.15).

Many copepods parasitize invertebrates. Unlike the species parasitizing fishes, the parasites of invertebrates are mostly marine and retain many morphological features similar to their free-living counterparts. Species of Lichomolgidae are parasites or symbionts of a wide variety of marine invertebrates, including corals, anemones, polychaetes, molluscs, echinoderms, and ascidians. These copepods have retained most of their free-living characteristics, including segmentation and swimming legs, although in some species, the latter are reduced. An excellent review of copepods parasitic on invertebrates can be found in Gotto (1979).

BRANCHIURA

Branchiurans are a relatively small group of crustaceans exclusively parasitic on marine and freshwater fishes, and occasionally amphibians (Fig. 8.5). Although they attach to the host by the antennal hooks and modified maxillules that form suckers in *Argulus* and hooks in *Dolops*, they are able to move over the surface of the host. They seem to prefer the gill chamber of fishes, but can also be found on the body surface and fins. Unlike most parasitic crustaceans, branchiurans exhibit low host specificity and are highly mobile, leaving the host to find mates, to locate new hosts, and to lay eggs in the substratum.

AMPHIPODA

Although many amphipods are associated with a number of marine organisms in some type of symbiotic or phoretic relationship, true parasitism is rare. Hyperiideans are pelagic amphipods that have established some type of symbiotic relationship with other pelagic organisms such as jellyfishes, ctenophores, molluscs, and tunicates.

Cyamids are a very unique, but still poorly known, group of amphipods parasitic on whales and dolphins (Fig. 8.3). Cyamids have several anatomical adaptations to parasitism. Their body is flattened dorso-ventrally, not laterally like other amphipods, and it is armed with strong ventral spines and powerful, hooked appendages. These are clearly adaptations to prevent dislodgment from the host and ensure attachment. Because of these adaptations, cyamids are unable to swim at any stage of their life cycle. Most cyamids attach to crevices, natural openings, or areas that provide some shelter from the currents produced by the swimming of the host. Attachment is such a critical factor, that the abundance of cyamids in a host is inversely proportional to its swimming speed. In slow-moving whales, for example, several thousand cyamids can be found on one host. In faster-moving cetaceans such as odontocetes, dolphins, orcas, etc., however, cyamid populations are much smaller (Raga, 1997). Because cyamids cannot swim, they lack dispersal capabilities, and the young individuals settle near their parents in

Fig. 8.16 (A) The isopod *Bopyrella macginitiei* in the gill chamber of the snapping shrimp *Alpheopsis equidactylus*. Note the presence of the small male. (B) Isopod removed from the gill chamber of the shrimp. Note the male on the lower left corner of the female. (Photographs courtesy of Todd Huspeni, University of California, Santa Barbara.)

crevices in the skin of the host. Transfer to new hosts probably occurs during host mating and lactation of the young. Even though cyamids cannot swim, they are able to move on the host and change locations, moving at an average rate of 4.5 m/h under experimental conditions (Rowntree, 1996).

ISOPODA

Even though parasitic isopods are not as abundant, in terms of numbers of species, as parasitic copepods, their morphological diversity, niches, and adaptations are extensive. Parasitic isopods include three major taxonomic groups, i.e., Gnathiidae, Cymothoidae, and Epicaridea. Gnathiid isopods parasitize fishes only during the gnathiid's larval stages; cymothoids are permanent ectoparasites of fishes, whereas epicarideans exhibit various degrees of morphological change and are exclusively parasites of crustaceans. The adults of Aegidae, Corallanidae, and Cirolanidae are also associated with fishes. The first two appear to be temporary parasites that often leave their host, whereas the Cirolanidae are actually predators or scavengers associated with restrained fishes, such as those in fish farms (Bunkley-Williams & Williams, 1998).

Gnathiid isopods are parasites of fishes only during their larval stage. The adults live in the benthos and do not do not feed, subsisting on the reserve nutrients accumulated during their larval stage. The larva and the adult show very little resemblance. Their morphology is so different that, before their life cycle was determined, the larvae and the adults were considered to be different species.

Cymothoids are permanent ectoparasites of both marine and freshwater fishes. Some species such as *Lironeca amurensis* burrow under a scale of the host and as the isopod grows it becomes completely surrounded by host tissue, communicating to the exterior by a small hole. Most cymothoids, however, are armed with strong claws used to cling to the skin, gills, operculum, and buccal cavity of the host, where they feed on whole blood or on oozings of plasma from wounds (Fig. 8.2). In many cases, the isopods cause pressure atrophy of the structures to which they attach, such as the gills and the tongue. A distinctive feature of cymothoids (discussed in section 8.1.6) is the phenomenon of protandrous hermaphroditism, where juveniles develop first into males and later into females. Most other crustaceans, both parasitic and free-living, are characteristically dioecious.

Epicaridean isopods are parasites of crustaceans during their entire lives. They are particularly interesting because their life cycles involve two hosts and sex determination is environmentally determined (as discussed in section 8.1.6). Two epicaridean families, Bopyridae and Dajidae, are ectoparasitic of the abdomen or gill chamber of marine crustaceans such as crabs and shrimps (Fig. 8.16). Those that parasitize the gill chamber may cause a large swelling in the host carapace. Isopods in the Entoniscidae are more modified and, although they are technically ectoparasites,

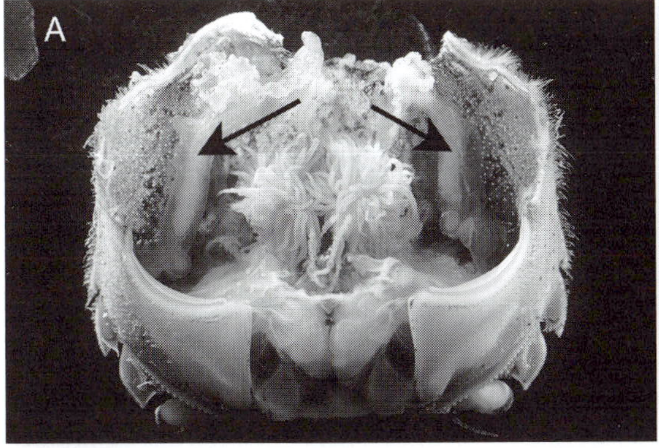

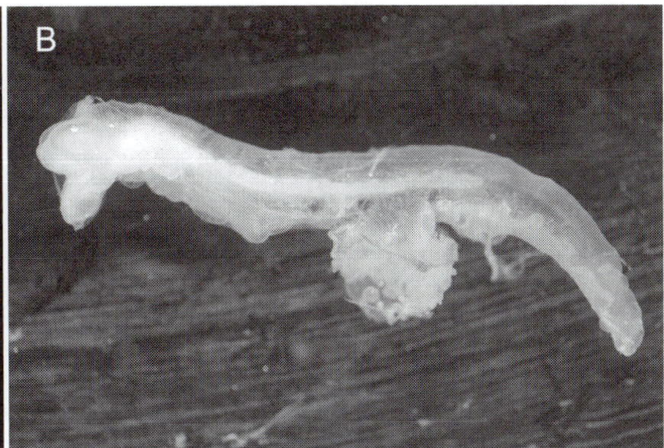

Fig. 8.17 (A) The entoniscid isopod *Portunion conformis* (arrows) inside the body of the shore crab *Hemigrapsus oregonensis*, after removal of the crab's carapace; (B) young female of *P. conformis* removed from the crab. (Photographs courtesy of Todd Huspeni, University of California, Santa Barbara.)

they are located inside the body of the host (Fig. 8.17). The female feeds on blood and becomes very large, whereas the males remain small and live on the surface of the female. The female broods the eggs and the larvae leave through a pore on the host's gill cavity. Species of *Danalia* and *Liriopsis* extend the concept of parasitism one step further. They are hyperparasites of rhizocephalan cirripedes parasitic on crustaceans. In an interesting turn of events, the isopod castrates the rhizocephalan host that, in turn, castrates its free-living crustacean host. Obviously, there is no such a thing as a free lunch!

CIRRIPEDIA

The rhizocephalan cirripedes are probably the most modified of all crustaceans and are often used as an example of extreme adaptation to parasitism. Although their larval stages resemble those of other crustaceans, the adults have sac-like bodies and lack all appendages, segmentation, and internal organs, except for the gonads and parts of the nervous system. The adult parasite consists basically of an external reproductive part, the externa, connected to an internal nutrient-absorbing structure, the interna.

The effects of rhizocephalans on their crab host are very dramatic. Because of their invasiveness, most of the internal tissues of the host are severely damaged. The parasite also affects the hormonal and reproductive activities of the host through a process known as parasitic castration. The host is not able to reproduce because the gonads atrophy and some sex characteristics change due to parasite-induced hormonal changes. The parasite takes full control of the host, to the point that the host is manipulated morphologically and behaviorally into accepting the parasite externa as its own egg mass by grooming, protecting, and aerating it. Moreover, when the parasite is ready to release its larvae, the crab performs its normal spawning behavior that involves moving into an open area where it then uses its appendages to create water currents for larval dispersal. The castration and hormonal interference in males is so profound that they also exhibit this kind of spawning behavior. Simply put, the parasite is a master of deception.

ASCOTHORACICA

The Ascothoracica is a group of ecto- and endoparasites of echinoderms and cnidarians that historically has been included within the Cirripedia. Recent studies, however, suggest that they should be considered as a distinct group. The body of most species is enclosed in a bivalved carapace, resembling an ostracod. The mouthparts of some species are arranged into an oral cone used to pierce the host's tegument and suck its body fluids. Other species live in pairs inside galls or cysts in corals and feed on the surrounding tissues.

TANTULOCARIDA

The tantulocarids are a group of recently discovered parasites. The first species was described in 1975 and, in 1983, they were recognized as a separate group within the Crustacea (Boxshall &

Lincoln, 1983). Tantulocarids are very small ectoparasites of other small crustaceans such as copepods, isopods, and ostracods, normally found in the deep sea. They have no cephalic appendages except for one pair of antennae on the sexual female. Their life cycle is rather unusual for a crustacean and includes alternating generations and a parthenogenetic phase. For further details on the life cycle of these unusual crustaceans see Huys *et al.* (1993) and Raibaut & Trilles (1993).

8.2 Chelicerata

8.2.1 General considerations

The Chelicerata includes such familiar organisms as horseshoe crabs, spiders, scorpions, mites, and ticks. Chelicerates include the Merostomata, Pycnogonida, and Arachnida. The Merostomata are a small group of free-living, marine, bottom-dwelling organisms, that includes the familiar horseshoe crabs. The Pycnogonida is also a small group of marine organisms known as 'sea spiders'. Pycnogonids are rather small, ranging from 1 to 10 mm; they are found in every ocean from the littoral zone to the deep sea. Most pycnogonids are carnivorous and feed on anemones, hydroids, soft corals, sponges, bryozoans, and polychaetes. A few species are commensal or ectoparasitic on other invertebrates, mainly molluscs. The Arachnida is the largest class of chelicerates. Most arachnids are terrestrial except for a few groups that have become aquatic. Spiders, scorpions, daddy longlegs, mites, and ticks, are the most familiar, but only the latter two, commonly referred to as Acari, or acarines, include parasitic organisms.

Mites include both free-living and parasitic forms, whereas ticks are all parasitic. Most of the parasitic mites are ectoparasites on both vertebrates and invertebrates, although some have become endoparasites through the infection of respiratory passages. Some mites are parasitic only as larvae, whereas others are parasitic during their entire life. Attachment, however, is not permanent and many are found on the host only when feeding. Ticks parasitize terrestrial vertebrates throughout their development and, in most cases, they remain attached to the host only when feeding.

The diseases and damage caused by ticks and mites can be of three general types, i.e., local damage, systemic damage, and transmitted infections. Local damage at, or around, the site of infection may include inflammation, swelling, ulceration, and itching. Sometimes, secondary infection by bacteria may occur. Ticks sometimes cause systemic damage. When ticks bite humans or other animals near the base of the skull, toxic secretions by the tick might cause a problem known as tick paralysis. Fortunately, prompt removal of the parasite reverses the condition. Heavy infections of ticks can result in high blood loss, with the debilitating effects of anemia. Ticks also are important vectors of a significant number of infectious and parasitic organisms such as viruses, bacteria, rickettsias, spirochaetes, protozoa, and microfilariae, including those causing Lyme disease and Rocky Mountain spotted fever. Mites are much less important as vectors and transmit just a handful of microorganisms. Some free-living mites are intermediate hosts for a number of helminths. Oribatid mites, for example, are intermediate hosts for *Moniezia expansa* and *M. benedeni*, two cestodes common in sheep, cattle, and goats. Other free-living mites, or even their remains, can be highly allergenic to some individuals, causing extreme discomfort when exposed to them (see Box 8.1).

The classification and taxonomy of ticks and mites is in a state of flux. We follow the system proposed by Evans (1992) that recognizes the Acari as a subclass of the Arachnida and divides it into seven orders (see Box 8.4).

8.2.2 The outside

Shape. Most mites are rather small, measuring between 0.2 and 0.8 mm; ticks, on the other hand, can reach up to 30 mm. The bodies of most arachnids can be divided into a prosoma or cephalothorax, and an opisthosoma or abdomen. In ticks and mites, most segmentation has disappeared and the abdomen has fused with the cephalothorax. In ticks and mites, the body can be divided in two regions, the **gnathosoma** or **capitulum**, and the **idiosoma** (Fig. 8.18). The capitulum carries all the mouthparts and feeding appendages (Fig. 8.19). These include a pair of **chelicerae** commonly used in piercing, tearing, or gripping

Box 8.1 Have you dusted lately?

Are you allergic to dust? If you are, be pleased to meet the mighty mites responsible for your ailment: *Dermatophagoides* spp. The genus *Dermatophagoides* includes a mixed batch of parasitic and non-parasitic mites. Most species live in the nest of birds and, of course, some of these occur on the birds themselves. Some species also occur on mammals where they feed on skin. A few free-living species, however, enjoy moist microhabitats where there is a good supply of sloughed skin or other similar products. These are the mites responsible for house-dust allergies and some types of asthma. Although they are not parasites, they are highly allergenic and, when the mites themselves, or parts of them, or their excrement products, are inhaled, they stimulate an allergic reaction or an asthma attack in sensitive individuals.

The house mites implicated in allergic reactions thrive in places like mattresses, pillows, sofas, and the ever-popular easy chairs, where people spend long periods of time and shed copious quantities of sloughed skin. Researchers in Brazil have recently discovered a new and unsuspected location for these mites, the scalps of children. Apparently, the conditions in the scalp are warm and humid and the mites are able to feast on skin flakes. So, previous efforts to control these mites, like vacuuming mattresses or putting pillows into the freezer, may have failed because nobody thought about vacuuming the heads of the children. Actually, vacuuming the head is not really necessary. A good antidandruff shampoo to reduce sloughed skin should be enough.

host tissues, and a pair of **pedipalps** also used in feeding. The coxae (first segment) of the pedipalps are fused and extend forward to form the **hypostome**. The hypostome, together with a labrum, forms the **buccal cone**, which, in some species, can be retracted. In other species, the pedipalps are greatly reduced and function as sensory organs. Most of the remaining body is formed by the idiosoma. It contains most internal organs and legs, and is covered dorsally with a single carapace or shield. Ticks and mites, like other arachnids, have four pairs of legs, but some may have fewer. A pair of claws is present at the tip of the legs in most Acari to aid in attachment.

The anus is near the posterior end of the body and the gonopore is normally located ventrally, in a genital plate, between the last two pairs of legs. Hairs, or setae of various shapes, many of which are sensory, cover the body and legs of most species.

Body wall. The cuticle of arachnids, as in other arthropods, is secreted by the epidermis and is formed by several layers containing, proteins, polysaccharides, and lipids. The epicuticle, the outermost layer of the cuticle, is made of **cuticulin**, a protein. Cuticulin is normally covered by a lipoidal layer that prevents water loss and by a kind of 'varnish' that protects the impermeable layer from abrasion. Immediately under the epicuticle is the procuticle that, in different species, has different degrees of sclerotization.

8.2.3 The inside

Digestive system. As in every other arthropod, the digestive system is divided into a foregut, midgut, and hindgut. The mouth continues into a muscular, sucking pharynx, and an esophagus. There is a pair of salivary glands that open to the buccal cone. The esophagus leads into a stomach with several ceca and continues into a sacculate intestine that terminates in the anus. In many acarines, the stomach becomes separated from the intestine and ends in a blind sac. In these species, the indigestible food is stored in specialized gut cells that accumulate in the ceca. When one of

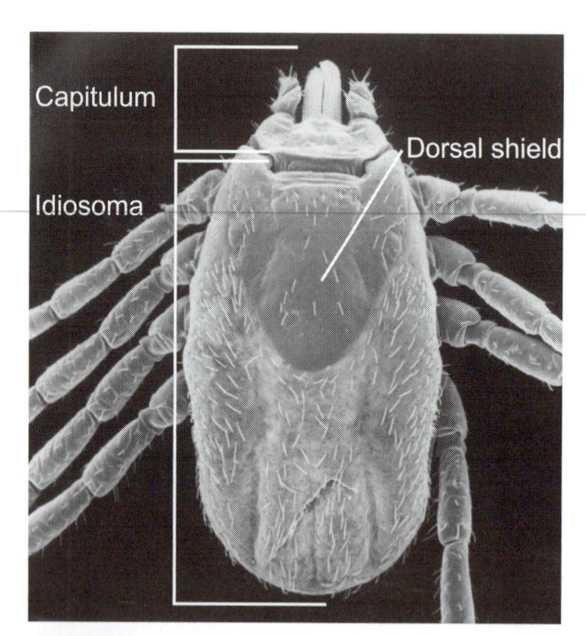

Fig. 8.18 External morphology of the tick *Boophilus microplus* (female) showing the two main body regions characteristic of ticks and mites: the gnathosoma or capitulum and the idiosoma. (Photograph courtesy of the US National Tick Collection, Georgia Southern University.)

these ceca is full, it detaches and is eliminated through a split in the dorsal cuticle.

Osmoregulatory and excretory system. The excretory organs in most ticks and mites are **coxal glands** and **Malpighian tubules**. Coxal glands open to the outside at the base (coxa) of certain legs. The Malpighian tubules lie free in the hemocoel and empty to the terminal part of the midgut. In those Acari whose stomach ends in a blind sac, the Malpighian tubules are connected to the hindgut and waste is still excreted through the anus. Uric acid, amino acids, and salts are taken up by the Malpighian tubules and secreted into their lumen. The main nitrogenous waste in ticks is guanine.

Circulatory system. The circulatory system of Acari is reduced in most groups and consists of a network of sinuses with colorless blood, although in some, a heart may be present. Movement of the blood probably is accomplished by the contraction of body muscles.

Respiratory system. Most ticks and mites possess **tracheae**, similar to those present in insects, for gas exchange. The tracheae are simple, chitin-lined tubes, branched or unbranched, through which gas diffuses. The tracheae open to the outside in one to four pairs of **spiracles**, all located on the anterior half of the body. Astigmatid mites, however, lack a tracheal system, have no spiracles, and respiration occurs through the tegument.

Reproductive system. Acari are dioecious and fertilization is internal. The male reproductive system consists of a pair of testes located in the mid region of the body. A vas deferens arises from each testis and both may join together to open through a gonopore, or through a chitinous penis projecting to the outside through the gonopore. In those species with a penis, transfer of sperm to the female is direct. In most Acari, however, sperm transfer is indirect. Sperm can be packed into a spermatophore that is delivered to the female or that may be injected by the chelicerae into special openings on the female body.

The female reproductive system consists of a single ovary that continues into an oviduct and uterus to open in the gonopore. During copulation, the spermatozoa are introduced into the uterus, although in some mites, a vagina is present. A seminal receptacle and accessory glands are also present.

Nervous system and sensory input. The nervous system of acarines is greatly compressed and includes a brain that is located above the esophagus. The brain is a ganglionic mass formed by two pairs of lobes, the protocerebrum and tritocerebrum, which innervate the eyes and anterior gut, respectively. Nerve trunks from the brain connect with the subesophageal ganglion forming a ring around the esophagus. Most, or all, of the ganglia originally present in the abdomen are fused with the subesophageal ganglia. Nerves arising from this anterior complex, or central nerve mass, innervate the appendages, organs, and sensory structures.

Sensory setae are probably the most important sensory organs in parasitic Acari. Innervated pits and slit sense organs are also present. Although many mites are blind, some possess simple eyes.

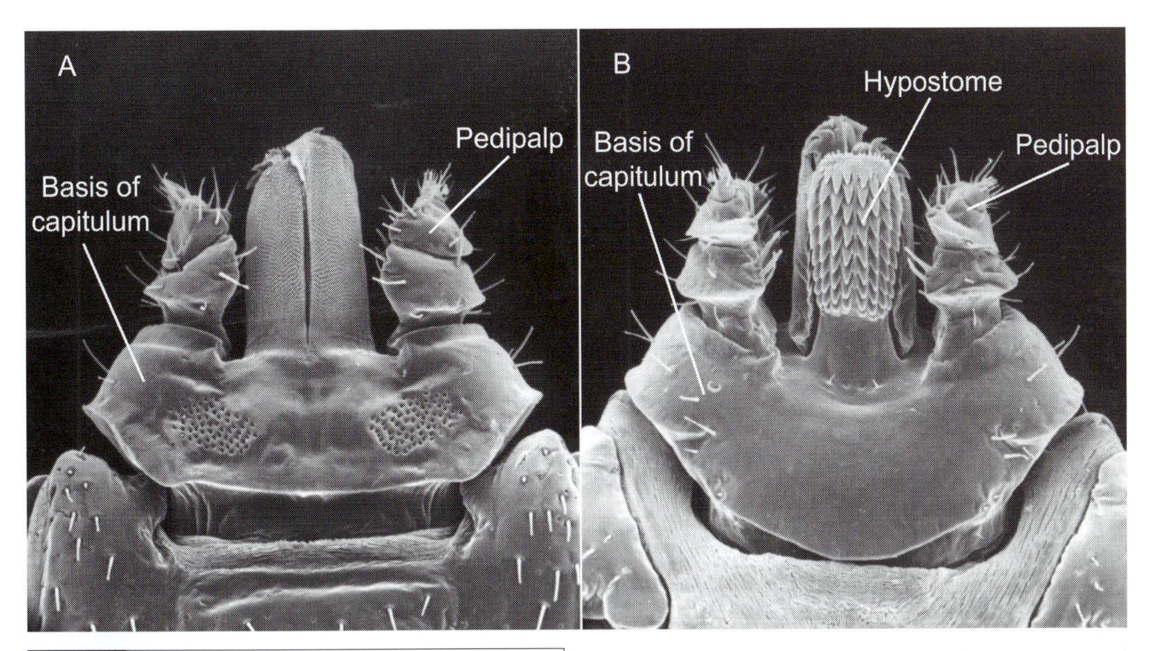

Fig. 8.19 External morphology of the capitulum of *Boophilus microplus*. (A) Dorsal view; (B) ventral view. (Photograph courtesy of the US National Tick Collection, Georgia Southern University.)

The Ixodida (all ticks) have a large set of different types of receptive sensilla clumped together on the first tarsi. This structure, known as **Haller's organ**, has an important aggregate of receptors, including mecano-, thermo-, hygro-, and chemoreceptors, which are used for host and mate location. Mites have a similar structure, but it lacks the modifications that characterize the Haller's organ in Ixodida (Lompen & Oliver, 1993).

8.2.4 Nutrient uptake

Ticks and mites ingest mostly fluids, but when feeding on solid foods, the food is digested externally and liquefied, then absorbed by the action of the sucking pharynx. Some feed by sucking blood and lymph released from tissues cut or damaged with the chelicerae. Others bite the skin of the host and feed on dermal tissue that is hydrolyzed externally by proteolytic enzymes. Ticks feed by pushing the hypostome into the skin of the host. A number of barbs present in the hypostome anchor the tick to the skin (Fig. 8.19B), although some species also produce a cement-like substance to help with the anchoring. Sharp teeth in the hypostome cut blood vessels under the skin causing the blood to form a pool. The tick then sucks this blood into its gut through the hypostome.

Tick saliva contains a number of compounds that aid in the feeding process, as well as in avoiding or neutralizing the host immune system. There are three mechanisms by which the host can stop bleeding. It can plug a hole in the capillaries using platelets, it can form a blood coagule, or it can pinch off the vessel (vasoconstriction). The tick saliva contains compounds able to block all three mechanisms. In addition, tick saliva contains the tick version of prostaglandins, which do everything from relaxing blood vessels to suppressing the host immune response. It also has compounds that can block the host chemicals that cause inflammatory responses, such as histamines and thromboxane. Still another compound in tick saliva breaks down bradykinin, a host chemical that causes pain at the bite site, allowing the tick to feed peacefully because the host does not feel the bite (Ribeiro *et al.*, 1985; Ramachandra & Wikel, 1992).

8.2.5 Development

Development of acarines is direct, although their life cycle may include one or more hosts. After mating, the female of most tick species requires a blood meal to initiate egg production. Following an incubation period of 2 to 6 weeks, a sexually

Fig. 8.20 Female *Dermacentor albipictus*, the winter tick of wild ungulates in North America, laying eggs. (Photograph courtesy of Bill Samuel, University of Alberta.)

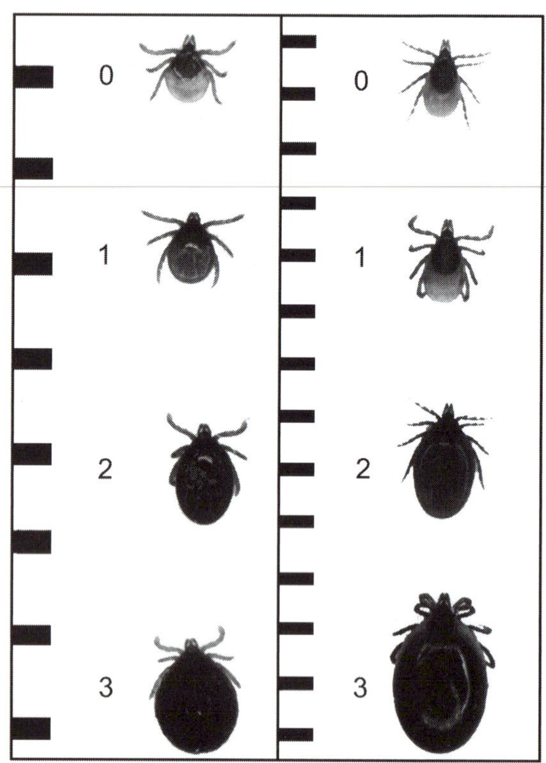

Fig. 8.21 Larvae (left) and nymphs (right) of *Ixodes persulcatus* at different stages of feeding. (0) Unfed; (1) partially fed for 1 day; (2) partially fed for 2 days; (3) fully fed for 3 days (engorged). Scale in mm. (From Nakao & Sato, 1996, with permission, *Journal of Parasitology*, **82**, 669–672.)

immature larva hatches from the egg. The larva has only three pairs of legs and looks like a miniature of the adult. In most acarines, the fourth pair of legs is acquired after the first molt, when the larva changes into a protonymph. In many acarines, successive molts transform the protonymph into a deuteronymph, a tritonymph, and, finally, an adult. During these changes, adult structures are gradually acquired. This pattern varies among the different groups. In the Ixodidae, or hard ticks, there is only one nymph; in the Argasidae, or soft ticks, there may be up to eight nymphal stages. Both males and females require a blood meal in every developmental stage in order to molt into the next developmental stage. In mites, the number is highly variable, ranging from zero to several.

8.2.6 Life cycle

The life cycle of ticks (Ixodida) may include one to many hosts, depending on the species. After mating and obtaining a blood meal, female ticks leave the host and lay the eggs in the soil (Fig. 8.20). A six-legged larva hatches from the egg, climbs into the vegetation, and waits for a suitable host. Once a host is found, the larva feeds, molts, and becomes a nymph (Fig. 8.21). Some ticks, like *Boophilus*, complete their development with just one host. Others may require two, three, or even more hosts. Most ixodids require three hosts, one for the larva, one for the nymph, and one for the adult, whereas most argasids use more than three hosts because they have more than one nymphal stage (Fig. 8.22).

The life cycle of parasitic mites is similar to that described for ticks, with a larva, one or more nymphal stages, and an adult, but many variations do occur. In many species, the life cycle includes free-living and parasitic stages. Some prostigmatid mites, like the chiggers, are parasitic only during their larval stage, and their nymphal and adult stages are free-living and prey on small arthropods.

Host specificity is variable in ticks and mites, ranging from highly specific to opportunistic. Both host specificity and the number of hosts needed to complete the life cycle have implications for their dispersal capabilities and for the potential to spread disease in those species that

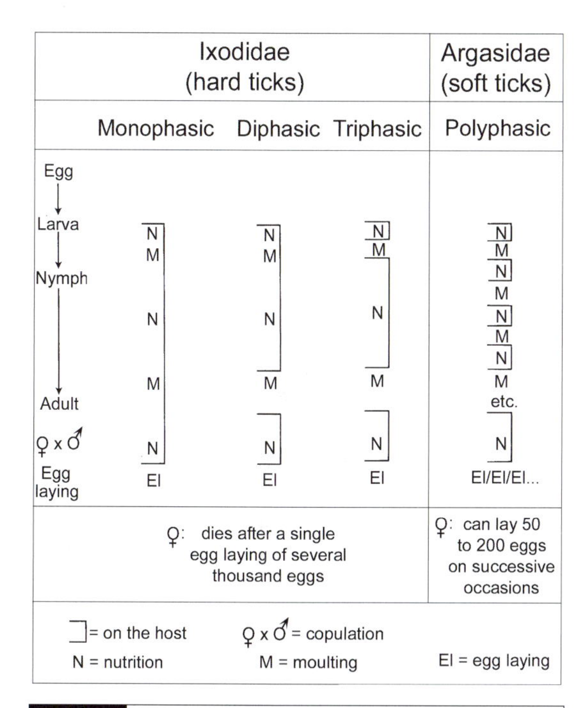

Fig. 8.22 Comparative life cycle of hard ticks (Ixodidae) and soft ticks (Argasidae) showing when feeding, molting, copulation, and egg laying occur. (Modified from Aeschlimann, 1991, with permission, Oxford University Press, New York.)

act as vectors. For example, in one-host ticks, the transmission of disease organisms from one animal to another is very low, but in multiple-host ticks, transmission among animals is very likely. Nonetheless, one-host ticks still carry disease-producing pathogens and, in this case, transmission is vertical, from the mother tick to her progeny, which, in turn, infect new hosts.

8.2.7 Diversity

Ticks and mites are commonly referred to as acarines. The diversity of Acari will be approached on a group-by-group basis following the taxonomic system proposed by Evans (1992); only the groups of Acari that have parasitic species in their ranks will be considered. What we commonly call ticks are all members of the Ixodida, whereas mites belong to a number of different taxonomic groups.

IXODIDA (= METASTIGMATA)

Ticks are highly specialized obligate parasites of a wide variety of terrestrial vertebrates, including amphibians, reptiles, birds, and mammals, plus a few marine snakes and lizards. About 800 species have been described, but most of the attention has been focused on the 80 or so species of medical or veterinary significance, either because of their own pathogenesis or because of the pathogens they carry.

Two kinds of ticks are normally recognized, i.e., hard ticks and soft ticks. *Nuttalliella namaqua*, a unique tick found in Africa, is somewhat intermediate between soft and hard ticks.

Hard ticks. Hard ticks belong to the family Ixodidae. In hard ticks, the capitulum is terminal and can be seen from above. They also have a sclerotized dorsal plate, or **scutum**, that covers the entire dorsum of the body in males, but only the leg-bearing portion of females. In females, the cuticle of the abdomen is extremely folded, allowing the abdomen to expand enormously when feeding and engorging on the host. Most hard ticks have only one nymphal stage between the larva and the adult. During their life span, hard ticks attach and feed on one, two, or more often, three hosts, which can be the same or different species. These ticks are called one-host, two-host, or three-host ticks (Fig. 8.22). Hard ticks feed for a long time. Each blood meal, depending on the developmental stage, can take 4 to 7 days to ingest. For most tick species, host specificity is very low and they attack any available host. In spite of their low host specificity, many species of tick show developmentally related patterns, where larvae and nymphs are found on small mammals and birds, whereas the adults are found on large mammals. After mating and engorging on the last host, the adult female leaves the host, lays a large clutch of eggs on the substratum, often in the thousands, and then dies.

Although most hard ticks have three hosts, species of *Boophilus* (Figs. 8.18, 8.19), including the American cattle tick *B. annulatus*, are one-host ticks, with all the developmental stages occurring on the same individual host. Several species of *Boophilus* are vectors for a number of viruses, rickettsias, and protozoans, including *Babesia bigemina*, the protozoan responsible for Texas cattle fever, also known as red-water fever. Although one-host ticks are, in general, poor vectors, pathogens

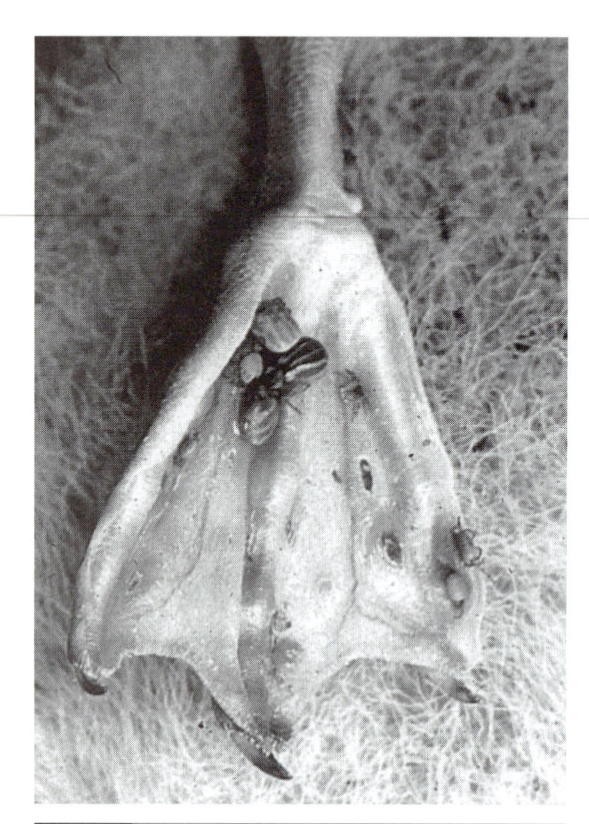

Fig. 8.23 The tick *Ixodes uriae* attached to the underside of the foot of an albatross chick from Bird Island, in the South Georgian Archipelago. (From Bergstrom *et al.*, 1999, with permission, *Journal of Parasitology*, **85**, 25–27.)

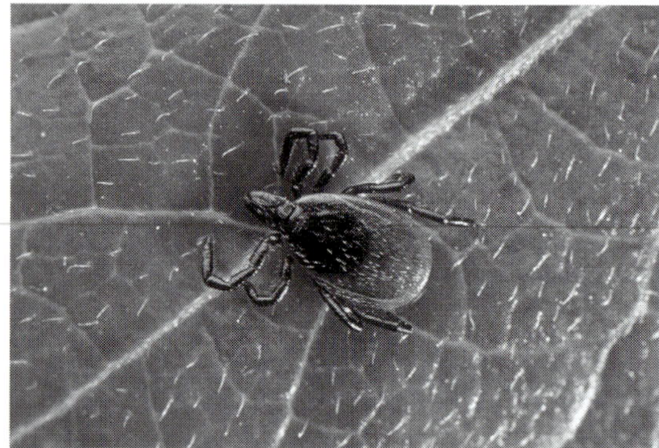

Fig. 8.24 The black-legged or deer tick *Ixodes scapularis*. (Photograph courtesy of the Agricultural Research Service, US Department of Agriculture.)

can be passed vertically, from mother to offspring, by transovarian transmission. Newly hatched ticks thus are already infected and ready to pass the disease onto new hosts when they disperse.

Hard ticks in the genus *Ixodes* are ubiquitous worldwide (Fig. 8.23). Although many species of *Ixodes* do not cause much damage to the host, several are vectors of the spirochete responsible for Lyme disease, *Borrelia burgdorferi* (for a comprehensive review of ticks and Lyme disease see Bennett, 1995; Oliver, 1996). *Ixodes scapularis* (previously known as *I. dammini*), the black-legged tick, is a common vector of Lyme disease in eastern North America (Fig. 8.24), whereas *I. pacificus* is the vector in western North America, *I. ricinus* in Europe, and *I. persulcatus* in Japan (Fig. 8.21). Several species of *Dermacentor*, such as the Rocky Mountain wood tick *D. andersoni*, the American dog tick *D. variabilis*, and the Pacific Coast tick *D. occidentalis*, are vectors for the pathogen causing Rocky Mountain spotted fever and for a number of other viral and bacterial diseases. The lone star tick *Amblyomma americanum*, is also a vector for the organism that causes Rocky Mountain spotted fever. Larvae and nymphs of this species usually feed on birds and almost any small terrestrial mammal available, whereas the adults, again, prefer larger hosts. This lack of specificity reaches as far as humans and, unlike most other ticks, all three developmental stages of *A. americanum* attack humans.

Dermacentor albipictus, also known as the winter tick, is a common species that overwinters on wild ungulates (moose, elk, white-tailed deer, mule deer) throughout the northern USA and southern Canada (Fig. 8.20). Moose, *Alces alces*, seems to be the preferred host, with some individuals harboring more than 100000 ticks. Moose that are heavily infected can be anemic, have low fat stores, and perform extensive grooming, which leads to the premature loss of winter hair. The lack of hair during winter is likely to affect thermoregulation and may affect survival and population size. Fortunately, moose have some help when ticks appear. Black-billed magpies (*Pica pica*) often land on moose and prey on their ticks, helping the moose get at least partially de-ticked whereas the magpies get a nutritious meal (for references see Samuel & Welch, 1991).

Although ticks do not parasitize many amphibians, *Amblyomma rotundatum* attacks reptiles and amphibians, especially the marine toad *Bufo marinus*, in Central and South America. This tick was introduced to Florida in the last 50 to 60 years with the toad host, and is now widely spread along both coasts of Florida (Oliver *et al*., 1993). Unlike other ticks, *A. rotundatum* is a parthenogenetic species, a factor that aids in the colonization of of new areas because, in theory, only one female is needed to start a new population.

The ticks' ability to endure environmental harshness is exemplified by species of *Hyalomma*. These hard ticks are found in desert areas of the Middle East, Asia, and Africa, where there is little shelter, very few hosts, and the ticks must endure periods of starvation. A possible adaptation to the scarcity of hosts is a reduction in the number of hosts required to complete their life cycle, and thus *Hyalomma* is a two-host tick. Many species of *Hyalomma* are vectors for a number of viruses, rickettsias, and protozoans, many of which are carried over long distances by migrating birds infected with ticks.

Soft ticks. Soft ticks belong to the Argasidae and their characteristics and biology are just the opposite of hard ticks. In soft ticks, the capitulum is subterminal and cannot be seen in dorsal view because it lies in a depression that forms a hood over it. Soft ticks have several nymphal stages and are called many-host ticks (Fig. 8.22). Unlike hard ticks, soft ticks do not engorge but feed repeatedly, mainly at night, with each feeding session lasting from minutes to 1 hour. After feeding, they leave the host and rest in their hideout between meals. They are normally found in nests, burrows, crevices, caves, loose soil of resting or rolling areas of big game, and other places often visited or inhabited by the hosts. In most species, the larva remains on the host until molting, whereas the nymphs and adults feed quickly and return to their resting place. Soft ticks parasitize the same individual or its family repeatedly. Adult females deposit eggs in their hiding places several times between feedings, and, unlike hard ticks, clutch sizes are small, of a few hundred eggs or fewer. Argasids prefer habitats with low humidity and are common in desert areas, where they are able to aestivate for months, or even years, without food.

Most species of soft ticks are found in areas of very low humidity and, whenever they happen to occur in humid or wetter climates, they seek dry microhabitats. Several species of *Ornithodoros* parasitize rodents and other non-flying mammals inhabiting burrows, dens, and caves; others parasitize a number of bats and a few are found on marine birds nesting, or resting, in large colonies on or near the ground. *Ornithodoros savignyi*, the sand tampan, is the most widely distributed species in this genus. It is common in dry regions of Africa, the Near East, and India where it feeds on camels, livestock, humans, and wildlife. It is a nocturnal species and most of its victims are asleep when the tick attacks. Although it appears not to vector any disease-causing organisms, the bite is quite painful. The extreme pain associated with the bite of *O. savignyi* seems to be a rather common phenomenon among most soft ticks. *Ornithodoros hermsi* is a common species found in forests up to 2500 m in altitude between the Rocky Mountains and the Pacific Coast of North America. The tick infests nests of chipmunks, wood rats, pine squirrels, and other rodents in hollow logs, tree stumps, and wood cabins, where tourists, woodsmen, hikers and residents are frequently bitten and may develop a severe relapsing fever caused by *Borrelia hermsi*.

A few species of *Ornithodoros* are associated with ground-nesting marine birds, including the Galapagos and Humboldt penguins (Fig. 8.25). The Humboldt penguin, *Spheniscus humboldti*, breeds on the barren coasts and offshore islands of Peru and Chile, where they normally cohabit with guano-producing bird colonies. *Ornithodoros spheniscus* is ubiquitous among penguins but they eagerly attack humans causing pruritus, slow-healing blisters, fever, and headaches. The tick is very annoying to the host, and the irritation produced by feeding ticks may induce the penguins to abandon their nests (Hoogstraal *et al*., 1985). Other related tick species also are known to cause nest desertion by terns and brown pelicans. Some species of Peruvian guano birds are prone to remain in flight longer than needed, staying away from their breeding grounds, just to avoid being bitten by the ticks. Eventually, they still may desert their nests (Feare, 1976; King *et al*., 1977; Duffy, 1983). Incidentally, guano farming was an

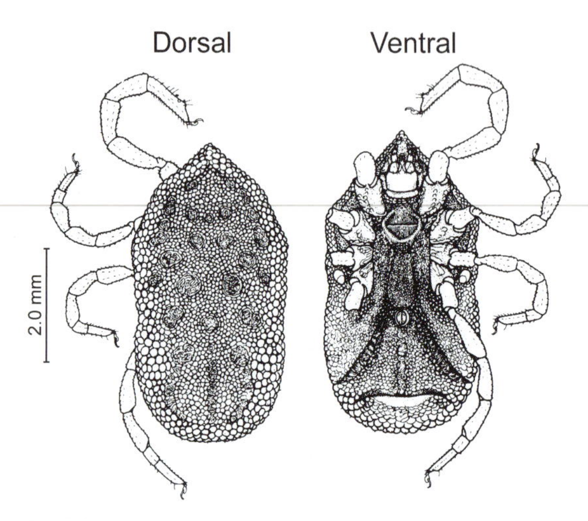

Fig. 8.25 Dorsal and ventral view of the soft tick *Ornithodoros spheniscus*, a parasite of the Humboldt penguin in Peru. (From Hoogstraal *et al.*, 1985, with permission, *Journal of Parasitology*, **71**, 635–644.)

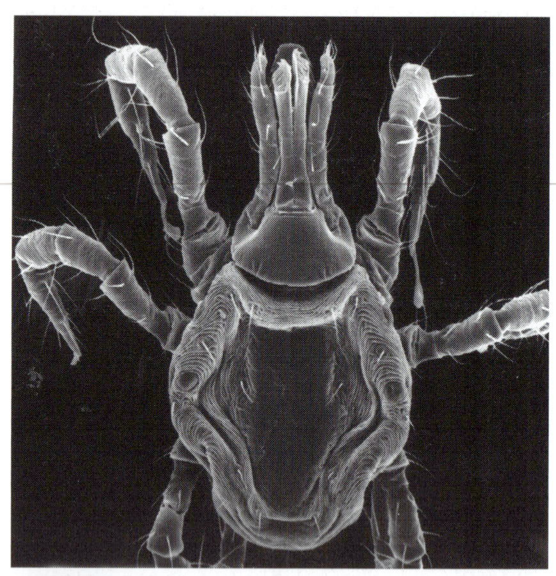

Fig. 8.26 Larval stage of the ear tick *Otobius megnini*. (Note that larval stages have only three pairs of legs.) (Photograph courtesy of the US National Tick Collection, Georgia Southern University.)

important source of revenue for Peru in the last century. It makes one itch just to think of the conditions the workers, mostly enslaved habitants of Easter Island, had to endure during the guano operation! Indeed, the discomfort associated with these ticks is legendary in Peru.

Soft ticks in the genus *Argas* are almost exclusively parasites of birds, both marine and terrestrial. Most species are nest parasites, feeding at night and hiding during the day in crevices, cracks, litter, or any other hiding place in or near the nest of the host. The fowl tick, *A. persicus*, is native to central Asia where it attacks arboreal nesting birds such as sparrows and crows, but has spread to many parts of the world where it has successfully colonized domestic chickens, turkeys, and other fowl. Under favorable conditions, this tick can form large populations in henhouses and become a serious pest. Humans that visit working, or even abandoned, chicken coops at night are usually attacked. *Argas cucumerinus* is the exception to this common nocturnal behavior. This species is found on barren cliffs facing the Pacific Ocean along the Peruvian coast and feeds on birds and humans that happen to stop and rest on the rock ledges where the ticks shelter. Because *A. cucumerinus* is not associated with a nest, host encounters are random and scarce. When a host is detected during the daytime, the tick races towards the host using its long, spider-like legs and feeds very rapidly. The racing legs, daytime behavior, and fast feeding are unique characteristics of this species, very different from the typical tick, and clearly constitute adaptations to the scarce and irregular presence of hosts.

Otobius megnini (Fig. 8.26) is a highly specialized soft tick that lives and feeds inside the ears of the host. Deer, mountain sheep, and pronghorn antelope are natural hosts, but the tick has adapted easily to cattle, other domestic animals, and humans. Unlike other soft ticks, *O. megnini* has a one-host life cycle, adult ticks do not feed, and mating occurs in the ground.

MESOSTIGMATA

Mesostigmatid mites attack a variety of vertebrates and invertebrates (Fig. 8.27), and many of them parasitize the respiratory system of their hosts. A number of species are associated with fowl and rats, and can become a nuisance to humans. The chicken mite *Dermanyssus gallinae* parasitizes fowl worldwide and causes severe dermatitis in humans (Fig. 8.28). The mites live near roosting places and feed at night. In heavy infections, they become so annoying that setting hens

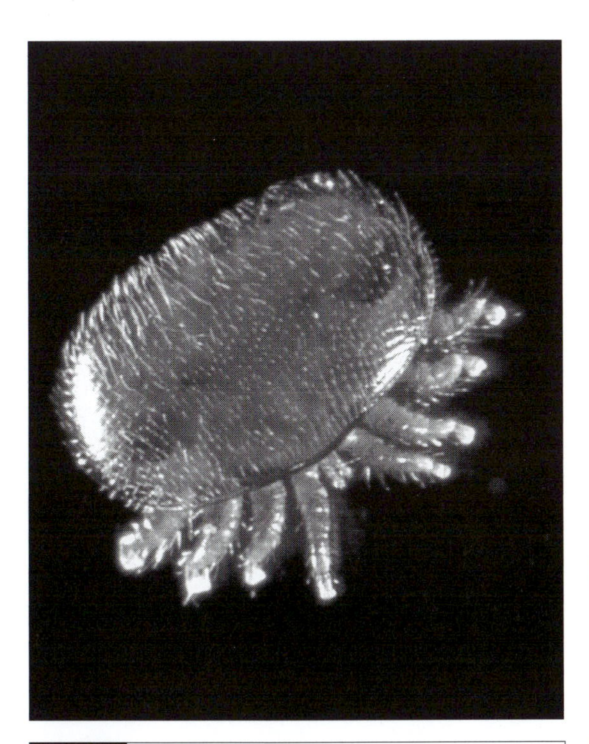

Fig. 8.27 *Varroa jacobsoni*, the honeybee mite. Notice the crab-like shape, with the antero-posterior axis being shorter than the side-to-side axis. (Photograph courtesy of Andrew Syred/Microscopix, © Andrew Syred/Microscopix.)

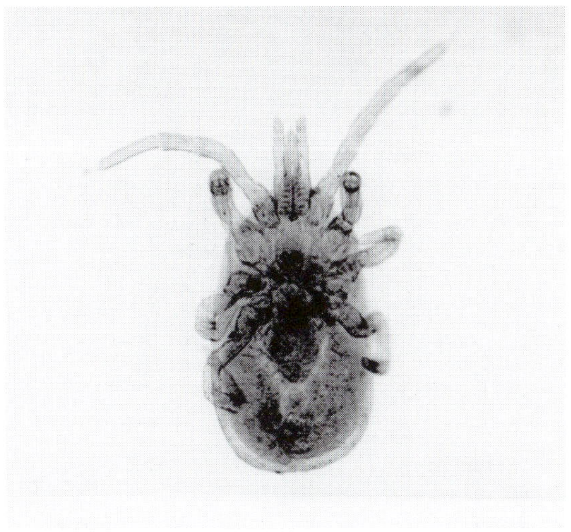

Fig. 8.28 The chicken mite *Dermanyssus gallinae*.

leave their nests and young chickens are likely to die. Pigeons may carry these mites near humans when roosting in the roofs or eaves of buildings. The northern fowl mite *Ornithonyssus sylviarum* is common in temperate regions and, although it is not pathogenic for the birds, it can be bothersome for people working in direct contact with chickens.

Echinolaelaps muris, the common rat mite, is the vector for a malaria-like protozoan among rats, whereas *Liponyssus sanguineus*, the house mouse mite, is the vector for the microorganism responsible for rickettsial pox in humans, a mild febrile condition that lasts about 2 weeks. The tropical rat mite *Ornithonyssus bacoti* can become a serious nuisance in research facilities where mouse colonies are maintained. In heavy infections, it can retard growth and even kill young mice, whereas in humans it may cause severe dermatitis in addition to the sharp pain produced by the bite.

Rhynonyssid mites parasitize the respiratory tracts of birds, feeding on blood and other tissues, and can be a significant agent of disease among wild birds. Halarachnids, on the other hand, are parasites of the respiratory system of other vertebrates. Species of *Orthohalarachne* parasitize the respiratory tract of marine mammals (Fig. 8.29). Males and females are relatively non-motile and have well-developed claws to maintain their position in the nares of the host. Transmission from host to host may occur by active larvae that crawl, or more likely, by larvae that are sneezed from one animal to another. There is a curious account of a human infection by *O. attenuata*, a common parasite of fur seals, sea lions, and walruses (Fig. 8.29). A single specimen was removed from the eye of a person who felt eye discomfort soon after visiting the walrus exhibit at Sea World in San Diego. Apparently, during the normal spitting and snorting behavior of walruses, one mite became dislodged from its walrus host and reached the eye of the innocent visitor (Webb *et al.*, 1985).

PROSTIGMATA

The Prostigmata includes both free-living and parasitic mites. The familiar chiggers, follicular mites, and several mites that infect insects belong to this group.

Chiggers are the larval forms of trombiculid mites. They are unusual in that only the larvae are

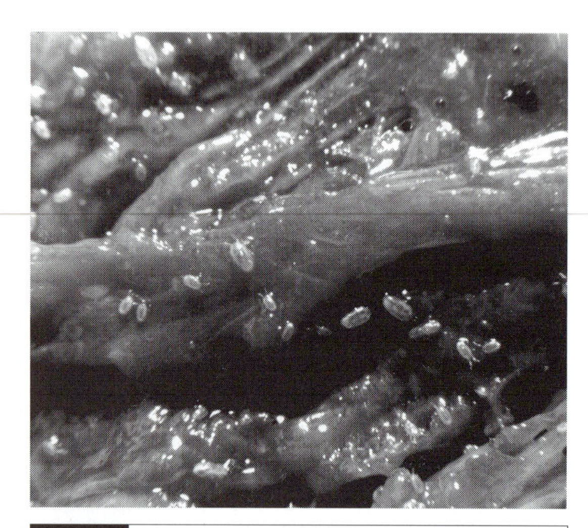

Fig. 8.29 *Orthohalarachne attenuata* (large mites) and *O. diminuata* (small mites) inside the nasal cavity of a Northern fur seal. (From Kim, K.C. [1985] Evolutionary relationships of parasitic arthropods and mammals. In *Coevolution of Parasitic Arthropods and Mammals*, ed. K. C. Kim, pp. 3–82. Copyright © 1985 John Wiley and Sons, Inc. Reprinted by permission of John Wiley and Sons, Inc.)

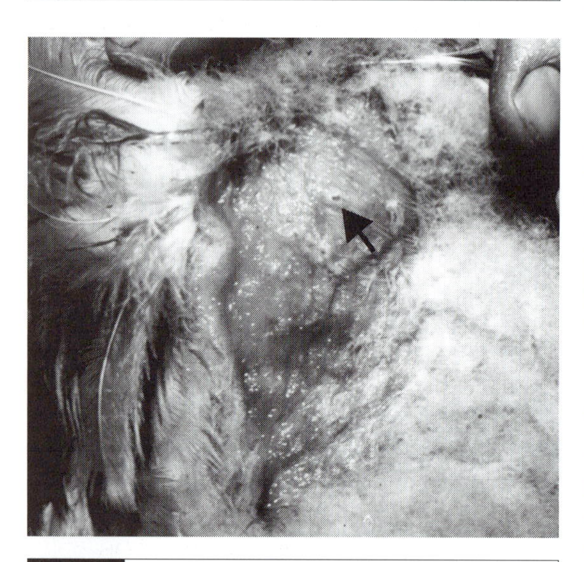

Fig. 8.30 Papules on the thigh of a Florida sandhill crane (*Grus canadensis pratensis*). Note the central depression (arrow) that contains one or more chiggers (*Blankaartia sinnamaryi*). (From Spalding *et al.*, 1997, with permission, *Journal of Parasitology*, **83**, 768–771.)

parasitic, whereas the nymphs and adults are free-living and prey on small terrestrial invertebrates and their eggs. Chiggers parasitize every major group of terrestrial vertebrates, and some aquatic ones too, but they rarely parasitize invertebrates.

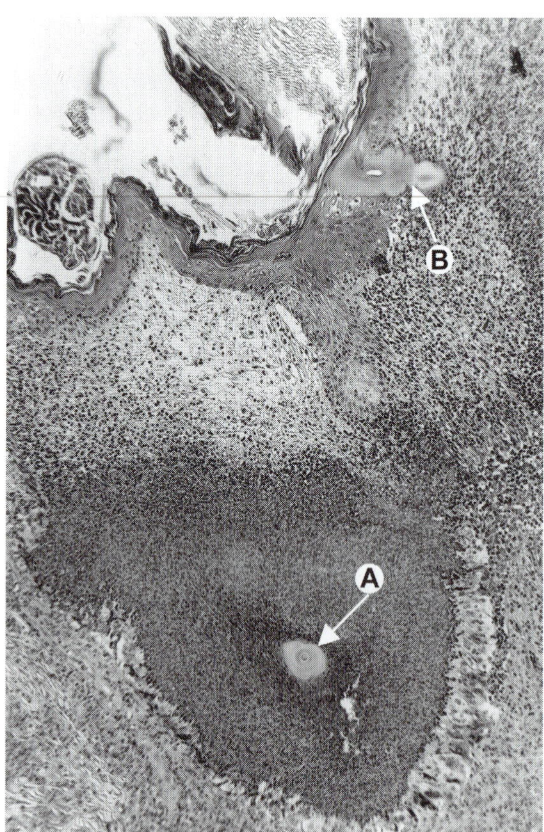

Fig. 8.31 Section of a papule containing the chigger *Blankaartia sinnamaryi*; within the central depression. A cross-section (arrow A) and a longitudinal section (arrow B) of the chiggers' stylostomes can be seen. (Modified from Spalding *et al.*, 1997, with permission, *Journal of Parasitology*, **83**, 768–771.)

Most chiggers cause some type of skin problem and some are vectors for pathogens. Contrary to popular belief, the skin condition caused by most chiggers is not due to burrowing of the larvae under the skin, but to their feeding activity. Chiggers do not burrow under the skin (Fig. 8.30); instead, after the mouthparts penetrate the skin, the larva injects a proteolytic salivary secretion that digests host cells, which are then sucked up, together with interstitial fluids. A specialized hollow feeding tube, or **stylostome**, made from hardened host and larva secretions, facilitates feeding, working as a drinking straw (Fig. 8.31). The larvae of *Blankaartia sinnamaryi* parasitize young Florida sandhill cranes and other birds and mammals in tropical America. These chiggers are

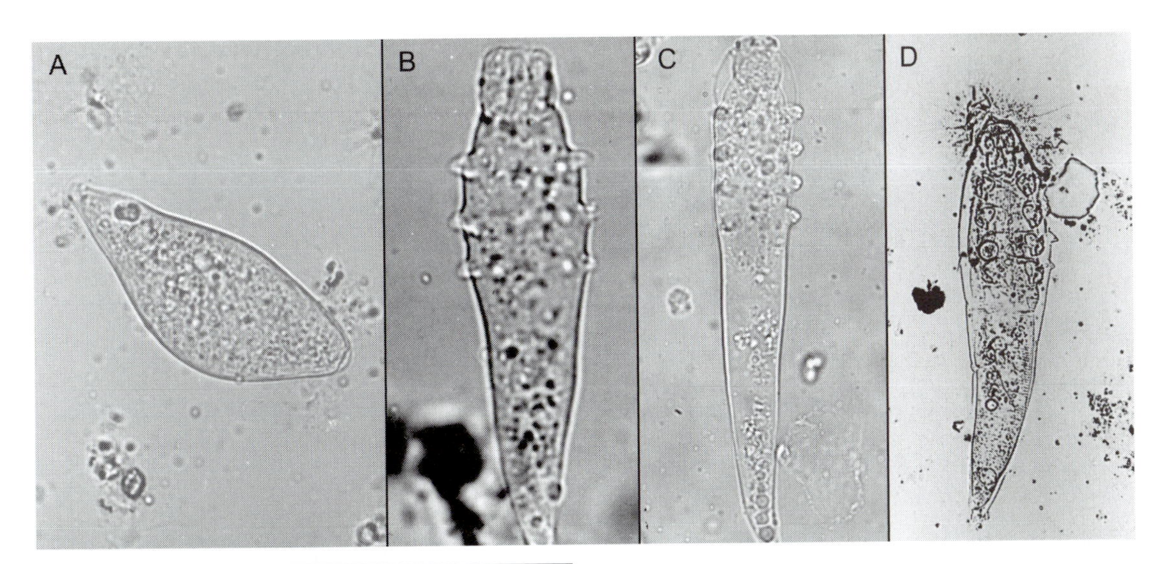

Fig. 8.32 Developmental stages of *Demodex canis*, the follicular mite of dogs. (A) Egg; (B) larva; (C) nymph; (D) adult. (From Caswell *et al.*, 1996, with permission, *Journal of Parasitology*, **82**, 911–915.)

bright orange and can be seen with the naked eye in the central depression of firm, raised papules. The papules are the result of a host reaction to the chigger, specifically to its stylostome, which triggers a localized granulomatous dermatitis (Fig. 8.30) (Spalding *et al.*, 1997).

Demodicid mites, commonly known as follicular mites, are very small, elongated parasites, with short stumpy legs. Follicular mites parasitize the hair follicles and sebaceous glands in the skin of mammals, where they feed on their secretions. Often, all the developmental stages, from egg to adult, are found in a single follicle. Demodicid mites are common in every group of mammals, including marsupials, and exhibit strict host specificity. *Demodex canis* is a common parasite of the hair follicles of dogs (Fig. 8.32). Sometimes, however, the mite population proliferates beyond normal levels, causing a clinical condition known as red mange. Heavily infected dogs have dry, scaly patches of hair loss, and may develop pustular lesions covered with a foul-smelling exudate, probably as a result of secondary infection by *Staphylococcus*. Young puppies, or older dogs suffering from malnutrition, parasitic infections, or debilitating diseases, are more susceptible; symptoms disappear gradually in older, healthy dogs, probably due to acquired immunity. There is some evidence that the host immune response is responsible for controlling the mite population.

Humans are no strangers to mites and are usually infected (unknowingly) by two species common in the follicles and sebaceous glands of facial hair. *Demodex folliculorum* measures almost 0.3 mm and inhabits the follicles of hairs above the level of the sebaceous gland, where three or more specimens can be found in a single follicle. *Demodex brevis* is slightly smaller, about 0.2 mm, and lives in the sebaceous glands of hairs, where a single individual, or at the most two (a female and her offspring), can be found (Desch & Nutting, 1972). *Demodex folliculorum* consumes cells of the follicular epithelium, whereas *D. brevis* feeds on cells from the sebaceous gland, a perfect example of habitat segregation and resource partitioning. Both species are rather benign, although some forms of rosacea and juvenile acne may be related to their presence.

Pyemotid mites are ubiquitous parasites of insects. Unlike other mites, the female retains the eggs in her abdomen and they complete all their developmental stages inside. Males emerge first and then wait to copulate immediately with the emerging females. *Pyemotes tritici*, the straw itch mite, is a common parasite of the grain moth, but it also feeds readily on humans, specially granary workers. The mites are so abundant that, in most cases, they kill their moth host. For this reason, the mite has been used for the biological control of several insects.

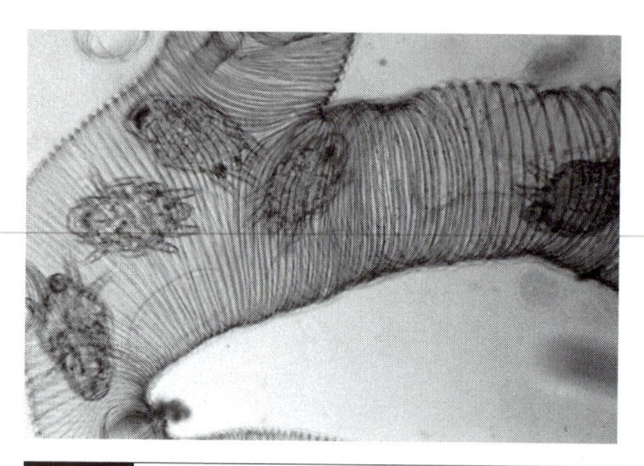

Fig. 8.33 Tracheal mites, *Acarapis* sp., in the tracheal system of a honeybee. (Photograph courtesy of the Agricultural Research Service, US Department of Agriculture.)

Mites in the genus *Acarapis* are highly host-specific as parasites of adult honeybees (*Apis* spp.). They live in the tracheal system (Fig. 8.33) where they perforate the wall of the trachea, reach into the body cavity of the honeybee, and feed on hemolymph. Tracheal infection affects gas exchange and impairs the bee's ability to fly because of the lack of oxygen. The bee eventually dies of starvation or suffocation. (For more details on this and other mites present in honeybees, see Box 8.2.)

Some mites parasitize aquatic insects. Several species of *Arrenurus* parasitize mosquito pupae while in the water; when the adult mosquito emerges, they transfer and parasitize the adults. In natural populations, female mosquitoes harbor more parasitic mites than males. The aquatic mites seem to detect and attach preferentially to female mosquito pupae, although males are also parasitized (Lanciani, 1988). Although several hypotheses may explain this preference, it seems that because the mites must return to the water to complete their life cycle, they are more likely to do so when parasitizing a female mosquito that also needs to return to the water to lay her eggs. The larva of the water mite *Hydrachnella virella* parasitizes the backswimmer *Buenoa scimitra*. It attaches to the external surface of the insect, punctures the tegument, and feeds on hemolymph. Infections by several mites may induce host mortality, but also, the number of mites on a host is inversely correlated with mite size. The more mites on a host, the smaller they are (Lanciani, 1984), i.e., too many mouths to feed and not enough food!

ASTIGMATA

Astigmatid mites lack a respiratory system. They are common parasites of mammals and birds, causing several types of **mange** and **scabies**. Mange is a generic term used to describe the dermatitis caused by various species of mites (Fig. 8.34). Scabies, or sarcoptic mange, on the other hand, is the allergic condition resulting from the burrowing of *Sarcoptes* spp. in the epidermis.

Sarcoptes scabiei (Fig. 8.35) is one of the best-known mites and is responsible for the scabies of humans and many wild and domestic animals. *Sarcoptes scabiei* burrows in the skin of the host, particularly in moist areas between fingers, behind knees and elbows, ankles, toes, and genital regions. Transmission is only by contact. Males live on the surface of the skin where they move seeking mates. Mature females, however, burrow and tunnel under the skin, where they live for about 2 months, during which time their tunnel advances between 0.5 and 5 mm per day. After 1 or 2 months of tunneling, humans develop an allergic reaction to eggs, mites, and mite droppings present in the

Box 8.2 *Varroa* and *Acarapis*, the bee's mite-y problem

When you think about bees, you normally think about honey, right? You also should think about fruits, vegetables, and other plants that rely on honeybees for pollination. About 30 crops in the USA, valued at 10 billion dollars annually, depend mostly on honeybees for pollination.

The mite-y problems these days are two species of mites, *Acarapis woodi* and *Varroa jacobsoni*, which are decimating the honeybee population and might ultimately affect agricultural yields.

Acarapis woodi (Fig. 8.33), the tracheal mite of honeybees, entered the USA from Mexico in 1984. The mite, however, has been recognized as a problem since 1917 in other parts of the world. The mites live in the tracheae of the respiratory system where they pierce the wall to feed on hemolymph; this also causes degeneration of the tissues around the feeding site. In high numbers, the mites also reduce the rate of air exchange in the tracheae, affecting the bee's capacity for flight. Although *A. woodi* does not necessarily kill the bees, the bees are weakened. Under adverse environmental conditions, the beehive may be affected. In warmer climates, where brood rearing is continuous, the constant production of new bees allows the colony to survive. In more temperate climates, however, the seasonal breeding is not enough to maintain the functionality of the colony. Unfortunately, the dynamics of the tracheal mite and bee population is not completely understood. Two other species of *Acarapis* have been documented as parasites on the external surfaces of honeybees, but parasitism by these does not seem to be debilitating.

Varroa jacobsoni (Fig. 8.27), the varroa mite, entered the USA in 1987; *V. jacobsoni* was originally a parasite of the Asian hive bee, *Apis cerana*. When the European honeybee *Apis mellifera*, a superior producer of honey, was introduced to Asia some 50 years ago, the varroa mite found a new and suitable host to colonize. Since then, varroa mites in honeybees have been spreading throughout the world. Varroa mites are external parasites of mature larvae and adult honeybees. The mites pierce the body wall and feed on hemolymph, weakening their victims. The adult females are reddish-brown in color, flattened, and measure about 1 to 1.5 mm in width. Females prefer to feed on adult bees where they attach and hide between the bee's abdominal segments. Female mites, ready to reproduce, abandon the bee inside the hive and enter a wax cell containing a bee larva 1 or 2 days before the cell is capped. The female mite is then sealed with the larva; the mite lays eggs, the eggs hatch, and the new mites feed on the developing honeybee larva. By the time the honeybee larva is mature and ready to emerge, several of the mites will have mated and reached adulthood; these are free to roam inside the hive and repeat the cycle. Mites spread from colony to colony by drifting workers and drones, or during swarming. Honeybees can also acquire the varroa mite when robbing other colonies that happen to be infected. Incidentally, infected colonies are weakened and are more likely to be robbed, facilitating the spread of the infection.

A solution to the honeybee crisis is not yet in sight. Currently, a few acaricides are available to control mite infection in hives but, unfortunately, mites are already showing signs of resistance. Several biological strategies also are being developed to fight these mites. One is to let natural selection runs its course until natural resistance increases in the bee population. It may take some time, but it has happened before with honeybees in Europe and South America, some of the earlier locations of the Asian mite invasion. Other strategies include the breeding of honeybees that groom, ridding each other of mites; the introduction of benign strains of varroa mites that could pre-empt the more virulent ones; the introduction of strains of honeybees with genetic resistance to mites; or the selective breeding of 'hygienic' bees, a naturally occurring behavioral trait that allows the bees to detect the presence of varroa mites in cells and evict them, reducing the mite load in the hive over time.

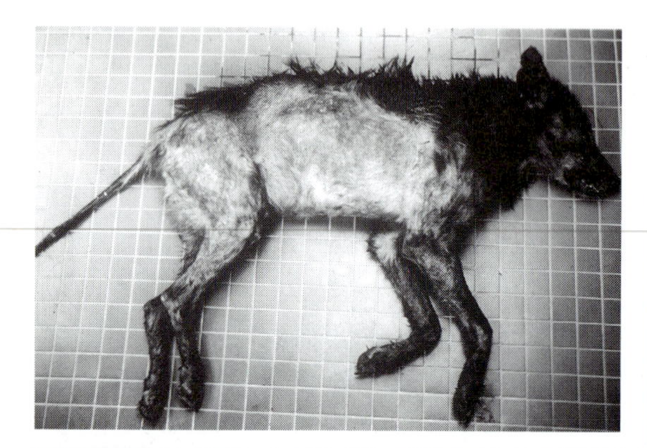

Fig. 8.34 Severe case of mange in a wolf *Canis lupus*. (Photograph courtesy of Bill Samuel, University of Alberta.)

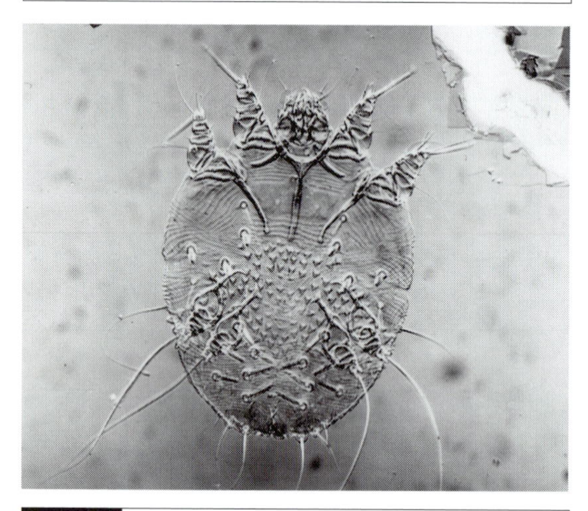

Fig. 8.35 *Sarcoptes scabiei*, the itch mite responsible for scabies or sarcoptic mange.

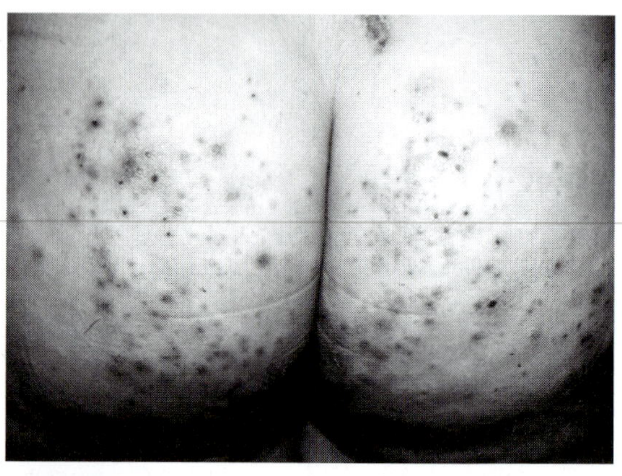

Fig. 8.36 Scabies in the buttocks of a Brazilian child. The area shows evidence of scratching and secondary bacterial infection. (Photograph courtesy of the Armed Forces Institute of Pathology.)

tunnels, which results in rashes and intense, maddening, itching (Fig. 8.36).

Sarcoptic mange in coyotes (also caused by *S. scabiei)*, however, becomes a chronic debilitating infection as evidenced by significant decreases in fat deposits and total body weight. Even though coyotes mount a significant humoral response, as indicated by marked increases in gamma globulin, the percentage of coyotes that recover completely from the infection is very low (Pence *et al.*, 1983). Secondary infections by fungi are very common, and it is not clear to what extent the fungi contribute to the pathology of mange in wild animals.

Scabies, and the mite causing it, have been known for a long time. Aristotle and Cicero were familiar with both. Because of the accurate record-keeping on this parasite in recent times, it seems that scabies follows epidemic cycles among humans, ranging from 20 to 30 years between peak levels of infection. Unfortunately, the reasons for this cycling phenomenon are not clear. Some correlate these peaks with periods of social unrest, such as wars. Others attribute them to the inability of physicians to recognize the disease because of lack of experience in periods of low prevalence, which results in a subsequent high level of transmission in the following years. A third explanation attributes the epidemic cycling to the development of 'herd immunity'. When herd immunity occurs, the proportion of the population that has been infected and overcomes the infection (the immune population) increases until it becomes so large that transmission decreases. Reduced transmission then results in a decline of prevalence and the disease may disappear for some time. Once a new susceptible population has grown up (a new host generation), a new wave of infection begins and prevalence increases (for references and further details about scabies, see Burgess, 1994).

Psoroptid mites, in particular species of *Psoroptes*, are responsible for a condition known as psoroptic mange or scabs. These mites do not burrow under the skin; instead, they pierce the skin at the base of hairs and suck on lymph and

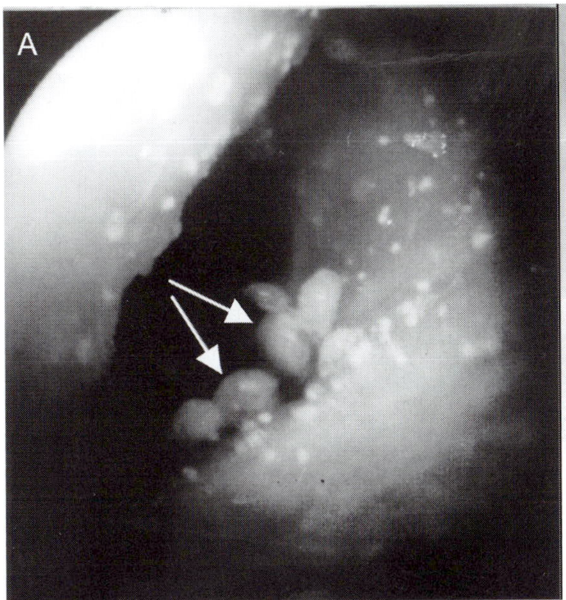

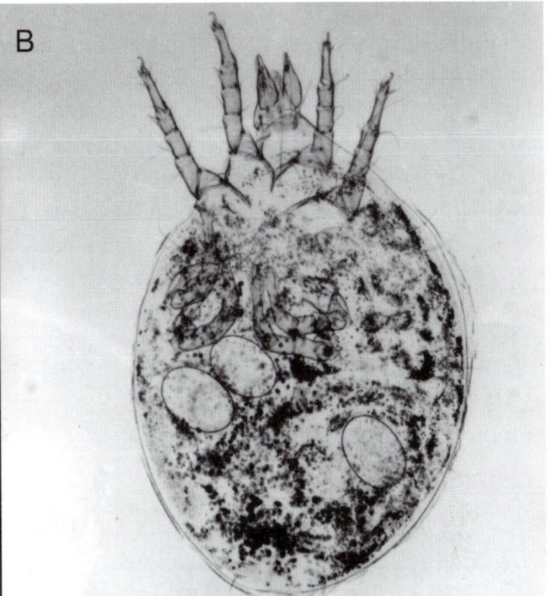

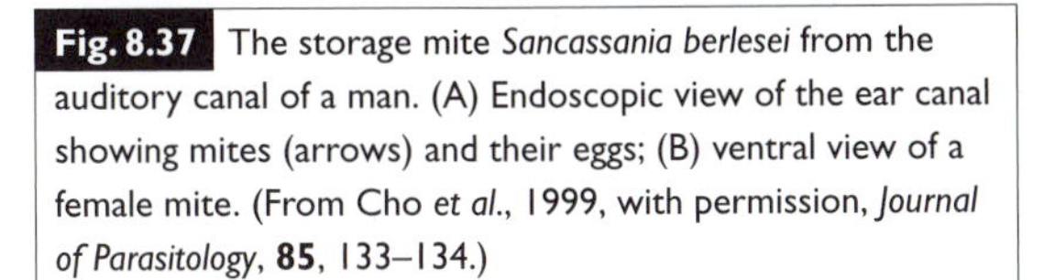

Fig. 8.37 The storage mite *Sancassania berlesei* from the auditory canal of a man. (A) Endoscopic view of the ear canal showing mites (arrows) and their eggs; (B) ventral view of a female mite. (From Cho *et al.*, 1999, with permission, *Journal of Parasitology*, **85**, 133–134.)

other exudates. The excess fluids left on the skin solidify, forming the characteristic scabs seen in infected animals. The scabs provide a protective surface for the mites, under which they reproduce and proliferate very rapidly. A related group of species in *Otodectes* attacks the ears and head of several canids and cats, causing much distress to the host.

Birds also are good hosts for astigmatid mites. Species of *Knemidokoptes* attack both feathered and scaly areas of the body. The depluming mite *K. gallinae* is common among chickens, pheasants, and geese. The mites embed in the skin at the base of the feathers and apparently cause intense itching; birds may actually be seen plucking their feathers in the affected area. Sometimes, however, the feathers may fall out spontaneously.

Some mites in the Acaridae are common in stored foods and may cause dermatitis in people handling the contaminated products. Some, however, have the potential to become established in other habitats and hosts. *Sancassania berlesei*, for example, is a common storage mite that sometimes may be found in cattle. This species was recently found in the ear canal of a man in Korea (Fig. 8. 37) where the mite maintained its life cycle for at least three generations, feeding probably on skin scales and wax (Cho *et al.*, 1999). Although this is an unusual situation, it exemplifies the habitat plasticity of many species of mites.

8.3 Uniramia

8.3.1 General considerations

The Uniramia includes the most common and familiar terrestrial invertebrates on earth and is divided into five classes, i.e, Chilopoda, Diplopoda, Symphyla, Pauropoda, and Insecta. The Chilopoda and Diplopoda are all terrestrial, free-living, flightless, and include the well-known centipedes and millipedes, respectively. The Symphyla and Pauropoda are rather small groups of free-living, terrestrial organisms that inhabit leaf mold and soil. The Insecta are by far the largest class of uniramians and the largest group of animals on earth; it includes about 75% of all animal species. Within the Uniramia, only the Insecta have species that have evolved a parasitic lifestyle (see Box 8.4 for a taxonomic account of parasitic insects).

Most insects are free-living and terrestrial, although a few have become parasitic. A number of insects are parasitic throughout their lives. Lice and bedbugs are bloodsuckers in all stages of their life cycles, but only lice are permanently associated with the host. Most insects are parasitic only during part of their life cycle. Fleas, mosquitoes, and a number of flies, for example, are bloodsucking parasites of birds and mammals only as adults, whereas some wasps and screwworm flies are parasitic only during their larval stage. In addition to their parasitic lives, some insects are vectors for a number of disease-causing organisms, including those responsible for sleeping sickness, malaria, elephantiasis, yellow fever, typhus, plague, typhoid fever, etc.

8.3.2 The outside

Shape. The body of all insects is divided into a head, thorax, and abdomen. The head bears one pair of compound eyes, one pair of antennae, and a mouth with three pairs of appendages that aid in feeding. Between the eyes and the antenna, one or more ocelli, or simple eyes, may be present. The mouth and mouthparts include a pair of mandibles, a pair of maxillae, and a single **labium**. The mandibles are the primary feeding appendages; the maxillae and labium may also have palps that aid in food handling. The mouth is covered by an extension of the head, forming an upper lip or **labrum**. Some insects also may have a tongue-like structure, or **hypopharynx**, in the floor of the mouth. Depending on the group of insects and their feeding behavior, these mouthparts may be modified or even lost. In chewing lice, the mouthparts conform to the basic pattern. In bloodsucking insects, however, some of the mouthparts are modified into **stylets**, used to penetrate the skin of the host in search of a blood vessel. In some groups such as mosquitoes, the stylets may be enclosed by a sheath that holds them in place. In sucking lice, the stylets are retracted into the body when not being used. Although biting flies also feed on blood, they do not have stylets. Instead, the mandibles produce a wound in the skin and blood is collected by a sponge-like labium and conveyed to the mouth.

The thorax consists of three segments each bearing a pair of legs. The second and third segments normally bear a pair of wings each. If wings are missing, their loss has been secondary. Some groups, such as the parasitic lice and fleas, lack wings completely, and others, like ants and termites, have wings only during certain periods of their life cycle. In the Diptera, which includes mosquitoes and flies, the second pair of wings is reduced to knobs, called **halteres**, which function as balancing organs during flight.

The abdomen contains most of the internal organs. Externally, it bears a pair of terminal sensory cerci that function as receptors for air currents, and the genital appendages or genitalia. Males possess a penis and females an ovipositor.

Several processes such as setae, scales, spines, etc., may be present on the surface of the body.

Body wall. The body wall of insects is very similar to that of arachnids. Three layers form it: an outer cuticle that contains chitin, polysaccharides, proteins, and often, pigments; an epidermis that lies beneath the cuticle and secretes it; and a thin, non-cellular layer beneath the epidermis, called the basement membrane. The cuticle can be divided into epicuticle and procuticle. The epicuticle, the thin outermost layer of the cuticle, is made of cuticulin, a protein. Cuticulin is normally covered by a wax layer that prevents water loss, and by a cement layer on top that prevents the abrasion of the wax and subsequent loss of water. Immediately under the epicuticle is the procuticle. The outer part of the procuticle is often dark and hard and is called the exocuticle; the remaining basal section is called the endocuticle.

The body wall has a number of internal processes or infoldings that strengthen the exoskeleton and serve as attachment points for the muscles.

8.3.3 The inside

Digestive system. The insect gut, as in other arthropods, is divided into three sections, i.e., foregut, midgut, and hindgut. Food taken into the mouth continues into the foregut, which is commonly divided into an anterior pharynx, an esophagus, a crop, and a proventriculus (Fig. 8.38). In insects that suck fluid, the pharynx is modified as a pumping organ. The crop is used for storage and may be absent. The function of the proventriculus depends on the type of food ingested. In insects that eat solid food, the proventriculus is a gizzard for triturating food, but in sucking insects it

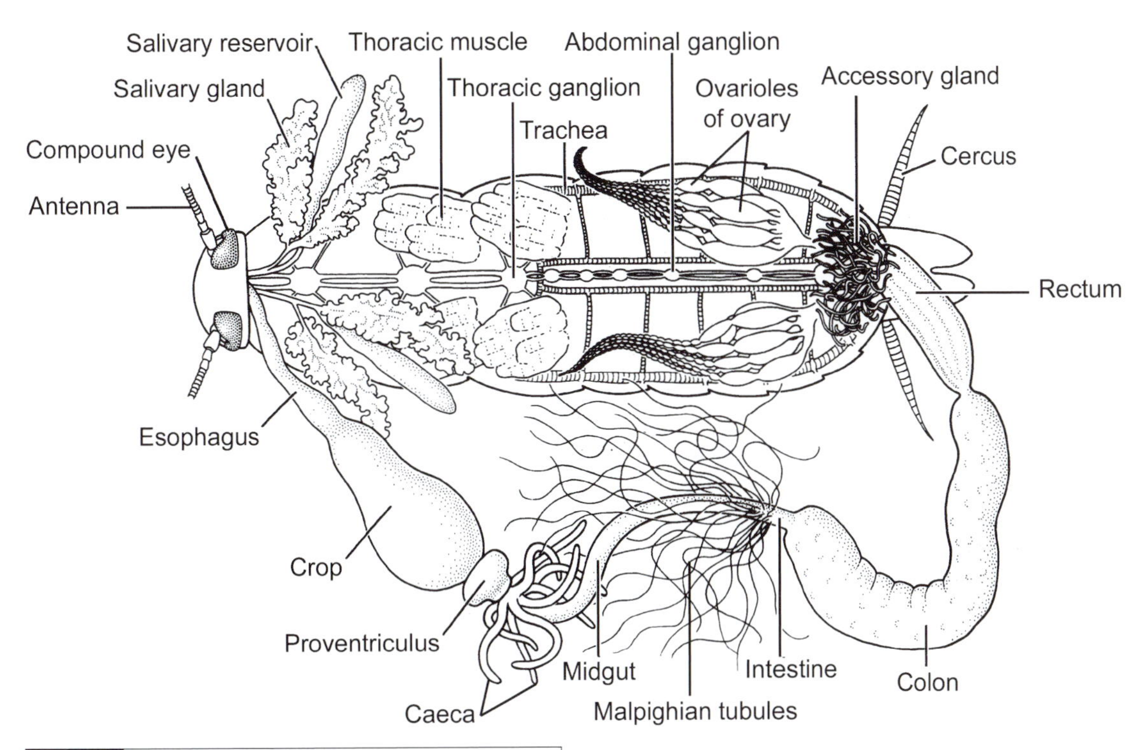

Fig. 8.38 Internal anatomy of a female insect. (Modified from Gullan, P. J. & Cranston, P. S. [1994] *The Insects: An Outline of Entomology.* Chapman & Hall, London. Copyright © P. J. Gullan & P. S. Cranston. Reprinted with kind permission from Kluwer Academic Publishers.)

becomes a valve to regulate the movement of food into the midgut. Most insects have a pair of salivary glands that open into the mouth (Fig. 8.38). The saliva produced by insects has a different composition depending on their feeding habits. Anticoagulants, for example, are present in saliva from blood-sucking insects.

The midgut, or stomach, is where enzyme production, digestion, and absorption occur. In many insects, the midgut secretes a thin membrane made of protein and chitin that surround the food as it moves along the gut. This structure, called the **peritrophic membrane**, seems to protect the gut from abrasion by the food mass while it remains permeable to enzymes and the products of digestion. Several gastric ceca located in the anterior end of the midgut are important in food and water absorption (Fig. 8.38). Following the midgut is the hindgut, divided into intestine and rectum, which are responsible for the elimination of waste and for regulating water and salts.

Osmoregulatory and excretory system. Malpighian tubules are the main structures used for excretion in insects (Fig. 8.38). They lie free in the hemocoel and open into the digestive system at the junction of the midgut and hindgut. Uric acid, amino acids, and salts are actively transported into the Malpighian tubules, but water enters passively. Water, some salts, and other compounds are later reabsorbed and are transferred back to the hemolymph in the body cavity. The uric acid within the Malpighian tubules is deposited into the hindgut where it crystallizes as water is reabsorbed. The crystallization of uric acid allows the insect to excrete solid waste with the feces, saving some valuable water in the process. Uric acid is the main nitrogenous waste in terrestrial insects, whereas aquatic insects excrete ammonia instead.

Some waste products are stored in specialized cells called nephrocytes. These cells are located on, or near, the heart and pick up waste to be degraded intracellularly.

Circulatory system. The circulatory system of insects is open and the body cavity functions as a hemocoel that is divided into several sinuses or compartments by perforated diaphragms. A dorsal and a ventral diaphragm divide the body into three compartments, i.e., a pericardial sinus, a perivisceral sinus, and a perineural sinus. A tubular heart that extends through most of the abdomen pumps blood from the posterior to the anterior end. The blood returns to the heart through the sinuses, aided by body movements. Blood enters the heart from the pericardial sinus through lateral openings, or ostia. The ostia are one-way valves that close when the heart contracts to force the blood anteriorly.

The blood, or hemolymph, is a clear fluid, usually colorless, but because of certain pigments it may be green, blue, or yellow. The red hemolymph of certain insects such as the gasterophilid bot flies and midges, owes its red color to hemoglobin. Hemolymph contains a number of distinct cells, or hemocytes, that function in phagocytosis, wound healing, encapsulation of foreign bodies, detoxification, coagulation, etc.

Respiratory system. Gas exchange between the tissues and the environment occurs through internal, air-filled tracheae that ramify throughout the body (Fig. 8.38; see also Fig. 8.33). The smallest branches, or tracheoles, contact all internal organs and tissues, and are especially numerous in tissues with high oxygen requirements. The tracheal system opens to the outside via spiracles. In many endoparasitic larvae, however, the spiracles are absent; in these species, the tracheae divide peripherally forming a network that covers the body surface and allows for cutaneous gas exchange

Reproductive system. The gonads of insects are located in the abdomen and their ducts open to the outside near the posterior end of the abdomen. The female reproductive system (Fig. 8.38) includes two ovaries and two oviducts that unite and continue into a vagina that opens to the outside. A seminal receptacle, or spermatheca, in which sperm is stored, and paired accessory glands, are connected to the vagina. The accessory glands secrete either an adhesive that cements the eggs to the substratum when they are laid or a material to form an egg capsule. In many insects, most of the eggs mature before they are laid.

The male reproductive system consists of a pair of testes that continue into a pair of vas deferens that fuse into a vas efferens or ejaculatory duct. The ejaculatory duct opens to the outside through a penis or aedeagus. A section of each vas deferens may be dilated into a seminal vesicle, where sperm is stored. Also associated with the ejaculatory duct is a pair of accessory glands that secrete seminal fluid to carry the sperm, or to form a sperm-containing capsule, the spermatophore. Most insects use spermatophores to transfer sperm, which are deposited directly into the female vagina during copulation.

Nervous system and sensory input. The nervous system of insects consists of a brain located in the head above the esophagus, a subesophageal ganglion connected to the brain by two commissures, and a ventral nerve cord that runs the length of the body. Three pairs of lobes, the protocerebrum, deuterocerebrum, and tritocerebrum, which innervate the eyes, antennae, and anterior gut, respectively, form the brain. The ventral nerve cord is double and connects the ganglia of the different body segments (Fig. 8.38). In some insects, these ganglia fuse and there are fewer ganglia than segments.

Sensory organs are abundant and well developed in insects. A pair of compound eyes and, normally, three ocelli are responsible for photoreception. Sensilla are sensory organs scattered over the body designed to monitor specific types of signals. In general, they function as mechanoreceptors, proprioceptors, chemoreceptors, and auditory receptors. Mechanoreceptors are used primarily for touch, orientation in space, and vibration reception. Most proprioceptors work as stretch receptors, and provide information about the tension of a particular muscle, or the forces being applied to the exoskeleton. Chemoreceptors are concentrated in the antennae, mouth, and around the ovipositor, and are involved in feeding behavior, mating, and habitat selection, including host selection. Auditory receptors are especially developed in sound-producing insects. In some mosquitoes and flies, males initially are attracted to the females by the sound of the wings.

8.3.4 Nutrient uptake

Insects have adapted to practically all types of food resources. Parasitic insects, however, feed mainly on blood, body secretions, feathers, skin,

Fig. 8.39 Feather louse from the breast feathers of the clapper rail *Rallus longirostris*. (From Overstreet, 1978, with permission from the Mississippi–Alabama Sea Grant Consortium.)

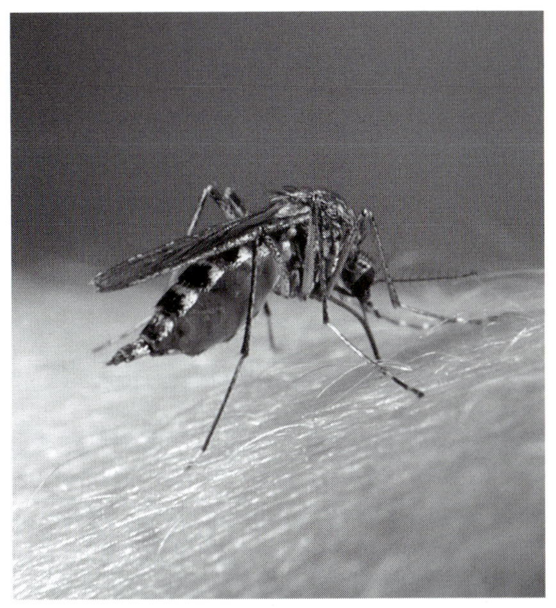

Fig. 8.40 Mosquito feeding on a human host. (Photograph courtesy of the Agricultural Research Service, US Department of Agriculture.)

and flesh. Most chewing lice feed by chewing the feathers, hair, and skin of their hosts (Fig. 8.39). The keratin present in these structures is digested with the aid of mutualistic bacteria. Some, however, may ingest blood if available. Scratching by the host or intense chewing of the skin may form small wounds, which provide chewing lice with blood. Some endoparasitic larvae are able to absorb food through their body surface, although most of them have fully functional digestive systems and consume tissues, blood, or other body fluids.

There are two basic types of blood-feeding insects, pool feeders and vessel feeders. Pool feeders use their strong mouthparts to rip and tear through tissue, causing abundant bleeding. The insect then just laps up the blood as it flows out of the wound. Vessel-feeding insects, on the other hand, take blood directly from small blood vessels. Their mouthparts are highly modified and function on the principle of a hypodermic syringe (Fig. 8.40). The mouthparts of mosquitoes, for example, include six stylets. Two of them are used to pierce the skin and the blood vessel, and the remaining four stylets form a straw-like tube through which blood is drawn. The saliva of vessel feeders usually contains anticoagulants, anticlotting agents, anesthetics, and even painkillers, so that the insect can feed in peace without being disturbed by the host.

Blood-sucking insects must contend with the problem of the excess liquid that is ingested with their meal and that needs to be removed from the food before it comes in contact with the digestive enzymes. In blood-sucking Hemiptera, for example, the blood meal is temporarily stored in the crop; the water is absorbed through the crop wall into the hemolymph and passes to the hindgut by way of the Malpighian tubules.

Although blood is a nutritious food item, it does not provide all the necessary compounds needed by these parasites to survive. Strict blood feeders cannot supplement their diet with other foods. Instead, they have mutualistic microorganisms that produce the compounds that are in short supply. These symbiotic organisms are housed in tissues and in special cells called **mycetomes**.

Parents normally transmit these symbionts to their offspring in several ways. In some species, the eggs are smeared with feces containing the symbionts and the shells are consumed by the hatching parasite. In sucking lice, the symbionts are incorporated directly into the eggshell, and in the viviparous tsetse flies, the growing larva inside the female feeds through a special gland that provides nutrients and the needed symbionts.

The metabolism of insects has been studied extensively and resembles that of mammals. A concise account of metabolism is given by Gillot (1995), but more detailed studies can be found in Gilmour's (1965) and Downer's (1981) texts on insect metabolism, and in the metabolism volume (vol. 10) of the series edited by Kerkut & Gilbert (1985).

8.3.5 Development

Most insects are oviparous, i.e., a larva or nymph hatches from the egg after being laid by the female. A few insects, however, some of them parasitic, are ovoviviparous. In these species, the developing eggs are retained and hatched within the body of the female. The number of eggs produced in this case may be small, but the protection offered by the female, and the fact that the young can commence feeding immediately, offsets the reduction in egg number. The sheep bot fly, for example, is ovoviviparous and the emerging larvae are deposited directly into the nostrils of sheep, increasing colonization success. A few other insects, such as tsetse flies, are viviparous. In viviparity, development also occurs within the female body but the embryos are not inside an egg. The female feeds the embryos inside her body until their development is complete, at which time they are born.

Once they hatch, insects exhibit one of three patterns of development, i.e., **ametabolous**, **hemimetabolous**, or **holometabolous**. In the ametabolous pattern, typical of the free-living Thysanura and Collembola, the newly hatched larva is a miniature of the adult. Growth occurs by successive molts and no significant morphological changes occur.

In hemimetabolous development, the newly hatched larva resembles the adult, with well-developed appendages, compound eyes, and the rudiments of the external genitalia. Only the wings and reproductive organs are undeveloped. The larval stages of hemimetabolous insects are called **nymphs** and have external wing buds that develop gradually through molting, until they reach the adult form. With some exceptions, the habits of the nymphs are usually similar to those of the adult, especially feeding. Several orders of insects, including some totally or partially parasitic such as Phthiraptera and Hemiptera, have hemimetabolous development.

In holometabolous development, there are three developmental stages, a feeding larva, a non-feeding pupa, and the adult. The newly hatched larva has no wings, and the antenna and eyes are often rudimentary. The larva usually goes through several instars, or developmental stages, before entering a non-feeding and quiescent stage, called a pupa. The pupal stage is normally passed in a protective location during which time adult structures are developed from embryonic buds and the organism is completely reorganized. When development is completed, a fully developed adult emerges from the pupa. The marked change from larva to adult is called metamorphosis. Many common groups that include parasitic species, such as Diptera, Coleoptera, Hymenoptera, and Siphonaptera, have holometabolous development. In holometabolous insects, the larva and the adult normally occupy different ecological niches, especially regarding their feeding preferences, a possible way to minimize intraspecific competition. In mosquitoes, for example, the larvae are aquatic and filter feed on microorganisms, whereas adult females suck blood and adult males feed on plant juices. In fleas, the larva is a scavenger with chewing mouthparts, whereas the adults feed on blood using piercing and sucking mouthparts.

8.3.6 Life cycle

Most parasitic insects require only one host to complete their life cycle. Some insects need a host during their larval stages, others only as adults, and still others need a host throughout their lives. Regardless of their host requirements, the developmental pattern remains basically the same. Hemimetabolous insects go from egg to nymph to adult, whereas holometabolous follow the egg, larva, pupa, and adult sequence.

Lice (Phthiraptera) are hemimetabolous and parasitic throughout their life. Adults lay eggs in the host and, after a short incubation, a nymph hatches. There are three nymphal instars, and the last one metamorphoses into the adult. The hemipteran bedbugs are also hemimetabolous and parasitic during their entire life, but, unlike lice, they do not live continually on the host. They attach only when feeding.

Fleas, and a number of dipterans, e.g., mosquitoes, black flies, deer flies, and tsetse flies, are holometabolous blood-sucking parasites only as adults. Eggs hatch away from the host and the larvae and pupa are free-living. Some others, such as screwworm flies, bot flies, and the parasitic hymenopterans, are also holometabolous, but they are parasitic only during their larval stages. The eggs are laid on the host or at least the hatched larvae manage to reach the host where they develop until it is time to pupate. The larvae then leave the host and pupate in the ground. A free-living adult emerges and completes the cycle.

8.3.7 Diversity

Although most insects are free-living, it comes as no surprise that many of them, about 15%, have adapted to a parasitic existence. The Insecta are divided into about 28 orders, a few more or a few less, depending on whose systematic scheme is followed. The Phthiraptera and Siphonaptera comprise only parasitic insects; the Diptera, Strepsiptera, Hemiptera, and Hymenoptera have a significant number of parasitic species; and the Coleoptera, Lepidoptera, and Dermaptera, are mostly free-living, with a few parasitic species in their ranks.

In a group as diverse as the Insecta, parasitism as a life style is also diverse, with many life-history strategies. Some insects are parasites throughout their lives, some only as adults, and some only during their larval stages. In those that are parasitic throughout their lives, some stay on the host at all times, whereas others visit the host only when feeding. Among those insects that are parasitic during their adult stage, only the females are parasites in some species, whereas both sexes are parasitic in other species.

The species parasitic only during their larval stages can be of two kinds, i.e., parasites in the traditional sense and parasitoids (see Chapter 1). Traditional parasites live and feed on the host and, under normal circumstances, they do not kill the host. Parasitoids, on the other hand, eventually kill their host. These specialized organisms begin as typical parasites in the sense that they are much smaller than the host and live in intimate association with it. However, they eventually grow, sometimes as much as the host, and the host is eventually killed. At this point, the original host–parasite association changes and becomes more like a prey–predator relationship. Because they are neither parasites in the traditional sense nor predators, they are referred to as parasitoids. For the sake of clarity (we hope) true parasites and parasitoids will be discussed in separate sections, with more emphasis on traditional parasites.

8.3.7.1 Parasitic insects

PHTHIRAPTERA

The Phthiraptera, or lice, are small ectoparasites of birds and mammals, rarely exceeding a few millimeters in length. Their body is flattened dorso-ventrally; they have reduced eyes, lack wings, and have well-developed grasping legs to attach to the host's hair or feathers. Lice are parasitic throughout their life cycles and do not survive away from the host. Development is hemimetabolous and the life cycle is completed on one individual host. The female glues the eggs to the feathers or hairs of the host and dispersal or transfer from one host to another occurs by direct contact during host mating, brooding, roosting, or lactation. Phthirapterans are usually divided into chewing lice (Mallophaga) and sucking lice (Anoplura) based on the structure of their mouthparts and the resulting feeding strategy.

The Mallophaga, or chewing lice, are ectoparasites of birds and mammals (Figs. 8.39, 8.41). Most chewing lice feed on feathers or hair, skin, and secretions from sebaceous glands. A few, however, ingest blood and serum. Most species of chewing lice are highly host specific, parasitizing only one species of host. One individual host, however, can host several species of chewing lice. When this occurs, the lice are often site specific and their morphology correlates with the sites preferred. Chewing lice found on the head and neck of birds have round bodies, large heads, and are slow moving. If they wander into other regions of the bird's body, they are likely to be removed during preening. Species living on the back and wings,

Fig. 8.41 The chewing louse *Plegadiphilus eudocimus*, from the feathers of the white ibis *Eudocimus albus*.

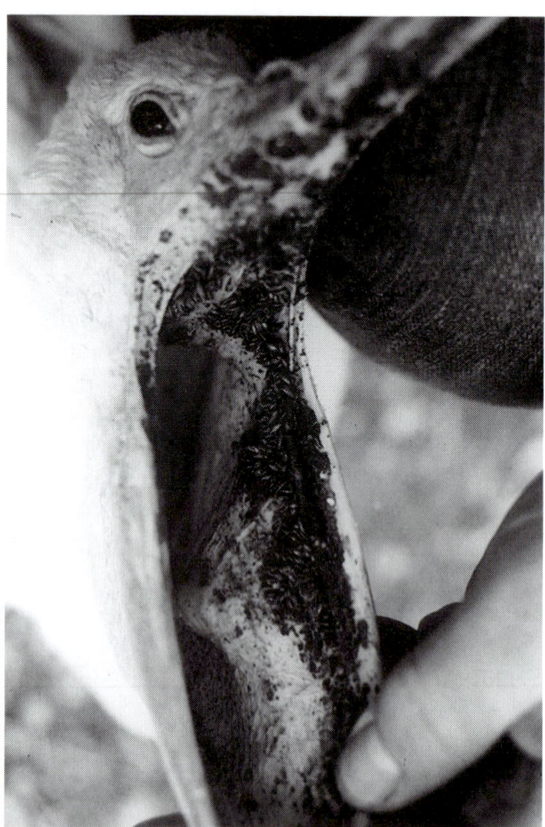

Fig. 8.42 Heavy infection by the louse *Piagetiella peralis* in the internal surface of the mandible of a white pelican *Pelecanus erythrorhynchus*. (From Samuel *et al.*, 1982, with permission, *Canadian Journal of Zoology*, **60**, 951–953.)

however, are flattened and elongated, and capable of moving very fast across the broad feathers of their habitat to avoid removal during preening. Chewing lice living in other body regions are normally intermediate between these forms. An interesting and highly site-specific louse is *Actornithophilus patellatus*. It lives inside the quill or shaft of the primary and secondary feathers of curlews, feeding on the pith of the shaft. It is rarely found on the outside, and all the developmental stages, from egg to adult, can be found in a single feather.

Several species of chewing lice are common in birds associated with humans. *Columbicola columbae* is the pigeon louse, *Anaticola crasicornis* and *A. anseris* are common in ducks, *Goniocotes gallinae* lives in the fluff at the base of the feathers in chickens, and *Menacanthus stramineus* is the yellow body louse of chickens and turkeys. Pelicans and cormorants are parasitized by species of *Piagetiella*. They live in the throat pouch of the birds (Fig. 8.42) and remain in place by attaching to the mucus membrane with their mandibles, where they apparently feed on blood and mucus. They only leave the pouch temporarily to lay eggs in the feathers of the same bird. The adults and nymphs are very mobile and, during the breeding season, they move rapidly from the adult birds to the newly hatched chicks (Samuel *et al.*, 1982).

Species of *Bovicola* are common on cattle, goats, sheep, deer, horses, and other ruminants. *Bovicola bovis* is one of the most common species present in cattle. Parasitized animals often try to eliminate the lice by biting the skin and rubbing themselves against posts, rocks, and other hard surfaces.

The Anoplura, or sucking lice, are exclusively ectoparasites of mammals. Their mouthparts are specialized for piercing the skin and sucking blood. Host and site specificity are also relatively high among sucking lice. Lice in the Enderleinellidae, for example, are parasites of squirrels, the Hematopinidae parasitize ungulates, and the Echithiriidae are found in the fur of seals and walruses. The latter are well adapted to the diving habits of their hosts; when in the water, the lice

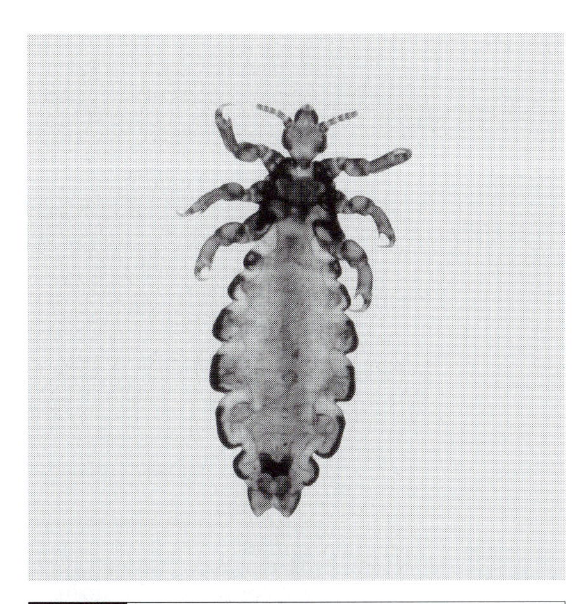

Fig. 8.43 *Pediculus humanus capitis*, the human head louse.

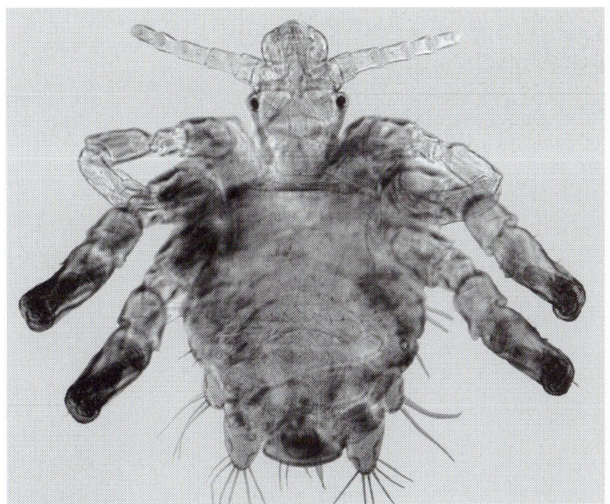

Fig. 8.44 *Phthirius pubis*, the human crab louse.

remain in the underfur and use the air trapped there as an oxygen source.

Humans are no strangers to sucking lice, and provide a good habitat for three types, i.e., the body louse, the head louse, and the crab louse. Body and head lice are closely related and usually are considered to be the same species, but different subspecies or varieties. Most people know the adults by some of their common names, e.g., cooties, graybacks, and 'mechanized dandruff', whereas the eggs that are laid and cemented to hair are called nits. The head louse, *Pediculus humanus capitis* (Fig. 8.43), is found mainly in the hair behind the ears and back of the neck. Transmission from one host to another occurs by physical contact and by stray hairs carrying lice or their eggs. The body louse, *P. humanus humanus*, is rather unusual in that it spends most of its time, and even lays its eggs, in the clothing of the host, visiting the host's body only to feed. Despite this habit, the lice must stay in close contact with the body of the host and are extremely sensitive to changes in temperature. If, during their nymphal stages, clothing is removed, development slows significantly and the lice may even die if the host does not wear the contaminated clothes for a few days. Although the species status of these two human lice is still under debate, it seems likely that the body louse is a relatively new form that evolved from a head louse ancestor when humans began wearing clothes. The third species found in humans, *Phthirius pubis*, the crab louse (Fig. 8.44), is found in the pubic region but, on hairy individuals, may occur almost anywhere in the body. They normally insert their mouthparts in the skin and, unlike the body and head lice, they remain in the same position for some time, causing intense itching. Transmission from host to host is by direct contact with the affected area, a potentially embarrassing souvenir as the result of a casual dalliance with an infected partner (for a comprehensive review of human lice and their management, see Burgess, 1995).

Lice are good vectors for prokaryotic pathogens. In humans, the body louse transmits at least three important diseases, epidemic typhus and trench fever caused by a rickettsial organism, and relapsing fever caused by a spirochaete. Epidemics of these diseases correlate with increases in lice populations, which, in turn, are favored when living quarters are crowded, sanitation is at a minimum, and people go for long periods without changing clothes. These conditions are commonly seen in times of social unrest and wars. A classic book by Hans Zinsser (1934) titled *Rats, Lice and History* gives detailed accounts of the impact that these diseases have had during the course of human history.

SIPHONAPTERA

The Siphonaptera, or fleas, are blood-sucking parasites only as adults. Fleas are small with laterally flattened bodies, have piercing–sucking mouths,

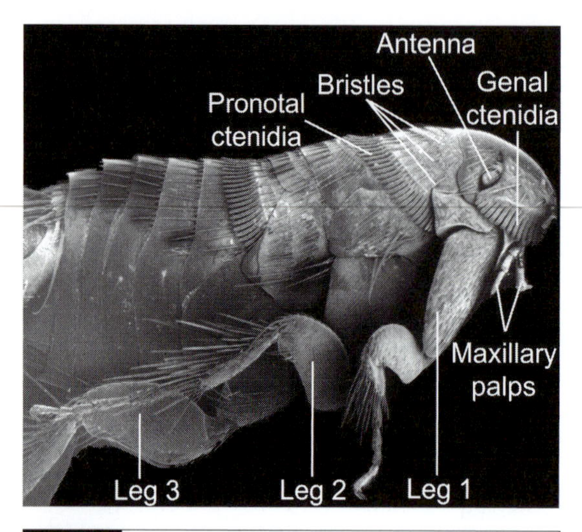

Fig. 8.45 The flea *Stenoponia americana* (male) from the white-footed mouse *Peromyscus leucopus*, showing details of the external morphology, in particular the presence of ctenidia. (Photograph courtesy of David Salmon, Ralph Eckerlin & Sherman Hendrix, Gettysburg College.)

A

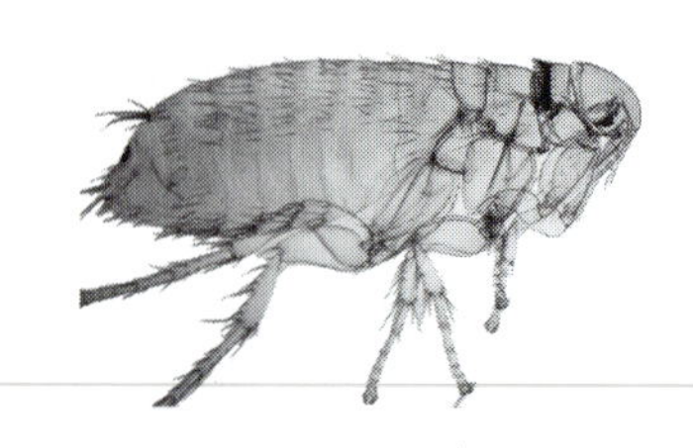

B

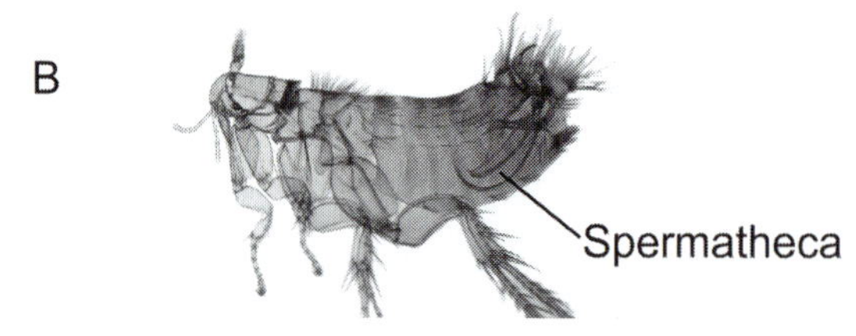

Fig. 8.46 The flea *Opisodasys pseudarctomis* from the southern flying squirrel *Glaucomys volans*. (A) Female; (B) male. (Photographs courtesy of Ralph Eckerlin, Northern Virginia Community College.)

and lack wings, probably to facilitate their movement between hairs on their hosts. Most fleas have characteristic bristles or spines directed backward, called **ctenidia**, for snagging hairs when the host attempts to remove them (Fig. 8.45). Development is holometabolous. Fleas lay their eggs in the nest of the host, or on the host, but because the eggs are not cemented to the host as in lice, they normally drop off and end up on the ground or in the host's nest. The larvae are maggot-like and feed on organic debris and flea feces accumulated in the host's nest. The pupae can remain dormant for a long period, until a host appears. The legs of most fleas are modified for jumping, and some can leap more than a hundred times their body length (Rothschild *et al.*, 1973). The jumping ability of most fleas is correlated with the type of host they parasitize. In general, fleas that parasitize nesting hosts (Fig. 8.46) are poor jumpers because the nest provides a restricted and safe place for transmission. Fleas that parasitize non-nesting hosts, however, have strong legs and are superb jumpers, being able to reach a host from the ground. Vibrations, heat, and high concentrations of carbon dioxide trigger the jumping or host-searching behavior of fleas.

Most adult fleas parasitize mammals, but birds are infected too. Some fleas are not very host specific, but they do have preferred hosts. For example, despite their common names, *Ctenocephalides canis* and *C. felis*, the dog and the cat flea, respectively, are widespread and parasitize cats, dogs, humans, and other mammals. Similarly, the human flea *Pulex irritans* (Fig. 8.47), has been found in a number of other hosts including pigs, dogs, coyotes, squirrels, etc.

Although most fleas can feed on a variety of hosts, some species require specific hosts to produce viable eggs. In other cases, such as the European rabbit flea *Spilopsyllus cuniculi*, and other related species parasitizing rabbits and hares, the host's hormones control reproduction in the flea (Rothschild & Ford, 1972). The European rabbit flea cannot mature sexually until the flea feeds on a pregnant rabbit. When the rabbits are born, the now sexually mature fleas move from the mother to the newborn where they feed, mate, and lay eggs. Then they return to the mother rabbit, leaving the newborn rabbits with a bumper crop of flea eggs. Hormonal changes in the mother rabbit during pregnancy trigger the sexual development of the fleas, and growth hormones in the newborn rabbits stimulate mating and egg laying.

The most annoying and insidious fleas are probably the 'stick-tight' fleas, a group of fleas that attach themselves to the host for extended

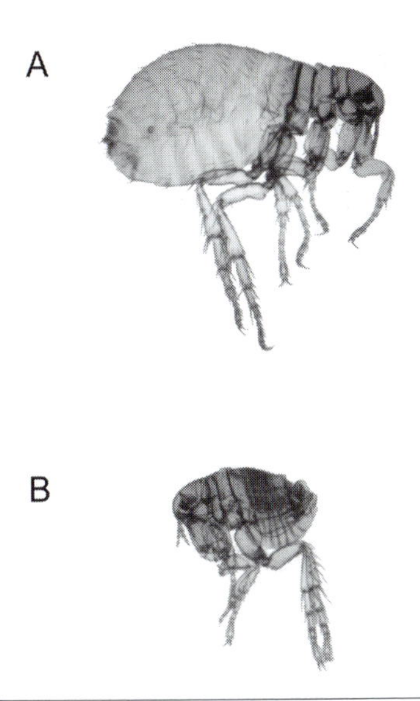

Fig. 8.47 The human flea *Pulex irritans*. (A) Female; (B) male.

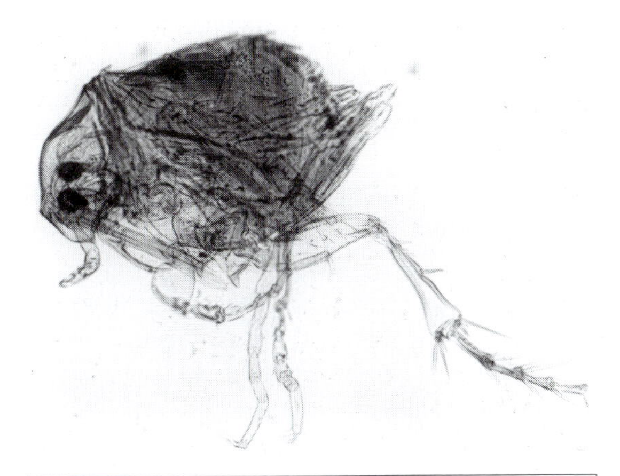

Fig. 8.48 A nongravid chigoe or sand flea *Tunga penetrans*. (Photograph courtesy of the Armed Forces Institute of Pathology.)

periods and are capable of swelling to the size of a pea as blood is ingested. *Echidnophaga gallinacea*, for example, is common in poultry and various other animals, and although the female remains attached to one host and rarely changes, the males move and feed on different individuals. One of the most extreme stick-tight fleas is *Tunga penetrans*, the chigoe, jigger flea, or sand flea, common in tropical areas, including the lovely Caribbean beaches (Fig. 8.48). It commonly parasitizes humans, although pigs and other mammals are often infected. Chigoes burrow under the skin (Fig. 8.49), but males remain on the outside and manage to copulate with the female through a small aperture in the skin of the host. Under the skin, the gravid female can reach the size of a pea. She then lays the eggs in her burrow and, after hatching, the larvae escape through the aperture in the skin and drop to the ground where they develop. Infection by this flea is rather painful and often results in gangrene and tetanus. Fleas beneath the nails are especially painful and appear to have given rise to the sailor's oath 'Well, I'll be jiggered'. Often, the female fleas have to be removed by minor surgery. Quite a memento from that relaxing trip to the Caribbean!

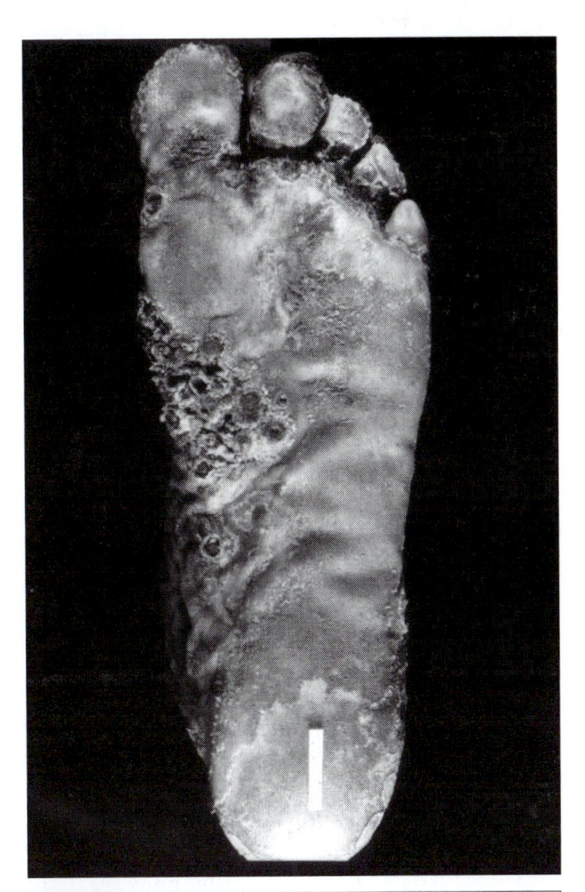

Fig. 8.49 A foot heavily infected with the sand flea *Tunga penetrans*. The foot shows multiple confluent craters. The non-pressure areas (the instep and between the toes) are most severely infected. The weight-bearing portion of the sole tends to be spared from infection. (Photograph courtesy of the Armed Forces Institute of Pathology.)

Fleas, like a number of other arthropods, are also vectors for a number of infectious diseases. A number of species, including the human flea *Pulex irritans*, but in particular the oriental rat flea

Xenopsylla cheopis, are the main vectors for plague and murine typhus. Plague, also known as black death, is caused by the bacterium *Yersinia pestis* and can affect humans, rats, mice, rabbits, dogs, etc. Epidemics of plague have had a profound effect on the history of humanity. The Black Death that swept Europe during the fourteenth century killed 25 million people. Similar epidemics in the sixth century, and the Great Plague of London in the seventeenth century also caused a heavy toll in human lives. Sporadic epidemics continued well into the twentieth century. About 7000 cases per year are reported but large-scale outbreaks are now prevented by insecticides (to limit flea proliferation) and antibiotics (for prompt treatment of infected individuals). Murine typhus, caused by *Rickettsia mooseri*, is a much less fatal disease affecting humans and other animals. A third disease transmitted by fleas is myxomatosis, a viral disease of rabbits transmitted by several blood-sucking arthropods, including the European rabbit flea *Spilopsyllus cuniculi*. The flea and the myxomatosis virus have been introduced in Australia to control the rabbit population.

Several species associated with humans and domestic animals are intermediate hosts for cestodes such as *Vampirolepis nana*, *Hymenolepis diminuta*, and *Dipylidium caninum*. So, when the hosts groom to rid themselves of the annoying fleas, and ingest them, they get rid of one parasite and acquire another, clearly a no-win situation.

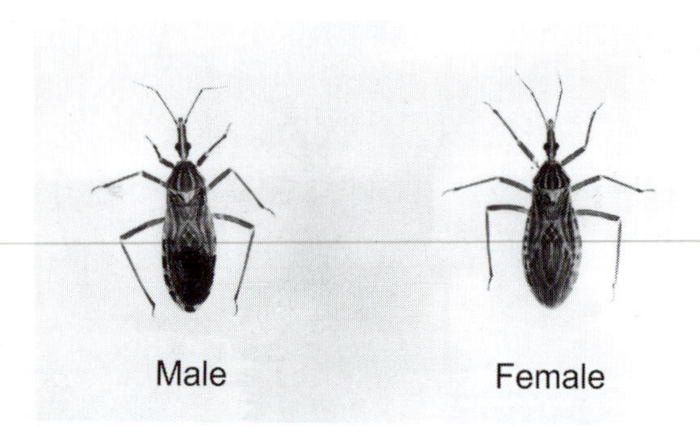

Fig. 8.50 Male and female adult kissing bugs *Rhodnius prolixus*. (Photograph courtesy of the Armed Forces Institute of Pathology.)

HEMIPTERA

Most Hemiptera, also known as true bugs, are free-living, hemimetabolous insects. Only three groups, the Cimicidae, or bedbugs, the Triatominae, or kissing bugs, and the Polyctenidae, are parasitic. Bedbugs and kissing bugs feed on host blood, but leave the host after each feeding; the polyctenids, however, are totally dependent on the host and die within 6–12 hours after removal from the host.

The cimicids are wingless, reddish brown, and measure up to 8 mm in length. Bedbugs are temporary parasites of birds and mammals, mainly bats. During the day, they hide in bird nests, crevices, cracks, etc., where they lay their eggs. At night, the adults and nymphs leave the hiding places, feed on their hosts until satiated, and then return to hiding. A complete blood meal can be obtained within 3 to 15 minutes. Most bedbugs are host specific. *Cimex lectularis*, *C. hemipterus* and *Leptocimex boueti* are found in human households where they hide in cracks in the walls, mattresses, beds, under wallpaper, etc. Because most cimicids are associated with cave-dwelling animals, it is thought that bedbugs evolved in caves (Askew, 1971). It is likely that humans became associated with cimicids during our cave-dwelling period. Bedbugs can certainly be annoying, but at least they are not known to be vectors of any disease-causing organisms.

The Triatominae, or kissing bugs, are temporary parasites of mammals, including humans. They can reach up to 35 mm in length, have wings, and two well-developed eyes (Fig. 8.50). Most species live in the burrows or nests of their hosts; those associated with humans are found in cracks, crevices, and roof thatching. Most kissing bugs are not host specific, and feed on multiple host species, mainly at night, biting any exposed parts. In the case of humans, they preferentially bite the face around the mouth; this site preference gives the insects their name, kissing bugs. Several species are vectors of *Trypanosoma cruzi*, the protozoan that causes Chagas' disease (see Box 3.5). Although the disease is common in South America, it also occurs in North America. Several wild and domestic mammals, including armadillos, opossums, squirrels, bats, dogs, cats, and rats are reservoir hosts. Although all kissing bugs seem to be suitable hosts for *T. cruzi*, their susceptibility varies. *Triatoma infestans*, *T. dimidiata*, *Rhodnius prolixus* (Fig. 8.50), and *Panstrongylus meg-*

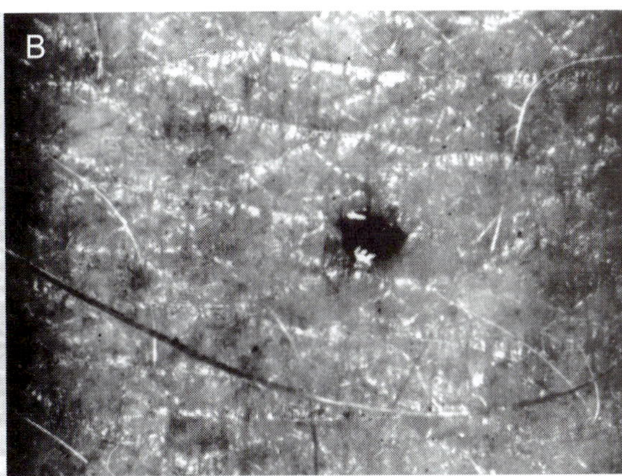

Fig. 8.51 The black fly *Simulium damnosum*. (A) Black fly in the biting position (head down); (B) site of bite of the black fly. The fly has just flown away leaving a distinctive hemorrhagic spot. Arm contributed by Dr. Harald Fuglsang, Medical Reasearch Council, London. (Photographs courtesy of the Armed Forces Institute of Pathology.)

istus are the most common and important vectors of *Trypanosoma cruzi* in the Americas.

The Polyctenidae are found exclusively in the micropteran bats of the tropics and subtropics. They are very well adapted to a totally parasitic life style. They lack eyes, the wings are reduced to a pair of large lobes, and the body has a number of combs similar to those present in fleas. Most of the bat species form small assemblages in tree holes and caves, which facilitates the transfer of parasites between individual bats (Marshall, 1982).

DIPTERA

The Diptera is a diverse group of insects with many parasitic and parasitoid species. Some are parasitic only during their larval stages, others only as adults, and some others throughout their lives. Furthermore, in many, only the females are parasitic, living permanently on the host, or just temporarily while feeding. Most Diptera have holometabolous development, including larval and pupal stages, with the larva and the adult usually occupying different ecological niches. Tsetse flies are exceptions; the larvae are retained and nourished viviparously within the female body until ready to pupate. The number of larvae produced by a single female is small, but survivorship is virtually guaranteed.

The Psychodidae (sand flies), Simuliidae (black flies), Ceratopogonidae (biting midges), Culicidae (mosquitoes), and Tabanidae (horse and deer flies) include species in which the females are blood-sucking parasites of vertebrates; in contrast, the males in these groups often feed on the nectar of flowers. The blood-sucking species of Psychodidae, or sand flies, are nocturnal; during the day they rest in moist, protected places such as caves, animal burrows, tree holes, and houses. Sand flies are vectors for a few pathogens, the most important being the protozoan responsible for leishmaniasis (see Box 3.6). Most species of sand flies belong to *Phlebotomus*, which includes *P. papatasi* and *P. sergenti*, two common species in the Old World responsible for transmitting the virus responsible for sand fly fever.

The parasitic species of Simuliidae, also known as black flies or buffalo gnats, have a worldwide distribution, but are more common in northern temperate and subarctic zones. They are daytime feeders and attack both birds and mammals (Fig. 8.51). Breeding and larval development occurs in running, well-oxygenated waters, mainly streams and rivers. Unfortunately, these are the same habitats where humans often go in search of nature and relaxation. Several species of *Simulium* are vectors of *Onchocerca* spp., the nematode responsible for onchocerciasis in humans and cattle, and of *Leucocytozoon* sp., a ubiquitous protozoan that causes a malaria-like disease in birds.

Only a few species in the Ceratopogonidae, or biting midges, feed on birds and mammals. They are very small, less than 1 millimeter, but they are quite bloodthirsty and ferocious. Species in four genera (*Culicoides*, *Forcipomyia*, *Austroconops*, and *Leptoconops*) attack mammals, including humans;

they can become quite annoying, especially near breeding areas that include aquatic and semi-aquatic habitats. Some biting midges are vectors for several filarial nematodes, protozoans, and viruses of domestic and wild animals.

Culicids, or mosquitoes, are probably the best known and most ubiquitous of all blood-sucking dipterans (Fig. 8.40). The larva and pupa of mosquitoes are aquatic; the larva normally feeds on algae, bacteria, and organic debris, although in a few species their larvae prey on other mosquito larvae. The pupa, even though motile, does not feed. Adult females feed on the blood of amphibians, reptiles, birds, and mammals. Some species are host specific, whereas others attack a variety of hosts. Most species of *Culex* are nocturnal and feed on birds, but they are not particularly host specific; some may also feed on humans. *Culex pipiens*, the common house mosquito, is a cosmopolitan species. Species of *Aedes*, in contrast, are diurnal or crepuscular feeders. Most species are fierce daytime eaters and attack in vast numbers. Mosquitoes are vectors of several disease-causing organisms. *Anopheles* spp. are vectors of *Plasmodium* spp., the protozoan that causes malaria; the viruses that cause yellow fever and dengue are transmitted by *Aedes* spp.; and the nematodes responsible for several types of filariasis, including elephantiasis, are transmitted chiefly by species of *Culex*.

The Tabanidae are robust flies, from 6 to 25 mm long, that feed during the day. Their bites are painful and continue bleeding after a meal is taken. Species of *Tabanus*, also known as horse flies, produce a loud buzz when flying; they attack cattle and horses, and, when available, humans. Species of *Chrysops* (Fig. 8.52), the deer flies, attack mainly deer. If humans are available, deer flies target the neck and head, whereas horse flies show some preference for the legs, another example of resource partitioning.

Unlike the groups discussed above, both males and females of Glossinidae, Hippoboscidae, Streblidae, and Nycteribiidae are obligate blood feeders. The Glossinidae, or tsetse flies, are restricted to Africa and parts of the Arabian Peninsula. Tsetse flies are not very host specific, but they are associated with specific habitats, which in turn determine their host availability. This pattern repeats itself for several other groups of insects (see below), where the habitat seems to be more important that the nature of the host. Most tsetse flies are diurnal and vision seems to play an important role in host location, mainly in open areas. Humans and cattle are common hosts (Fig. 8.53), and the flies are widely known because they are vectors for *Trypanosoma* spp., the protozoans that cause African sleeping sickness.

Fig. 8.52 A deer fly *Chrysops* sp. feeding on a human. (Photograph courtesy of the Armed Forces Institute of Pathology.)

Hippoboscid flies are ectoparasites of birds and mammals; only a few species are host specific and, like tsetse flies, the range of hosts is mostly determined by ecological factors related to the habitat. Keds and louse flies are the most common hippoboscids. The sheep ked, *Melanophagus ovinus*, spends its entire life associated with the host and, consequently, has lost its wings and even the halteres. The Streblidae and Nycteribiidae are very specialized flies that feed exclusively on bats. The latter are the most specialized, and superficially they do not resemble flies. They lack wings, the thorax is reduced, and the small head and the insertion of the legs are displaced to a dorsal position, giving the flies a spider look. Bat flies are not very host specific and their host ranges are determined mostly by the habitat of the bats. Nycteribiids are intimately associated with their hosts, taking fresh blood meals frequently. They die within 1 or 2 days if separated from the host.

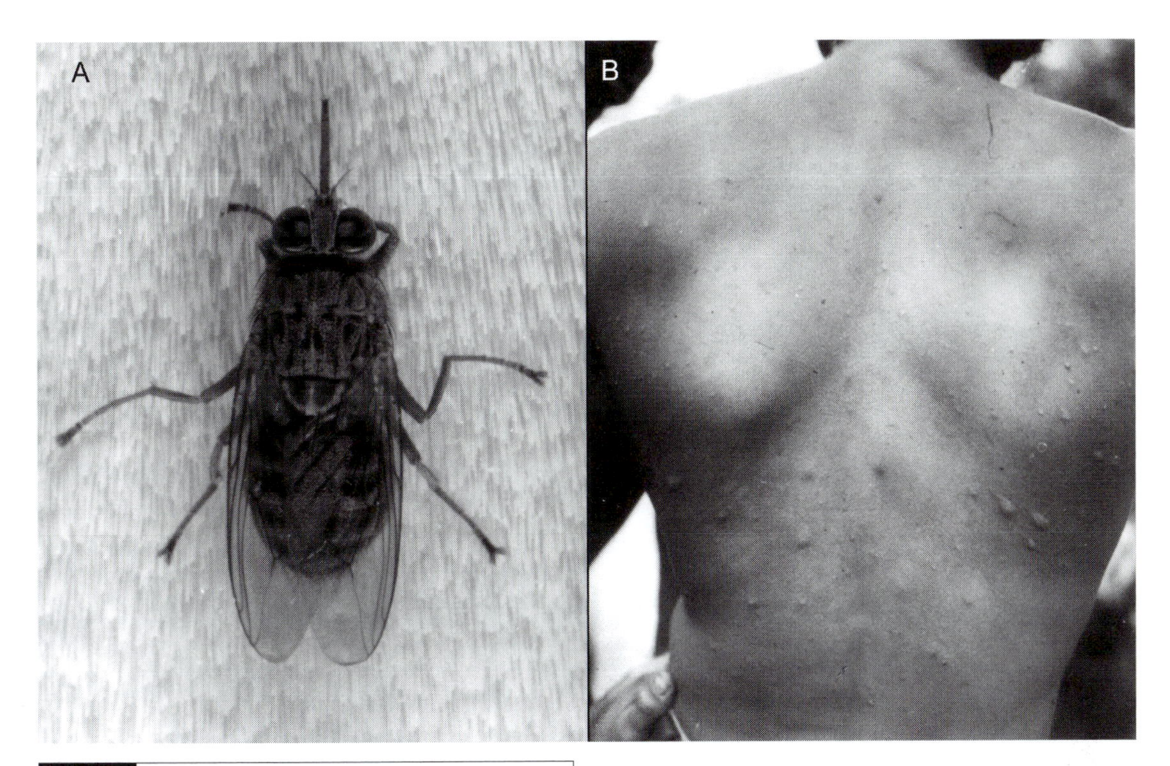

Fig. 8.53 The tsetse fly *Glossina morsitans*. (A) Adult fly; (B) bites of a tsetse fly on the back of an adult male. (Photographs courtesy of the Armed Forces Institute of Pathology.)

As mentioned earlier in this section, parasitism in dipterans is not restricted to the adults. In a number of families, the adult flies are free-living, whereas the larva is the parasitic stage. The infection of vertebrate hosts by fly larvae, or maggots, is often referred to as **myiasis**. Parasitic larvae can feed on host tissue, body fluids, secretions, or food ingested by the host. The two most important families with larval forms parasitic on vertebrates are Calliphoridae and Oestridae.

Adult calliphorids, known as blow flies, are about the size of a house fly, and frequently are metallic green or blue. Free-living species lay their eggs on dead organic compounds, but parasitic species require the presence of a wound or a sore to lay their eggs. The screwworm fly (Fig. 8.54A) *Cochliomyia hominivorax*, is a common species that lays eggs in the wounds of mammals. After hatching, the screwworm larva (Fig. 8.54B) feeds on living tissue and, when fully developed, drops to the ground and pupates.

The Oestridae include the unpopular bot flies as well as the terrifying warble flies. The adult flies are robust, hairy, and resemble bees. Unlike the screwworm flies, the larvae of oestrids do not require a pre-existing wound to lay eggs and gain entrance to the host. Skin bot flies, such as *Cuterebra* spp., lay their eggs on or near natural orifices of the host. After hatching, the larva enters the body, tunnels under the skin, and lives in a cyst that communicates to the exterior through a small hole. The fully mature larva then leaves the cyst and pupates on the ground. Head maggots such as *Oestrus ovis*, the sheep bot fly, and *Cephenemyia* sp., a bot fly of deer, deposit their larvae in the nostrils. Once there, the larva finds its way into the sinuses and nasal passages where it attaches to the mucosa and feeds.

Warble flies are primarily parasites of bovids and cervids. The presence of adult flies nearby truly terrifies cattle, which initiate intense evasive maneuvers attempting to avoid the flies. The female lays eggs in the hairs of the host and, on hatching, the larvae penetrate the skin. What terrifies the host is probably the extensive subcutaneous migration that lasts several months and that takes the larvae first to the front end of the host

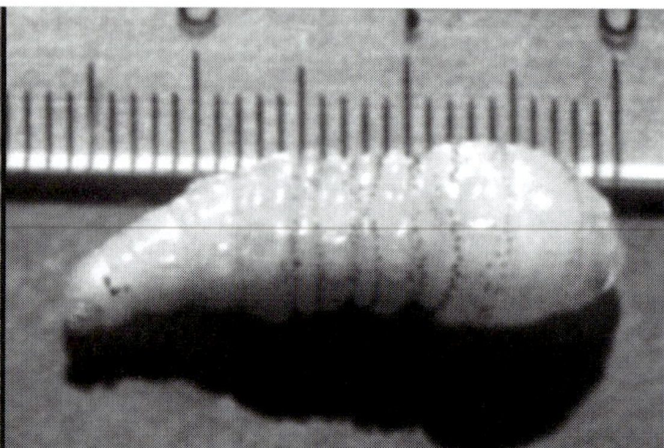

Fig. 8.54 (A) Screwworm fly *Cochliomyia hominivorax*; (B) screwworm larva extracted from the nose of a patient in French Guiana. ((A) Photograph courtesy of the Agricultural Research Service, US Department of Agriculture; (B) Photograph courtesy of Jean-Phillipe Chippaux, Institut de Recherche pour le Développement, Paris, France.)

and finally to the dorsal area, where they open a small hole in the skin. When development is completed, the larva drops to the ground and pupates.

Several species of bot flies parasitize the stomach of equids, elephants, and rhinoceroses, causing gastric myiasis. Each stomach bot fly has a different infection strategy and occupies different sites in the alimentary canal of the host. Species of *Gasterophilus* deposit their eggs on grass or on the skin of horses from where they are ingested when eating or licking themselves, respectively. The larva hatches in the mouth, penetrates the mucosa and migrates through the host's body to its preferred site; depending on the species, this may be the stomach, intestine, or rectum. Eventually, the larva detaches, passes out with the feces, and pupates in the soil. One species, *G. intestinalis,* has achieved a cosmopolitan distribution and is common in horses, donkeys, and mules worldwide.

Not all myiasis-causing flies deposit their eggs directly on the host (or on the grass the host eats). The human skin bot, *Dermatobia hominis*, is a common species found in the jungles of Central and South America (Fig. 8.55). Its behavior is rather unusual because the females seek other blood-feeding insects and glue their eggs to that insect's abdomen. When the carrier insect finds a warm creature to feed on, the skin bot eggs hatch immediately, the larvae drop onto the new host, penetrate the skin and settle there. Over the next two months, the larva grows reaching about 2.5 cm in length. When the larva is ready to pupate, it emerges from the skin, drops to the ground, and pupates. The human skin bot also can develop in the skin of almost any warm-blooded animal, including birds. Attempts to remove this larva can cause more damage than allowing it to complete its development. Removal normally requires a small incision in the skin that allows the larva to be expressed. A very effective treatment involves the layering of several strips of raw bacon over the lesion housing the larva. In most cases, the larva moves out of the skin and into the bacon in a matter of hours. Parasites obviously know what is good after all. For an in-depth discussion of myiasis in humans and domestic animals, see Hall & Wall (1995).

STREPSIPTERA

The Strepsiptera is a small group of minute insects, most of which are parasitic in other insects, especially Orthoptera, Hemiptera, Homoptera, Hymenoptera, and Thysanura. They are sexually dimorphic. The males are free-living and winged, whereas the females are wingless and, in the parasitic species, never leave the host. The parasitic females are highly modified; they usually lack legs, antennae, and eyes, and the head and thorax are fused, resulting in a vermiform body. Females inside the host produce large numbers of small larvae, which escape from her body and the body of the host, dropping to the ground. These larvae have well-developed eyes and legs and actively seek and enter the body of an insect, where they feed and develop into pupae. When

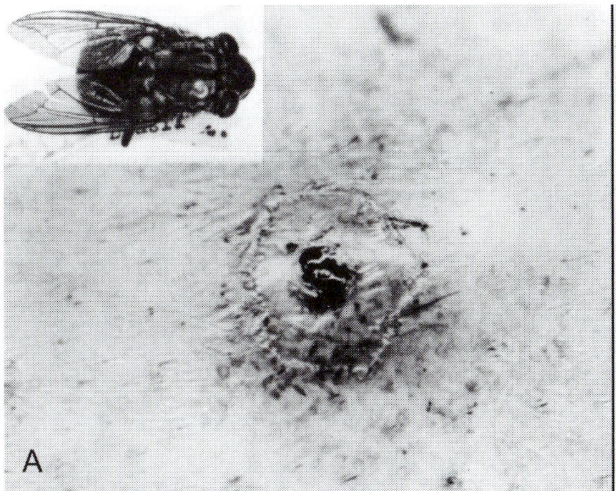

Fig. 8.55 The human skin bot *Dermatobia hominis*. (A) Myiasis caused by *D. hominis* in the leg of a patient in Panama. The infection is about 1 month old and the shiny spot in the center is the posterior end of the larva. (Insert: adult *D. hominis*.) (B) Myiasis caused by *D. hominis* in the brain of a child in Panama. There were several warbles on the scalp. The larva in the cavity entered through a 4-mm hole that the larva had bored through the anterior fontanel. (Photographs courtesy of the Armed Forces Institute of Pathology.)

development is completed, the adult males emerge from the host, whereas the females remain inside, with only the anterior part of their body protruding through the abdominal segments of the host. After emerging, the male seeks out and copulates with a female. Males usually cause extensive damage to the host when emerging and, although the host is not always killed, the host may be castrated, show unusual coloration patterns, or malformation of the abdomen.

COLEOPTERA, LEPIDOPTERA, AND DERMAPTERA

Very few species in each of these groups are parasitic and, in most cases, they are rather unusual and very host specific. Species of *Hemimerus*, a dermapteran or earwig, are strict parasites of pouched rats in tropical Africa. They lack eyes, are viviparous, and all the developmental stages are parasitic. They live in the fur of the rodents where they feed on epidermis and a fungus that grows on the skin of the rats.

The Lepidoptera are better known as moths and butterflies, and it is not easy to think of them as parasites. However, just as another example of the great adaptability of insects, a few have become parasites. The best-known are the eye moths, which suck lachrymal secretions and pus from the eyes of their hosts, mostly cattle. Sometimes, they may even obtain blood from the conjunctiva and cornea. As bizarre as this may seem, this type of feeding can be interpreted just as another variation on the more generalized liquid feeding of lepidopterans that includes fruits and other moist foods.

One particularly interesting species is the coleopteran *Platypsyllus castoris*. This is a small louse-like beetle measuring 3–5 mm in length that parasitizes beavers across their geographical distribution. The beaver beetle is blind, and is a permanent parasite of beavers as a larva and adult, feeding on the fatty skin secretions produced by the host. The only time the beetles leave their host is to pupate and possibly to lay eggs, but the life cycle is still completed within the beaver lodge. This species is one of the very few holometabolous insects that are parasitic on the same host both as larva and adult. In most holometabolous species, as discussed previously, larval and adult stages tend to have different feeding strategies, a possible way of avoiding intraspecific competition.

The Leptinidae is a group of beetles found in the nests and fur of small mammals (Fig. 8.56). Their eyes are reduced, or absent, and although their food requirements are poorly understood, they seem to feed on host hair and skin secretions. Species of *Leptinus* are common in the nests and fur of mice, moles, and shrews.

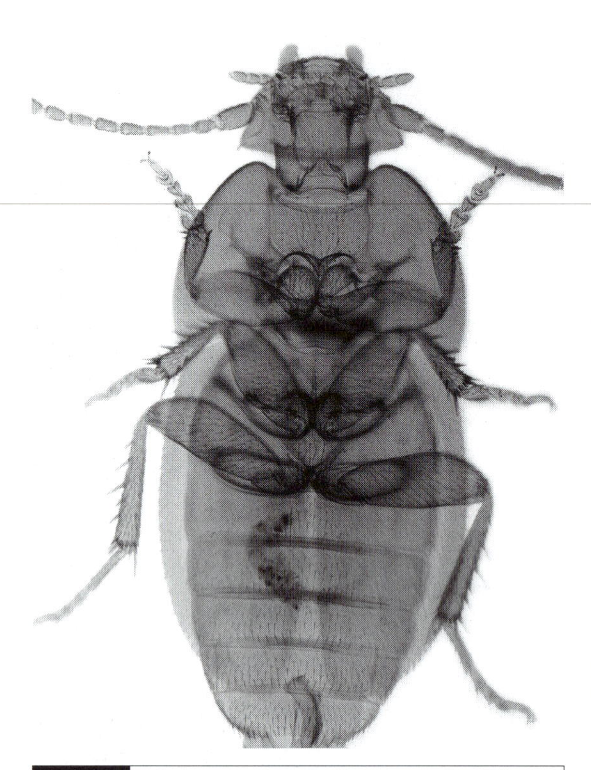

Fig. 8.56 The coleopteran *Leptinus orientamericanus* from the fur of the star-nosed mole *Condylura cristata*. (Photograph courtesy of Ralph Eckerlin, Northern Virginia Community College.)

Fig. 8.57 The female parasitoid wasp *Catolaccus grandis* homes in on a boll weevil larva *Anthonomus grandis*. The wasp immobilizes the larva with the ovipositor and withdraws enough nourishment to complete the development of her eggs. Later, she, or another female wasp, will revisit the boll weevil larva and will deposit an egg next to it. When the wasp larva hatches, it consumes the boll weevil larva. (Photograph courtesy of the Agricultural Research Service, US Department of Agriculture.)

8.3.7.2 Parasitoids

The free-living adults of parasitoids look just like any free-living relative. They may feed on nectar and sap, but many also feed on potential hosts. Parasitoids parasitize other insects, although spiders and centipedes may be used as hosts too. The adult female locates a host, in most cases the larval stage or egg of another insect, and lays her eggs directly on or in it. Sometimes, the eggs are laid in the vicinity of the host and, when they hatch, the larvae actively seek and infect the host. In some cases, the female lays the eggs on the plant where the host feeds and the host becomes infected by ingesting the eggs. Some parasitoids develop within the body of their hosts, feeding from the inside, and are called endoparasitoids. Ectoparasitoids on the other hand, live externally, but their mouthparts are buried in the body of the host. Some parasitoids are highly specific, attacking one or very few species of insects. Host specificity is higher among endoparasitoids, where a most intimate host–parasitoid relationship exists. In most cases, only a single individual parasitoid can exist within a host. Most parasitoids can recognize and reject hosts that are already parasitized. Parasitoids normally begin feeding on non-vital tissues until their development is nearly complete. Later, they may consume vital tissues. In many cases, parasitoids slow the development of the host to ensure that food will be available throughout their own development.

The Hymenoptera is one of the largest groups of insects and includes such familiar forms as bees, ants, and wasps. About half of the hymenopterans are parasitoids. Hymenopteran parasitoids have highly specialized ovipositors that are used both to lay eggs and to sting the host (Fig. 8.57). The sting temporarily (and sometimes permanently) immobilizes the host victim, making it easier for the female parasitoid to lay her eggs, either on the surface of the host or in the hemocoel. Parasitoids that attack hosts living on wood, underground, or other concealed places often have long ovipositors (Fig. 8.58). Cutting ridges at the end of the ovipositor allow parasitoid wasps to penetrate through plant tissues, including wood, to locate hidden hosts. In some cases, the ovipositor can be up to eight times the length of the body. In a number of hymenopteran and dipteran parasitoids, the abdomen is compressed so that it can slide into narrow openings, such as between the gills of fungi, to locate hosts living there. Most of the hymenopteran parasitoids belong to the Ichneumonidae, Braconidae, and Chalcidae, and parasitize the eggs and larvae of Lepidoptera, Coleoptera, Diptera, sawflies, and spiders (see Box 8.3 for an account of a unique parasitoid of tarantulas).

The Diptera are the second most important group of insects in terms of the number of parasitoids. Unlike most Hymenoptera, the ovipositor of the dipterans is inconspicuous; accordingly, they are not able to pursue hosts protected in galleries in wood, leaf-mines, galls, etc. As discussed in the previous paragraph, however, the abdomen of some is secondarily modified into a flat structure to reach into narrow openings. Although the parasitoid dipterans are not as diverse as the

Fig. 8.58 The wasp *Diapetimorpha introita* is preparing to lay an egg in the pupal tunnel of a corn earworm pupa *Heliothis zea*. Notice the long ovipositor. (Photograph courtesy of the Agricultural Research Service, US Department of Agriculture.)

Box 8.3 Providing for the children

In the deserts of Mexico and the southwestern United States lives the largest spider in North America, the tarantula (*Aphonopelma chalcodes*). Tarantulas are fierce-looking spiders with a leg span of approximately 10 cm. Despite their ferocious appearance, they are prey to many creatures ranging from amphibians to reptiles and from birds to mammals.

An interesting, and deadly, relationship they have is with the tarantula hawk (*Hemipepsis* sp). Despite their name, tarantula hawks are not birds, belonging instead to a group of parasitoids known as spider wasps. Tarantula hawks are large, approximately 2.5 to 3.5 cm long, and colorful with a metallic blue–green or blue–black body and bright orange wings. As adults, they are nectivores and feed commonly on milkweed. As larvae, they feed on the mighty tarantula!

When the female tarantula hawk is ready to lay an egg, she spends much of her time walking along the desert floor touching her antennae repeatedly to the surface searching for a tarantula. She may find a tarantula out in the open searching for prey or in its burrow. In the latter case, she will continually disturb the tarantula's web and entice it to leave its burrow. In either case, once the tarantula is in the open, the wasp attacks it attempting to sting the larger beast. These battles may last for some time and, usually, the wasp wins. Interestingly, the sting of the wasp does not kill the spider but paralyzes it instead. The paralysis is permanent. The wasp then digs a burrow, flips the spider over so that the ventral surface faces up, and drags it into the burrow. Once in the burrow, the wasp lays an egg on the ventral surface of the tarantula, which, although paralyzed, is still very much alive. When the young wasp larva hatches, it begins to feed on the tarantula. The paralyzed tarantula is the sole source of food for the larva and remains alive for most, if not all, of the wasp's growth. In several weeks, depending on temperature, an adult wasp emerges from the burrow as a nectivore, leaving behind what remains of the tarantula's carcass and the wasp's period of carnivory. This relationship is interesting because the female wasp is forced to engage a larger and poisonous spider to provide sufficient food for its larva. By paralyzing the tarantula, the tarantula hawk ensures 'fresh meat' for its young!

hymenopterans, their hosts are certainly more varied. Insects are still the likely hosts, but slugs, earthworms, snails, and centipedes are also attacked.

For a comprehensive view of parasitoids from an ecological, behavioral, and evolutionary perspective, see Godfray (1994) and Hawkins & Sheenan (1994).

8.4 Phylogenetic relationships

The arthropods are the most diverse group of animals, and one of the most studied. However, despite their species richness, a rich fossil record going back to the Cambrian, and much intensive study, their phylogenetic relationships remain unclear. This is true for both the relationships among the different arthropod groups and between the arthropods and other phyla. Moreover, the basic question regarding the origin of arthropods is still debated. It is not clear if all the arthropod groups derive from a common ancestor (monophyletic origin) or, if the arthropod groups derive from more than one common ancestor (polyphyletic origin).

Historically, monophyly has been the prevailing view because all arthropods share a number of peculiar traits, such as a cuticle made of chitin and proteins, segmentation with at least some segments bearing paired, articulated limbs, and similar patterns of cephalization. However, extensive work by Sidney Manton on the comparative morphology and embryology of the different groups of arthropods suggested that the three major groups, Chelicerata, Crustacea, and Uniramia, should be recognized as separate phyla having no common ancestor (Manton, 1973). Supporters of the polyphyletic origin indicate that the similarities between the groups should be regarded as extensive convergence, in which the possesion of an exoskeleton leads, by necessity, to an array of changes that always result into an organism recognized as an arthropod, regardless of the starting point. It all boils down to a sort of chain reaction that results in what Willmer (1990) calls a 'syndrome of arthropodization'.

It appears, however, that at the present time, most authors support the hypothesis regarding a monophyletic origin of arthropods. They base their views on the fact that all arthropods are more closely related to each other than to members of other phyla, that their nearest common ancestor itself looked like an arthropod, and that their similarities are too great to be accounted for by anything but shared ancestry. Most importantly, perhaps, cladistic analyses of arthropods, including Cambrian fossils, strongly indicate that the primitive crustaceans are the living forms that come closest to the ancestral arthropod (Briggs & Fortey, 1989; Nielsen, 1995). This evidence is also supported by the fact that most crustaceans are aquatic (and marine), whereas the uniramians and chelicerates are, for the most part, terrestrial, and all their embryological and anatomical specializations are related to the change in habitat, from aquatic to terrestrial. In this context, then, the Uniramia can be considered as a very specialized group of terrestrial crustaceans (Nielsen, 1995).

The phylogenetic relationships of the arthropods with other phyla are not clear either. Evidence based on the nature of the nervous system and the segmentation of the body suggested that arthropods derived from one or more ancestral annelids. The Annelida, then, was thought to be the sister group of arthropods. However, a number of studies using morphological, paleontological, and molecular data, question the arthropod–annelid relationship. Recently, phylogenetic analysis of 18S ribosomal DNA sequences indicates a close relationship between arthropods, nematodes, tardigrads, onycophorans, nematomorphs, kinorhynchs, and priapulids, but not with annelids (Aguinaldo *et al.*, 1997). The common bond between the arthropods and their close allies is that they all undergo molting during their growth and development.

Box 8.4 Classification of Arthropoda

A unique, widely accepted scheme of classification for the arthropods does not exist. Several authors have proposed classifications for the many groups of arthropods but there is no consensus yet as to which one is a better reflection of their phylogeny. Despite their differences, most classifications are very similar and differ mainly in the hierarchical level of the groups. For example, some consider the Crustacea, Chelicerata, and Uniramia as classes within the phylum Arthropoda, whereas others assign them to subphylum level. The same is true for many of the minor divisions. Regardless of their exact hierarchical level, the relationships among the groups are, for the most part, the same. The classification of Crustacea, Chelicerata, and Uniramia follows Bowman & Abele (1982), Evans (1992), Ruppert & Barnes (1994), and Gillot (1995). Only the groups of arthropods that include parasitic species are considered. For a complete classification of the arthropods, including all taxa, see the texts cited above.

Subphylum Crustacea
Head appendages consisting of one pair of antennules, one pair of antennae, one pair of mandibles, and two pairs of maxillae. Cephalothorax usually with a dorsal carapace; apppendages, except antennules (first antennae) biramous (primitive condition); development includes a nauplius stage.

Class Maxillopoda
Thorax with no more than six segments; abdomen with no more than four segments, plus the telson; a naupliar eye with a unique structure, called a maxillopodan eye, may be present.

Subclass Copepoda
Small, with cylindrical bodies; cephalothorax formed by the head and one or two thoracic segments; remainder of the thorax bearing five pairs of biramous appendages; abdomen without appendages; thorax and abdomen form a trunk; antennules longer than the antennae; compound eyes absent but a naupliar eye is often present. Many parasitic forms are highly modified and often do not conform to the group description.

Order Cyclopoida
Includes marine and freshwater, planktonic and benthic copepods, both

free-living and parasitic. Antennules short, with only 10 to 16 articles; antennae uniramous.

Representative genera: *Afrolernaea, Ascidicola, Enterocola, Lernaea, Notodelphys*

Order Monstrilloida

Nauplii free-swimming but subsequent larval stages parasitic in polychaetes and gastropods. Adults planktonic; antennae and mouthparts missing; functional gut absent.

Representative genus: *Haemocera*

Order Poecilostomatoida

Marine copepods parasitic on invertebrates and fishes. Adults often lack segmentation; antennules reduced and small; antennae often modified into attachment organs; base of mandibles sickle-shaped; maxillules and maxillae reduced; thoracic appendages modified and reduced.

Representative genera: *Bomolochus, Chondracanthus, Clausia, Colobomatus, Ergasilus, Lichomolgus, Mytilicola, Philichthys, Sarcotaces, Taeniacanthus*

Order Siphonostomatoida

Freshwater and marine copepods parasitic on fishes and invertebrates. Body segmentation often reduced or lost; antennules modified (reduced or elongated), may end in a claw used for attachment to the host; mandibles enclosed in a siphon or tube formed by the fusion of the labrum and labium; maxillae may be modified for attachment to the host; thoracic appendages may be normal and used for swimming, or highly modified.

Representative genera: *Caligus, Cecrops, Clavella, Dichelesthium, Lepeophtheirus, Pandarus, Pennella, Phrixocephalus, Salmincola, Sphyrion*

Subclass Tantulocarida

Small copepod-like ectoparasites on deep-water crustaceans. Only a median ventral stylet is present in the cephalic area; other cephalic appendages missing; thorax with six segments, each with a pair of appendages; appendages one through five are biramous, sixth pair uniramous.

Representative genera: *Basipodella, Deoterthron*

Subclass Branchiura

Crustaceans ectoparasitic mainly on fishes. Head and thorax covered by a shield-like carapace; thorax with four pairs of biramous appendages; abdomen small, unsegmented, without appendages; compound eyes present; antennules and antennae reduced, the former possess claws for attachment; maxillules often modified to form suctorial disks.

Representative genera: *Argulus, Dolops*

Subclass Cirripedia

Barnacles. Sessile free-living and parasitic crustaceans. Head reduced and abdomen rudimentary; body does not show signs of segmentation; the carapace becomes a mantle in non-parasitic epizoic forms; a nauplius larva hatches from the egg but develops into a bivalved cypris larva.

Order Acrothoracica

Barnacles without exoskeleton that bore into calcareous substrata.

Four pairs of thoracic appendages present; males very small and parasitic on the female.
Representative genera: *Berndtia*, *Trypetesa*
Order Ascothoracica
Barnacles without exoskeleton, parasitic on echinoderms and corals. Thorax with up to six pairs of appendages.
Representative genera: *Ascothorax*, *Gorgonolaureus*, *Petrarca*, *Synagoga*
Order Rhizocephala
Barnacles without exoskeleton, parasitic primarily on decapod crustaceans. Adults lack appendages, segmentation, and digestive system; root-like processes extend through the host and absorb nutrients.
Representative genera: *Chthamalophilus*, *Clistosaccus*, *Lernaeodiscus*, *Peltogasterella*, *Ptychascus*, *Sacculina*, *Typhosaccus*
Order Thoracica
Free-living and commensal barnacles with six pairs of thoracic appendages; mantle usually covered with calcareous plates.
Representative genera: *Conchoderma*, *Coronula*, *Xenobalanus*
Class Ostracoda
Small crustaceans with the body enclosed within a hinged, bivalved carapace; trunk reduced, with no more than two pairs of appendages. A few species might be symbiotic in the gills of marine fishes.
Class Malacostraca
Thorax with eight segments; abdomen with six segments (in most species) plus the telson; all the segments bear appendages; telson and uropods form a tail fan; female gonopore opens onto the fifth thoracic segment and the male gonopore onto the eighth; compound eyes present in most species.
Order Amphipoda
Amphipods. Marine, freshwater and terrestrial crustaceans. Body compressed laterally; sessile compound eyes present; body lacks a carapace; ventral brood pouch present.
Suborder Hyperiidea
Marine amphipods, most species are symbiotic on medusae and tunicates. Head and eyes very large; head fused with first thoracic segment.
Representative genera: *Phronima*, *Hyperia*, *Primmo*
Suborder Caprellidea
Marine amphipods that include the skeleton shrimp and the whale lice. Head fused with first two thoracic segments; abdomen reduced, with vestigial appendages.
Representative genera: *Aeginella*, *Caprella*, and *Cercops*, the skeleton shrimps; *Cyamus*, *Paracyamus*, and *Syncyamus*, the whale lice
Order Isopoda
Marine, freshwater, and terrestrial crustaceans. Body flattened dorsoventrally, sessile compound eyes present; body without a carapace; ventral brood pouch present.
Suborder Gnathiidea
Marine isopods. Manca larvae ectoparasitic on marine fishes; adults do not feed; abdomen much narrower than the thorax.

Representative genus: *Gnathia*

Suborder Flabellifera

Mostly marine isopods including free-living and parasitic species. Uropods fan-shaped, form a tail fan together with the telson.

Representative genera: *Cymothoa, Nerocila, Lironeca*

Suborder Epicaridea

Marine isopods parasitic on other crustaceans. Females modified for parasitism, some lack segmentation and appendages; males smaller but less modified.

Representative genera: *Bopyrus, Danalia, Entoniscus, Pinnotherion, Portunion, Liriopsis*

Subphylum Uniramia

All appendages uniramous; head appendages include one pair of antennae, one pair of mandibles, and one or two pairs of maxillae.

Class Insecta

Insects. Body divided into a well-differentiated head, thorax, and abdomen; head with one pair of antennae; thorax with three segments, three pairs of legs and usually two pairs of wings, although some have one pair or none at all.

Order Coleoptera

Beetles. Body hard with two pairs of wings; forewings modified as strong protective covers (elytra); hindwings membranous.

Representative genus: *Platypsyllus*

Order Dermaptera

Earwigs. Elongated, resemble beetles; chewing mouthparts; compound eyes present; forceps-like cerci at the posterior end; most species with fan-shaped wings and elytra. A few are ectoparasites on mammals.

Representative genera: *Arixenia, Hemimerus*

Order Diptera

Flies and mosquitoes. Only one pair of functional wings present; forewings membranous and functional; hindwings reduced and knob-like (halteres).

Representative genera: *Aedes, Anopheles, Chrysops, Cochliomyia, Conops, Culex, Cuterebra, Dermatobia, Gasterophilus, Glossina, Hippelates, Hippobosca, Lucilia, Lutzomyia, Melophagus, Oestrus, Phlebotomus, Simulium, Tabanus*

Order Hemiptera

True bugs; kissing bugs and bed bugs are parasitic. Mouthparts adapted for piercing and sucking; basal section of the forewings thick and leathery, distal section membranous; hindwings entirely membranous; some species lack wings.

Representative genera: *Cimex, Leptocimex, Panstrongylus, Rhodnius, Triatoma*

Order Hymenoptera

Ants, bees, wasps. Two pairs of transparent, membranous wings, connected together with small hooklets; wings with only a few veins; some lack wings.

Representative genera: *Aphelinus, Coeloides, Eupelmus, Macrocentrus, Phytodietus, Rhyssella, Tetrastichus*

Order Lepidoptera

Butterflies and moths. Soft-bodied insects; body, appendages and wings covered with scales; mouthparts modified for sucking nectar and juices (and lachrymal fluid).

Representative genera: *Arcyophora, Calyptra, Lobocraspis*

Order Phthiraptera

Biting and sucking lice. Body flattened and without wings; eyes reduced or absent; all thoracic segments fused; mouthparts modified for chewing or piercing–sucking.

Representative genera: *Actornithophilus, Anaticola, Bovicola, Columbicola, Haematopinus, Menacanthus, Menopon, Pediculus, Phthirius, Piagetiella, Trichodectes*

Order Siphonaptera

Fleas. Body small, wingless, flattened laterally; legs very long, adapted for jumping; mouthparts modified for piercing and sucking.

Representative genera: *Ctenocephalides, Echidnophaga, Pulex, Spilopsyllus, Tunga, Xenopsylla*

Order Strepsiptera

Very small, beetle-like; only males possess wings; females extremely modified, lack wings, eyes, legs, and antennae.

Representative genera: *Corioxenos, Elenchus, Eoxenos, Stylops*

Subphylum Chelicerata

Body divided into prosoma (cephalothorax) and opisthosoma (abdomen), usually unsegmented; prosoma with four pairs of legs and without antennae.

Class Arachnida

Opisthosoma without appendages.

Subclass Acari

Ticks and mites. Body usually divided into an anterior proterosoma and a posterior hysterosoma; appendages of the mouth and its segments form a differentiated area called the capitulum, the rest of the body is called the idiosoma; usually with four pairs of legs, but some may have fewer.

Order Astigmata

Small mites without tracheal system; respiration occurs across the body surface.

Representative genera: *Knemidocoptes, Megninia, Otodectes, Psoroptes, Sarcoptes*

Order Cryptostigmata (Oribatida)

Beetle mites. Free-living mites common in soil and leaf mold. Body without spiracles. Although none of the species is parasitic, many are intermediate hosts for cestodes.

Representative genera: *Galumna, Oppia*

Order Ixodida

Ticks. Large acari; teeth modified for piercing and for attachment to the host; a specialized sensory organ, Haller's organ, present on the first pair of legs; respiratory system with only one pair of spiracles opening near the base of the fourth pair of legs.

Representative genera: *Amblyomma, Argas, Boophilus, Dermacentor, Hyalomma, Ixodes, Nuttalliella, Ornithodoros, Otobius, Rhipicephalus*

Order Mesostigmata
Mites with several sclerotized plates on their dorsal and ventral surface; tracheal system with one pair of spiracles between the base of the second and fourth pairs of legs.
Representative genera: *Dermanyssus, Echinolaelaps, Lyponyssus, Ornithonyssus, Orthohalarachne, Pneumonyssus, Sternostoma*
Order Prostigmata
Free-living and parasitic mites; one or two pairs of spiracles located near the mouthparts.
Representative genera: *Acarapis, Arrenurus, Blankaartia, Demodex, Hydrachnella, Pyemotes, Trombicula*

References

Aguinaldo, A. M. A., Turbeville, J. M., Linford, L. S., Rivera, M. C., Garey, J. R., Raff, R. A. & Lake, J. A. (1997) Evidence for a clade of nematodes, arthropods and other moulting animals. *Nature*, **387**, 489–493.

Askew, R. R. (1971) *Parasitic Insects*. New York: Elsevier.

Bennett, C. E. (1995) Ticks and Lyme disease. *Advances in Parasitology*, **36**, 343–405.

Bowman, T. E. & Abele, L. G. (1982) Classification of the recent Crustacea. In *The Biology of the Crustacea*, vol. 1, *Systematics, The Fossil Record, and Biogeography*, ed. L. G. Abele, pp. 1–27. New York: Academic Press.

Boxshall, G. A. & Lincoln, R. J. (1983) Tantulocarida, a new class of Crustacea ectoparasitic on other crustaceans. *Journal of Crustacean Biology*, **3**, 1–16.

Bresciani, J. (1986) The fine structure of the integument of free-living and parasitic copepods. A review. *Acta Zoologica (Stockholm)*, **67**, 125–145.

Bresciani, J. & Dantzer, V. (1980) Fine structural localization of acid phosphatase in the root system of the parasite *Clistosaccus paguri* (Crustacea, Rhizocephala). *Electron Microscopy*, **2**, 290–291.

Briggs, D. E. & Fortey, R. A. (1989) The early radiation and relationships of the major arthropod groups. *Science*, **246**, 241–243.

Bunkley-Williams, L. & Williams, E. H. Jr. (1998) Isopods associated with fishes: a synopsis and corrections. *Journal of Parasitology*, **84**, 893–896.

Burgess, I. (1994) *Sarcoptes scabiei* and scabies. *Advances in Parasitology*, **33**, 235–292.

Burgess, I. F. (1995) Human lice and their management. *Advances in Parasitology*, **36**, 271–342.

Cho, J. H., Kim, J. B., Cho, C. S., Huh, S. & Ree, H. I. (1999) An infestation of the mite *Sancassania berlesei* (Acari: Acaridae) in the external auditory canal of a Korean man. *Journal of Parasitology*, **85**, 133–134.

Desch, C. & Nutting, W. B. (1972) *Demodex folliculorum* (Simon) and *D. brevis* (Akbulatova) of man: redescription and reevaluation. *Journal of Parasitology*, **58**, 169–177.

Downer, R. G. H. (ed.) (1981) *Energy Metabolism in Insects*. New York: Plenum Press.

Duffy, D. C. (1983) The ecology of tick parasitism in densely nesting Peruvian seabirds. *Ecology*, **64**, 110–119.

Evans, G. O. (1992) *Principles of Acarology*. Wallingford: CAB International.

Feare, C. J. (1976) Desertion and abnormal development in a colony of sooty terns *Sterna fuscata* infested by virus-infected ticks. *Ecology*, **118**, 112–115.

Gillot, C. (1995) *Entomology*, 2nd edn. New York: Plenum Press.

Gilmour, D. (1965) *The Metabolism of Insects*. Edinburgh: Oliver & Boyd.

Godfray, H. C. J. (1994) *Parasitoids, Behavioral and Evolutionary Ecology*. Princeton: Princeton University Press.

Gotto, R. V. (1979) The association of copepods with marine invertebrates. *Advances in Marine Biology*, **16**, 1–109.

Hall, M. & Wall, R. (1995) Myiasis of humans and domestic animals. *Advances in Parasitology*, **35**, 257–334.

Hawkins, B. A. & Sheenan, W. (eds.) (1994) *Parasitoid Community Ecology*. New York: Oxford University Press.

Hoogstraal, H., Wassef, H. Y., Hays, C. & Keirans, J. E. (1985) *Ornithodoros (Alectorobius) spheniscus* n. sp. [Acarina: Ixodoidea: Argasidae: *Ornithodoros (Alectorobius) capensis* group], a tick parasite of the Humboldt penguin in Peru. *Journal of Parasitology*, **71**, 635–644.

Huys, R., Boxshall, G. A. & Lincoln, R. J. (1993) The tantulocaridan life cycle: the circle closed? *Journal of Crustacean Biology*, **13**, 432–442.

Kabata, Z. (1974) Two new features in the morphology of Caligidae (Copepoda). *Proceedings of the Third International Congress of Parasitology*, **3**, 1635–1636.

Kabata, Z. (1981) Copepoda (Crustacea) parasitic on fishes: problems and perspectives. *Advances in Parasitology*, **19**, 2–71.

Kannupandi, T. (1976*a*) Cuticular adaptations in two parasitic copepods in relation to their mode of life. *Journal of Experimental Marine Biology and Ecology*, **22**, 235–248.

Kannupandi, T. (1976*b*) Occurrence of resilin and its significance in the cuticle of *Pennella elegans*, a copepod parasite. *Acta Histochemica*, **56**, 73–79.

Kerkut, G. A. & Gilbert, L. I. (eds.) (1985) *Comprehensive Insect Physiology, Biochemistry and Pharmacology*. Elmsford, NY: Pergamon Press.

Khan, R. A. (1988) Experimental transmission, development, and effects of a parasitic copepod, *Lernaeocera branchialis*, on Atlantic cod, *Gadus morhua*. *Journal of Parasitology*, **74**, 586–599.

Khan, R. A., Lee, E. M. & Barker, D. (1990) *Lernaeocera branchialis*: a potential pathogen to cod ranching. *Journal of Parasitology*, **76**, 913–917.

King, K. A., Keith, J. O., Mitchell, C. A. & Keirans, J. E. (1977) Ticks as a factor in nest desertion of California brown pelicans. *Condor*, **79**, 507–509.

Lanciani, C. A. (1984) Crowding in the parasitic stage of the water mite *Hydrachna virella* (Acari: Hydrachnidae). *Journal of Parasitology*, **70**, 270–272.

Lanciani, C. A. (1988) Sexual bias in host selection by parasitic mites of the mosquito *Anopheles crucians* (Diptera: Culicidae). *Journal of Parasitology*, **74**, 768–773.

Lompen, J. S. H. & Oliver, J. H. Jr. (1993) Haller's organ in the tick family Argasidae (Acari: Parasitiformes: Ixodida). *Journal of Parasitology*, **79**, 591–603.

Manton, S. M. (1973) Arthropod phylogeny – a modern synthesis. *Journal of Zoology, London*, **171**, 111–130.

Marshall, A. G. (1982) The ecology of the bat ectoparasite *Eoctenes spasmae* (Hemiptera: Polyctenidae) in Malaysia. *Biotropica*, **14**, 50–55.

Moser, M. & Taylor, S. (1978) Effects of the copepod *Cardiodectes medusaeus* on the lanternfish *Stenobrachius leucopsarus* with notes on hypercastration by the hydroid *Hydrichthys* sp. *Canadian Journal of Zoology*, **56**, 2372–2376.

Nielsen, C. (1995) *Animal Evolution: Interrelationships of the Living Phyla*. Oxford: Oxford University Press.

Oliver, J. H. Jr. (1996) Lyme borreliosis in the southern Unites States: a review. *Journal of Parasitology*, **82**, 926–935.

Oliver, J. H. Jr., Hayes, M. P., Keirans, J. E. & Lavender, D. R. (1993) Establishment of the foreign parthenogenetic tick *Amblyomma rotundatum* (Acari: Ixodidae) in Florida. *Journal of Parasitology*, **79**, 786–790.

Pence, D. B., Windberg, L. A., Pence, B. C. & Sprowls, R. (1983) The epizootiology and pathology of sarcoptic mange in coyotes, *Canis latrans*, from South Texas. *Journal of Parasitology*, **69**, 1100–1115.

Perkins, P. S. (1985) Iron crystals in the attachment organ of the erythrophagous copepod *Cardiodectes medusaeus* (Pennellidae). *Journal of Crustacean Biology*, **5**, 591–605.

Perkins, P. S. (1994) Ultrastructure of the holdfast of *Phrixocephalus cincinnatus* (Wilson), a blood-feeding parasitic copepod of flatfishes. *Journal of Parasitology*, **80**, 797–804.

Raga, J. A. (1997) Parasitology of marine mammals. In *Marine Mammals, Seabirds and Pollution of Marine Systems*, ed. T. Jauniaux, J. M. Bouquegneau & F. Coignoul, pp. 67–90. Liège: Presses de la Faculté de Médecine Vétérinaire de l'Université de Liège.

Raibaut, A. & Trilles, J. P. (1993) The sexuality of parasitic crustaceans. *Advances in Parasitology*, **32**, 367–444.

Ramachandra, R. N. & Wikel, S. K. (1992) Modulation of host immune response by ticks (Acari: Ixodidae): effect of salivary gland extracts on host macrophages and lymphocyte cytokine production. *Journal of Medical Entomology*, **29**, 818–826.

Ribeiro, J. M. C., Makoul, G. T., Levine, J., Robinson, D. R. & Spielman, A. (1985) Antihemostatic, anti-inflammatory, and immunosuppressive properties of the saliva of a tick, *Ixodes dammini*. *Journal of Experimental Medicine*, **161**, 332–344.

Rothschild, M. & Ford, B. (1972) Breeding cycle of the flea *Cediopsylla simplex* is controlled by breeding cycle of host. *Science*, **178**, 625–626.

Rothschild, M., Schlein, Y., Parker, K., Neville, C. & Sternberg, S. (1973) The flying leap of the flea. *Scientific American*, **229 (5)**, 92–100.

Rowntree, V. J. (1996) Feeding, distribution, and reproductive behavior of cyamids (Crustacea: Amphipoda) living on humpback and right whales. *Canadian Journal of Zoology*, **74**, 103–109.

Ruppert, E. E. & Barnes, R. D. (1994) *Invertebrate Zoology*, 6th edn. Fort Worth: Saunders College Publishing.

Samuel, W. M. & Welch, D. A. (1991) Winter ticks on moose and other ungulates: factors influencing their population size. *Alces*, **27**, 169–182.

Samuel, W. M., Williams, E. S. & Rippin, A. B. (1982)

Infestations of *Piagetiella peralis* (Mallophaga: Menopodidae) on juvenile white pelicans. *Canadian Journal of Zoology*, **60**, 951–953.
Spalding, M. G., Wrenn, W. J., Schwikert, S. T. & Schmidt, J. A. (1997) Dermatitis in young Florida sandhill cranes (*Grus canadensis pratensis*) due to infestation by the chigger, *Blankaartia sinnamaryi*. *Journal of Parasitology*, **83**, 768–771.
Webb, J. P., Furman, D. P. & Wang, S. (1985) A unique case of human ophthalmic acariasis caused by *Orthohalarachne attenuata* (Banks, 1910) (Acari: Halarachnidae). *Journal of Parasitology*, **71**, 388–389.
Willmer, P. (1990) *Invertebrate Relationships: Patterns in Animal Evolution*. Cambridge: Cambridge University Press.
Zinsser, H. (1934) *Rats, Lice and History*. Boston: Little, Brown.

Chapter 9

Parasitism in other metazoan groups

Parasitism is a ubiquitous life style. A quick survey of the animal phyla (Table 9.1) shows that practically every major phylum has some species that have adapted to a parasitic existence. In addition to the major groups of parasitic organisms discussed in the previous chapters, there are other phyla that are completely parasitic, i.e., Orthonectida, Dicyemida, or that have some parasitic species, e.g., Annelida, Mollusca, or that are parasitic in some stage of their life cycle, e.g., Nematomorpha. Unfortunately, very little is known about some of these species and, for many of them, a fine line separating commensalism, mutualism, and parasitism (as a type of exploitation) is not clear. For a discussion of these concepts, see Chapter 1.

9.1 Porifera: the sponges

Poriferans are mostly sedentary, aquatic animals, predominantly marine, whose bodies contain numerous, small external apertures. They are unique because each cell functions independently from the others and because their functional morphology is constructed in order to create an internal water current. Although none of the sponges known can be considered true parasites, a few species of the marine sponge *Cliona* bore into the shells of molluscs, causing them to become brittle. This condition is commonly known as 'spice bread disease' and mass mortalities of oysters have been attributed to it. Although these boring sponges do not acquire nutrients directly from the host, some consider them parasites, in a loose sense, because of the extra energetic cost incurred by the host during shell repair, and because most of these sponges seem to be host specific.

9.2 Cnidaria: hydras, jellyfishes, sea anemones, corals

The Cnidaria includes the familiar hydras, jellyfishes, sea anemones, and corals. Cnidarians are almost exclusively marine organisms, with radial symmetry. They have a mouth and gastrovascular cavity, a diffuse nervous system, a **planula larva**, and specialized cells called **cnidocytes**. Cnidocytes possess a stinging structure called a **nematocyst**, used for food capture, defense, and sometimes attachment. Another important characteristic of cnidarians is the **alternation of generations** during the life cycle, i.e., switching between a mostly asexual polyp form (a kind of sea anemone) and a mostly sexual medusa stage (a form of jellyfish). One of the two forms, however, may be reduced or totally absent in some groups. The cnidarians have traditionally been divided into four classes, Hydrozoa, Anthozoa, Scyphozoa, and Cubomedusae, of which the first two include a few parasitic species.

Among the hydrozoans, *Hydrichthys* is a colonial cnidarian that attaches to the body of marine fishes. The polyp has lost the characteristic tentacles and feeds on blood and tissues by means of a root-like stolon that penetrates the body of the fish. *Hydrichthys* can also parasitize parasitic copepods, a case of hyperparasitism (Fig. 9.1). Moser &

Table 9.1 Estimated number of parasitic species among various groups in the animal kingdom

Phylum	Subgroups	Total number of species	Number of parasite species
Protozoa *sensu lato*		70000	*ca.* 12000
Porifera		5000	few
Cnidaria		8900	few
Ctenophora		80	few
Placozoa		1	0
Dicyemida		40	40
Orthonectida		15	15
Platyhelminthes		>9300	>7700
	Turbellaria	>1600	>80
	Digenea	>4000	>4000
	Monogenea	>2200	>2200
	Cestoda	1500	1500
Gnathostomulida		80	0
Priapulida		3	0
Entoprocta		60	0
Nemertea		750	10
Rotifera		1500	20
Gastrotricha		150	0
Nematoda		>10000	>5000
Nematomorpha		<100	<100
Kinorhyncha		100	0
Acanthocephala		500	500
Annelida		7000	*ca.* 420
Onycophora		70	0
Arthropoda			
	Xiphosura	5	0
	Arachnida	30000	4300
	Pycnogonida	350	<350
	Crustacea	20000	2500
	Myriapoda	10500	0
	Insecta	many millions	many millions[a] 20000[b]
Pentastomida		75	75
Tardigrada		180	1?
Phoronida		18	0
Bryozoa		>4000	0
Brachiopoda		280	0
Mollusca		112000	>100 1300[c]

Table 9.1 (cont.)

Phylum	Subgroups	Total number of species	Number of parasite species
Echiura		70	few (intraspecific parasites)
Sipunculida		250	0
Hemichordata		250	0
Echinodermata		*ca.* 6000	few (≥2)
Pogonophora		47	0
Chaetognatha		50	0
Chordata		62 000	few

Notes:

[a] Millions include all the herviborous insects.

[b] According to Toft (1991) there are about 20 000 parasitic insects and about 117 500 parasitoids. Only animal parasites are included in her count.

[c] According to Toft (1991) there are 1300 species of parasitic molluscs. Her count includes 100 gastropods and 1200 freshwater bivalves.

Source: With permission, modified from Rohde, K. (1993) *Ecology of Marine Parasites*, 2nd edn. Wallingford, UK: CAB International.

Taylor (1978) found this hydrozoan in the copepod *Cardiodectes medusaeus*, a parasite, in turn, of mesopelagic lanternfishes. The attachment organ of the copepod is embedded in the bulbus arteriosus of the fish, while most of its body remains outside the fish (see sections 8.1.4 and 8.1.7). In this tripartite relationship, the copepod seems to castrate the fish host and *Hydrichthys*, in turn, seems to castrate the copepod, an interesting case of hypercastration. The life cycle of hydrozoans such as *Hydrichthys* includes a free-living medusa stage, whereas the polyp is the parasitic form.

The planula larva of some anemones parasitizes other cnidarians, in particular the gastrovascular cavity of medusae. The planula larva of *Peachia quinquecapitata*, for example, seems to be ingested together with other food items by the medusa host. After ingestion, the larva of *P. quinquecapitata* grows and remains endoparasitic for about 11 days, feeding on food particles present in the host's gastrovascular cavity. Then, the larva becomes ectoparasitic on the medusa and feeds on the gonads and other body tissues of the host. Indeed, a larval anemone can consume all the host gonads in just two days. Finally, after about a month of being ectoparasitic, the anemone has developed most of the adult characteristics, drops off the medusa host, and becomes a free-living anemone. Quantitative infection data for the medusa *Phialidium gregarium* infected with *Peachia quinquecapitata* shows that peak prevalences of 62% can be reached by late spring at Friday Harbor, Washington (Spaulding, 1972).

Probably one of the most studied, but not necessarily better-known, cnidarians is the freshwater hydrozoan *Polypodium hydriforme*, a parasite that infects the oocytes of sturgeons and polyodontid fishes in Eurasia and North America (sturgeons are the main source of caviar). Because of the heavy losses inflicted on the caviar industry in the former Soviet Union, a significant amount of research has been done on this species but, to date, the complete life cycle of the parasite has not been elucidated. The earliest parasitic stages found inside the fish oocytes are binucleate cells (Fig. 9.2). These cells move through different developmental stages, including a morula-like phase and a planula larva until they reach the stolon stage, all of which takes several years. When the fish spawns, the parasite's stolon in the fish oocytes bursts and releases free-living polyps. These polyps swim in the water column, feed, multiply by longitudinal fission, and develop gonads. The final fate of these free-living polyps, or how they infect new fish oocytes, is not known (Raikova, 1994).

Fig. 9.1 The cnidarian *Hydrichthys* sp. (arrow) parasitizing the copepod *Cardiodectes medusaeus* (not visible), a parasite of the lanternfish *Stenobrachius leucopsarus*. (Photograph courtesy of Mike Moser, University of California, Berkeley.)

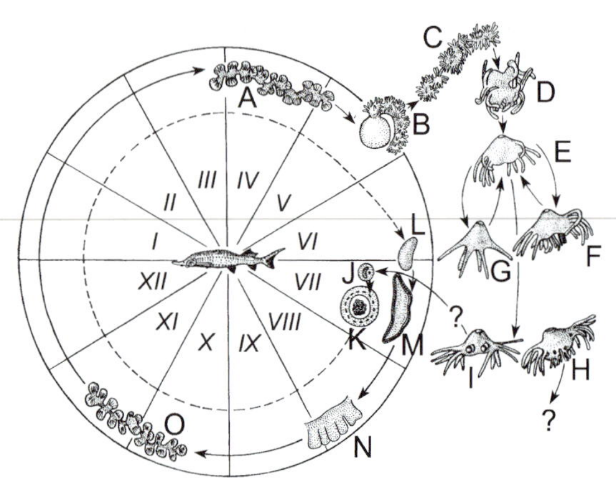

Fig. 9.2 Life cycle of *Polypodium hydriforme* in sturgeons. Parasitic stages are within the circle; free-living stages are outside. (A) Mature stolon with tentacles inside the egg before the fish spawns; (B) stolon with external tentacles emerging from egg at the time of spawning; (C) free stolon in water; (D) fragment of a stolon; (E) 12-tentacled specimen; (F) 24-tentacled specimen; (G) 6-tentacled specimen; (H) 24-tentacled specimen with four 'female' sexual complexes (the question mark at H indicates that the fate of these female animals is unknown); (I) 12-tentacled specimen with four 'male' gonads (the question mark between I and J indicates that the mode of infection of the fish oocytes is unknown); (J) binuclear parasitic cell inside a young oocyte; (K) morula encircled by a trophamnion; (L) planula (trophamnion not shown here or at any further parasitic stage); (M) budding planula; (N) stolon with tentacles; (O) stolon with internal tentacles. (From Raikova, 1994, with permission, *Journal of Parasitology*, **80**, 1–22.).

9.3 Myxozoa

The Myxozoa includes a group of parasitic organisms for which our basic knowledge regarding phylogenetic affinities and life cycles has changed dramatically in the last two decades.

Historically, the myxozoans were included among the unicellular protozoans. In the widely accepted classification of the Kingdom Protozoa proposed by Levine *et al.* (1980), the myxozoans are treated as the phylum Myxozoa. However, recent evidence clearly indicates that myxozoans are true metazoans (Smothers *et al.*, 1994; Katayama *et al.*, 1995; Siddall *et al.*, 1995; Schlegel *et al.*, 1996). For example, their spores are multicellular, not unicellular as in protozoans, and cell junctions, similar to those found in all other metazoans, hold some of the cells in the spores together. Also, they reportedly produce collagen, a protein characteristic only of metazoan phyla.

The definite position of myxozoans among the metazoans, however, is not yet clear. Some studies suggest that myxozoans have close affinities with the cnidarians, whereas others are not able to pinpoint their exact affinities among the metazoans. Although many researchers had long remarked on the similarities between the myxozoan polar capsules and the cnidarian nematocysts, only recent phylogenetic analyses combining molecular (18S ribosomal RNA gene sequences) and morphological data have provided sufficient evidence indicating affinities between Myxozoa and Cnidaria (Siddall *et al.*, 1995). In this phylogenetic analysis, the myxozoans are included within the cnidarian clade and are closely related to *Polypodium hydriforme*, the sturgeon parasite discussed previously. Additional phylogenetic analyses based on molecular data alone (Katayama *et al.*, 1995; Schlegel *et al.*, 1996), however, do not show close affinities between the Myxozoa and Cnidaria. Future research should help clarify the exact relation of Myxozoa with the Cnidaria and with the other metazoan phyla.

The other significant advance in myxozoan biology has been the determination of a complete life cycle (Markiw & Wolf, 1983) involving an invertebrate and a vertebrate host. Previously, the stages found in the vertebrate and the invertebrate hosts had been treated as different species. So different are they, that they were included in different taxonomic classes within the Myxozoa when, in reality, they are simply different developmental stages of the same species. Since the first life cycle was completed, life cycles for several other species have been studied, and all require a vertebrate and an invertebrate host, including the two morphological forms. The challenge now is to pair up the two morphological types of spores that had been described as different species. Molecular techniques, especially the comparison of 18S ribosomal RNA between the developmental stages found in the vertebrate and the invertebrate hosts, have facilitated the completion and confirmation of a number of life cycles (Andree *et al.*, 1997; Bartholomew *et al.*, 1997).

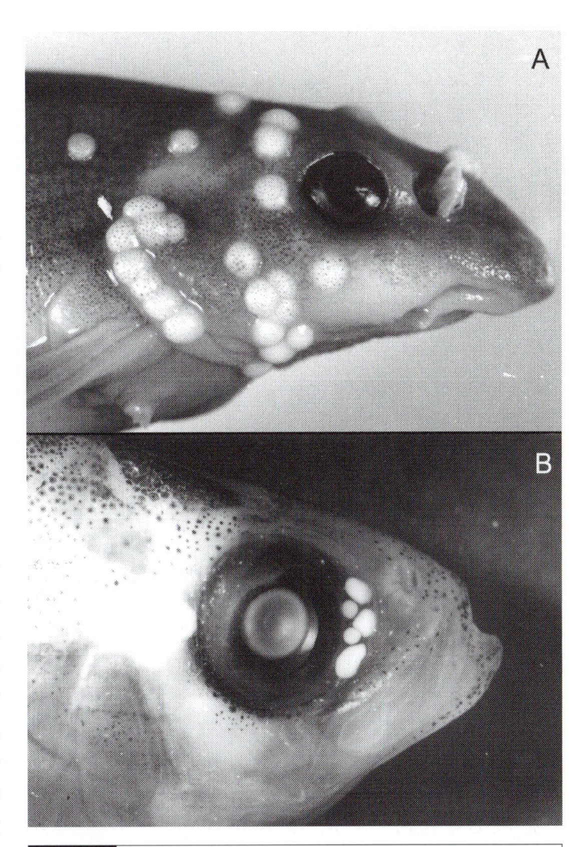

Fig. 9.3 (A) Head of a longnose dace *Rhinichthys cataractae*, infected with the myxozoan *Myxobolus rhinichthidis*; (B) head of a threadfin shad *Dorosoma petenense*, showing cysts of *Myxobolus penetenses* in the anterior margin of the eye. ((A) From Cone & Raesly, 1995, with permission, *Canadian Journal of Aquatic Sciences*, **52** (Supplement 1), 7–12; (B) from Frey et al., 1998, with permission, *Journal of Parasitology*, **84**, 1204–1206.)

BIOLOGY

Myxozoans are extracellular parasites of vertebrates (mainly fishes, and a few amphibians and reptiles) and invertebrates (mostly annelids). In fishes, they are found in cavities such as the gall bladder, urinary bladder, and ureters, or in tissues such as cartilage, muscle, gills, and skin (Fig. 9.3). In annelids, they occur in the intestinal epithelium. Myxozoans feature multicellular spores consisting of two or more shell valves that join at a sutural plane, a **sporoplasm** that is infective to the host, and **polar filaments** coiled within one or more **polar capsules** (Fig. 9.4). When an appropriate host ingests the spores, the polar filaments within the polar capsules are expelled and apparently are used for anchoring.

Myxobolus cerebralis is probably one of the best-known species because of the significant losses inflicted in salmonid hatcheries. *Myxobolus cerebralis* affects the cartilage and nervous systems of many salmonids worldwide (see Box 9.1). The life cycle of most myxozoans follows essentially the same pattern. In the case of *M. cerebralis*, spores released by the fish host are ingested by a tubificid oligochaete, the polar filaments are everted, and spores release the infective sporoplasma. Development in the oligochaete involves a form of sexual reproduction as well as the production of infective **triactinomyxon** spores (Fig. 9.4E, F) that are released into the water with the feces. (These spores were previously thought to be a different species, i.e., *Triactinomyxon gyrosalmo*.) When the salmonid fish ingests the spores, the polar capsules evert, the infective sporoplasm emerges, and migrates to the infection site (spine and head cartilage in the case of *M. cerebralis*). In the tissues, the sporoplasm grows and its nucleus divides many times, forming a multinucleate mass (or **trophozoite**) that feeds on the surrounding tissues. As development proceeds, some nuclei inside the trophozoite, i.e., generative nuclei, become surrounded by cytoplasm and form a unique type of cell, a **pansporoblast**, which will produce the spores. It is not clear how these spores reach the water to gain access to the invertebrate host, but it appears that when a predator consumes the salmonid host, the spores are released during digestion and are eliminated with the feces of the predator (Taylor & Lott, 1978). The spores also could be released after the death and decomposition of the infected fish.

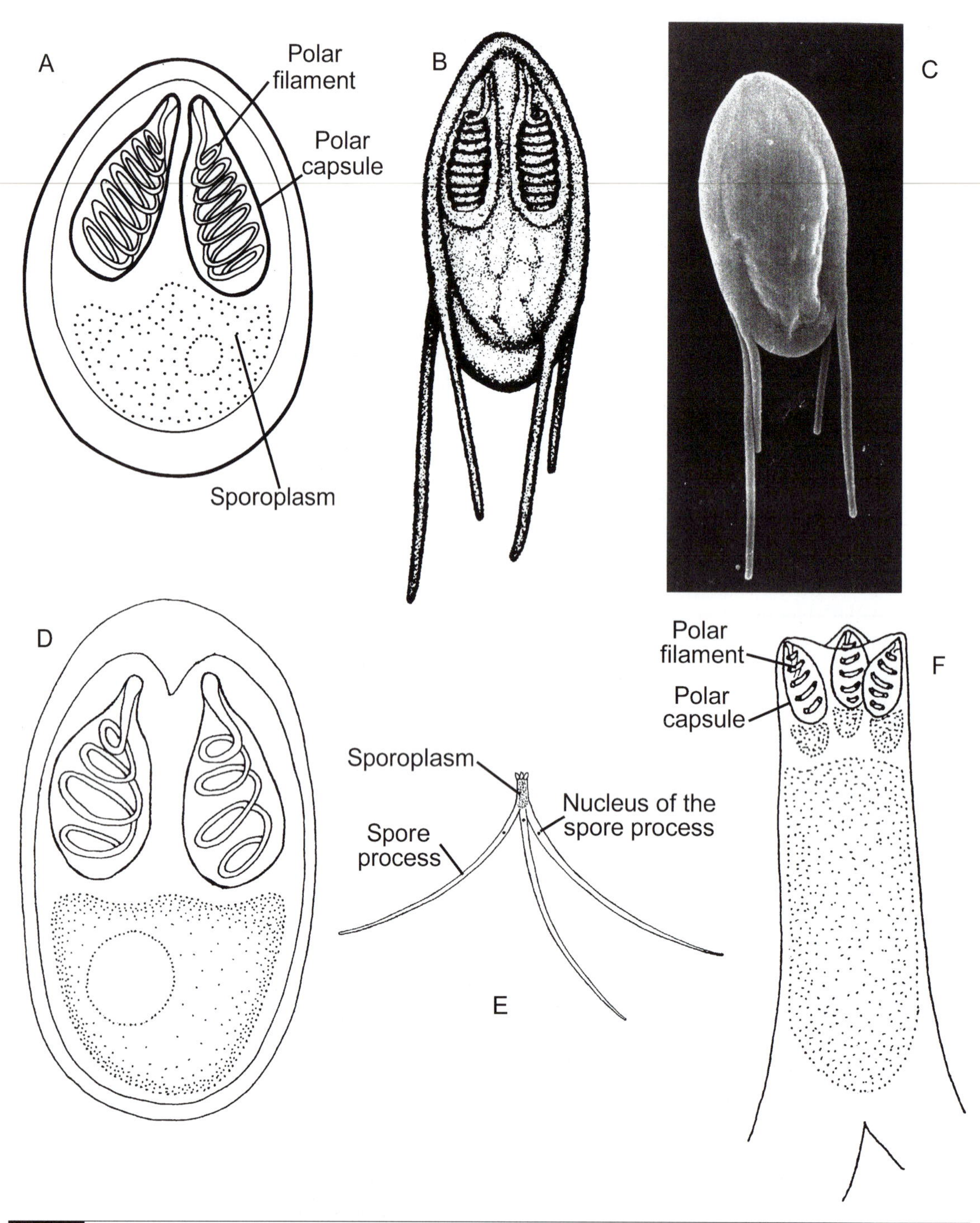

Fig. 9.4 Spores of various myxozoans. (A) Spores of *Myxobolus rhinichthidis* from the head of a longnose dace *Rhinichthys cataractae*; (B) Spore of *Tetrauronema desaequalis* from the ventral fin of the fish *Hoplias malabaricus*; (C) scanning electron micrograph of the spore of *T. desaequalis*; (D) spore of *Myxobolus cultus* from the cartilage of the goldfish *Carassius auratus*; (E) triactinomyxon spore of *Myxobolus cultus* from the intestinal epithelium of the tubificid oligochaete *Branchiura sowerbyi*; (F) detail of the upper portion of the triactinomyxon spore *of M. cultus* showing the polar capsules. ((A) Modified from Cone & Raesly, 1995, with permission, *Canadian Journal of Aquatic Sciences*, **52** (Supplement 1), 7–12; (B, C) modified from Azevedo & Matos, 1996, with permission, *Journal of Parasitology*, **82**, 288–291; (D, E, F) modified from Yokohama *et al.*, 1995, with permission, *Journal of Parasitology*, **81**, 446–451.)

Box 9.1 *Myxobolus cerebralis* and whirling disease

If you are serious about fly-fishing, you are probably familiar with the tragic consequences of whirling disease. The disease affects mainly young salmonids with as yet uncalcified skeletons and where cartilage is destroyed or consumed by the trophozoites of *Myxobolus cerebralis* (see section 9.3). When the parasite attacks the cartilaginous tissue of the auditory and equilibrium organs, it interferes with coordination and equilibrium, the fish loses its sense of balance and tumbles erratically (or whirls) as it attempts to feed. As the disease progresses, the parasite invades the vertebral column, mainly that section from the anus to the tail, and prevents normal ossification, causing the tail and trunk to curve. Simultaneously, the sympathetic nerves that control the melanocytes in that region become impaired, producing a permanent dark coloration of the caudal region known as 'black tail'.

Malformation of the skeleton is often seen among older fishes that manage to overcome and survive the infection. These malformations include retraction of the operculum, curvature of the vertebral column, permanently open or twisted jaws, and deformed heads. In reality, however, very few individuals are likely to survive because the lack of coordination, distinctive coloration, and skeletal deformities are serious handicaps that greatly increase the risk of predation in natural habitats.

As with many other pathogens and parasites, *M. cerebralis* is a good example of what sometimes happens when organisms colonize new hosts or habitats. *Myxobolus cerebralis* was originally endemic to brown trout in central Europe and southeast Asia, where it causes no visible symptoms to its natural host. The disease was first reported in central Europe at the turn of the century soon after the introduction of the rainbow trout to Europe. In North America, the disease was first observed in 1956, probably after being introduced with imported frozen trout from Europe. Presently, *M. cerebralis* has a near-cosmopolitan distribution, causing high mortalities and significant losses in farm- and hatchery-reared brook and rainbow trout. Other salmonids, such as coho and chinook salmon, also are affected.

The disease had not been of great significance in natural bodies of water in North America until it was detected in wild rainbow trout in the Madison River of Montana in late 1994. Because of whirling disease, the wild rainbow trout population in this river has plummeted from about 2100 fish per km in 1990 to about 200 fish per km in 1994. Current efforts aimed at controlling the expansion and even eliminating this parasite have focused on the alternate invertebrate host in the life cycle, the aquatic oligochaete *Tubifex tubifex*. Some individuals of *T. tubifex* seem to be resistant to infection by *M. cerebralis*. If resistant tubificids could displace susceptible ones in nature, the life cycle of *M. cerebralis* would be interrupted, and this should be enough to control the disease. An alternative approach, also under study, is directed at the temperature-dependent release of infective spores from the tubificids. At 15 °C, the tubificids release huge numbers of spores into the water, but at 5 °C, very few spores are released. Generally, rainbow trout (the most susceptible species to infection) hatch in May when the temperatures are warm and water is filled with infective spores. If earlier spawning in rainbow trout could be induced through artificial selection, then the synchrony of trout hatching and spore production could be disrupted and parasite transmission reduced.

9.4 Ctenophora: the comb jellies

Ctenophorans or comb jellies are almost exclusively free-swimming marine organisms that resemble cnidarian medusae. Unlike cnidarians, however, ctenophorans have a complete digestive system, lack cnidocytes, have comb-like plates, and do not have the alternation of generations, with polyp and medusa stages, of cnidarians. However, like some of the cnidarian species, the larva of at least one species is parasitic, whereas the adults are free living. The larva of *Gastrodes parasiticum*, for example, bores into the mantle of the planktonic tunicate *Salpa fusiformis* where it develops into an adult. This adult stage leaves the host, becomes free-living, matures, and reproduces, producing new larvae to repeat the cycle.

9.5 Mesozoa

Mesozoa is a generic name given to the phyla Orthonectida and Dicyemida. Both phyla include only about 60 species, all are parasitic in a variety of marine invertebrates. The phylogenetic position of the mesozoans is uncertain and many doubt that the two phyla are even related. One hypothesis regarding their origins suggests that mesozoans are living relicts of an ancient group of metazoans that have not advanced beyond the cell aggregate stage. Another hypothesis proposes that mesozoans evolved their simple form by degenerating from more complex ancestors such as the Platyhelminthes (flatworms). Developmental as well as phylogenetic studies using 18S ribosomal RNA of dicyemids seem to support the

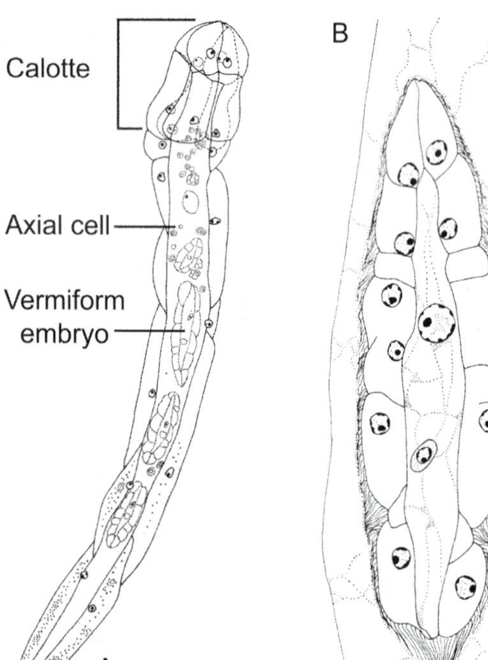

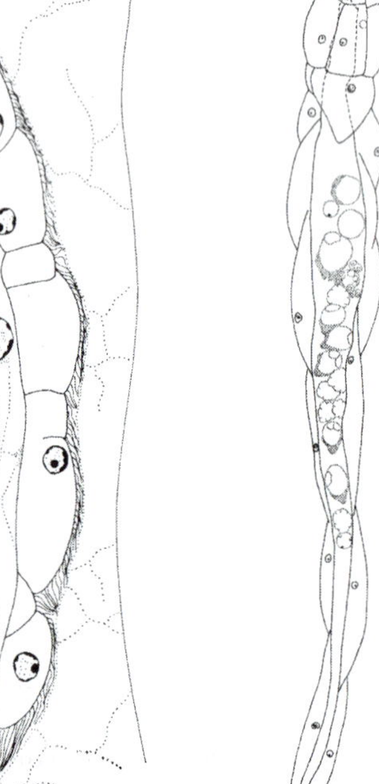

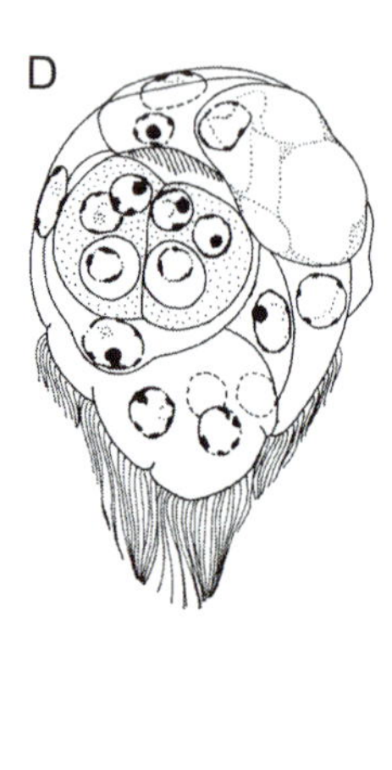

Fig. 9.5 Some developmental stages of *Dicyemennea kaikouriensis* (cilia not shown) from the kidneys of *Octopus maorum*. (A) Nematogen; (B) vermiform embryo within the axial cell of the nematogen in (A); (C) rhombogen; (D) infusoriform larva. (Modified from Short & Hochberg, 1969, with permission, *Journal of Parasitology*, **55**, 583–596.)

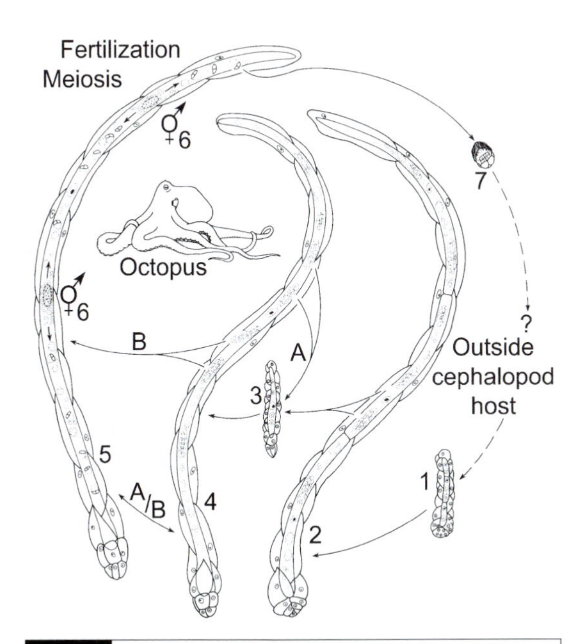

Fig. 9.6 Generalized life cycle of a dicyemid mesozoan in *Octopus* sp. (1) Larval stem nematogen; (2) stem nematogen; (3) vermiform embryo; (4) nematogen; (5) rhombogen; (6) infusorigen; (7) infusoriform larva released from parent. (A) Low density of parasites; (B) high density of parasites. (From Hochberg, 1982, with permission, *Malacologia*, **23**, 121–134.)

latter hypothesis, suggesting that dicyemids may share a common ancestry with flatworms or that they were derived from a more complex, free-living metazoan (Furuya *et al.*, 1992, 1994; Katayama *et al.*, 1995).

DICYEMIDA (OR RHOMBOZOA)

Members of the Dicyemida (=Rhombozoa) live inside the kidneys of squids, cuttlefishes, and octopods. The development of dicyemids inside the host involves many stages, of which the nematogen and rhombogen stages are the most common. Their bodies are composed of a polar cap, or **calotte**, in the anterior end and a trunk formed by one or more large axial cells surrounded by a single layer of ciliated somatic cells (Fig. 9.5). They are cylindrical, elongated, and measure between 0.5 and 9 mm in length.

The life cycle of dicyemids is still incompletely known despite intensive study (Fig. 9.6). It is not known how the cephalopod becomes infected. The earliest developmental stage, found in juvenile cephalopods, is a ciliated stage called the **larval stem nematogen** (Fig. 9.6). This larva grows to become a **stem nematogen** (Figs. 9.5A, 9.6). As development continues, **vermiform embryos** are produced asexually inside the axial cells of the stem nematogen (Figs. 9.5B, 9.6). The vermiform embryos then escape the body of the stem nematogen and attach to the kidney tissues of the host. These vermiform embryos develop and grow into **nematogens** (Fig. 9.6). These nematogens, in turn, give rise to more generations of asexually produced vermiform embryos resulting in a massive build-up of nematogens that fills the kidney of the cephalopod. When the population density of nematogens reaches a critical level (option B in Fig. 9.6), or when the cephalopods mature sexually (or become older), the nematogens transform into **rhombogens** (Figs. 9.5C, 9.6). It is not known which one of these two factors, or maybe a combination of them, is responsible for the transformation from nematogen to rhombogen. The axial cell of the rhombogen produces agametes that divide to become a mass of reproductive cells called an **infusorigen**. The infusorigen produces male and female gametes that fuse to form zygotes. The zygotes become **infusoriform larvae** (Figs. 9.5D, 9.6) that escape the parent rhombogen and leave the host with the urine (Hochberg, 1982). Anatomically, the infusoriform larva is the most complex stage of the life cycle and is the only one that can survive in seawater. It is not known what happens to the infusoriform larva once it leaves the host. It may be infective to the next cephalopod host, or it may require an additional host.

Nematogens inside the kidney (Fig. 9.7) feed by consuming particulate and dissolved nutrients from the host urine via phagocytosis or pinocytosis; they do not appear to damage the host. Because the cephalopod urine is mostly anoxic, dicyemids living in the kidney are anaerobes that derive their energy from lactate fermentation. Infusoriform larvae that escape from the host, in contrast, are probably metabolically aerobic.

ORTHONECTIDA

Orthonectids are much smaller than dicyemids and rarely reach more than 0.3 mm. They parasitize the gonads and body cavities of various marine invertebrates including brittle stars,

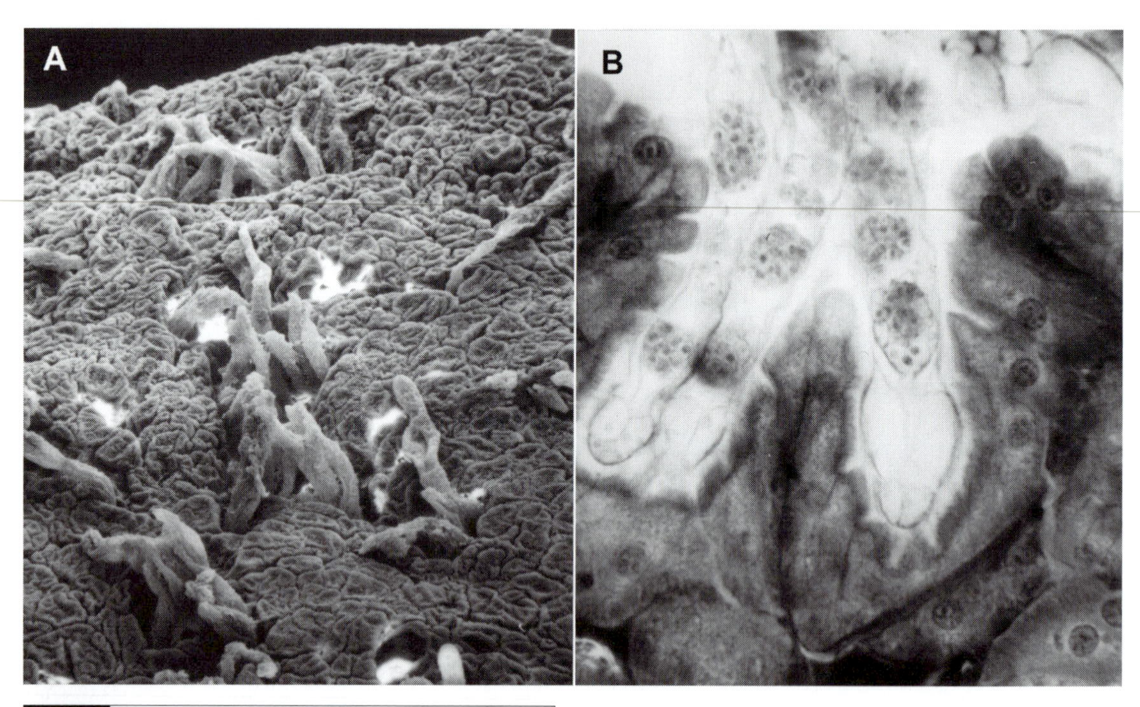

Fig. 9.7 *Dicyema* sp. in the renal appendages of *Octopus* sp. (A) Scanning electron micrograph; (B) section through the renal appendages. (From Hochberg, F. G. [1990]. In *Diseases of Animals*, Vol. 3, ed. O. Kinne, pp. 21–227. Hamburg: Biologische Anstalt Helgoland.)

nemerteans, polychaetes, turbellarians, and bivalves (Fig. 9.8). Most species are known from the Mediterranean Sea or the Atlantic Coast of Europe. Only a few species have been found in the Atlantic Coast of North America or in the Pacific. Their life cycle also includes an alternation of sexual and asexual generations. Asexual reproduction takes place in an amoeboid **plasmodium** inside the host and gives rise to free-swimming, sexual individuals. These sexually mature forms are released from the host simultaneously. Sperm from male individuals penetrate the bodies of females and fertilize their eggs. Cleavage of the eggs results in a ciliated larva that is released from the female. This larva seeks and infects a new host by entering through the host's reproductive openings. Inside the new host, the larva becomes a plasmodium, which is a multinucleate amoeboid stage that feeds on host tissues, often castrating the host in the process. The plasmodium reproduces asexually, forming new cell masses that dissociate and spread the parasite to other parts of the host. These asexually produced stages develop into the free-swimming, sexually mature individuals that leave the host and reproduce sexually.

Rhopalura ophiocomae is probably the best-known orthonectid to date (Fig. 9.8). It lives in the gonadal tissues of brittle stars castrating its host. It is relatively common in European waters; in Roscoff, France, up to 6% of the hosts examined, the ophiuroid *Amphipholis squamata*, are parasitized. The same orthonectid has been found in the same host on the Pacific coast of North America, but apparently is not as common (Kozloff, 1969).

9.6 Nemertea: the ribbon worms

Most nemerteans are intertidal marine animals with soft, flat, unsegmented bodies, complete digestive system, closed circulatory system, and simple reproductive structures. The most distinctive feature of the group is the presence of a remarkable protrusible **proboscis** used both in food capture and defense. Although most nemerteans are marine predators or scavengers, a few are parasites, or commensals, or both, of marine invertebrates. Species of *Malacobdella* are found in the mantle cavity of clams and snails, but they

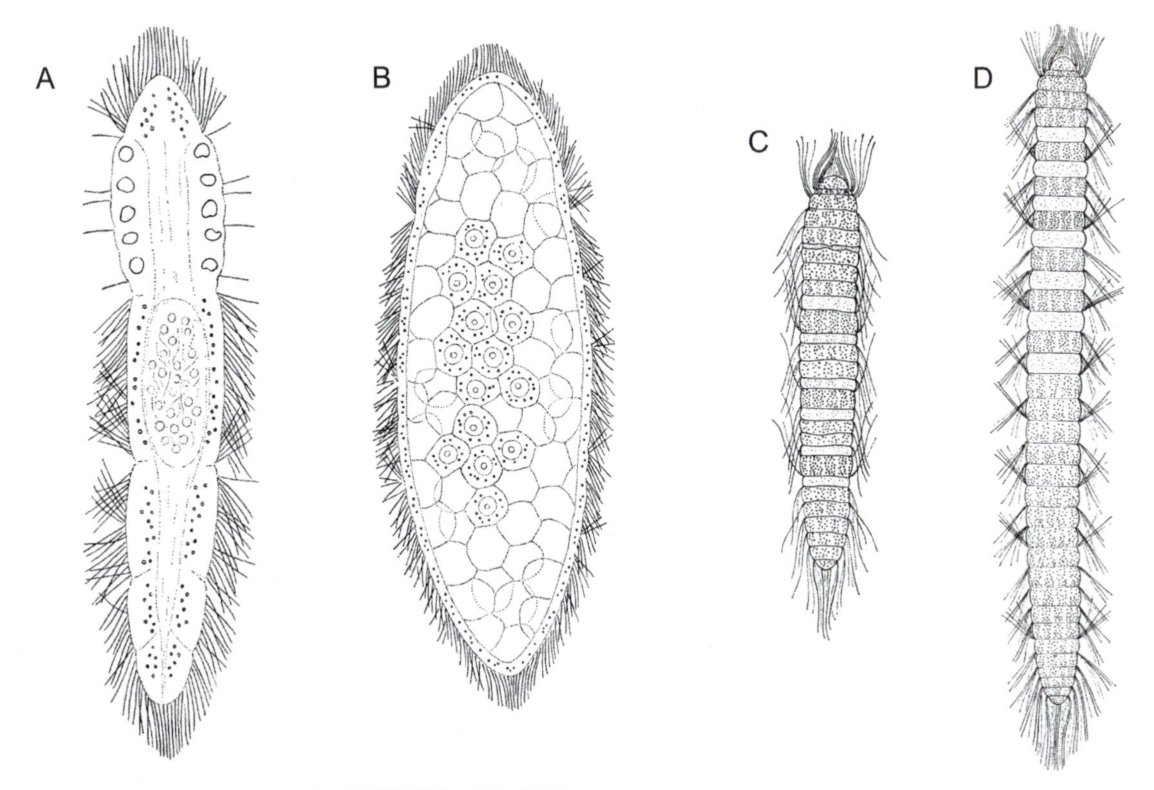

Fig. 9.8 Orthonectids. (A and B) Male and female *Rhopalura ophiocomae* from the ophiuroid *Amphipholis squamata;* (C and D) male and female *Ciliocincta sabelariae* from the polychaete *Sabellaria cementarium.* ((A, B) From Kosloff, 1969, with permission, *Journal of Parasitology,* **55**, 171–195; (C, D) from Kosloff, 1965, with permission, *Journal of Parasitology,* **51**, 37–44.)

seem to be commensals rather than parasites. Detailed studies of *M. grossa* from bivalves in Europe indicate that the nemerteans have no measurable effect on the hosts and that their diet consists of free-living organisms (Gibson, 1968; Gibson & Jennings, 1969). *Uchidana parasita,* however, also lives in the mantle cavity of clams in Japan, but feeds extensively on the gill tissues of its hosts (Iwata, 1966).

Species of *Carcinonemertes* are considered to be specialized egg predators or brood parasites of crabs (Fig. 9.9). Although their ecological impact is that of a predator because they kill individual embryos, they do not kill their host. They do, however, affect their host's fitness in terms of offspring production. During their life cycles, the nemertean planktonic larvae settle on both juvenile and adult crabs. Sexually immature nemerteans are found in the gills or arthrodial membranes of the host, particularly between the legs, in the joints, and eye sockets, where they feed on dissolved organic material leaked by the crab (Crowe *et al.*, 1982). When the female crab oviposits her eggs for brooding under her abdomen, the juvenile nemerteans migrate to the egg mass where they feed on developing crab embryos and yolk from within the eggs. There, they grow, reach sexual maturity, and reproduce, laying their eggs among the eggs of the crab host (Fig. 9.10). After reproduction, the nemerteans move back to the gills of the crab.

This group of nemerteans may have a significant impact on the population dynamics of their crustacean hosts. Epizootics of *Carcinonemertes errans* in the Dungeness crab *Cancer magister*, along the Pacific Coast of North America, have been implicated in the collapse of this commercial crab fishery. For example, up to 99% of Dungeness crabs on the California coast are infected with *C. errans*, and each crab may harbor as many as 100 000 worms that consume about 70 eggs per worm during the brooding season. Heavily

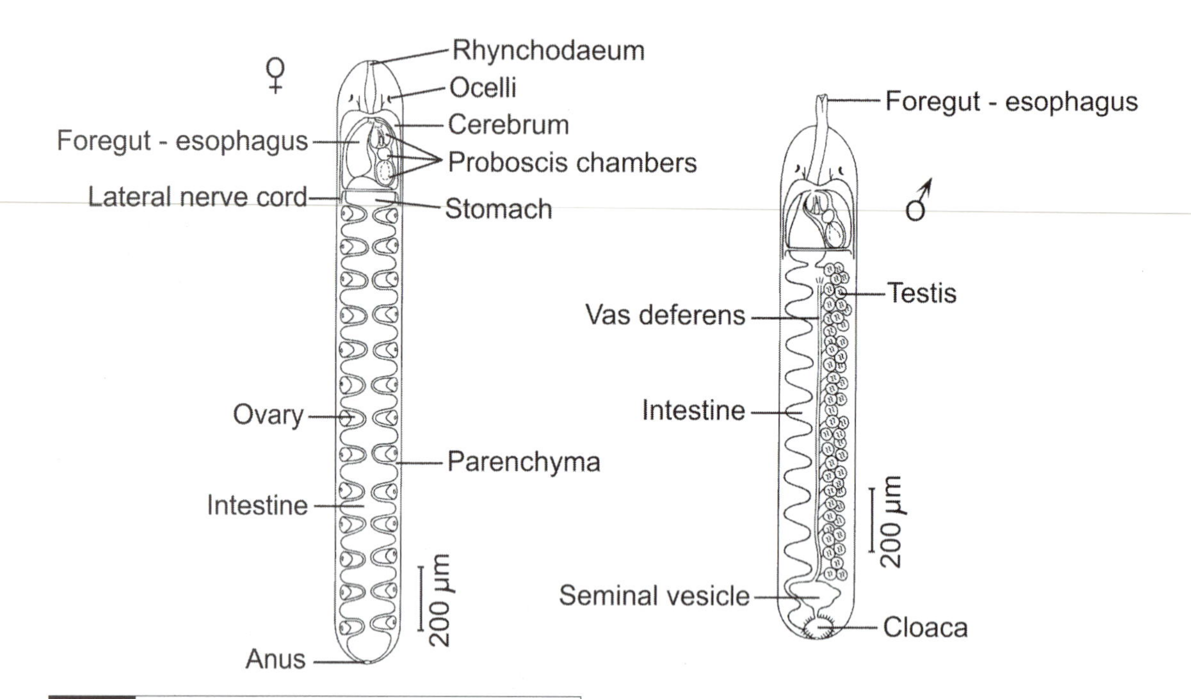

Fig. 9.9 Drawing of female and male *Carcinonemertes regicides* from the red king crab *Paralithodes camtschatica*. The foregut of the male is everted for comparison. Also, the reproductive system of the male is shown on the right half of the body, while the digestive system is figured on the left half. (Modified from Shields *et al.*, 1989, with permission, *Canadian Journal of Zoology*, **67**, 923–930.)

infected crabs may lose more than 50% of their eggs (Wickham, 1979, 1980, 1986). Significant infections by other nemertean species also have been reported for red king crabs and American lobsters (Wickham, 1986; Kuris *et al.*, 1991). For recent reviews on the biology of nemerteans associated with crustacean hosts, see Kuris (1993) and McDermott & Gibson (1993).

9.7 Nematomorpha: the horsehair worms

Adult nematomorphs are free-living in aquatic habitats, but the larvae are obligatory parasites, mainly of arthropods. Nematomorphs are long (up to 1 m in length), cylindrical, dioecious worms, with a non-functional digestive system. Adult males and females are found in shallow puddles, marshes, streams, and ponds (Fig. 9.11). Copulation takes place in the summer and several million fertilized eggs, in long gelatinous strings, are released into the water or among plant roots along the shore (Fig. 9.12). Males die after copulation and females die later, after laying their eggs. A larva with an armed proboscis hatches from the egg (Fig. 9.13).

It is not yet clear how the larva, after emerging from the egg, reaches its definitive host. One possibility, involving a direct cycle, is that the larva actively seeks and penetrates a definitive host. It is also possible that the definitive host might become infected by consuming a larva that has encysted in the environment. A second scenario suggests that the definitive host becomes infected by consuming a larva encysted in an intermediate host.

In general, after the larva hatches from the egg, it infects an arthropod host living in the water or along the water's edge. If the life cycle is indirect, the larva encysts in the intermediate host (Fig. 9.14). When this host is ingested by a suitable definitive host, the larva excysts, bores through the intestine, and enters the host's hemocoel. Inside the arthropod definitive host, the larva develops and, just prior to becoming sexually mature, it leaves the host, but only when the host is near water. Experimental studies in the life

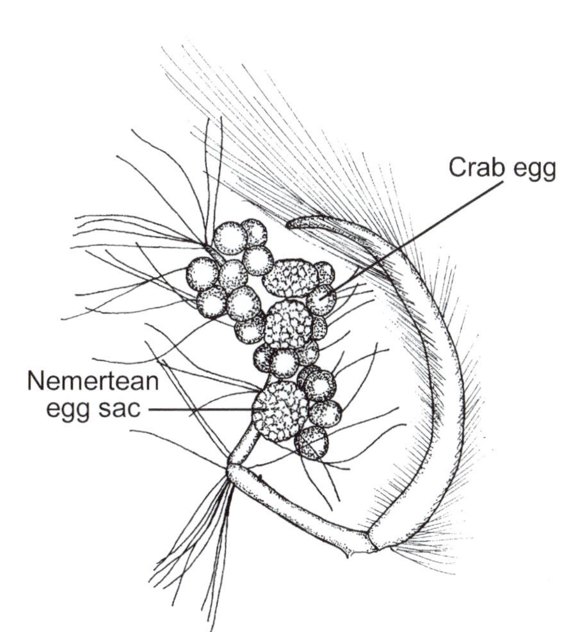

Fig. 9.10 Second abdominal appendage of a female crab *Pinnixa chaetopterana*, infected with *Carcinonemertes pinnotheridophila*, showing the nemertean egg sacs attached to the appendage's hairs along with the crab eggs. (Modified from McDermott, J. J. & Gibson, R. [1993] *Carcinonemertes pinnotheridophila* sp. nov. (Nemertea, Enopla, Carcinonemertidae) from the branchial chambers of *Pinnixa chaetopterana* (Crustacea, Decapoda, Pinnotheridae): description, incidence and biological relationships with the host. *Hydrobiologia*, **266**, 57–80. Reprinted with kind permission from Kluwer Academic Publishers.)

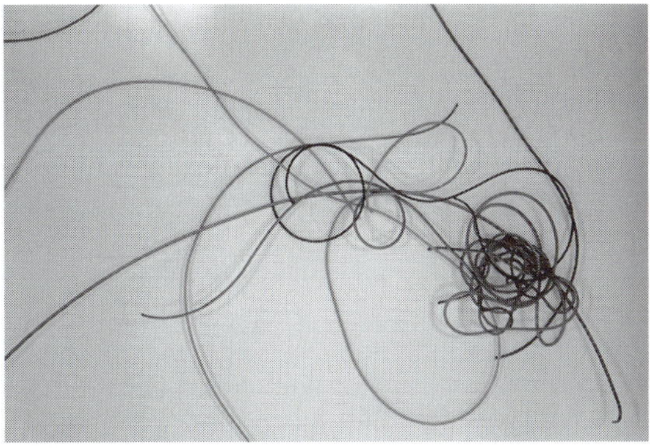

Fig. 9.11 Free-living adults of the nematomorph *Gordius robustus*. (Photograph courtesy of Ben Hanelt, University of Nebraska.)

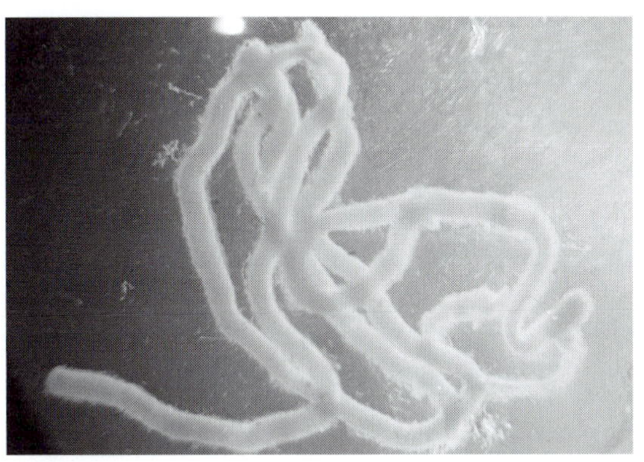

Fig. 9.12 Egg sacs of the nematomorph *Gordius robustus*, in the water. (Photograph courtesy of Ben Hanelt, University of Nebraska.)

cycle of *Gordius robustus*, a common species in North America, indicate that an intermediate host is required for infection of the definitive host (Fig. 9.14) (Hanelt & Janovy, 1999).

It is not clear if the larva manipulates the host's behavior and entices it to move towards the water when mature, or if it just waits until the host nears water by its own means. Poinar (1991) suggests that the host may be 'driven' to water when the parasite completes its development and is ready to emerge, but no experimental data are available to support such an hypothesis. Nematomorphs attain sexual maturity shortly after leaving the host and reaching the water.

Beetles, cockroaches, crickets, grasshoppers, centipedes, millipedes, and leeches are common hosts for nematomorphs. Although the free-living adult nematomorph does not feed and relies on stored energy reserves (mainly glycogen), the larva secretes digestive enzymes through its tegument and absorbs host nutrients directly through the body. In New Zealand, *Euchordodes nigromaculatus* parasitizes several species of orthopterans. A quantitative study of this host–parasite association (Poulin, 1995) revealed that only female crickets were infected, with each normally harboring only a single horsehair worm. One female cricket, however, had two worms, but they were smaller than average. This study raises some interesting questions about the mechanism involved in host selection and the potential for intraspecific competition.

Species of *Nectonema* are exclusively marine. Adults are free-living, pelagic, and their larvae parasitize crustaceans such as crabs and shrimps

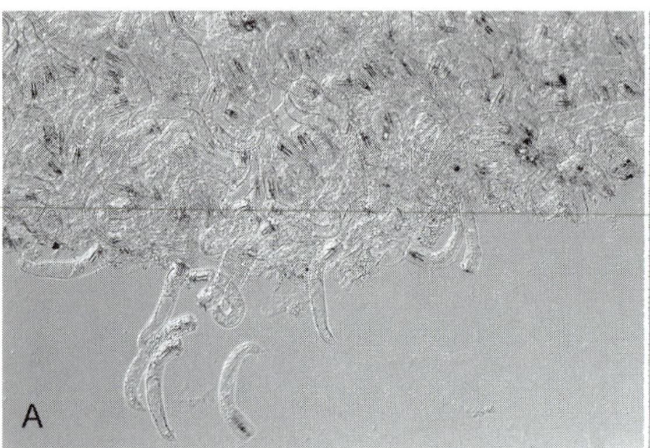

Fig. 9.13 (A) Mass of larvae emerging from egg sacs of the nematomorph *Gordius robustus*; (B) larva of *Gordius robustus*. (Photographs courtesy of Ben Hanelt, University of Nebraska.)

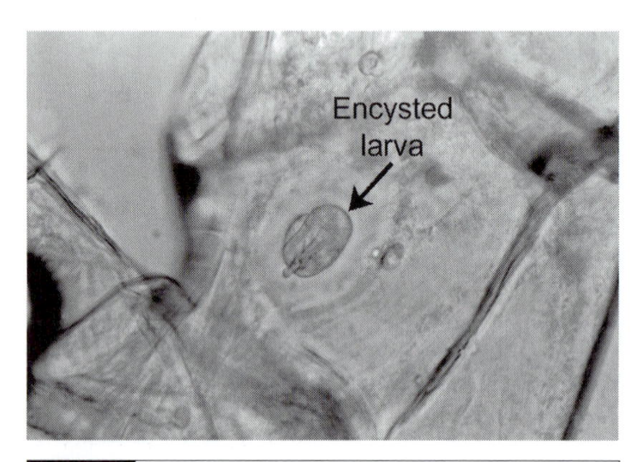

Fig. 9.14 Walking appendage of the freshwater amphipod *Hyalella azteca* with an encysted larva of the nematomorph *Gordius robustus*. (Photograph courtesy of Ben Hanelt, University of Nebraska.)

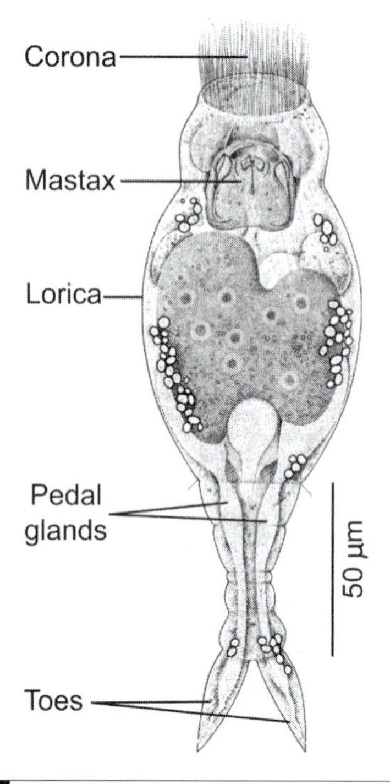

Fig. 9.15 The rotifer *Proales paguri* from the gills of a hermit crab. (Modified from Thane-Fenchel, 1966, with permission, *Ophelia*, **3**, 93–97.)

(Nielsen, 1969). *Nectonema agile,* for example, is found in the hemocoel of shrimps along the northeastern coast of North America where it reaches at least 10 cm in length. The ovaries of infected females shrink and become opaque. Although the host is not killed, its reproductive capabilities are severely constrained. A similar effect on the host was observed by Hanelt & Janovy (1999), where crickets infected with *G. robustus* lacked gonads and fat bodies.

9.8 Rotifera: the wheel animals

Rotifers are small (less than 1 mm long), mostly free-living, and abundant in fresh water, although a few species are marine and a few others terrestrial. Only a few species are parasitic. Rotifers are dioecious, have a characteristic ciliary organ, or **corona**, in the anterior end of the body, a pharynx armed with movable jaws called a **mastax**, and a thickened cuticle that forms a conspicuous encasement known as the lorica. The posterior end of the body, or foot, contains pedal glands that produce an adhesive substance for temporary attachment. In the parasitic species, the corona is reduced, and either the foot, or the mastax, is modified as an attachment organ. They are called

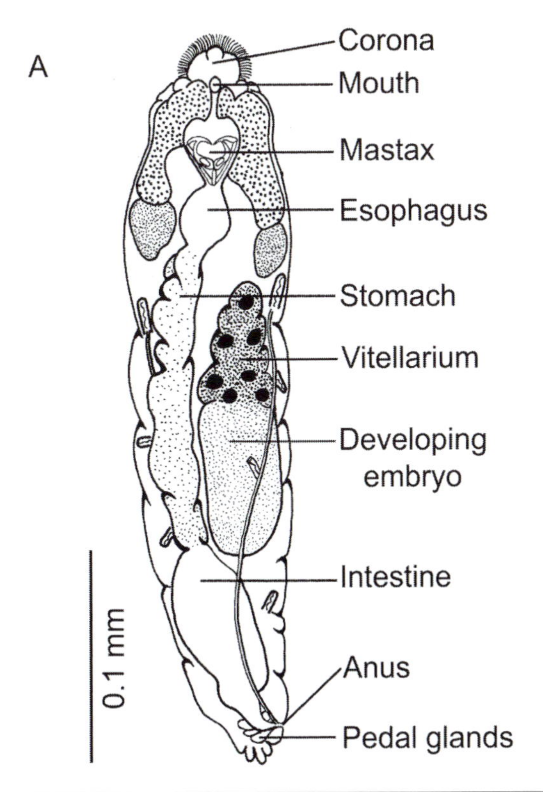

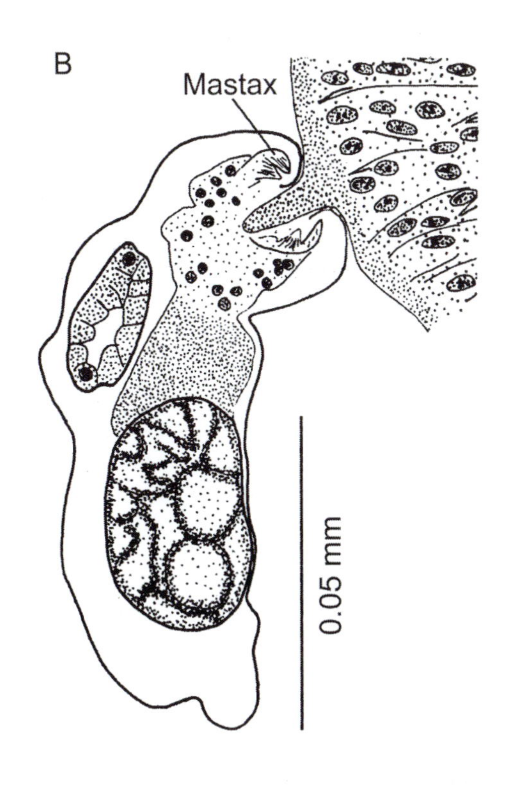

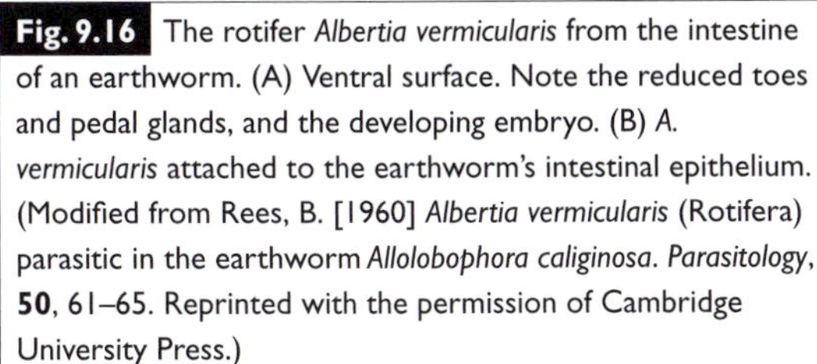
Fig. 9.16 The rotifer *Albertia vermicularis* from the intestine of an earthworm. (A) Ventral surface. Note the reduced toes and pedal glands, and the developing embryo. (B) *A. vermicularis* attached to the earthworm's intestinal epithelium. (Modified from Rees, B. [1960] *Albertia vermicularis* (Rotifera) parasitic in the earthworm *Allolobophora caliginosa*. *Parasitology*, **50**, 61–65. Reprinted with the permission of Cambridge University Press.)

wheel animals because, when the cilia of the corona are beating, the corona resembles a set of spinning wheels.

Proales paguri is a marine rotifer that lives on the gills of hermit crabs and feeds on epithelial cells from the gills of the host (Fig. 9.15). The rotifer measures about 0.2 mm and its pedal glands are unusually large, an adaptation to parasitism. They adhere to the gills and suck the gill epithelium with their buccal cavity. Up to 40 *P. paguri* can be found on a single host; their prevalence increases with the size of the crab, probably because larger crabs molt less often, providing a more stable and persistent environment for colonization (Thane-Fenchel, 1966). Other species of *Proales* parasitize heliozoan protozoans, feeding on the host's protoplasm; some infect freshwater snails, feeding on their eggs.

Some species of *Albertia* are endoparasites in the intestine of annelids, both terrestrial and marine. *Albertia vermicularis,* for example, is a parasite of the intestine and body cavity of earthworms (Fig. 9.16) (Rees, 1960). It attaches to the host's intestinal epithelium by pinching off a small papilla of the intestinal mucosa with its mastax (Fig. 9.16B), in a manner similar to that of the intestinal nematodes *Uncinaria* and *Ancylostoma*. Because attachment to the host is accomplished by the mastax, the foot and pedal glands of *A. vermicularis* are greatly reduced. These animals are viviparous and embryos can often be seen inside the large individuals (Fig. 9.16A).

9.9 Annelida: polychaetes, earthworms, and leeches

The annelids are segmented, soft-bodied worms, found in the oceans, freshwater, and moist soils; they range in length from 1 mm to 3 m. Annelids are metameric coelomates, with a complete digestive system and metameric nervous, excretory, and circulatory systems. Generally, three major groups

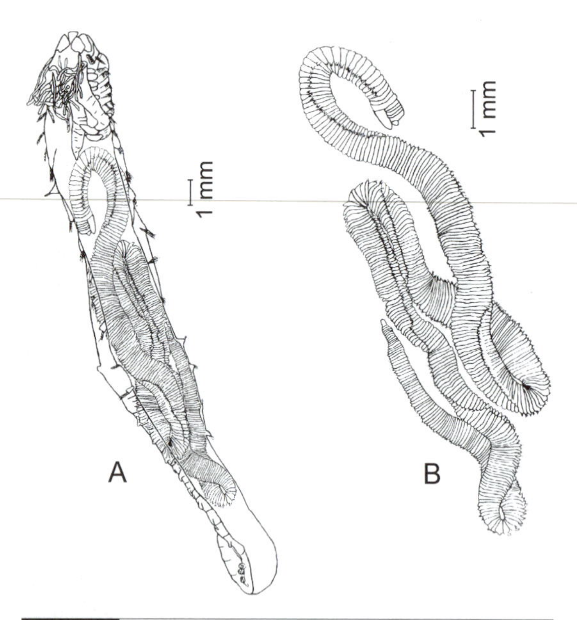

Fig. 9.17 (A) The parasitic polychaete *Labrorostratus zaragozensis* within the body cavity of its host, the polychaete *Terebellides californica*; (B) the parasitic polychaete *L. zaragozensis* outside the host. (From Hernández-Alcántara & Solís-Weiss, 1998, with permission, *Journal of Parasitology*, **84**, 978–982.)

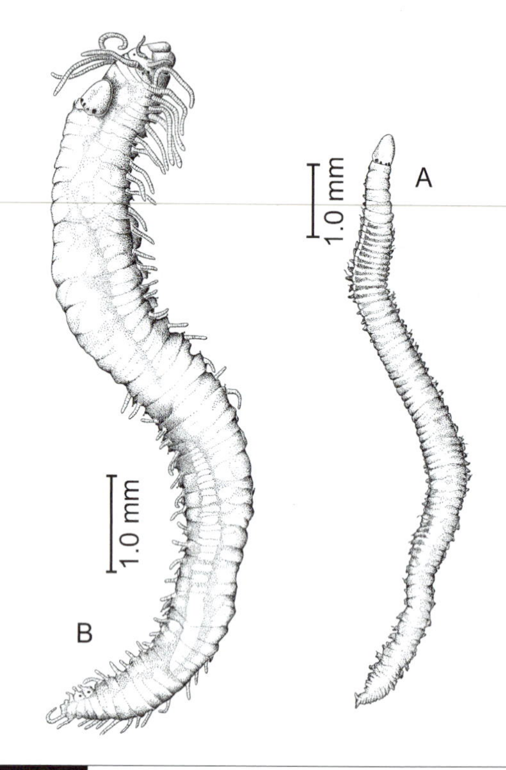

Fig. 9.18 (A) The parasitic polychaete *Labrororostratus luteus*, a parasite of the polychaete *Haplosyllis spongicola*; (B) the parasitic polychaete shown inside the body of the host. Note that the head of the parasite is protruding on the anterior, dorsal side of the host; usually the parasite is completely enclosed in the host's body. (From Uebelacker, 1978, with permission, *Journal of Parasitology*, **64**, 151–154.)

are recognized, i.e., Polychaeta, Oligochaeta, and Hirudinea, all of which have some parasitic species.

POLYCHAETA

Polychaetes are mostly marine, segmented organisms, characterized by a well-defined head region, and by the presence on each segment of a pair of paddle-like appendages bearing setae, called **parapodia**. The parapodia are used for locomotion, for building tubes (in the tube- or burrow-dwelling forms) and in moving water around the organism. Many species in the families Histriobdellidae and Ichthyotomidae are only found in the branchial chambers of lobsters and crabs, but it is not clear if they are cleaning symbionts or true parasites. The ichthyotomid *Ichthyotomus sanguinarius*, however, is a blood-sucking ectoparasite of marine eels and other fishes. The polychaete attaches to the host by burrowing in a hole near the fins and proceeds to suck blood from the wound with the help of an anticoagulant secreted by the salivary glands.

Most endoparasitic polychaetes belong to the Oenonidae, which also includes free-living forms. These polychaetes are found, at least during part of their lives, within the body cavity of other polychaetes, and rarely in echiurans and bivalves. Little is known about their biology, but it seems that they enter their host at an early stage, grow, and then leave the host before reaching reproductive maturity (Pettibone, 1957). The endoparasitic forms do not show much tendency toward specialization when compared with the free-living species, suggesting that parasitism is a secondary and rather recent development among these polychaetes. Moreover, parasitism in this group is not usually accompanied by a reduction in size, and the size of the host and parasite species are very similar. Parasites such as *Labrorostratus zaragozensis*, for example, are much longer than their hosts (Fig. 9.17) (Hernández-Alcántara & Solís-Weiss, 1998). The oenonid *Labrorostratus luteus* is found in

Fig. 9.19 The leech *Stibarobdella macrothela* on the anal fin of a large bull shark. Note the round area where the large posterior sucker had attached to the fin. (From Overstreet, 1978, with permission from the Mississippi–Alabama Sea Grant Consortium.)

the body cavity of another polychaete, the syllid *Haplosyllis spongicola* that lives in sponges near Grand Bahama Island (Fig. 9.18). Each syllid host is parasitized by a single polychaete, which is totally enclosed within the host and occupies most of its body cavity. The individual depicted in Fig. 9.18B, with the head protruding through the anterior dorsal wall of the host, may be an anomaly due to accidental tearing, or may also be the beginning of the parasite emergence from the host (Uebelacker, 1978).

The oenonid *Pholadiphila turnerae* is rather unique. Unlike other oenonids, this species lives in the mantle cavity of deep-sea (3600 m), wood-boring bivalves. Even though *P. turnerae* is not an endoparasite, its mouthparts and setae are reduced in a manner similar to other endoparasitic polychaetes. Likely food sources for this species include gill filaments, the mucus coating the filaments, or even partially digested material within the host's wood-storing cecum (Dean, 1992).

Unlike the poorly modified oenodids, the myzostomes, sometimes placed in a separate group, are highly modified. Their body is oval and flattened, with only five pairs of parapodia present on the ventral side. All myzostomes are either commensal or parasitic on, or in, echinoderm hosts, mainly crinoids (feather stars). Some species burrow into their hosts and form cysts within which a pair of myzostomes is found.

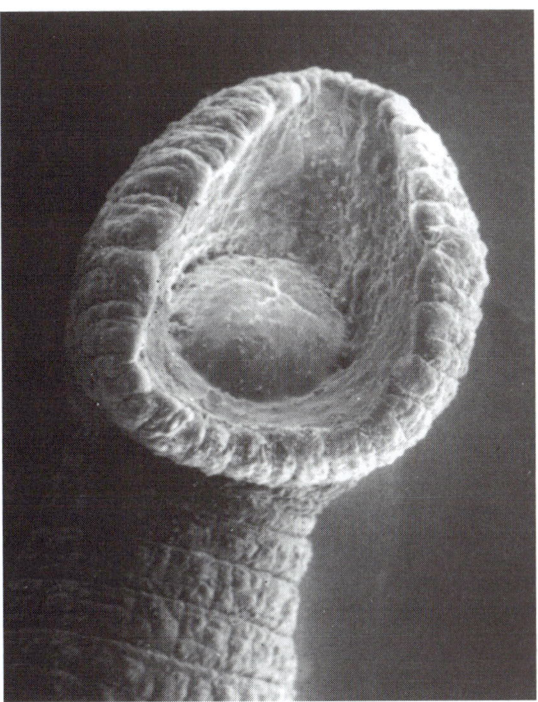

Fig. 9.20 Anterior sucker of *Johanssonia arctica*, a common leech on fishes of the Atlantic Northwest. (Photograph courtesy of David Cone, Saint Mary's University.)

Myzostomum pulvinar is commonly found in the intestine of crinoids, whereas *Protomyzostomun polynephris* lives in the coelomic cavity of brittle stars where it feeds on the genital organs, causing partial host castration. Interestingly, myzostomes, together with parasitic snails, are the oldest known fossil parasites. Galls produced by myzostomes have been found in the arms of crinoids from the Silurian and Devonian periods.

OLIGOCHAETA

Unlike polychaetes, oligochaetes, such as the common earthworms, have no parapodia, relatively few and inconspicuous setae, and a reduced head. They have, however, a prominent glandular zone, called the **clitellum**, that secretes a cocoon to house and protect the eggs. Several oligochaete species are associated with other oligochaetes, molluscs, crustaceans, fishes, and amphibians but, in most cases, their relationship seems to be more commensalistic than parasitic. Branchiobdellidans, for example, are obligate symbionts of Holarctic freshwater crustaceans such as

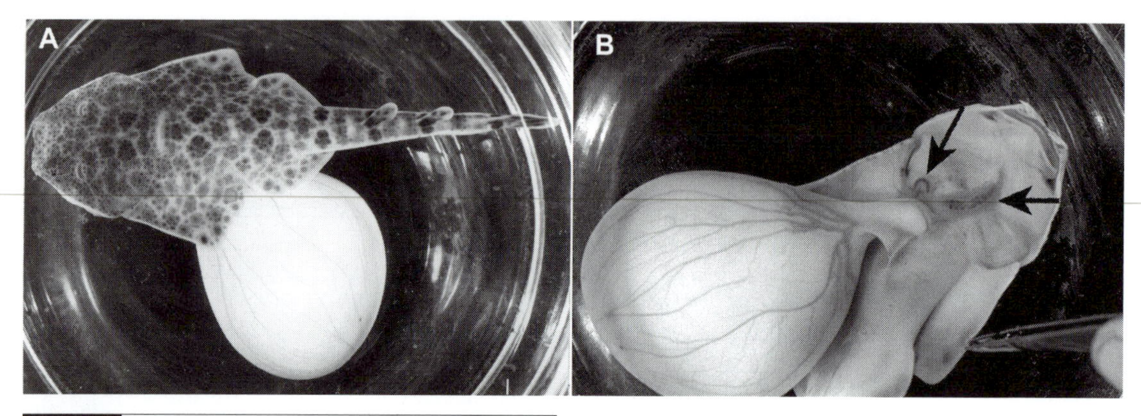

Fig. 9.21 The leech *Branchelion lobata* from embryos of the Pacific angel shark *Squatina californica*. (A) Dorsal view of an angel shark embryo; (B) ventral surface of the embryo with two leeches attached (see arrows). (From Moser & Anderson, 1976, with permission, *Canadian Journal of Zoology*, **55**, 759–760.)

crayfishes, crabs, shrimps, and isopods. Some species are ectoparasites on the gills of these crustaceans, whereas others live on the outer surface of the exoskeleton and graze on accumulated organic debris and microorganisms. All species, however, seem to deposit their egg-containing cocoons in their host's body (Holt, 1986). Species of *Acanthobdella*, oligochaetes that sometimes are considered to be primitive leeches, are true parasites on the fins of freshwater fishes. Species of *Dero* live in the ureters of toads and tree frogs (Harman, 1971).

HIRUDINEA

The hirudineans or leeches have elongated, dorsoventrally flattened bodies and, like oligochaetes, have a clitellum, and lack parapodia and head appendages. Some of the body segments at both extremities have been modified to form suckers, used both for locomotion and feeding. The majority of leeches are blood-sucking ectoparasites on a variety of hosts, but about 25% of the species are predators of small invertebrates. Most predatory leeches feed on worms, snails, and insect larvae. Their prey is either swallowed whole or the leech may simply suck the soft parts.

Most blood-sucking leeches attack vertebrate hosts, although some species feed only on invertebrates such as oligochaetes, snails, crustaceans, and insects. Depending on the concept of parasitism adopted (see Chapter 1) some leeches can be treated as parasites and others as micropredators. For purely simplistic reasons, they all will be called parasites with the understanding that this is a very loose use of the term. Most parasitic leeches in the Glossiphonidae feed on amphibians, turtles, snakes, alligators, and crocodiles, but species of *Theromyzon* attach to the nares and nasal cavities of aquatic birds and have been known to kill young waterfowl. Piscicolid leeches are common ectoparasites of both marine and freshwater fishes, including sharks and rays (Figs. 9.19, 9.20). Their body is cylindrical, ranging from a few mm to more than 30 mm; during their life cycle, they feed on their fish host, but leave, temporarily, to deposit their cocoons on rocks, oyster beds, or, more commonly, benthic crustaceans. The role of the crustaceans in the life cycle of piscicolid leeches is to provide a substrate for cocoon deposition and an alternative means of dispersal. *Johanssonia arctica* (Fig. 9.20), for example, feeds on the blood of various marine fishes, including cod and plaice, along the Atlantic Coast of North America, then deposits its egg-containing cocoon on the spider crab *Chionoecetes opilio*. It takes between 6 to 8 months for the eggs to hatch, depending on the ambient temperature, at which time the newly emerged leeches seek their fish host and proceed to feed (Khan, 1982).

Many species of freshwater and terrestrial leeches parasitize mammals, including humans. Adults of *Hirudo medicinalis*, the European medicinal leech, feed primarily on warm-blooded animals. Interestingly, the feeding habits of this leech, and of many others, vary with development. Young leeches begin consuming insect blood, later they move to frogs and fishes, and finally to homeothermic animals. *Hirudo medicinalis* normally sucks blood from the skin and, occasionally, from the nasal membranes or lining of the buccal cavity.

Although most leeches are considered to be ectoparasites, some species are found in unusual places. The marine piscicolid *Branchelion lobata* is normally found in the spiracles on the head and the opening of the cloaca of the Pacific angel shark. However, Moser & Anderson (1977) found several individuals of *B. lobata* on the ventral surface of embryos inside the uterus of this shark (Fig. 9.21). Given the rough surface of the skin of the host, it seems logical to find these parasites in softer external areas such as the cloaca and spiracles. It is likely that the leeches found in the embryos moved through the cloaca, entered the uterus, and attached to the embryos. *Limnatis africana* is a freshwater species that can enter the body of its hosts through the vagina and urethra, usually when bathing, and causes severe bleeding. It seeks humans, canines, and monkeys, and is common in parts of Africa and Asia.

Feeding is not a simple task for blood-sucking leeches. Three different types of pharyngeal glands aid and facilitate the feeding process. One gland secretes a local anesthetic, another causes vasodilatation of the host's capillaries, and a third produces **hirudin**, an anticoagulant that induces blood to flow freely from the host to the leech intestine. Following ingestion of a blood meal, water is extracted, and the blood is digested very slowly, probably because the gut secretes only a small number of enzymes, mainly exopeptidases. Although leeches lack important digestive enzymes such as amylases, lipases, and endopeptidases, they have an important symbiotic bacterial flora that aids in the breakdown of high-molecular-weight proteins, fats, and carbohydrates. Many blood-sucking leeches feed infrequently, but when they do, they can ingest an enormous amount of blood. For example, during a single feeding period, a terrestrial leech such as *Haemalipsa* may ingest ten times its own weight, and *Hirudo* two to five times its own weight. Then, it may take the leeches up to 200 days to completely digest this large meal.

The duration of attachment to the host is highly variable in leeches. Some species of *Hirudo*, for example, attach to their host only when feeding. Others, such as most piscicolids of fishes, are relatively permanent, leaving their hosts only when breeding and just long enough to attach their cocoons to other organisms.

Leeches are vectors for a number of protozoans and other microorganisms. Piscicolid leeches, for example, transmit many trypanosomes of marine and freshwater fishes, and species of *Helobdella* are vectors for amphibian trypanosomes. Their success as vectors depends on their host specificity and feeding habits. Thus, leech species that are host specific, and that remain permanently attached to their hosts, are poor vectors. However, leeches such as *Johanssonia arctica*, found on many marine fishes in Newfoundland, are extremely good vectors because they feed on a variety of fishes, depending on their availability. *Johanssonia arctica* appears to be the main vector of several blood protozoa, including trypanosomes, hemogregarines, piroplasms, and trypanoplasms, parasitic on fishes in the Newfoundland area (Khan, 1982).

9.10 Echiura: the spoonworms

Echiurans are a small group of vermiform, bottom-dwelling, marine organisms. Echiurans are dioecious and some species, such as *Bonellia viridis*, display extreme sexual dimorphism in which males are diminutive and live as parasites within the female's kidneys. Females are about 80 mm long, whereas the dwarf males are only 1 to 3 mm in length. This is an interesting example of intraspecific parasitism with a unique method of sexual determination. If a larva of *B. viridis* develops independently in the environment, it becomes a female. But if a larva comes in contact with an adult female, a hormone secreted by the proboscis of the female induces the larva to develop into a dwarf male (Jaccarini *et al.*, 1983). After a few days in the proboscis, the larva moves into the female esophagus and eventually into the kidneys, where it matures into a male in just 2 weeks. Females, on the other hand, take over a year to reach sexual maturity. A single female may be parasitized by up to 20 males.

9.11 Mollusca: mussels, clams, snails, squids, and the like

Molluscs, together with arthropods, are among the most conspicuous and familiar of all invertebrate animals and include such organisms as chitons, clams and oysters, squids and octopods, snails and slugs, and many others. Molluscs have well-developed digestive, circulatory, respiratory, excretory, and nervous systems. Their body is divided into three distinct regions, the head–foot, the visceral mass, and the mantle, all of which is normally contained within a hard, protective shell. Most molluscs are large, conspicuous, and well known, but the ones that have become parasitic are very small and poorly known. Molluscs are divided into eight classes, i.e., Bivalvia, Gastropoda, Cephalopoda, Polyplacophora, Scaphopoda, Aplacophora, Caudofoveata, and Monoplacophora, but only the bivalves and gastropods have parasitic species.

BIVALVIA

Some adult bivalves have evolved commensal relationships with a number of invertebrates such as echinoderms, crustaceans, annelids, and other molluscs, but a few species, such as *Entovalva mirabilis*, which lives in the gut of sea cucumbers, have become true parasites. Although parasitism by adult bivalves is rare, larval stages of many freshwater clams and mussels are modified for a parasitic existence. Freshwater bivalves incubate their eggs in their gills, where an egg will develop into a small, shelled larva called a **glochidium** (Figs. 9.22, 9.23). This larva is highly modified and, when mature, leaves the gills of the bivalve and becomes parasitic on fishes (Fig. 9.23. 9.24). Some species swim by clapping the shell valves together and disperse into the surrounding water while seeking a suitable host, whereas others lie in wait for a host, at the bottom, with the valves open wide. Within a few days, the glochidium attaches to the gills, fins, or body surface of a suitable fish host by clamping the valves together on the host tissue (Fig. 9.24). Host epithelium grows around the glochidium, forming a cyst where development continues. During this obligate ectoparasitic period that may last 10 to 30 days in species of the freshwater mussel *Anodonta*, phagocytic cells from the glochidium feed on host tissues, many of the larval structures disappear, and the adult organs begin development. Subsequently, the glochidium breaks out of the cyst, sinks to the substratum, and gradually transforms into an adult bivalve. The glochidia of some bivalve species require a specific species of fish, whereas others have a wide range of potential hosts. Adult fish may carry as many as 3000 glochidia, without signs of harm. Young fish fry, however, may die from secondary infections.

GASTROPODA

Parasitism among adult gastropods seems to be more common than among bivalves. Gastropods manifest an interesting adaptive series, ranging from free-living forms, to epizoic and ectoparasitic, to highly modified endoparasites that are, literally, nothing more than sacs of reproductive organs. Many of the anatomical changes evolved by parasitic gastropods are associated with the buccal region and digestive system. For example, snails in the Pyramidellidae have chitinous jaws, stylets, and a pumping pharynx for sucking blood from bivalve, polychaete, echinoderm, and ascidian hosts. Moreover, the characteristic radula of gastropods is not even present (Fretter & Graham, 1949; Robertson & Mau-Lastovicka, 1979). The pyramidellid *Boonea impressa* parasitizes the oyster *Crassostrea virginica* and significantly reduces the growth rate of its host. The presence of large numbers of *B. impressa* may have a significant impact on the health, population structure, and productivity of commercial oyster beds.

Most parasitic gastropods belong to a group of small, related families that live on or within echinoderms. Species of *Thyca*, infecting starfish in the Indian Ocean, retain the typical gastropod morphology, but their foot is greatly reduced and the oral area is enlarged to form a large sucker

Fig. 9.22 Frontal section through the brooding demibranch of the female freshwater clam, *Hyridella depressa*, showing glochidia. (Photograph courtesy of Ronald V. Dimock, Wake Forest University.)

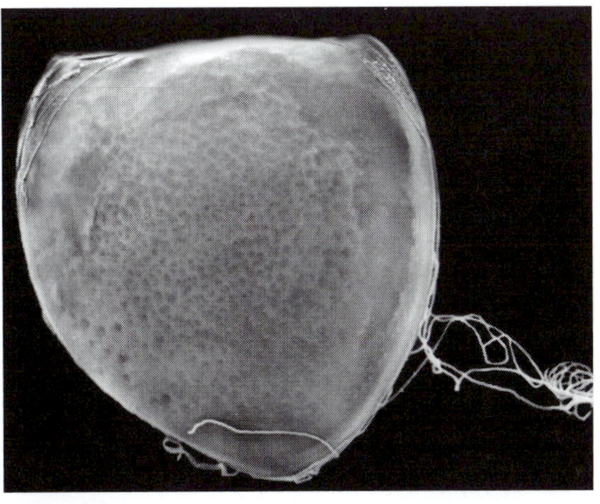

Fig. 9.23 Glochidium of the freshwater mussel *Pyganodon cataracta* with its larval thread. (Photograph courtesy of Richard Tankersley, Florida Institute of Technology)

used for attachment. They feed by means of a proboscis that pierces the host's tegument and sucks tissue fluids. *Thyca crystallina* parasitizes the starfish *Linckia laevigata* in Indonesian waters where prevalence may exceed 60%. Interestingly, prevalence is correlated with the degree of water movement, and the position of the snails on the surface of the starfish is related to the size of the snail. The smallest snails are found on the aboral surface and side of the arms, whereas the largest are on the oral surface, facing the starfish's mouth. Small snails are able to move very slowly over the surface of the starfish, but the larger ones, reaching up to 12 mm, seem to be permanently attached. As in many other parasitic organisms, the males of this species are dwarfs, living under the female's mantle (Elder, 1979).

Species of *Stylifer* have very thin shells and their foot is reduced. They are permanently embedded in cysts or galls in the surface of starfishes and, in most cases, a male and a female pair are found in each gall. The capsule-like gall remains open to the exterior by means of a small pore allowing water to flow in and out of the gall for respiration and the removal of waste material from the snails. Feeding in these species is accomplished by means of a long proboscis that reaches the host's coelom and absorbs body fluids. Some species, such as *S. linckiae*, a parasite of the starfish *Linckia multiflora* in Hawaii, can prevent an infected starfish from self-cutting and releasing its arms (autotomy), thus thwarting the host in its effort to eliminate the parasite-infested arms (Davis, 1967).

Finally, the entoconchids are highly modified endoparasites of sea cucumbers. They have lost all the external characteristics of their free-living snail ancestors, including the shell, and only their larval stages show any indication of their affinities. Females of *Enteroxenous* are reduced to vermiform-like morphs measuring 100 to 150 mm in length, whereas males are dwarfs embedded in the female as small clumps of testicular material (Lutzen, 1979).

An unusual snail among parasitic gastropods is *Cancellaria cooperi*, a blood-sucking snail found in the body surface of the California electric ray *Torpedo californica*. These snails are normally buried in the sand but, with appropriate chemostimulation, they move rapidly to locate their host. In laboratory trials, the snails are able to move 14 cm/min; in the field, snail trails of up to 24 m have been detected as the snail approaches its host. Electric rays appear to be convenient hosts because they normally remain on the substratum, partially

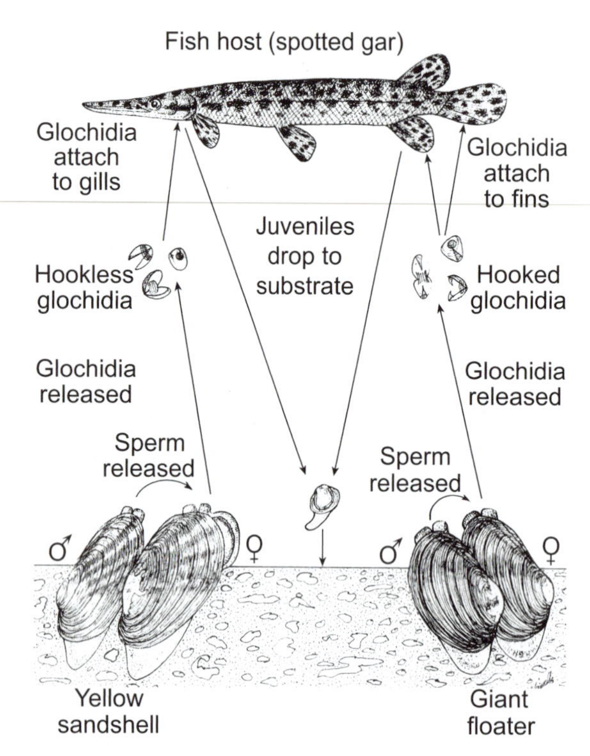

Fig. 9.24 Generalized life cycle of two freshwater mussels, the yellow sandshell *Lampsilis teres* and the giant floater *Pyganodon grandis*. The glochidia of the yellow sandshell lack hooks and attach to the gills of the spotted gar *Lepisosteus oculatus*, whereas the glochidia of the giant floater have hooks and attach to the fins of the fish. (Drawing courtesy of Robert G. Howells, Texas Parks and Wildlife Department.)

buried, for long periods of time and show very little responsiveness to mechanical disturbances. Once the snail reaches the host, it makes a small cut with the radula on the ventral surface of the fish, inserts its proboscis, and proceeds to suck blood. The snails may also insert the proboscis into wounds, mouth, gill slits, and even the anus (O'Sullivan *et al.*, 1987).

Most parasitic gastropods parasitize echinoderms, which raises some interesting questions about the long-term nature of their association. Fossil gastropods, very similar to the present day *Thyca*, are known from crinoids and starfishes from the Silurian–Triassic era, indicating that the relationship between molluscs and echinoderms is an old one.

9.12 Chordata: from lampreys to bats

Parasitism, at least, 'conventional parasitism' (see Box 9.2) is a rare phenomenon among chordates, and vertebrates in particular. Moreover, many of the examples fall in the gray area between parasitism and micropredation (see Chapter 1). Most cases of parasitism, or near parasitism, are associated with fishes and are qualified as such based on size alone. Parasitic lampreys of the genus *Petromyzon*, for example, attach and suck the blood of fishes by means of a specialized buccal apparatus and the secretion of an anticoagulant known as **lamphredin**. According to some, lampreys are simply specialized, or 'prudent' predators; but because they do not always kill their 'prey', others consider them to be parasites. However, the most appropriate description is that of a micropredator. Survival of the host is common as evidenced by the presence of external scars left on fishes to which lampreys had once attached. Survival seems to depend on the host's weight, with heavier fishes having higher survival probabilities. Lampreys such as *Petromyzon marinus* show host selectivity based on the host's body mass. Thus, if all potential hosts are small, lampreys choose the heavier hosts; but, if all potential hosts exceed a minimum size threshold, then they are not selective. Apparently if the host body mass is above the threshold, the lamprey's energy intake from a blood meal is about the same regardless of the host size (Swink, 1991; Cochran & Jenkins, 1994).

Lampreys are not the only blood-feeding fishes. A group of small South American catfishes, the Trichomycteridae, enter the gill cavities of other fishes, attach to the fish host with the aid of spines in their opercula, and feed on gill filaments and blood. One species, especially feared in some parts of Brazil, is *Vandellia cirrhosa*, a tiny fish known locally as candiru (Fig. 9.25). The candiru is attracted to nitrogenous wastes such as those that are normally produced by fish gills, but may mistakenly enter the urinary opening of animals and humans bathing in rivers, with painful consequences. Indeed, the people who live in that region wear special sheaths made of palm fibers to protect themselves and avoid this most painful

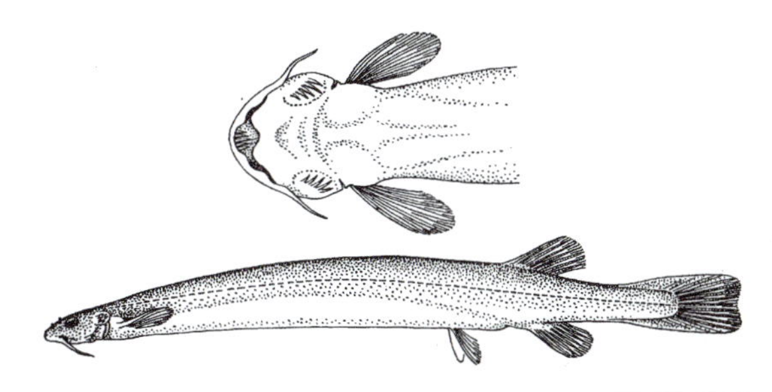

Fig. 9.25 The candiru *Vandellia cirrhosa*, a parasite of the gill cavity of freshwater fishes. It may mistakenly enter the urinary opening of mammals, including man. (From Norman & Greenwood, 1963, with permission, Ernest Benn Ltd, London.)

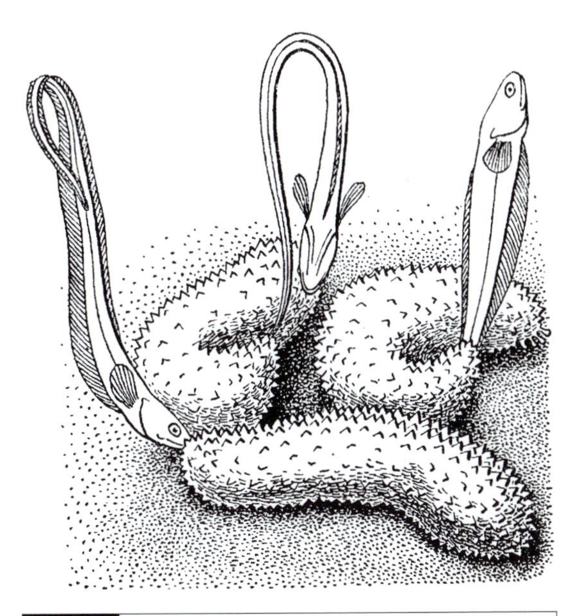

Fig. 9.26 The pearlfish *Carapus acus*, and its sea cucumber host. (From Norman & Greenwood, 1963, with permission, Ernest Benn Ltd, London.)

encounter. Removal of the fish can only be accomplished surgically because of the presence of recurved spines on the opercula.

Not all the fishes that fall in this gray area of parasitism are blood feeders. Most pearlfishes in the Carapidae live in a close association with some marine invertebrates. Some species live inside the body cavity of sea cucumbers (holothuroideans) where they feed on the gonads (Fig. 9.26). The host survives, but is unable to reproduce. The fish enters the sea cucumber through the anus,

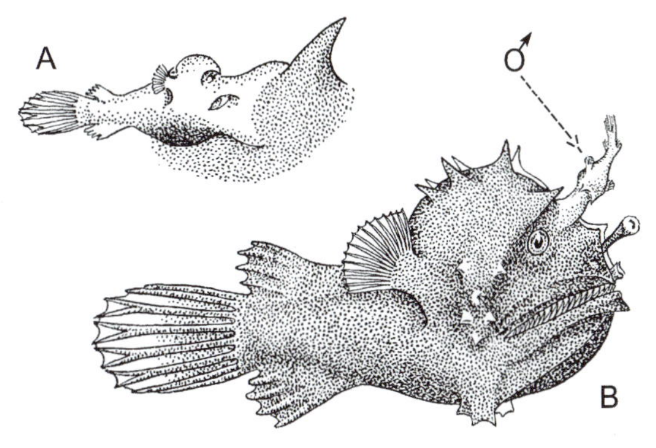

Fig. 9.27 (A) Detail of a dwarf male anglerfish *Photocorynus spiniceps*; (B) female anglerfish with a dwarf parasitic male. (From Norman & Greenwood, 1963, with permission, Ernest Benn Ltd, London.)

tail first, and wiggles backward until completely inside the host. Interestingly, the anus of the pearlfish is located close to the head, allowing the fish to void its waste without having to completely emerge from the holothuroidean host.

The presence of dwarf males, parasitic on females, is not restricted to invertebrates. In the deep reaches of the oceans, a region of perpetual darkness, life is constrained by two main factors, i.e., finding food and finding a mate. The deep-sea anglerfishes (Ceratioidei) have solved the feeding problem by modifying one of their fin rays into a fishing lure that usually bears a luminous flap or bulb at its tip; this acts as a lamp or beacon to attract unsuspecting prey. The mating problem also has been resolved in a clever manner, in some families, by the evolution of dwarf, parasitic males that attach permanently to the female with their jaws (Fig. 9.27). In most cases, the skin and blood vessels around the mouth of the male fuse with the skin of the female. The male, then, is totally dependent on the female for nourishment, and its body is so poorly developed that, in fact, it is no more than an attached gonad (Pietsch, 1976).

Among mammals, blood-sucking bats of the genus *Desmodus* are common in the American tropics and subtropics where they attack wild and domestic warm-blooded animals. When feeding, they insert their well-developed canine teeth into the flesh of the victim, rupture a large blood vessel, and proceed to lick the blood with their tongue. Unlike lampreys that sometimes cause mortality of

Box 9.2 Surrogate mothers and free childcare

Imagine being able to perpetuate your genes with neither the rigors, nor energetic costs, associated with raising your offspring. What a deal! Actually, this is precisely what happens in the symbiotic relationship known as 'brood parasitism', sometimes also called 'nest parasitism'. Brood parasites are birds that lay their eggs in the nest of a different species; once this is done, the brood parasites are, quite literally, finished with their offspring. Brood parasitism is widespread (obligate brood parasites are found in six families of birds) but not particularly common (only about 1% of all species of birds are obligate brood parasites). Some brood parasites are host specialists, meaning that they will only lay their eggs in the nests of one, or a few, species of hosts; others are host generalists, meaning that just about any nest will do, thank you! Also, some brood parasites lay eggs that mimic the eggs of the host (this trait would obviously be more likely in host specialists) whereas others lay eggs that may, or may not, show morphological differences from the host's eggs (a trait likely to be found in host generalists). Not surprisingly, host birds parasitized by brood parasite specialists tend to discriminate amongst the eggs in their nest and to reject, or remove, eggs dissimilar to their own. In contrast, host birds parasitized by generalist brood parasites are mostly non-discriminatory and will accept any eggs laid within their nests.

Brood parasitism, whether by specialists or generalists, is clearly a distinct energetic benefit to the parasite. But, what of the costs to the host? These can be extreme! For example, first elucidated by Aristotle is the fact that, when Old World cuckoos hatch from their eggs, they remove the host's offspring from the nest – kind of a parasitic sibling rivalry at its worst! More subtle, but still costly, is the situation when the parasitic nest mate does not remove the host's offspring physically but, instead, being larger and noisier, it outcompetes the other nest mates for food. The overall cost to host species due to brood parasitism can be large. For example, Kirtland's warblers are an endangered species in the United States. Part of the recovery plan for this species was to trap, and kill, brown-headed cowbirds that were parasitizing the nests of the warblers (Brewer, 1994). Paton (1992) notes that brood parasites tend to be more costly to birds nesting in edge habitats. Winfree (1999) provides an interesting discussion on why hosts tolerate nest parasitism. Indeed, nest parasitism is interesting in, and of, itself and it might prove useful in addressing some ideas on more conventional parasites. For example, 'From whom do nest parasites acquire their more conventional parasites?' Do they share the same parasites as their hosts? Do they have specialists?

their smaller prey, blood-sucking bats do not kill their prey. Their feeding strategy is very similar to the one used by leeches that detach from the host after feeding. Although hematophagous leeches are commonly referred to as parasites, hematophagous bats normally are described as specialized predators or micropredators. Vampire bats can consume up to 50 ml of blood in a single feeding. They normally return and 'visit' the same animal night after night, even if the victim belongs to a large herd. This peculiar behavior ensures a close relationship between the bat and the victim, making their relationship more akin to parasitism than to predation.

References

Andree, K. B., Gresoviac, S. J. & Hedrick, R. P. (1997) Small subunit ribosomal RNA sequences unite alternate myxosporean stages of *Myxobolus cerebralis*, the causative agent of whirling disease in salmonid fish. *Journal of Eukaryotic Microbiology*, **44**, 208–215.

Bartholomew, J. L., Whipple, M. J., Stevens, D. G. & Fryer, J. L. (1997) The life cycle of *Ceratomyxa shasta*, a myxosporean parasite of salmonids, requires a freshwater polychaete as an alternate host. *Journal of Parasitology*, **83**, 859–868.

Brewer, R. (1994) *The Science of Ecology*, 2nd edn. New York: Saunders College Publishing.

Cochran, P. A. & Jenkins, R. E. (1994) Small fishes as hosts for parasitic lampreys. *Copeia*, **1994**, 499–504.

Crowe, J., Crowe, L., Roe, P. & Wickham, D. E. (1982) Uptake of DOM by nemertean worms: association of worms with arthrodial membranes. *American Zoologist*, **22**, 671–682.

Davis, L. V. (1967) The suppression of autotomy in *Linckia multiflora* (Lamarck) by a parasitic gastropod, *Stylifer linckiae* Sarasin. *Veliger*, **9**, 343–346.

Dean, H. K. (1992) A new arabellid polychaete living in the mantle cavity of deep-sea wood boring bivalves (Family Pholadidae). *Proceedings of the Biological Society of Washington*, **105**, 224–232.

Elder, H. Y. (1979) Studies on the host parasite relationship between the parasitic prosobranch *Thyca crystallina* and the asteroid starfish *Linckia laevigata*. *Journal of Zoology*, **187**, 369–391.

Fretter, V. & Graham, A. (1949) The structure and mode of life of the Pyramidellidae, parasitic opisthobranchs. *Journal of the Marine Biological Association UK*, **28**, 493–532.

Furuya, H., Tsuneki, K. & Koshida, Y. (1992) Development of the infusoriform embryo of *Dicyema japonicum* (Mesozoa: Dicyemidae). *Biological Bulletin*, **183**, 248–257.

Furuya, H., Tsuneki, K. & Koshida, Y. (1994) The development of the vermiform embryos of two mesozoans, *Dicyema acuticephalum* and *Dicyema japonica*. *Zoolological Science*, **11**, 235–246.

Gibson, R. (1968) Studies on the biology of the entocommensal rhynchocoelan, *Malacobdella grossa*. *Journal of the Marine Biological Association UK*, **48**, 637–656.

Gibson, R. & Jennings, J. B. (1969) Observations on the diet, feeding mechanism, digestion and food reserves of the entocommensal rhynchocoelan, *Malacobdella grossa*. *Journal of the Marine Biological Association UK*, **49**, 17–32.

Hanelt, B. & Janovy, J. Jr. (1999) The life cycle of a horsehair worm, *Gordius robustus* (Nematomorpha: Gordioidea). *Journal of Parasitology*, **85**, 139–141.

Harman, W. J. (1971) A review of the subgenus *Allodero* (Oligochaeta: Naididae: *Dero*) with a description of *D. (A.) floridiana* n. sp. from *Bufo terrestris*. *Transactions of the American Microscopical Society*, **90**, 225–228.

Hernández-Alcántara, P. & Solís-Weiss, V. (1998) Parasitism among polychaetes: a rare case illustrated by a new species: *Labrororostratus zaragozensis*, n. sp. (Oenonidae) found in the Gulf of California, Mexico. *Journal of Parasitology*, **84**, 978–982.

Hochberg, F. G. (1982) The 'kidneys' of cephalopods: a unique habitat for parasites. *Malacologia*, **23**, 121–134.

Holt, P. C. (1986) Newly established families of the order Branchiobdellida (Annelida: Clitellata) with a synopsis of the genera. *Proceedings of the Biological Society of Washington*, **99**, 676–702.

Iwata, F. (1966) *Uchidana parasita* nov. gen. et nov. sp., a new parasitic nemertean from Japan with peculiar morphological characters. *Zoologischer Anzeiger*, **178**, 122–136.

Jaccarini, V., Agius, L., Schembri, P. J. & Rizzo, M. (1983) Sex determination and larval sexual interaction in *Bonellia viridis*. *Journal of Experimental Marine Biology and Ecology*, **66**, 25–40.

Katayama, T., Wada, H., Furuya, H., Satoh, N. & Yamamoto, M. (1995) Phylogenetic position of dicyemid mesozoa inferred from 18S rDNA sequences. *Biological Bulletin*, **189**, 81–90.

Khan, R. A. (1982) Biology of the marine piscicolid leech *Johanssonia arctica* (Johansson) from Newfoundland. *Proceedings of the Helminthological Society of Washington*, **49**, 266–278.

Kozloff, E. N. (1969) Morphology of the orthonectid *Rhopalura ophiocomae*. *Journal of Parasitology*, **55**, 171–195.

Kuris, A. M. (1993) Life cycles of nemerteans that are symbiotic egg predators of decapod Crustacea: adaptations to host life histories. *Hydrobiologia*, **266**, 1–14.

Kuris, A. M., Blau, S. F., Paul, A. J., Shields, J. D. & Wickham, D. E. (1991) Infestation by brood symbionts and their impact on egg mortality of the Red King Crab, *Paralithodes camtschatica*, in Alaska: geographic and temporal variation. *Canadian Journal of Fisheries and Aquatic Sciences*, **48**, 559–568.

Levine, N. D., Corliss, J. O., Cox, F. E. G., Deroux, G., Grain, J., Honigberg, B. M., Leedale, G. F., Loeblich,

A. R. III, Lom, J., Lynn, D., Merinfeld, E. G., Page, F. C., Poljansky, G., Sprague, V., Vavra, J. & Wallace, F. G. (1980) A newly revised classification of the protozoa. *Journal of Protozoology*, **27**, 37–58.

Lutzen, J. (1979) Studies on the life history of *Enteroxenous* Bonnevie, a gastropod endoparasitic in aspidochirote holothurians. *Ophelia*, **18**, 1–51.

Markiw, M. E. & Wolf, K. J. (1983) *Myxosoma cerebralis* (Myxozoa: Myxosporea) etiologic agent of salmonid whirling disease requires tubificid worms (Annelida: Oligochaeta) in its life cycle. *Journal of Protozoology*, **30**, 561–564.

McDermott, J. J. & R. Gibson. (1993) *Carcinonemertes pinnotheridophila* sp. nov. (Nemertea, Enopla, Carcinonemertidae) from the branchial chambers of *Pinnixa chaetopterana* (Crustacea, Decapoda, Pinnotheridae): description, incidence and biological relationships with the host. *Hydrobiologia*, **266**, 57–80.

Moser, M. & Anderson, S. (1977) An intrauterine leech infection: *Branchellion lobata* Moore, 1952 (Piscicolidae) in the Pacific angel shark (*Squatina californica*) from California. *Canadian Journal of Zoology*, **55**, 759–760.

Moser, M. & Taylor, S. (1978) Effects of the copepod *Cardiodectes medusaeus* on the lanternfish *Stenobrachius leucopsarus* with notes on hypercastration by the hydroid *Hydrichthys* sp. *Canadian Journal of Zoology*, **56**, 2372–2376.

Nielsen, S.-O. (1969) *Nectonema munidae* Brinkman (Nematomorpha) parasitizing *Munida tenuimana* G.O. Sars (Crust. Dec.) with notes on host parasite relations and new host species. *Sarsia*, **38**, 91–110.

O'Sullivan, J. B., McConnaughey, R. R. & Huber, M. E. (1987) A blood sucking snail: the Cooper's nutmeg, *Cancellaria cooperi* Gabb, parasitizes the California electric ray, *Torpedo californica* Ayres. *Biological Bulletin*, **172**, 362–366.

Paton, P. W. (1992) The effect of edge on avian nest success: how strong is the evidence? *Conservation Biology*, **8**, 17–26.

Pettibone, M. H. (1957) Endoparasitic polychaetous annelids of the family Arabellidae with descriptions of new species. *Biological Bulletin*, **113**, 170–187.

Pietsch, T. W. (1976) Dimorphism, parasitism, and sex: reproductive strategies among deep sea ceratoid anglerfishes. *Copeia*, **1976 (4)**, 781–793.

Poinar, G. O. Jr. (1991) Hairworm (Nematomorpha: Gordioidea) parasites of New Zealand wetas (Orthoptera: Stenopelmatidae). *Canadian Journal of Zoology*, **69**, 1592–1599.

Poulin, R. (1995) Hairworms (Nematomorpha: Gordioidea) infecting New Zealand short-horned grasshoppers (Orthoptera: Acridiidae). *Journal of Parasitology*, **81**, 121–122.

Raikova, E. V. (1994) Life cycle, cytology, and morphology of *Polypodium hydriforme*, a coelenterate parasite of the eggs of acipenseriform fishes. *Journal of Parasitology*, **80**, 1–22.

Rees, B. (1960) *Albertia vermicularis* (Rotifera) parasitic in the earthworm *Allolobophora caliginosa*. *Parasitology*, **50**, 61–66.

Robertson, R. & Mau-Lastovicka, T. (1979) The ectoparasitism of *Boonea* and *Fargoa* (Gastropoda: Pyramidellidae). *Biological Bulletin*, **157**, 320–330.

Rohde, K. (1993) *Ecology of Marine Parasites*, 2nd edn. Wallingford: CAB International.

Schlegel, M., Lom, L., Stechmann, A., Bernhard, D., Leipe, D., Dyková, I. & Sogin, M. L. (1996) Phylogenetic analysis of complete small subunit ribosomal RNA coding region of *Myxidium lieberkuehni*: evidence that Myxozoa are Metazoa and related to the Bilateria. *Archiv für Protistenkunde*, **147**, 1–9.

Siddall, M. E., Martin, D. S., Bridge, D., Desser, S. S. & Cone, D. L. (1995) The demise of a phylum of protists: phylogeny of Myxozoa and other parasitic Cnidaria. *Journal of Parasitology*, **81**, 961–967.

Smothers, J. F., Von Dohlen, C. D., Smith, L. H. Jr. & Spall, R. D. (1994) Molecular evidence that the myxozoan protists are metazoans. *Science*, **265**, 1719–1721.

Spaulding, J. G. (1972) The life cycle of *Peachia quinquecapitata*, an anemone parasitic on medusae during its larval development. *Biological Bulletin*, **143**, 440–453.

Swink, W. D. (1991) Host-size selection by parasitic sea lampreys. *Transactions of the American Fisheries Society*, **120**, 637–643.

Taylor, R. L. & Lott, M. (1978) Transmission of salmonid whirling disease by birds fed trout infected with *Myxosoma cerebralis*. *Journal of Protozoology*, **25**, 105–106.

Thane-Fenchel, A. (1966) *Proales paguri* sp. nov., a rotifer on the gills of the hermit crab *Pagurus bernhardus* (L.). *Ophelia*, **3**, 93–97.

Toft, C. A. (1991) An ecological perspective: the population and community consequences of parasitism. In *Parasite–Host Associations, Coexistence or Conflict?*, ed. C. A. Toft, A. Aeschlimann & L. Bolis, pp. 319–343. New York: Oxford University Press.

Uebelacker, J. M. (1978) A new parasitic polychaetous annelid (Arabellidae) from the Bahamas. *Journal of Parasitology*, **64**, 151–154.

Wickham, D. E. (1979) Predation by the nemertean *Carcinonemertes errans* on eggs of the Dungeness crab *Cancer magister*. *Marine Biology*, **55**, 45–53.

Wickham, D. E. (1980) Aspects of the life history of *Carcinonemertes errans* (Nemertea: Carcinonemertidae), an egg predator of the crab *Cancer magister*. *Biological Bulletin*, **159**, 247–257.

Wickham, D. E. (1986) Epizootic infestations by nemertean brood parasites on commercially important crustaceans. *Canadian Journal of Fisheries and Aquatic Sciences*, **43**, 2295–2302.

Winfree, R. (1999) Cuckoos, cowbirds and the persistence of brood parasitism. *TREE*, **14**, 338–343.

Chapter 10

Population concepts

10.1 Background

If the study of the ecology of parasitic organisms is considered in its broadest form, then it has a long history, extending back to the middle of the nineteenth century when the connection between cysticerci and adult taeniid cestodes was identified by Dujardin in 1845, when the life cycle of the canine tapeworm, *Taenia pisiformis*, was experimentally completed by Kuchenmeister in 1852, or when the life cycle of *Fasciola hepatica* was mostly elucidated by Leuckart and Thomas in 1882/3 (Box 1.1). These discoveries were matched, if not exceeded, by Sir Patrick Manson who described the life cycle of *Wuchereria bancrofti* in 1877 and by Ross who reported the life cycle of avian malaria in 1895; Sir Ronald Ross eventually was to receive one of the first Nobel Prizes in physiology in recognition of this exceptional research achievement.

Modern parasite ecology can be traced to V. A. Dogiel and his colleagues (Dogiel, 1964). Many of these early studies had a decided 'natural history' flavor, but they were nonetheless ecological in their scope and approach. The research trends in parasite ecology and epidemiology have continued to become more and more complex with time and the development of new technologies. However, it was not until Holmes (1961, 1962*a*, *b*) published his now classical studies on the interspecific competitive interactions between the rat tapeworm *Hymenolepis diminuta* and the acanthocephalan *Moniliformis dubius* that a quantitative perspective became solidly entrenched in the parasite literature. These papers also became the modern standard for parasite community ecology. Ten years later, Crofton (1971*a*, *b*) established a similar standard for a quantitative approach in the study of parasite population dynamics. MacDonald (1965) was the first to extensively employ mathematical modeling in the study of host–parasite relationships. Over the last 30–35 years, investigations on the ecology and epidemiology of parasitic organisms and their hosts have expanded significantly.

10.2 General definitions

Ecology can be considered within the context of two, quite different methodologies. **Autecology** is concerned with the study of individual organisms or species. The approaches taken here can be varied but, plainly, population biology is autecological in character. A population can be defined as a group of organisms of the same species occupying a given space in time and comprising a single gene pool. Each population can be described by several parameters. Some of these include birth rate, death rate, age distribution, biotic potential (reproductive potential), dispersion, growth pattern, and density. Investigations on the population biology of human parasites contribute to the discipline of **epidemiology**, whereas studies on the population biology of parasites in other animals are in the realm of **epizootiology**. **Synecology** deals collectively with groups of organisms of different species that live together. For synecology, the approach is at the community level. A **community** of organisms is defined as a group of populations of different species occupying a similar habitat or ecosystem. An **ecosystem** encompasses the community of organisms, plus their physical surroundings.

Most of these concepts were developed for free-living organisms, but there are certain problems when the concepts are applied to parasites. For example, do all members of a given parasite species within a single host constitute a population? Or, do all members of a given parasite species within all hosts in an ecosystem represent a population? Another difficulty with the concept as applied to parasites is that populations of free-living organisms increase in number through birth, or immigration, or both, whereas most adult helminth parasites in a host increase only through immigration, or **recruitment**. In an effort to clarify these issues, Esch *et al.* (1975*a*) proposed that parasites within a single host and those within all hosts in an ecosystem be separated and considered independently. They developed the concepts of the **infrapopulation** and **suprapopulation** to address these issues. An infrapopulation was described as one that includes all of the parasites of a single species in one host. A suprapopulation was defined to include all of the parasites of a given species, in all stages of development, within all hosts in an ecosystem. Margolis *et al.* (1982) reiterated this terminology and expanded on it.

Subsequently, Bush *et al.* (1997) modified the terminology and applied the term **component population** to describe all of the infrapopulations of a species of parasite within all hosts of a given species in an ecosystem. The numbers of studies on infrapopulation and component populations are substantial and far exceed those at the suprapopulation level. This is mainly because the logistics are far too complex in the latter case to cover all stages of a parasite's life cycle in a single ecosystem at the same time. Several investigations have been attempted on suprapopulations (Hairston, 1965; Dronen, 1978; Jarroll, 1979,1980), or at least partially attempted (Riggs *et al.*, 1987; Marcogliese & Esch, 1989), and these will be discussed subsequently. Since the development of this scheme for parasite population ecology, a similar hierarchical approach has been adopted by those working in the area of parasite community ecology (see Bush & Holmes, 1986; Holmes & Price, 1986; Esch *et al.*, 1990*a*; Bush *et al.*, 1993).

It should be noted here that a terminology issue for parasite ecology was precipitated by several of the earlier works and the initial resolution was provided by Margolis *et al.* (1982). However, with the increase in ecological research in the following years, the need to revisit the issue became apparent. Toward this end, Bush *et al.* (1997) re-examined the use of terms and concepts as they applied to the community ecology of parasitic organisms. They attempted to clarify some of the old questions and to raise some new issues as well. Students are advised to refer to both the Margolis *et al.* (1982) and Bush *et al.* (1997) papers, as they will be useful in providing a preamble to the terminology employed in ecological parasitology.

The conceptualization of parasite population structure can be viewed in terms of a nested hierarchy (Fig. 10.1). Consider, for example, the three-host life cycle of a hypothetical digenean, one that involves a snail as the first intermediate host, a fish as the second intermediate host, and a bird as the definitive host. For this parasite, active recruitment (solid lines in Fig. 10.1) is associated with free-swimming miracidia (asterisks) and cercariae (plus signs). The transmission of metacercariae to the definitive host is a passive process, almost always depending on predator–prey interactions (dashed line in the figure). A spatial aspect to the transmission process is introduced with the positioning of the symbols for miracidia and cercariae. The large box that circumscribes the system represents the suprapopulation; observe that it includes all of the component and infrapopulations, as well as the free-living stages. The large triangle represents the component population of parasites in snail hosts that, in turn, is composed of a series of individual snails (smaller triangles), each with its own infrapopulation of parasites. The same kind of nesting is associated with the large rectangle and the large circle. The latter represent component populations of metacercariae in fish second intermediate hosts and the rectangles are definitive hosts. Infrapopulations within individual fish are smaller circles and individual birds are the smaller rectangles. In considering

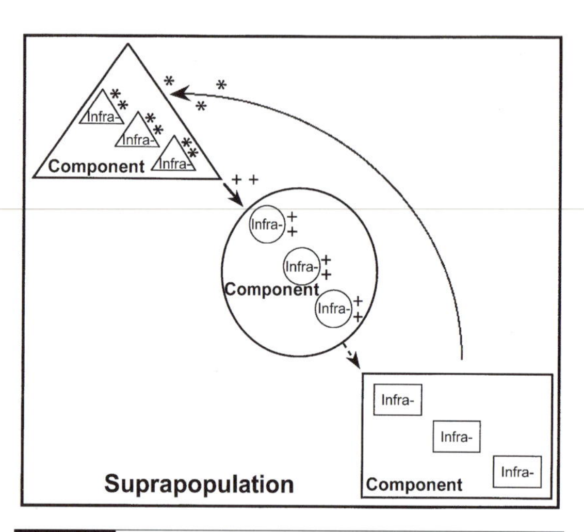

Fig. 10.1 A schematic representation for the hierarchical organization of parasite infra-, component, and suprapopulations. Whereas it represents a hypothetical digenean, the scheme applies to virtually any host–parasite system. The small asterisks represent miracidia in the vicinity of snails; the small triangles are infrapopulations within molluscs; the large triangle represents a component population of several molluscs. The small + signs are cercariae in the vicinity of the second intermediate host; the small circles are infrapopulations in second intermediate hosts. The large circle is a component population of metacercariae. The small rectangles are infrapopulations of the parasite in the definitive host; the large rectangle is the component population of the adult parasite. Solid lines represent active transmission and dashed lines are passive transmission. The large box circumscribing the entire figure represents the suprapopulation of parasites.

the suprapopulation as a nested hierarchy, the ideas of parasite flow, and of recruitment and turnover, can also be visualized. More importantly, even though the nature of the processes that impact on the dynamics of the system have not been illustrated, the idea of **gene flow** within the suprapopulation is also implicit.

10.3 Factors affecting parasite populations

The diagram in Fig. 10.1 is a simplified way of illustrating the idea of parasite transmission through a series of hosts within the framework of a suprapopulation. Parasite recruitment and turnover are, however, influenced by a range of factors both external and internal to the host (see also Chapter 13). Recruitment into a host can be passive, requiring no expenditure of energy by the parasite, or it can be active in which case an expenditure of energy is required. Table 10.1 provides an idea of the wide-ranging tactics that parasites have evolved in achieving recruitment success, by either active or passive means. The list is somewhat distorted, however, since the process of active recruitment is almost entirely restricted to parasites that require water to complete their life cycles. Exceptions here would include certain nematode species and, even then, their free-living larvae require special soil conditions (moisture, shade, etc.) in which to survive. The turnover of parasite infrapopulations will result from natural host and parasite mortality, inter- and intraspecific interactions, and from the effects of host immune responses. In some cases, the turnover processes will be directly affected by external as well as internal environmental factors.

10.3.1 External environmental factors

Free-living stages are directly affected by abiotic environmental factors. For helminth parasites, such stages would include, among others, miracidia and cercariae of digeneans, coracidia of many pseudophyllidean cestodes, and the rhabditiform and filariform larvae of some rhabditid and strongylid nematodes. The eggs of helminth parasites and the cysts of many parasitic protozoans are also affected by the external environment. If terrestrial, then photoperiod, temperature, and humidity, are but a few of the environmental factors that will affect their development, survivorship, etc. In freshwater habitats, temperature and water current are important elements in the biology of ectoparasites, e.g., copepods. In marine environments, salinity and temperature have been shown to affect the biology of ectoparasitic organisms. In terrestrial, freshwater (Jokela & Lively, 1995), and marine (Curtis, 1997) habitats, a prime feature affecting the risk of infection is the spatial distribution of infective agents and of hosts shedding infective agents.

Consider the host-finding capability of free-swimming miracidia and the combination of physical and chemical factors that influence this

Table 10.1 Examples of parasite species using passive or active recruitment strategies for infecting the intermediate (I) or definitive (D) hosts

Passive (stage involved – host type)	Active (stage involved – host type)
Protozoans	
Entamoeba histolytica (cyst–D)	*Trypanosoma cruzi* (metacyclic trypomastigote–D)
Giardia lamblia (cyst–D)	*Ichthyophthirius multifiliis* (tomite–D)
Plasmodium spp. (gametocyte–D)	
Cestodes	
Taenia pisiformis (cysticercus–D)	*Proteocephalus ambloplitis* (coracidium–I)
Moniezia expansa (egg–I)	*Diphyllobothrium latum* (coracidium–I)
Hymenolepis diminuta (cysticercoid–D)	*Ligula intestinalis* (coracidium–I)
Digeneans	
Clonorchis sinensis (egg–I)	*Schistosoma japonicum* (cercaria–D)
Uvulifer ambloplitis (metacercaria–D)	*Crepidostomum cooperi* (miracidium–I)
Dicrocoelium dendriticum (cercaria–I)	*Benedenia melleni* (oncomiracidium–D)
Alaria canis (mesocercaria–D)	*Paragonimus westermani* (miracidium–I)
Nematodes	
Ascaris lumbricoides (egg–D)	*Necator americanus* (filariform larva–D)
Wuchereria bancrofti (microfilaria–I)	*Strongyloides stercoralis* (filariform larva–D)
Pneumostrongylus tenuis (L_1 larva–I)	*Uncinaria lucasi* (filariform larva–D)

behavior (Kennedy, 1975; Christensen *et al.*, 1978; Callinan, 1979; Kearn, 1980; Esch, 1982; see also Chapter 13). Finding hosts involves light and gravity stimuli, insuring that the miracidia are spatially distributed in such a way as to enhance the probability of contact between the molluscan host and the parasite. Then, once the miracidia are swimming in the vicinity of their first intermediate hosts, they are further stimulated by appropriate chemical stimuli released from snails or clams, causing them to seek out and penetrate these hosts. The chemical stimuli in these cases are not complicated, consisting mainly of simple amino acids and fatty acids. The processes of host-finding by some digenean miracidia are thus not random, but occur in a definite and predictable sequence.

Callinan (1979) has emphasized 'an understanding of the ecology of the free-living stages of sheep nematodes is essential if maximum use is to be made of the natural processes controlling nematode populations'. Free-living, 3rd stage larvae of many nematode species are negatively geotropic and positively thermotropic, both of which are involved in host location. Evidence suggests that optimum soil conditions, including humidity and temperature, are absolutely necessary for the development and transmission of free-living larval nematodes to their normal definitive hosts (Callinan, 1979). Deviations from optimum microhabitat conditions will disrupt life cycles and, in turn, affect both the local and the global (zoogeographic) distributions of these parasites and the diseases they produce. On the other hand, a changing climate may also affect microhabitat conditions in such a way as to be conducive to colonization by new parasites.

The patterns of dispersal and transmission of digeneans, and of nematodes having free-living larval stages, are influenced by many microhabitat physical and chemical forces, some of which are very subtle in character. These same microgeographic factors also may affect the dispersal of parasites that have stages transmitted passively. Consider the digenean *Halipegus occidualis*, as an example. This parasite has received considerable attention in recent years and much is known regarding its life cycle, population biology, and transmission dynamics (Esch *et al.*, 1997). The life-cycle pattern of this hemiurid digenean is much more complicated than that of most other digeneans as it has three obligate hosts, and a paratenic host (Goater *et al.*, 1990; Zelmer & Esch, 1998*a*). Adults live under the tongue of the green frog *Rana clamitans*, throughout temperate areas of North America. Eggs are swallowed and then shed by the frog when it defecates. The snail intermediate host *Helisoma anceps* must then eat the eggs. Once inside the snail gut, the eggs hatch and a motile larval stage penetrates the gut wall and migrates to the hepatopancreas where sporocysts develop. After a sporocyst generation, rediae are released from the sporocysts and eventually spill over into the gonads. These voracious larvae consume the gonads and cause host castration. Cercariae are then produced by the rediae and are released from the snail. One of the unique characteristics of this parasite is the unusual cystophorous cercaria, also called a cercariocyst. These cercariocysts are non-motile, but they have a relatively extended life span (Fig. 10.2). It has been speculated that long-term survival by the cystophorous cercariae has evolved as a temporal mechanism for dispersal in time as opposed to the spatial dispersal strategies employed by free-swimming, but short-lived, cercariae of most other species. Studies by Wetzel & Esch (1995) have shown that infectivity of the cystophorous cercaria for the next host (an ostracod, a microcrustacean) is not related to the ability of the parasite to excyst. Excystment is a purely physical phenomenon apparently related to rupture of the outer cyst wall, followed by changes in the internal osmotic pressure which forces eversion of a delivery tube from the body of the cercariocyst, then by the forceful ejection of the body of the parasite through the delivery tube (Zelmer & Esch, 1998*b*). Infectivity of the parasite is a strictly temporal factor.

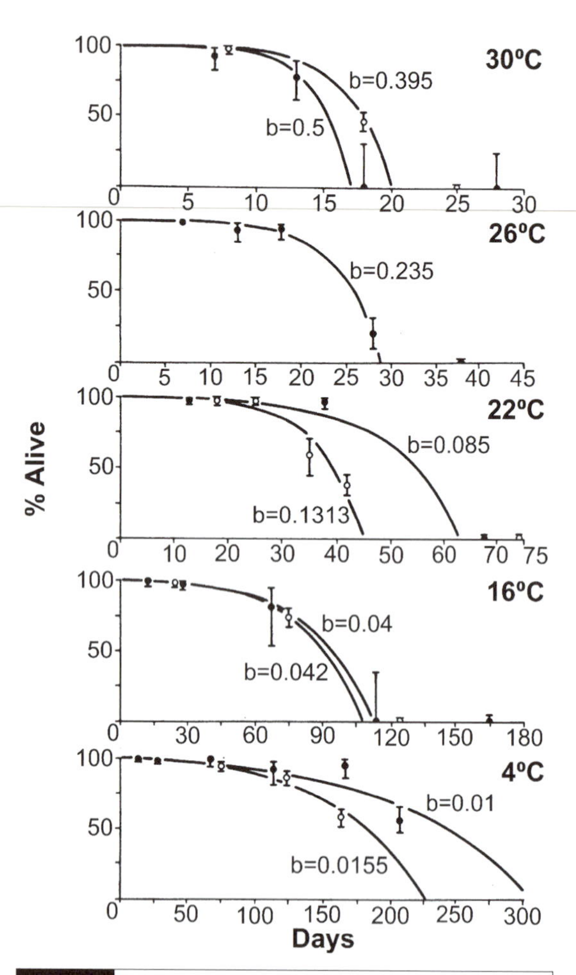

Fig. 10.2 Survival of cercariae of *Halipegus occidualis* at constant temperature. Values are percent and 95% confidence limits. Solid and open circles show results from two experiments. (Modified from Shostak & Esch, 1990, with permission, *International Journal for Parasitology*, **20**, 95–99.)

The cystophorous cercariae are accidentally ingested by benthic ostracods in which the parasites penetrate the gut wall and develop into infective metacercariae within the hemocoel. However, transfer of the metacercariae from an ostracod to a frog definitive host is unlikely to occur since frogs do not normally feed on ostracods. Instead, an alternate host, the odonate naiad, is used to bridge the ecological, or trophic, gap (Zelmer & Esch, 1998*a*). If development of the metacercaria is incomplete when the ostracod is consumed by the odonate, development will continue inside the midgut of the naiad until it becomes infective for the frog.

The distribution of *Halipegus*-infected snails was examined from a microhabitat perspective in a small, 2-ha pond in the Piedmont area of North Carolina, USA (Williams & Esch, 1991). They found that infected snails were not randomly distributed in the littoral zone of the pond, but were concentrated in very shallow water, <5.0 cm,

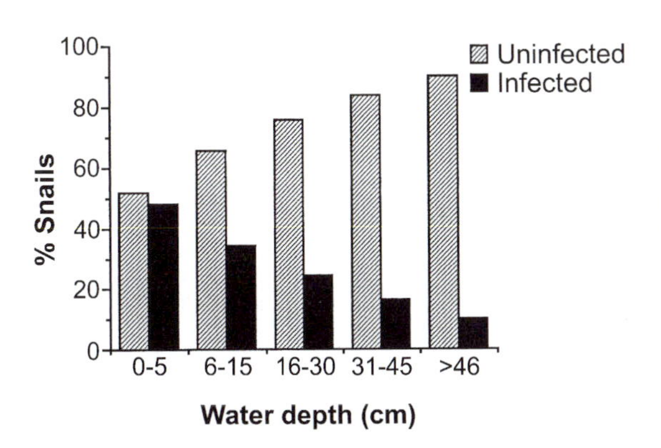

Fig. 10.3 The prevalence of *Halipegus occidualis* in the pulmonate snail *Helisoma anceps*, as a function of water depth. (Modified from Williams & Esch, 1991, with permission, *Journal of Parasitology*, **77**, 246–253.)

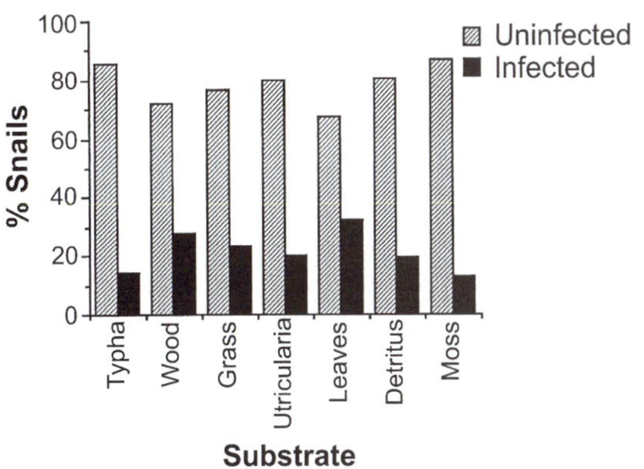

Fig. 10.4 The prevalence of *Halipegus occidualis* in the pulmonate snail *Helisoma anceps*, with respect to the substrate from which the snails were captured. (Modified from Williams & Esch, 1991, with permission, *Journal of Parasitology*, **77**, 246–253.)

immediately adjacent to the shoreline (Fig. 10.3) and were found significantly more often on leaf substrata in the shallow waters (Fig. 10.4). These observations on the microgeographic distribution of *H. occidualis* serve to emphasize the concept of 'scale' in considering the overall population dynamics of different species of digeneans and probably other parasites as well. They explained the distribution of the infected snails by correlating it with the behavior of the definitive host. Green frogs are ambush predators, spending most of their time submerged in very shallow water along the pond's edge. It is here that they defecate and shed eggs in high concentrations. Snails such as *H. anceps* graze the leaf litter and other substrata in the littoral zone. As they do, they encounter aggregations of parasite eggs, primarily in shallow water and on leaf litter that, in effect, function as enhanced transmission foci, or 'hot spots', for the parasite.

The idea of these enhanced transmission foci was extended to encompass the concept of landscape ecology/epidemiology. According to Beeby & Brennan (1997), landscape ecology is the 'study of the spatial arrangement of ecosystems and the large-scale processes that unite them'. **Landscape epidemiology** would include application of the same ideas of scale processes to the transmission of parasites. Zelmer *et al.* (1999) examined the distribution of infected green frogs in the North Carolina farm pond mentioned above. They determined that these so-called 'hot spots' (Fig. 10.5) were primarily associated with areas of the pond which favored all facets of the parasite's transmission success, i.e., the frog's predatory and reproductive activities, an appropriate substratum conducive for the life histories of snails, ostracods, and odonates, and emergent vegetation which would permit odonate naiads to successfully crawl out of the pond and onto the vegetation, then transform from an aquatic to an aerial stage. The concept of landscape ecology/epidemiology has broad implications for the transmission dynamics of a wide range of protozoan and helminth parasites, including many of those infecting humans, and will be discussed, as appropriate, in other sections of the text.

Of the external factors affecting infrapopulation biology of endoparasitic organisms, diet is probably the most basic. For example, many of the enteric parasites within a definitive host are present because the host ingested an infective stage of some parasite. Dogiel (1964) made a strong case for the importance of diet in influencing the parasite fauna within hosts when he said, 'the parasitological indicators of diet are among those clues that allow us to make deductions from the type of parasite fauna about various aspects of the ecology of the host'. In other words, parasitologically, you are what you eat! He illustrated his point by noting that herbivorous cyprinids within an aquatic system were virtually devoid of enteric hel-

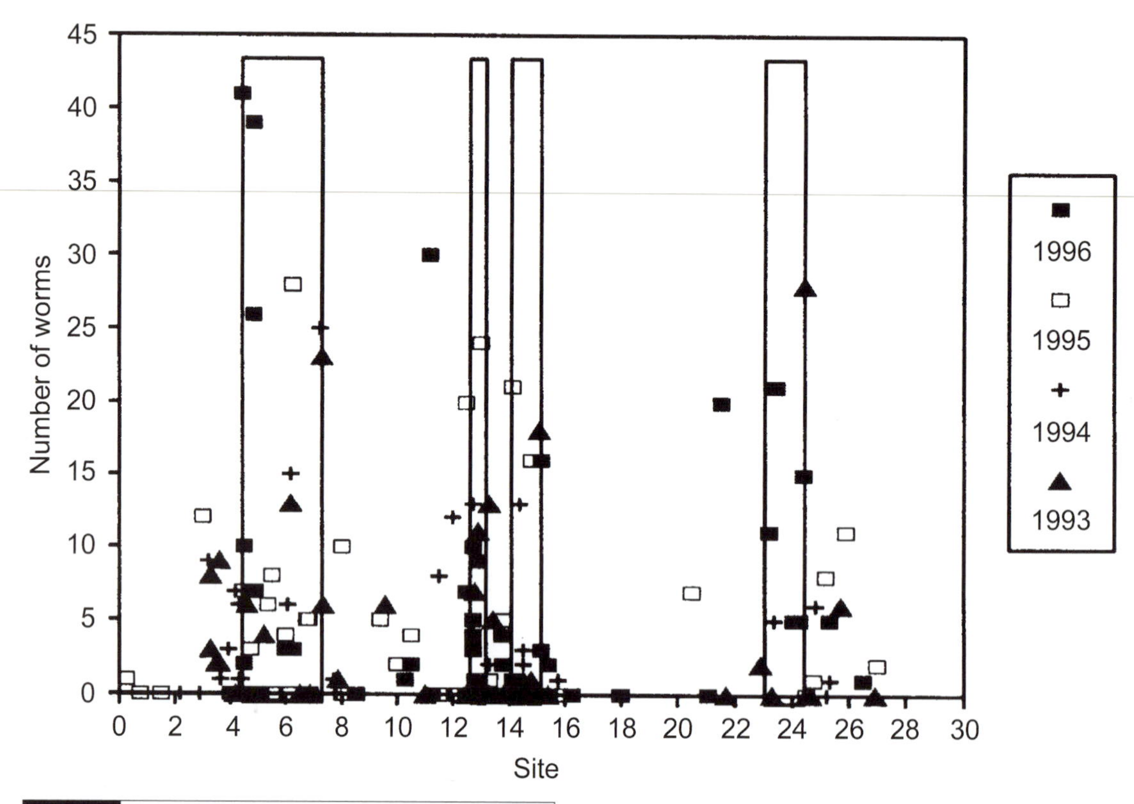

Fig. 10.5 Distribution of *Halipegus occidualis*-infected *Rana clamitans* in a North Carolina (USA) farm pond showing the parasite's heterogeneous distribution and illustrating the concept of 'landscape epidemiology'. (Modified from Zelmer *et al.*, 1999, with permission, *Journal of Parasitology*, **85**, 19–24.)

minths, whereas carnivorous cyprinids within the same system possessed rich helminth faunas. The implication is that carnivory is more likely to result in recruitment of parasites than herbivory, but there are many exceptions to this pattern. For example, slider turtles *Trachemys scripta* are typically omnivorous, but feed extensively on aquatic vegetation. It is the herbivory that results in the recruitment of an extensive acanthocephalan fauna of the genus *Neoechinorhynchus* (Esch *et al.*, 1979*a*, *b*; Jacobson, 1987). These parasites use various species of ostracods as first intermediate hosts. The ostracods are part of the epiphytic fauna associated with aquatic vegetation and are accidentally consumed as the turtles feed. Whereas turtles are accidentally exposed to infected ostracods, it is the herbivorous diet that in reality causes them to become infected with species of *Neoechinorhynchus*. Horses are herbivores and they also often harbor rich faunas of strongylid nematodes in their gut. These parasites are accidentally acquired as filariform larvae while the horses graze. Some detritivores such as mullets (Mugilidae) have rich digenean faunas that are acquired as the fish feed on the benthos of estuarine and marine coastal areas, accidentally ingesting metacercariae of the parasites present in the detritus (Paperna & Overstreet, 1981; Fernández, 1987). Bush *et al.* (1993) describe what they call 'source communities' in intertidal crabs. In these cases, the crabs have been shown to recruit several species of larval parasites and then, when ingested by appropriate predators, the infective larval forms are transferred to the new hosts where they establish as adults. As described by Lotz *et al.* (1995), 'these source communities are transmitted in packets of infective propagules to target communities in definitive hosts'. They went on to note that such transfers would result in greater numbers of positive associations among parasite species within some definitive hosts than what might otherwise be expected.

Also of course, a strong case can be made for the relationship between diet and parasites in humans. Many of the enteric protozoans of humans are, for example, acquired through the accidental ingestion of cysts that occur externally on contaminated food or in water, e.g., *Giardia lamblia, Entamoeba histolytica, E. coli, Balantidium coli.* The same route of infection is used by eggs of many enteric helminths, e.g., *Ascaris lumbricoides, Enterobius vermicularis,* and *Trichuris trichiura*. Human parasites are acquired by ingestion of a wide range of foods, e.g., *Fasciola hepatica* (watercress), *Diphyllobothrium latum* (fish), *Taenia solium* (pork), *Taenia saginata* (beef), and *Trichinella spiralis* (pork). For a number of these latter parasites, transmission depends on long-standing, local 'culinary' customs, e.g., the consumption of poorly cooked or raw fish, beef, and pork, or on poor socioeconomic conditions, or a combination of diet and wealth (or lack thereof). Finally, there is an inextricable link between nutrition (not just diet) and parasitic infection (Crompton, 1993); this will be discussed more fully subsequently.

Only a few of the external environmental factors known to affect the infrapopulation biology of parasites have been identified here. We hope, however, that these examples will serve as an introduction. Other factors will be considered more fully in subsequent chapters. These include seasonal changes in photoperiod and light intensity, ecological succession, the many ramifications of nutrient loading and pollution in aquatic habitats, ecosystem stability, dispersion, and other spatial distribution patterns such as those discussed above for *H. occidualis*. The zoogeographic factors influencing the global distribution of both parasites and their hosts will be considered as well.

10.3.2 Internal environmental factors

The range of internal environmental factors affecting the infrapopulation biology of parasites is probably not as great as the external ones, but, in some ways, internal factors are more complex than those external to the host. This is true primarily because phenomena such as behavior, host and parasite genetics, natural and acquired resistance, ontogenetic factors associated with the aging process, and host sex may be involved; moreover, several of these frequently act in concert. Inter- and intraspecific interactions among infrapopulations of certain species may also affect parasite densities and fecundity. It is well known, for example, that egg production by some enteric helminths will rise with increasing parasite density, but will reach a threshold or even decline if parasite densities increase above a certain point (see Chapters 11 and 14 for a more extensive discussion of competition).

The two most complex internal factors affecting parasite infrapopulation biology are associated with the host's immune system and with the genetics of both the parasite and the host. This complexity is compounded since the genetics of the host also impacts directly on its immune capabilities. One of the best-studied cases of such an interaction involves the deer mouse *Peromyscus maniculatus* and the tapeworm *Hymenolepis citelli*. Wassom *et al.* (1973, 1974, 1986) reported that these tapeworms were aggregated within the definitive host and that the prevalence was low, less than 5%. They attributed these observations to the heterogeneous distribution of infected intermediate hosts (camel crickets, *Ceuthophilus utahensis*) within the habitat of the deer mice. Moreover, they presented evidence that about 75% of the deer mice are endowed with natural resistance. That natural resistance is related to a single, autosomal dominant gene. The result is rejection of the tapeworms before they can become sexually mature and this contributes to both the low prevalence and the aggregated distribution of the parasite.

Host behavior, age, and sex are factors that may influence the infrapopulation biology of some parasitic organisms. In a few cases, all three factors can be inextricably linked. A good example of this linkage is associated with aspects of the infrapopulation biology of the cestode *Proteocephalus ambloplitis*. Definitive hosts for this parasite are smallmouth bass *Micropterus dolomieui* and largemouth bass *Micropterus salmoides*. After eggs are released from the adult tapeworm, they are shed in the feces. Eggs are freed from the fecal material, float in the water column, and are consumed by several species of planktonic copepods. An oncosphere emerges from the egg and penetrates the gut wall of the copepod; it enters the hemocoel and develops into a procercoid. Planktivorous fishes then eat the copepods. It should be emphasized that, for the most part, only species of smaller fishes and fingerlings that will grow to much larger sizes will normally feed on copepods. We thus have an example of an age-related or, perhaps more precisely, a size-related transmission tactic. Within the gut of the small fish, the procercoid is freed from the copepod by digestion and the liberated parasite penetrates the gut wall where it develops into a parenteric (outside the intestine) plerocercoid. The plerocercoids of *P. ambloplitis* are able to migrate throughout the visceral mass, e.g., liver, gonads, and spleen, of infected centrarchids using a metalloproteinase for digestion of host tissues, and are capable of inducing considerable tissue damage in the process (Esch & Huffines, 1973; Coggins, 1980*a*, *b*; Polzer *et al.*, 1994).

Another age-related component of the parasite's life cycle is also involved. In order to complete the cycle, the plerocercoid must first reach a parenteric site within the definitive host, almost always a sexually mature smallmouth or largemouth bass. This can be accomplished in one of two ways. First, a larger bass can eat the plerocercoid-infected small fish or fingerling. If this happens, the plerocercoid will be digested from the flesh of the intermediate host and then migrate through the gut wall of the bass into the body cavity. Once inside, it will migrate into various organs such as the gonads, spleen, and liver. Apparently, it continues to wander in these organs, growing in size until an appropriate stimulus is received.

In the second route, the copepod containing the procercoid is consumed by a fingerling bass, in which case the parasite migrates into a parenteric site within the fingerling and changes into a plerocercoid. It wanders through various abdominal organs, growing in size until the bass becomes sexually mature. The infrapopulation of plerocercoids within a bass that becomes sexually mature for the first time will thus consist of a mix of parasites acquired directly from procercoids in copepods and those acquired through predation on small fishes and fingerlings infected with plerocercoids that were obtained from copepods infected with procercoids.

The final step in the cycle involves several phenomena about which not much is understood. In the spring of each year, when water temperatures rise, adult tapeworms begin to accumulate in the pyloric ceca of sexually mature bass (all adult worms are lost each fall and are replaced the following spring by a new infrapopulation). Under laboratory conditions, Fischer & Freeman (1969) were able to demonstrate that rising temperature was apparently the stimulus for migration of parenteric plerocercoids from sites within the abdominal cavity into the lumen of the pyloric ceca. However, this migration also coincides with increases in the level of circulating spawning hormones released from the gonads and pituitary gland in the spring. The spawning act in smallmouth bass is precisely determined by water temperature. An interesting feature in this process is that, when sexually mature bass begin to acquire new plerocercoids from plerocercoid-infected fishes as the summer progresses, the new plerocercoids migrate into parenteric sites and do not develop into adult tapeworms until the following spring. In other words, all plerocercoids apparently must spend at least part of one annual cycle in parenteric sites within a sexually mature bass before migrating back into the pyloric ceca and developing into adults within that bass. This could be interpreted as a mechanism to minimize competition between an already established adult infrapopulation and new, potential adults. It may also be a way to separate gene pools on an annual basis. Plerocercoids occurring in other centrarchid species such as the bluegill, *Lepomis macrochirus*, will not migrate out of parenteric sites after the fishes become sexually mature even though subjected to rising spring temperatures. This clearly implies that the stimulus for migration from parenteric sites in bass is not rising temperature alone, but is something specifically associated with the appropriate definitive host.

The precise nature of the stimulus in smallmouth bass is not known. Fischer & Freeman (1969) and Esch *et al.* (1975*b*) suggested that it was a combination of rising temperatures and increasing levels of hormones since migration coincided with the spawning act and because juvenile bass are not infected with adults even though they usually possess plerocercoids. The problem with this hypothesis is that in South Carolina (USA), adults of *P. ambloplitis* are present in largemouth bass during winter and disappear in the spring at about the time of spawning (Eure, 1976). This is opposite to the pattern that occurs in the north. Whatever the migration cues, this system is an elegant one to highlight the complex relationships between parasite infrapopulation biology and host behavior, age, and sexual maturation.

Another excellent example of a hormonal effect is associated with the synchronization of the life cycles of the monogenean *Polystoma integerrimum* and its frog host *Rana temporaria*. *Polystoma integerrimum* enters the frog bladder before the tadpoles metamorphose and juvenile frogs become terrestrial. Therefore, transmission of the parasite to new hosts can occur only when frogs invade temporary bodies of water to reproduce. When frogs are preparing to enter the water prior to copulation (after 3 years on land), the reproductive system of the monogenean develops. As the frog spawns for the first time, maturation of *P. integerrimum* occurs and eggs are released. In this way, the synchronized mechanism ensures that when the monogenean eggs hatch, tadpoles will be available for infection. The high correlation between the host's and the parasite's reproductive cycles suggests that reproductive development in *Polystoma* is controlled by the hormonal activity of the frog. Indeed, experiments have demonstrated that maturation and stimulation of gamete production in *P. integerrimum* can be elicited by injecting the frog host with a pituitary extract. However, it is not known yet whether the effect is a direct one through the injected pituitary hormones, or via the host gonadal hormones that are produced in response to those released from the pituitary. A similar synchrony in life cycles has been demonstrated for *Polystoma stellai* and the tree frog *Hyla septentrionalis* (Stunkard, 1955). In contrast, however, polystomatids parasitic in more aquatic amphibians, e.g., *Protopolystoma xenopodis* in *Xenopus* spp., exhibit development that is not influenced by season of the year, reproductive condition of the host, or by experimental treatment with sex hormones. In these cases, the uninterrupted reproductive activity of the parasite is in some manner related to the continuous availability of its aquatic host (Tinsley & Owen, 1975).

One of the most unusual sequences of internal migration and responsive parasite adaptation is associated with the monogenean *Pseudodiplorchis americanus*. This parasite passes through a wider range of external and internal habitats and is exposed to more diverse physiological conditions than any other monogenean (Cable & Tinsley, 1992). It reaches sexual maturity in the urinary bladder of the desert toad *Scaphiopus couchii*. Following a very brief, free-living oncomiracidium stage in an ephemeral desert pond, the parasite enters the toad's nares and migrates to the lungs. Within the lungs, there are ontogenetic changes in morphology that, presumably, prepare the worms for migration through the entire length of the intestine. The majority of the parasites migrate through the gut when their hosts are gorging themselves in order to replenish body fats necessary to withstand the upcoming estivation. This migration presents a formidable barrier with which to cope, e.g., the parasites are subjected to radical changes in pH, secretions of bile acids from the liver, digestive enzymes produced in the stomach and pancreas, and anaerobiasis. In partial response to these conditions, the pre-adults in the lungs accumulate PAS-positive vesicles in the tegument, which are then released to form a carbohydrate-rich glycocalyx as the parasites enter the gut. Presumably, the glycocalyx serves to isolate the surface of the fluke from the extreme vagaries of the gut during migration. The entire gut migration lasts for about 30 minutes, after which the production of the PAS-positive vesicles stops; they are then replaced by vesicles that are morphologically distinctive of the adult stage that now occupies the urinary bladder, and the glycocalyx disappears.

10.4 The dispersion concept

Before considering further any basic ideas regarding parasite population biology and regulatory interactions, it is first necessary to identify and discuss several additional population terms and concepts. The sample **mean**, $\bar{x}$, is defined as the sum of all measurements in the sample divided by the number of measurements in the sample. It is one of the most important measures of what statisticians refer to as measures of central tendencies. The **variance**, (s^2), of a sample mean is a measure of mathematical variability within the population. It is important to note that the mean and variance, as we use them here, represent values derived from a *sample* as opposed to the entire population. They are *estimates* of the true population mean (μ) and variance (σ^2).

Since the sample mean and the variance for a given population are only estimates, it becomes critical that sampling be conducted both accurately and randomly. Some species are clearly more difficult than others to enumerate and, accordingly, a variety of sampling protocols have been developed for various plants and animals (Tanner, 1987).

Parasite infrapopulations present special problems. For all but a few species, hosts must be killed before parasites can be counted. The problem with this approach is that by killing the hosts, both the hosts and the parasites they carry are removed from both their overall populations and gene pools. In some cases, this does not create a serious problem because the population size of the host is sufficiently large that the removal of a few individuals will not significantly affect the reproductive capabilities of the remaining hosts or their parasites. It is, therefore, necessary to obtain certain basic information on the population biology of both the host and the parasite and evaluate it carefully before undertaking a 'kill and count' procedure.

We might note here that *Halipegus occidualis*, the hemiurid fluke that occupies the buccal cavity of the green frog *Rana clamitans*, is unusual in that it is one of the few endoparasitic helminths that can be enumerated without killing the host and therefore offers a number of unique advantages over other species. *Halipegus eccentricus* occurs in the eustachian tube of the same host and is also readily enumerated without killing its host. Extensive studies using these two parasites have shown that the dynamics of infrapopulation recruitment and turnover may be more volatile than previously expected or shown, and that further efforts in this area are clearly warranted. Wetzel & Esch (1996), for example, observed large and rapid increases in infrapopulation densities, reminiscent of the 'source community' structural changes alluded to by Bush *et al.* (1993) and Lotz *et al.* (1995). These increases were frequently followed by sudden and sharp declines in infrapopulation densities, not unlike those reported by Jackson & Tinsley (1994) for *Gyrdicotylus gallieni*, a monogenean that occupies the buccal cavity and pharynx of *Xenopus laevis*. Both Wetzel & Esch (1996) and Jackson & Tinsley (1994) ascribed the swift infrapopulation declines to host reactions in response to high parasite densities.

The component parasite population's variance is an important estimate when assessing dispersion of parasites among the hosts. Consider the six distinct component population frequency distributions in Fig. 10.6. In each of these cases, the mean is identical, but note how the component populations differ in their dispersion characteristics. If an assessment is made regarding the degree to which dispersion patterns of various component populations differ from each other, then substantially more information about the component populations could be generated. Such a measure would provide an idea of the spread, or variability, within the component population, which, in space, typically occurs in three distinct patterns called **random**, **aggregated** (or **clumped**), and **uniform** (or **regular**) (Fig. 10.7). (See Box 10.1 for semantic issues associated with these terms in the fields of ecology and parasitology.) Random distributions occur when the position of one individual is completely independent of any other individual and when each segment of the habitat has the same probability of being colonized. Stated differently, random distributions suggest that no interactions occur among organisms. As such, random distributions are the appropriate 'null model' against which to test observed patterns. Aggregated patterns of distribution, on the other hand, indicate possible social interactions, a need to be together for some reason (mutual defense, cooperative feeding, mating purposes), or the presence of a suitable resource. Aggregated distributions are the most common pattern found in nature. In uniform distributions the organisms are evenly distributed or spaced in an area, imply-

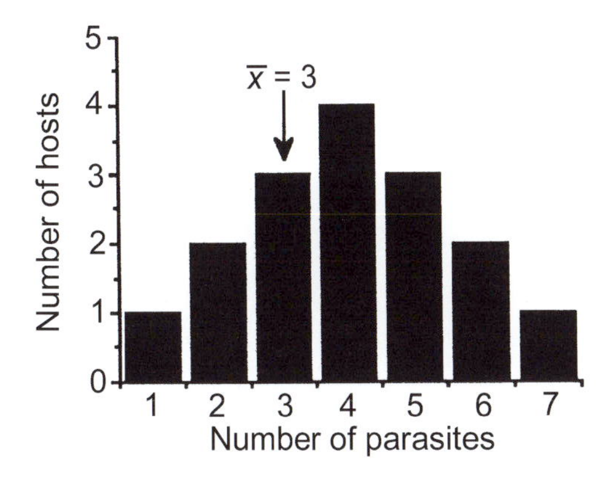

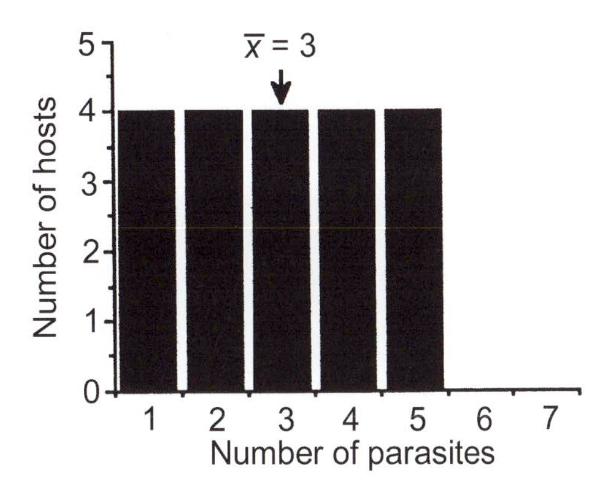

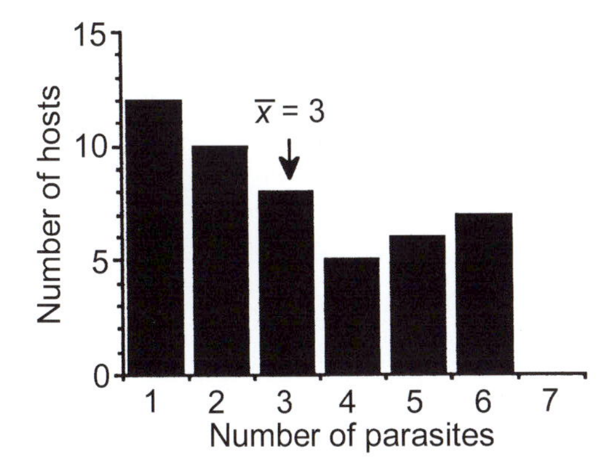

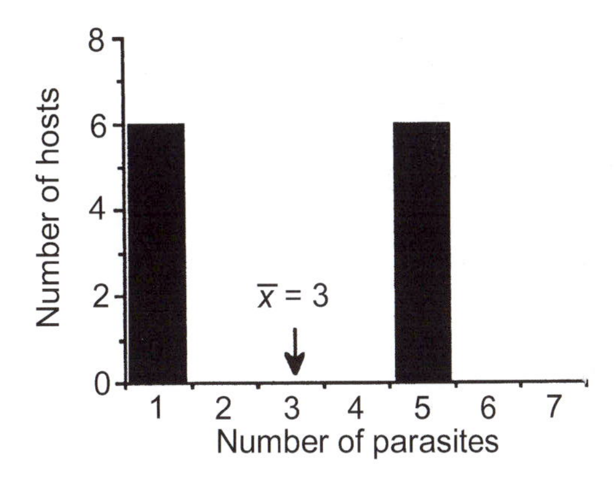

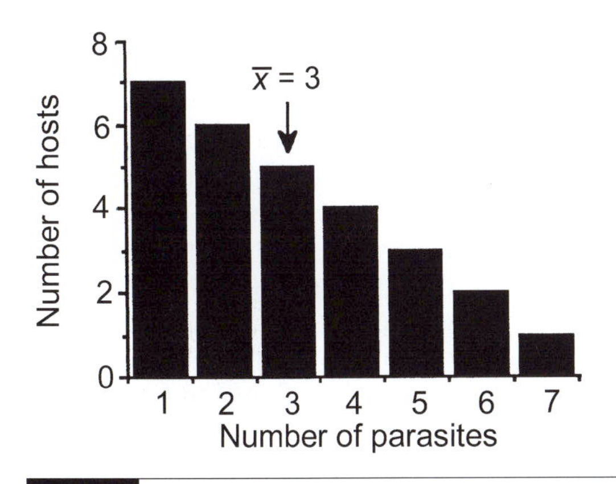

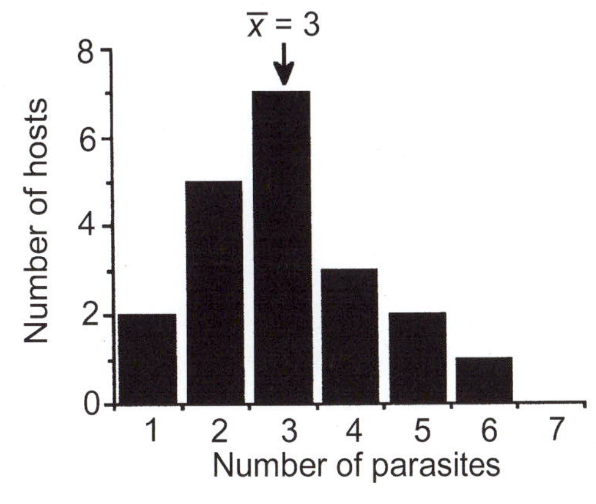

Fig. 10.6 An example of the variability with which different populations can be dispersed around the same mean. In each case, the mean is the same ($\bar{x} = 3$), but the distribution of the observed numbers is quite different.

ing the existence of strong antagonistic interactions between them.

The importance of knowing the variances and means of parasite component populations is apparent when these two estimates are computed for the three hypothetical component populations given in Table 10.2. For each, the mean ($\bar{x}$), the variance (s^2), and the variance/mean ($s^2/\bar{x}$) ratio, or coefficient of dispersion has been calculated. Note that for each component population, the mean ($\bar{x}$) is the same ($\bar{x} = 6.0$). The statistical significance for goodness of fit can be assessed using a simple χ^2 test. In component population I, the variance (s^2) is not significantly different from the mean ($\bar{x}$); thus, $s^2 \approx \bar{x}$). In this case, the component population is randomly distributed. Whenever the variance/mean ratio is close to 1, a component population is randomly distributed. In

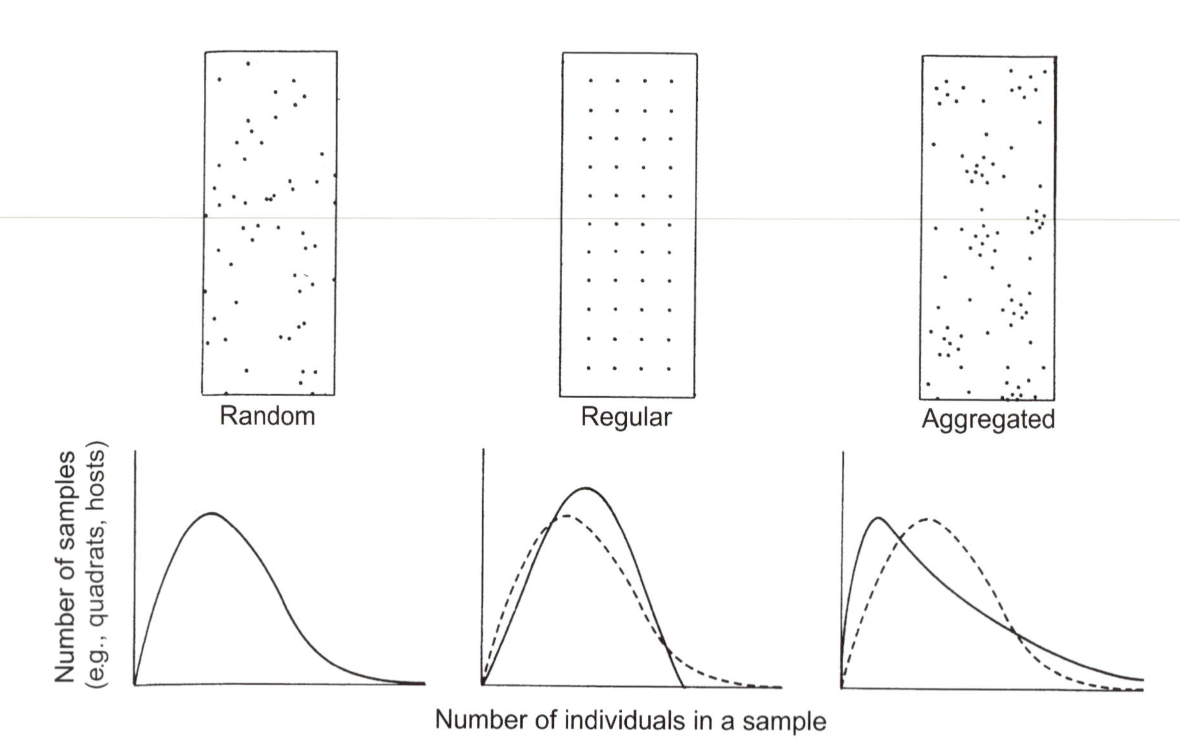

Fig. 10.7 Distributional patterns in space (above) for populations that are random, regular and aggregated. Below are the frequency curves that describe these distributions; the dashed line is the same random distribution for comparison.

component population II, the variance/mean ratio ($s^2/\bar{x}$) is greater than unity ($s^2 >> \bar{x}$) and it is aggregated. In component population III, the variance/mean ratio ($s^2/\bar{x}$) is less than unity ($s^2 << \bar{x}$) and this component population is uniformly distributed. In other words, the distribution of values around the mean in each component population is quite distinct and each distribution can be described mathematically; they will be random, aggregated, or uniform. The significance of understanding the frequency distribution of parasite component populations will become more apparent when Crofton's (1971*a, b*) definition of parasitism is discussed in greater detail. Since Crofton's seminal work, the concept of aggregation (he used the term overdispersion) has become a central component for paradigms involving parasite population biology and of models employed in describing regulatory interactions between host and parasite populations (Anderson & May, 1979; May & Anderson, 1979).

Uniform frequency distributions are uncommon among parasite infrapopulations. Cestodarians of the genus *Gyrocotyle*, the digenean *Deretrema philippinensis*, and the copepod *Leposphilus labrei* are among the few exceptions. Studies of different species of *Gyrocotyle* in holocephalan hosts such as *Chimaera monstrosa*, *Hydrolagus colliei*, *Callorhynchus millie*, and *C. callorhynchus*, reveal some striking similarities. In most instances, prevalence is very high (90%–100%) and parasite densities are low, with infrapopulations composed mostly of two individuals (Halvorsen & Williams, 1967; Dienske, 1968; Simmons & Laurie, 1972; Allison & Coakley, 1973; Fernández *et al.*, 1983; Williams *et al.*, 1987; Donnelly & Reynolds, 1994). Four possible mechanisms have been proposed to explain why uniform distributions occur among these species of parasites. The first suggests a reduction in the probability of infection caused by ontogenetic changes in host biology, e.g., in food selection or habitat. The second proposes that parasite recruitment rates are equal to death rates, thereby maintaining constant population size. A third explanation suggests that heavier infections are lethal for the host. Finally, it is also possible that further infections are prevented through intraspecific competition, or the

Box 10.1 Patterns of distribution: a matter of terminology and tradition

Since the early nineteenth century, 'free-living' ecologists have recognized that organisms are, for the most part, not randomly distributed in space. Most scientists now recognize that the distributions of both free-living and parasitic organisms follow one of three basic patterns. Random patterns can be described mathematically by the expression $s^2 \approx \bar{x}$. A second pattern, variously called regular, uniform, spaced, negative contagion, or overdispersed can be defined mathematically by the expression $s^2 < \bar{x}$. The third pattern has also been variously described as aggregated, clustered, clumped, contagious, patchy, positive contagion, or underdispersed. This distribution is described mathematically as $s^2 > \bar{x}$.

The difficulty with some of these terms is that 'free-living' ecologists and parasitologists use 'overdispersion' and 'underdispersion' to describe opposite patterns of distribution. Unfortunately, when Crofton (1971*a*) first looked at the distribution of parasites, he used the mathematical term 'overdispersed' as a synonym for aggregated or clumped distributions and 'underdispersed' as a synonym for uniform or regular distributions, exactly the opposite to the accepted manner in which most 'free-living' ecologists had applied the same terminology. Some parasitologists, including one of the most oft-cited papers in parasite population ecology, i.e., that of Anderson and May (1979), follow Crofton's lead and continue to use overdispersion to describe aggregated or clumped distributions.

To avoid further confusion, and in an attempt to standardize the usage of these terms, we will use the more intuitive terms random, aggregated (or clumped), and uniform (or regular) to describe the patterns of parasite distribution. Because the terms 'overdispersion' and 'underdispersion' have had general acceptance among some parasite population ecologists, we think that any student, new to parasitology, should be aware of this semantic problem. In short, much of the primary literature in parasitology refers to aggregated or clumped distributions as overdispersed, and to uniform or regular distributions as underdispersed. Beware!

host's immune response, or both. Although not much is known about the biology of either hosts or parasites, the evidence available favors explanations based on the last mechanism, with regulation mediated by antagonistic, parasite–parasite interactions (Williams *et al.*, 1987). Another example of a uniform distribution is *Deretrema philippinensis*, a gallbladder digenean in the flashlight fish, *Anomalops katopron*. Burn (1980) and Beverley-Burton & Early (1982) consistently found two digeneans per gallbladder, with a prevalence of almost 100%. Donnelly & Reynolds (1994) found that 1922 of 1924 infected corkwing wrasse (*Crenilabrus melops*) in Mulroy Bay, on the coast of Ireland, were infected with a single individual of the copepod *Leposphilus labrei*, and that the other two hosts had just two copepods. Again, however, the mechanism responsible for these highly remarkable uniform distributions has not been determined.

Neither random nor uniform frequency distributions are common in nature. By far, the most common form of frequency distribution is one that is aggregated. This pattern is as common for parasite component populations as it is for free-living organisms. Several different theoretical

Table 10.2 Mean ($\bar{x}$), variance (s^2) and variance/mean ratios for three hypothetical populations

Population I	Population II	Population III
$x_i = 3, 7, 10, 5, 8, 7, 2$	$x_i = 10, 11, 2, 3, 11, 3, 2$	$x_i = 6, 5, 8, 7, 6, 5, 5$
$N = 7$	$N = 7$	$N = 7$
$\bar{x} = 6.0$	$\bar{x} = 6.0$	$\bar{x} = 6.0$
$s^2 = 8.0$	$s^2 = 19.3$	$s^2 = 1.33$
$s^2/\bar{x} = 1.33$	$s^2/\bar{x} = 3.22$	$s^2/\bar{x} = 0.22$
$s^2 \approx \bar{x}$	$s^2 > \bar{x}$	$s^2 < \bar{x}$
Random distribution	Aggregated distribution	Uniform distribution

Note:
x_i = number of parasites in individual hosts, N = number of hosts sampled in each population.

models, including the lognormal, log series, Neyman type A, and the negative binomial, can describe aggregated frequency distributions. The significance of some of these models and their application to parasite population dynamics will be discussed in Chapter 11.

10.5 Dynamics of population growth

Generally speaking, there are two patterns of growth exhibited by populations of free-living organisms. In an ideal environment that has unlimited resources, growth may be exponential; that is to say, numbers per unit of space will increase 1, 2, 4, 8, 16, 32, etc. (Fig. 10.8A). The curve in Fig. 10.8A that illustrates this growth can be described by the following equation:

$$\frac{dN}{dt} = rN$$

where N = the number of individuals at time t, and r = the per capita rate of natural increase, or biotic potential. The per capita rate of natural increase, r, is the difference between the instantaneous birth rate and instantaneous death rate. Thus, r is said to 'represent in a single number all of the physiological responses of all members of a population to a given set of environmental factors' (Hairston *et al.*, 1970). Species with growth curves that are exponential in character are said to be ***r*-selected** (Pianka, 1970; Esch *et al.*, 1977). Species that employ such a growth strategy are also frequently described as opportunistic in the sense that they tend to maximize their reproductive capacities and do not direct their energies toward enhancing the survivorship of their offspring.

Most species do not exhibit exponential growth except under ideal laboratory conditions or during initial colonization. Instead, they follow a logistic pattern (Fig. 10.8B); only when these species are colonizing a new habitat, or perhaps at the beginning of a new reproductive season, will they approximate an exponential pattern of growth. Even then, however, it is not truly exponential because of the resistance created by many environmental factors. The logistic growth curve can be described by the equation:

$$\frac{dN}{dt} = rN\left(\frac{K-N}{K}\right)$$

where K = the carrying capacity of the habitat, or that point in time when $dN/dt = 0$. The term $[(K-N) / K]$ means that as N increases, dN/dt decreases. It is a measure of environmental resistance or crowding. When $N = K$, the term will be 0 and dN/dt also will be 0. The term $[(K-N)/K]$ is the simplest method of expressing the manner in which a population can expand up to equilibrium K. If N exceeds K, the term will become negative and N will approach K from the other direction.

The shape of a logistic curve is characteristically sigmoid (Fig. 10.8B). When a population ini-

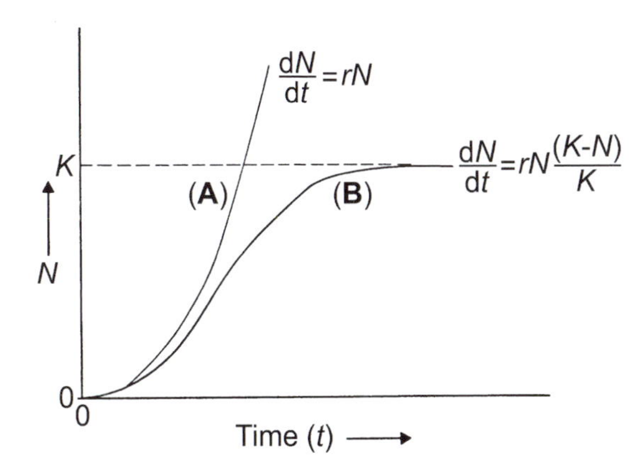

Fig. 10.8 Population growth curves. (A) Exponential growth; (B) logistic or sigmoidal growth. N = number of individuals; K = carrying capacity of the environment; r = rate of increase.

tially colonizes a habitat, growth is slow. This is a time for physiological adjustment to new surroundings, or perhaps new mates are being sought. This period of time is followed by a near-exponential increase in population size. The difference between the shape of the growth curve at this point in time and a true exponential curve is said to represent the biotic resistance of the environment. Eventually, however, environmental pressures begin to slow population growth. Resources such as nutrients or space may become limiting, or toxic materials produced by members of the population may begin to accumulate. Finally, the population growth stabilizes; at this time, birth and death rates become equal. Once the carrying capacity is reached, population size will remain reasonably constant except for erratic or irregular fluctuations in response to short-term shifts in various resources. Competition, predation, and parasitism may also influence population fluctuations. There are some populations that oscillate after reaching their carrying capacities; in these cases, there are regular changes in density over fixed periods of time. The classic case is the 10-year lynx/hare cycle recorded by the Hudson Bay Company since the middle of the nineteenth century. In this type of oscillation, population densities of the lynx lag those of the hare with a high degree of fidelity. Some have speculated that the population density of the lynx is directly related to that of the hare, although others have challenged this conclusion.

If limiting factors are increased, then the initial phase of the growth curve may be repeated until some new carrying capacity is reached and then it will level off again. In some cases, the population will undergo senescence and become locally extinct. In these situations, perhaps the population has become too specialized for a changing environment, or perhaps a competitor enters the habitat and is more successful than the previously established population. Species that exhibit a logistic growth curve are said to be **K-selected**.

In concluding this brief primer on population biology, a number of questions regarding parasite population dynamics should come to mind. For example, how do parasites fit into the exponential–logistic schemes of population growth, or do they fit at all? Are parasite population dynamics regulated primarily by abiotic factors or by density-dependent forces? If the latter, are they consistent over time? If the former, are most parasite species of the 'boom–bust' variety? Are growth patterns of most parasite species influenced by the stability or instability of their physical environment? At what point(s) is parasite population growth most likely to be checked and which factors are most likely to impact on this growth? Can these check points be exploited by intervention of drug therapies for the control of parasite population dynamics? Most parasite component populations are aggregated within their host populations, but what is the significance of this sort of dispersion pattern within the framework of parasite population dynamics? May & Anderson (1979) claimed that parasitic organisms are highly effective in regulating host populations. Indeed, they stated, 'parasites (broadly defined) are probably as important as the more usually-studied predators and insect parasitoids in regulating natural populations'. In general terms, there may be some, but very little, validity to this statement if one includes viruses and bacteria as parasites (and they did). On the other hand, if one includes only what the traditional parasitologists consider as parasites (mostly protozoans, helminths, and arthropods), then this proposition would be questioned.

References

Allison, F. R. & Coakley, A. (1973) Two species of *Gyrocotyle* in the Elephant fish, *Callorhynchus milii* (Bory). *Journal of the Royal Society of New Zealand*, **3**, 381–392.

Anderson, R. M. & May, R. M. (1979) Population biology of infectious diseases: Part I. *Nature*, **280**, 361–367.

Beeby, A. & Brennan, A.-M. (1997). *First Ecology*. London: Chapman & Hall.

Beverley-Burton, M. & Early, G. (1982) *Deretrema philippinensis* n. sp. (Digenea: Zoogonidae) from *Anomalops katoptron* (Beriformes: Anomalopidae) from the Philippines. *Canadian Journal of Zoology*, **60**, 2403–2408.

Burn, P. R. (1980) Density dependent regulation of a fish trematode population. *Journal of Parasitology*, **66**, 173–174.

Bush, A. O. & Holmes, J. C. (1986) Intestinal parasites of lesser scaup ducks: patterns of association. *Canadian Journal of Zoology*, **64**, 132–141.

Bush, A. O., Heard, R. W. Jr. & Overstreet, R. M. (1993) Intermediate hosts as source communities. *Canadian Journal of Zoology*, **71**, 1358–1363.

Bush, A. O., Lafferty, K. D., Lotz, J. M. & Shostak, A. W. (1997) Parasitology meets ecology on its own terms. Margolis *et al.* revisited. *Journal of Parasitology*, **83**, 575–583.

Cable, J. & Tinsley, R. C. (1992) Unique ultrastructural adaptations of *Pseudodiplorchis americanus* (Polystomatidae: Monogenea) to a sequence of hostile conditions following host infection. *Parasitology*, **105**, 229–241.

Callinan, A. P. L. (1979) The ecology of free-living stages of *Trichostrongilus vitrinus*. *International Journal for Parasitology*, **9**, 133–136.

Christensen, N. O., Nansen, P. & Fransen, F. (1978) The influence of some physico-chemical factors on the host finding capacity of *Fasciola hepatica* miracidia. *Journal of Helminthology*, **52**, 61–67.

Coggins, J. R. (1980*a*) Apical end organ structure and histochemistry in plerocercoids of *Proteocephalus ambloplitis*. *International Journal for Parasitology*, **10**, 97–101.

Coggins, J. R. (1980*b*) Tegument and apical end organ fine structure in the metacestode and adult of *Proteocephalus ambloplitis*. *International Journal for Parasitology*, **10**, 409–418.

Crofton, H. D. (1971*a*) A quantitative approach to parasitism. *Parasitology*, **62**, 179–193.

Crofton, H. D. (1971*b*) A model for host–parasite relationships. *Parasitology*, **63**, 343–364.

Crompton, D. W. T. (1993) Human nutrition and parasitic infection. *Parasitology* (Supplement), **107**, S1–S203.

Curtis, L. A. (1997). *Ilyanassa obsoleta* (Gastropoda) as a host for trematodes in Delaware estuaries. *Journal of Parasitology*, **83**, 793–803.

Dienske, H. (1968) A survey of the metazoan parasites of the rabbit-fish, *Chimaera monstrosa* L. (Holocephali). *Netherlands Journal of Sea Research*, **4**, 32–58.

Dogiel, V. A. (1964) *General Parasitology*. Edinburgh: Oliver & Boyd.

Donnelly, R. E. & Reynolds, J. D. (1994) Occurrence and distribution of the parasitic copepod *Leposphilus labrei* on corkwing wrasse (*Crenilabrus melops*) from Mulroy Bay, Ireland. *Journal of Parasitology*, **80**, 331–332.

Dronen, N. O. (1978) Host–parasite population dynamics of *Haematoloechus coloradensis* Cort, 1915 (Digenea: Plagiorchidae). *American Midland Naturalist*, **99**, 330–349.

Esch, G. W. (1982) Abiotic factors: an overview. In *Parasites – Their World and Ours*, ed. D. F. Mettrick & S. S. Desser, pp. 279–288. Proceedings of the Fifth International Congress of Parasitology (ICOPA V), Toronto, Canada. Amsterdam: Elsevier Biomedical Press.

Esch, G. W. & Huffines, W. J. (1973) Histopathology associated with endoparasitic helminths in bass. *Journal of Parasitology*, **59**, 306–313.

Esch, G. W., Gibbons, J. W. & Bourque, J. E. (1975*a*) An analysis of the relationship between stress and parasitism. *American Midland Naturalist*, **93**, 339–353.

Esch, G. W., Johnson, W. C. & Coggins, J. R. (1975*b*) Studies on the population biology of *Proteocephalus ambloplitis* (Cestoda) in smallmouth bass. *Proceedings of the Oklahoma Academy of Sciences*, **55**, 122–127.

Esch, G. W., Hazen, T. C. & Aho, J. M. (1977) Parasitism and *r*- and K-selection. In *Regulation of Parasite Populations*, ed. G. W. Esch, pp. 9–62. New York: Academic Press.

Esch, G. W., Gibbons, J. W. & Bourque, J. E. (1979*a*) Species diversity of helminth parasites in *Chrysemys s. scripta* from a variety of habitats in South Carolina. *Journal of Parasitology*, **65**, 633–638.

Esch, G. W., Gibbons, J. W. & Bourque, J. E. (1979*b*) The distribution and abundance of enteric helminths in *Chrysemys s. scripta* from various habitats on the

Savannah River plant in South Carolina. *Journal of Parasitology*, **65**, 624–632.
Esch, G. W., Bush, A. O. & Aho, J. M. (eds.) (1990*a*) *Parasite Communities: Patterns and Processes*. New York: Chapman and Hall.
Esch, G. W., Shostak, A. W., Marcogliese, D. J. & Goater, T. M. (1990*b*) Patterns and processes in helminth parasite communities: an overview. In *Parasite Communities: Patterns and Processes*, ed. G. W. Esch, A. O. Bush & J. M. Aho, pp. 1–19. New York: Chapman & Hall.
Esch, G. W., Wetzel, E. J., Zelmer, D. A. & Schotthoefer, A. M. (1997) Long-term changes in parasite population community structure: A case history. *American Midland Naturalist*, **137**, 369–387.
Eure, H. (1976) Seasonal abundance of *Proteocephalus ambloplitis* (Cestoidea: Proteocephalidae) from largemouth bass living in a heated reservoir. *Parasitology*, **73**, 205–212.
Fernández, J. (1987) Los parásitos de la lisa, *Mugil cephalus* L., en Chile: sistemática y aspectos poblacionales (Perciformes: Mugilidae). *Gayana, Zoología*, **51**, 3–58.
Fernández, J., Villalba, C. S. & Albiña, A. (1983) Parásitos del pejegallo, *Callorhynchus callorhynchus* (L.), en Chile: aspectos biológicos y sistemáticos. *Biología Pesquera*, **15**, 63–73.
Fischer, H. & Freeman, R. S. (1969) Penetration of parenteral plerocercoid of *Proteocephalus ambloplitis* (Leidy) into the gut of smallmouth bass. *Journal of Parasitology*, **55**, 766–774.
Goater, T. M., Browne, C. R. & Esch, G. W. (1990) The structure and function of the cystophorous cercariae of *Halipegus occidualis* (Trematoda: Hemiuridae). *International Journal for Parasitology*, **20**, 923–934.
Hairston, N. G. (1965) On the mathematical analysis of schistosome populations. *Bulletin of the World Health Organization*, **33**, 45–62.
Hairston, N. G., Tinkle, D. W. & Wilbur, H. M. (1970) Natural selection and the parameters of population growth. *Journal of Wildlife Management*, **34**, 681–698.
Halvorsen, O. & Williams, H. H. (1967) Studies on the helminth fauna of Norway, IX. *Gyrocotyle* (Platyhelminthes) in *Chimaera monstrosa* from Oslo Fjord, with emphasis on its mode of attachment and regulation in the degree of infection. *Nytt Magasin for Zoologi*, **15**, 130–142.
Holmes, J. C. (1961) Effects of concurrent infections on *Hymenolepis diminuta* (Cestoda) and *Moniliformis dubius* (Acanthocephala). I. General effects and comparison with crowding. *Journal of Parasitology*, **47**, 209–216.
Holmes, J. C. (1962*a*) Effect of concurrent infections on *Hymenolepis diminuta* (Cestoda) and *Moniliformis dubius* (Acanthocephala). II. Effects on growth. *Journal of Parasitology*, **48**, 87–96.
Holmes, J. C. (1962*b*) Effect of concurrent infections on *Hymenolepis diminuta* (Cestoda) and *Moniliformis dubius* (Acanthocephala). III. Effects in hamsters. *Journal of Parasitology*, **48**, 97–100.
Holmes, J. C. & Price, P. W. (1986) Communities of parasites. In *Community Ecology: Patterns and Processes*, ed. D. J. Anderson & J. Kikkawa, pp. 187–213. Oxford: Blackwell Scientific Publications.
Jackson, J. A. & Tinsley, R. C. (1994) Infrapopulation dynamics of *Gyrdicotylus gallieni* (Monogenea: Gyrodactylidae). *Parasitology*, **108**, 447–452.
Jacobson, K. C. (1987) Infracommunity structure of enteric helminths in the yellow-bellied slider, *Trachemys scripta scripta*. MSc. thesis, Wake Forest University, Winston-Salem, North Carolina, USA.
Jarroll, E. L. (1979) Population biology of *Bothriocephalus rarus* Thomas (1937) in the red-spotted newt, *Notophthalmus viridescens* Raf. *Parasitology*, **79**, 183–193.
Jarroll, E. L. (1980) Population dynamics of *Bothriocephalus rarus* (Cestoda) in *Notophthalmus viridescens*. *American Midland Naturalist*, **103**, 360–366.
Jokela, J. & Lively, C.M. (1995) Spatial variation in infection by digenetic trematodes in a population of freshwater snails (*Potamopyrgus antipodarum*). *Oecologia*, **103**, 509–517.
Kearn, G. C. (1980) Light and gravity responses of the oncomiracidium of *Entobdella soleae* and their role in host location. *Parasitology*, **81**, 71–89.
Kennedy, C. R. (1975) *Ecological Animal Parasitology*. Oxford: Blackwell Scientific Publications.
Lotz, J. M., Bush, A. O. & Font, W. F. (1995). Recruitment driven, spatially discontinuous communities: a null model for transferred patterns in target communities of intestinal helminths. *Journal of Parasitology*, **81**, 12–24.
MacDonald, G. (1965) The dynamics of helminth infections with special reference to schistosomes. *Transactions of the Royal Society of Tropical Medicine and Hygiene*, **59**, 489–506.
Marcogliese, D. J. & Esch, G. W. (1989) Alterations in seasonal dynamics of *Bothriocephalus acheilognathi* in a North Carolina cooling reservoir over a seven year period. *Journal of Parasitology*, **75**, 378–382.
Margolis, L., Esch, G. W., Holmes, J. C., Kuris, A. M. & Schad, G. A. (1982) The use of ecological terms in parasitology (report of an ad hoc committee of the American Society of Parasitologists). *Journal of Parasitology*, **68**, 131–133.

May, R. M. & Anderson, R. M. (1979) Population biology of infectious diseases. II. *Nature*, London, **280**, 455–461.

Paperna, I. & Overstreet, R. M. (1981) Parasites and diseases of mullets (Mugilidae). In *Aquaculture of Grey Mullets*, ed. O. H. Oren, pp. 411–494. International Biological Program 26. Cambridge: Cambridge University Press.

Pianka, E. R. (1970) On *r*- and K-selection. *American Naturalist*, **104**, 592–597.

Polzer, M., Overstreet, R. M. & Taraschewski, H. (1994) Proteinase activity in the plerocercoid of *Proteocephalus ambloplitis* (Cestoda). *Parasitology*, **109**, 209–213.

Riggs, M. R., Lemly, A. D. & Esch, G. W. (1987) The growth, biomass and fecundity of *Bothriocephalus acheilognathi* in a North Carolina cooling reservoir. *Journal of Parasitology*, **73**, 893–900.

Simmons, J. E. & Laurie, J. S. (1972) A study of *Gyrocotyle* in the San Juan Archipelago, Puget Sound, USA, with observations on the host, *Hydrolagus colliei* (Lay and Bennett). *International Journal for Parasitology*, **2**, 59–77.

Stunkard, H. W. (1955) Induced gametogenesis in a monogenetic trematode, *Polystoma stellai* Vigueras, 1955. *Journal of Parasitology*, **45**, 389–394.

Tanner, J. T. (1987) *Guide to the Study of Animal Populations*. Knoxville: University of Tennessee Press.

Tinsley, R. C. & Owen, R. W. (1975) Studies on the biology of *Protopolystoma xenopodis* (Monogenoidea): the oncomiracidium and life cycle. *Parasitology*, **71**, 445–463.

Wassom, D. L., Guss, V. M. & Grundmann, A. W. (1973) Host resistance in a natural host–parasite system. Resistance to *Hymenolepis citelli* by *Peromyscus maniculatus*. *Journal of Parasitology*, **59**, 117–121.

Wassom, D. L., DeWitt, C. W. & Grundmann, A. W. (1974) Immunity to *Hymenolepis citelli* by *Peromyscus maniculatus*: genetic control and ecological implications. *Journal of Parasitology*, **60**, 47–52.

Wassom, D. L., Dick, T. A., Arnason, N., Strickland, D. & Grundmann, A. W. (1986) Host genetics: a key factor in regulating the distribution of parasites in natural host populations. *Journal of Parasitology*, **72**, 334–337.

Wetzel, E. J. & Esch, G. W. (1995) Effect of age on infectivity of cercariae of *Halipegus occidualis* (Digenea: Hemiuridae). *Invertebrate Biology*, **114**, 205–210.

Wetzel, E. J. & Esch, G. W. (1996) Seasonal population dynamics of *Halipegus occidualis* and *Halipegus eccentricus* (Digenea: Hemiuridae) in their amphibian host, *Rana clamitans*. *Journal of Parasitology*, **82**, 414–422.

Wetzel, E. J. & Esch, G. W. (1997) Infrapopulation dynamics of *Halipegus occidualis* and *Halipegus eccentricus* (Digenea: Hemiuridae): temporal changes within individual hosts. *Journal of Parasitology*, **83**, 1019–1024.

Williams, H. H., Colin, J. A. & Halvorsen, O. (1987) Biology of gyrocotylideans with emphasis on reproduction, population ecology and phylogeny. *Parasitology*, **95**, 173–207.

Williams, J. A. & Esch, G. W. (1991) Infra- and component community dynamics in the pulmonate snail, *Helisoma anceps*, with special emphasis on the hemiurid trematode, *Halipegus occidualis*. *Journal of Parasitology*, **77**, 246–253.

Zelmer, D. A. & Esch, G. W. (1998*a*) Bridging the gap: the odonate naiad as a paratenic host for *Halipegus occidualis* (Trematoda: Hemiuridae). *Journal of Parasitology*, **84**, 94–96.

Zelmer, D. A. & Esch, G. W. (1998*b*) The infection mechanism of the cystophorous cercariae of *Halipegus occidualis* (Digenea: Hemiuridae). *Invertebrate Biology*, **117**, 281–287.

Zelmer, D. A., Wetzel, E. J. & Esch, G. W. (1999) The role of habitat in structuring *Halipegus occidualis* metapopulations in the green frog. *Journal of Parasitology*, **85**, 19–24.

Chapter 11

Factors influencing parasite populations

11.1 Density-independent factors

11.1.1 Introduction

According to Scott & Dobson (1989), '**regulation** occurs through the action of processes that tend to reduce the per capita survival or reproduction of the members of a population as the density of that population increases'. In other words, regulation implies the influence of **density-dependent** forces. Probably, however, the dynamics of most parasite infrapopulations are not regulated by density-dependent constraints. Instead, they are mostly affected by **density-independent** factors such as temperature or other general climatic conditions external to the host and by some biotic factors, primarily those associated with host behavior.

There are several generalizations that have been made regarding density-independent factors and the role they play in the dynamics of parasite infra-, component, and suprapopulations (Bradley, 1972). These factors are almost always extrinsic to both the hosts and parasites. Many of these species may be influenced by a large number of abiotic factors in the environment. Other species may be linked to an aspect of host behavior that the parasite exploits in order to enhance the probability of transmission. In some of these latter cases, the parasite would capitalize on predator–prey interactions in such a way that the host is more likely to be eaten by the next host in the life cycle (Moore & Gotelli, 1990).

Much available evidence suggests that the mortality rates of most parasites in hosts under natural conditions are independent of infrapopulation density (Esch & Fernández, 1994). Instead, parasite density or prevalence tends to be a function of seasonal changes in temperature and other physical parameters in the environment, or of natural senescence and mortality within the host population. For example, many freshwater snail species are the equivalent of annual plants, with a new cohort replacing an existing one each year. This means that as the snails die, the parasite community associated with them is also lost. Therefore, the replacement of an existing digenean fauna within a snail population having an annual cycle would also be an annual event, and would occur independent of any density-related effects.

The densities of certain parasite species under the influence of density-independent factors may fluctuate erratically over the long term. In reality, this statement is more of a prediction than a fact. The problem here is that there are far too few long-term studies to make acceptable generalizations. Kennedy & Rumpus (1977) argued that annual persistence in the prevalence levels of some parasites is not the result of density-dependent regulation, but rather is due to long-term stability within the ecosystem in which the parasite is found. Even though the parasite populations are non-equilibrial, they appear to be stable because the habitat is stable. In these situations, if the ecosystem is perturbed, the population biology of the parasite will be affected in some unpredictable fashion. Certainly, the findings of Marcogliese *et al.* (1990) in their study of the allocreadid fluke *Crepidostomum cooperi* in a eutrophic

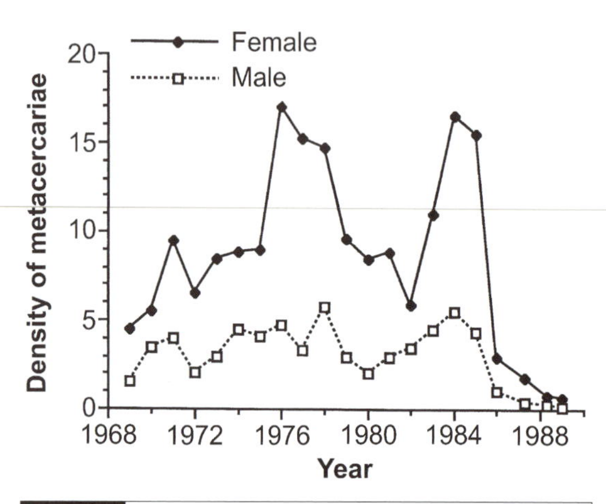

Fig. 11.1 Densities of *Crepidostomum cooperi* metacercariae in female and male *Hexagenia limbata* subimagoes from 1969 to 1989 in Gull Lake, Michigan, USA. (Modified from Marcogliese *et al.*, 1990, with permission, *American Midland Naturalist*, **124**, 309–317.)

lake in Michigan (USA) would support such a contention. This parasite uses centrarchid fishes as definitive hosts and sphaeriid clams as the first intermediate host. Cercariae released from the clams encyst as metacercariae in nymphs of the burrowing mayfly *Hexagenia limbata*. When nymphs emerge from the lake as subimagoes, they are highly vulnerable to predation by many species of fish. After 48 hours, the mayflies molt for the last time, mate, and the females return to the lake to oviposit. They are again highly vulnerable to predation. If an appropriate fish eats an infected mayfly, the life cycle of *C. cooperi* will be completed.

Several hundred subimagoes were collected each August over a 20-year period and the metacercariae counted (Marcogliese *et al.*, 1990). For the first 8 years, parasite densities increased significantly (Fig. 11.1). They declined slightly for the next 6 years and then increased again, sharply. In 1986, the component population density of metacercariae crashed and continued to decline over the next 3 years, reaching a 20-year low in the summer of 1989. The increasingly higher densities over the first 16 years corresponded with an environmental perturbation, namely progressive, cultural eutrophication. A sewer system for the lake's drainage basin was constructed, made operational in 1984, and the eutrophication process was reversed. Esch *et al.* (1986) earlier had speculated that mayfly nymphs were driven from deeper anoxic areas of the lake over the first 16 years (1969–84) into parts of the lake where transmission of the parasites to the mayflies was enhanced. Then, as eutrophication was reversed, they predicted that nymphs would move back into deeper parts of the lake and away from higher densities of infected sphaeriid clams. The result would be a diminution in the transmission of cercariae to the mayfly nymphs and a decline in metacercariae prevalence and density. Based on the follow-up study by Marcogliese *et al.* (1990), this appears to have happened in the way Esch *et al.* (1986) had predicted. One of the serious problems with such investigations, however, is that there are too few long-term studies with which their results could be compared. Moreover, without adequate controls or the ability to set up and conduct experimental protocols over the long term, many of the conclusions, or speculations, or both, must necessarily be inferential in nature. This makes the 'science' of long-term studies, whether at the population or community level, more difficult to assess and evaluate.

Another characteristic of parasite species that is influenced mainly by density-independent factors is that they are frequently subject to local extinction. Again, however, in order to know if extinction has occurred, the system must be followed over a period of time and, as just alluded to, such studies are virtually non-existent. In general, local extinction is most likely to occur in a perturbed ecosystem, or one in which colonization by infected definitive hosts is an annual, but erratic, event. An example of the latter situation is with the digenean *Hysteromorpha triloba* in a lake in South Dakota, USA (Hugghins, 1956). The definitive hosts for this parasite are double-crested cormorants *Phalacrocorax auritus*, that used the lake as a nesting area but, in 1955, following the opening of a public recreation area nearby, they abandoned the habitat. With the disappearance of the birds, the parasite also disappeared. These concepts of local colonization in combination with local extinction, and the nature of contributing factors, have not received the attention they deserve. Conversely, the sudden appearance of a new parasite in an ecosystem is also a characteristic of many species whose densities are primarily controlled by density-independent factors; it is, in effect, a colonization phenomenon (see Chapter 13).

Many parasites that are influenced by density-independent phenomena are found in short-lived hosts. Most of these parasites are highly seasonal in their transmission patterns, especially since their short-lived hosts are usually seasonal with their own life-cycle patterns. Indeed, many of their hosts are also affected by density-independent factors (and accordingly may be considered as *r*-strategists). In these cases, it is absolutely necessary for the life cycles of the parasites to be highly synchronized with those of their hosts in order for them to be successful. In many situations, both hosts and parasites are responding to the same external stimuli, probably because they coevolved, or coadapted, and could have responded to some of the same or similar selection pressures over evolutionary time. Parasites in short-lived hosts generally will not build up large numbers because the life spans of the hosts are not long enough to permit the establishment of large infrapopulations. The result is that density-independent constraints may not influence population size because the parasite infrapopulation densities do not reach a size where nutrient or other resources could become restraining. There are, however, several exceptions to such a possibility and these will be considered subsequently.

Finally, many parasites that are near the limit of their geographic ranges will be affected by density-independent factors. This is because their physiological tolerances are usually quite narrow and they are simply not highly adaptable to extreme environmental vagaries such as would occur at the edges of their range. These species, and perhaps even their hosts, if similarly affected by a range limitation, would be easily subject to such vagaries and, therefore, to limitations conferred for both colonization and extinction.

11.1.2 *Caryophyllaeus laticeps* and temperature

Caryophyllideans include a relatively small group of monozoic, or unsegmented, cestodes. Adults of *C. laticeps* are found within the intestines of freshwater fishes. Eggs of the parasite are shed from definitive hosts and are ingested by a tubificid oligochaete. The egg hatches, an oncosphere emerges and develops into a procercoid in the hemocoel. When a bottom-feeding fish eats the infected tubificid, the parasites develop into what many consider as **neotenic plerocercoids** that mature sexually and produce eggs, completing the life cycle. This life cycle is representative of most other caryophyllideans.

Caryophyllaeus laticeps has a fairly cosmopolitan distribution in the Northern Hemisphere. It has a wide range of definitive hosts and probably intermediate hosts as well. The ease with which definitive hosts can be collected and the relative simplicity of the parasite's life cycle make it an attractive model for study. It has, therefore, been the focus of fairly intense research efforts over the years. The choice of *C. laticeps* for illustrating the effects of temperature on the population biology of a given parasite species was made with a clear purpose. A review of the literature will show not only a striking correlation between water temperature and the parasite's population biology, it will also demonstrate how differently the parasite behaves from one host species to another, even within a relatively narrow geographic range.

Consider first the study by Kennedy (1968) on *C. laticeps* in dace *Leuciscus leuciscus*, in the River Avon, Hampshire, England. His investigation was conducted for 12 consecutive months. Two sets of data from his paper are instructive with respect to the influence of temperature on the population biology of the parasite. The first, in Table 11.1, shows the clear seasonal characteristics of both the prevalence and density (mean worm burden) of *C. laticeps* in dace. The second, in Fig. 11.2, shows the changing pattern of maturation by the parasites after recruitment from infected tubificids. Several conclusions can be drawn from these data. Recruitment of juvenile *C. laticeps* began in December and was continuous through June when it stopped completely. Mature, egg-shedding adults were present from January through June when they disappeared completely. Highest prevalences of all worms, disregarding the state of maturity, occurred from January through June. The highest densities were from January through May, with a few present in December and June. An

Table 11.1 Seasonal variation in the degree of infection of dace *Leuciscus leuciscus* by the cestode *Caryophyllaeus laticeps*

Year and month		Number of dace		Percent infection	Number of worms found	Mean worm burden	Water temperature (°C)	
		Examined	Infected				Mean	Range
1966	March	27	16	59.2	144	9	7.8	6.6–8.9
	April				no samples taken		9.25	5.0–11.1
	May	30	18	60.0	153	8.5	11.5	8.3–14.4
	June	27	6	22.2	7	1.1	14.6	13.3–15.5
	July				no samples taken		14.9	14.3–17.6
	August	30	0	0	0	0	15.5	13.3–17.7
	September	21	0	0	0	0	14.9	11.2–17.7
	October	28	1	3.6	1	1	11.1	8.3–13.3
	November	27	0	0	0	0	8.6	7.2–10.0
	December	23	2	8.7	3	1.5	6.8	4.4–8.9
1967	January	26	6	23.1	35	5.8	6.9	4.4–10.2
	February	25	14	56.0	262	18.9	7.3	4.4–10.2

Source: From Kennedy, 1968, with permission, *Journal of Parasitology*, **54**, 538–543.

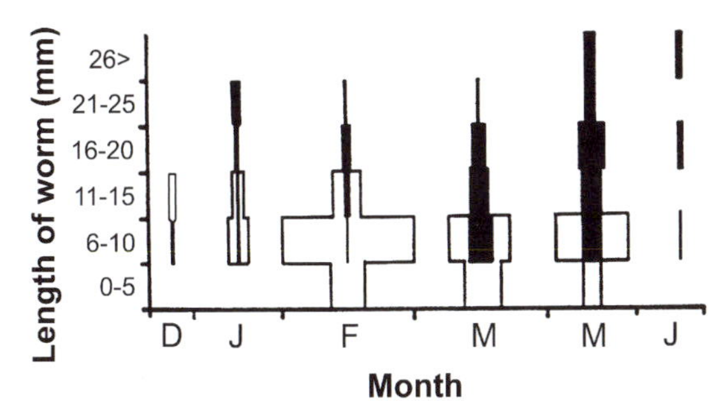

Fig. 11.2 Monthly pattern of changes in the length and state of maturity of *Caryophyllaeus laticeps* in dace *Leuciscus leuciscus*. The data are expressed as actual numbers of mature (not shaded) and gravid (shaded) worms. (Modified from Kennedy, 1968, with permission, *Journal of Parasitology*, **54**, 538–543.)

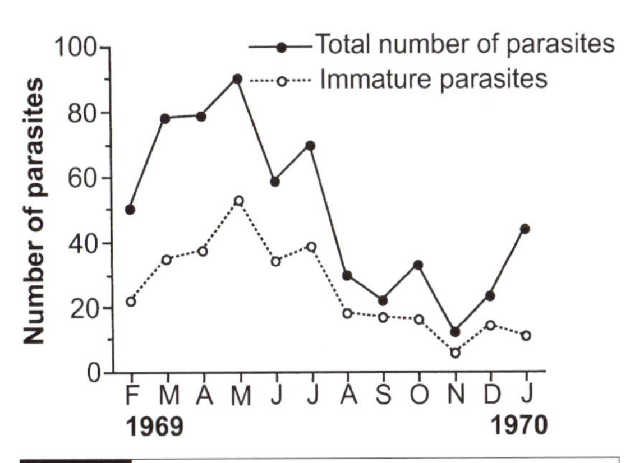

Fig. 11.3 Seasonal variation in total parasite numbers and immature parasites (*Caryophyllaeus laticeps*) per sample of 30 fish (*Abramis brama*). (Modified from Anderson, 1974, with permission, *Journal of Animal Ecology*, **43**, 305–321.)

examination of the data in Table 11.1 shows an almost perfect relationship between rising temperature and both the recruitment and maturation of the parasite. Despite these strong correlations, Kennedy (1968) proposed that a direct relationship did not exist, arguing instead that it was spawning hormones that were probably triggering maturation. He also suggested that elimination of the parasite from dace in early summer was due to some sort of rejection response that was stimulated by the rising temperature. Kennedy & Walker (1969) subsequently presented evidence that purported to support their contention.

One facet of the transmission process not examined by the 1968 Kennedy study was infection in the tubificid intermediate host *Psammoryctes barbatus*. Subsequently, however, Kennedy (1969) found that tubificids were infected throughout the year and that the feeding behavior of dace was unaltered from one season to the next. The implication was that dace were constantly being exposed to infected tubificids but, during certain times of the year, the parasites were incapable of becoming established even though they were still being actively recruited. Based on this observation and the one made by Kennedy & Walker (1969), he asserted that the overriding controlling factor was the temperature-dependent rejection response in the fish host. Subsequently, Kennedy (1971) conducted a series of experimental studies on *C. laticeps* using the orfe *Leuciscus idus*, a species of fish that is closely related to dace. Based on these results, he concluded that circulating antibodies were not involved in rejection, but he still could not characterize the exact nature of the response.

Anderson (1974, 1976) also studied the component population biology of *C. laticeps*, but the definitive host for his investigation was the bream *Abramis brama*. The results he obtained were similar in some ways to those of Kennedy (1969, 1971), but different in others. For example, Anderson (1974) observed a clear seasonal pattern, with the density of parasites declining sharply during the summer months (Fig. 11.3). Indeed, Anderson (1974) observed a significant relationship between temperature and survivorship of *C. laticeps* in bream (Fig. 11.4). A striking difference in the results of Kennedy (1968) and Anderson (1974), however, was that in dace the parasite disappeared completely after June, not to reappear until the following December. In bream, on the other hand, gravid adults remained throughout the summer and the parasite was present throughout the year. The occurrence of gravid adults in the summer months was one of the key elements of Anderson's (1974) analysis regarding the dynamics of *C. laticeps*. Thus, it was during the summer months that the largest proportion of gravid worms was present, at the time when the overall densities of the worms were also the lowest. He speculated that this strategy by the parasite prevented a large number of intermediate hosts from becoming infected at the same time. In

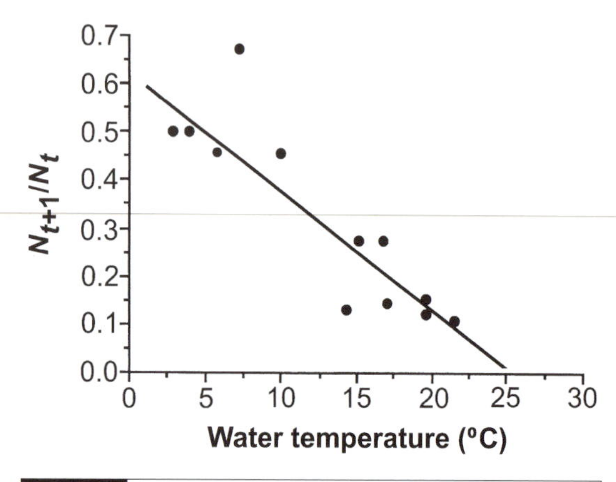

Fig. 11.4 The relationship between the ratio N_{t+1}/N_t and water temperature, where N_t is the number of immigrant parasites at time t and N_{t+1} is the number of mature parasites at time $t+1$. (Modified from Anderson, 1974, with permission, *Journal of Animal Ecology*, **43**, 305–321.)

turn, he suggested this would keep the size of the adult population in the fish host at a low level while minimizing the potential for intraspecific competition and maximizing the efficiency of reproductive effort.

Another aspect of Anderson's (1976) study involved the spatial distribution of infected intermediate hosts in combination with the feeding behavior of bream and the frequency distribution of parasites within the definitive hosts. The parasites were aggregated within the fish hosts and the distribution was best described by the negative binomial model (see Chapters 10 and 12 for additional discussion of this concept). Additionally, he noted that the age structure of the fish population was a significant factor affecting the component population dynamics of *C. laticeps*. In short, both the age distributions and the feeding behaviors in different size classes of fishes varied and, in turn, affected the population biology of the parasite.

There are strong seasonal similarities between the results obtained by Anderson (1976) and those of Kennedy (1968, 1969, 1971). On the other hand, a basic difference was that *C. laticeps* was present in bream throughout the year, but not in dace or orfe. This would suggest that the latter species may not be the normal hosts for the parasite and could provide an explanation for the reported differences.

In summary, then, these studies serve to illustrate the impact of a ubiquitous abiotic factor, namely temperature, on the seasonal infrapopulation dynamics of the cestode *C. laticeps*. Moreover, the efforts of Anderson (1976) also indicate how the spatial distribution of infected tubificids may influence the dispersion pattern of adult parasites in their definitive hosts.

11.1.3 *Bothriocephalus acheilognathi*: the role of temperature, competition and predation

The Asian tapeworm *Bothriocephalus acheilognathi* was first described from cyprinid fishes in Japan. It was introduced into the USA in the mid-1970s (Granath & Esch, 1983*a*) and into Britain at about the same time (Andrews *et al.*, 1981). The parasite has several unique characteristics. First, it has been reported from at least 40 different species of definitive hosts, so its host specificity is much broader than that of almost any other species of cestode. This, in combination with its broad specificity at the copepod intermediate host level, unquestionably accounts for its wide geographical distribution and for its ability to colonize rapidly once it is introduced into a new habitat. Another unusual feature of the parasite is its capacity for what appears to be indeterminate growth. For example, in the mosquitofish *Gambusia affinis*, it will grow to a maximum length of approximately 3 cm, whereas in the common carp *Cyprinus carpio*, it may reach 15–20 cm. The size of most helminth parasites at maturity is fixed within some narrow range, indicating determinate growth patterns for such species, but not for *B. acheilognathi*.

The life cycle of *B. acheilognathi* is simple. Eggs are produced in the piscine definitive host and shed in the feces. They hatch in the water and the coracidium that emerges is consumed by cyclopoid copepods. In the hemocoel of the copepod, the parasite develops into a procercoid. When the infected copepod is eaten by the definitive host, the procercoid develops first to the plerocercoid stage in the small intestine of the fish and, when provided an appropriate stimulus, into an adult parasite.

Most of the investigations that comprise the

present case study were conducted in Belews Lake, a 1563-ha cooling reservoir, located in the Piedmont region of North Carolina (USA). There are three reasons for focusing on these particular investigations. First, they were conducted over a 7-year period, giving them a reasonably long time frame. Second, because the lake is a cooling reservoir, there are ambient areas where temperatures are 32–33 °C during summer, whereas in thermally-altered areas, water temperature rises to nearly 40 °C. These differences provided an ideal opportunity to compare the effects of warm, and even warmer, temperatures on the population biology of the parasite. Finally, the lake was established in 1971 and had a normal fauna of fishes until 1974 when the effects of selenium pollution were first observed. Then, over a period of 2 years, all but two of the original 26 piscine species became locally extinct; a cyprinid species was introduced into the lake in 1978, increasing the number of species present to three. By the time this series of studies was begun in 1980, those species present were the mosquitofish *Gambusia affinis*, the fathead minnow *Pimephales promelas*, and the red shiner *Notropis lutrensis*. Carp *Cyprinus carpio*, channel catfish *Ictalurus punctatus*, and green sunfish *Lepomis cyanellus* were occasionally seen, but were apparently unsuccessful in spawning. Near the headwaters of the lake, there is an exceedingly strong chemocline that has effectively created two distinct, but contiguous, bodies of water, the main one polluted by selenium and another that is not polluted. At the latter location, there is a fish community that is identical to the one that was present throughout the reservoir before the selenium pollution; it is also typical, in terms of its fish community, for the southeastern part of the USA. Among the other species present at the unpolluted end of the reservoir is the carnivorous largemouth bass *Micropterus salmoides*. The tapeworm is present in fishes in the unpolluted part of the reservoir. This unique physical boundary and the resulting biological structure in the lake also created an opportunity for comparing the presence and absence of piscine predation on the population biology of the parasite.

The studies, begun in 1980, were initially designed to compare the prevalence, density, and recruitment patterns of the parasite in mosquitofish from thermally altered and ambient locations in the polluted part of Belews Lake (Granath & Esch, 1983*a*, *b*, *c*). The effects of temperature were striking (Figs. 11.5, 11.6). In both areas, parasite densities and prevalences were lowest during the summer months. Densities at both locations then increased sharply in the fall, peaked in early winter, and declined again in late winter. At the thermal location, however, recruitment began 2 weeks earlier, lasted 2 weeks longer, and was interrupted for several weeks in the summer when water temperatures exceeded 35 °C. This observation illustrates how a parasite's component population biology can be affected by seasonal changes in water temperature and that excessively high temperatures can fundamentally alter normal patterns of parasite recruitment.

Laboratory investigations confirmed the field observations regarding the relationship between the recruitment of *B. acheilognathi* and water temperature (Granath & Esch, 1983*b*). The studies also analyzed the infrapopulation structure with respect to the influence of water temperature on the parasite's developmental pattern. Beginning in October and continuing into May, more than 98% of the worms present were non-segmented. When water temperatures rose above 25 °C, segmentation proceeded at a rapid rate (Fig. 11.7). The segmentation was accompanied by a sharp and significant decline in infrapopulation densities, an observation not unlike that of Anderson (1976) for *C. laticeps* in bream. Granath & Esch (1983*b*) speculated that a temperature-dependent rejection response could be responsible for the decline. However, they favored the notion that intraspecific exploitative competition for limited spatial and nutrient resources was most likely responsible for the decline in infrapopulation densities within the mosquitofish during the summer months. In effect, they proposed that whereas a density-independent factor (temperature) was the driving force for the seasonal changes in infra- and component population biology, a density-dependent factor (competition) was involved in regulating (*sensu* Scott & Dobson, 1989) infrapopulation densities.

Subsequently, a follow-up study of the *Bothriocephalus*–mosquitofish system in Belews

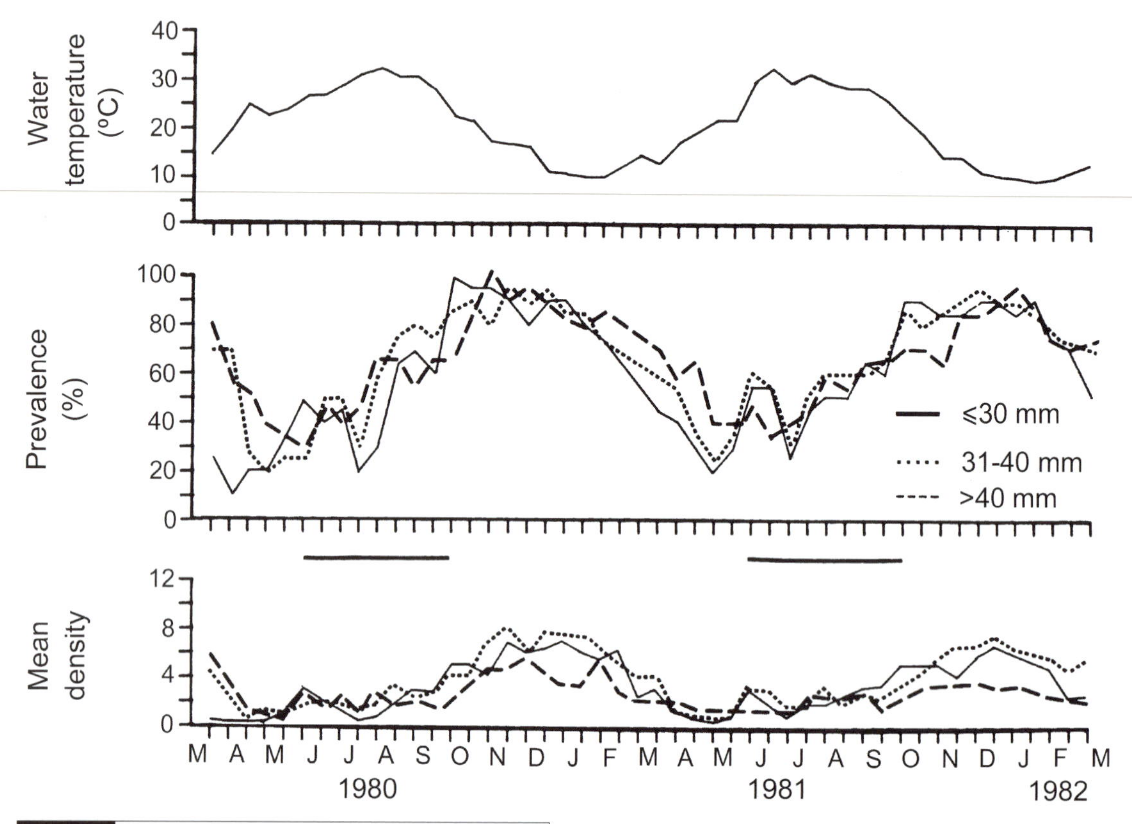

Fig. 11.5 Biweekly changes in temperature, prevalence, and density of *Bothriocephalus acheilognathi* within three size classes of mosquitofish *Gambusia affinis* from an ambient-temperature site in Belews Lake, North Carolina, USA. The horizontal bar indicates when recruitment of the cestode occurred. (Modified from Granath & Esch, 1983, with permission, *Proceedings of the Helminthological Society of Washington*, **50**, 205–218.)

Lake found that a striking change in the population biology of the parasite had taken place from the 1980–2 period to 1984–6 (Marcogliese & Esch, 1989). Both the prevalence and densities of the parasite in mosquitofish were much lower in 1984–6. Moreover, maximum prevalence and density in mosquitofish shifted from the fall months in 1980–2 to summer in 1984–6. These observations would, at first glance, tend to raise questions regarding the conclusions of Granath & Esch (1983*a*, *b*, *c*) as to the influence of temperature on the population biology of the parasite. However, as mentioned earlier, a third species of fish was introduced into the lake in 1978. Between 1976 and 1980, the dominant species was the fathead minnow *Pimephales promelas*, a detritivore. Mosquitofish were also quite abundant in the early 1980s, ranging far from their normal littoral zone habitat into limnetic parts of the lake while foraging, probably because of the absence of both piscine predators and potential competitors. The introduced species was the red shiner *Notropis lutrensis*, which is a highly efficient planktivore and a competitor with mosquitofish. Three copepods are known to transmit the parasite in Belews Lake, but the primary copepod host for *B. acheilognathi* in the lake at the earlier time was *Diacyclops thomasi*. In 1980–2, it was abundant in both late spring and early summer, but underwent diapause from early summer to fall. In 1984–6, only the copepod *Tropocyclops prasinus* was abundant and *D. thomasi* densities were sharply lower than in the earlier period. Marcogliese & Esch (1989) proposed that, with the introduction of the red shiner in 1978, the diversity of the copepod community was substantially altered between then and the 1984–6 period. Moreover, the population biology of the parasite was significantly changed as well. Although *T. prasinus* is a suitable intermediate host for the parasite, it is a much smaller

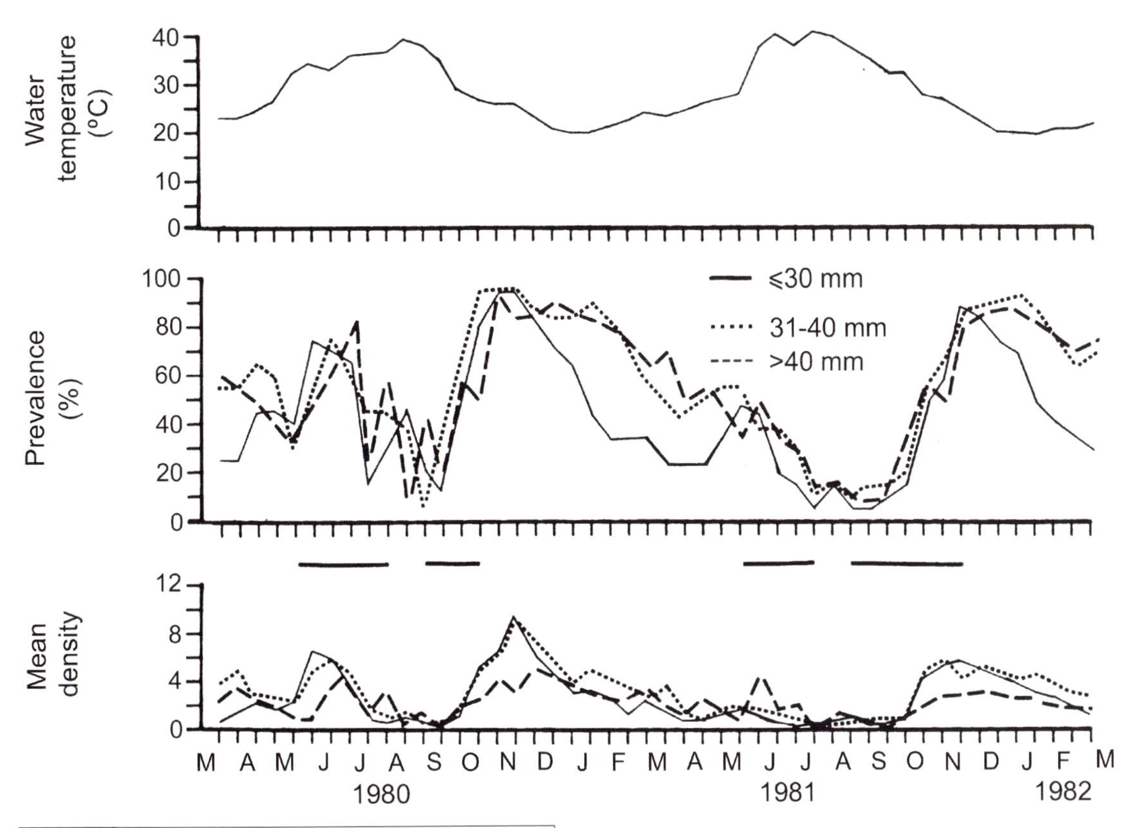

Fig. 11.6 Biweekly changes in temperature, prevalence, and density of *Bothriocephalus acheilognathi* within three size classes of mosquitofish *Gambusia affinis* from a thermally altered site in Belews Lake, North Carolina, USA. The horizontal bar indicates when recruitment occurred. (Modified from Granath & Esch, 1983, with permission, *Proceedings of the Helminthological Society of Washington*, **50**, 205–218.)

species than *D. thomasi*, about half its size. As a result, it is less vulnerable to predation by visually oriented fish such as the dominant, planktivorous red shiner. According to Marcogliese & Esch (1989), '*T. prasinus* would not be as effective as *D. thomasi* in transmitting the parasite to the definitive host, simply because the former is smaller and suffers less predation. As a consequence, fewer parasites were transmitted to mosquitofish, with lower autumnal abundance (density) and prevalence in 1984–6 than in 1980–2.' Indeed, Riggs (1986) and Marcogliese & Esch (1989) reported that the diversity of the copepod community was strikingly altered by size-selective predation over the 10-year period from 1976 through 1986 and that, presumably, changes in the parasite's population biology within the reservoir reflect these changes.

Bothriocephalus acheilognathi is a unique cestode for several reasons, not the least of which is its lack of specificity at both the intermediate and definitive host levels. As previously noted, this accounts, in part, for its wide dispersal throughout Asia, Europe, and North America. It is clear that the population biology of the cestode in its definitive host is affected by normal seasonal changes in water temperatures. However, in Belews Lake its population biology also has been greatly impacted through the long-term alterations in the community structure of the copepod intermediate hosts brought on via selenium pollution in the reservoir and, in turn, by the selective predation of planktivorous fishes such as the red shiner.

11.1.4 *Eubothrium salvelini*, a case for temperature and host foraging behavior

A 15-year study was conducted on the component population biology of the pseudophyllidean cestode *Eubothrium salvelini*, in sockeye salmon *Oncorhynchus nerka*, from Babine Lake, British

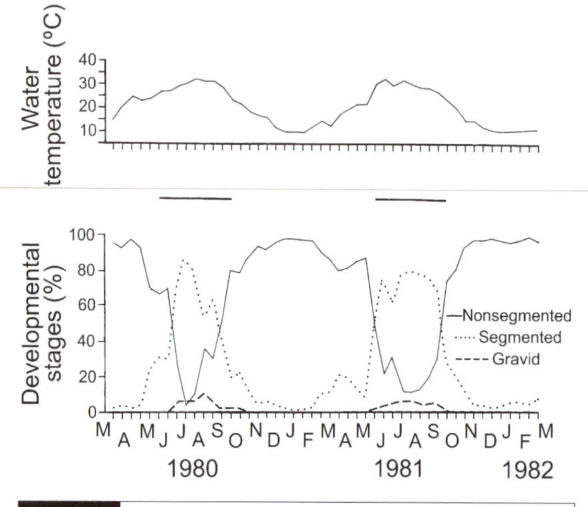

Fig. 11.7 Seasonal changes in water temperature and developmental stages of *Bothriocephalus acheilognathi* within mosquitofish *Gambusia affinis* from an ambient site in Belews Lake, North Carolina, USA. Each percent refers to the proportion of the total infrapopulation of the cestode at that stage of development. The horizontal bar indicates when recruitment of the cestode occurred. (Modified from Granath & Esch, 1983, with permission, *Journal of Parasitology*, **69**, 1116–1124.)

Columbia, Canada (Smith, 1973). Salmon spawn in tributaries that feed the lake. Fry of the salmon then migrate into the lake in May and June where they remain for 1 year as underlings. By the age of 1, they mature into smolts and are ready for their migration to the ocean. There are two populations within the lake, one residing in the northern basin and one in the southern basin. Migration to the sea is bimodal, with the northern population moving a few weeks sooner than the southern population. The stimulus for migration is apparently 'ice-off', or spring turnover.

Adult *E. salvelini* occur in the intestines of underlings and smolts. Eggs released from the fish are ingested by the copepod *Cyclops scutifer*. On hatching, the parasites penetrate the gut wall and enter the hemocoel where they develop into procercoids. These are ingested by planktivorous smolts and underlings where procercoids are freed upon digestion of host tissue; they then mature sexually. On migration to the ocean, the adult parasites are lost from the salmon.

The parasite recruitment pattern generally was quite consistent over the 15 years of study. As smolts pass through the lake in May and June, they shed eggs that are eaten by *C. scutifer*. At about the same time, fry move into the lake from tributaries where they were spawned and then become underlings. As underlings, they begin to consume copepods and become infected in the process. Prevalence in the underlings reached between 30% and 40%. Interestingly, the prevalence did not go higher even though infected copepods remained in the lake. The explanation for the cessation in recruitment is related to a complete shift in prey preference from copepods to cladocerans in July; as a result, the fish stop recruiting *E. salvelini*. Moreover, *C. scutifer* is most abundant in May and June and then the population density declines rather precipitously at about the time the cladocerans become more abundant. This seasonal succession in the plankton community in Babine Lake is typical for most lake systems in temperate parts of the world.

The shift in diet from copepods to cladocerans is most important for the population biology of the parasite and probably the host as well since it is known from hatchery studies that high parasite densities will kill smolts. Thus, continued feeding of underlings on copepods beyond June could easily induce death of the fish in Babine Lake. Moreover, not only is there synchrony in dietary preference and transmission of the parasite, synchrony is also important to the survival of both the host and the parasite. The patterns of parasite recruitment and the level of parasite prevalence were consistent over the 15-year period with two exceptions, one lasting for two summers and another for three. During these years, the prevalence of the parasite was substantially lower than normal. The change was attributed to the asynchronous appearance of potential intermediate hosts (the copepods) and the presence of infected definitive hosts. No reasons for the asynchrony were suggested.

The component population biology of *E. salvelini* in salmon is certainly influenced by seasonal temperature changes. However, there is also a behavioral element involved in affecting the population biology of the parasite when the underlings change their dietary preferences from one planktonic species to another. The dietary shift may be related to a natural, seasonal decline in the copepod densities, but more likely it is due to alterations in size-selective predation by the underlings. In other words, the underlings are better off foraging for larger cladocerans than for smaller copepods from the standpoint of energy expenditure and return for the effort (a 'getting the most bang for the buck' idea). This is supported by the observation that the population density of *C. scutifer* is seasonally bimodal with a large increase in numbers in August as well as in the early summer. The underlings do not return to copepods as a dietary preference during August, foraging instead on the larger cladocerans.

The effects of temperature on the component population biology of *E. salvelini* are clearly illustrated by the long-term study of Smith (1973). However, this abiotic influence has been modified by a shift in foraging behavior by underlings during mid-summer away from copepod intermediate hosts to cladocerans that do not serve as hosts for the parasite. For this system, then, there is a combination of abiotic and ontogenetic biotic forces that affect the component population biology of the cestode in the fish definitive host.

11.1.5 *Diphyllobothrium sebago*, spatial and temporal transmission

Diphyllobothrium sebago is a common tapeworm in herring gulls *Larus argentatus* in the Rangeley Lakes area of northern Maine, USA (Meyer, 1972); the parasite has a three-host life cycle. Eggs are shed from herring gulls in their feces and, from these, coracidia hatch and emerge in the water. The planktonic copepod *Diacyclops thomasi* consumes the coracidia, which develop into procercoids in the hemocoel. In turn, planktivorous smelt fry *Osmerus mordax* eat infected copepods. The procercoids then enter the tissues of the new host where they develop into plerocercoids. When a gull eats an infected smelt, the cycle is complete.

An unusual aspect in the life cycle of *D. sebago* is the elegant and complex temporal and spatial synchrony exhibited by the parasite and its hosts. Smelt infected with plerocercoids migrate from the lakes into feeder streams and spawn. Then they return to the lake where they die. Subsequently, the young fry return to the lake where they migrate into deeper, hypolimnetic locations. They assume a planktivorous mode of feeding before they move back into the streams to spawn two years later.

Gulls become infected when they feed intensively on dead and dying smelt as they return to the lake after spawning. This corresponds with the time when gulls are raising their young so that young birds also become infected. Adults mostly favor the littoral zone in their early foraging. However, when the young birds fledge, both the fledglings and the parents move to the deeper, limnetic parts of the lake to feed. By this time, *D. sebago* has matured sexually and is producing eggs. Simultaneously, smelt fry are intensively feeding on infected copepods and thus become infected; in doing so, they enable the parasite to complete its life cycle.

In summary, the success of the parasite in this setting rests, in part, on the temporal sequence of smelt spawning and on the reproductive activity of the definitive host. There is also a spatial component in which infected birds move from the near-shore, littoral zone to the limnetic areas of the lakes. This ensures that eggs released from adult parasites are shed in open water; eggs sink into the hypolimnion before hatching, providing a much greater chance for the free-swimming larval stages to come into contact with appropriate intermediate hosts which are cold-water stenotherms. The population biology of the parasite appears to be directly influenced by normal seasonal events that involve the behavior of both definitive and intermediate hosts and, to a much lesser extent, by other density-independent abiotic factors.

11.2 Density-dependent factors

11.2.1 Introduction

Any density-dependent factor that influences the population biology of a free-living organism automatically implies regulation. Among others, such factors may involve predation, competition, or parasitism. For most parasite infrapopulations, however, predation can be excluded as a major regulator; the only predatory parasites are some digeneans that have rediae as a part of their life cycle (Kuris, 1990; Sousa, 1990). Indeed, Kuris (1990) has developed an elaborate scheme of dominance hierarchies that applies to digenean infracommunities within the horn snail *Cerithidea californica*. For some of these digeneans, predation by parasitic rediae is an important mechanism for maintaining their position within the dominance hierarchy.

The immune responsiveness of some hosts is certainly a density-dependent phenomenon in some systems. However, a serious problem in these cases is that field evidence, except in a few instances, is unavailable and conclusions can only be inferred from laboratory-based, experimental observations. Generally, immune capacities are only partially successful in regulating parasite infrapopulations; complete or **sterile immunity**, such as the kind induced by measles, has been reported for only a very few species. The complex regulatory system described by Wassom and his co-workers (1973, 1974, 1986; see section 11.2.3) for *Hymenolepis citelli* and *Peromyscus maniculatus* is unusual in that it is based on a combination of both host genetics and immunity.

Competitive interactions among parasites appear to result in both exclusion, or at least partial exclusion, and reductions in fecundity. The effects of competition have been viewed largely in terms of parasite infra- and component community dynamics and less so from an infra- or component population standpoint (Holmes, 1973; Bush & Holmes, 1986; Esch *et al.*, 1990). In the present chapter, only a few of the more classic cases of competition will be reviewed; most of the examples will be reserved for separate consideration within the context of parasite community dynamics (see Chapter 14).

The phenomenon of developmental arrest in nematodes (Schad, 1977) is, in part, influenced by density-dependent factors. Arrest, as defined by Michel (1974), is a temporary cessation in the development of parasites at a precise point in their early development, 'where such an interruption contains a facultative element, occurring only in certain hosts, certain circumstances, or at certain times of the year, and often affecting only a proportion of the worms'. Schad (1977) also emphasized that arrest is a temporary cessation of development. He pointed out that in some situations, it is a natural mechanism by the parasite to avoid harsh external environmental conditions and, in this sense, could be a density-independent process. However, he also noted that 'it is equally apparent that the entry of newly acquired parasitic nematodes into arrest, often at precisely the time when adult worm burdens are becoming particularly dense, is a highly adaptive mechanism for regulating [infra]populations of adult worms'. It is, therefore, an unusual mechanism for avoiding competition and clearly has density-dependent ramifications.

An interesting example of developmental arrest has long been associated with *Strongyloides stercoralis* in humans (Pelletier, 1982). A number of prisoners of World War II in southeastern Asia were diagnosed with strongyloidiasis some 40 years later and without any chance of reinfection from extraneous sources in the intervening years. Apparently, the infection became patent after the 40 years when previously arrested female worms resumed reproduction and the hosts became autoinfected. According to Schad *et al.* (1997), autoinfection-based turnover of adult worms of this sort may imply that adult worms (all females) in occult infections can return to fecundity. They provided experimental evidence to support this hypothesis and, more interestingly, suggest that the progeny of rejuvenated worms can quickly initiate an autoinfective burst. Moreover, they suggested that the phenomenon is probably related to some sort of immune responsiveness rather than a direct effect on the parasite itself. The implications regarding autoinfection, immunity, and parasite population biology are clear and far-reaching.

Finally, there is the interaction between parasite density and the potential for parasite-induced host mortality. Beginning with Crofton (1971*a*, *b*), the relationship between aggregation of parasite infrapopulations and regulation has been examined by a large number of investigators, both in the field and in the laboratory. The contention is that parasites are able to cause host mortality under certain circumstances. If this could be established, then there would be a basis for inferring regulation. There is also the potential for mutual, regulatory interaction between host and parasite populations. This occurs when parasite density becomes high enough to cause mortality of a host either directly or indirectly. By causing the host to die, the parasite clearly is affecting the host population density. Simultaneously, the parasite is affecting its own density at the component level and possibly at the suprapopulation level. If the stage that is eliminated through host mortality is reproductive, then the rate at which new propagules are introduced into the environment could be affected and this would also affect parasite numbers. The extent of density-dependent regulatory interaction under natural conditions is still under debate and will be considered further in Chapter 12.

11.2.2 Host immunity and the regulation of parasite infrapopulations

The impact of the immune response on parasite infrapopulation biology within humans is known but, from an experimental and strictly quantitative point of view (*sensu* Crofton, 1971*a*), it is not well understood or studied. In natural populations of other animals, much less is known. In the laboratory, the tapeworm *Hymenolepis nana* has been found to stimulate sterile immunity in its mouse host lasting for up to 120 days (Heyneman, 1962; Weinmann, 1966). Esch (1983) speculated that the duration of such an immune response under natural conditions could be enough to provide a density-dependent regulatory influence, but there are no data from field studies to support the assertion. A second tapeworm, *Taenia pisiformis*, has a tissue-dwelling cysticercus in its rabbit intermediate host *Oryctolagus cuniculus*. In natural populations of rabbits in New Zealand, prevalence of the parasite increased with increasing host age, but mean density remained constant once a certain density threshold was reached (Bull, 1964). The number of eggs that were initially consumed by the rabbit apparently determined the long-term size of the infrapopulation since subsequent exposure to eggs of the parasite did not produce an expansion of the infrapopulation size. While the immune responsiveness in rabbits was not evaluated in the laboratory, the evidence certainly suggests density-dependent regulation via a strong immune response. Sterile immunity is also known for the hemoflagellate *Leishmania tropica*, the causative agent for oriental sore (see Fig. 3.9). This parasite produces an open skin lesion on humans; after it has healed, the individual cannot be re-infected with the parasite.

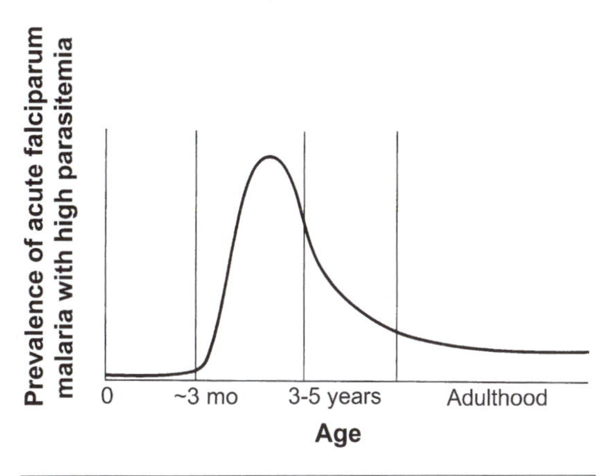

Fig. 11.8 An age–prevalence curve for the malarial parasite *Plasmodium falciparum*, in a hypothetical African village. Newborns have transplacentally acquired immunity to infection. After 3 to 6 months of age, individuals become highly susceptible to infection. After several years of exposure, immunity is acquired and parasitemias are low.

Whereas this parasite is not regulated in terms of the classic density-dependent process, it is nonetheless regulated by the host immune system.

The prevalence of the malarial parasite *Plasmodium falciparum* within a human population where the parasite is being transmitted at a high rate produces a time-curve that suggests transplacentally acquired immunity in newly born infants (Fig. 11.8). The immunity lasts for about 3 months before disappearing. Then, for approximately the next 5 years, children are totally susceptible to infection by the parasite; in Africa alone, at least 1 million die from malaria each year. Humans are apparently capable of acquiring immunity to the parasite because, after 5 years, the prevalence of the disease declines. It is impossible to assign any real significance to the effects of the immune response in humans on the overall population biology of the parasites or the hosts since field studies are not generally designed to answer such questions.

Schistosomiasis is another disease in which a relationship between immunity and parasite density is known to occur and for which there are some field data available. If the prevalence or density of schistosomes is plotted against age in a village where the parasite is endemic, peak numbers in both parameters will be seen in the

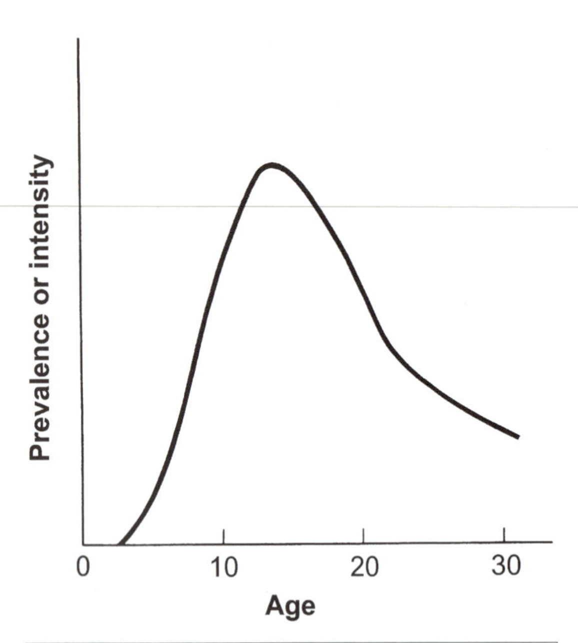

Fig. 11.9 An age–prevalence curve for *Schistosoma* spp. in a hypothetical African village. The decline in prevalence with age may be attributable to a slow, spontaneous mortality of adult worms combined with acquired resistance, or a reduced level of exposure, or both.

early teenage years (Fig. 11.9). After the peak, both prevalence and parasite density decline. The exact explanation for the decline is not known, but it may be due to any of several factors. It could be related to natural senescence and mortality of older worms. Perhaps it is affected by acquired resistance, or by a change in behavior of the potential host with age so that there is a reduction in exposure to the parasite with passing time. Based on animal models in laboratory settings, it is known that **concomitant immunity** will keep parasite infrapopulations at a lower level by effectively preventing the superimposition of new infections on pre-existing ones.

The immune capacity of an individual definitive host obviously plays a significant role in the infrapopulation biology of many parasites. However, as we have noted, documentation of this effect in natural systems is not extensive and, therefore, requires substantial additional study. Moreover, it would be of interest to extend these efforts to the level of the intermediate host and, in particular, to molluscan intermediate hosts since there is mounting laboratory evidence that snails have a reasonably powerful immune capacity.

11.2.3 Host genetics, immunity, and *Hymenolepis citelli*

One of the most unusual forms of host–parasite interaction occurs between the deer mouse *Peromyscus maniculatus* and the tapeworm *Hymenolepis citelli*. Adult parasites are found within the intestines of deer mice where they produce eggs that are shed to the outside in feces. Eggs are then accidentally ingested by camel crickets *Ceuthophilus utahensis*. They hatch, larvae emerge, penetrate the gut wall, and migrate to the hemocoel of the cricket. In the hemocoel, the parasite develops into a cysticercoid. When eaten by a deer mouse, the cysticercoid develops into an adult cestode and the cycle is complete.

Lab-reared deer mice were 100% susceptible to infection with *H. citelli* (Wassom *et al.*, 1973). However, most of the animals eliminated the worms before they were able to produce mature proglottids. When mice were subsequently challenged with a new infection, they demonstrated a marked resistance to the parasite. In a few mice, however, there was no resistance to either the stimulating or the challenge infection. The resistance established in mice was density-dependent, being directly related to the dose of cysticercoids presented to the mice in the challenge infection.

Subsequently, it was shown that the acquired resistance by *P. maniculatus* to *H. citelli* was under the influence of a single autosomal dominant gene and that susceptible hosts were genetically homozygous for the recessive gene (Wassom *et al.*, 1974). Moreover, the immunity could be transferred to uninfected hosts via what they termed 'immune' lymphoid cells, but not with serum from infected hosts. The susceptibility of outbred deer mice to infection was compared between the Utah (USA) strain of *H. citelli* and one isolated from California (USA) deer mice. The Utah mice responded much more rapidly to the Utah strain of the parasite than to the one from California. Field research revealed that the parasite also was patchily distributed, occurring only in small foci that favored the coexistence of both the definitive and the intermediate hosts. It was speculated that the parasite was sustained in such foci by susceptible hosts that represent about 25% of the total population.

In field populations of deer mice, Wassom *et al.* (1973, 1974) found that the parasite was highly aggregated and suggested that the aggregation was influenced by resistance-related genetic factors. Wassom *et al.* (1986) tested and confirmed this hypothesis using controlled laboratory experiments. By using such an approach, most of the ecological variables that otherwise could have been invoked in contributing to the parasite's aggregated frequency distribution were eliminated. It was stressed, and correctly so, that host population genetics should be considered more frequently in evaluating other parasite frequency distributions.

The component population biology of *H. citelli* also was examined in the white-footed deer mouse *Peromyscus leucopus*, in Wisconsin (USA) (Munger *et al.*, 1989). They also found *H. citelli* in very low prevalences (2%–3%), but the parasite in *P. leucopus* behaved much differently than in *P. maniculatus*. For example, although up to 100% of *P. leucopus* could become infected with *H. citelli* under laboratory conditions, the parasites also developed to maturity before being rejected 28 days after infection. Some of the laboratory-infected mice were returned to the field. These mice retained their infections much longer than in the lab, with some animals still shedding eggs up to 100 days post-infection. When *P. leucopus* were challenged with new infections, the majority resisted infection. Another interesting observation was that when lab-infected mice were introduced into the field, they apparently stimulated transmission of the parasite to mice that had been previously uninfected, thereby increasing prevalence of the parasite in the natural population. Whereas genetically based expulsion of worms may be important in the Utah system, the rejection phenomenon was not the overriding regulatory factor at the Wisconsin study area. Aggregation and low prevalence at the Wisconsin location were attributed to heterogeneous distributions and low densities of the intermediate host in conjunction with the abbreviated life span of the parasite.

Whereas the Wassom *et al.* (1973, 1974, 1986) and Munger *et al.* (1989) studies differ in conclusions regarding parasite population dynamics, they serve to illustrate two important points regarding regulation. First, in the future, the population genetics of hosts must be assessed in more depth than it has been. Second, the population biology in two closely related systems, whereas seemingly very similar in pattern, is not necessarily affected by similar forces. This provides a clear warning that the conclusions reached in a single study, or even a series of studies, cannot be extrapolated to other systems without great care in the analysis of all events and factors associated with the population biology of both the parasites and their hosts.

11.2.4 Competition as a regulatory effector

Competition occurs whenever two or more organisms use the same resources and when those resources are in short supply. **Exploitative competition** is the interaction between two species or individuals that is mediated indirectly through the utilization of a common resource. **Interference competition** occurs when two individuals or species are attempting to acquire a common resource and there is direct confrontation or interaction that reduces access to the resource by one or more of the individuals involved.

The outcome of competition can be considered in three ways. The first is exclusion, which forms the basis for the so-called Gausian Principle. In essence, it states that two species cannot coexist simultaneously in the same space at the same time. According to Holmes (1973), competitive exclusion among helminth parasites is a common phenomenon. **Interactive site segregation** refers to the specialization or segregation of niches by two species in which the **realized niche** of one, or both, is reduced by the presence of a second species (Holmes, 1973). **Selective site segregation** occurs within an evolutionary context. It is non-interactive and implies the absence of current competition. This suggests that, within an evolutionary time frame, genetic changes occurred among one or both formerly competing species and the resource requirements of the competitors diverged.

As discussed earlier, most of the investigations on competition that followed those of Holmes (1961, 1962*a*, *b*) have focused largely on the structure and organizing forces involving parasite infra-, component, and compound communities. We will review these studies subsequently within the framework of parasite community ecology (Chapter 14). The present chapter will review

Holmes's reports (1961, 1962*a*) because these were seminal efforts in this area and because these can be most closely identified with the concept of parasite population regulation.

The cestode *Hymenolepis diminuta* has a two-host cycle. Adults are found in the small intestine of rats, *Rattus rattus* or *R. norvegicus*. Eggs are shed in the feces and then consumed by a variety of beetles (*Tribolium confusum* or *Tenebrio molitor* are used most commonly for laboratory infections). On hatching, an oncosphere emerges, penetrates the gut wall, and enters the hemocoel where it develops into a cysticercoid. After consumption by the rat definitive host, the parasite matures into a sexually mature adult. A second parasite employed by Holmes was the acanthocephalan *Moniliformis dubius*. This parasite also uses rats as a definitive host and, in fact, occupies the same region of the small intestine as does *H. diminuta*. Eggs are released from sexually mature females and are shed with the host feces. On ingestion by cockroaches *Periplaneta americana*, eggs hatch and the spined acanthor penetrates the gut wall; it enters the hemocoel and develops into an infective cystacanth. When a rat eats the infected cockroaches, the parasite matures sexually, completing its life cycle.

Holmes (1961) conducted a series of experimental infections in laboratory rats. In one protocol, the rats were intubated with known numbers of cysticercoids and, after a period of maturation, the rats were killed and the precise linear location of the attachment site of the adult's scolex was noted. The attachment site by the tapeworms was always in the first 35% of the intestine. Infections of *M. dubius* were obtained by intubating rats with known numbers of acanthocephalan cystacanths. The acanthocephalans likewise became attached in the first 35% of the intestine. When the two parasites were introduced simultaneously, however, *M. dubius* continued to occupy the same site, whereas *H. diminuta* was forced posteriad in the intestine into a region that it did not normally occupy. Though an elegant series of experiments, Simberloff (1990) argues correctly that such experiments cannot rule out that the niche shift by *H. diminuta* in the presence of *M. dubius* was due simply to an increase in the total number of worms.

A different set of experiments was designed to determine the effects of single and concurrent species infections on several physical parameters of the worms (Holmes, 1961). These included weight, length, and weight:length ratios. The data (Table 11.2) show that the number of individual parasites that became established was not affected by the presence of the other species, but that weight, length, and weight:length ratios of *H. diminuta* were substantially affected in the rats with double infections. *Moniliformis dubius* also was influenced by the presence of *H. diminuta*, but not to the extent that the cestode was affected.

He further assessed the impact of crowding and concurrent infection by *M. dubius* on the linear intestinal distribution of *H. diminuta*. The reverse of the experiment was also attempted for the two parasites. As the numbers of *H. diminuta* increased, the effects of crowding could be seen clearly, with the points of attachment being forced more and more posteriorly in the intestine. In single-species infections with *M. dubius*, basically the same pattern was observed, with the site of attachment moving further back in the intestine as parasite densities increased. In concurrent infections though, the impact was not as great as in crowding from heavy infections with a single species. Based on the role of carbohydrate in crowding and competition, Read (1959) and Holmes (1961) concluded that their observations were consistent with the hypothesis that competition occurred because of limited nutrient resources.

As has been noted, the role of competition in influencing the infrapopulation dynamics of parasites in natural systems has not been well documented. However, there is evidence that infracommunity structure is affected by interspecific competition in certain host species, indicating that parasite infrapopulation densities could also be affected.

11.2.5 Developmental arrest

Referred to earlier in this chapter, developmental arrest is a widespread form of **diapause** that frequently occurs among several nematode species infecting mammals (Schad, 1977; Adams, 1986) (see also Chapter 5). Schad (1977) identified three sets of factors that are believed to induce arrest.

Table 11.2 Effects of concurrent infection on the number, wet weight, length, and weight:length ratios of *Hymenolepis diminuta* and *Moniliformis dubius*

	Single infection					Concurrent infection					F^a	Significance[a]
	Rats	$\bar{x}$	±	S.E.	Range	Rats	$\bar{x}$	±	S.E.	Range		
H. diminuta												
Numbers	14	4.0	±	0.3	2–5	14	4.7	±	0.2	4–5	2.7	NS
Weight (mg)	14	664.6	±	30.7	496–918	14	251.4	±	28.0	122–434	134.0	HS
Length (mm)	14	529.4	±	21.3	388–662	14	323.5	±	14.6	261–458	23.9	HS
Weight:length	14	1.272	±	0.056	1.03–1.77	14	0.766	±	0.055	0.46–1.04	38.3	HS
M. dubius (aggregate)												
Number	14	7.3	±	0.6	3–10	14	6.4	±	0.6	2–9	0.8	NS
Weight	—		—		—	—		—		—	19.9^b	HS[b]
Length	—		—		—	—		—		—	3.7^b	B[b]
Weight:length	—		—		—	—		—		—	20.2^b	HS[b]
M. dubius males												
Weight	14	79.5	±	2.3	65–88	14	59.1	±	2.4	40–74	[b]	[b]
Length	13	93.8	±	1.2	85–103	13	88.4	±	2.4	77–101	[b]	[b]
Weight:length	13	0.862	±	0.098	0.64–1.04	13	0.697	±	0.038	0.51–0.90	[b]	[b]
M. dubius females												
Weight	14	363.7	±	15.1	285–430	13	235.8	±	14.7	141–332	[b]	[b]
Length	14	207.4	±	3.8	188–231	13	188.4	±	4.7	151–208	[b]	[b]
Weight:length	14	1.751	±	0.052	1.34–1.97	13	1.248	±	0.072	0.88–1.77	[b]	[b]

Notes:

[a] F = variance ratio; NS = not significant at the 10% level; HS = significant at the 1% level; B = significant at the 10% level but not the 5% level.

[b] Data on *M. dubius* males and females were combined in the analysis of variance (with segregation of a highly significant variance due to worm sex) to give a single test of significance that takes into account the difference in each sex due to concurrent infection, and also the concordance of direction of change between the sexes.

Source: From Holmes, 1961, with permission, *Journal of Parasitology*, **47**, 209–216.

The first includes external environmental factors, causing a condition in the parasite that resembles diapause after it reaches the host. Second, there are factors associated with the host that will influence its capacity to either stimulate the development of the parasite or cause it to stop development when external conditions become adverse. And third, there are genetic and density-dependent factors related to the parasite that are involved with the induction of arrest in some species.

Environmentally induced developmental arrest appears to be fairly common among certain species of gastrointestinal nematodes in mammals from temperate regions of the world. Arrest among these species typically occurs during seasons of the year when external environmental conditions would be most detrimental to the survival of free-living larval stages. The primary stimuli for arrest thus vary from species to species. Examples of stimuli include low temperature, declining temperature, changing photoperiod, or combinations thereof (Fernando *et al.*, 1971; Hutchinson *et al.*, 1972; McKenna, 1973; Michel *et al.*, 1975*a*, *b*; Eysker, 1981).

Host-induced developmental arrest also may be related to either natural or acquired resistance (Dineen *et al.*, 1965). Arrest under these conditions is additionally influenced by host age, sex, or species. Adams (1986) suggested that 4th stage larvae of the sheep nematode *Haemonchus contortus* included a number of different subpopulations in a given host. He speculated, and then confirmed, that some of these subpopulations could be stimulated to undergo arrest in response to immunological stimuli while, in others, arrest was genetically inherent to the parasite. In the canine hookworm *Ancylostoma caninum* larvae may remain in a state of arrest in the musculature of bitches until parturition, at which time the 3rd stage larvae migrate to the mammary glands and are transferred to suckling pups via milk (Stoye, 1973) (see also Box 5.2).

It is also clear that arrest in some species, even among strains of the same species, is probably genetically controlled, as is the case for *H. contortus* (Adams, 1986). Nawalinski & Schad (1974) suggested that a strain of *Ancylostoma duodenale* (a human hookworm), isolated from a single female worm in West Bengal, was genetically fixed with respect to arrest. Frank *et al.* (1988) switched populations of the nematode *Ostertagia ostertagi* between northern Ohio (USA) and southern Louisiana (USA) and examined responses of the parasites to new environmental stimuli. They reported that 'the transplanted northern isolate (Ohio) had not adapted to respond to environmental stimuli in the south (Louisiana), whereas the southern isolate continued to respond to spring stimuli in both the north and the south, with no adaptation to autumn stimuli in the north'. This is clearly an indication of genetically based arrest.

Developmental arrest is a density-dependent phenomenon among species of *Ostertagia*, *Cooperia*, *Nematodirus*, *Graphidium*, *Obeliscoides*, and *Haemonchus* (Schad, 1977). The nematode *Obeliscoides cuniculi* was intensively studied in rabbits by Russell *et al.* (1966) and Michel *et al.* (1975*b*). Russell *et al.* (1966) infected rabbits with doses of larvae ranging from 2500 to 25 000. Subsequently, all rabbits were killed and the larvae were counted. They observed a direct relationship between dose size and arrest in the larval infrapopulation up to a certain point; the density of arrested larvae then reached a plateau and leveled off. This plateau was referred to as a 'biomass or immunogenic threshold'.

Michel *et al.* (1975*a*, *b*) extended these observations and concluded that both environmental conditioning and density-dependent factors were involved in many of the observed dose effects. These conclusions were based on a combination of laboratory and field observations, which clearly add a dimension of credibility.

11.3 Suprapopulation dynamics

11.3.1 Introduction

The complexity of abiotic and biotic factors involved in regulating or controlling infrapopulation densities should be apparent by this point. The complexity of these factors at the suprapopulation level is obviously compounded by the multi-host life-cycle patterns of many parasitic organisms. The occurrence of free-living life-cycle stages complicates the situation even further. Moreover, many parasitic organisms have more than one species of intermediate or definitive host and this places even greater demands on the investigator who chooses to undertake the study of a suprapopulation. The numbers of investigations at the suprapopulation level are, therefore, very limited. For the most part, however, whenever such studies have been attempted, the results that were generated were well worth the effort.

11.3.2 *Schistosoma japonicum* and the importance of reservoir hosts

Hairston (1965) conducted the first comprehensive study on the suprapopulation biology of a helminth parasite, *Schistosoma japonicum*, in the Philippines. This digenean occurs in the superior mesenteric venous system that drains the small intestine of its definitive host (see Box 4.1).

One of the critical observations made by Hairston (1965) relates to the importance of reservoir hosts in maintaining the parasite infrapopulations at high levels. He stated that if the parasite could be eliminated completely from a given village, within a year it would return to the same levels it had been at before its elimination. This assertion was based on its high prevalence in reservoir hosts such as dogs, pigs, and field rats. Indeed, it was suggested that component populations of parasites in humans or dogs could not be maintained except by the parasite flow through field rats. This high flow rate was accounted for by the intimate contact between rats and snails; he even argued that, in the Philippines, field rats are the primary hosts, not humans. This was despite the fact that in humans, egg-producing females had both their longest life span and highest total egg output. In field rats, the prevalence was equal to that in humans, but parasite densities were only about a third of that in humans and the egg output was less than 0.1% of that in humans.

The consequences of a study such as Hairston's (1965) are clear from an epidemiological point of view. Studies can be designed to generate baseline data that can then be used in developing strategies for appropriate control measures. Without certain basic information regarding critical aspects of the suprapopulation biology of the parasite, however, the design of control efforts for *S. japonicum* would have been totally inadequate.

11.3.3 *Metechinorhynchus salmonis* and flow rates

The suprapopulation dynamics of the acanthocephalan *Metechinorhynchus salmonis* was examined in several species of fishes within Cold Lake, Alberta, Canada and in the parasite's intermediate host, the amphipod *Pontoporeia affinis* (Holmes *et al.*, 1977). On the surface, this would seem like a reasonably easy system for a comprehensive study at the suprapopulation level because the parasite has a simple two-host life cycle. This is not the case, however, because at least 10 species of fishes in the lake can become infected by the parasite. Some species are infected directly by ingesting infected amphipods, as is the case for whitefish *Coregonus clupeaformis*. Others, such as lake trout *Salvelinus namaycush*, or coho salmon *Oncorhynchus kisutch*, are infected indirectly; they acquire the parasite by feeding on smaller, infected fishes that, in effect, become transport (or paratenic) hosts. The complexity of the 10 definitive host system became somewhat less complicated when two species of sucker (*Catostomus* spp.), northern pike (*Esox lucius*), stickleback (*Pungitius pungitius*), burbot (*Lota lota*) and walleye (*Stizostedion vitreum*) were eliminated from the study because, although they can recruit the parasite, it seldom or never matures sexually in these hosts.

The prevalence, density, and percent of gravid worms were determined in several hundred individual salmonid hosts from Cold Lake (Holmes *et al.*, 1977). The proportion of gravid worms is important for the assessment of relative parasite flow rates through the different species of definitive host and, therefore, for the suprapopulation dynamics of the parasite in the lake. According to their results, the parasite densities in whitefish are constrained by density-dependent forces. This conclusion was based on observations that: (1) the mean number of gravid worms remained constant with increasing density of worms in older age classes of fishes, (2) the mean number of gravid worms remained the same seasonally, despite radical and irregular shifts in overall parasite densities from month to month, and (3) there was a significant negative correlation between the percentage of gravid females and the density of infrapopulations within individual fish.

In cisco, *Coregonus artedii*, the pattern was completely different than in whitefish. Cisco are planktivores and the acquisition of infected benthic amphipods is purely by chance. There would, therefore, be little opportunity to evolve effective feedback mechanisms for regulating *M. salmonis* infrapopulations in cisco. Lake trout acquire the acanthocephalans by preying on fishes that are already infected with the parasite. Whereas the percentage of gravid females in lake trout was relatively low, Holmes *et al.* (1977) indicated that there was a negative correlation between gravid *M. salmonis* and densities of the cestode *Eubothrium salvelini*. This suggested the possible influence of intra- or interspecific competition on parasite maturation and, therefore, parasite fecundity. Coho salmon were introduced into Cold Lake in the summers from 1970 to 1972; results of the introduction were difficult to assess as far as the parasite's suprapopulation biology was concerned. They suggested, however, that the acanthocephalan may have been killing large numbers of young coho salmon and thereby could have had a significant impact on parasite flow within the system as well as on the population biology of the salmon.

Using data on parasite prevalence, densities, numbers of gravid worms, and the relative abundance of different fish species in the lake, the relative output of eggs from each host species was computed. The resulting estimate indicated that most of the flow through the system was via whitefish. It was hypothesized that regulation of flow through whitefish was 'sufficient to regulate the entire system'. After modeling the system mathematically, Holmes *et al.* (1977) concluded that regulation at the suprapopulation level might operate through individual infrapopulations in only one of several host species. Moreover, it was emphasized that the host species in which regulation occurs need not to have the highest numbers or even be the one through which the largest parasite flow is occurring. They admonished 'those working with the population dynamics of parasites having alternative definitive hosts investigate relative flow rates through those hosts, and keep their populations in perspective when studying potential regulatory mechanisms'.

11.3.4 *Haematoloechus complexus* and transmission efficiencies

In a very elegant study, Dronen (1978) examined the suprapopulation dynamics of *H. complexus* (= *H. coloradensis*) focusing on the flow rates and transmission efficiencies at different steps in the parasite's life cycle rather than on the nature of potential regulatory processes. The parasite's definitive host is the Rio Grande leopard frog *Rana berlandieri*, where it occurs in the lungs. Eggs produced by adults are released in the feces. Eggs must be eaten by the pulmonate snail *Physa gyrina*, where they hatch, releasing miracidia that migrate to the hepatopancreas and give rise to sporocysts. Cercariae emerge from snails and penetrate nymphs of three different species of odonates where they encyst as metacercariae. When odonates are consumed by *R. berlandieri*, the metacercariae migrate to the lungs and develop into adults.

For two years, Dronen (1978) assessed the density and prevalence of *H. complexus* in different size classes in each host of the life cycle. The parasite's seasonality was examined in all hosts in the life cycle and the efficiencies with which each step in the life cycle could be completed were determined. The dynamics of the parasite's suprapopulation were monitored in 11 permanent ponds located in Sierra County, New Mexico (USA).

Haematoloechus complexus was highly seasonal, increasing rapidly in both prevalence and density in early summer. Overall prevalence of the parasite was lowest in snails (5%–7%), next in odonates (15%–17%), and highest in frogs (69%–76%). The parasites were aggregated in both odonates and frogs, and the negative binomial model could describe the distributions. No firm conclusions were reached regarding the manner in which the distributions were generated although Dronen (1978) noted that the frequency distributions did conform to the Poisson distribution in some of the smallest frogs. This suggested that perhaps the compounding of Poisson variates was producing aggregation. There was a positive correlation between nymph size and parasite prevalence throughout the aquatic phase of the dragonfly life cycle.

Density estimates (numbers per m^3) were made for each host in the cycle. Using data from estimates of egg and cercariae production in the laboratory, data regarding parasite prevalence and density in the field, and data for host densities, flow diagrams were developed to characterize efficiencies for each step in the life cycle. In 1972, these efficiencies were 0.02% for eggs becoming successful miracidia, 4.4% for cercariae production from successful eggs, 0.08% for metacercariae from successful cercariae, and 0.9% for new adults from metacercariae. The next year, these efficiencies were 0.01%, 3.3%, 0.1%, and 0.2%, respectively.

Based on these numbers, not surprisingly, it was concluded that transmission efficiency was very low, but that reproductive efficiency was very high. As pointed out by both Holmes *et al.* (1977) and Dronen (1978), there is a very delicate balance between reproductive and transmission efficiencies in parasite life cycles. On the one hand, if transmission efficiency drops too low, the chances of local extinction are increased. If the parasite's transmission efficiency becomes too great, there is the risk a host will recruit too many parasites, which, in turn, could produce host morbidity and even mortality. As Dronen (1978) noted, 'it should be kept in mind that this equilibrium between reproductive potential and efficiency has evolved under the influence of selective pressures on the gene pools of both the parasite and its host; and although these selective pressures were probably numerous, host density and the selective pressures that determined the expression of these densities probably played a major role'.

11.3.5 *Bothriocephalus rarus*, a case for transmission efficiencies and reservoir hosts

The definitive host of the pseudophyllidean cestode *Bothriocephalus rarus* is the red-spotted newt *Notophthalmus viridescens*. Eggs are shed in the feces of newts and hatch releasing coracidia. The coracidium is ingested by a copepod in which the parasite develops to the procercoid stage in the hemocoel. When newts feed on copepods, the life cycle is completed. Both larval and adult newts may acquire the parasite. The latter are also known to cannibalize their larvae. If an adult eats an infected larval newt, its parasites will be successfully transferred to the predator. Larval newts migrate from the pond for periods of time lasting from 2 to 7 years, carrying adult worms with them when they leave the pond and when they return.

The suprapopulation dynamics of *B. rarus* was studied in a small, permanent pond in Ritchie County, West Virginia (USA) (Jarroll, 1980). All parts of the parasite's reproduction and transmission were examined experimentally in the field or in the laboratory. The seasonal dynamics of the parasite in copepods and newts was also assessed. Based on the results of his investigation, the probability of an egg successfully surviving to the procercoid stage was 2.2%. The probability of a procercoid surviving to become a worm in an adult newt was 3.8% and to a worm in a larval newt, the probability was 2.9%.

He indicated that the period of greatest loss was from the egg to the procercoid stage. This pattern was similar to that seen by Dronen (1978) for *H. complexus*, but not by Hairston (1965) for *S. japonicum* where the greatest losses in the life cycle were in the cercariae to definitive host phase of the cycle. Jarroll (1980) observed a much higher efficiency in going from the procercoid to the adult parasite. This was attributed to two factors, one associated with the strong spatial overlap between the copepods and newts and the second associated with a strong predator–prey relationship between larval newts and copepods. It was suggested, although no evidence was provided, that infected copepods might in some way be more vulnerable to predation.

Finally, larval newts had the largest parasite densities and exhibited the highest recruitment rates. However, when they became efts and left the pond, there was a tendency to lose worms, so much so that mean infrapopulation densities came to resemble those of adult newts in the pond. Jarroll speculated, 'If one assumes that a dynamic equilibrium exists in the newt and worm populations, then the returning efts and their worms may serve to replace the lost newts and their worm infrapopulations.' In this way, the returning infected efts would serve as a reservoir for the parasite and act to prevent its local extinction. This is also an excellent mechanism for the dispersal of the parasite to other ponds.

11.3.6 *Bothriocephalus acheilognathi*, interaction of season, location, and diet in transmission

The life cycle of *B. acheilognathi* includes a free-swimming coracidium, a procercoid in several species of copepods, and an adult in at least 40 species of fish. Certain aspects of the parasite's suprapopulation biology were investigated in Belews Lake, a large cooling reservoir in the Piedmont area of North Carolina (USA) (Riggs & Esch, 1987; Riggs *et al.*, 1987). As was described elsewhere (section 11.1.3), the lake had undergone severe selenium pollution, to the extent that the main body of water was completely devoid of piscine predators, such as largemouth bass. Indeed, at the beginning of the study, only three fish species were represented in the lake. These included mosquitofish *Gambusia affinis*, fathead minnows *Pimephales promelas*, and red shiners *Notropis lutrensis*, all of which could be infected by *B. acheilognathi*. Over a 30-month period beginning in 1980 and continuing into 1983, the density, prevalence, and dispersion patterns of the parasite were followed at three distinct locations in the reservoir. One was in the polluted part of the lake (without piscivorous predators). A second was in the headwaters of the lake where no pollution had occurred, but where there was a diverse fish community, including piscivorous largemouth bass. The third was in an interface zone where predation was probably occurring, but its effects were clearly less than at the unpolluted location. There was some pollution at the intermediate locality, but selenium concentrations were comparatively low.

The approach taken by Riggs & Esch (1987) was different from those of Holmes *et al.* (1977), Dronen (1978), and Jarroll (1980). The *Bothriocephalus* study was designed to examine the interaction of season, locality, host size, and host diet on changes in parasite prevalence, density, dispersion, and fecundity in three species of definitive host. There were seasonal changes in prevalence of the parasite at all three locations, among all three definitive host species, and in three, arbitrarily established, host size classes. Interestingly, the variations observed were strongly related to seasonal changes in species richness within the copepod community, to size-related changes in prey preferences among the three host species, and to the influence, or absence, of predation pressure.

They also observed striking differences in the ability of the parasites to mature sexually in the three host species. Holmes (1979) defined three classes of hosts in terms of their capacity to allow successful maturation of a parasite. The so-called 'required hosts' are those in which the majority of gravid worms are found. In 'suitable hosts', the parasite will mature sexually, but will not possess gravid individuals in large numbers. 'Unsuitable hosts' may recruit the parasites, but the parasites will not become gravid. In Belews Lake, both fathead minnows and red shiners are clearly required hosts whereas, for the most part, mosquitofish fall into the suitable category. In mosquitofish, worms appeared to be stunted and occupied only a very narrow region of the gut. Moreover, Granath & Esch (1983*b*) observed, and Riggs (1986) confirmed, that growth of even a few worms to sizes larger than 40–50 proglottids caused mortality in mosquitofish.

One of the more striking observations was the difference in fecundity and egg viability among the worms from polluted and unpolluted areas, as well as between red shiners and fathead minnows. Fecundity and egg viability were higher in fathead minnows; the same two parameters were also higher at the unpolluted locations. The former observation is a clear indication of differences in physiological compatibility of the parasite for these two host species. The latter finding suggests that the power-plant effluent was definitely having an impact on the parasite's fecundity throughout the polluted part of the reservoir. Selenium toxicity halts reproduction in fishes and was responsible for the negative impact on the fish community in Belews Lake (Cumbie & Van Horn, 1978). The selenium concentration was examined in the tissues of all three fish species in the polluted and unpolluted parts of the reservoir as well as from tapeworms removed from the same fishes. The results were clear. The concentration of selenium in tapeworm tissues containing gravid proglottids was 10 times higher than in muscle from the respective host fish and was 6 to

8 times higher than in the scolices of the same worms. The study also showed that selenium was being concentrated in the host gonadal tissue, but was still far less than half what it was for the gravid proglottids from the tapeworms. When the concentration of selenium in host and parasite tissues from the unpolluted localities was examined, it was found to be inconsequential and up to 4 orders of magnitude less than comparable tissues from fishes and parasites from polluted areas. It was apparent that the parasites were acting as selenium 'sinks' and that both fecundity and egg viability were being affected as a result.

Riggs & Esch (1987) concluded their study by saying, 'factors that affect, directly and indirectly, the dynamics of the intermediate hosts and the frequency and intensity of predation on them by definitive hosts seem to be of primary importance' in influencing the prevalence, density, and aggregation of *B. acheilognathi* in Belews Lake. They emphasized, however, that the one variable in the transmission process not associated with predator–prey relationships was parasite fecundity in required hosts. In the end, then, it is the unique combination of biotic and abiotic components within the host's ecosystem that affects the suprapopulation dynamics of *B. acheilognathi*.

11.3.7 *Cystidicoloides tenuissima*, the importance of habitat preference

Aho & Kennedy (1987) examined the circulation pattern and transmission dynamics of the nematode *Cystidicoloides tenuissima*, in all its intermediate and definitive hosts in the River Swincombe in Devon, England. Definitive hosts for the parasite include brown trout *Salmo trutta*, and juvenile Atlantic salmon *Salmo salar*. Intermediate hosts included 18 species of insects, but the parasite was able to develop into infective 3rd stage larvae only in the mayfly *Leptophleba marginata*.

Even though trout and salmon were suitable definitive hosts, the former species was the primary host because the mayfly was a major component in its diet. It was estimated that 99% of the parasite's egg production originated in the trout, but that insects ingested only 10% of the eggs. Most of the eggs were apparently ingested by two species of insects; nearly 80% of these intermediate hosts were infected by *C. tenuissima*, whereas only 10% of *L. marginata* were infected. The overall transmission rate of eggs to larval *L. marginata* varied between 0.25 and 0.87%; transmission rates from insect to fish were much higher, ranging from 10.8 to 39.8%.

Whereas unsuitable hosts apparently consumed most of the parasite eggs, the parasite could not become established, perhaps because of physiological incompatibility or because of an immunological response. It was concluded that the differences between trout and salmon as a potential definitive host were not related to a physiological/immunological response, but to an ecological factor. In the River Swincombe, trout were restricted primarily to pools and salmon to riffles. During periods of the year when parasite transmission occurred, *L. marginata* was also confined to the pools thereby bringing the mayfly into the foraging arena of the trout. The parasite could not, therefore, be transmitted to salmon since they did not typically feed on the mayfly host. Also of interest was the comparison made by Aho & Kennedy (1987) of the suprapopulation dynamics of *C. tenuissima* with those of the acanthocephalans, *Pomphorhynchus laevis* and *Metechinorhynchus salmonis* (Hine & Kennedy, 1974; Holmes *et al.*, 1977). All three parasites have a two-host life cycle, with the definitive hosts being fishes. Aho & Kennedy (1987) pointed out that there was a strict specificity at both the intermediate and definitive host levels for all three species of helminths. However, in the case of the nematode, large numbers of eggs were consumed by insects in which the parasite could not develop properly, whereas, apparently, only hosts in which proper development could occur consumed the acanthocephalan eggs. Conversely, acanthocephalan intermediate hosts were consumed by a wide range of potential fish hosts, but could mature in only one.

11.3.8 Conclusions

As noted earlier, the suprapopulation dynamics of parasitic organisms have not received extensive attention. However, in those studies which have been conducted, there is ample indication of their worth. Thus, Hairston (1965) used the approach in identifying a critical epidemiological factor for the potential control of schistosomiasis in the Philippines. Other investigators (Holmes *et al.*, 1977; Dronen, 1978; Jarroll, 1980; Granath & Esch, 1983*a*, *b*; Aho & Kennedy, 1987; Riggs *et al.*, 1987) employed several different host–parasite systems to characterize reproductive and transmission efficiencies at various life-history steps and to emphasize the delicate balance between the two processes in parasite life cycles. Studies such as these are quite valuable and should be extended to other systems as well.

References

Adams, D. B. (1986) Developmental arrest of *Haemonchus contortus* in sheep treated with corticosteroid. *International Journal for Parasitology*, **16**, 659–664.

Aho, J. M. & Kennedy, C. R. (1987) Circulation pattern and transmission dynamics of the suprapopulation of the nematode *Cystidicoloides tenuissima* (Zeder) in the River Swincombe, England. *Journal of Fish Biology*, **31**, 123–141.

Anderson, R. M. (1974) Population dynamics of the cestode *Caryophyllaeus laticeps* (Pallas, 1781) in the bream (*Abramis brama* L.). *Journal of Animal Ecology*, **43**, 305–321.

Anderson, R. M. (1976) Seasonal variation in the population dynamics of *Caryophyllaeus laticeps*. *Parasitology*, **72**, 281–305.

Andrews, C., Chubb, J. C., Coles, T. & Dearsley, A. (1981) The occurrence of *Bothriocephalus acheilognathi* Yamaguti, 1934 (*B. gowkongensis*) (Cestoda: Pseudophyllidea) in the British Isles. *Journal of Fish Diseases*, **4**, 89–93.

Bradley, D. J. (1972) Regulation of parasite populations. A general theory of the epidemiology and control of parasitic infections. *Transactions of the Royal Society of Tropical Medicine and Hygiene*, **66**, 697–708.

Bull, P. C. (1964) Ecology of helminth parasites of the wild rabbit *Oryctolagus cuniculus* (L.) in New Zealand. *New Zealand Department of Scientific and Industrial Research, Bulletin*, **158**, 1–147.

Bush, A. O. & Holmes, J. C. (1986) Intestinal parasites of lesser scaup ducks: an interactive community. *Canadian Journal of Zoology*, **64**, 142–152.

Crofton, H. D. (1971*a*) A quantitative approach to parasitism. *Parasitology*, **62**, 179–193.

Crofton, H. D. (1971*b*) A model for host–parasite relationships. *Parasitology*, **63**, 343–364.

Cumbie, P. M. & Van Horn, S. L. (1978) Selenium accumulation associated with fish mortality and reproductive failure. *Proceedings of the Annual Conference of the Southeastern Fish and Wildlife Agencies*, **32**, 612–624.

Dineen, J. K., Donald, A. D. & Wagland, B. M. (1965) The dynamics of the host–parasite relationship. III. The response of sheep to primary infection with *Haemonchus contortus*. *Parasitology*, **55**, 515–525.

Dronen, N. O. (1978) Host–parasite population dynamics of *Haematoloechus coloradensis* Cort, 1915 (Digenea: Plagiorchidae). *American Midland Naturalist*, **99**, 330–349.

Esch, G. W. (1983) The population and community ecology of cestodes. In *Biology of the Eucestoda*, ed. P. Pappas & C. Arme, pp. 81–137. New York: Academic Press.

Esch, G. W. & Fernández, J. C. (1994) Snail–trematode interactions and parasite community dynamics in aquatic systems: a review. *American Midland Naturalist*, **131**, 209–237.

Esch, G. W., Hazen, T. C., Marcogliese, D., Goater, T. M. & Crews, A. E. (1986) Long-term studies on the population biology of *Crepidostomum cooperi* (Allocreadidae) in the burrowing mayfly, *Hexagenia limbata* (Ephemeroptera). *American Midland Naturalist*, **116**, 304–314.

Esch, G. W., Bush, A. O. & Aho, J. M. (eds.) (1990) *Parasite Communities: Patterns and Processes*. New York: Chapman & Hall.

Eysker, M. (1981) Experiments on inhibited development of *Haemonchus contortus* and *Ostertagia circumcincta* in the Netherlands. *Research in Veterinary Science*, **30**, 62–65.

Fernando, M. A., Stockdale, P. H. G. & Ashton, G. C. (1971) Factors contributing to the retardation of development of *Obeliscoides cuniculi* in rabbits. *Parasitology*, **63**, 21–29.

Frank, G. R., Herd, R. P., Marbury, K. S., Williams, J. C. & Willis, E. R. (1988) Additional investigations on hypobiosis of *Ostertagia ostertagi* after transfer between northern and southern U.S.A. *International Journal for Parasitology*, **18**, 171–177.

Granath, W. O. & Esch, G. W. (1983*a*) Temperature and other factors that regulate the composition and infrapopulation densities of *Bothriocephalus acheilognathi* (Cestoda) in *Gambusia affinis* (Pisces). *Journal of Parasitology*, **69**, 1116–1124.

Granath, W. O. & Esch, G. W. (1983*b*) Survivorship and parasite-induced host mortality among mosquitofish in a predator-free, North Carolina cooling reservoir. *American Midland Naturalist*, **110**, 314–323.

Granath, W. O. & Esch, G. W. (1983*c*) Seasonal dynamics of *Bothriocephalus acheilognathi* in ambient and thermally altered areas of a North Carolina cooling reservoir. *Proceedings of the Helminthological Society of Washington*, **50**, 205–218.

Hairston, N. G. (1965) On the mathematical analysis of schistosome populations. *Bulletin of the World Health Organization*, **33**, 45–62.

Heyneman, D. (1962) Studies on helminth immunity. II Influence of *Hymenolepis nana* (Cestoda: Hymenolepididae) in dual infections with *H. diminuta* in white mice and rats. *Experimental Parasitology*, **12**, 7–18.

Hine, P. M. & Kennedy, C. R. (1974) Observations on the distribution, specificity and pathogenicity of the acanthocephalan *Pomphorhynchus laevis* (Muller). *Journal of Fish Biology*, **6**, 521–535.

Holmes, J. C. (1961) Effects of concurrent infections on *Hymenolepis diminuta* (Cestoda) and *Moniliformis dubius* (Acanthocephala). I. General effects and comparison with crowding. *Journal of Parasitology*, **47**, 209–216.

Holmes, J. C. (1962*a*) Effect of concurrent infections on *Hymenolepis diminuta* (Cestoda) and *Moniliformis dubius* (Acanthocephala). II. Effects on growth. *Journal of Parasitology*, **48**, 87–96.

Holmes, J. C. (1962*b*) Effect of concurrent infections on *Hymenolepis diminuta* (Cestoda) and *Moniliformis dubius* (Acanthocephala). III. Effects in hamsters. *Journal of Parasitology*, **48**, 97–100.

Holmes, J. C. (1973) Site selection by parasitic helminths: interspecific interactions, site segregation, and their importance to the development of helminth communities. *Canadian Journal of Zoology*, **51**, 333–347.

Holmes, J. C. (1979) Parasite populations and host community structure. In *Host–Parasite Interfaces*, ed. B. B. Nickol, pp. 27–46. New York: Academic Press.

Holmes, J. C., Hobbs, R. P. & Leong, T. S. (1977) Populations in perspective: community organization and regulation of parasite populations. In *Regulation of Parasite Populations*, ed. G. W. Esch, pp. 209–245. New York: Academic Press.

Hugghins, E. J. (1956) Ecological studies on a strigeid trematode at Oakwood Lakes, South Dakota. *Proceedings of the South Dakota Academy of Science*, **35**, 204–206.

Hutchinson, G. W., Lee, E. H. & Fernando, M. A. (1972) Effects of variations in temperature on infective larvae and their relationship to inhibited development of *Obeliscoides cuniculi* in rabbit. *Parasitology*, **65**, 333–342.

Jarroll, E. L. (1980) Population dynamics of *Bothriocephalus rarus* (Cestoda) in *Notophthalmus viridescens*. *American Midland Naturalist*, **103**, 360–366.

Kennedy, C. R. (1968) Population biology of the cestode *Caryophyllaeus laticeps* (Pallas, 1781) in dace, *Leuciscus leuciscus* L., of the River Avon. *Journal of Parasitology*, **54**, 538–543.

Kennedy, C. R. (1969) Seasonal incidence and development of the cestode *Caryophyllaeus laticeps* (Pallas), in the River Avon. *Parasitology*, **59**, 783–794.

Kennedy, C. R. (1971) The effect of temperature upon the establishment and survival of the cestode *Caryophyllaeus laticeps* in orfe, *Leuciscus idus*. *Parasitology*, **63**, 59–66.

Kennedy, C. R. & Rumpus, A. (1977) Long term changes in the size of the *Pomphorhynchus laevis* (Acanthocephala) population in the River Avon. *Journal of Fish Biology*, **10**, 35–42.

Kennedy, C. R. & Walker, P. J. (1969) Evidence for an immune response in dace, *Leuciscus leuciscus*, to infections by the cestode *Caryophyllaeus laticeps*. *Journal of Parasitology*, **55**, 579–582.

Kuris, A. M. (1990) Guild structure of larval trematodes in molluscan hosts: prevalence, dominance and significance of competition. In *Parasite Communities: Patterns and Processes*, ed. G. W. Esch, A. O. Bush & J. M. Aho, pp. 69–100. New York: Chapman & Hall.

Marcogliese, D. J. & Esch, G. W. (1989) Alterations in seasonal dynamics of *Bothriocephalus acheilognathi* in a North Carolina cooling reservoir over a seven year period. *Journal of Parasitology*, **75**, 378–382.

Marcogliese, D. J , Goater, T. M. & Esch, G. W. (1990) *Crepidostomum cooperi* (Allocreadidae) in the burrowing mayfly, *Hexagenia limbata* (Ephemeroptera) related to trophic status of a lake. *American Midland Naturalist*, **124**, 309–317.

McKenna, P. B. (1973) The effect of storage on the infectivity and parasitic development of third-stage *Haemonchus contortus* larvae in sheep. *Research in Veterinary Science*, **14**, 312–316.

Meyer, M. C. (1972) The pattern of circulation of *Diphyllobothrium sebago* (Cestoda: Pseudophyllidea) in an enzootic area. *Journal of Wildlife Diseases*, **8**, 215–220.

Michel, J. F. (1974) Arrested development of nematodes and some related phenomena. *Advances in Parasitology*, **12**, 279–366.

Michel, J. F., Lancaster, M. B. & Hong, C. (1975*a*)

Arrested development of *Ostertagia ostertagi* and *Cooperia oncophora*: effect of temperature at the free-living third stage. *Journal of Comparative Pathology*, **85**, 133–138.

Michel, J. F., Lancaster, M. B. & Hong, C. (1975*b*) Arrested development of *Obeliscoides cuniculi*: the effect of size of inoculum. *Journal of Comparative Pathology*, **85**, 307–315.

Moore, J. & Gotelli, N. J. (1990) A phylogenetic perspective on the evolution of altered host behaviours: a critical look at the manipulation hypothesis. In *Parasitism and Host Behaviour*, ed. C. J. Barnard & J. M. Behnke, pp. 193–233. New York: Taylor & Francis.

Munger, J. C., Karasov, W. H. & Chang, D. (1989) Host genetics as a cause of overdispersion of parasites among hosts: how general a phenomenon? *Journal of Parasitology*, **75**, 707–710.

Nawalinski, T. A. & Schad, G. A. (1974) Arrested development in *Ancylostoma duodenale*: course of a self-induced infection in man. *American Journal of Tropical Medicine and Hygiene*, **23**, 895–898.

Pelletier, L. L. (1982) Chronic strongyloidiasis in World War II Far East prisoners of war. *American Journal of Tropical Medicine and Hygiene*, **33**, 55–61.

Read, C. P. (1959) The role of carbohydrates in the biology of cestodes. VIII. Some conclusions and hypotheses. *Experimental Parasitology*, **8**, 365–382.

Riggs, M. R. (1986) Community dynamics of the Asian fish tapeworm, *Bothriocephalus acheilognathi*, in a North Carolina cooling reservoir. PhD dissertation, Wake Forest University, Winston-Salem, North Carolina, USA.

Riggs, M. R. & Esch, G. W. (1987) The suprapopulation dynamics of *Bothriocephalus acheilognathi* in a North Carolina cooling reservoir: abundance, dispersion and prevalence. *Journal of Parasitology*, **73**, 877–892.

Riggs, M. R., Lemly, A. D. & Esch, G. W. (1987) The growth, biomass and fecundity of *Bothriocephalus acheilognathi* in a North Carolina cooling reservoir. *Journal of Parasitology*, **73**, 893–900.

Russell, S. W., Baker, N. F. & Raizes, G. S. (1966) Experimental *Obeliscoides cuniculi* infections in rabbits: comparison with *Trichostrongylus* and *Ostertagia* infections in cattle and sheep. *Experimental Parasitology*, **19**, 163–173.

Schad, G. A. (1977) The role of arrested development in the regulation of nematode populations. In *Regulation of Parasite Populations*, ed. G.W. Esch, pp. 111–167. New York: Academic Press.

Schad, G. A., Thompson, F., Talham, G., Holt, D., Nolan, T. J., Ashton, F. T., Lange, A. M. & Bhopale, V. M. (1997) Barren female *Strongyloides stercoralis* from occult chronic infections are rejuvenated by transfer to parasite-naive recipient hosts and give rise to an autoinfective burst. *Journal of Parasitology*, **83**, 785–791.

Scott, M. E. & Dobson, A. (1989) The role of parasites in regulating host abundance. *Parasitology Today*, **5**, 176–183.

Simberloff, D. (1990) Free-living communities and alimentary tract helminths: hypotheses and pattern analyses. In *Parasite Communities: Patterns and Processes*, ed. G. W. Esch, A. O. Bush & J. M. Aho, pp. 289–319. New York: Chapman & Hall.

Smith, H. D. (1973) Observations on the cestode *Eubothrium salvelini* in juvenile sockeye salmon (*Oncorhynchus nerka*) at Babine Lake, British Columbia. *Journal of the Fisheries Research Board of Canada*, **30**, 947–964.

Sousa, W. P. (1990) Spatial scale and the processes structuring a guild of larval trematode parasites. In *Parasite Communities: Patterns and Processes*, ed. G. W. Esch, A. O. Bush & J. M. Aho, pp. 41–67. New York: Chapman & Hall.

Stoye, M. (1973) Untersuchungen über die Möglichkeit pränataler und galaktogener Infektionen mit *Ancylostoma caninum* Ercolani, 1859 (Ancylostomidae beim Hund). *Zentralblatt Veterinärmedizin, B*, **20**, 1–39.

Wassom, D. L., Guss, V. M. & Grundmann, A. W. (1973) Host resistance in a natural host–parasite system. Resistance to *Hymenolepis citelli* by *Peromyscus maniculatus*. *Journal of Parasitology*, **59**, 117–121.

Wassom, D. L., DeWitt, C. W. & Grundmann, A. W. (1974) Immunity to *Hymenolepis citelli* by *Peromyscus maniculatus*: genetic control and ecological implications. *Journal of Parasitology*, **60**, 47–52.

Wassom, D. L., Dick, T. A., Arnason, N., Strickland, D. & Grundmann, A. W. (1986) Host genetics: a key factor in regulating the distribution of parasites in natural host populations. *Journal of Parasitology*, **72**, 334–337.

Weinmann, C. J. (1966) Immunity mechanisms in cestode infections. In *Biology of Parasites*, ed. E. J. L. Soulsby, pp. 301–320. New York: Academic Press.

Chapter 12

Influence of parasites on host populations

12.1 Introduction to regulation

In the previous chapter, we made an effort to assess the nature of factors, both biotic and abiotic, that are known to affect the biology of parasite populations. The purpose of the present chapter is to reverse that thrust and to determine if, and under what circumstances, parasites can affect host population dynamics.

In order to identify the nature of regulatory influence on host population dynamics, it is first necessary to define what is meant by regulation. Scott & Dobson (1989) described it as a phenomenon involving processes that will 'reduce the per capita survival or reproduction' within a population as that population's density increases. The key element in the definition is the reduction in host survivorship, or fecundity (reproduction), as a function of parasite density. In other words, there must be the clear implication of a density-dependent effect on host mortality or reproductive fitness for there to be regulation. As the authors point out, however, the measurement of density-dependent regulation is not easily accomplished since it is not always feasible, or even possible, to manipulate host populations in the field or in the laboratory. For this reason, some of the examples included in this chapter as being regulatory in character must be inferential only.

12.2 Crofton's approach

Before considering Crofton's (1971*a*) analysis of the host–parasite system with which he worked, or the mathematical model he developed (Crofton, 1971*b*), it is instructive to know why he undertook such an effort. He was disenchanted with the various definitions of parasitism, mainly because they were too qualitative in character. He was, however, evidently impressed with the approaches taken by Kostitzin (1934, 1939) and Lotka (1934) who were familiar with frequency distributions, but lacked the sophisticated computational technologies necessary for fitting theoretical models to observed distributions. After considering several applications to the problem, Crofton (1971*a*) became convinced that a quantitative methodology could best be applied through the use of the **negative binomial model**.

The negative binomial model is an empirical model which is defined by a positive exponent k, and μ, the population mean, given by $\mu = pk$. The variance σ^2 of the distribution is $\sigma^2 = pkq$; k is an index of aggregation. As k approaches 0, the aggregation becomes greater and as k increases, the distribution becomes more random. Frequency distribution curves that can be described by the negative binomial model for various values of k and μ (the population mean) are shown in Fig. 12.1. (Remember that, in the real world, it is almost impossible to determine means and variances for an entire population; instead, we rely on samples of the population and we use, therefore, $\bar{x}$ as an estimate of the sample mean and s^2 as an estimate of the sample variance.)

Crofton (1971*a*) stated, 'the Negative Binomial distribution is a 'fundamental model' of parasitism in so far as it describes the distribution of parasites among hosts'. The elegance of the negative binomial rests with the observation that

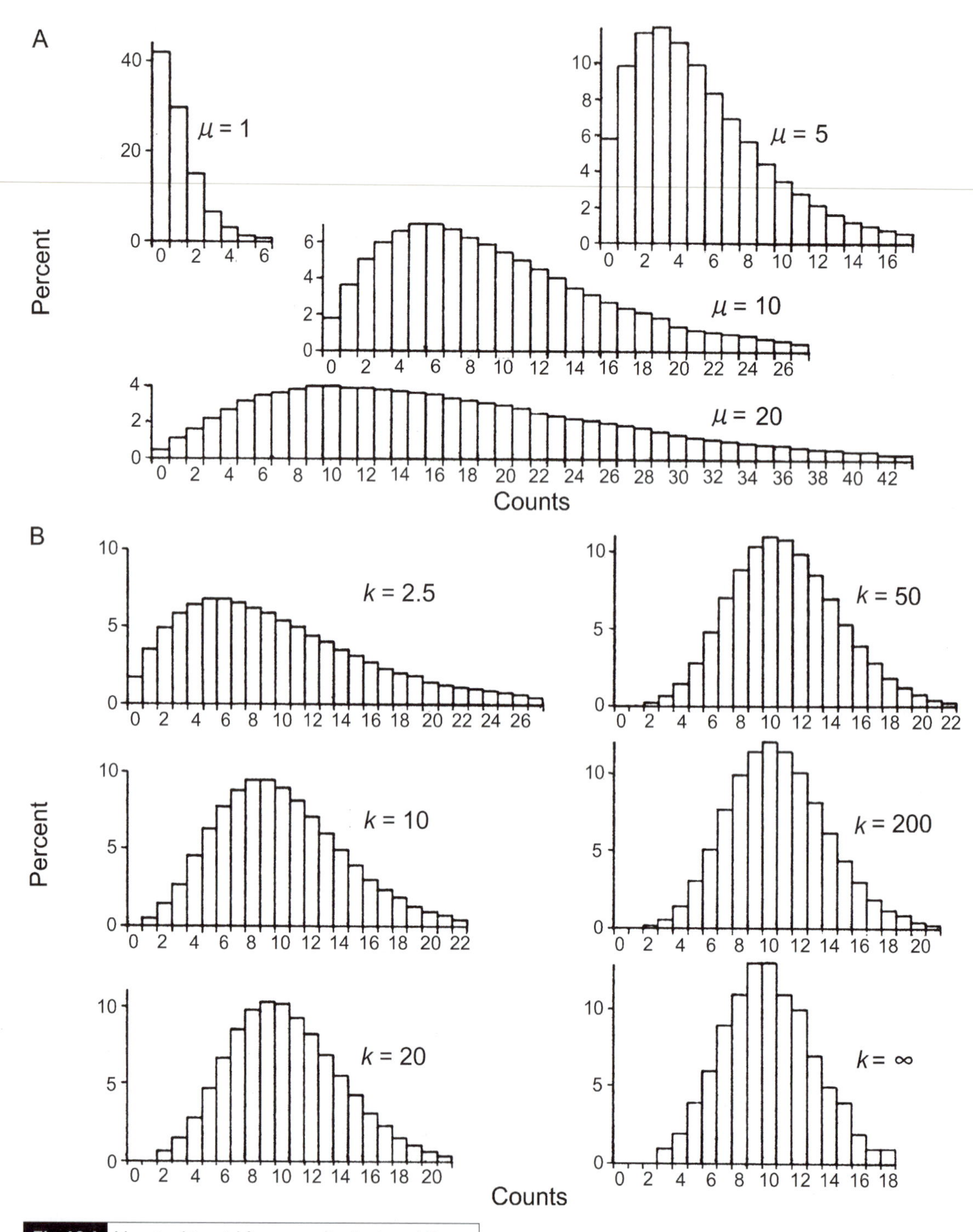

Fig. 12.1 Negative binomial frequency distributions. (A) Frequency distributions for $k = 2.5$ and various values of μ, from 1 to 20; (B) frequency distributions for $\mu = 10$ and various values of k, from 2.5 to infinity. Frequency of each count is expressed as percentage of total count. (Modified from Elliot, 1983, with permission, *Freshwater Biological Association, Scientific Publication No. 25.*)

distributions, which can be described by the model, have hypothetical bases for inferring the manner in which they can arise. With a hypothesis in place, it then can be tested and either supported or rejected. Most other models do not offer such an advantage. For example, a distribution that can be described by the negative binomial might be generated through the compounding of Poisson variates. In other words, if a host acquires parasites through a series of random waves of infection so that the chances of infection by each succeeding wave is independent of the previous ones, then the final aggregated distribution may be described by the negative binomial model.

Despite its many advantages, the negative binomial model is not applicable to all host–parasite systems. Sporocysts and rediae of digeneans in snails, for example, will not generate aggregated frequency distributions. This is because, once infected, intramolluscan larval stages reproduce asexually thereby generating frequency distributions that are highly skewed. Several species of cestodes, e.g., *Taenia crassiceps, Echinococcus multilocularis,* also reproduce asexually and their infrapopulations are not aggregated in their intermediate hosts. The same observation holds for most protozoans in, or on, their hosts.

If the infective stages of a parasite are distributed in an aggregated pattern in space, there is a likelihood that the parasite will be aggregated within the population of the next host in the life cycle in two ways. First, intermediate hosts may acquire parasites that become aggregated. These aggregated parasite component populations may then be transferred through predation, more or less *in toto*, to their definitive hosts, creating 'instantaneous' aggregation in the definitive host population. A number of parasite transfers are believed to occur in just such a manner, e.g., *Crepidostomum cooperi* metacercariae via mayflies, *Halipegus occidualis* metacercariae via dragonflies, *Dicrocoelium dendriticum* metacercariae via ants, etc. The second has to do with microgeographic foci and, in effect, with so-called 'landscape epidemiology'. Eggs of *Ancylostoma duodenale, Necator americanus, Trichuris trichiura, Ascaris lumbricoides, Ascaris suum*, etc., are all released in feces. These eggs, or 3rd stage larvae which emerge from the eggs, are initially distributed in the soil as patches, or microgeographic foci. Exposure to a host may, as a result, occur in 'packets', and may produce aggregated distributions of parasites in the human hosts.

Some parasites have the ability to alter a host's behavior. If infection increases the transmission probability for a parasite, then aggregation may be an outcome. If, within a host population, the chances for infection are unequal because of some form of variation in individual hosts, then aggregation also may occur. The individual variability in hosts may be related to genetic differences in susceptibility–resistance traits, to physiological differences, to age differences within a host population, etc. Similarly, individuals within a population change over time and such changes could also create conditions that could lead to aggregated frequency distributions.

The reason for emphasizing the nature of aggregation is that, because of it, the majority of parasites within a host population will be found in only a few hosts. This suggests that, if the parasite has any lethal qualities, then mortality may be restricted to but a few hosts, those that are heavily infected. The risk of parasite-induced mortality, and **morbidity**, is thus spread unevenly within a host population. Presumably, during the course of **coevolution** in the various host–parasite systems, efficiencies in life-cycle patterns have developed to the point that transmission strategies involving parasite-induced host mortality are such that even local extinction is a low probability except under very exceptional conditions. There are other evolutionary implications regarding the dispersion patterns of parasites, but these will be discussed in subsequent chapters dealing with evolution and biogeography (see Chapters 15 and 16).

12.2.1 The *Polymorphus–Gammarus* system

The database for Crofton's (1971*a*) quantitative approach was provided by Hynes & Nicholas (1963) who examined the component population biology of the acanthocephalan *Polymorphus minutus* in its amphipod intermediate host *Gammarus pulex*. The definitive hosts for the parasite are ducks. In their study area, ducks were kept in cages next to a stream into which feces (and parasite eggs) from the ducks were washed. The amphipods were collected at several stations, each located at a distance further and further away from the source of eggs. The amphipods were returned to the lab where they were dissected and the cystacanths counted. Not surprisingly, parasite densities were highest at the station immediately adjacent to the duck cages. At each station downstream from the duck pens, the parasite densities declined in the amphipods. It is worth noting here that the database generated in this study was entirely observational in character. Hynes & Nicholas (1963) had no a priori indication that it would subsequently be used by a theoretician to create a foundation for modern parasite population ecology, as well as the basis for our current approaches to the epidemiology of the enteric helminths of mankind.

These data were analyzed by Crofton (1971*a*); both the Poisson and negative binomial models were used to describe parasite frequency distributions in the amphipods, with the latter clearly giving the best fit. However, he also observed that the fits for zero and one parasite per host classes were unsatisfactory, especially at locations closest to the duck pens, e.g., there were too many individuals in these two classes. He reasoned that an explanation for the poor fit was that heavily infected hosts had been killed by the parasite and that by removing these hosts from the population the resulting distribution had become skewed.

This hypothesis was then tested using a truncated negative binomial model to describe the distributions. According to Crofton (1971*a*), 'truncation is usually applied when the negative and sometimes the zero terms of a frequency distribution are rejected or accumulated at the zero frequency'. It was concluded that there was little or no mortality at stations located the greatest distance from the egg source. At stations closer and closer to the duck pens (and the source of parasite eggs), parasite losses due to host mortality appeared to become greater and greater. Losses at stations closest to the duck pens were estimated at between 5% and 60%.

The real significance of Crofton's (1971*a*) analysis was that it combined the concepts of parasite frequency distributions and parasite-induced host mortality in a practical manner for the first time. However, it was vigorously emphasized that the precise form of the frequency distribution is unimportant as long as the distribution is aggregated. This point, alluded to by Crofton (1971*a*), needs to be reiterated. Aggregation (note that he used the term 'overdispersion'; see Box 10.1) is the key element in his concept, not whether the observed distribution can be described by the negative binomial model. 'It is the overdispersion and the relationship of parasite density to lethal factors that produce a disparity in parasite and host deaths. This disparity, offset by the greater reproductive rate of the parasite, can produce the dynamic equilibrium of host and parasite populations that is essential to the continued association of the host and parasite species. In effect, the parasite acts as a regulator of the host population, the intensity of the regulatory function being related to both host and parasite population dynamics' (Crofton, 1971*a*).

12.2.2 Crofton's model

In conjunction with his study of the *Polymorphus–Gammarus* system, Crofton (1971*b*) developed a mathematical model to describe some of the dynamic qualities of host–parasite interactions. The deterministic model that was developed employed several basic assumptions regarding certain characteristics of the host and parasite populations. For example, the host population was considered to have varying rates of logistic growth. In the model, the population never reached its carrying capacity.

The parasite was described as always having a higher rate of reproduction than the host, a component of his quantitative definition of parasitism (Crofton, 1971*a*). In the model, the reproductive rate of the parasite and its potential to infect a host were combined into a single term called the Achievement Factor. Since the probability of infecting a host is always less than 1, the Achievement Factor for the parasite was always less than its reproductive rate. The frequency distribution of the parasites within the host population was aggregated and could be described by the negative binomial model, another essential element in his quantitative approach to parasitism. The concept of the Lethal Level (*L*) was also incorporated into the model. It was 'defined in terms of the number of parasites required to kill the host before it can reproduce' (Crofton, 1971*b*). After using several different approaches, he was concerned about the nature of transmission rates in relation to host density, and decided to make the model's transmission rates correlate directly with host density. The efficiency of the model was expressed in terms of equilibrium between densities of hosts and parasites. Equilibrium was assumed to have been reached when constant levels of parasites and hosts were attained or when there were persistent oscillations in parasite and host densities.

After developing the basic model, simulations were conducted in order to assess the effects of changing various parameters. For example, *k* is a measure of aggregation in the negative binomial distribution (note that the term here is *k*, not the same term, *K*, used to describe the carrying capacity in logistic population growth). As *k* was increased in the model, the amount of aggregation decreased. The influence of *k* was found to be greatest on the equilibrium levels when it was <2; when it was >2, only small changes in densities were observed. Population densities of both hosts and parasites were a function of the Achievement Factor. As the Achievement Factor increased, population densities of both parasites and hosts fell because of host mortality. The effects of three different *L* values were then simulated. With low pathogenicity, the equilibrium levels of the parasite population were high. As *L* increased, the mean density of parasites also increased. Crofton (1971*b*) pointed out, 'this is due not only to an increase in the number of parasites but, perhaps surprisingly, to a decrease in the number of hosts'. Apparently, as transmission success increased, more and more hosts acquired parasite numbers that were greater than *L* and were thus eliminated. Host sterilization by parasites was found to be similar to host death in terms of the equilibria that could be established. Another observation worth noting is that, when dispersion was reduced, a low *L* produced instability, causing the entire system to collapse. Finally, when the effects of immunity were introduced, the results were inconclusive for the most part. When aggregation was high, immunity produced a stable, oscillating system. If immunity was strong enough to increase host survivorship and produce higher host population densities, parasite densities will also be higher, not lower as might be expected.

Based on this modeling effort, Crofton (1971*b*) concluded that his quantitative description of parasitism (Crofton, 1971*a*) was a 'functional' one because parasite pathogenicity was found to be a primary factor in establishing host–parasite population equilibria. The significance of the model is not so much that it attempted to replicate with precision the 'real world', but that it provided the basic hypothesis from which subsequent modelers could develop more sophisticated methods for interpreting host–parasite interactions. In effect, it became both the baseline and the initial stimulus for future effort in this area of study.

12.2.3 Crofton revisited

As pointed out by May (1977), the Crofton (1971*b*) model for the dynamics of host–parasite relationships was overly simplistic, but it 'retains pedagogical value as *the* (his emphasis) basic model'. According to May (1977), there were two essential flaws in the original model. First, Crofton (1971*b*) used a fixed value for his Lethal Level (*L*). In other words, all hosts with a parasite density greater than *L* parasites were eliminated. This feature of Crofton's model was regarded as too deterministic. Lethality should have been considered in more probabilistic terms. Thus, some hosts with $>L$ parasites might die and others would not, a much more plausible assumption, especially given the high degree of genetic variability in many host populations. When the deterministic effects of parasite-induced host mortality were exchanged for probabilistic expressions, May (1977) found that the key parameters in stabilizing the system were *k*, which describes parasite aggregation, and α, which refers to the growth potential of the host population. Another objection was that Crofton assumed parasite transmission was a direct function of host density. Accordingly, May included 'saturation effects in the parasite transmission factor' in his modification of the model. He indicated that Crofton devised a parasite transmission factor (*F*) that was 'linearly in proportion to the number of hosts'. The problem with this assumption is that transmission will not increase linearly with host density indefinitely; instead 'it must saturate to unity at high host population levels' (May, 1977).

A basic similarity also was noted between the stabilization of the host–parasite relationship by aggregation and related interactions involving prey–predator and host–parasitoid systems (May, 1977). He emphasized further that relationships involving predators and prey 'can be stabilized by differential aggregation of predators or by explicit refuges for the prey'.

12.3 Aggregation and regulation

Keymer (1982) noted that the statistical distribution of parasites within a host population is not, in and of itself, a regulatory factor. The effects of aggregation may be influenced by a number of other factors, e.g., diet, stress, host age, immunity, etc. In some of these situations, aggregation may be exacerbated and, in others, it may be constrained.

The effects of aggregation may be measured in two different ways. First, it may induce host mortality, either directly or indirectly. Second, some parasites have the capacity to cause a reduction in host fecundity, even castration, and this too is regulatory. The effects of aggregation and other factors on host mortality and fecundity will now be examined through consideration of a series of case studies.

12.3.1 *Hymenolepis diminuta* and its intermediate host

Keymer (1981, 1982) examined the effects of the cestode *Hymenolepis diminuta*, on its intermediate host, the common grain beetle *Tribolium confusum*. Experimental protocols were designed to assess various aspects of the relationships between parasite establishment, growth, and infectivity, and the host's population biology.

Results show that there were no density-dependent constraints on the establishment of cysticercoid infrapopulations within the grain beetles, at least within the range of exposures employed. However, there was an inverse relationship between infrapopulation densities and cysticercoid size, suggesting intraspecific competition for food or space, or both. No relationship was observed between infrapopulation densities and infectivity although there was a marked decline in cysticercoid infectivity with the passage of time. However, Evans *et al.* (1992, 1998) showed that the fecal contents of rats infected with *H. diminuta* are attractive in some manner to beetles, and this is the first evidence that a tapeworm is able to enhance its transmission chances by influencing the foraging of its intermediate host. The chemical attractant(s) is a volatile agent, but its precise nature is not known.

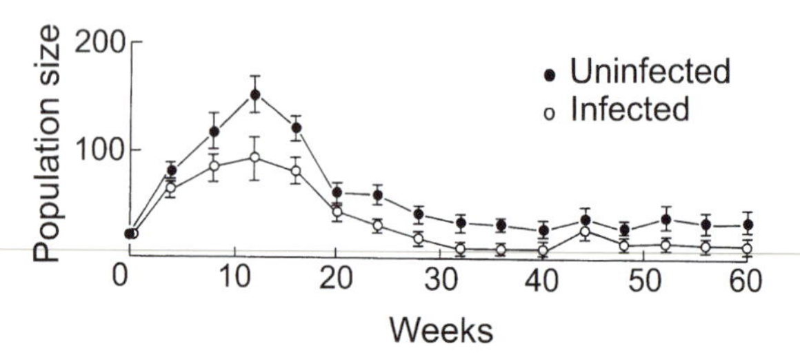

Fig. 12.2 The influence of infection with *Hymenolepis diminuta* on the population growth of *Tribolium confusum*. The points represent observed values of total population size; the vertical bars indicate the 95% confidence limits of the means. (Modified from Keymer, 1981, with permission, *Journal of Animal Ecology*, **50**, 941–950.)

The transmission of the cysticercoid to the definitive host is a much more complex process than of an egg to the intermediate host. Transmission dynamics in both cases are, in part, related to host feeding behavior, but in the case of the definitive host there is also an apparent link between infrapopulation size and aggregation of cysticercoids in the beetles.

Changes in beetle population size were followed over time, with and without the parasite. The populations were begun with 400 beetles of uniform size, half of which were exposed to parasite eggs and half of which were not. Beetles were counted and re-exposed to additional parasite eggs at regular intervals. The results are illustrated in Fig. 12.2. Under these conditions, densities of infected beetle populations were reduced up to 50% as compared with uninfected populations. Keymer (1981) pointed out that while the parasite is regulatory under laboratory conditions, it is probably ineffective as a regulator under field conditions where predation and competition are more likely to be the primary constraining factors for the grain beetle. On the other hand, Evans (1983) examined the cestode *Hymenolepis tenerrima* that uses ostracods as an intermediate host and ducks as definitive hosts. In a natural pond in West Sussex, England, the cysticercoid was found to cause substantial mortality in heavily infected hosts and fecundity in the ostracod population was reduced by nearly 10%.

In summary, Keymer (1981) indicated that

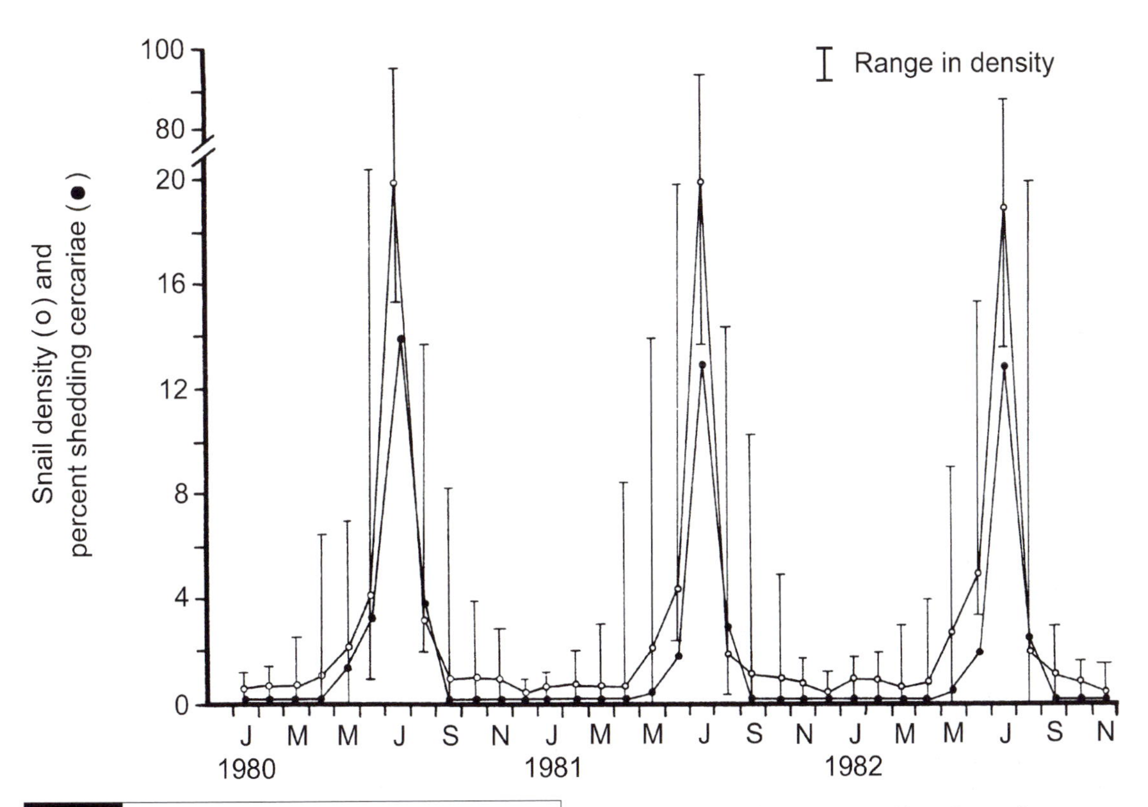

Fig. 12.3 The density (number of individuals per square meter) of the snail *Helisoma trivolvis*, in the littoral zone of Reed's Pond, North Carolina, USA, and the percentage of snails shedding cercariae of *Uvulifer ambloplitis*. (Modified from Lemly & Esch, 1984, with permission, *Journal of Parasitology*, **70,** 461–465.)

adult parasites are probably incapable of inducing mortality of rat definitive hosts, except under conditions of nutritional stress (see also Goodchild & Moore, 1963; Dunkley & Mettrick, 1977). Regulation of *H. diminuta* dynamics at the suprapopulation level is probably via density-dependent effects on fecundity and survival of adult worms in the definitive host.

12.3.2 *Uvulifer ambloplitis* and bluegill sunfishes

The metacercaria of the digenean *Uvulifer ambloplitis* causes 'blackspot' disease in centrarchid fishes. Kingfishers, *Ceryle alcyon,* are definitive hosts, the pulmonate snail *Helisoma trivolvis* is the first intermediate host, and centrarchid fishes are second intermediate hosts. The component population biology of *U. ambloplitis* was studied within the snail and fish intermediate hosts in a combined laboratory and field effort that lasted for approximately 3 years (Lemly & Esch, 1983, 1984*a*, *b*, *c*, 1985).

Snail densities and parasite prevalences had a striking annual peak in July of each year, followed by a decline (Fig. 12.3). It was speculated that a combination of the life cycle of the snail and the visitation periodicity of the kingfishers to the pond (most frequently in spring and early summer) were responsible for the seasonal pattern of parasite prevalence in snails. The snail thus recruited the parasite when kingfishers were foraging in the pond in the spring and early summer. The decline in prevalence then corresponded with the annual period of senescence and mortality within the snail population and the recruitment of a new snail cohort in mid-summer.

Seasonal changes in prevalence and density of metacercariae were observed in juvenile (<70 mm) bluegill *Lepomis macrochirus* and largemouth bass *Micropterus salmoides* in the pond over a 3-year period. Prevalences in three arbitrarily established size classes of bluegill showed a marked seasonal pattern; highest percentages occurred

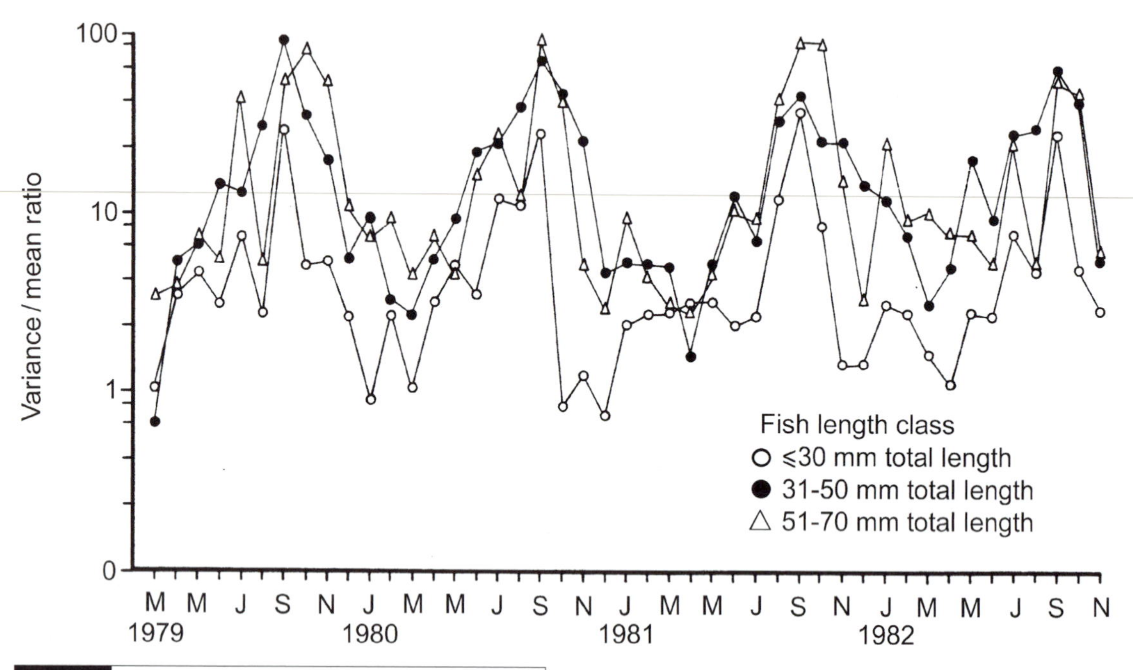

Fig. 12.4 Variance-to-mean ratios for metacercariae of *Uvulifer ambloplitis* in bluegill sunfishes *Lepomis macrochirus*, from Reed's Pond, North Carolina, USA, from March, 1979 to November, 1982; $n = 10$ fish per data point. (Modified from Lemly & Esch, 1984, with permission, *Journal of Parasitology*, **70**, 466–474.)

from spring to the mid-winter. There were then decreases in prevalence. Peak metacercariae densities were in September, followed by declines into the following spring and then again by increases until the next September. Fishes were also maintained in liveboxes within the littoral zone and the rates of parasite recruitment were followed throughout the year. Acquisition of parasites began in April, peaked in July, and stopped completely by October. All of the parasite recruitment coincided with those periods in which snail shedding also occurred. The parasite frequency distributions were highly aggregated in each of the years and could be described by the negative binomial model.

The variance-to-mean ratio ($s^2/\bar{x}$) for *U. ambloplitis* in bluegills increased continuously in all three size classes from early spring to September and then stopped (Fig. 12.4). During that time period, the parameter appeared to be tracking parasite recruitment by the fish hosts. Lemly & Esch (1984*c*) reasoned that if host or parasite mortality did not occur and if no fish moved in or out of the pond, then the $s^2/\bar{x}$ ratios should remain constant from September until the following spring, but they did not. Instead, a huge decrease occurred so that within 3 months following the September peak, the ratio was 10- to 100-fold smaller. Under similar circumstances, other investigators had suggested parasite-induced host mortality as an explanation for changes in these ratios (Crofton, 1971*a*; Gordon & Rau, 1982). However, direct observation of host mortality was not made in any of these studies.

Lemly & Esch (1984*c*) hypothesized that blackspot was causing mortality of heavily infected fishes in the pond. A series of experiments was then designed to examine the relationship between parasite density and host mortality. Bluegills with a wide range of parasite densities were kept in unheated outdoor aquaria and within liveboxes in the pond during the winter months. A set of aquaria was placed in a controlled-temperature room and fishes were exposed to temperature and light regimes that simulated those of a typical winter, while another set was placed in a controlled-temperature room where temperature was held constant throughout winter. When fishes died in any of the aquaria or in the pond, they were dissected and the metacercariae counted. In

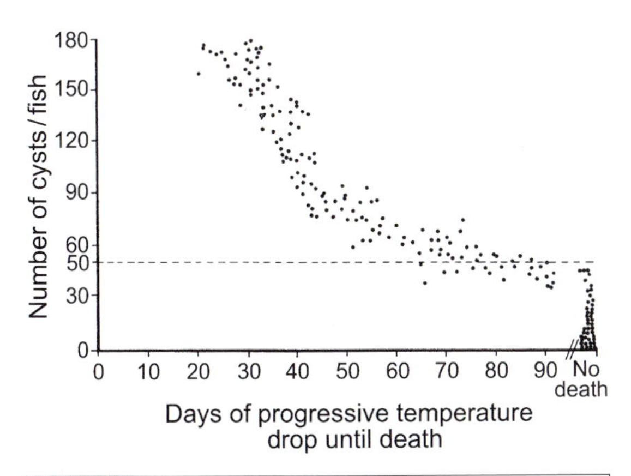

Fig. 12.5 Association between parasite density and survival, following onset of declining water temperature for bluegill sunfishes *Lepomis macrochirus*, from Reed's Pond, North Carolina, USA, during the autumn and winter of 1980, 1981, and 1982. Data are for fish held in outdoor aquaria at ambient temperature ($n=18$) or in liveboxes in the littoral zone ($n=174$). All fish were naturally infected with *Uvulifer ambloplitis* and were <70 mm total length. Correlation coefficient $r_s=-0.94$ ($p<0.001$). The dotted line indicates the maximum density observed for bluegill that overwintered successfully in Reed's Pond. (Modified from Lemly & Esch, 1984, with permission, *Journal of Parasitology*, **70**, 475–492.)

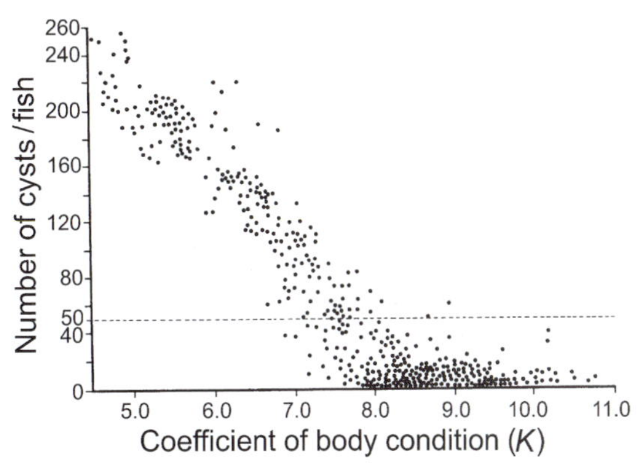

Fig. 12.6 Association between parasite density and body condition of bluegill sunfishes *Lepomis macrochirus*, from Reed's Pond, North Carolina, USA, March 1979 to November, 1982. All fish (31–50 mm total length) were naturally infected with *Uvulifer ambloplitis*, $n=440$, correlation coefficient $r_s=-0.93$ ($p<0.001$). The dotted line indicates the maximum density observed for bluegill that successfully overwintered in Reed's Pond. (Modified from Lemly & Esch, 1984, with permission, *Journal of Parasitology*, **70**, 475–492.)

several experiments, fishes that died were also frozen for subsequent lipid analysis.

In most of the field and laboratory experiments, the coefficient of body condition was calculated for each fish in which metacercariae were counted. The coefficient is expressed as:

$$K=\frac{100 \times \text{weight (g)}}{[\text{standard length (cm)}]^3}$$

Body condition is a standard method by which fisheries biologists measure individual robustness within a given population. (Again, K is not to be confused with k in the negative binomial model, or with K in the logistical growth curve.)

All of the heavily infected juvenile fishes (>50 metacercariae), in both the field and the laboratory experiments, died when water temperatures fell below 10 °C (Fig. 12.5). If fishes were >70 mm, high parasite densities had no effect. Similarly, high parasite densities had no effect on fishes held under constant temperature (22 °C) and fed *ad libitum* in the laboratory. Based on these experiments, in combination with the field observations on changes in $s^2/\bar{x}$ ratios, it was concluded that parasite-induced host mortality was occurring among juvenile bluegill in the pond. Moreover, it was proposed that it was also directly involved with density-dependent regulation of the bluegill population. Based on population estimates for juvenile bluegill and on the prevalence of heavily infected fishes in the pond during the late fall, it was suggested that 10%–20% of the bluegill were eliminated annually by parasitism.

During the course of the study, metacercariae were counted in a large sample of juvenile bluegill taken from the pond. A striking and significant correlation was observed between body condition and parasite density. As parasite densities increased, body conditions declined (Fig. 12.6). It was reasoned that, in some way, the decline in body condition contributed to a fish's inability to survive the winter months. Measurements of total body lipid revealed that it was positively correlated with body condition and negatively correlated with parasite density. Finally, in one last set of experiments, a group of uninfected fishes was starved in order to reduce their body conditions to levels similar to those caused by heavy infections.

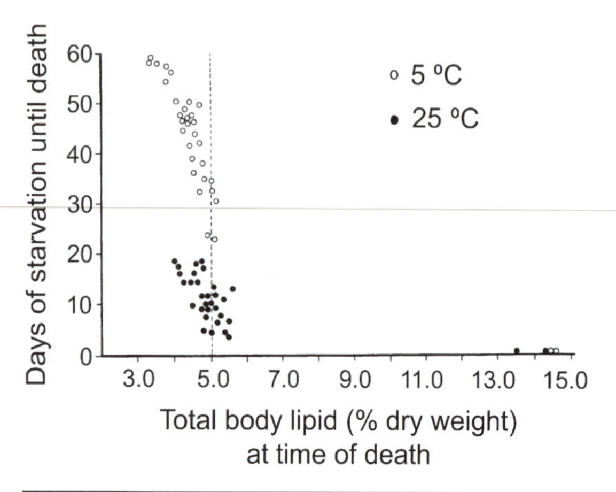

Fig. 12.7 Total body lipid present at the time of death for uninfected bluegill sunfishes *Lepomis macrochirus*, from Reed's Pond following starvation; $n = 32$ for each water temperature; all fishes were <70 mm total length. Correlation coefficient $r_s = -0.211$ for 5 °C ($p > 0.05$); $r_s = -0.364$ for 25 °C ($p > 0.05$). (Modified from Lemly & Esch, 1984, with permission, *Journal of Parasitology*, **70**, 475–492.)

When these fishes were subjected to temperatures that simulated winter conditions (<10 °C), they all died (Fig. 12.7).

It is clear that the energetic demands on hosts during the infection process by certain parasites are substantial. When strigeid cercariae enter the flesh of a fish, for example, there is a striking host reaction that includes inflammation, localized bleeding, cell destruction, and edema (Davis, 1936; Lewert & Lee, 1954*a*, *b*; Erasmus, 1960). With *U. ambloplitis*, the reaction is similar initially but, as it progresses, it actually becomes more and more intense. Thus, in addition to the localized tissue response, the fish host produces a large fibrous capsule around the parasite; this is followed by the migration of melanocytes into the wall of the cyst creating the characteristic appearance of the 'blackspot'. Lemly & Esch (1984*c*) concluded that the metabolic demand involved with the encystment of metacercariae significantly decreased a fish's lipid reserves and because bluegills do not feed during the winter months, these lipid reserves are critical to their survival through the winter.

There are two important points that should be emphasized by this series of studies on *U. ambloplitis*. First, parasite-induced host mortality was inferred from changes in $s^2/\bar{x}$ ratios of parasite component populations within fish from field studies. Hypotheses regarding mortality of heavily infected hosts were then proposed, tested, and confirmed or rejected through both field and laboratory experiments. Second, further analysis of the data provided a framework for interpreting the physiological basis for the parasite-induced host mortality. It was thus not surprising to find that the mortality was related to a massive pathological response of the host to the parasite. Among most tissue-dwelling helminth parasites, it is the trauma of the tissue reaction and the parasite-induced pathology that produces the serious morbidity problems for the host. A somewhat similar conclusion was reached by Gordon & Rau (1982) in their study of the strigeid digenean *Apatemon gracilis* in brook sticklebacks *Culaea inconstans*. Although this parasite does not encyst, the metacercariae are quite large and capable of inducing considerable damage if they occur in large numbers. Moreover, Gordon & Rau (1982) also invoked the notion of parasite-induced host mortality based on seasonal changes in the $s^2/\bar{x}$ ratios, but they did not conduct experimental studies to confirm or reject the hypothesis.

12.3.3 *Heligmosomoides polygyrus* and regulation

Heligmosomoides polygyrus is a common intestinal nematode in mice. It has a direct life cycle in which eggs shed in the feces hatch and free-living larvae become infective within 7 days. The larvae are accidentally ingested by mice where they migrate into the intestinal wall, stay in the area of the serosa for about a week, and then migrate back into the lumen of the intestine before maturing sexually. Scott (1987*a*) designed a series of experiments in which the aim was to compare the population dynamics of infected (with *H. polygyrus*) and uninfected mouse colonies. The colonies were established using outbred animals of the same strain.

In the uninfected colony, reproduction and mortality stabilized at a point where densities were about 320 mice/m^2. At these high densities, there was density-dependent mortality among young mice, some cannibalism, and a decline in reproductive success over time. As Scott (1987*a*) indicated, such mechanisms are considered as 'intrinsic regulatory mechanisms in mouse populations where emigration is impossible (MacKintosh, 1981)'.

On introduction of the parasite into immunologically naive mouse populations, the impact was dramatic (Fig. 12.8). In the three arenas housing the exposed colonies, mouse densities were reduced by 50% within 7 weeks after introducing the parasite and, by 9 weeks, 90% of the mice were dead. Parasite densities in necropsied mice from these three arenas averaged from 213 to 892 per animal, with some mice harboring up to 1500 nematodes. Mice in one of the three arenas were then transferred into a habitat in which high transmission rates were prevented. Mortality of mice continued at a high rate. However, when these mice were treated with an anthelmintic, the population growth rate increased significantly.

Based on these results, it is clear that *H. polygyrus* is a potential pathogen of considerable magnitude for mice. The study also demonstrated that the parasite is capable of regulation in the sense that it reduced population equilibria to levels lower than those that existed without the parasite. Scott (1987*a*) noted, however, that the experiments were conducted using naive populations and that the high mortality rates she observed could have been a result of the lack of prior exposure to the parasite. She concluded by emphasizing that 'the relative importance of parasitism in relation to other regulatory factors including predation, competition and dispersal will need to be assessed'. This admonition is certainly a clear warning for ecologists working on free-living organisms and that is, parasites can kill.

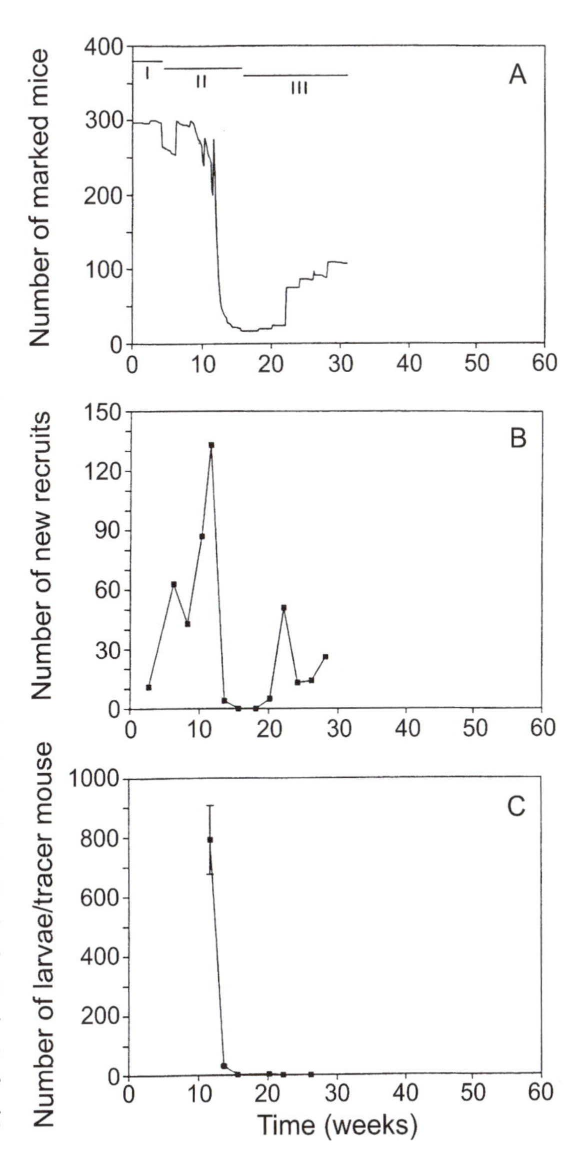

Fig. 12.8 Effect of the introduction of *Heligmosomoides polygyrus* into an immunologically naive mouse population. (A) Change with time in the number of marked mice (>2 weeks old) during three phases of the experiment. Phase I represents the mouse population prior to parasite transmission. Phase II represents the mouse population during the initial period of *H. polygyrus* transmission. Phase III represents the mouse population during continued low-level transmission of *H. polygyrus*. (B) Numbers of newly recruited mice (2 weeks old) at each census. (C) Mean (±S.E.) numbers of *H. polygyrus* larvae in tracer mice exposed to contaminated arena for 24 h following each census. (Modified from Scott, M. E. [1987] Regulation of mouse colony abundance by *Heligmosomoides polygyrus*. *Parasitology*, **95**, 111–124. Reprinted with the permission of Cambridge University Press.)

12.3.4 *Gyrodactylus turnbulli*, guppies, and aggregation

The component population biology of *G. turnbulli*, a monogenean of guppies *Poecilia reticulata*, was examined in a series of studies by Scott (1982, 1985*a*, *b*, 1987*b*), Scott & Anderson (1984), Scott & Nokes (1984), Scott & Robinson (1984), and Madhavi & Anderson (1985). In addition to evaluating the parasite's component population biology, the report by Scott (1987*b*) is of particular interest because it provides a useful discussion of the value k in the negative binomial model. Perhaps more importantly, it makes a careful comparison between the use of k and the variance-to-mean ($s^2/\bar{x}$) ratio in evaluating parasite aggregation.

The system on which she based her assessment was described in detail by Scott & Anderson (1984). In brief, *G. turnbulli* is an ectoparasitic monogenean that lives on the fins and other body surfaces of guppies. It has a short generation time, a direct life cycle, and transmission between fishes is rapid. Another attractive feature of the system is that fishes can be anesthetized with ease and the parasites can be counted using an ordinary dissecting microscope. In other words, the more traditional 'kill and count' method of enumerating parasite infrapopulation density was not necessary.

The aims of Scott's (1987*b*) investigation were to examine the temporal aggregation of the parasite and to determine if changes in the level of aggregation would provide any insights into the biological processes that produced the aggregation. The results of several carefully designed experiments revealed that the parasite undergoes repetitive **epizootic** cycles on guppies. In that part of the epizootic where parasite numbers are increasing, parasites became more aggregated presumably because of direct reproduction by the parasites on individual fish. Then, as parasite prevalences and densities peaked, aggregation began to fall and continued downward as the prevalence and densities declined. According to Scott (1987*b*), 'this is thought to be a function of density-dependent death of infected hosts, and density-dependent reduction in parasite survival and reproduction on hosts that recover from infection'.

Scott & Anderson (1984) had previously discussed and evaluated the same data set and concluded 'that the degree of aggregation decreased while the prevalence and abundance of infection increased, based on the positive correlation between k and abundance'. Scott (1987*b*) subsequently rejected that interpretation of their data. Thus, whereas k did rise as density increased, it would not necessarily also indicate that aggregation was simultaneously decreasing since both the density and the percentage of infected fish were changing. She stated, 'it is suggested that this relationship is simply a function of the interaction between mean burden (density) and k in the negative binomial distribution, and does not indicate anything *per se* about the changes in the degree of aggregation'. The term k was considered a 'mathematical convenience' and it was urged that it should not be interpreted outside of that consideration. The $s^2/\bar{x}$ ratio is the appropriate term of biological sensitivity, especially if prevalence or density, or both, are in flux, or if the tail of the distribution is the focus of evaluation. While k can be useful in describing a given frequency distribution, it must be considered with caution when drawing conclusions having any sort of biological implications. As she declared, the value of k in the negative binomial is a theoretical description, whereas the variance-to-mean ratio is an absolute measure of the degree of aggregation.

12.4 Epidemiological implications

12.4.1 Aggregation and predisposition

Despite the availability of suitable anthelminthic drugs, the human intestinal helminths continue to represent an undesirable and unnecessary global scourge. Most are concentrated within the so-called Third World countries where their prevalences continue to increase. *Ascaris lumbricoides*, for example, was estimated to occur in more than 1 billion individuals in 1986 (Crompton, 1989, 1999). Stoll (1947) had placed the number with ascariasis at 645 million, meaning that the numbers of humans infected by this parasite increased by about 50% in just under 40 years, and definitely is still increasing. The prevalences of several of the other debilitating parasite diseases of man are also still rising (see Table 1.1).

The life cycles of these parasites and many of their epidemiological characteristics have been known for years. However, it was not until the early 1980s that the emergent ecological doctrine regarding the aggregation of parasites, developed by Crofton (1971*a*, *b*), was applied to human helminths. At about the same time, however, some epidemiologists and ecologists recognized that there was a problem in Crofton's assertion that parasites have the capacity to kill. We know that many parasites, probably most, do not cause host mortality. Therefore, Anderson & Gordon (1982) qualified Crofton's mortality position by stating 'that the probability of a host dying in a given time interval is some function of its parasite burden'. Their revised approach appears much closer to reality, certainly as it applies to the large group of enteric helminths affecting man. However, it must be emphasized that, whereas these parasites may not cause extensive mortality, they remain as one of humankind's most serious morbidity problems. This is because morbidity (incapacitation) translates into time lost from work or school, relatively high costs in terms of treatment and hospitalization, and just plain ill health.

The capacity of the enteric helminths to cause morbidity or mortality is clearly influenced by an enormous variety of environmental factors that operate in conjunction with nutritional status, immunocompetency, and the physiological condition of the host. These three factors frequently are so intertwined as to be virtually inseparable. Consider hookworm disease as an example and, in doing so, recall that hookworms infect in excess of 1 billion humans. A single adult hookworm attached to the intestinal wall of a human host will consume approximately 0.2 ml of blood per day. Assume that a host has an infrapopulation density of 100 worms. In a single day, the host will lose 20 ml of blood. If this individual is on a high-protein diet, the probability of morbidity is low and mortality is nil. On the other hand, if the host is living under exceedingly poor socioeconomic conditions and the diet is meager, typical in some Third World countries, the loss of 20 ml of blood per day by itself may very well produce morbidity and perhaps even contribute to an individual's death.

The insidious nature of the disease is what exacerbates the hookworm problem. The loss of blood from the individual living under poor socioeconomic conditions is compounded by a low-protein diet, meaning that lost hemoglobin cannot be readily replaced. The immunocompetency of the individual is further compromised through the loss of gamma globulin and this translates into antibody loss. Thus, the poor diet will influence the host's immunological capacity, with the worms affecting both the O_2-carrying capacity of the host and its immune system. The host's size, age, sex, genetics, and cultural background also will influence the situation regarding morbidity or mortality. These sorts of host–parasite relationships are exceedingly complex and understanding the epidemiological characteristics of the diseases caused by these parasites requires multivariate approaches.

For the epidemiologist, there are two generally accepted ways of estimating infrapopulation densities of enteric helminths in human populations. One is to count parasites in the feces immediately after treating the individual with an appropriate anthelmintic. This procedure has its problems, one of which is the cost of the treatment itself. Another difficulty is that parasite expulsion can be extended over a period of several days so that the reliability of feces collections may be questionable, especially when dealing with children (Kightlinger *et al.*, 1995). The second procedure for estimating parasite densities in humans relies on fecal egg counts; however, a primary difficulty with the technique is related to the tremendous variability in the number of eggs shed per day over a given time interval. Moreover, there are also density-dependent constraints created by intraspecific parasite competition, causing even greater variability in daily egg output. In many of these diseases, egg output will increase as parasite densities increase, but only until a threshold is reached and then it may decline or fluctuate erratically.

A third quantification procedure that has been pursued by some epidemiologists is the use of prevalence, or the percentage of hosts infected in a given population, in predicting high parasite densities and the potential for morbidity. As just indicated however, with many enteric helminths, egg counts are very unreliable in estimating worm densities. For some species, e.g., *Schistosoma mansoni*, a reliance on prevalence alone has been suggested as a viable alternative since, if effective, it might prove useful in the design of rational treatment programs (Guyatt *et al.*, 1994). Unfortunately, such a proposal has been shown to be impractical, in part at least, because of variations in area-specific transmission of the parasite, both in the past and at present, and to variations in levels of local immunity. A number of investigators have attempted to resolve these sorts of problems and the student is referred to these papers for a more thorough review of the issues and solutions (Croll *et al.*, 1982; Anderson & Schad, 1985; Holland *et al.*, 1988; Elkins & Haswell-Elkins, 1989; Guyatt *et al.*, 1994; Kightlinger *et al.*, 1996).

Croll & Ghadarian (1981) made the earliest application of the approach developed by Crofton (1971*a*, *b*) to the epidemiology of human helminth infections. They worked in a number of Iranian villages generating data on the frequency distributions of several enteric helminths, including the nematodes *Ancylostoma duodenale*, *Necator americanus*, *Ascaris lumbricoides*, and *Trichuris trichiura*. The first two (both hookworms) employ soil-dwelling, 3rd stage larvae that penetrate the skin of their human hosts. The latter two species of worms are transmitted directly, through the accidental ingestion of eggs. All four parasites were aggregated in their frequency distributions within the village populations. All of the villagers were then de-wormed using appropriate anthelmintics. After about a year, the same people were re-examined for parasites. The investigators found that those individuals who had been wormy the first time were not wormy the second time. In other words, 'wormy persons' from the initial screening were not predisposed to re-infection. As Croll & Ghadarian (1981) stated, 'in this population of mixed ages, sexes, and social histories, the 'wormy persons' prior to treatment were not reliable predictors of subsequent intensities of infection'.

These findings suggested two important conclusions. First, host genetic predisposition did not appear to play a role in the acquisition of these helminth parasites, or in creating aggregation. Second, the worms did not appear to have stimulated the host's immune responsiveness to the extent that immunity precluded re-infection. They concluded that 'risk factors are much more subtle than the classic categories we are considering or that 'wormy persons' in these communities result from the superimposition of otherwise random events'.

The study by Croll & Ghadarian (1981) was a watershed in modern epidemiology as applied to the enteric helminths of humans. Their results and conclusions regarding the absence of genetic predisposition were, however, viewed somewhat skeptically by some investigators. Since the time of their report, there has been a large number of studies in different parts of the world that have attempted to determine if predisposition is, or is not, a corner-stone of helminth epidemiology (Bundy *et al.*, 1987; Haswell-Elkins *et al.*, 1988; Anderson, 1989; Kightlinger *et al.*, 1995, 1996, 1998). In other words, is an individual within a population, or a subgroup within a population, more vulnerable to the acquisition of a given parasite than other individuals or groups and, if so, why?

Generally, the notion of predisposition is considered within one of three contexts. The first relates to non-genetic factors such as personal hygiene, religious practices, socioeconomic status, water resources and their possible contamination, etc. Any of these factors could influence transmission dynamics in such a manner as to create predisposition, as well as help produce parasite distributions that are aggregated. The second context relates parasite frequency distributions with host or parasite genetics, or both. For example, there could be genetically determined resistance that would create differential levels of immunity within a community and thus contribute to parasite aggregation. Genetically based resistance has been documented for parasites in a number of animal models, such as mice, cattle, and guinea pigs, but its influence in humans is less well understood. Finally, differential levels of nutrition within a population could also affect the presence or absence of predisposition. This could be a confounding factor in assessing the resistance/susceptibility and, therefore, the genetic component in parasite transmission dynamics.

The significance of aggregation and predisposition as they affect the enteric helminths of humans has several practical implications when considered within the framework of mathematical models (Anderson & May, 1979; May & Anderson, 1979; Schad & Anderson, 1985). The latter authors, especially, argued that treatment of only the heavily infected (and presumably predisposed) humans would be effective in reducing overall infection within a community, or possibly in eliminating the parasite entirely. They based their assertions on predictions regarding transmission dynamics generated from mathematical models and on the numbers of parasites necessary to sustain a given parasite species within a host population. In other words, it may be necessary to apply therapy to only a few people within a community in order to effectively control the parasite. As they suggested, cost-effective treatment becomes a clear possibility except for the expense involved in identifying the predisposed individuals within a population. Keymer & Pagel (1990) reported that fewer than 30% of paired human hosts had the same ranked infrapopulation density order at the first and final sampling times. In other words, there were no significant changes in parasite infrapopulation densities following treatment and re-recruitment of new infrapopulations. This caused Guyatt & Bundy (1990) to suggest 'fixed, long-term predisposition is modified by transient changes during the relatively short course of the observation (usually less than two years). Equally it may indicate that the phenomenon of predisposition is a reflection only of short-term factors that remain constant for some individuals during the study period.' In any event, the idea of predisposition through genetic mechanisms continues as a contentious issue at the present time.

12.4.2 Non-genetic predisposition

Defecation practices are among the most important factors in creating the differential frequency distribution patterns of certain human enteric helminths. This behavior is related directly to the transmission patterns associated with these parasites. Haswell-Elkins *et al.* (1988), for example, observed in an Indian coastal village that adult females and children were much more heavily infected with hookworms because they used shaded tree groves in which to defecate, whereas older males used the sandy beaches. Hookworm larvae require special soil and shade conditions in order to develop successfully; such conditions were optimal in the shaded tree groves, but not on the sandy beaches. Sixty to 70 years ago, hookworm disease was much more common than it is now in the southeastern USA and there was a distinct, age-related pattern of infection. Teenagers were likely to have more hookworms than either young children or adults because they would commonly return to a previously established outdoor location to defecate, whereas young children tended to defecate more randomly and adults would use indoor facilities when they were available. Kightlinger *et al.* (1995, 1996, 1998), in a series of studies from Madagascar, reported prevalences of *Ascaris lumbricoides* of up to 94% among children 4 to 10 years old, with significantly higher levels among females than males. The overall high prevalences were attributed to poor socioeconomic conditions, but gender-related behavior was said to account for the differences between the sexes in these villages. At very early ages, boys were required to begin tending cattle and participate in the cultivation of rice, away from the source of infection within the villages. Young girls stayed 'at home' and were, therefore, continuously exposed to infection.

The wearing of shoes prevents larvae of hookworms and *Strongyloides stercoralis* from penetrating the skin and is a considerable deterrent to transmission in certain parts of the world. Holland *et al.* (1988) reported that Panamanian children living under poverty conditions, such as in houses with dirt floors, were much more likely to have enteric helminths than children living in concrete block houses. Kightlinger *et al.* (1996) affirmed this observation in their study of enteric helminthiases on the island of Madagascar.

12.4.3 Genetic predisposition

Long before the study by Croll & Ghadarian (1981), a strong basis was indicated for genetic predisposition to parasite infection. For example, in the southeastern USA, Keller *et al.* (1937) demonstrated that the frequency of hookworm disease could be separated along racial lines, with blacks having significantly lower prevalences of the disease than whites. Cram (1943) reported the same racial trend for the benign human nematode *Enterobius vermicularis*.

Schad & Anderson (1985) examined hookworm infections in several villages in West Bengal, India, prior to, and after, drug therapy was administered. They were able to show that the frequency distributions were aggregated both before and after drug therapy and, moreover, that certain individuals appeared to be predisposed to infection. They stated, 'we suspect that both heterogeneity in exposure to infection within human communities and genetically determined host resistance mechanisms play important roles as determinants of parasite aggregation and host predisposition to infection'. Efforts have been made to correlate resistance to infection with certain types of known genetic markers, but the results were not very conclusive (Bundy, 1988). However, several animal models have been successfully employed to establish relationships between genetically based predisposition and infection (for a review, see Wakelin, 1996). For example, Sher *et al.* (1975) provided a clear indication of a single-locus control in the protective immunity against the digenean *Schistosoma mansoni*, in laboratory mice. Enriques *et al.* (1988) reported that resistance to infection of *Heligmosomoides polygyrus* (= *Nemotaspiroides dubius*) in rats could be under the control of several genes, some of which may belong to the major histocompatibility complex.

12.4.4 The nutritional aspects of parasitism, predisposition, and epidemiology

There is a wide range of signs and symptoms associated with specific viral, bacterial, protozoan, and helminth pathogens of humans. Perhaps the most devastating of these conditions is attributed to the diarrheal diseases and their impact on children, especially in Third World countries. It has been estimated that fully ⅓ of the world's mortality of children less than 5 years of age is related to diarrhea. This translates into about 4 million deaths annually. According to Hodges (1993), the protozoan and helminth parasites most associated with this problem include *Entamoeba histolytica*, *Balantidium coli*, *Cryptosporidium parvum*, *Capillaria philippinensis*, *Strongyloides* spp., and *Trichuris trichiura*. In four of these six species, infection is hand-to-mouth, a clear indictment of contaminated food or water, substandard sanitary conditions, and dismal socioeconomic circumstances. In many of the diseases caused by these, and other, enteric parasites, the problem is also exacerbated by low-protein diets (Kightlinger *et al.*, 1996). These, and other, parasites frequently aggravate an already difficult nutritional situation by causing blood loss, anemia, etc. Hookworm disease and schistosomiasis are two other examples of human helminth parasites contributing to such problems, although neither of these parasites is transmitted from hand to mouth. Iron-deficiency anemia caused by *Ancylostoma duodenale* and *Necator americanus* has been known since antiquity (see Box 1.1).

For most of these parasite-induced problems, poor diet, socioeconomic conditions, and behavioral predisposition may be considered as synergistic. Disregarding genetic considerations for the moment, you should recall that the transmission of most of these enteric parasites is direct, with no intermediate hosts required. If each of these predisposing factors is considered independently, however, the underlying basis for the parasites' success must be considered within a socioeconomic framework, the so-called 'bottom line'. The life style of the poor is an overt 'invitation' to parasite transmission. Among such populations, the subsistence diet is minimally nutritious and immune capabilities may, accordingly, be affected in a detrimental fashion. The absence of adequate sanitation increases the probability of contact with infective agents of the parasites. Finally, human behavior may then augment the likelihood of parasite transmission and insure continuation of the inexorable pattern. Despite the availability of anthelminthic drugs and the development of sophisticated treatment strategies (see Box 12.1), the most effective way of neutralizing the impact of these devastating diseases is to deal with the basic problem of poor socioeconomic conditions and all that this entails. Only when this is accomplished will the mortality and morbidity associated with these parasitic diseases begin to pale. For a broad perspective on these sorts of diseases and related problems, see Crompton (1993, 1999).

12.4.5 The use of molecular technologies in epidemiological studies

Contemporary molecular technologies have the potential to play significant roles in understanding and interpreting the epidemiological tendencies of many important protozoan and helminth parasitic diseases of humans. Starch-gel electrophoresis is a relatively old tool that depends on the differential mobility of enzymes of varying molecular weights, size, and charge when subjected to an electrical

Box 12.1 Control of enteric helminth parasites in humans

As shown in Table 1.1, the extent of enteric helminth infection in humans, or prevalence, is staggering, with most of the parasites occurring in Third World countries. Morbidity due to disease induced by helminth infection is likewise at an unacceptable level. Over the years, strategies involved in dealing with these problems have revolved around three approaches. The first is via mass treatment with anthelminthic drugs. In this case, all of the potential hosts in a given locality are treated without regard to who may be heavily infected, or predisposed to infection. A second scenario in the treatment of these worm problems is to target individuals within a population who, for one reason or another, are predisposed to infection. These individuals may be prone to infection through some genetic, behavioral, socioeconomic, or other factor operating to increase the likelihood of becoming infected. A third approach is to identify individuals who are carrying the heaviest infections. The frequency distributions of most of the enteric helminths in humans are aggregated, which means that a few individuals in a given population carry a large proportion of the worms within a given locality.

Current control programs are aimed at reducing the infection load and transmission potential as a way of minimizing morbidity and stopping mortality rather than eliminating the parasites (Albonico *et al.*, 1998). These efforts include a program of sustained improvement in sanitation and health education, combined with periodic chemotherapy. This strategy can be reasonably cost-effective and successful.

Current drug therapies for the nematode parasites of humans include albendazole, ivermectin, levamisole, mebendazole, piperazine, pyrantel, and thiabendazole. In some situations, these drugs are not recommended during certain periods of pregnancy and for some age groups. Even though most of these drugs are effective, there are certain side-effects that have been reported.

As noted above, however, a necessary element in the long-term treatment of these enteric helminths includes effective sanitation improvement. This is simply because all of these enteric helminths are passed from their human hosts via the feces. Effective sanitation must include the construction of latrines and septic systems that will not leak, thereby contaminating nearby water supplies with infective agents of the parasites. Note that improved sanitation alone does not ensure meaningful results unless it is combined with effective and continuous drug therapy. Moreover, improved sanitation may be more effective against helminths such as *Ancylostoma* and *Necator* since these parasites have larvae as the infective agent and these larvae are not long lived, whereas the eggs of a worm such as *Ascaris lumbricoides* may survive in the soil for several years.

Control programs directed at certain of the human helminths, while of absolute necessity, are not the end-all in the elimination of all parasitic disease in Third World countries, or even in developed countries with supposedly high standards of living and superior sanitation. Serious problems include schistosomiasis, malaria, amoebiasis, leishmaniasis, Chagas' disease, neurocysticercosis, etc., in some Third World countries. In developed countries, cryptosporidiosis and toxoplasmosis, coupled with increasing use of immunosuppressant drugs, are considered as emergent diseases with public health authorities. Giardiasis and vaginal trichomoniasis are also on the increase in these same developed countries.

Fifty years ago, soon after World War II, Stoll (1947) pointed out the monumental extent of parasitic diseases, in some part due to the turmoil following the war. Crompton (1999) has revisited this worldwide situation and determines that virtually all of the major parasitic diseases of humans are on the increase, this despite all of the new epidemiological information that has emerged and the availability of new drugs known to be effective against many of these parasites. Albonico *et al.* (1998) provide a very thorough and thoughtful review dealing with the control strategies of human intestinal nematode infections. The last line of their review states, however, 'The challenge for the next century will be to develop sustainable approaches for controlling intestinal helminth infections and associated diseases, being aware of the growing demographic trends and increasing urbanization that will be occurring, particularly in developing countries.' This is a wonderful sentiment, but is it a realistic expectation in view of the inexorable trend for poverty, illiteracy, and political upheaval associated with humankind? When Stoll presented his presidential address to the American Society of Parasitologists in 1947, who knew anything about cryptosporidiosis or AIDS? Can these 'new' diseases, plus all those that were extant 50 years ago, be controlled in the next 50 years, let alone during the entire century? It will be a most interesting exercise!

field. The success of Linus Pauling and Harvey Itano in identifying the differences between normal and sickle-cell hemoglobin depended on this technique as early as 1949. Polymerase chain reaction (PCR) is a newer procedure that relies on the synthesis of DNA primers used in the amplification of specific DNA segments. Restricted fragment length polymorphisms (RFLPs) depend on restriction enzymes to cut DNA into small pieces that can then be compared with each other in terms of length and nucleotide sequence. Hybridization techniques are employed to copy specific DNA fragments. Thus, cDNA can be synthesized from a mRNA transcript and a hybrid DNA–RNA molecule created by reverse transcription. Other enzymes can split the hybrid molecule and a new, double-stranded cDNA molecule can then be generated. All of these techniques have been, and are being, used to determine and compare the genetic character of individual parasites and parasite populations. From an epidemiological perspective, they are useful in understanding the effects of genetic isolation and the significance of gene flow between **zymodemes** (strains). The range in applications of the various molecular probes and techniques is limited only by the imagination of those using them.

Studies that have employed molecular techniques have focused variously on Chagas' disease (D'Oro *et al.*, 1993), the African trypanosomes (Truc & Tibyrenc, 1993; Pencheiner *et al.*, 1997), giardiasis (Meloni *et al.*, 1995), trichinosis (Zarlenga *et al.*, 1996; Pozio *et al.*, 1997), ascariasis (Nadler, 1995), echinococcosis (Bowles & McManus, 1993; McManus & Bryant, 1995), and several other parasites. These efforts have drawn attention to the genetic variations in parasite populations, both locally and on a more global scale, and, as will be seen, they have exceptional heuristic value. Despite the enormous contribution of molecular biology to the study of the epidemiology of these and many other communicable diseases, investigators none the less must continue to seek an even greater understanding of the subtleties in parasite transmission and the nature of host–parasite interactions which influence the cellular/molecular and the organismic/behavioral characteristics of disease and disease processes.

Whereas the molecular approaches taken in the study of parasite epidemiology are wide ranging, various forms of enzyme electrophoresis have been quite productive. For example, D'Oro *et al.* (1993) examined enzyme patterns, using polyacrylamide-gel electrophoresis, in 95 isolates of *Trypanosoma cruzi* from domesticated and wild animals, insect vectors, and humans in several different geographic localities throughout Argentina. Their findings indicate that sexual reproduction is, not surprisingly, restricted or entirely absent in these *T. cruzi* isolates and that the isolates were essentially clonal in character. Thus, there appears to be little gene flow between geographic localities. This finding, in combination with certain genetic correlations and clinical signs of the disease caused by the parasite, may account for the wide and persistent variability in the manner in which Chagas' disease is manifested in endemic areas of Central and South America. Pencheiner *et al.* (1997) similarly employed electrophoresis and enzyme polymorphisms to examine the population genetics of *Trypanosoma brucei* isolates from pigs and humans in Côte d'Ivoire, Africa. Their approach was to isolate the parasite from pigs and humans and then to compare genetic loci. Results demonstrated clearly the clonal nature of the various isolates and confirmed previous hypotheses that pigs are the primary reservoir for the parasite in West Africa. Meloni *et al.* (1995) used eletrophoresis to examine 97 isolates of *Giardia lamblia* from throughout Australia and several overseas locations, including a number in North America. The parasites were taken from humans and a wide range of other animals, including cats, dogs, sheep, beaver, cattle, and rats. Similar to the work on *T. brucei* and *T. cruzi* just considered, Meloni *et al.* (1995) reported a clonal population structure for *G. duodenalis*, but also with occasional indications of genetic exchange between zymodemes. They interpreted the occurrence of constant genetic identity between isolates from humans and various animal isolates as evidence for the **zoonotic** transmission of the parasites.

Two other parasites on which a great deal of work has been conducted are *Ascaris suum* and *Ascaris lumbricoides*, the former species long considered a parasite of pigs and the latter of humans (Anderson *et al.*, 1993; Nadler *et al.*, 1995). The transmission biology of the two species is apparently identical, i.e., accidental ingestion of eggs by swine or humans. However, except for a number of anecdotal reports, cross-infections of humans with swine *Ascaris* and vice versa were thought to be uncommon. Considering that opportunities regularly exist for cross-infection in many rural areas of the world where pigs and humans live in very close proximity, if not virtually together, the question arises as to why there are not more exchanges of the parasites between the two hosts. Anderson *et al.* (1993) used mtDNA sequence analysis to pursue this question and found that mtDNA haplotypes fell into two distinct clusters, indicating that cross-infections between the two species were uncommon. However, they emphasized that with a clear phylogenetic relationship between *A. lumbricoides* and *A. suum*, the pattern of low cross-transmission in Guatemala may not be representative of other geographic areas. Subsequently, Anderson (1995) screened a number of *Ascaris* infections from humans in North America using an rDNA probe and determined that the probable sources of the parasite were swine, indicating *A. suum* was the agent of disease. Murrell *et al.* (1997) examined the migratory route of *A. suum* in experimentally infected pigs and re-emphasized the possibility of more extensive cross-infection than had been previously suspected, affirming that a reported outbreak of visceral larval migrans in Japan was due to infection by *A. suum* (Muruyama *et al.*, 1996). Although they did not use these kinds of molecular probes, the possibility of cross-infection by *A. suum* into humans in China was also raised by Weidong *et al.* (1996).

There are at least five recognized and accepted species of *Trichinella* worldwide. The so-called 'domestic' species and the one most likely to be encountered by humans is *T. spiralis*. Typically, this species cycles through swine and rats, with humans becoming involved when undercooked pork is consumed. *Trichinella britovi*, *T. nelsoni*, and *T. nativa* are considered as sylvatic species since they employ wild mammals and more classical predator–prey interactions for transmission. The fifth species is *T. pseudospiralis* and it cycles through avian hosts, although the biology of this species certainly is not well understood. When the epidemiological characteristics of trichinosis were first being considered, there was much confusion since only one species, *T. spiralis*, was then recognized. The use of a broad spectrum of biological approaches and biochemical/genetic probes has since reduced the earlier chaos and created at least a modicum of order for the group, although much work remains before a clearer understanding of the epidemiology of trichinosis is obtained.

Hydatid disease, caused by *Echinococcus granulosus*, is a serious zoonosis throughout the world. Not surprisingly, it has received a substantial amount of research attention and from a variety of perspectives (for a review, see McManus & Bryant, 1995). One of the special problems still attached to *E. granulosus*, however, concerns the nature and status of strains within the species. As previously noted (Box 4.2), the primary final hosts for *E. granulosus* are canines, usually wolves and domesticated dogs. However, it is the definitive host/intermediate host life-cycle patterns that have created the greatest consternation. Nuclear and mitochondrial DNA probes indicate, for example, that in the UK, there are distinct dog–horse and dog–sheep strains and, moreover, that each strain is genetically uniform throughout the world. DNA sequence data also suggest that sheep and equine strains do not cross even though they may employ the same definitive host and are thus sympatric in their distribution within a single definitive host.

All of these molecular techniques have contributed to significant progress in understanding the population genetics and epidemiology of the most dangerous and widespread parasitic diseases in humans and their domesticated animals. The same molecular probes also have been useful in establishing basic phylogenetic relationships among several of the protozoan and helminth groups. From an epidemiological and a systematic perspective, we are in a major transition in the study of parasitic organisms, and it is being driven, in large part, by some very specialized and sophisticated molecular technologies.

12.4.6 Concluding remarks on epidemiology

It should be clear that in the last 50–60 years of study on the epidemiology of parasitic diseases, an enormous amount of new information has been generated, despite the fact that the life cycles of most of the parasites causing these diseases were elucidated in the late nineteenth and early twentieth centuries. Despite all of this new information, however, the incidence of most of these diseases still is on the rise in many Third World countries (Crompton, 1999; see Table 1.1). There are several reasons for this discouraging observation. In part, the increase is the result of continuing divisive social and political pressures in many of these countries which, in turn, have created even more social and economic repression. The widespread migration of the Kurds in northern Iraq following war in the Persian Gulf or the vast movement of populations during recent tribal conflicts in Ruanda and the neighboring Congo are indications that dangerous problems still exist and that potentially greater difficulties may arise. For example, a 1998 press release from WHO reported an increase of 439% in visceral leishmaniasis from one Sudanese treatment center during a single 4-month period. They reported that this particular outbreak was 'exacerbated by the traditional seasonal migration of workers AND (emphasis ours) the resettlement of returnees and refugees in highly endemic areas'. These sorts of observations and the explanations for them are the rule, not the exception, in many parts of the world.

Another problem is the disproportionate emphasis for research on diseases afflicting much smaller segments of the world's population. Many of these diseases are horribly unpleasant and deadly and, because of this, they attract enormous public attention and financial treasure, especially in developed nations. This has produced the apparently intractable paradox of spending disproportionately larger sums of money on some of the 'lesser', though no less deadly, diseases of mankind. But this should not be surprising. The priorities for disease research vary from nation to nation and, in a general sense, there is a strong dichotomy in the population needs of those living in developed versus undeveloped areas of the world. This is a reality of wealth and power versus poverty and impotence.

Finally, the levels of funding allocated by individual governments, for malaria research as an example, almost always increase during the time of war when their own service personnel are exposed to the prospects of acquiring a given disease. However, all too often, after the war is over and the soldiers have returned home, the money spent on these research programs declines, leaving many such programs underfunded and unfinished. In effect, national research priorities change rapidly and this too is most unfortunate.

Whether through governmental efforts, or the private sector, the basic issue in the development of treatments for many of the tropical diseases is one of economics. There is, quite simply, no sense of urgency to spend money because there is no public outcry (except from those affected) to do anything about the problem, or there is no money to be made by corporate investors. This is a sad commentary on the realities of public health problems in Third World countries.

Guyatt & Bundy (1990) have urged, 'future research should focus on longitudinal studies that monitors an individual's infection and variables related to exposure or susceptibility through time'. This issue was also raised by Quinnell *et al.* (1990) who commented on the ethics of long-term studies as applied to human populations. As they indicated, 'There are, however, noticeable gaps in the available data [regarding long-term epidemiology]. These arise mainly from the ethical requirement to treat people where possible, thus making long-term observational studies unacceptable and the necessity to spend considerable effort persuading subjects to collaborate, thus reducing sample sizes.'

To reiterate, even though the epidemiologies of disease produced by enteric helminths are well known, they are still incompletely understood. For example, the present concept of genetic predisposition to infection is relatively new and, therefore, the basic or subtler nuances of the concept have not yet been evaluated clearly. If genetic predisposition to infection can be demonstrated, it may be necessary to treat only the 'wormy persons' in order to effectively reduce the prevalence and density of most enteric helminths (Schad & Anderson, 1985). The problem with such an approach, however, is that it depends on the validity of the genetic predisposition concept. A question persists, is genetic predisposition the core of a realistic paradigm, or is it simply a smaller segment of a much larger landscape?

12.5 Models

'It will be acknowledged that a predictive mathematical model of the epidemiology of any disease is desirable, both from the standpoint of intellectual satisfaction and from the standpoint of the usefulness in planning measures to control the disease' (Hairston, 1965). With this statement, the *raison d'être* for the mathematical model in epidemiology is certainly made clear. There are, however, two other important issues that need to be raised with respect to the broader development and use of mathematical models. First, any model is only as good as the assumptions required to generate it and, moreover, the assumptions are only as good as the quality of the data provided (in many cases) to make them. Second, any model, as a research tool, is a mathematical hypothesis. Models are attempts to predict reality in simplistic terms. In biology, one of the greatest problems with modeling is associated with the vagaries of biological systems; another is the extraordinary complexity of the same systems. Despite these almost intractable difficulties, the mathematical models developed for host–parasite population dynamics have become valuable tools in understanding how these systems function, and for developing important and useful applications in the treatment and control of parasitic diseases.

Hairston's (1965) study was a clever effort to apply the life-table concept to understanding the efficiencies of transmission dynamics in the life cycle of the digenean *Schistosoma japonicum*, in the Philippines. Two important findings emerged from his investigation. First, he found that the then-existing information regarding the epidemiology of *S. japonicum* was inadequate for the approach he wanted to use in developing his mathematical model. This observation serves as a critical object lesson in that one cannot assume that an existing database, even though it may be accurate, will provide the sort of information necessary to develop a realistic model. Second, he made the determination that reservoir hosts were much more important to the transmission dynamics of *S. japonicum* than had been previously suspected. This led him to the conclusion that humans were not the primary hosts for the schistosome in the Philippines. As previously noted, if the parasite were eliminated from a local population, it would quickly return because of its high prevalence in many reservoir hosts.

The use of modeling for host–parasite interactions was extended when Anderson (1974) focused attention on the biology of the cestode *Caryophyllaeus laticeps*. He developed an immigration–death model to assess various components of the parasite's life cycle, but found the model inadequate because of the complexity of the parasite's life cycle. Specifically, the problems were associated with too many time lags, the age structure of the host population, and the periodicity associated with the maturation of the parasite.

Subsequently, Anderson (1978) developed a more sophisticated model for the analysis of host–parasite interactions. In it, he suggested that stability in host populations would be affected by parasite aggregation, by density-dependent restrictions on the growth of parasite infrapopulations, and by non-linear, parasite-induced host mortality. On the other hand, he also determined that host and parasite infrapopulation growth could be destabilized when parasites are able to reduce the host's reproductive potential, as with the castration of molluscs by larval digeneans. Thus, some parasites that reproduce within their hosts can have a destabilizing effect on host population growth without causing mortality. Finally, time delays can affect population growth as well as the stability between host and parasite interactions.

Anderson & May (1979) and May & Anderson (1979) presented a most comprehensive treatment of mathematical models as applied to what they called microparasites (viruses, bacteria, and protozoans) and macroparasites (helminths and parasitoids). They related the patterns of disease caused by microparasites to: (1) the host providing a suitable environment, (2) the extent of pathogenicity caused by the parasite or its effectiveness in reducing the host's reproductive capacity, (3) the parasite's capacity for the induction of acquired immunity, and (4) the requirement that a parasite be transmitted from one host to the next.

Macroparasites were considered within two contexts, those with direct life cycles and those with indirect life cycles. In the former group, pathogenicity, host resistance, and production of transmission stages all affected parasite densities within the host population. The phenomenon of parasite aggregation also was re-emphasized. They predicted that aggregation would influence both pathogenicity and parasite transmission dynamics. For macroparasites with indirect life cycles, they emphasized the difficulty in modeling systems with multi-host life cycles. Based on these models, however, they provided several generalizations regarding the evolution of host–parasite relationships. Assuming no genetic change in the parasite, for example, it was suggested that selection in an evolutionary time frame would ultimately lead to a reduction in the numbers of susceptible hosts. In this way, if a host population is initially regulated by a parasite, it was predicted over the long term that it may escape regulatory influences altogether.

They also predicted that parasites with indirect life cycles could be identified with a number of distinctive characteristics in their host populations. For example, pathogenicity in the definitive host would be low, prevalence high, and the expected life span of the host would be long. Conversely, for the first intermediate host, pathogenicity should be high, prevalence low, and the host's life span should be short. To support this contention, they cited as evidence the *Haematoloechus complexus* study of Dronen (1978) (described earlier in Chapter 11). In that system, prevalence was highest in frogs, but no apparent pathology was produced. In dragonflies, where prevalence was moderate, there was likewise only moderate mortality caused by the parasite. In snails, prevalence was low, but mortality was high. Also consider the *Halipegus occidualis*–ranid frog combination described in Chapter 10. This parasite occurs in approximately 60% of the frog definitive hosts and very little pathology is apparently produced. On the other hand, up to 60% of the molluscan first intermediate hosts can have patent infections by the end of their life span and, in all cases, the snails are totally castrated. As has been stressed, whereas castration is not death, it is the equivalent from the standpoint of host reproductive fitness.

A long-standing discussion concerns the extent of parasite-induced host mortality in natural populations. Empirical evidence to support this notion is not extensive, although some does exist. The best case for supporting the idea of mortality caused by parasites is provided by several of the digenean–molluscan combinations, several of which were just cited (Dronen, 1978; Crews & Esch, 1986; Goater, 1989; Fernández & Esch, 1991). Other intermediate hosts are also known to be adversely affected by certain protozoan and helminth parasites (Crofton, 1971*a*; Keymer, 1980; Evans, 1983). In many of these systems, one of the overriding factors supporting the case for parasite-induced host mortality is the aggregated frequency distribution. Anderson & Gordon (1982), in a series of Monte Carlo simulations, examined parasite dispersion patterns within the context of probability models for increasing and decreasing host and parasite populations. They showed that, 'for certain types of host–parasite associations, convex curves of mean parasite abundance in relation to age (age-intensity curves), concomitant with a decline in the degree of dispersion in the older classes of hosts, may be evidence of host mortality by parasite infection'. In a very real sense, the laboratory studies by Scott (1987*b*) on the monogenean *Gyrodactylus turnbulli*, tend to confirm the interpretation of Anderson & Gordon (1982) regarding the effects of aggregation on host mortality within a temporal framework (see section 12.3.4).

Holmes & Bethel (1972), Moore (1983*a*, *b*), Barnard & Behnke (1990), and Lafferty (1992) have reviewed some of the consequences of parasite-induced changes in host behavior. Dobson (1985) attempted to quantify the population dynamics of hosts exhibiting behavioral modifications resulting from parasitism by developing a set of mathematical models. In each case, he estimated the basic reproductive rate (R_0) of the parasite at the outset of its introduction into the host population. The R_0 was then used to predict the threshold values for intermediate and definitive hosts that were required to maintain the parasites at levels where local extinction would not occur. In each system examined, when the parasite caused changes in host behavior, the R_0 would increase and the threshold numbers necessary to maintain the parasite were reduced. He showed that intermediate host and parasite infrapopulations had a tendency to oscillate over time. When intermediate host densities were low, changes in host behavior increased the frequency of oscillations. In contrast, when intermediate host densities were high, parasite densities were affected more by interactions with definitive hosts and with intermediate host behavior becoming less significant in influencing parasite infrapopulation sizes. He concluded by emphasizing that, since host–parasite systems are patchily distributed in space, any change in host behavior would clearly benefit the parasite. It would act in this manner by serving to reduce the density threshold necessary to sustain the host over time and increase the basic reproductive rate of the parasite. These adaptive permutations were viewed as highly advantageous in exploiting host populations that occur in small pockets or in situations when hosts are erratic in their visits to a particular habitat.

Lafferty (1992) generated a model indicating that the 'energetic cost of parasitism for a predator against the energetic value of prey items that transmit the parasite to the predator suggests that there is no selective pressure to avoid parasitized prey'. He went on to suggest that predators may even benefit if the energetic costs of being parasitized are moderate and if prey capture is facilitated by the parasites.

In a review regarding the extent of regulation by parasite species in nature, Scott & Dobson (1989) identified several host–parasite systems where regulation was clearly demonstrated, or at least could be strongly inferred. They even referred to the possibility that parasites may function as 'keystone species' at the community level. They quickly qualified this statement by saying 'we should curb our enthusiasm and understand that just because parasites were largely ignored in ecological contexts in the past, an understanding of host–parasite interactions will not explain everything about host population regulation and community structure'. This exhortation regarding less heat and more light should provide a strong indication of where the areas of parasite population biology and mathematical modeling are at the present time. As the literature indicates over and over, there is a severe shortage of appropriate field and laboratory data and, moreover, not enough are of a long-term nature. When more of these studies are completed and more information regarding the epidemiological or epizootiological nature of parasitic disease is acquired, then, and only then, will there be a reasonable basis for effectively considering parasitic diseases from the standpoint of mathematical models.

References

Albonico, M., Crompton, D. W. T. & Salvioli, L. (1998) Control strategies for human intestinal nematode infections. *Advances in Parasitology*, **42**, 277–341.

Anderson, R. M. (1974) Population dynamics of the cestode *Caryophyllaeus laticeps* (Pallas, 1781) in the bream (*Abramis brama* L.). *Journal of Animal Ecology*, **43**, 305–321.

Anderson, R. M. (1978) The regulation of host population growth by parasitic species. *Parasitology*, **76**, 119–157.

Anderson, R. M. (1989) Transmission dynamics of *Ascaris lumbricoides* and the impact of chemotherapy. In *Ascariasis and its Prevention and Control*, ed. D. W. T. Crompton, M. C. Nesheim & Z. S. Pawlowski, pp. 253–273. London: Taylor & Francis.

Anderson, R. M. & Gordon, D. M. (1982) Processes influencing the distribution of parasite numbers within host populations with special emphasis on parasite-induced host mortalities. *Parasitology*, **85**, 373–398.

Anderson, R. M. & May, R. M. (1979) Population biology of infectious diseases: Part I. *Nature*, **280**, 361–367.

Anderson, R. M. & Schad, G. A. (1985) Hookworm burdens and faecal egg counts: an analysis of the biological basis of variation. *Transactions of the Royal Society of Tropical Medicine and Hygiene*, **79**, 812–825.

Anderson, T. J. C. (1995) *Ascaris* infections in humans from North America: molecular evidence for cross-infection. *Parasitology*, **110**, 215–219.

Anderson, T. J. C., Romero-Abal, M. E. & Jaenike, J. (1993) Genetic structure and epidemiology of *Ascaris* populations: patterns of host affiliations in Guatemala. *Parasitology*, **107**, 319–334.

Barnard, C. J. & Behnke, J. M. (eds.) (1990) *Parasitism and Host Behaviour*. New York: Taylor & Francis.

Bowles, J. & McManus, D. P. (1993) Molecular variation in *Echinococcus*. *Acta Tropica*, **53**, 291–305.

Bundy, D. A. P. (1988) Population ecology of intestinal helminth infections in human communities. *Philosophical Transactions of the Royal Society of London, B*, **321**, 405–420.

Bundy, D. A. P., Cooper, E. S., Thompson, D. E., Didier, J. M., Anderson, R. M. & Simmons, I. (1987) Predisposition to *Trichuris trichiura*. *Epidemiology and Infection*, **98**, 65–71.

Cram, E. B. (1943) Studies on oxyuriasis. XXVIII. Summary and conclusions. *American Journal of Diseases of Children*, **65**, 46–59.

Crews, A. & Esch, G. W. (1986) Studies on the population biology of *Halipegus occidualis* (Hemiuridae) in the snail host *Helisoma anceps* (Pulmonata). *Journal of Parasitology*, **72**, 646–651.

Crofton, H. D. (1971*a*) A quantitative approach to parasitism. *Parasitology*, **62**, 179–193.

Crofton, H. D. (1971*b*) A model for host–parasite relationships. *Parasitology*, **63**, 343–364.

Croll, N. A. & Ghadarian, E. (1981) Wormy persons: contributions to the nature and patterns of overdispersion with *Ascaris lumbricoides*, *Ancylostoma duodenale*, *Necator americanus* and *Trichuris trichiura*. *Tropical and Geographical Medicine*, **33**, 241–248.

Croll, N. A., Anderson, R. M., Gyorkos, T. W. & Ghadarian, E. (1982) The population biology and control of *Ascaris lumbricoides* in a rural community in Iran. *Transactions of the Royal Society of Tropical Medicine and Hygiene*, **76**, 187–197.

Crompton, D. W. T. (1989) Prevalence of ascariasis. In *Ascariasis and its Prevention and Control*, ed. D. W. T. Crompton, M. C. Nesheim & Z. S. Pawlowski, pp. 45–69. London: Taylor & Francis.

Crompton, D. W. T. (1993) Human nutrition and parasitic infection. *Parasitology* (Supplement), **107**, S1–S203.

Crompton, D. W. T. (1999) How much human helminthiasis is there in the world? *Journal of Parasitology*, **85**, 397–404.

Davis, D. J. (1936) Pathological studies on the penetration of the cercariae of the strigeid trematode *Diplostomum flexicaudum*. *Journal of Parasitology*, **22**, 329–338.

Dobson, A. P. (1985) The population dynamics of competition between parasites. *Parasitology*, **91**, 317–347.

D'Oro, D., Gardenal, C. N., Perret, B., Crisci, J. V. &

Montamat, E. E. (1993) Genetic structure of *Trypanosoma cruzi* populations from Argentina estimated from enzyme polymorphism. *Parasitology*, **107**, 405–410.

Dronen, N. O. (1978) Host–parasite population dynamics of *Haematoloechus coloradensis* Cort, 1915 (Digenea: Plagiorchidae). *American Midland Naturalist*, **99**, 330–349.

Dunkley, L. C. & Mettrick, D. F. (1977) *Hymenolepis diminuta*: migration and the host intestinal and blood plasma glucose levels following carbohydrate intake. *Experimental Parasitology*, **41**, 213–228.

Elkins, D. B. & Haswell-Elkins, M. (1989) The weight/length profiles of *Ascaris lumbricoides* within a human community before mass treatment and following reinfection. *Parasitology*, **99**, 293–299.

Enriques, F. J., Zidian, J. L. & Cypess, R. H. (1988) *Nematospiroides dubius*: genetic control of immunity to infections of mice. *Experimental Parasitology*, **67**, 12–19.

Erasmus, D. A. (1960) The migration of cercaria X Baylis (Strigeida) within the fish intermediate host. *Parasitology*, **49**, 173–190.

Esch, G. W., Wetzel, E. J., Zelmer, D. A. & Schotthoeffer, A. M. (1997) Long-term changes in parasite population community structure: a case history. *American Midland Naturalist*, **137**, 369–387.

Evans, N. A. (1983) The population biology of *Hymenolepis tenerrima* (Linstow, 1882) Fuhrmann, 1906 (Cestoda, Hymenolepididae) in its intermediate host *Herpetocypris reptans* (Ostracoda). *Zeitschrift für Parasitenkunde*, **69**, 105–111.

Evans, W. S., Hardy, M. C., Singh, R., Moodie, G. E. & Cote, J. J. (1992) Effect of the rat tapeworm, *Hymenolepis diminuta*, on the coprophagic activity of its intermediate host, *Tribolium confusum*. *Canadian Journal of Zoology*, **70**, 2311–2314.

Evans, W. S., Wong, A., Currie, R. W. & Vanderwel, D. (1998) Evidence that the factor used by the tapeworm, *Hymenolepis diminuta*, to direct the foraging of its intermediate host, *Tribolium confusum*, is a volatile attractant. *Journal of Parasitology*, **84**, 1098–1101.

Fernández, J. & Esch, G. W. (1991) Guild structure of larval trematodes in the snail *Helisoma anceps*: patterns and processes at the individual host level. *Journal of Parasitology*, **70**, 528–539.

Goater, T. M. (1989) The morphology, life history, ecology and genetics of *Halipegus occidualis* (Trematoda: Hemiuridae) in molluscan and amphibian hosts. PhD dissertation, Wake Forest University, Winston-Salem, North Carolina, USA.

Goodchild, C. G. & Moore, T. L. (1963) Development of *Hymenolepis diminuta* in mice made obese by aurothioglucose. *Journal of Parasitology*, **49**, 398–402.

Gordon, D. M. & Rau, M. E. (1982) Possible evidence for mortality introduced by the parasite *Apatemon gracilis* in a population of brook sticklebacks (*Culaea inconstans*). *Parasitology*, **84**, 41–47.

Guyatt, H. L. & Bundy, D. A. P. (1990) Are wormy people parasite prone or just unlucky? *Parasitology Today*, **6**, 282–283.

Guyatt, H. L., Smith, T., Gryseels, B., Lengler, C., Mshinda, H., Siziya, S., Salanave, B., Mohome, N., Makwala, J., Ngimbi, K. P. & Tanner, M. (1994) Aggregation in schistosomiasis: comparison of the relationships between prevalence and intensity in different endemic areas. *Parasitology*, **109**, 45–55.

Hairston, N. G. (1965) On the mathematical analysis of schistosome populations. *Bulletin of the World Health Organization*, **33**, 45–62.

Haswell-Elkins, M. R., Elkins, D. B., Manjula, K., Michael, E. & Anderson, R. M. (1988) An investigation of hookworm infection and reinfection following mass anthelmintic treatment in the South Indian fishing community of Vairavankuppam. *Parasitology*, **96**, 565–577.

Hodges, M. (1993) Diarrheal disease in early childhood; experiences from Sierra Leone. *Parasitology* (Supplement), **107**, S37–S51.

Holland, C. V., Taren, D. L., Crompton, D. W. T., Nesheim, M. C., Sanjur, D., Barbeau, I., Tucker, K., Tiffany, J. & Rivera, G. (1988) Intestinal helminthiases in relation to the socioeconomic environment of Panamanian children. *Social Science and Medicine*, **26**, 209–213.

Holmes, J. C. & Bethel, W. M. (1972) Modification of intermediate host behaviour by parasites. In *Behavioural Aspects of Parasite Transmission*, ed. E. U. Canning & C. A. Wright. *Journal of the Linnean Society of London*, Supplement **1**, 123–149.

Hynes, H. B. N. & Nicholas, W. L. (1963) The importance of the acanthocephalan *Polymorphus minutus* as a parasite of domestic ducks in the United Kingdom. *Journal of Helminthology*, **37**, 185–198.

Keller, A. E., Leathers, W. S. & Knox, J. C. (1937) The present status of hookworm infestation in North Carolina. *American Journal of Hygiene*, **26**, 437–454.

Keymer, A. E. (1980) The influence of *Hymenolepis diminuta* on the survival and fecundity of the intermediate host, *Tribolium confusum*. *Parasitology*, **81**, 405–421.

Keymer, A. E. (1981) Population dynamics of *Hymenolepis diminuta* in the intermediate host. *Journal of Animal Ecology*, **50**, 941–950.

Keymer, A. E. (1982) Density-dependent mechanisms in

the regulation of intestinal helminth populations. *Parasitology*, **84**, 573–587.

Keymer, A. & Pagel, M. (1990) Predisposition to infection. In *Hookworm Disease: Current Status and New Directions*, ed. G. A. Schad & K. S. Warren, pp. 177–209. London: Taylor & Francis.

Kightlinger, L. B., Seed, J. R. & Kightlinger, M. B. (1995) The epidemiology of *Ascaris lumbricoides, Trichuris trichiura,* and hookworm in children in the Ranomafana Rainforest, Madagascar. *Journal of Parasitology*, **81**, 159–169.

Kightlinger, L. B., Seed, J. R. & Kightlinger, M. B. (1996) *Ascaris lumbricoides* aggregation in relation to child growth status, delayed cutaneous hypersensitivity, and plant anthelmintic use in Madagascar. *Journal of Parasitology*, **82**, 25–33.

Kightlinger, L. B., Seed, J. R. & Kightlinger, M. B. (1998) *Ascaris lumbricoides* intensity in relation to environmental, socioeconomic, and behavioral determinants of exposure to infection in children from southeast Madagascar. *Journal of Parasitology*, **84**, 480–484.

Kostitzin, V. A. (1934) *Symbiose, Parasitisme et Evolution*. Paris: Hermann.

Kostitzin, V. A. (1939) *Mathematical Biology*. London: Harrap.

Lafferty, K. D. (1992) Foraging on prey that are modified by parasites. *American Naturalist*, **140**, 854–867.

Lemly, A. D. & Esch, G. W. (1983) Differential viability of metacercariae of *Uvulifer ambloplitis* (Hughes, 1927) in juvenile centrarchids. *Journal of Parasitology*, **69**, 746–749.

Lemly, A. D. & Esch, G. W. (1984*a*) Population biology of the trematode *Uvulifer ambloplitis* (Hughes, 1927) in the snail intermediate host, *Helisoma trivolvis*. *Journal of Parasitology*, **70**, 461–465.

Lemly, A. D. & Esch, G. W. (1984*b*) Population biology of the trematode *Uvulifer ambloplitis* (Hughes, 1927) in juvenile bluegill sunfish, *Lepomis macrochirus*, and largemouth bass, *Micropterus salmoides*. *Journal of Parasitology*, **70**, 466–474.

Lemly, A. D. & Esch, G. W. (1984*c*) Effects of the trematode *Uvulifer ambloplitis* on juvenile bluegill sunfish, *Lepomis macrochirus*: ecological implications. *Journal of Parasitology*, **70**, 475–492.

Lemly, A. D. & Esch, G. W. (1985) Black-spot caused by *Uvulifer ambloplitis* (Trematoda) among juvenile centrarchids in the Piedmont area of North Carolina. *Proceedings of the Helminthological Society of Washington*, **52**, 30–35.

Lewert, R. M. & Lee, C. L. (1954*a*) Studies on the passage of helminth larvae through host tissues. I. Histochemical studies on the extracellular changes caused by penetrating larvae. *Journal of Infectious Diseases*, **95**, 13–35.

Lewert, R. M. & Lee, C. L. (1954*b*) Studies on the passage of helminth larvae through host tissues. II. Enzymatic activity of larvae *in vivo* and *in vitro*. *Journal of Infectious Diseases*, **95**, 36–51.

Lotka, A. J. (1934) *Théorie Analytique des Associations Biologiques*. Paris: Hermann.

Madhavi, R. & Anderson, R. M. (1985) Variability in the susceptibility of the fish host, *Poecilia reticulata*, to infection with *Gyrodactylus bullatarudis* (Monogenea). *Parasitology*, **91**, 531–544.

May, R. M. (1977) Dynamical aspects of host–parasite associations: Crofton's model revisited. *Parasitology*, **75**, 259–276.

May, R. M., & Anderson, R. M. (1979) Population biology of infectious diseases. Part II. *Nature*, **280**, 455–461.

McManus, D. P. & Bryant, C. (1995) Biochemistry, physiology, and molecular biology of *Echinococcus* and hydatid disease: In *Echinococcus and Hydatid Disease*, ed. R. C. A. Thompson & A. J . Lymbery, pp. 135–171. Wallingford, UK: CAB International.

Meloni, B. P., Lymbery, A. J. & Thompson, R. C. A. (1995) Genetic characterization of isolates of *Giardia duodenalis* by enzyme electrophoresis: implications for reproductive biology, population structure, taxonomy, and epidemiology. *Journal of Parasitology*, **81**, 368–384.

Moore, J. (1983*a*) Responses of an avian predator and its isopod prey to an acanthocephalan parasite. *Ecology*, **64**, 1000–1015.

Moore, J. (1983*b*) Altered behavior in cockroaches (*Periplaneta americana*) infected with an archiacanthocephalan, *Moniliformis moniliformis*. *Journal of Parasitology*, **69**, 1174–1176.

Murrell, K. D., Eriksen, L., Nansen, P., Slotved, H.-C. & Rasmussen, T. (1997) *Ascaris suum*: a revision of its early migratory path and implications for human ascariasis. *Journal of Parasitology*, **83**, 255–260.

Muruyama, H., Nawa, Y., Noda, S., Mimori, T. & Choi, W.-Y. (1996) An outbreak of visceral larval migrans due to *Ascaris suum* in Kyusha, Japan. *Lancet*, **347**, 1766–1767.

Nadler, S. A. (1995) Microevolution and the genetic structure of parasite populations. *Journal of Parasitology*, **81**, 395–403.

Nadler, S. A., Lindquist, R. L. & Near, T. J. (1995) Genetic stucture of midwestern *Ascaris suum* populations: a comparison of isozyme and RAPD markers. *Journal of Parasitology*, **81**, 385–394.

Pencheiner, L., Mathieu-Daudé, F., Brengues, C., Bañuls, A. & Tibayrenc, J. (1997) Population structure of *Trypanosoma brucei s.l.* in Côte d'Ivoire by multilocus enzyme electrophoresis: epidemiological and taxonomical considerations. *Journal of Parasitology*, **83**, 19–22.

Pozio, E., Serrano, F. J., La Rosa, G., Reina, D., Pérez-Martin, E. & Navarrette, I. (1997) Evidence of potential gene flow in *Trichinella spiralis* and in *Trichinella britovi* in nature. *Journal of Parasitology*, **83**, 163–166.

Quinnell, R. J., Medley, G. F. & Keymer, A. E. (1990) The regulation of gastrointestinal helminth populations. *Philosophical Transactions of the Royal Society of London, B*, **330**, 191–201.

Schad, G. A. & Anderson, R. M. (1985) Predisposition to hookworm infection in humans. *Science*, **228**, 1537–1539.

Scott, M. E. (1982) Reproductive potential of *Gyrodactylus bullatarudis* (Monogenea) on guppies (*Poecilia reticulata*). *Parasitology*, **85**, 217–236.

Scott, M. E. (1985*a*) Experimental epidemiology of *Gyrodactylus bullatarudis* (Monogenea) on guppies (*Poecilia reticulata*): short- and long-term studies. In *Ecology and Genetics of Host–Parasite Interactions*, ed. D. Rollinson & R. M. Anderson, pp. 21–38. New York: Academic Press.

Scott, M. E. (1985*b*) Dynamics of challenge infections of *Gyrodactylus bullatarudis* (Monogenea) on guppies, *Poecilia reticulata* (Peters). *Journal of Fish Diseases*, **8**, 495–503.

Scott, M. E. (1987*a*) Regulation of mouse colony abundance by *Heligmosomoides polygyrus*. *Parasitology*, **95**, 111–124.

Scott, M. E. (1987*b*) Temporal changes in aggregation: a laboratory study. *Parasitology*, **94**, 583–595.

Scott, M. E. & Anderson, R. M. (1984) The population dynamics of *Gyrodactylus bullatarudis* (Monogenea) within laboratory populations of the fish host *Poecilia reticulata*. *Parasitology*, **89**, 159–194.

Scott, M. E. & Dobson, A. (1989) The role of parasites in regulating host abundance. *Parasitology Today*, **5**, 176–183.

Scott, M. E. & Nokes, D. J. (1984) Temperature-dependent reproduction and survival of *Gyrodactylus bullatarudis* (Monogenea) on guppies (*Poecilia reticulata*). *Parasitology*, **89**, 221–227.

Scott, M. E. & Robinson, M. A. (1984) Challenge infections of *Gyrodactylus bullatarudis* (Monogenea) on guppies, *Poecilia reticulata* (Peters), following treatment. *Journal of Fish Biology*, **24**, 581–586.

Sher, A. N., Smithers, S. R. & Mackenzie, P. (1975) Passive transfer of acquired resistance to *Schistosoma mansoni* in laboratory mice. *Parasitology*, **70**, 347–357.

Stoll, N. R. (1947) This wormy world. *Journal of Parasitology*, **33**, 1–18.

Truc, P. & Tibyrenc, M. (1993) Population genetics of *Trypanosoma brucei* in Central Africa: taxonomic and epidemiological significance. *Parasitology*, **106**, 137–149.

Wakelin, D. (1996) *Immunity to Parasites: How Parasitic Infections are Controlled*, 2nd edn. Cambridge: Cambridge University Press.

Weidong, P., Xiamin, Z., Xiaomin, C., Crompton, D. W. T, Whitehead, R. R., Jiangqin, X., Haigeng, W., Jiyuan, P., Yang, Y., Weixing, W., Kaiwu, X. & Yonxing, Y. (1996) *Ascaris*, people and pigs in a rural community of Jiangxi Province, China. *Parasitology*, **113**, 545–557.

Zarlenga, D., Aschenbrenner, R. & Lichtenfels, J. R. (1996) Variations in microsatellite sequences provide evidence for population differences and multiple ribosomal gene repeats within *Trichinella pseudospiralis*. *Journal of Parasitology*, **82**, 534–538.

Chapter 13

From reproduction and transmission to colonization

One of the most fundamental elements in the overall success of any organism, free-living or otherwise, is the ability to complete its life cycle. The life cycles of parasitic and free-living organisms are similar in many ways. They include an array of processes, ranging from reproduction and the generation of reproductive propagules, to the release and dispersal of their various developmental forms and their subsequent maturation to adult stages. In most free-living organisms, this sequence of events is completed without the association of another organism. For a parasite, however, the presence of a host in at least one phase of its life cycle is a necessity.

Among parasitic organisms, reproduction of some type usually occurs in, or on, a host, resulting in the production of offspring. At some point in time and space, the reproductive propagules are usually released from the host; these need to disperse, spreading away from the host in which reproduction occurred. For many parasites, the stages involved in dispersal must seek out and secure a new host. In other cases, a host finds the dispersing stages. In any event, finding a host, or somehow being able to reach a host, is one of the most critical stages in the life cycle of an organism. Moreover, even though many potential organisms may be available to serve as hosts, not all are equally suitable for a given species of parasite.

The successful reproduction and transmission of a parasite generally assures that its genome will be perpetuated in time. Even though successful transmission of a parasite can be equated with perpetuation of their genetic information, most organisms also have the intrinsic ability to expand their distribution and colonize, or pioneer, new habitats away from the immediate area were reproduction took place. For parasites, the intrinsic ability to pioneer may involve the colonization of new individual hosts, new host populations, and even new host species.

Since the life cycles of many parasites have already been discussed in some detail elsewhere, the phenomena associated with reproduction, transmission and host finding, and colonization, will be addressed in the present chapter in more general terms. First, we will consider some of the reproductive schemes employed by various parasitic groups. Next, we will examine how transmission to an appropriate host is accomplished, with an emphasis on various adaptations of both hosts and parasites. Finally, we will consider the colonization of individual hosts, host populations, and host species, and how colonization might relate to the idea that hosts are nothing but islands embedded in an abiotic matrix. This will lead to some considerations of hosts as islands within the context of the Theory of Island Biogeography.

13.1 Reproduction and fecundity

For most species of parasites, the probability that a single propagule from an individual parent will find a host is exceedingly small, seemingly infinitesimal. Parasites have solved this problem using various approaches that require a number of morphological and/or physiological adaptations, ranging from high offspring production to changing the behavior of parasitized hosts. In whatever manner these successes have been achieved, they can be classified in some ways as striking, and always as fascinating.

13.1.1 Asexual strategies

Asexual reproduction is the production of new organisms without the participation of gametes. The resulting offspring is genetically identical to the parent organism, forming a clone. Theoretically, asexual reproduction is most advantageous in stable, favorable environments because it perpetuates successful genotypes exactly. Asexual reproduction is most commonly associated with protozoans, the intramolluscan stages of digeneans, and the larval stages of some cestodes.

In a protozoan, asexual reproduction is accomplished through simple division of the single cell that makes up the organism. Binary fission is the division of one cell into two daughter cells and it is the most basic form of multiplication. Because the axis of division in the different groups of protozoans is fixed, cell division can be accomplished in several different ways. Among microsporidians, e.g., *Nosema*, division is irregular and the cytoplasm is divided irregularly, resulting in the production of daughter cells of different size. In sarcomastigophorans, e.g., *Entamoeba*, the division is also somewhat irregular because the axis of cell division is not fixed. It is sometimes referred to as 'amoeba-like' division. Among ciliates, e.g., *Ichthyophthirius*, the axis of cytoplasmic division runs perpendicular to the axis of the nuclear spindles; it is therefore transverse. In zoomastigophoreans, e.g., *Trichomonas* and *Trypanosoma*, the divisional axis is longitudinal. In trichomonads, however, the axis is longitudinal only at the beginning of the division, when the basal bodies and nuclei are duplicated; later, the divisional axis changes to an oblique, or even a transverse, angle. In opalinids, e.g., *Opalina*, the axis is also longitudinal, but becomes parallel to the characteristic oblique rows of cilia, being intermediate between the longitudinal division of most flagellated forms and the transverse pattern seen in ciliated forms.

Multiple fission of a mother cell occurs in some apicomplexan and microsporidian protozoans. In multiple fission, the nucleus and other essential organelles divide several times before cytokinesis, the division of the cytoplasm, resulting in the simultaneous production of many daughter cells. The process of multiple fission can also be referred to as schizogony, merogony, or sporogony, depending on the type of cells involved. During schizogony and merogony, the mother cell is called a schizont or meront and the resulting daughter cells are merozoites, which are of the same life-cycle stage as the mother cell. During sporogony, however, the mother cell is a zygote and the resulting cells are typically a different life-cycle stage.

Endopolyogeny is a form of multiple fission sometimes referred to as internal budding. In this process, the daughter cells begin to form before the nucleus has divided. In most cases, however, the nucleus of the original cell has grown significantly and presents a lobulated surface. During the final stages, the daughter cells simultaneously incorporate portions of the lobulated giant nucleus. Endodyogeny is similar to endopolyogeny, but only two daughter cells are produced. This form of multiple fission occurs among some of the coccidia that produce tissue cysts.

Among parasitic multicellular organisms, digeneans exhibit what some consider as the most remarkable process of asexual reproduction. Recall that often, the eggs of many digeneans hatch when released from their definitive host. A free-swimming miracidium emerges and eventually enters an intermediate host, always a mollusc, where it develops into a sac-like stage, the sporocyst. In some species, the miracidium remains inside the egg after release from the definitive host and the mollusc becomes infected when it ingests the egg. In either case, a single miracidium transforms into one sporocyst, two miracidia into two sporocysts, etc. A single sporocyst may give rise to one or more daughter sporocysts, which, in turn, may each produce a generation of rediae. These rediae may then give rise to many daughter rediae, which finally produce cercariae. The cercariae usually leave the molluscan host and are the infective stage for the next host in the life cycle. Depending on the species of digenean, the daughter sporocyst, and one, or both, rediae stages, may be omitted from the life-cycle pattern. Regardless of how many sporocyst and redia stages occur, a single miracidium and its intramolluscan progeny are capable of producing enormous numbers of cercariae. In this remarkable process, known as polyembryony, cells that divide by mitosis form germ balls, which give rise to embryos (sporocysts, rediae, or cercariae) and to a line of new germ cells that are incorporated into these embryonic stages. As strange as this may sound in these days of 'high tech' research, the exact processes involved in polyembryony remain unknown. Among multicellular organisms, some consider digeneans as the masters of reproduction because of this magnificent amplification process. Perhaps more important than mere numbers, the clones represent the magnification of a winning genome, i.e., that of a miracidium which was successful. However, success as a reproductive machine does not necessarily translate into success in transmission and in reaching adulthood. There are many 'trade-offs' along the way, and this highly efficient reproductive process means that something must 'give in' somewhere else in the life cycle.

The larval stages of some cestodes can also reproduce asexually in their intermediate host. They increase their numbers by external or internal budding, or both. In its rodent intermediate host, a cysticercus of *Taenia crassiceps* (or bladder worm), for example, can produce numerous external buds, which disengage from the parent and form daughter cysticerci. The significance of such a process becomes clear when a canine definitive host consumes the infected rodent. Instead of becoming infected with maybe just one cestode, the canine acquires a much larger number of worms. In other cestodes, such as *Echinococcus granulosus*, budding in the hydatid cyst is internal, with the potential of producing hundreds of thousands of protoscolices that remain inside the hydatid cyst. The hydatid cyst of *E. multilocularis*, on the other hand, buds both internally and externally, metastasizing and spreading within the rodent intermediate host. (For a comprehensive treatment of asexual reproduction among parasitic platyhelminths see Whitfield & Evans, 1983.)

13.1.2 Sexual reproduction

The majority of parasitic arthropods, nematodes, and all acanthocephalans are dioecious (separate sexes), whereas most digeneans, monogeneans, and cestodes are monoecious (hermaphrodites). Many protozoans, primarily among the Apicomplexa, also are capable of sexual reproduction. Unlike asexual reproduction, where the offspring are genetically identical to the parent organism (clones), in sexual reproduction the offspring have a unique combination of genes inherited from two 'parents'. These offspring are somewhat different from their parents and siblings, and each one is genetically unique. Here we will not deal with the details of sexual reproduction, instead we will emphasize the variations or peculiarities of some reproductive processes that may increase the biotic potential of some parasites and the chances of perpetuating their genes.

One of the problems that adult parasites face is finding each other in, or on, a host in order to mate. Hermaphroditism has certainly solved the problem of finding a mate for many parasites. However, despite the advantage of housing both sexes in a single body, apparently most hermaphrodites still seek another partner and cross-fertilize. If a partner cannot be found, then self-fertilization, or selfing, will occur. Most cestodes also have increased their reproductive potential through the continuous production of new reproductive segments, or proglottids, a process known as modular iteration. In most cestodes, each proglottid houses both male and female reproductive organs, with each female system having the potential of being fertilized and, therefore, producing eggs. Sperm transfer can occur between different proglottids in the same worm or between proglottids of two different worms. Because of the combination of hermaphroditism and serial repetition of reproductive structures, 'egg factory' is a well-deserved epithet for most adult cestodes.

13.1.3 Fecundity

Because of the routinely high mortality of many larval stages and the rich supply of nutrients available to most parasites inside a host, high fecundity is generally considered to be one of the most characteristic features of parasites. Moreover, the fact that many parasites can undergo asexual reproduction, as indicated in previous sections, serves to reinforce the view that parasites are highly effective reproductive machines.

From an evolutionary perspective, it has been generally assumed that the fecundity of parasites tends to evolve toward greater and greater egg

Table 13.1 Fecundity of some parasitic helminths

Parasite	Eggs/day	Life span[a]
Acanthocephala		
Macracanthorhynchus hirudinaceus	260000	10 months
Cestoda		
Taenia saginata	700000	10 years
Taenia solium	300000	—
Taenia hydatigena	60000	—
Hymenolepis diminuta	250000	2–3 years
Hymenolepis microstoma	45600	2 years
Diphyllobothrium latum	36000	—
Trematoda		
Fasciola hepatica	25000	—
Schistosoma japonicum	1200	2 years
Schistosoma mansoni	300	15 years
Nematoda		
Haemonchus contortus	6000	—
Ascaris lumbricoides	200000	9 months
Ancylostoma duodenale	20000	5 years
Necator americanus	10000	5 years
Toxocara canis	200000	4 months
Trichinella spiralis[b]	1500	—
Litomosoides carinii[c]	15000	—

Notes:
[a] The dashes under life span indicate that data are unknown or are uncertain.
[b] Average number of larvae produced per female.
[c] Microfilariae released from a female.

output because of the heavy losses incurred by the infective stages during transmission or colonization. High fecundity has been seen as a way to compensate for these losses, ensuring that at least a few offspring would be able to complete their life cycle. In ecological terms, many parasites also are viewed as classic examples of *r*-selected organisms since they produce large numbers of eggs containing low energy reserves. (For further discussion of parasites and *r*/*K* strategies see Jennings & Calow, 1975; Esch *et al.*, 1977; Sibly & Calow, 1985.)

Although it is true that most parasites may produce astonishing numbers of eggs, or other larval stages through various reproductive processes (Table 13.1), it is also true that many free-living organisms also have high reproductive capacities. Some parasites have very low fecundities, yet they are still able to complete their life cycle. This leads to the justifiable conclusion that they must be doing something 'right' somewhere in their life cycle! The truth is that fecundity is but one of many life-history traits of parasites. The relationships, or trade-offs, among many demographic traits is what determines the ultimate success of each organism, parasitic or otherwise. These demographic traits include body size, number of offspring produced, life span, age at maturity, etc. (Poulin, 1996).

Fecundity and body size are positively correlated in a wide range of free-living invertebrates, as well as in a variety of parasites, including monogeneans, cestodes, nematodes, copepods, bopyrid isopods, and ticks (for references see Poulin, 1996). Similarly, egg size and egg number tend to be inversely related. Parasitic organisms

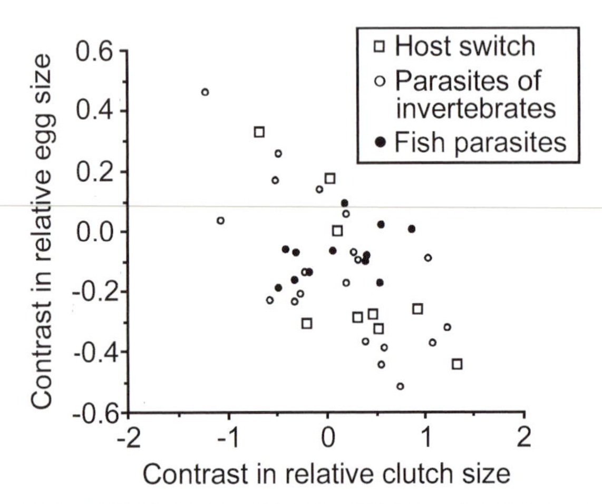

Fig. 13.1 Relationship between egg size and clutch size in parasitic copepods. Each point represents a contrast between two sister branches in the copepod phylogeny. Both variables were corrected for parasite body size, which covaries with the number and size of eggs produced. The negative relationship suggests that there is a trade-off between egg size and clutch size. Host switches represent cases in which one taxon is parasitic on invertebrates and its sister taxon is parasitic on fish. (Modified from Poulin, R. [1998] *Evolutionary Ecology of Parasites.* London: Chapman & Hall. © R. Poulin. Reprinted with kind permission from Kluwer Academic Publishers.)

may produce many small eggs, or a few large eggs, or somewhere in between. Parasitic copepods, as well as schistosomes, range throughout this spectrum (Figs. 13.1, 13.2 top right). Comparisons among copepods parasitic on invertebrates and on fishes indicate that, after 'controlling' for the effects of phylogeny and body size, parasitic copepods on fishes produce relatively more, and smaller, eggs than their counterparts on invertebrates. Some copepods parasitic on invertebrates may produce as few as one or two eggs per clutch, but these eggs are relatively large compared to the body size of the parent (Poulin, 1995). This difference is probably related to the degree of difficulty involved in finding a host. Most invertebrate hosts tend to be sessile or at least less mobile than most fishes, a circumstance which would seem to facilitate encounter between hosts and parasites. To latch on to a more mobile fish host is a different challenge. It seems clear that selection has favored an investment in offspring numbers rather than in offspring size with the purpose of covering a greater spatial area.

Trade-offs also appear between the production of eggs and the production of asexual stages such as the cercariae of digeneans and the larval forms of cestodes. Loker (1983) observed that, among mammalian schistosomes, species producing fewer eggs also produce larger eggs; larger eggs usually produce larger miracidia; and larger miracidia generate larger numbers of cercariae in the snail intermediate host (Fig. 13.2). The rate of egg production in these schistosomes is inversely correlated with the number of cercariae produced in the snail intermediate host, suggesting that a trade-off exists between the production of eggs and the production of cercariae (Poulin, 1998). Moore (1981) observed a similar pattern for taeniid cestodes. The adult stages of those species of taeniids capable of reproducing asexually by budding in the intermediate host are relatively small and short-lived, whereas the adult stages of those species without asexual reproduction are large and long-lived.

13.1.4 Factors affecting fecundity

A number of host-related factors are known to affect the fecundity of parasites. Among others, these include host species, host age, host diet, site in the host, the host's immune response, and the presence of other parasites.

The extent of host specificity for different hosts in a parasite's life cycle varies greatly. Some parasites show an exceptionally high level of host specificity, whereas others are capable of parasitizing a wide range of hosts. For most generalist parasite species, the different host species provide a variety of biochemical, immunological and physiological challenges that can affect the parasite's fecundity (Berrie, 1960; Dixon, 1964). For example, *Schistosoma bovis* has a broad specificity for is snail host. Mouahid & Théron (1987) infected snails of three different species with miracidia of *S. bovis* and then monitored cercariae production. There were significant differences in both the pattern and number of cercariae released by the three different hosts (Fig. 13.3).

Even within a single host population, not all hosts are 'created' equal. For example, the size of a snail intermediate host can affect the production of cercariae. For the common sheep liver fluke, *Fasciola hepatica*, the size of the snail host is positively correlated with cercariae production

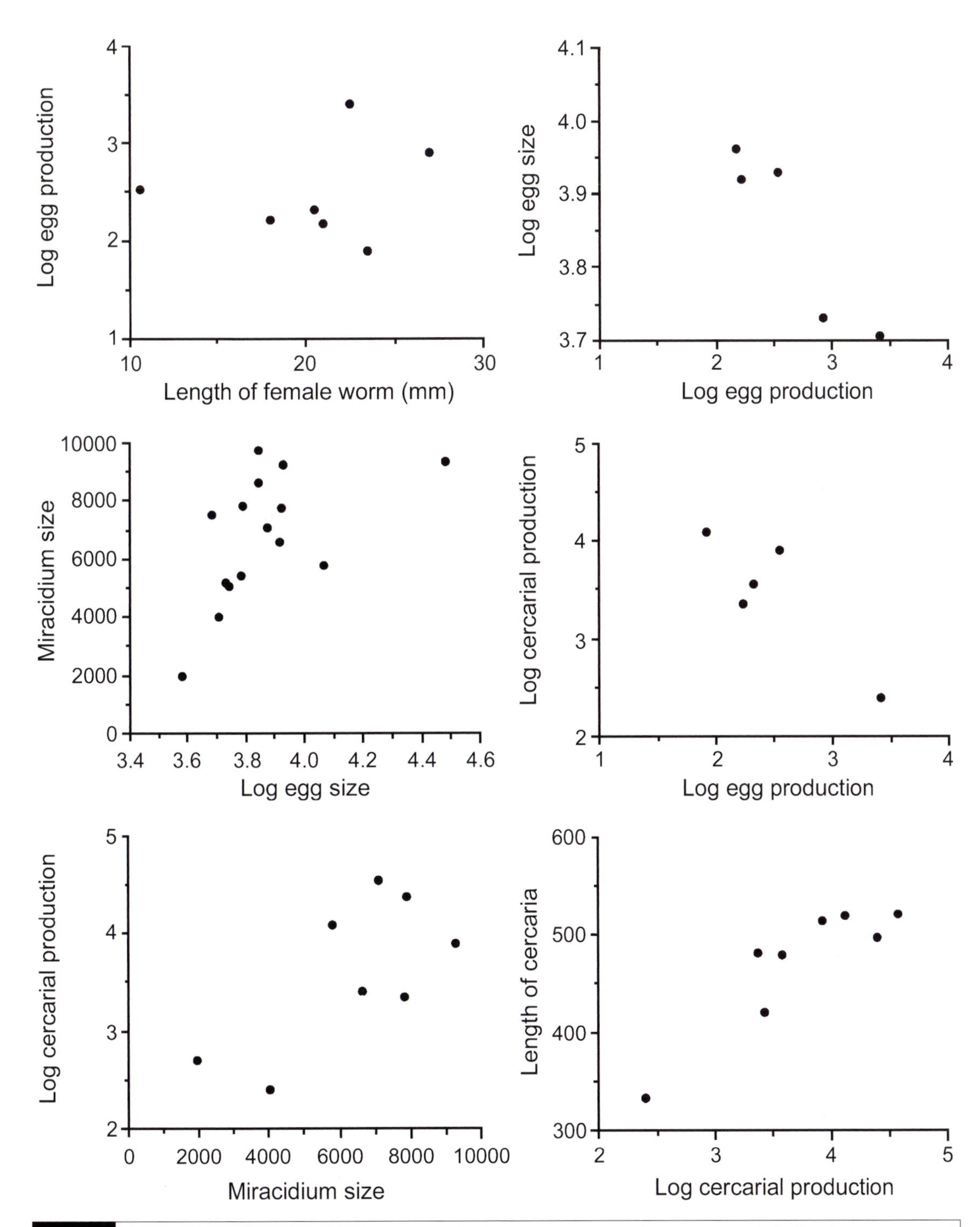

Fig. 13.2 Relationships among various life-history traits of schistosome species parasitic on mammals. Each point represents a different parasite species. Egg production is the number of eggs produced per female per day; egg size is the product of egg length (μm) and egg width (μm); miracidium size is the product of miracidium length (μm) and miracidium width (μm); cercarial production is the total number of cercariae produced per infected snail intermediate host; cercarial length includes the tail (μm). (Modified from Poulin, R. [1998] *Evolutionary Ecology of Parasites*. London: Chapman & Hall. © R. Poulin. Reprinted with kind permission from Kluwer Academic Publishers.)

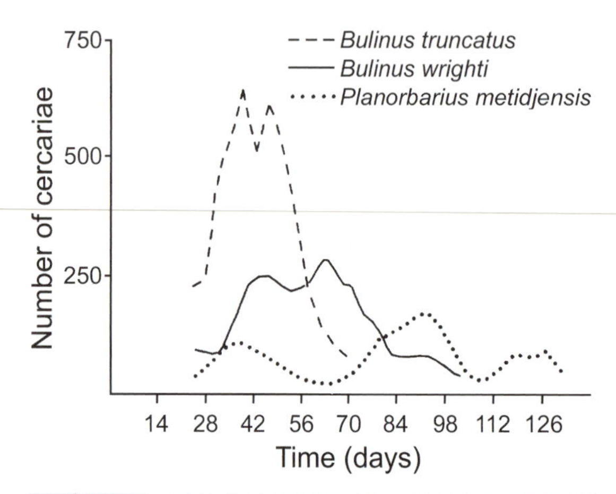

Fig. 13.3 Dynamics of cercariae production of *Schistosoma bovis* in three different snail hosts. (Modified from Mouahid & Théron, 1987, with permission, *International Journal for Parasitology*, **17**, 1431–1434.)

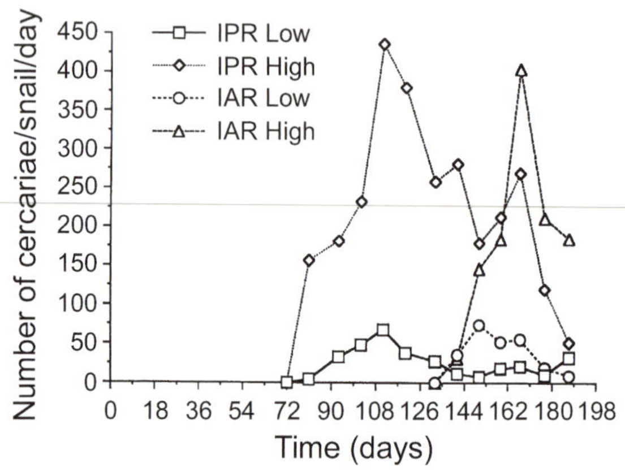

Fig. 13.4 Mean number of cercariae of *Halipegus occidualis* shed by infected *Helisoma anceps* raised on two diet qualities (Low and High) and two infection conditions (IPR = infected prior to reproductive maturity, and IAR = infected after reproductive maturity). (Modified from Keas & Esch, 1997, with permission, *Journal of Parasitology*, **83**, 96–104.)

(Rondelaud & Barthe, 1987). The ability of a host to respond immunologically to a parasite is well known. Both parasite size and fecundity, however, can be affected by the host immune response. Ito *et al.* (1988*a*, *b*) have provided evidence that the immune response provoked by *Hymenolepis nana* in mice is not only capable of reducing fecundity and biomass of the parasite, but it prevents internal autoinfection as well. Among some protozoans such as the African trypanosomes, the importance of the host immune response has been demonstrated many times by experimentally immunosuppressing the host. In immunosuppressed hosts, parasitemias are often higher and are maintained for a longer period of time.

One of the premises behind the assumption that parasites tend to evolve towards high fecundity is that parasites live in a nutrient-rich environment. This assumption is probably true for most endo- and ectoparasites. In many cases, however, the quantity and quality of nutrients available to intestinal parasites for growth and reproduction will depend on what the host eats. Several studies have shown that the quality of the host diet can have a negative impact on the fecundity of certain cestodes under laboratory conditions (Read, 1959; Keymer *et al.*, 1983). Most of these dietary studies have focused on the effects created by a decrease in the intake of carbohydrates. Crompton (1987) reviewed the literature regarding the responses of 21 species of protozoan and helminth parasites to experimental manipulations in host diet and feeding programs. The effects were varied and included retardation in parasite growth rate, gametogenesis, fecundity, and asexual reproduction. Cercariae production in snails can also be affected by the nutritional status of the host. Keas & Esch (1997) showed that experimentally infected *Helisoma anceps* raised on a high-quality diet (rich in protein) produced more cercariae than snails raised on a low-quality diet (with only 25% the nutritional value of the high-quality diet) (Fig. 13.4).

Intraspecific interactions among parasites may have a negative impact on fecundity. This phenomenon has long been recognized and was initially termed the 'crowding effect' by Read (1951). The presence of conspecific parasites in a host increases competition for carbohydrates or other essential nutrients; it also may increase the intensity of the host immune response, thereby affecting the body size and fecundity of the parasites. Boray (1969), for example, exposed sheep to various numbers of the liver fluke *Fasciola hepatica*; in heavy infections, the percentage of patent parasites declined sharply and so did their individual fecundity. Fleming (1988) noted the same kind of density-dependent effects in lambs infected with

Haemonchus contortus, a particularly pathogenic nematode. Whereas carbohydrate shortage traditionally has been identified as an important component in the 'crowding effect', the reduction in growth and fecundity that results from high infrapopulation densities has also been attributed to growth inhibitors, spatial constraints, and immunological factors (Bush & Lotz, 2000; Roberts, 2000). It is likely that in most cases, a combination of factors, perhaps acting synergistically, is truly responsible for the decrease in fecundity.

Interspecific competitive interactions such as those described between the cestode *Hymenolepis diminuta* and the acanthocephalan *Monilifomis dubius* in laboratory rats result in the displacement of the cestodes into a less desirable position in the gut, with a concomitant reduction in their biomass and probably fecundity as well (Holmes, 1961, 1962). Changes in the optimum position of the worms in the intestine of their hosts affect their fecundity. The fecundity of the nematode *Trichinella spiralis*, for example, is related to the position of the worm in the intestine (Sukhdeo, 1991).

The factors affecting parasite fecundity are multiple and, in most cases, interrelated. The host is clearly an active force in the phenotypic plasticity exhibited by most parasites.

13.2 Transmission and dispersal

Transmission traditionally refers to the transfer of a parasite to a host, regardless of whether the host is already parasitized by the same, or a different, species of parasite. Dispersal, on the other hand, implies less directionality and relates to the process of spreading away from the point of origin. Transmission is more like 'moving to', whereas dispersal is more like 'moving away from'. Dispersal away from the host where reproduction occurs is generally desirable to avoid additional infections and therefore reducing the potential for overcrowding. Infection of the same host leads to overcrowding and, potentially, higher intraspecific competition with its deleterious effects on life history traits such as parasite body mass and fecundity. Repeated infection also is likely to increase the probability of inbreeding in the parasite population and decrease its genetic diversity. The life-cycle stages involved in dispersal are usually the same ones involved in transmission to a host; dispersion, then, can be viewed as an extension of the transmission process.

In most species of parasites with free-living larval stages that actively seek the next host, dispersal is generally not random, but directed at finding a host. Even among those species with free-living larval stages that disperse passively, the larvae often have adaptations allowing them to reach particular areas where they are more likely to be encountered by a host. It should be emphasized that whereas dispersal may seem secondary to transmission, the need to disperse is an intrinsic characteristic of most organisms.

For practical purposes, parasite transmission (and dispersal) can be considered in terms of passive, or active, processes. **Passive transmission** implies that the larval stage does not require an expenditure of energy in the transfer to a host. Nonetheless, there will be an energy cost by the parasite in preparation for transmission, e.g., an eggshell or a cyst wall may be produced to protect the infective agent from desiccation. Some consider this to be a physiological trade-off in the life history of parasites, a kind of 'pay now or pay later' strategy. In contrast, during **active transmission**, larvae deliberately seek a host and a direct energy cost is incurred. Whereas dispersal in space is an important tactic to assure a wider distribution of many parasites, some species can extend their dispersal abilities through time. For example, if current environmental conditions are not favorable for development or are not conducive for dispersal, inactive transfer stages may be produced which will allow them to delay their life cycles until conditions are more favorable for successful transmission.

13.2.1 Passive transmission and dispersal by non-motile stages

Free-living larval stages that disperse passively cannot move on their own volition, however, many of them have adaptations that facilitate their dispersal into the microenvironments where the required host is likely to be found. Eggs and cysts are commonly dispersed passively. They are often light and small and are easily moved by water and wind. The eggs or cysts of many parasites infecting terrestrial hosts are usually released with the host feces. They then can be dispersed in space by the host as it moves, by wind once the feces have dried up, or by moving water. Mobile hosts that defecate at their convenience facilitate dispersion by spreading the egg-containing feces over the area in which they roam. Water runoff may be an effective mechanism of dispersion over larger areas, if the runoff reaches a larger body of moving water such as a creek or river. Dispersion of the eggs away from the fecal pile by water or wind increases their spatial distribution and may increase their chance of coming in contact with an appropriate host. Occasionally, however, as in the case of the acanthocephalan *Macracanthorhynchus hirudinaceus*, remaining with the fecal mass will enhance transmission success since the next hosts in the life cycle of this parasite are dung beetles, or other dung-consuming insects.

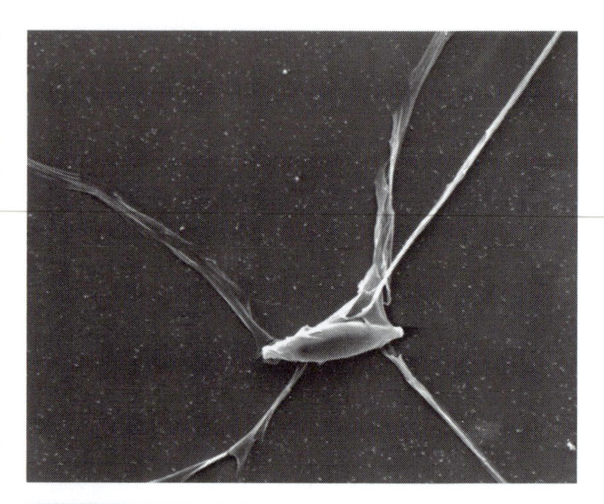

Fig. 13.5 Scanning electron micrograph of the egg of the acanthocephalan *Leptorhynchoides thecatus*. Sometime after the egg reaches water, it loses the outer membrane and the fibrillar coat tears and unwraps to form filaments. (From Barger & Nickol, 1998, with permission, *Journal of Parasitology*, **84**, 534–537.)

The eggs of species parasitizing aquatic organisms have different shapes and densities that influence and determine their distribution in aquatic systems. These adaptations appear to be more important for increasing transmission than in effecting dispersal. For example, the buoyancy of the eggs of certain species of cestodes using copepods as intermediate hosts correlates with the habitat of the target host. Eggs that sink require benthic copepods as their intermediate hosts, whereas eggs that float infect planktonic copepods (Jarecka, 1961). The buoyancy and sinking characteristics of eggs of the sealworm *Pseudoterranova decipiens* are such that they maximize transmission, while reducing their dispersal outside the habitats where transmission to the next host is less likely. Sealworms parasitize the stomach of seals; eggs passed in the feces hatch after a number of days and free-living larvae adhere to the bottom. Bottom-dwelling invertebrates are first intermediate hosts whereas bentho-demersal fish are second intermediate hosts and carry the stage infective to seals. Calculations of the sedimentation rate of eggs and their horizontal transport due to ocean currents on the Scotian Shelf off Nova Scotia, Canada, indicate that they will likely be deposited in the bottom before leaving the continental shelf, the area where transmission to the invertebrate and bentho-demersal fish hosts occurs (McConnell *et al.*, 1997). Under current oceanographic conditions, a slower sedimentation rate would likely lead to excessive dispersion in space and eggs would be lost off the continental shelf.

Similar morphological adaptations also are seen in many acanthocephalans. For example, when the eggs of *Leptorhynchoides thecatus* are released into the water, they 'unwrap' or 'unravel' one of their egg layers, forming ribbon-like filaments (Fig. 13.5). These filaments become entangled in submerged vegetation upon which amphipod intermediate hosts are known to feed, thereby enhancing the probability of contact between the egg and a potential host (Barger & Nickol, 1998). The outer membrane of the egg of *Paulisentis nagpurensis*, on the other hand, swells in contact with water causing the egg to float, which facilitates transmission to the planktonic amphipod intermediate host.

Another transmission strategy involves offering a nutritional reward to a potential host to increase the chances of being preyed on. Cestodes with benthic-feeding intermediate hosts have large, heavy eggs, with high concentrations of various nutrients such as glycogen, fats, and proteins; these provide a nutritional reward, as well as an uninvited guest, to the predator. Instead of offering nutritional rewards, the eggs of some parasites mimic the food items of the next host, in essence tricking the host into eating a parasite instead of their regular prey. The eggs of some hymenolepids, for instance, mimic the diatoms normally eaten by the ostracod intermediate host (Jarecka, 1961).

The egg is the most common non-motile larval stage that is dispersed passively. However, in many cases, the egg itself is not the infective stage to the next host. In many species, after a period of passive dispersal, the egg hatches and a motile stage actively seeks the next host. Usually, hatching is not random. Egg hatching can follow circadian rhythms, where hatching occurs at a time coinciding with some host activity, increasing the odds of encounter. Hatching can also be triggered by the presence of the host, once again increasing the chances of encounter.

13.2.2 Active transmission and dispersal by motile stages

Active transmission and dispersal requires an expenditure of energy on the part of the parasite. Active transmission can be successfully achieved in two ways. In the first, a parasite's mobile larval stage moves until it is placed in a position where it can actively penetrate the surface of a host directly. Or, second, the larval stage moves into a position in space where it is likely to be encountered and then consumed by a host.

The forms of locomotion used by parasites are as varied as those used by free-living organisms. Interestingly, however, the locomotory organelles with which parasitic protozoans can be most easily identified, i.e., pseudopodia, cilia, flagella, are generally present only during their obligate parasitic stages. They do not occur in the free-living phases of their life cycles and are not, therefore, used in dispersal or host finding. Indeed, most parasitic protozoans do not have mobile free-living stages, and transmission, for the most part, involves either passive non-motile stages or vectors. An exception to this pattern is the ciliate *Ichthyophthirius multifiliis*, an ectoparasite of fishes (see also Chapter 3). Trophozoites of *I. multifiliis* are released from the fish, but they are poor swimmers and settle quickly to the substratum or attach to aquatic vegetation, where they encyst. The parasite then undergoes a series of transverse fissions, producing up to 1000 infective cells, called tomites; tomites, by the way, are excellent swimmers. These forms can survive up to 96 hours before finding and infecting a host.

The parasitic flatworms (Platyhelminthes) employ several different free-living larval stages for transmission and dispersal. Oncomiracidia are produced by monogeneans, coracidia by cestodes, and digeneans employ miracidia and cercariae. None of these free-living larval stages feed, relying entirely on their stored energy reserves. Ultimately, survival and infectivity depend on the finite energy reserves of the larvae. Most rely on stored glycogen, although lipids also may be used as energy reserves (Furlong *et al.*, 1995). As a result, their life spans are quite short, generally measured in hours, with a high limit of about 72 hours. Environmental temperatures also affect life span. Under experimental conditions, life span is temperature-dependent; maximum life expectancy usually occurs at those temperatures resembling the normal conditions of their environment. Among monogeneans, for example, the oncomiracidia of *Discocotyle sagittata* has one of the longer survival rates among temperate species. It can survive for up to 96 hours at 6 °C, but only 26 hours at 22 °C. Swimming activity usually decreases as the oncomiracidia age; they become progressively less active, and spend less and less time in the water column (Gannicot & Tinsley, 1998). Infectivity of the larval stage also decreases with age. Even though the larvae might still be alive, it may not be infective. Finding a host is only half of the adventure. The other half is penetrating the host, and this event requires a significant amount of energy. If a larval stage finds a host 'late in life', it may not have enough energy left to effectively invade the host. Similar patterns of temperature-dependent survival, swimming activity, and infectivity, have been observed in most larval stages.

The eggs released by most digeneans settle onto the substratum. When they hatch, therefore, miracidia are already in proximity to their mostly benthic-dwelling molluscan hosts. The miracidia then swim close to the substratum until impacted, in some manner, by their molluscan hosts. It appears that specific chemicals released by their hosts stimulate and attract both miracidia and oncomiracidia. The oncomiracidia of monogeneans are attracted by sugar-containing substances such as glycoproteins, proteoglycans, and polysaccharides, all of which are present in the epithelium of the piscine host (most monogeneans parasitize fishes). The chemical stimuli for many miracidia of digeneans appear to be chemicals of snail origin, such as, magnesium and hydrogen ions, amino acids, peptides like glutathione, short-chain fatty acids, N-acetyl neuraminic acid, and ammonia.

Most cercariae, on the other hand, seem to locate their hosts using a two-step process that requires a number of behavioral adaptations. In the first step, which seems to be independent of the host, the cercariae exhibit a number of behaviors that allow them to localize and reach the microgeographic area ('host space'), or time window ('host time'), most likely to be occupied by the next host in the parasite's life cycle (Combes *et al.*, 1994). Cercariae may respond to several different environmental stimuli, such as light intensity, gravity, temperature, and water currents, to reach a given 'host space'. Success from the standpoint of 'host time' usually requires a non-random temporal pattern of cercarial emergence from the molluscan host. In many species, emergence occurs once or twice a day and is timed to correspond with the behavior of the next host in the life cycle. For example, the cercariae-shedding patterns of the various species of *Schistosoma* coincide with the time when the definitive host is most likely to visit the aquatic habitat where transmission will occur (Fig. 13.6). The rhythmic emergence of the cercariae is synchronized by exogenous factors to which the cercariae respond. The nature of these factors, or synchronizers, varies with the species involved, but photoperiod and thermoperiod are most commonly involved. In some marine species, the chronobiology of the host, as it relates to changes in ocean tides, for example, may be a most important feature in regulating cercarial release.

Once cercariae are positioned in 'host time' or 'host space', they still must encounter an individual host. In the second step of their host-finding behavior, cercariae respond to a number of physical and a few chemical stimuli emanating from the host. Physical factors include shadow patterns created by the swimming behavior of a host, or the turbulence created by a fin or the opening and closing of a host's gill opercula. Unlike miracidia and oncomiracidia, most cercariae are not very responsive to chemical stimuli of host origin (Combes *et al.*, 1994). Among the few exceptions are cercariae of *Echinostoma revolutum*, which are attracted to chemicals present in water conditioned by the snail *Biomphalaria glabrata*, the second intermediate host of the parasite (Fried & King, 1989). It is interesting that the only cercariae known to respond to chemical stimuli also employ a snail as second intermediate host. Perhaps the relatively low vagility of this host, a mollusc, encourages an efficient use of chemical cues, which are likely to form a chemical gradient in the water. If a host is relatively more mobile, it may not allow the cercariae to efficiently track chemicals, because the host mobility may distort gradients required for chemotactic responses.

Instead of actively seeking and infecting a host, some cercariae utilize a reverse approach, in which the next host in the life cycle becomes infected by preying upon the cercariae. In these cases, the cercariae attract their hosts by mimicking the behavior or morphology of common prey items. The cercariae of *Azygia lucii*, for example, are unusually large (7 mm) and their swimming behavior closely resembles the movement of mosquito larvae, upon which the potential hosts are likely to prey. Some marine cercariae cluster together in large numbers (up to 700) forming an attractive prey item for the curious host (Beuret & Pearson, 1994).

The free-living larval stages of nematodes are long-lived compared to those of the platyhelminths, but they do not appear to disperse as far. Among terrestrial nematodes, in particular those infecting domestic animals, most of the larvae and eggs are found in the vicinity of the feces in which the eggs were deposited. Eggs and larvae are highly aggregated in space and dispersal is local. When they escape the fecal mass, they frequently climb blades of grass or other prominent objects. The mechanisms controlling this type of response are not, as yet, well known. The larvae of

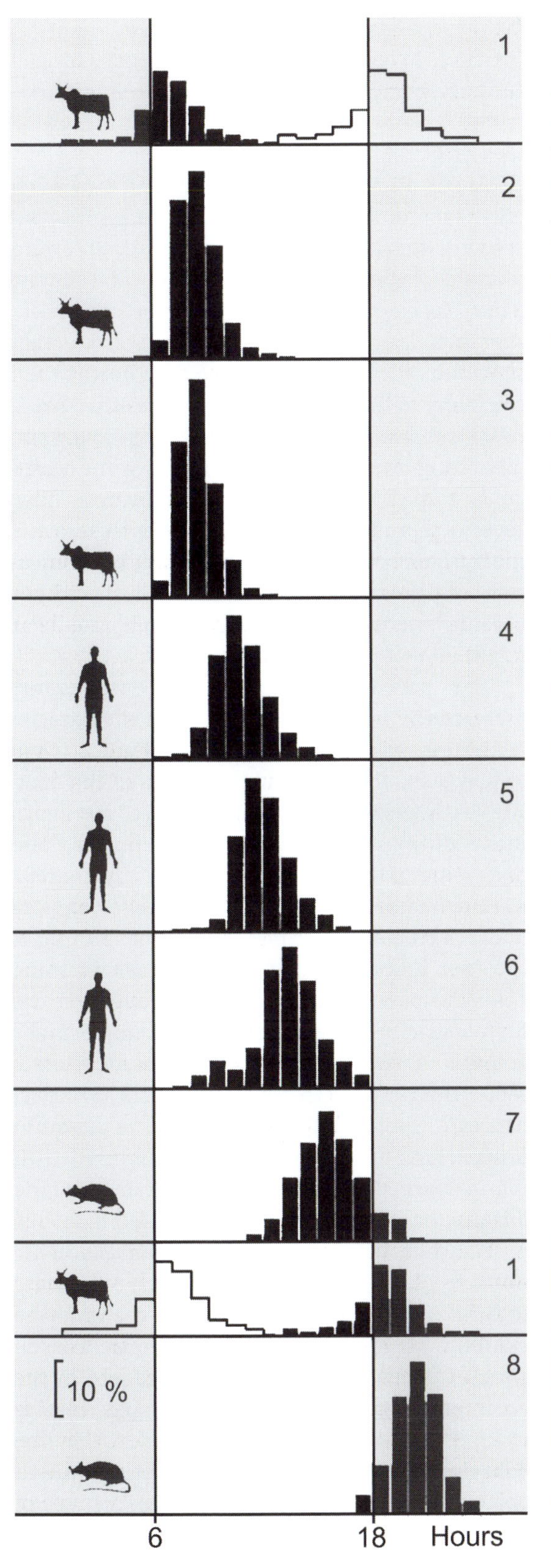

some species of hookworm are known to be highly sensitive to thermal stimuli. Some also seem to respond to light, although it is not clear if the response is to the light or the heat associated with it. When the environmental conditions are favorable, infective larvae of species such as *Ancylostoma caninum*, *Nippostrongylus braziliensis*, *Necator americanus*, etc. may enter a resting, presumably energy-saving, posture. When a host is near, carbon dioxide from the host, at concentrations as low as 0.16%, coupled with radiant heat, activates the questing response of these larvae. The nematodes then begin to undulate their anterior end from side to side as a means of increasing their chances of physically contacting the host to which they attach and penetrate directly.

13.2.3 Transmission and dispersal by vectors and other hosts

Vectors, intermediate hosts, and paratenic hosts, are probably the most important means of transmission and dispersion for most parasites. In most cases, the movement of vectors and other hosts will greatly facilitate the dispersal of parasites on a geographical scale because they range much greater distances than most free-living larval stages. For example, the embryonated eggs of the acanthocephalan *Polymorphus minutus* can spread passively up to 2500 m downstream from the source of infection. However, infected *Gammarus pulex*, the amphipod intermediate host, can be found 900 m upstream and 5000 m downstream from the source of infection (Hynes & Nicholas, 1963). The movement of infected ducks, the definitive hosts of *P. minutus*, is even greater, dispersing the acanthocephalan's eggs over continental areas.

Fig. 13.6 Daily shedding patterns of cercariae of different species of *Schistosoma*. Note the correspondence between the time of shedding and the likely time of host activity. Scale: percentage of daily cercarial production. (1) *S. margrebowiei*; (2) *S. curassoni*; (3) *S. bovis*; (4) *S. mansoni*; (5) *S. haematobium*; (6) *S. intercalatum*; (7) *S. mansoni* of murine origin; (8) *S. rodhaini*. (Modified from Combes, C., Fournier, A., Moné, H. & Théron, A. [1994] Behaviours in trematode cercariae that enhance parasite transmission: patterns and processes. *Parasitology* (Supplement), **109**, S3–S13. Reprinted with the permission of Cambridge University Press.)

Parasites that use vectors, and some that employ paratenic hosts for transmission and dispersal, possess a number of adaptations to ensure success. Vector-transmitted parasites often synchronize their presence in a given area of the host with the daily or annual cycle of their vectors. Daily cycles of synchronization between parasites and the biting behavior of their vectors have been confirmed in many host–parasite combinations. The vector for *Wuchereria bancrofti*, for example, is a nocturnally feeding mosquito. The microfilariae of *W. bancrofti* in the blood of the human definitive host show a marked periodicity, with peak numbers of parasites in the peripheral blood vessels at night. Several strains of *W. bancrofti* show, however, an interesting geographical adaptation to variations in the feeding behavior of their mosquito hosts. In the South Pacific the vectors are day-biting mosquitoes, and as expected, the microfilariae are most abundant in the peripheral blood during the daylight hours. The movement of microfilariae into the peripheral blood is an active process and clearly an adaptation to enhance transmission success. By moving in and out of the peripheral circulation, the parasites increase their chances of being acquired by the appropriate vector, while excluding or reducing the chances of being picked up by hematophagous insects that do not serve as vectors. Microfilariae that inhabit tissues other than blood also exhibit site preferences that correlate with the biting habits of their vectors. For example, microfilariae of *Onchocerca gutturosa* in cattle occur primarily in subcutaneous regions of the umbilical area, where the vector, the black fly *Simulium ornatum*, prefers to take its blood meal. Microfilariae of the congener *O. cervicalis* in horses migrate to the midline region of the abdomen, which is the preferred biting area for the vector *Culicoides nubeculosus*.

Annual rhythms of synchronization between the appearance of blood parasites and the activity of the insect vector have also been documented. In some haemosporidians of birds, for example, the seasonal peak of the protozoans in the peripheral blood is apparently triggered by hormonal changes in the host that occur in the spring. These peaks coincide with the reappearance of the vector after overwintering. Similarly, microfilariae of the dog heartworm, *Dirofilaria immitis*, increase in the peripheral circulation in late summer, coinciding with increased ambient temperatures, and increased numbers, of the mosquitoes that vector the parasite.

Anderson & Bartlett (1994) suggest two novel ideas about how reproductive strategies can enhance the probability of transmission of bird filarioids by mallophagan vectors. On the one hand, species of *Eulimdana* produce long-lived, skin-inhabiting microfilariae; after reproduction the adults die and are resorbed. The host is then refractory to further infection. On the other hand, *Pelecitus fulicaeatrae* also produces long-lived, skin-inhabiting microfilariae, and, although the adults do not die subsequent to reproduction, they become reproductively senescent. For either case, the authors speculate that ephemerality or senescence ensures that the skin of the host will not become over-saturated with microfilariae that might prove detrimental to the vector.

The parasite adaptations just mentioned increase the transmission success of the parasite from host to vector. The vector, however, is likely to play a small role in the dispersion of the parasite on a geographical scale because the movement of most vectors is rather limited. The dispersion of the parasites is more likely to occur as an infected host moves around allowing the 'local' vectors to pick up the infection and pass it to other hosts in that area. Moreover, in some cases the parasites have a negative effect on the flight capacity of their vectors. For example, flight activity of the mosquito *Anopheles stephensi* is reduced when infected with *Plasmodium cynomolgi* or *P. yoelii*. Similarly, flight activity in the mosquito *Aedes aegypti* also is reduced when infected with filarial worms such as *Brugia pahangi* or *Dirofilaria immitis* (for references, see Moore & Gotelli, 1990).

Parasites using intermediate and paratenic hosts to reach the next host in their life cycles exploit predator–prey relationships and often exhibit a number of adaptations to increase the chances that the infected prey will be taken by the right predator. These adaptations may involve changes in the external appearance, or in the behavior, of the infected hosts. Parasite-induced changes in the normal appearance of a host are likely to make the infected prey more susceptible

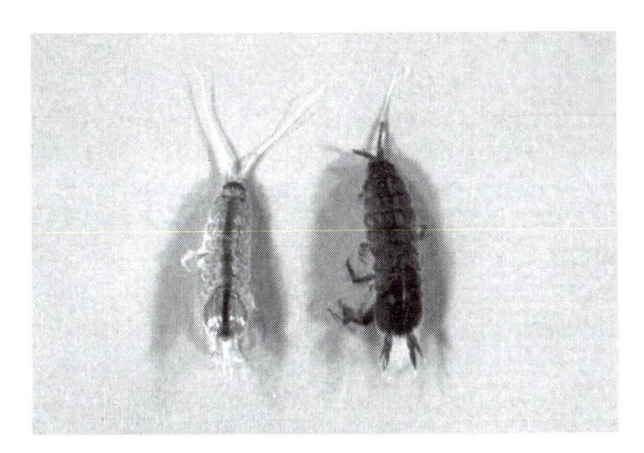

Fig. 13.7 Pigment changes in the isopod *Asellus intermedius* induced by the acanthocephalan *Acanthocephalus dirus*. The isopod on the left is infected; the specimen on the right is not infected and shows the normal pigmentation. (Photograph courtesy of Joseph W. Camp, Purdue University–North Central, Westville.)

to visually oriented predators. Several of these cases have been observed in larval acanthocephalans infecting invertebrate intermediate hosts. Most cystacanths are colorless. The cystacanths of *Corynosoma constrictum*, however, are bright orange and can easily be seen inside the amphipod *Hyalella azteca*, making the amphipod very conspicuous to its avian predator. Cystacanths of several species of *Polymorphus* and *Pomphorhynchus* also have enough color to invite differential predation of infected intermediate hosts by an unsuspecting host. The pulsating sporocysts of *Leucochloridium paradoxum* in the tentacles of its molluscan host also invite predation by birds. Another tactic used by some parasites is to induce changes in the normal pigmentation of the intermediate host. *Acanthocephalus dirus*, for example, disrupts the formation of pigment in the isopod *Asellus intermedius*, producing a striking color contrast with uninfected isopods (Fig. 13.7) (Camp & Huizinga, 1979). A comprehensive investigation on the feeding behavior of perch and sticklebacks, both visual predators, showed that these hosts were more likely to consume infected, rather than uninfected, isopods.

After a successful transmission process, the mobility of the host over a geographical range is likely to be the most important means of dispersal for most of the parasites (Sousa & Grosholz, 1991). Although dispersal of a parasite does not ensure its success in terms of completing the life cycle and reproducing, it increases the chances of colonizing new host individuals and new geographic areas.

13.2.4 Host behavior and transmission

It is becoming increasingly clear that many parasites have evolved mechanisms for altering host behavior to ensure transmission between host species. Most of these behavioral modifications appear to result from mechanical damage or chemical interference inflicted by the parasite on the physiology of its host (Helluy & Holmes, 1990; Holmes & Zohar, 1990). In some host–parasite systems, these effects are relatively subtle, whereas in others they are clear and direct. For example, rats infected with *Toxoplasma gondii* show an increase in their exploratory behavior, thereby increasing, at least indirectly, the likelihood of predation by feline definitive hosts (Berdoy *et al.*, 1995). Another protozoan, the sporozoan *Eimeria vermiformis*, seems to induce just the opposite kind of behavior. Under normal conditions, uninfected rodents display a variety of defensive behaviors, e.g., wariness, fear, 'anxiety', when confronted with a potential predator. However, in rodents infected with *E. vermiformis*, this behavior becomes subdued, suggesting the possibility that the parasite is increasing the likelihood of the host's predation. It appears that the abnormal behavior of the host is due, in part, to parasite-induced changes in the serotonin neuromodulatory system, and to an opioid-mediated analgesia, both induced by the parasite (Kavaliers & Colwell, 1994; Thompson & Kavaliers, 1994).

Parasites may also modify the behavior of their vectors, particularly their feeding behavior. For example, tsetse flies infected with trypanosomes appear to be more active, and seem to probe and feed more frequently than uninfected flies, which would increase the spread of the parasite within a host population (Jenni *et al.*, 1980). Similar behaviors associated with increased probing have been observed in other biting flies and mosquitoes; the opposite behavior, however, has also been observed (for a review see Molyneux & Jefferies, 1986; Moore, 1993)

Parasite manipulation of host behavior to increase the visibility of the host seems to be common among invertebrate intermediate hosts. The amphipod *Gammarus lacustris* is the intermediate host for the acanthocephalan *Polymorphus paradoxus*, and mallard ducks are the final hosts. Uninfected amphipods avoid light and do not cling to floating vegetation or other floating substrata. Infected amphipods, however, spend most of their time in the photic zone of a lake, clinging to floating vegetation. When dislodged from vegetation, they skim the water surface. These behaviors make infected amphipods significantly more vulnerable to predation by mallards (Holmes & Bethel, 1972; Bethel & Holmes, 1977). A similar type of behavior is shown by the terrestrial isopod *Armadillidium vulgare* when infected by the acanthocephalan *Plagiorhynchus cylindraceus*. Infected, dark-colored isopods are found more frequently on light-colored backgrounds and in less humid locations, both of which result in increased predation by starlings *Sturnus vulgaris* (Moore, 1984).

Perhaps one of the best-known examples of parasite-induced behavior is the remarkable effect produced by the metacercariae of *Dicrocoelium dendriticum* on the behavior of its ant intermediate host. When an ant becomes infected by cercariae of *D. dendriticum*, most encyst as metacercariae in the hemocoel. One cercaria, however, migrates to, and encysts in, the subesophageal ganglion of the ant. The presence of this metacercaria prevents the ant from opening its jaws when temperatures fall at dusk. If the ant happens to be grazing on a blade of grass, it will stay attached for the night in a fixed position, with its jaws tightly closed until temperatures rise the next morning. Since sheep and other definitive hosts graze extensively during the early evening and early morning hours, an infected ant caught on a blade of grass is therefore highly vulnerable to accidental ingestion. This sort of behavioral modification would appear to involve an intricate manipulation of the nervous system of the host and would seem to represent an extreme evolutionary novelty. However, the evolutionary steps leading to this peculiar behavior are much simpler than presumed. Nonformicine hymenopterans, the postulated ancestors of ants, sleep in the same way as do infected ants, e.g., attached to a plant. It would appear then, that when a metacercaria induces an ant to cling to a plant, the parasite might be 'simply' capitalizing on an ancient behavior by reactivating it (Wickler, 1976; Curio, 1988).

Within the framework of transmission success, parasite-induced changes in host behavior are likely to carry some sort of cost for the parasite. To be favored by natural selection, the fitness benefits of such an effort by the parasite must outweigh the cost incurred. Although some parasites may cause alterations in host behavior by their chance presence in a critical organ, most changes appear to result from interference with host neurochemistry through the secretion and release of chemicals by the parasite (Helluy & Holmes, 1990; Holmes & Zohar 1990; Hurd, 1990; Thompson & Kavaliers, 1994; Maynard *et al.*, 1996). In terms of energy allocation by the parasite, therefore, energy invested in manipulating the host's behavior may not be available for growth, reproduction, countering the host's immune system, or other energy-dependent processes that individually, or in some combination, ensure survival of the parasite. The trade-off between energy allocation to manipulate the host's behavior and other needs of the parasite suggests that **natural selection** will likely favor an optimal value of energy allocation for modification of host behavior, and not necessarily the maximum investment (Poulin, 1994, 1998).

Poulin (1994) developed a theoretical framework for the evolution of parasite manipulation of host behavior based on a cost–benefit analysis of their investment in such endeavor. He proposed that the energy investment of a parasite in manipulation of host behavior should increase when parasite prevalence increases, because of the need to stand out among many others showing the same kind of altered behavior (Fig. 13.8A). When prevalence is low, a minimum investment should result in a small change in host behavior that will be enough to make it stand out among all the other uninfected hosts. On the other hand, energy investment by a parasite manipulating host behavior should decrease as infrapopulation size, parasite longevity, host longevity, passive transmission rates, or parasite fecundity, increase (Fig. 13.8B). When a parasite is alone in a host, with no conspecifics, this individual has to assume the full cost of manipulating the host's behavior. However, if many parasites of the same species are found in a host, they all can share the cost of

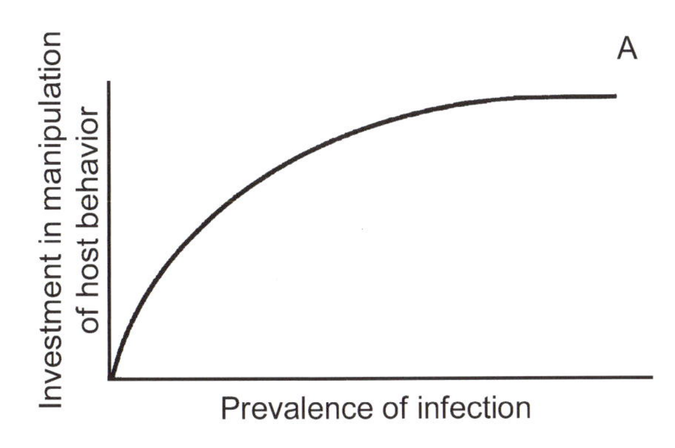

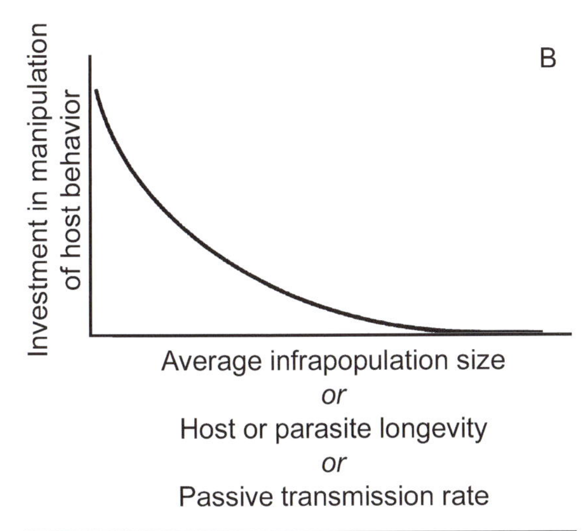

Fig. 13.8 Predicted relationship between the parasite investment in the manipulation of host behavior and some variables that may affect transmission success or manipulation costs. (A) Predicted relationship between parasite investment in the manipulation of host behavior and parasite prevalence. (B) Predicted relationship between the parasite investment in the manipulation of host behavior and three similar variables: the average infrapopulation size, the longevity of either the parasite or the host following infection, and the rate of passive transmission. (Modified from Poulin, R. [1994] The evolution of parasite manipulation of host behavior: a theoretical analysis. *Parasitology* (Supplement), **109**, S109–S118. Reprinted with the permission of Cambridge University Press.)

modifying the host's behavior, decreasing their individual investment. If the life span (longevity) of the parasite or the host is short, then the parasite investment in manipulating the host behavior should increase; this is because the parasite is under pressure to leave a host before it, or its host, dies, without achieving transmission. But, if its life expectancy is high, then there is no 'rush' in effecting transmission by modifying the host behavior, and the parasite can afford being 'patient', and wait for that chance event when a predator may find the infected host. A similar principle applies when passive transmission rates are high. Thus, if rates are high, there is no need for the parasite to improve over an already efficient process. Only when transmission rates are low will manipulation of host behavior result in an improvement in the transmission rate. Finally, life-history theory suggests that only one subset of all life-history characteristics can be maximized in an organism. It is unlikely, therefore, that a parasite would maximize its investment in host manipulation *and* fecundity at the same time. It is important to remember that investment trade-offs do not usually reach extreme levels of maximum and minimum investment. A balance is often achieved without reaching extreme values. Similarly, these predictions regarding the trade-offs and potential investment of parasites in manipulation of host behavior are subject to exceptions and a number of, as yet, unstudied facets of the host–parasite relationship. Factors such as the possibility that infected hosts may oppose the manipulative efforts of the parasite, that parasites may be able to assess the local conditions and gage their investment on an individual basis (and not on a species-specific basis), or that predators may avoid prey showing unusual behaviors, may each affect the theoretical predictions of parasite investment in host manipulation and their outcome (Poulin, 1994). With respect to the latter, Lafferty (1992) has suggested that, since most adult parasites do not cause 'harm' and because hosts with behaviors modified by parasites might be easier to catch, there is no penalty and therefore final hosts need not ignore, or avoid, infected prey.

13.3 Dispersion, colonization, and the Theory of Island Biogeography

13.3.1 The colonization process

Transmission and dispersal are two processes in the life cycle of a parasite in which the ultimate objective is to reach a suitable host; both result in perpetuation of the species. In strict terms, colonization refers to the establishment of a parasite in a population where none of that species was present; it results in pioneering by the species. Simply put, this means that an uninfected host is colonized when it acquires an infection. However, if the host is already infected by other parasites of the same species, then the process is transmission, not colonization. The idea of colonization can be applied to an individual host, to a host population, or to a host species. The opposite of colonization is extinction, and this also can occur at the level of an individual host, population, or community (Bush *et al.*, 1997).

Colonization of new hosts and repeated infection of previously infected hosts is achieved by many of the stages that occur in the life cycles of parasites. Thus, both colonization and infection can be achieved by ingesting eggs, resistant cysts, or larval stages of the parasites, or by the infection of any of the free-living larval stages. Dispersal and colonization lead intuitively to the idea of pioneering and range extension of the parasite species.

Colonization within the same host population ensures the local survival of the parasite population. Colonization of new host populations, and new host species, implies that new geographical areas are being reached by the parasite. Dispersal, over relatively large geographical areas, is most likely to be achieved by long-lived eggs or cysts transported by wind or water, or by dispersal of larval stages and adults in wide-ranging intermediate, paratenic, and definitive hosts. Dispersal by free-living larval stages such as miracidia, cercariae, oncomiracidia, or the motile larval stages of some nematodes and insects, is limited to local areas due to their short life span, or limited vagility, or both.

Dispersal of a parasite into a new habitat or geographic area is usually a chance event and does not guarantee success in colonizing a new host or in becoming established reproductively. If the abiotic conditions of the new area are inappropriate for the free-living stages, or if the appropriate intermediate or definitive hosts are not present, then success will be nil. Parasite species that exhibit low host specificity are more likely to succeed but, even then, the right combination of hosts must be present to assure completion of the life cycle (at least for those species with indirect life cycles). In general, parasites with one-host life cycles are more likely to succeed than those with multiple hosts.

Large numbers of waterfowl, as well as some marine organisms, including fishes, birds and mammals, are known to migrate extensively on an annual basis. This means that parasites' reproductive propagules being released from these hosts may be dispersed over great distances. However, not all the parasites present at the beginning of the hosts' migration will be retained. For example, alterations in the physiology of the digestive system of these hosts induced by dietary changes affect many parasites, with some being lost; others gained. In Pacific salmonids, migration from freshwater to marine habitats results in both dietary and extreme physiological changes, preventing parasites of one environment from successfully invading the other. The dispersal of parasites and the colonization of new hosts may not be successful at the opposite ends of a home range, but they may work over contiguous areas, forming a gradient.

13.3.2 Colonization and the Theory of Island Biogeography

In the 1960s, Robert MacArthur and Edward O. Wilson (1963, 1967) developed a general theory of island biogeography aimed primarily at explaining how, and why, **species diversity** changes on an island. The theory introduced the idea that some variables associated with rates of extinction and colonization should explain the variability in **species richness** among communities on islands. For example, the number of species present on an island should be determined by the rate at which new species colonize the island and the rate at which species become extinct (Fig. 13.9A). Colonization and extinction rates, in turn, should be affected by the size of the island and its distance from the mainland. (Here, the mainland is used to imply a source of potential colonists.) Large islands should have lower extinction rates than smaller ones because they are likely to have more resources and more diverse habitats, which translates into the potential for higher species

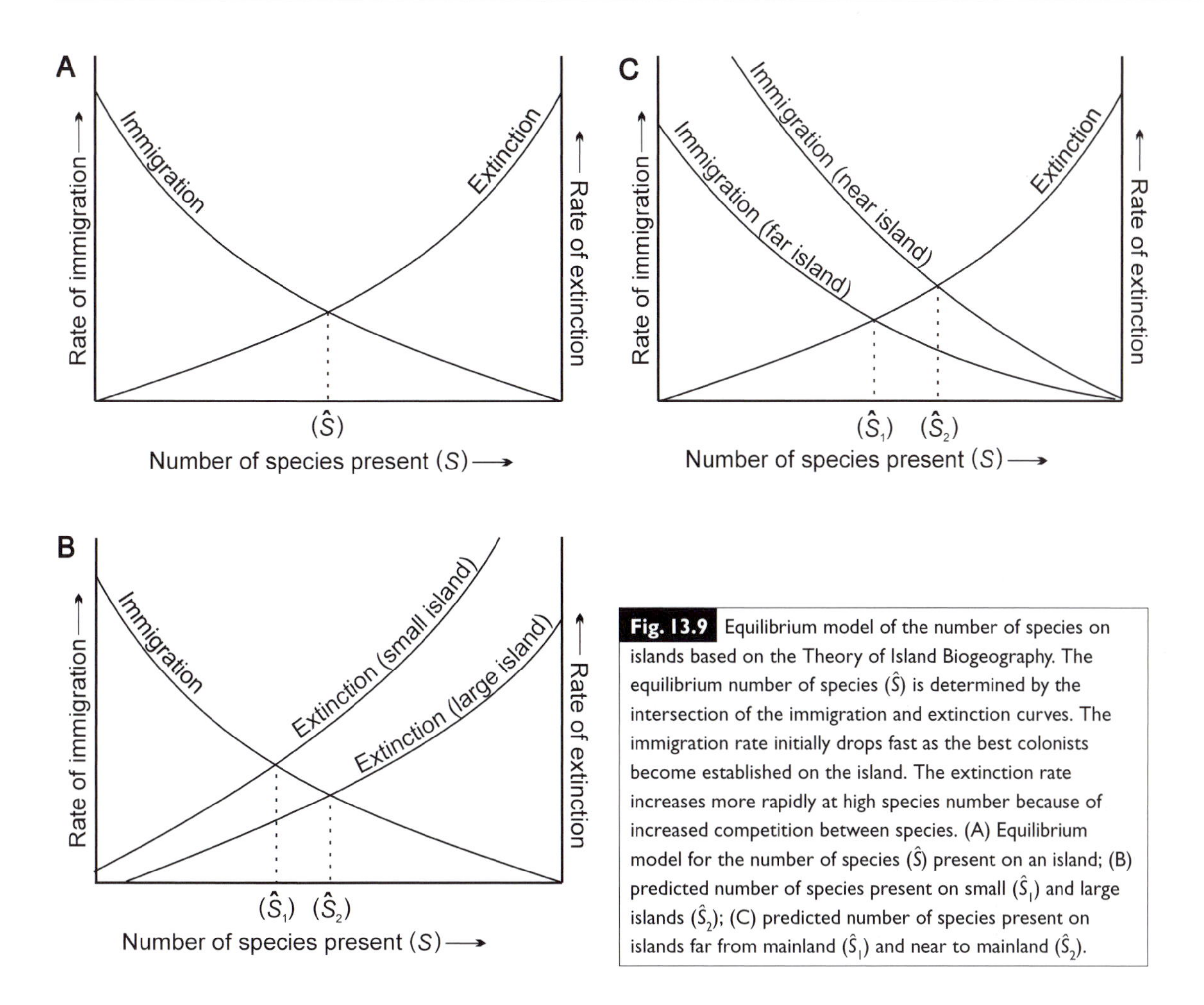

Fig. 13.9 Equilibrium model of the number of species on islands based on the Theory of Island Biogeography. The equilibrium number of species $(\hat{S})$ is determined by the intersection of the immigration and extinction curves. The immigration rate initially drops fast as the best colonists become established on the island. The extinction rate increases more rapidly at high species number because of increased competition between species. (A) Equilibrium model for the number of species $(\hat{S})$ present on an island; (B) predicted number of species present on small $(\hat{S}_1)$ and large islands $(\hat{S}_2)$; (C) predicted number of species present on islands far from mainland $(\hat{S}_1)$ and near to mainland $(\hat{S}_2)$.

richness (Fig. 13.9B). Larger islands should have higher colonization rates compared to smaller ones because their larger sizes will make it easier for potential colonizers to locate. As for distance from the mainland, islands close to a mainland should have higher colonization rates and, therefore, more species than islands at a greater distance (Fig. 13.9C). Over time, an equilibrium should be reached between the rates of colonization and extinction, and the number of species present at this equilibrium point should correlate with island size and distance from the mainland.

Since the theory of Island Biogeography was introduced, it has been applied to 'habitat islands' as well, e.g., lakes, mountaintops surrounded by lowlands, etc. It has also been applied to the colonization of plants by herbivorous insects and to the colonization of hosts by parasites. In the case of host–parasite systems, the individual host, host population, or host species may be considered as an 'island'.

Among parasites, body size and geographic range of the host species have been suggested to correlate with parasite richness (see Chapter 14). Larger hosts usually have more parasites because they should provide more habitats and more space to occupy; generally, they also consume more prey that may harbor more infective stages and, thereby, acquire larger numbers and a greater diversity of parasites. Larger hosts usually live longer, providing a more permanent habitat for colonization and allowing for the accumulation of parasite species over time. Hosts with wider geographical ranges also may be more heavily parasitized because they have more chances of encountering parasites and, therefore, of being colonized. In addition to body size and host range, a number of other characteristics, including host population density, type of diet, behavior, and the phyletic diversity of the host taxon, have been used as determinants of parasite richness (Chapter 14).

The Theory of Island Biogeography provides a solid set of working hypotheses with which to examine those factors that might contribute to colonization and succession. There will be few, if any, Krakataus in the life of most ecologists. But, consider, for example, the hatching of some precocial bird, e.g., a duck, in nature. Are the ducklings not replicate Krakataus? Indeed, this is a prime example where colonization, and succession, can be examined naturally; it is also the basis for using sentinel hosts to study parasite acquisition. Clearly, however, some caution must be applied when analogizing a living organism with an island. Except for some ephemeral habitat islands, such as a temporary clearing in a rainforest, most hosts have a shorter life span than most islands. Size can also be ambiguous. For example, the physical size of a host individual might be an important parameter, but so too could the host population size be important, and perhaps even the host population density (see Chapter 14). Likewise, distance can mean several things. It might relate simply to physical distance; for the example of ducklings say, on a lake, prior to their taking flight, the lake defines distance. Once they take flight, that measure of 'distance' is no longer meaningful. But distance, from a parasite's point of view might also include phylogeny (the number of related hosts they might be able to infect), or even the genetics (particularly with respect to the immune status) of an individual host. In summary, there are a number of caveats that must be judged when considering parasites and hosts in the context of the Theory of Island Biogeography. None seem insurmountable however and few, if any, other systems afford such a natural, replicable opportunity to investigate colonization and succession in the absence of anthropogenic factors.

References

Anderson, R. C. & Bartlett, C. M. (1994) Ephemerality and reproductive senescence in avian filarioids. *Parasitology Today*, **10**, 33–34.

Barger, M. & Nickol, B. (1998) Structure of *Leptorhynchoides thecatus* and *Pomphorhynchus bulbocolli* (Acanthocephala) eggs in habitat partitioning and transmission. *Journal of Parasitology*, **84**, 534–537.

Berdoy, M., Webster, J. P. & MacDonald, M. W. (1995) Parasite-altered behavior: is the effect of *Toxoplasma gondii* on *Rattus norvegicus* specific? *Parasitology*, **111**, 403–409.

Berrie, A. D. (1960) The influence of various definitive hosts on the development of *Diplostomum phoxini* (Strigeida, Trematoda). *Journal of Helminthology*, **34**, 205–210.

Bethel, W. M. & Holmes, J. C. (1977) Increased vulnerability of amphipods to predation owing to altered behavior induced by larval acanthocephalans. *Canadian Journal of Zoology*, **55**, 110–115.

Beuret, J. & Pearson, J. C. (1994) Description of a new zygocercous cercaria (Opisthorchioidea: Heterophyidae) from prosobranch gastropods collected at Heron Island (Great Barrier Reef, Australia) and a review of zygocercariae. *Systematic Parasitology*, **27**, 105–125.

Boray, J. C. (1969) Experimental fascioliasis in Australia. *Advances in Parasitology*, **7**, 96–210.

Bush, A. O. & Lotz, J. M. (2000) The ecology of crowding. *Journal of Parasitology*, **86**, 212–213.

Bush, A. O., Lafferty, K. D., Lotz, J. M. & Shostak, A. W. (1997) Parasitology meets ecology on its own terms: Margolis *et al.* revisited. *Journal of Parasitology*, **83**, 575–583.

Camp, J. W. & Huizinga, H. W. (1979) Altered color, behavior and predation susceptibility of the isopod, *Asellus intermedius*, infected with *Acanthocephalus dirus*. *Journal of Parasitology*, **65**, 667–669.

Combes, C., Fournier, A., Moné, H. & Théron, A. (1994) Behaviours in trematode cercariae that enhance parasite transmission: patterns and processes. *Parasitology* (Supplement), **109**, S3–S13.

Crompton, D. W. T. (1987) Host diet as a determinant of parasite growth, reproduction and survival. *Mammalogical Review*, **17**, 117–126.

Curio, E. (1988) Behavior and parasitism. In *Parasitology in Focus*, ed. H. Mehlhorn, pp. 149–160. New York: Springer-Verlag.

Dixon, K. E. (1964) The relative suitability of sheep and cattle as hosts for the liver fluke, *Fasciola hepatica* L. *Journal of Helminthology*, **38**, 203–212.

Esch, G. W., Hazen, T. C. & Aho, J. M. (1977) Parasitism and r- and K-selection. In *Regulation of Parasite Populations*, ed. G. W. Esch, pp. 9–62. New York: Academic Press.

Fleming, M. W. (1988) Size of inoculum dose regulates in part worm burdens, fecundity, and lengths in ovine *Haemonchus contortus* infections. *Journal of Parasitology*, **74**, 975–978.

Fried, B. & King, B. W. (1989) Attraction of *Echinostoma revolutum* cercariae to *Biomphalaria glabrata* dialysate. *Journal of Parasitology*, **75**, 55–57.

Furlong, S. T., Thibault, K, S. Morbelli, L. M., Quinn, J. J. & Rogers, R. A. (1995) Uptake and compartmentalization of fluorescent lipid analogs in larval *Schistosoma mansoni*. *Journal of Lipid Research*, **36**, 1–12.

Gannicott, A. M. & Tinsley, R. C. (1998) Larval survival characteristics and behaviour of the gill monogenean *Discocotyle sagittata*. *Parasitology*, **117**, 491–498.

Helluy, S. & Holmes, J. C. (1990) Serotonin, octopamine, and the clinging behaviour induced by the parasite *Polymorphus paradoxus* (Acanthocephala) in *Gammarus lacustris* (Crustacea). *Canadian Journal of Zoology*, **68**, 1214–1220.

Holmes, J. C. (1961) Effect of concurrent infections on *Hymenolepis diminuta* (Cestoda) and *Moniliformis dubius* (Acanthocephala). I. General effects and comparison with crowding. *Journal of Parasitology*, **47**, 209–216.

Holmes, J. C. (1962) Effect of concurrent infections on *Hymenolepis diminuta* (Cestoda) and *Moniliformis dubius* (Acanthocephala). II. Effects on growth. *Journal of Parasitology*, **48**, 87–96.

Holmes, J. C. & Bethel, W. M. (1972) Modification of intermediate host behavior by parasites. In *Behavioural Aspects of Parasite Transmission*, ed. E. U. Canning & C. A. Wright, pp. 123–149. London: Academic Press.

Holmes, J. C. & Zohar, S. (1990) Pathology and host behavior. In *Parasitism and Host Behavior*, ed. C. J. Barnard & J. M. Behnke, pp. 34–64. London: Taylor & Francis.

Hurd, H. (1990) Physiological and behavioural interactions between parasites and invertebrate hosts. *Advances in Parasitology*, **29**, 271–318.

Hynes, H. B. N. & Nicholas, W. L. (1963) The importance of the acanthocephalan *Polymorphus minutus* as a parasite of the domestic ducks in the United Kingdom. *Journal of Helminthology*, **37**, 185–198.

Ito, A., Lightowlers, M. W., Rickard, M. D. & Mitchell, G. F. (1988*a*) Failure of auto-infection with *Hymenolepis nana* in seven inbred strains of mice initially given beetle-derived cysticercoids. *International Journal for Parasitology*, **18**, 321–324.

Ito, A., Onitake, K. & Andreassen, J. (1988*b*) Lumen phase-specific cross immunity between *Hymenolepis microstoma* and *H. nana* in mice. *International Journal for Parasitology*, **18**, 1019–1027.

Jarecka, L. (1961) Morphological adaptations of tapeworm eggs and their importance in the life cycles. *Acta Parasitologica Polonica*, **9**, 409–426.

Jenni, L., Molyneux, D. H., Livesey, L. & Galun, R. (1980) Feeding behavior of tsetse flies infected with salivarian trypanosomes. *Nature*, **283**, 383–385.

Jennings, J. B. & Calow, P. (1975) The relationship between high fecundity and the evolution of entoparasitism. *Oecologia*, **21**, 109–115.

Kavaliers, M. & Colwell, D. D. (1994) Parasite infection attenuates nonopioid-mediated predator-induced analgesia in mice. *Physiology and Behaviour*, **55**, 505–510.

Keas, B. E. & Esch, G. W. (1997) The effect of diet and reproductive maturity on the growth and reproduction of *Helisoma anceps* (Pulmonata) infected by *Halipegus occidualis*. *Journal of Parasitology*, **83**, 96–104.

Keymer, A. E, Crompton, D. W. T. & Singhvi, A. (1983) Mannose and the 'crowding effect' of *Hymenolepis* in rats. *International Journal for Parasitology*, **13**, 561–570.

Lafferty, K. D. (1992) Foraging on prey that are modified by parasites. *American Naturalist*, **140**, 854–867.

Loker, E. S. (1983) A comparative study of the life-histories of mammalian schistosomes. *Parasitology*, **87**, 343–369.

MacArthur, R. H. & Wilson, E. O. (1963) An equilibrium theory of insular zoogeography. *Evolution*, **17**, 373–387.

MacArthur, R. H. & Wilson, E. O. (1967) *The Theory of Island Biogeography*. Princeton: Princeton University Press.

Maynard, B. J., DeMartini, L. & Wright, W. G. (1996) *Gammarus lacustris* harboring *Polymorphus paradoxus* show altered patterns of serotonin-like immunoreactivity. *Journal of Parasitology*, **82**, 663–666.

McConnell, C. J., Marcogliese, D. J. & Stacey, M. W. (1997) Settling rate and dispersal of sealworm eggs (Nematoda) determined using a revised protocol for myxozoan spores. *Journal of Parasitology*, **83**, 203–206.

Molyneux, D. H. & Jefferies, D. (1986) Feeding behavior of pathogen-infected vectors. *Parasitology*, **92**, 721–736.

Moore, J. (1981) Asexual reproduction and environmental predictability in cestodes (Cyclophyllidea: Taeniidae). *Evolution*, **35**, 723–741.

Moore, J. (1984) Altered behavioral responses in intermediate hosts: an acanthocephalan parasite strategy. *American Naturalist*, **123**, 572–577.

Moore, J. (1993) Parasites and the behavior of biting flies. *Journal of Parasitology*, **79**, 1–16.

Moore, J. & Gotelli, N. J. (1990) Phylogenetic perspective on the evolution of altered host behaviours: a critical look at the manipulation hypothesis. In *Parasitism and Host Behaviour*, ed. C. J. Barnard & J. M. Behnke, pp. 193–233. London: Taylor & Francis.

Mouahid, A. & Théron, A. (1987) *Schistosoma bovis*: variability of cercarial production as related to the snail hosts: *Bulinus truncatus*, *B. wrighti* and *Planorbarius metidjensis*. *International Journal for Parasitology*, **17**, 1431–1434.

Poulin, R. (1994) The evolution of parasite manipulation of host behavior: a theoretical analysis. *Parasitology* (Supplement), **109**, S109–S118.

Poulin, R. (1995) Clutch size and egg size in free-living and parasitic copepods: a comparative analysis. *Evolution*, **49**, 325–336.

Poulin, R. (1996) The evolution of life history strategies in parasitic animals. *Advances in Parasitology*, **37**, 107–134.

Poulin, R. (1998) *Evolutionary Ecology of Parasites*. London: Chapman & Hall.

Read, C. P. (1951) The 'crowding effect' in tapeworm infections. *Journal of Parasitology*, **37**, 174–178.

Read, C. P. (1959) The role of carbohydrates in the biology of cestodes. VIII. Some conclusions and hypotheses. *Experimental Parasitology*, **8**, 365–382.

Roberts, L. S. (2000) The crowding effect revisited. *Journal of Parasitology*, **86**, 209–211.

Rondelaud, D. & Barthe, D. (1987) *Fasciola hepatica* L.: étude de la productivité d'un sporocyste en fonction de la taille de *Lymnaea truncatula* Müller. *Parasitology Research*, **74**, 155–160.

Sibly, R. M. & Calow, P. (1985) Classification of habitats by selection pressures: a synthesis of life-cycle and *r*/K theory. In *Behavioral Ecology*, ed. R. M. Sibly & R. H. Smith, pp. 75–90. Oxford: Blackwell Scientific Publications.

Sousa, W. P. & Grosholz, E. D. (1991) The influence of habitat structure on the transmission of parasites. In *Habitat Structure: The Physical Arrangement of Objects in Space*, ed. S. S. Bell, E. D. McCoy & H. R. Muskinsky, pp. 300–324. London: Chapman & Hall.

Sukhdeo, M. V. K. (1991) The relationship between intestinal location and fecundity in adult *Trichinella spiralis*. *International Journal for Parasitology*, **21**, 855–858.

Thompson, S. N. & Kavaliers, M. (1994) Physiological bases for parasite-induced alterations of host behaviour. *Parasitology* (Supplement), **109**, S119–S138.

Whitfield, P. J. & Evans, N. A. (1983) Parthenogenesis and asexual multiplication among parasitic platyhelminths. *Parasitology*, **86**, 121–160.

Wickler, W. (1976) Evolution-oriented ethology, kin selection and altrustic parasites. *Zeitschrift für Tierpsychologie*, **42**, 206–214.

Chapter 14

Communities of parasites

14.1 General considerations

Most of you probably have a sense of what the term community means. Indeed, unless you were born to parents living on a remote farm, or perhaps to parents who were lighthouse keepers, you probably were born and educated in a village, a town or a city, i.e., in a community. However, in most cases, the term community is ambiguous. Where does the 'community' begin and end? If your house is just across the city limits, are you excluded from that community? What is included in the community? Does it include all peoples of all ethnic, religious, and political backgrounds? Does it include the dogs and cats, the rats and mice, the birds and trees? Our point is that, although we all have a sense of 'community', it is really a vague and artificial concept. What constitutes a community may be too large and encompassing to some, or too small and exclusive to others. In a real sense, a community is in the eyes of the beholder.

The same problems exist when considering communities of free-living organisms, divorced from a human perspective. From an ecological viewpoint, a community includes all of the organisms living in a prescribed area. Although the current literature is rich in community studies, to our knowledge, there exist no thorough studies of communities. The reasons are several. First, similar to the argument we present above, where does the community begin and end? The 'community' in most community-level studies on free-living organisms is determined by the whims of the investigator. These whims may be influenced by geographic factors, by the interests of the investigators or, more commonly, by logistics. The result is that the boundaries of the community are artificial – often in terms of area or habitat and always in terms of organisms. The problem of area or habitats can sometimes be alleviated in studies of free-living organisms by, for example, studying discrete units of habitats such as ponds or islands. But, even then, additional problems may arise such as seasonally resident organisms (e.g., consider migratory birds that may use a pond, lake, or island for breeding purposes but are absent for the remainder of the time). The problem of 'what organisms to include (or exclude)' cannot be easily remedied. A proper study of a community would include an investigation of virtually all representatives, of all kingdoms, found within the area or habitat of interest. Who has such expertise? Even assembling a multidisciplinary team with sufficient credentials would be an awesome task unto itself!

With all of these apparent problems associated with the study of communities, you might well ask – why bother? There are many reasons actually, but we will highlight just a few. Humans have an inherent interest in classifying things and determining how they work. The current interest in 'biodiversity' should be ample evidence of our desire to know and classify. Although some communities appear relatively constant over time, most are dynamic, with ever-changing numbers and identities of organisms. About such communities, we wonder, what makes them change, do they change randomly or predictably, will they persist or change yet again? Many who study populations of one organism branch out, in time, to investigate other species. Perhaps this is due to the potential boredom of looking at only one type of organism. More likely, they recognize that few species are independent of other species. It is this *potential* interaction between species, whether positive or negative, that seems to drive community-level studies. On a more practical note, and tied in with biodiversity questions, is the idea of preservation. As the world population grows, and habitat becomes more precious, many are concerned with setting up parks or reserves as a refuge for organisms. How large must these parks or reserves be to insure the survival of the target organisms? What other organisms, and in what numbers, should be included in these areas to foster a sustainable community?

In addition to the above, studies on communities of parasites can, sometimes, tell us much about the hosts in which they are found. For example, because many parasites exploit food webs, their presence in a host can tell us some of the things that the host has eaten. The presence/absence of some parasites can be used to identify different populations of hosts (see section 15.4.1) and such analyses have even been used in settling legal questions that have arisen due to poaching! These are just some of the applied questions whose underpinnings are found in community studies.

14.2 The kinds of communities of parasites

Similar to populations of parasites, communities of parasites can be viewed hierarchically, or in a nested manner. Bush & Holmes (1986*a*) coined the term 'infracommunity' to include all of the parasite infrapopulations in a single host individual. This is the most fundamental level of parasite communities and it is at this level that all data on parasite communities begin. Holmes & Price (1986) extended the hierarchical terminology by adopting the ideas proposed by Root (1973) for free-living communities. They suggested the term 'component community' to include all of the infrapopulations of parasites associated with a host species (or, more realistically, a subset of a host species). The compound community, the most encompassing level, includes all suprapopulations of all parasites in the host(s) of interest. Bush *et al.* (1997) suggested two changes to these ideas. First, they acknowledge that a component community might also include all of the free-living phases of parasites associated with the abiotic environment (or, more realistically, some subset of the abiotic environment). Second, they suggest that the term 'supracommunity' be used in place of compound community for parallelism with the terminology used for parasite populations. Figure 14.1 is a stylized representation of a nested series of communities with one supracommunity, four component communities, and many infracommunities. Needless to say, this is highly simplified!

Still another hierarchical way of looking at parasite communities might be to focus on the identity of the hosts. This results in what are called '**guild** communities': the parasite community found in a guild of hosts. This approach has been pioneered for parasites by Zander (1998) and provides a level somewhat beyond the component community level. Few studies have been conducted, explicitly, with this level in mind, but such studies would allow for an additional suite of questions to be asked. It is at the guild community level that one might ask questions such as, from whom does one obtain their parasites? This too is a level where one might examine 'host switching' or 'host capture'.

Parasite communities can also be viewed on spatio-temporal scales that go beyond their hierarchical nature. For example, it is not uncommon to see studies on 'parasites of . . . on their breeding grounds' or 'parasites of female . . .'. The study of the same host species that occurs in different habitats seasonally, or that is stage-structured, seems to go beyond the concept of a 'component community', although the host individuals still represent infracommunities and they are all part of the supracommunity. It is at different spatio-temporal scales that one might address questions related to the stability (or resilience) of the community, to replacement patterns within the community (succession), or to the influence of gender or age.

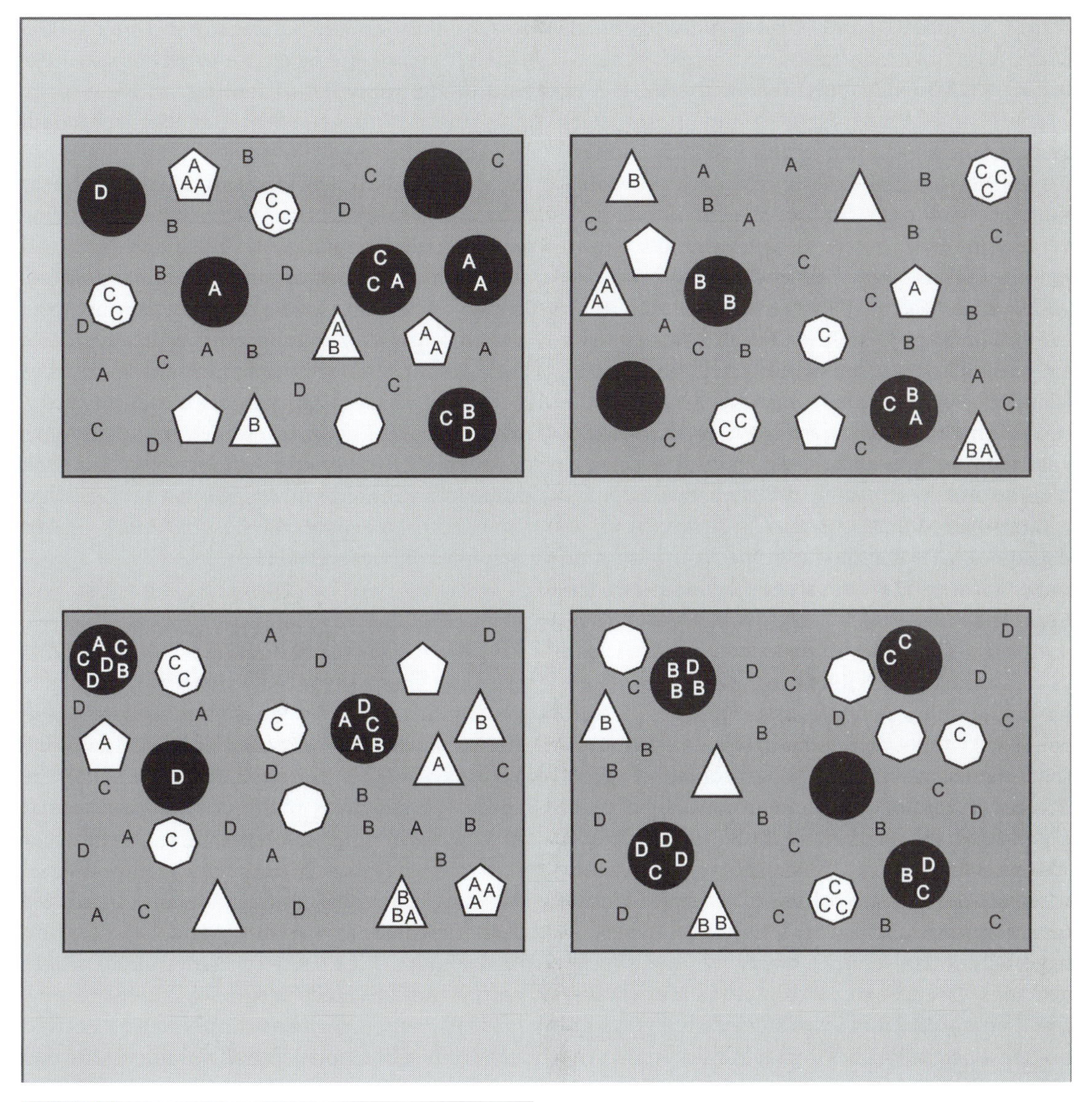

Fig. 14.1 A highly simplified supracommunity. The large square represents the supracommunity, the four rectangles represent the only four component communities comprising the supracommunity, and the circles, triangles, octagons, and pentagons represent infracommunities. Circles are definitive hosts whereas triangles, octagons, and pentagons are intermediate hosts. Letters refer to different species of parasites; those within the rectangle, but not within the infracommunities, represent free-living stages. Note that species D has a direct life cycle and therefore it is not found in any intermediate host infracommunities; it is absent from the upper right component community, perhaps an historical accident. Species A is absent in the component community in the lower right; note too that one of its intermediate hosts, the pentagon, is also absent.

These community-level terms provide us with categories for labeling our studies but, just as was true for studies on free-living communities and those on free-living, or parasite, populations, these too are fraught with problems of scale (both in a geographical, and in a taxonomic sense). Indeed, almost all community-level studies are conducted on some subset of the infracommunity. For example, it is not uncommon to see analyses of, say, the 'ectoparasites of . . .', the 'gastrointestinal parasites of . . .', the 'digeneans of . . .', or the 'macroparasites of . . .'. Depending on the nature of the question(s) asked, using a subset of the overall infracommunity is not, necessarily, a bad thing. In other words, comparative studies, perhaps comparing digenean species richness between different populations of the same host species in different areas, or different host species in the same area, can be quite insightful. Where caution must be advised, is when the question(s) is(are) mechanistic. For example, many studies, using some subset of the infracommunity, seek to explain the distribution of parasites, between, or within, hosts based on deterministic functions. We already know that the presence of a parasite can alter the host's physiology or immune status (Chapter 2) and, it is quite possible for a parasite, not within the subset under investigation, to influence the presence or absence (or, perhaps even distribution) of one, or more, of the parasites under investigation. Figure 14.2, is a diagrammatic representation of a *sample* (12 infracommunities) of some hypothetical component community. The investigation seeks to explain patterns of species' presence and absence in the 'gastrointestinal parasites'. Each black ellipse represents the entire host (and thus, the infracommunity) and each white rectangle represents the gastrointestinal tract within each host. An examination of all subsets of interest (the gastrointestinal tracts) of the infracommunites shows that species C and species D can co-occur with each other and with either species A or B. However, species A and species B never co-occur. This might lead one to suggest that the presence of one of these species precludes the presence of the other. Indeed, this could be the case. Considering all of the parasites in the host, additional possibilities arise. For example, species F, which lives outside of the gastrointestinal tract, can co-occur with any of the other species and species E, which also lives outside the gastrointestinal tract, can co-occur with all but species B. As above for species A and B, perhaps the presence of species E precludes the presence of species B, or vice versa. Whether considering just the 'subsets' or the entire infracommunities, there exists another possible (likely?) explanation. With such a small sample size, the 'pattern' observed may simply reflect an artifact due to sampling error! In other words, had more subsets or infracommunities been examined, all parasite species might be found concurrently in the same host individual.

One possible way to overcome some of the problems of analyzing only a subset of the infracommunity would be to consider guilds of parasites (Bush & Holmes, 1986*b*). This approach would be particularly useful when the analysis seeks to explain distributions of parasites. The rationale is that, if organisms interact (e.g., compete with one another), it is likely that any such interactions will occur between species that use similar resources in a similar fashion. Using this rationale, and considering Fig. 14.2, it is highly unlikely that species E or F would belong to the same guild as parasites A through D. This does not mean that the extra-enteric parasites could not influence the presence or absence of the enteric species, only that such influences would likely have to be mediated through host mechanisms (e.g., immunological or even physiological responses).

14.3 Describing and quantifying parasite communities

Studies of parasite communities mostly have one of two objectives, although these need not be mutually exclusive. On the one hand, are what we will call 'descriptive' studies and, on the other hand, are what we will call 'mechanistic' studies. The goals of the former are many and published studies on parasite communities mostly fall into this category. It is these types of studies where comparisons are made between different populations of hosts (i.e. presence/absence or derived data from all sampled infracommunites in one population are compared to presence/absence or derived data from all sampled infracommunities in one or more different populations), usually addressing some form of similarity (or dissimilarity) between different component or guild communities. These studies allow many questions to be addressed ranging from the importance of parasite/host phylogeny to assessing the impact of pollution; from providing baseline data on parasitism in a host of interest to assessing the impact of an introduced host or parasite; from evaluating the importance of host gender to evaluating the importance of host size. In short, given the large number of ideas that might be addressed, it is no wonder that most published

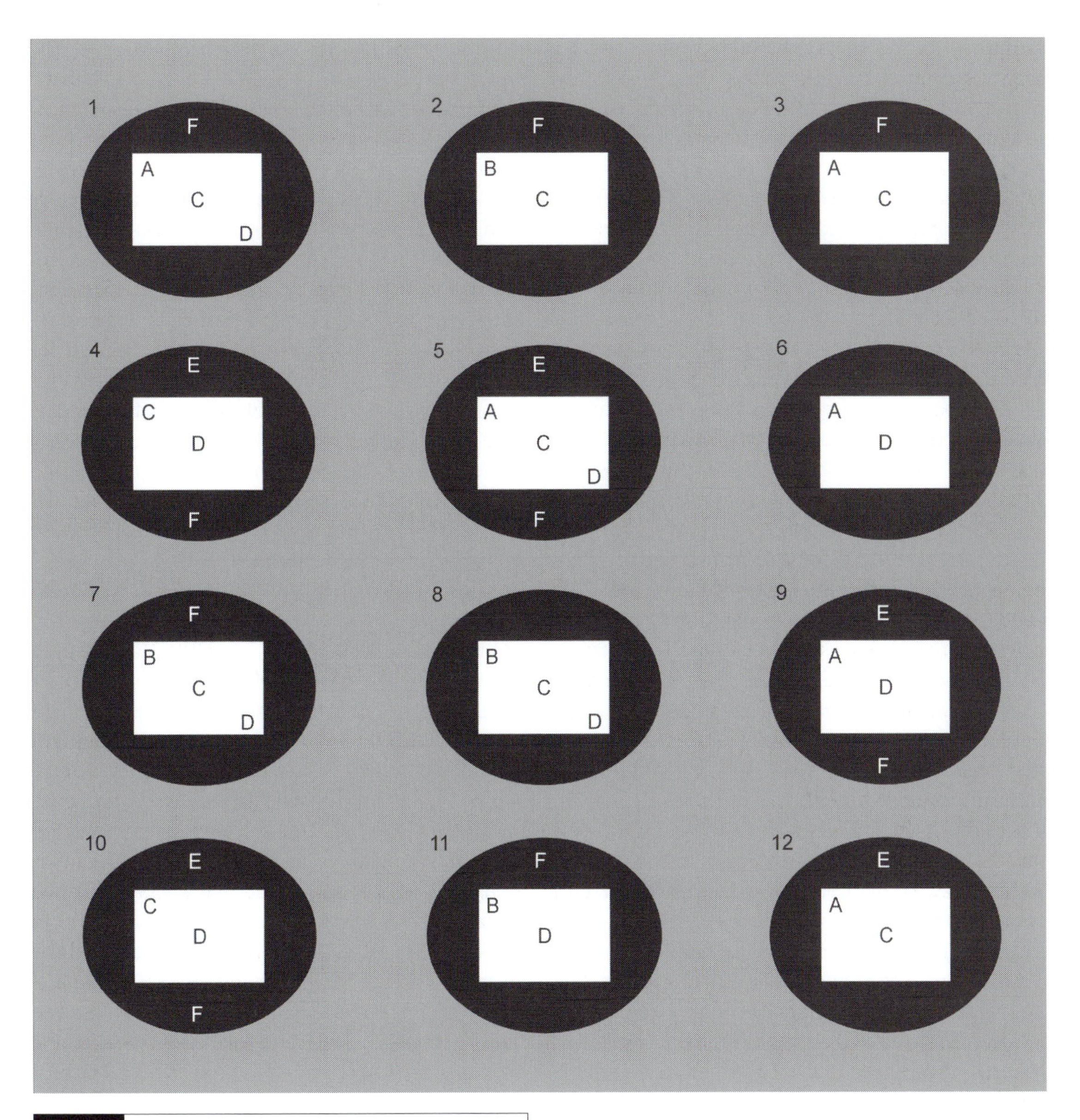

Fig. 14.2 A stylized component community (the large square) containing 12 infracommunities (ellipses). The white rectangles within the infracommunities represent the small intestine. Letters differentiate different species of parasites. Species A–D are found in the gastrointestinal tract whereas species E and F are found elsewhere in the body. Note that species C, D, and F appear to co-occur with any other parasite species. Species A and E can co-occur with each other and with any other species but neither co-occurs with species B.

community-level studies on parasites are descriptive.

Minimally, for a descriptive study, we need a list of the parasites present. Ideally, however, we would want to know a bit more than just presence/absence; as a minimum, some measure (and its variance) of the mean species richness and some measure (and its variance) of the mean abundance of each species. Although calculating species richness would seemingly be a simple task, it is not. The problem arises because, when the intent is to compare species richness between different samples, there is likely to be uneven sampling effort. In other words, more species might be found in one sample simply because more hosts were examined in that sample. Judging from the recent papers focusing solely on that question, e.g., Poulin (1998*a*), Walther & Morand (1998), Zelmer & Esch (1999), there is no clear consensus. Often, ecologists calculate a measure of species diversity which is a single datum incorporating not only the actual number of species but also a measure of their evenness in distribution. The intent is to weight those species having more individuals present. Debating the relative merits of diversity indices, estimators of species richness, and different measures of abundance, as they apply to parasite communities, is beyond the scope of this book and can be found elsewhere (Bush *et al.*, 1997). However, for any descriptive study, it is important that the analyses be appropriate to the questions asked. For example, if the goal was to compare individual infracommunities within some host population, it would be appropriate to analyze, say the variance in species richness (or diversity), between individual infracommunities. On the other hand, if the goal was to compare between component communities, it would be appropriate to analyze the variance in mean species richness (or diversity) between component communities. Recently, Rózsa *et al.* (2000) suggest that the minimum data required are the host sample size, the prevalence for each parasite, and, if at all possible, a frequency distribution for each parasite species. This latter feature obviates the need for deriving data (i.e., means, medians, geometric means), meaning that no information is lost. Unfortunately, if there were more than just a few parasite species, such a proposal would result in a very lengthy paper that few editors would likely accept. Above all else, Rózsa *et al.* (2000) advise that the biological interpretation of statistical descriptors and comparisons must be considered.

Mechanistic studies, as the name implies, focus on how communities are organized. It is in this type of study where comparisons are usually made between infracommunities, i.e., the infracommunities are used as replicates. The types of questions asked in these studies are fewer than for descriptive studies and generally focus on biotic variables, e.g., how the presence, absence, or *location* of one parasite population might influence the presence, absence, or *location* of one, or more, of a different parasite's population. It is in mechanistic studies that one can meaningfully ask whether biotic forces such as predation or competition (either intra- or inter-) structure communities. For the latter, we need to know, minimally, some measure of the distribution of each parasite's population with respect to some resource axis (the parasite's realized niche) and we need to know how much of that resource axis can be used by the parasite in the absence of other parasite species (the parasite's **fundamental niche**). Determining the fundamental niche is best accomplished experimentally by infecting an appropriate host with varying population sizes of a particular parasite and determining the range of the resource axis used. Often, however, such experimental infections are intractable and the fundamental niche must be inferred. Bush & Holmes (1986*b*) suggested that, by summing the realized niches of a parasite species from a sample of infracommunities, a measure of the fundamental niche could be derived (Fig. 14.3). In addition, where the resource axis is linear or radial, a measure of central tendency, such as the medians (and their variances), and a measure of the mean endpoints (and their variances) of the distribution are necessary. The ideal, of course, would be to present the actual distributions, either by way of figures or tables, for each parasite species in each host. But, just like the problem of providing a frequency histogram for each species of parasite in a descriptive study, providing such data would result in a very lengthy manuscript.

14.4 The kinds of parasites found in communities

Often, authors wish to focus attention on certain parasites that comprise the community. The goal seems mostly to identify parasites having some common attribute(s). Such attributes may include a range of diverse features such as common life-history patterns, common sites of occupation, common food resources, and so forth. In the fol-

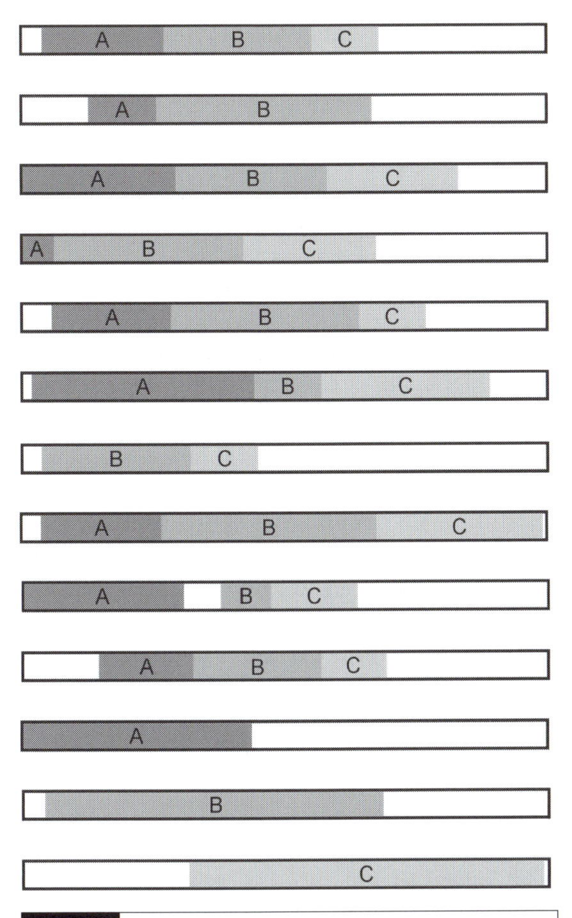

Fig. 14.3 Realized and fundamental niches of three parasites along a linear gradient in infracommunities. The first 10 infracommunities represent the actual (realized) observed distributions of each of three different parasites along a linear gradient. The bottom three 'artificial infracommunities' show the fundamental (potential) distributions of each species.

lowing subsections, we discuss some of the criteria, and the implications, used most frequently to try and create subsets of parasite species within the community milieu.

14.4.1 Generalists and specialists

Perhaps the most widely used descriptor of the kinds of parasites in a community is to classify them as being host generalists or as host specialists. Most commonly, the terms are applied to a single life-history stage. For example, it might be practical to consider a species, say a digenean, which is specific for a particular first intermediate host, a second intermediate host, or for a final host. It is not, however, beyond the realm of possibility to consider a species that might be a host specialist at virtually all life-history stages.

The implication for a host generalist is that it will do equally well, in terms of Darwinian fitness, in a variety of host species. By this, we mean that 'just about any host will do'. For parasites that have indirect life cycles, there do not appear to be too many parasites that are generalists in virtually all of the hosts required in their life cycle, but there are certainly some with very catholic host use at a particular stage during their life cycle. For example, the plerocercoid of the tapeworm *Diphyllobothrium mansonoides* can be found in a wide variety of amphibians, reptiles, and mammals. At the final host level, one can also find some very extreme examples of host generalists. As we noted in Chapter 11, the cestode *Bothriocephalus acheilognathi* has been reported to mature in >40 species of fishes; taking it even a step further is the digenean *Zygocotyle lunata* which is reported from a wide range of waterfowl species as well as several species of mammals.

The other side of the coin to the host generalist is the host specialist. As their name implies, host specialists will survive, and be able to perpetuate, in only one, or a very few, host species. Like host generalists, host specialists can be found at various points in the life cycle of a particular parasite. In the southeastern USA, the only intermediate host for the digenean *Catatropis johnstoni* is the snail *Cerithidea scalariformis*; along the Pacific Coast, a similar situation exists with the cognate species *Catatropis martini* and *Cerithidea californica*. The only known final hosts for *Taenia solium* and *Wuchereria bancrofti* are humans.

There are several caveats to note when considering the application of 'generalist' or 'specialist' to a parasite. First, a parasite might be considered a host specialist at the intermediate host level but a host generalist at the final host level (or vice versa). For example, in endemic areas, adult *Schistosoma japonicum* can be found in a wide variety of mammals; presumably, any mammal that comes into contact with water will suffice. In contrast, the miracidia of *S. japonicum* are highly specific. Molluscan hosts for *S. japonicum* can only be *Oncomelania hupensis*. Further, specificity may even extend to strains of intermediate hosts. In the Philippines, only certain strains of *Oncomelania hupensis* will serve as a gastropod host for *S. japonicum*. The potential importance of strains of parasites is demonstrated nicely by the work of Shostak *et al.* (1993) who examined metacercariae of *Zygocotyle lunata* from five different individuals of the snail *Helisoma trivolvis*. When mice were infected experimentally, the resulting adult worms exhibited different patterns of growth and survivorship attributable to the different sources of metacercariae. They conclude: 'variation associated with the intermediate host may confound studies of specificity to the definitive host'. Second, to identify a parasite as being a host generalist or host specialist means that we need to know much about the parasites in all potential hosts. In other words, not only is the sample size of the host species under investigation important, so too is the breadth of knowledge about what parasites are in other hosts. Third, and related, is that many times a parasite is designated as being a 'generalist' or 'specialist' based on an examination of host–parasite checklists. Rarely do such checklists provide information on the number of individual parasites found in a particular host species and, therefore, checklists equate common species with rare species. Further, they seldom tell us if the parasites are mature. The result from such a **meta-analysis** might indicate that a parasite is a generalist species because it was found in 25 different host species. However, on closer examination, the actual data used to compile the checklist may reveal that the parasite matured, and had a high prevalence and abundance, in only one of those hosts; in the remaining 24 hosts, it was very rare and did not mature. Seemingly, such a parasite would be a host specialist and, its presence in the other 24 host species represents accidental infections. Finally, just as we noted the vagaries for defining a community, where does one draw the line between a generalist and a specialist? Consider a parasite in a final host. Even given that original data are available (i.e., the distinction is not based on a meta-analysis), is a parasite that occurs in essentially equal numbers and is found to develop equally well in two host species a specialist or generalist? What about three species of final host? What about four species of final host?

Depending on the nature of the questions asked, one might also wish to distinguish between phylogenetic versus ecological specificity. In phylogenetic specificity, a particular parasite is likely to be found in a range of related host taxa, whereas, in ecological specificity, some abiotic, e.g., hosts live in the same type of habitat, or biotic, e.g., hosts eat the same type of food, variable might determine who parasitizes whom. It is probably true that phylogenetic specificity is the template upon which ecological specificity might be superimposed. The numbers of studies considering specificity at higher host taxonomic levels and addressing community-level ideas are few. Kennedy & Bush (1994) used such an approach to ask whether knowledge of parasites of a host in its heartland, and where it had been introduced, provide insight into community structure. They found that rainbow trout *Oncorhynchus mykiss* in its heartland had parasite communities composed predominantly (52%–72%) of salmonid specialists, with the remaining species in the communities being host generalists. As the distance from the heartland increased, i.e., where the trout had been introduced to other geographical regions, the proportion of specialists also decreased to the point that in Chile and New Zealand, areas with no native salmonids, the parasite communities were composed entirely of host generalists. Poulin (1997*a*) asked a related question, but more broadly. Is there a relationship between species richness in a community and host specificity? Based on an analysis of 116 fish species, he concluded that rich parasite communities, i.e., communities with many different species of parasites, were composed of both host specialists and host generalists. In contrast, species-poor communities were composed mostly of host generalists.

We conclude our consideration of host specificity with an evolutionary question of sorts. What drives parasites to specialize? Stated differently, what limits the host range of a parasite? First, consideration must be given to phylogeny. An oft-cited mechanism accounting for specialists among parasites is that of descent. Here, the argument is that current-day host specialists are derived from ancestors that, for whatever reason, parasitized a particular host species or host group. Therefore, the answer to our question is simple – history. However, parasites evolve. In other words, they can, and do, change. One purported advantage (perhaps the only one?) is the notion that specialists are very good at what they do; the implication being that the specialist could 'win' over a generalist. Presumably, 'winning' means that the specialist is better adapted, perhaps a better colonizer, or competitor and, should resources (hosts in this case) be limiting, the specialist would survive to the detriment of the generalist. For parasites, at least, we have no convincing evidence if this is true or not. Consider, however, the disadvantages of specialization. The fate of any parasite requiring a single, specific host at any stage in its life history is inextricably linked to the fate of that host. (We already know that when the passenger pigeon *Ectopistes migratorius* went extinct, two species of highly specific bird lice went extinct also [Stork & Lyal, 1993]. What then of any host specialists of dodos and dinosaurs?)

Evolutionary ecology provides for two potential reasons why parasites might specialize. First, a parasite might be restricted to a single host species simply because it does not come into contact with other hosts. Second, a parasite might come into contact with other hosts, but colonizing such a host would be maladaptive to the parasite. Recently, Tompkins & Clayton (1999) devised a rather nifty experiment to test the maladaptive hypothesis using the parasitic chewing lice on swiftlets. First, through a very large-scale survey of several species of swiftlets and their chewing lice, they noted that some lice were mostly restricted to one host species, whereas other lice species could be found on several species of swiftlets. Importantly, they noted that host-specific lice were found, occasionally, on other swiftlet species. This, and the related observation that some species of chewing lice were generalists on up to three species of swiftlet, would seem to argue against the suggestion that specificity is due to the lice not coming into contact with alternative hosts. To address whether the lice might be specialists on a particular host species because it is maladaptive, they removed lice and placed them on novel hosts. They compared survivorship of these lice on novel hosts with survivorship of lice that had transplanted to different individuals of the normal host. They found that survivorship of the lice on novel host species was very poor.

14.4.2 Site specialists

Once parasites have reached an appropriate host, they have two options. First, if they enter a host by being ingested, they may develop at the site where they are released, e.g., where they excyst, exsheath, or hatch. If they penetrate directly, once inside the host epithelium, they can develop. If they are ectoparasites, they can select where, on a host, they will attach. Second, the parasites, once on, or inside, the host, may migrate to specific sites far removed from where they attach initially, or where they excyst, exsheath, or hatch. This latter scenario, which is most common, suggests that parasites show a degree of specialization that goes far beyond the entire host. For example, as we note in Box 1.1, Galen, long ago, first alluded to site selection of two nematodes in the gut of humans. (Probably, this was also the first observation of 'resource partitioning'.) Findings that some parasites are always found in the same organ, the same site within an organ, the same cells, or the same sites on a host, are legendary. Indeed, such observations are so common today they almost seem trivial. Despite the common reference to site selection that one now finds in the literature about parasite populations and communities, several issues about those observations are relevant.

Adamson & Caira (1994) state, emphatically, 'Parasites are first and foremost specific to microhabitat.' We do not know if this is true, but site specificity often transcends host specificity. What we mean by this is that a given parasite might be found in a wide spectrum of host species (i.e., it is a host generalist) yet it is always found in the same organ and, perhaps, even at the same site within that organ. For example, although the generalist *Zygocotyle lunata* can be found in such diverse hosts as waterfowl and mammals, it is always found in the cecae or cecum.

Most parasitic protozoa are highly specific for cells or tissues within their hosts. For example, *Plasmodium* spp. are found in a wide variety of vertebrate hosts yet they are characteristically found in the erythrocytes or hepatic cells; *Leishmania* spp. also are found in a wide variety of vertebrates but, within the vertebrate, they are restricted to macrophages. Many ectoparasites, such as lice on birds or monogeneans on fishes, are highly specific; some of the former are specific for certain feather barb diameters or feather types, some of the latter are restricted to a particular gill arch. Cestodes and acanthocephalans are parasites of the gut tube. Although some may be found in organs with very disjunct boundaries (e.g., *Gastrotaenia cygni* under the gizzard lining of waterfowl, *Corynosoma cetaceum* in the stomach of dolphins), most are found in the intestinal tract. Although we can recognize broad regions within the small intestines of many vertebrates, e.g., duodenum, jejunum, and ileum, they still appear to be rather homogenous units. Apparently, however, there are clear differences since most cestodes and acanthocephalans occupy very discrete sites even within these broad subdivisions of the small intestine. Most digeneans and nematodes exhibit more catholic use of the vertebrate body. It is true that many use the gut tube, and these may show high site fidelity, extending even to ontogenetic habitat shifts (Aznar *et al.*, 1997) within the gut, but it is not uncommon to find them in other organs such as eyes, nasal fossa, muscle cells, the pancreas, the liver, the lungs, the epidermis, etc. Still, whether in or out of the gut tube, most of these parasites show a preference for certain organs, or parts of organs (Fig. 14.4). In short, most parasites are highly predictable in their distribution within infracommunities.

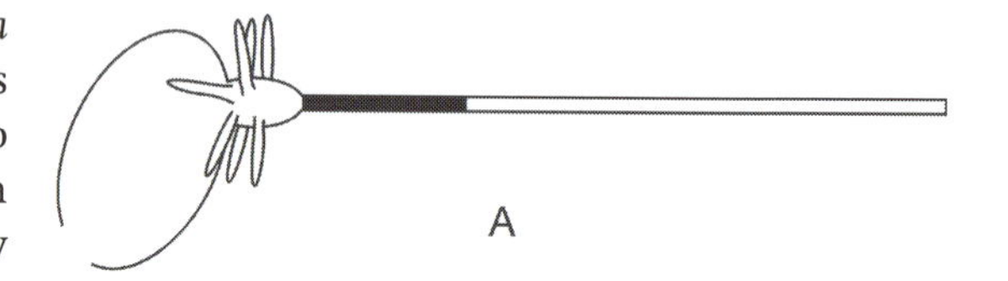

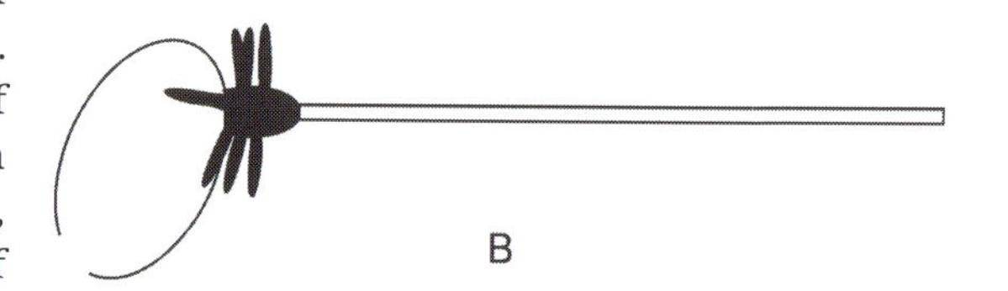

Fig. 14.4 An example of an ontogenetic shift in the site of a parasite showing a change from the anterior part of the intestinal tract (A) to the pyloric ceca (B). (Based on ideas presented by Richardson, D. J. & Nickol, B. B. [1999] Physiological attributes of the pyloric ceca and anterior intestine of green sunfish (*Lepomis cyanellus*) potentially influencing microhabitat specificity of *Leptorhynchoides thecatus* (Acanthocephala). *Comparative Biochemistry and Physiology Part A*, **122**, 375–384. With permission from Elsevier Science.)

Just as we could ask 'why be host specific?', so too can we ask 'why be site specific'? Perhaps the best answer is that, being adapted to a specific site should increase the fitness of the parasite in a particular site over the parasite's fitness in some other site. Clearly, being organ specific, or site specific, should be adaptive. For example, a number of nematodes are ingested while their herbivorous host grazes. These nematodes are ingested as infective 3rd stage larvae and are protected by the cuticle of the former 2nd stage. To develop further, these larvae must exsheath and, often, there are very specific requirements necessary to stimulate the exsheathment process. For intestinal dwelling nematodes, the specific requirements necessary to stimulate exsheathment are often found only after the sheathed larvae have passed from the stomach into some region of the small intestine. Perhaps the exsheathment-specific requirements ensure that the sheath-protected parasite passes through what is perceived to be the harsh environment of the stomach, until it reaches the more benign conditions of the small intestine. The importance of specific cues that parasites use has been demonstrated well by Sukhdeo & Croll (1981). These authors examined the distribution of the nematode *Heligmosomoides polygyrus* along the small intestine of its rat final host in which the bile duct had been surgically relocated. They found that the incoming larvae tend to establish in that region of the intestine where bile enters (Fig. 14.5). Richardson & Nickol (1999) provide very detailed information on the physiological attributes of the anterior intestine and pyloric cecae in green sunfish *Lepomis cyanellus* with respect to site specificity of the acanthocephalan *Leptorhynchoides thecatus*. They note that there exist 'subtle spatial and temporal complexities intrinsic to the environment of helminths inhabiting the teleost enteric system'. Perhaps it is these 'complexities', and pos-

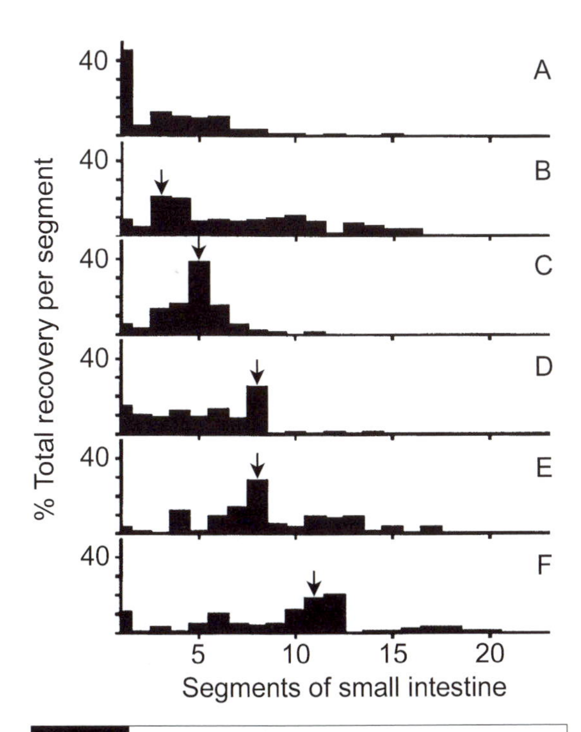

Fig. 14.5 The distribution of larvae of *Heligmosomoides polygyrus* in the small intestine of rats following the surgical re-location of the bile duct. The distributions are 6 days after intragastric infection with 300 larvae. (A) represents a sham-operated control. In (B)–(F), the arrows represent the new location of the entry of bile. (Modified from Sukhdeo, M. V. K. & Sukhdeo, S. C. [1994] Optimal habitat selection by helminths within the host environment. *Parasitology* (Supplement), **109**, S41–S55. Reprinted with the permission of Cambridge University Press.)

sibly their predictability, upon which parasites can cue and recognize an appropriate site. Sukhdeo & Sukhdeo (1994; see also Adamson & Caira, 1994) elaborate on site and organ specificity by parasites with three important observations. First, hosts, at least internally, tend to be very predictable, obviating the need for an elaborate 'orientation' when a parasite enters an appropriate host. In other words, parasites do not need to evaluate the appropriate habitat; instead, they need only identify the habitat. Second, as the habitat is predictable, so too are the behaviors of the parasites and these behavior patterns ensure that the parasites maintain their site specificity. Finally, Sukhdeo & Sukhdeo (1994) go on to suggest that migrating to a site, or organ, and maintaining their positions at those sites, may become a fixed behavior of parasites during the course of evolution. Such observations demonstrate clearly that site specificity is not an accident; they provide information on how some parasites might recognize appropriate sites, but they only hint as to what selective pressure(s) might have led to such specificity.

14.4.3 Core and satellite species

Caswell (1978) suggested 'Perhaps a community consists of a core of dominant species, which interact strongly enough among themselves to arrive at an equilibrium, surrounded by a larger set of non-equilibrium species playing their roles against the background of the equilibrium species.' In 1982, Hanski proposed the core/satellite species dichotomy, attempting to explain mechanisms that might produce two empirical patterns of species distributions. Those patterns are a positive relationship between distribution and abundance and a bimodal distribution of species. We should note, at this point, that the core/satellite species dichotomy is somewhat controversial (Hanski, 1991, Nee *et al.*, 1991) and other mechanisms can produce similar patterns. Poulin (1998*b*) provides a further caution when the dichotomy is applied specifically to parasite communities. He notes that a species considered to be a satellite species with respect to abundance may be a core species with respect to biomass. Indeed, Bush (1990) noted that too few authors consider the importance of biomass when assessing species of parasites in infracommunities. However, depending on the questions asked, providing that alternative mechanisms are considered, and heeding the cautionary notes of Bush and Poulin, it seems a reasonable way to label, and thus focus attention on, particular subsets of species in a community.

Applied to parasites, the implication of the dichotomy is as follows. If there is an inverse relationship between the abundance of a parasite species and the probability of that species' extinction or if there is random variation in the transmission or extinction, or both, then each parasite in a community will tend towards one of two states. Core species are regionally common and locally abundant, i.e., they colonize most host individuals and are found in high numbers. Satellite species, in contrast, are regionally uncommon and locally rare, i.e., they colonize few host individuals and are found in low numbers. Bush & Holmes (1986*a*) applied Caswell's and Hanski's dichotomy to the almost 1 million parasites, representing 52 species, found in 45 lesser scaup ducks *Aythya affinis*. They used prevalence to represent regional dispersion and intensity to represent abundance. They tested explicitly for a positive correlation between prevalence and intensity and examined the modes of distribution before assigning species to categories. Based on these criteria, they identified a trimodal, not bimodal, pattern. Core species were few, eight of 52 species, and abundant, 91.7% of all individual parasites. What they identified as a second mode was termed 'secondary species'; these were also few, eight of 52 species, and were not particularly abundant, 7.3% of all individual parasites. Finally, they noted that most species (satellite species) were many, 36 of 52 species, but rare, 1% of all individual parasites. They then used these categories to explore mechanistic questions. Others have also used the Caswell–Hanski dichotomy profitably as a first step in data exploration, i.e., using it to identify predictably co-occurring species and not to infer mechanism. Unfortunately, far too many authors use the terms 'core' and 'satellite' to reflect, respectively, high or low prevalence. It is also true that differentiating between core and satellite species can be arbitrary, particularly if there is no clear separation between modes on a frequency histogram.

14.4.4 Guilds of parasites

We have already used the term guild as it might apply to communities and, to some extent, its application to groups of parasites within a community. Here, we emphasize the use of guilds as ways of partitioning data into what may be more meaningful subsets. Guilds separate those species in a community that exploit resources in a similar fashion (Root, 1967, 1973). It is important to note that taxonomy is irrelevant to membership in a guild. The guild concept is applied frequently to the study of putative interspecific competitive interactions between parasites and to predation between larval digeneans in their molluscan hosts. Both of these antagonisms have been discussed previously and here we only elaborate on how guild membership might be determined. Depending on the nature of the questions asked, a particular species could be viewed as a member of one type of guild on the one hand, or a member of a different type of guild, on the other. For example, consider some hypothetical gut community in a vertebrate comprised of two nematodes, two cestodes, and an acanthocephalan, all of which are similar in size. If the questions address the use of space along some linear axis, the guild might be considered a single enteric guild and would include all five species. If the questions address acquisition of nutrients, two guilds would be meaningful: an engulfing guild (the two nematode species) and an absorbing guild (the two cestodes and the acanthocephalan). As a last example, if the questions address niche restriction based on mating processes, two guilds would again be meaningful: a gonochorist guild (the two nematode species and the acanthocephalan) and a monoecious guild (the two cestodes). A final comment on the use of guilds is that the term 'community' is often used to describe a group of species that would more meaningfully be described as a guild or infraguild (Bush *et al.*, 1997).

14.4.5 Other 'kinds' of parasites

Sometimes, it is instructive to consider parasites based on how they colonize hosts or how they maintain themselves within hosts. Esch *et al.* (1988) proposed the terms **autogenic** and **allogenic** species as part of an effort to examine and explain the helminth colonization and transmission patterns found in freshwater fishes in Great Britain. One of the important observations made by these authors was that the colonization and transmission abilities of allogenic species were much greater than those of autogenic species. Allogenic species mature in birds, or mammals, or both. In contrast, autogenic species mature in fishes. Since birds and mammals are unlikely to be confined to a particular body of water, as is often the case with freshwater fishes, it follows that allogenic parasites have a much greater potential for dispersal than do autogenic parasites.

Parasites in three selected groups (**anadromous**, **catadromous**, and **stenohaline**) of British fishes were characterized according to the allogenic/autogenic dichotomy (Esch *et al.*, 1988). The first included the salmonids, which were dominated by autogenic species. These autogenic species also accounted for a substantial portion of the similarity within and between the various study locations. In contrast, cyprinids were dominated by allogenic species; this also accounted for much of the similarity within, and between, collecting areas. Anguillids were intermediate, with neither allogenic nor autogenic species showing clear dominance. They conclude by stating, 'our conclusions relating to the different contributions of autogenic and allogenic species to community structure in the three groups of fish hosts are novel and unexpected, and are not simply logical consequences of the dispersal abilities of the helminths themselves. An understanding of colonization strategies, therefore, including the separation of autogenic and allogenic species and recognition of the different roles of both transient/resident and **euryhaline**/stenohaline host species, provides important clues for evaluating the stochastic nature of parasite community structure.'

It is common to find, in a list of parasites reported from one or more host species, the notation that some parasites are 'accidentals'. The use of the 'accidental' label takes two forms. The first is when a parasite does not mature in the host(s) of interest and the second is when the parasite is rare, at least relative to other species in the community. As the name suggests, the implication is, of course, that accidental parasites do not belong to the community under investigation; it further implies they are unimportant. The implication that a parasite failing to mature in a host is therefore unimportant can be very misleading. It is most definitely true that a parasite failing to mature in a particular parasite community cannot respond, in an evolutionary sense, to the presence of any other species in that community. Does that mean they are unimportant? We think the answer is no and our reason is that the other parasites in the community may have to respond to the presence of the 'accidental'. For example, Bush & Holmes (1986*a*, *b*), in their study of the parasite communities in lesser scaup ducks, found an undescribed species of hymenolepidid cestode to be frequent (>50% of all hosts infected) and to have high mean intensities ($\bar{x} = 2194$, S.D. $= 3692$, range $= 27$–17375). A total of 52 species of parasite was found in all hosts examined and this hymenolepidid ranked 9th in prevalence and 4th in mean abundance of the almost 1 million parasites found. Is this an accident? As we note above, this parasite (which was found to mature in redhead ducks *Aythya americana*) cannot evolve to the presence of sympatric species. Nonetheless, it has clearly solved any colonization problems associated with infecting lesser scaup. Furthermore, it had a very specific distribution along the linear aspect of the small intestine. Although this cestode species cannot evolve in lesser scaup, there is no reason to believe that the other parasites along the linear aspect of the small intestine do not have to respond to its presence.

Simply because of the connotation of the word 'accidental', its use as a synonym for rarity is also unfortunate and may be misleading. Often, when authors use accidental as a synonym for rarity, their goal is to reduce 'noise' in their data. However, the notion that rarity of a parasite in a community is an accident has no implicit logic. Rarity can take several forms. Rare parasites can be infrequent, i.e., they have low prevalences in the sampled communities; they can have low intensities, i.e., there are few individuals in the sampled communities; or they can exhibit both patterns. None of these reasons seem to be logical grounds for considering the parasites as accidents. Indeed, there are many parasites that are rare (using any connotation of the word; just why they are rare, and how they circumvent extinction, would be an interesting avenue to explore!).

14.5 Niche restriction in parasite communities

Rohde (1994) noted 'There is no 'universal' parasite which infects all hosts and all sites on the hosts equally, occurs in all macrohabitats and geographical regions, in all seasons, *etc.*' Point well taken! Indeed, all of the organisms to which we ascribe a parasitic life style exhibit various degrees of specialization. And, as we note above, no form of specialization is more apparent than specialization to a specific site in, or on, a host organism. Since the site occupied by a parasite will represent at least some portion of that parasite's niche, it seems perfectly legitimate to ask 'What mechanism might we invoke to explain niche restriction (site selection) in parasites?' Although there might be a number of answers, they mostly fall into one of five categories: phylogeny, mating, adaptation, predation, or competition.

14.5.1 Niche restriction by descent

Phylogenetic niche restriction (Brooks, 1980) asserts that a parasite's niche in contemporary time might simply reflect descent. In other words, the parasite's present-day niche is simply a reflection of its ancestor's niche in an ancestral host species. An hypothetical example of this scenario is depicted in Fig. 14.6A. Here, the distributions of the descendent parasites in the descendent hosts simply mirror the distribution of the ancestral parasite in the ancestral host. This provides us with no information since it begs the question: 'What force, or forces, restricted the niche in the ancestral host in the first place?'. Recalling the ideas of Sukhdeo & Sukhdeo (1994), this situation provides the parasites with no means of orientation other than 'history'. A more plausible explanation would be that seen in Fig. 14.6B. Here, the only difference is that some 'cue', indicated by the arrow pointing to the resource gradient, provides a focal point for the establishment and subsequent development of the parasites. In other words, niche restriction does not represent simple descent; instead it is mediated by orientation to some morphological or physiological parameter. In Fig. 14.6C, we extend this hypothetical example to suggest that the precise location of the parameter may differ in descendants of the ancestral host and thus there will be a niche shift by the parasite in that descendent lineage.

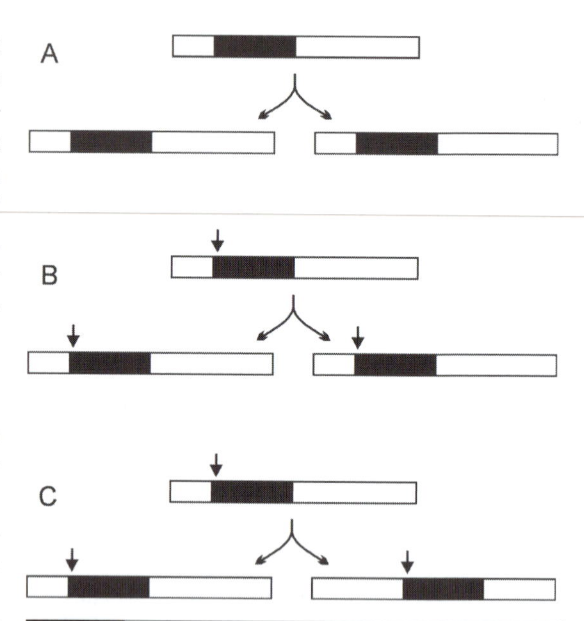

Fig. 14.6 Niche restriction by descent. In (A), the descendent parasites mirror the ancestral niche position in the descendent species. In (B), the descendent parasites mirror the ancestral niche position in the descendent species but here, the arrow indicates the location (perhaps the entry of the bile duct? – see Fig. 14.5) of a cue. In (C), the niche position of the descendent parasite in one descendent daughter host differs from that in the ancestral host but it seems to track the same cue (which has changed position in the daughter host).

14.5.2 Niche restriction for mating

The notion that niche restriction might be a way to reinforce reproductive isolation was first proposed by Sogandares-Bernal (1959) and, subsequently, by Martin (1969). This basic idea has since been extended to include the suggestion by Rohde (e.g., 1979; 1982; 1994; and references therein) that niche restriction might also enhance, or facilitate, sexual reproduction to account for the pattern of niche exploitation seen mostly in monogeneans on fishes. Rohde has based his argument that selection has favored niche restriction to facilitate mating on the following evidence: (1) narrow host ranges and microhabitats lead to increased intraspecific contact, (2) adult stages often have fewer hosts and narrower microhabitats than sexually immature and larval stages, (3) microhabitats of sessile and rare species often are narrower than those of more motile and common species, and (4) microhabitats of some species can be shown to become more restricted at the time of mating. In most of his work, Rohde suggests the mating hypothesis as a contrast to competition as the mechanism promoting niche restriction. He indicates that invoking competition as a mechanism to explain current niche restriction relies on the evolutionary argument that parasites competed for a site in the past and have now sorted things out (Rohde, 1979, 1982). A second suggestion is that many parasite communities are unsaturated, that is, they have fewer species than they could conceivably have and, therefore, competition cannot be important as an organizing force. We see no reason why competition cannot occur between two niche-specific parasite species, regardless of how much unoccupied space remains in, or on, the host. In fact, when we discuss competition below, we will provide examples of competition occurring when communities are unsaturated. Adamson & Caira (1994) refute many of Rohde's assumptions, suggesting that the same patterns could instead be produced by alternative mechanisms; they note also that aggregated patterns seen frequently in parasites may obviate any problems associated with finding a mate. Despite these caveats, the mating hypothesis is one of several mechanisms that might restrict niches in some species.

14.5.3 Niche restriction by adaptation

Many parasites show specific adaptations, often morphological, to their niche. This led Price (1980) to suggest that specialization for one niche precludes occupation of any other niches. Williams *et al.* (1970) noted that the bothridia of the cestode *Echeneibothrium maculatum* were perfectly adapted to the mucosal folds found in the spiral valve of the ray *Raja montagui*, but not to the mucosal folds in the spiral valves of related rays. This means that *E. maculatum* is not only host-specific to *R. montagui*, but that it is also specific to certain regions of the spiral valve. Rohde (1982) referred to this morphological adaptation as a 'lock-and-key'. The idea behind morphological adaptations is that, once a parasite has become specialized, it will no longer 'fit' into, or on, another niche. Adaptations to physiological parameters are also known, and would include such things as the cues required by parasites to establish discussed previously under site specialists. Adamson & Caira (1994) state that adaptations, whether morphological or physiological, are best viewed as a result, rather than a determinant, of niche restriction. Nonetheless, once such an adaptation has occurred during the evolutionary history of a parasite, there is no reason to believe that it cannot be inherited.

14.5.4 Niche restriction due to predation

Predation is rare on, or by, animal parasites. It occurs in what are called 'cleaning symbioses' and it occurs in the larval, intramolluscan stages of some digeneans. A systematically diverse group of organisms ranging from invertebrates to vertebrates are known to prey, selectively, on ectoparasites. Cleaning symbioses are known to exist where the cleaner is an invertebrate preying on ectoparasites of other invertebrates or vertebrates and where the cleaner is a vertebrate preying on ectoparasites of invertebrates or other vertebrates. There is little information on what impact cleaners might have on site specificity by ectoparasites. There are, however, some suggestions that it might be important. Potts (1973) noted that the cleaner *Crenilabrus melops* cleaned the pelvic fin region and cloaca of fishes >50% of the time, the sides of the bodies 17% of the time and the head region about 10% of the time. Clearly, ectoparasites that avoided the pelvic fin/anal region in this system should increase their fitness. Although Rohde (1993) concludes that there is little evidence for population control of parasites by cleaners (but see Grutter, 1996 for arguments to the contrary) he suggests 'Hence, the possibility exists that certain parasite species, as a result of selection processes to avoid predation by cleaners, have become adapted to certain protected habitats on the host.' (Rohde, 1994).

The only other documented form of predation within parasites involves the predation, by rediae of some digeneans, on sporocysts of other digeneans, within the molluscan host. Intramolluscan interactions among digeneans in snails have been observed for many years, e.g., Kuris (1990), Sousa (1990, 1993, 1994), Kuris & Lafferty (1994), Curtis (1995, 1997), Jokela & Lively (1995), Curtis & Tanner (1999), and references therein. Interactions have also been found in the digeneans of clams (Rantanen *et al.*, 1998). From these studies, two dominant themes have emerged. The first is that multiple species (double and triple) infections in most snails occur at a rate that is significantly less than would be expected by chance alone. The second relates to the explanation for the depauperate infracommunities in molluscs. Some studies indicate that antagonistic interactions involving larval digeneans in snails include predation, and possibly competition, but other investigations have led to the conclusion that interactions are not important factors in terms of niche restriction in intramolluscan infracommunities.

Most of the information regarding intramolluscan antagonism in natural snail populations comes from studies on one host–parasite system (see reviews by Kuris, 1990; Kuris & Lafferty, 1994; Sousa, 1990, 1993, 1994). These investigations mainly have focused on the parasite faunas of the salt-marsh snail *Cerithidea californica* in two geographically separate locations on the coast of California (USA).

Kuris' (1990) analysis was based on collections from Upper Newport Bay, California; 12 995 snails (all older than 2 years of age) were collected and checked for shedding cercariae. Seventeen digenean species were identified. The focus of his effort was a comparison of expected and observed frequencies of double infections among digeneans with different combinations of intramolluscan larval stages (redia–redia, redia–sporocyst, sporocyst–sporocyst). Intramolluscan development begins with either the penetration of a free-swimming miracidium or the ingestion of an egg. By far, the majority of species use the first method. After gaining access to the snail, the larva migrates to a specific organ or tissue where it develops into a sporocyst; morphologically, the sporocyst is little more than a shapeless sac. In some species, sporocysts then give rise to rediae, all of which have a mouth and a primitive gut; this larval stage is able to ingest host tissue as well as prey on the larvae of other digenean species. (An interesting footnote here is that there is no evidence of cannibalism among the redial stages, suggesting that digenean rediae have evolved mechanisms of self-recognition that must be rather sophisticated.) Some species do not produce rediae, remaining in the snail as sporocysts. Niche restriction within the snail is pronounced, at least at the beginning of the infection. Later, when space or nutrients become limited, rediae, or sporocysts, or both, may spill over into adjacent tissues.

The various combinations of two-species digenean infections that occurred in *C. californica* were identified and compared with the type of developmental stage that occurred in each species (Kuris, 1990). The aim of this analysis was to determine if either competition or predation was a structuring force at the infracommunity (infraguild) level. In order for antagonistic interaction to be a significant determinant in organizing the infraguilds, Kuris contended that three criteria must be met: (1) prevalence within the parasite community must be high, (2) a dominance hierarchy must exist, and (3) interference competition should be more important than exploitation competition. All three criteria were satisfied for the digeneans in the salt-marsh snail. The process of structuring the infracommunities occurred through intense interspecific interaction such that death or complete suppression of inferior competitors was the outcome. Kuris (1990) concluded that the infraguilds were strongly influenced by antagonistic interactions, so much so that he was able to construct an elaborate dominance hierarchy for many of the species in *C. californica*. On the other hand, among most other digenean communities, antagonism is unlikely to affect infraguild interactions because parasite prevalences are generally too low. When antagonism does occur, it causes site displacement, total exclusion, or crowding effects such as size reduction, delayed maturity, or reduced fecundity of subordinate species (Kuris, 1990). Lie *et al.* (1966) reported similar guild structures in the freshwater snail *Lymnaea rubiginosa*. In that assemblage, however, dominance was established through priority of occupancy, conferring a strong temporal component to the processes that organize digenean infraguilds.

Sousa (1990) described the results of a 7-year study on the infraguild structures in *C. californica* at two locations (different from Kuris', 1990 location) on the California coast. From a total of 4462 snails examined, he identified 15 species of digeneans. Sousa (1990, 1993) made three important observations with respect to negative antagonistic interactions for the infraguilds in his investigation. First, double infections were rare, occurring with much less frequency than would be expected by chance alone. Second, direct evidence was presented for the existence of hierarchical antagonism among co-occurring species. Third, he recorded hierarchical species replacement within a temporal framework. There were, therefore, antagonistic interactions that affected infraguilds in the snail populations he studied, confirming some of the observations and conclusions of Kuris (1990).

Interestingly, Bush *et al.* (1993) screened a number of *C. scalariformis* (an eastern cognate of *C. californica* and with many cognate parasites species) from Cedar Key, Florida (USA), for cercarial shedding and found the number of multiple infections to be much higher (>16%) than that reported in either California location of *C. californica* (3% reported by Kuris, 1990; 2% reported by Sousa, 1990). They also examined *C. scalariformis* from several other locations in Florida and found no mixed infections; indeed, one population of snails was completely uninfected. Why these differences occur is unclear. It may be related simply to the samples; Bush *et al.* (1993) examined only about 100 snails from each location and on a single date.

14.5.5 Niche restriction due to competition

We have discussed competition, at least in part, in several earlier chapters and sections. We have noted that organisms can interact intra- and interspecifically (and competition is definitely an 'interaction'), that competition can be exploitative (in which the interaction is mediated by utilization of common resources) or interference (in which the interaction takes the form of direct confrontation), and that the result of present-day competition can be niche restriction or exclusion. Competition, in any form, is most definitely the 'black sheep' in the family of interactions that might lead to niche restriction. Indeed, most who study niche restriction contrast their proposed mechanism(s) with competition and then go on to inveigh that competition is unimportant and rare as an organizing force. Actually, we are pleased to note that competition is alive and well in parasite infracommunities! Although competition is notoriously difficult to quantify, it can be demonstrated by experimental manipulation and it can be inferred from rigorous field studies.

The best-known and most often demonstrated form of competition is intraspecific, exploitative competition. Commonly called the 'crowding effect', when there are too many parasites, packed into a limited, finite area, the result is often stunting, reduced fecundity, or both. Experimentally, this form of competition has been demonstrated many times (Bush & Lotz, 2000). It has also been inferred from field data (e.g., Goater & Bush, 1988).

Almost as common are demonstrations of interspecific, exploitative competition. We have discussed already (see section 11.2.4), the seminal papers by Holmes (1961, 1962*a*) who was perhaps the first to demonstrate competition between two phyla of animals and who (1962*b*) demonstrated that space was not a limiting factor. Others (e.g., Lang, 1967; Courtney & Forrester, 1973) have subsequently demonstrated experimentally interspecific, exploitative competition between members of different phyla. Patrick (1991), using manipulative studies, demonstrated ecological release by one nematode parasite when interspecific competitors were removed. Initially, he examined the occurrence and distribution of helminth parasites in the intestine of naturally infected flying squirrels *Glaucomys volans*. He found three species of nematode, *Strongyloides robustus, Capillaria americana*, and *Citellinema bifurcatum*, to co-occur in the anterior 30% of the small intestine of infected squirrels. He then live-trapped squirrels, dosed them with an anthelmintic to remove all nematodes, and subsequently reinfected the squirrels with *S. robustus*. In the absence of competitors, *S. robustus* expanded its niche significantly. This evidence, plus the following, (1) as the population size increased, the mean linear range of the parasite increased, and (2) in natural infections, there was no overlap in the distribution between parasite species, led Patrick to conclude that there was strong evidence for competition between *S. robustus* and the other two nematodes. Since 70% of the gut was unoccupied, his study also negates the notion that 'empty space' in a community means that competition does not occur.

There are many studies that infer interspecific, exploitative competition based on data acquired from field studies. By their very nature, these can neither be as demonstrative nor unequivocal as experimental studies. However, when appropriate data are analyzed rigorously, they can provide prima facie evidence that competition can play a role in restricting some aspect of a parasite's niche. Most often, the data used to infer interspecific, exploitative competition in field studies revolve around the assumption that the gut of a host is a linear, or radial, gradient and that it represents a meaningful niche axis for a parasite. Such studies usually demonstrate that the distributions of the parasites differ from random, e.g., distributions are tested against a null model assuming random distributions as in Bush & Holmes (1983), that the parasites are concordant in their distribution across infracommunities, that the parasites co-occur frequently, and that the realized niche of a parasite is reduced significantly from that parasite's fundamental niche. The assumption of this latter observation is that, in the absence of competitors, a parasite's linear, or radial, distribution would equal its fundamental distribution, i.e., ecological release would occur, as has been demonstrated in experimental studies. This latter observation is paramount because it allows competition to be differentiated from other plausible mechanisms restricting niches such as mating.

One of the most detailed studies invoking interspecific exploitative competition from field data is that of Bush & Holmes (1986*b*). As noted earlier (section 14.4.3) they used the core/satellite hypothesis to screen, initially, those parasites that were regionally common and locally abundant. In that analysis, they noted a trimodal, rather than bimodal pattern and considered the intermediate mode to represent what they called 'secondary species'. The core species and secondary species, collectively, occupied all parts of the small intestine and exhibited an even dispersion along the small intestine (Fig. 14.7). Note from Fig. 14.7 that six of the core and secondary species use the same intermediate host *Hyalella azteca* and that another four species use *Gammarus lacustris.* Further, note that species identified as specialists occupy all but the very anterior of the small intestine. When these patterns of dispersion were tested against a model assuming random distributions, the distribution of core species was significantly more even than expected by chance, the secondary species fit into gaps between core species, and the satellite species (not shown in the figure) were distributed randomly. They concluded that the interactive determination of distributions appeared to be mediated by asymmetrical niche shifts. Focusing on two species of small absorbers, e.g., members

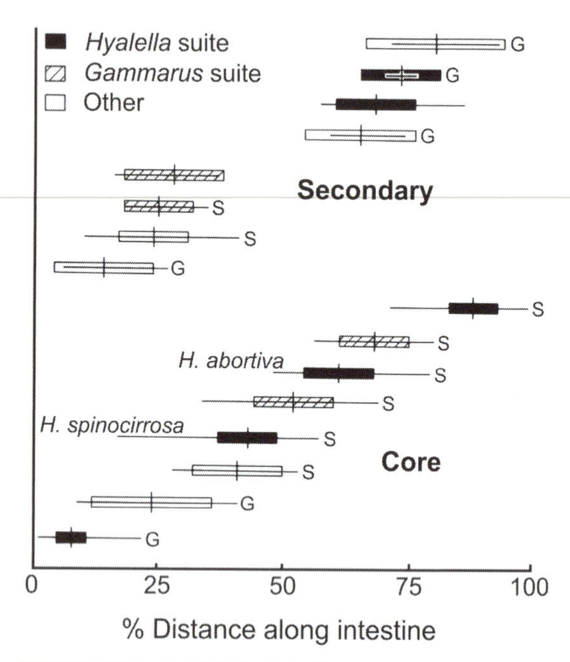

Fig. 14.7 The average realized distribution of core and secondary parasites along the small intestine of lesser scaup ducks. The vertical lines represent the mean of the median distributions of each parasite in each infracommunity; boxes enclose one standard deviation of the mean, and horizontal lines connect the mean anterior and mean posterior endpoints in each infracommunity. Note the even distribution of core species; secondary species are similar although none regularly occurs in the middle of the gut. S and G refer, respectively, to parasites that are specialists in lesser scaup ducks and parasites that are generalists in waterfowl. The status of two species, with respect to this dichotomy, is uncertain. The identity of only two species is presented – see Fig. 14.8. (Modified from Bush & Holmes, 1986, with permission, *Canadian Journal of Zoology*, **64**, 142–152.)

of the same guild both in their use of intermediate hosts and in their mode of using the small intestine, they provide evidence that the niche of *Hymenolepis abortiva* increases symmetrically in response to higher populations but that, under the same conditions, the niche of *H. spinocirrosa* shifts anteriorly (Fig. 14.8). This study represents one of the most detailed investigations available on the distribution of parasites along a linear gradient and concludes that both inter- and intraspecific exploitative competition is important in niche restriction. Several caveats are in order, however. First, competition assumes that a re-

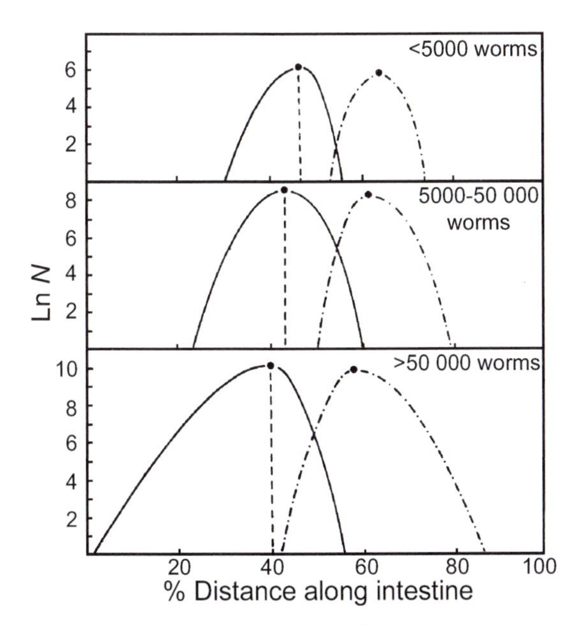

Fig. 14.8 A detailed examination of the linear distributions of two purported exploitative competitors, *Hymenolepis spinocirrosa* (solid line on the left) and *H. abortiva* (dot/dashed line on the right) along the small intestine of lesser scaup ducks. Each of the three panels represents the average distribution of the two species under different total numbers of all helminths present. The distributions are based on the mean anterior, median, and posterior individuals. The mean of the medians is indicated by a dot along each curve. The dashed line within the distribution of *H. spinocirrosa* shows the significant anteriad movement of the mean medianth individual. (Modified from Bush & Holmes, 1986, with permission, *Canadian Journal of Zoology*, **64**, 142–152.)

quired resource must be in short supply. Based on the nature of the data, the authors in the above study had no information on the availability of resources. Second, even though the authors could show a dramatic niche shift by one species in the presence of a putative competitor, they had no evidence that fitness, of either species, was either increased or decreased. Finally, just because competition is inferred in this particular community does not mean that interspecific exploitative competition is a dominating force in niche restriction.

Studies demonstrating interspecific interference competition are not as common. In the absence of predation, the observation that double and triple infections of digenean larvae in snails, e.g., Kuris & Lafferty (1994), or clams, e.g., Rantanen *et al.* (1998), are rare is often attributed to interference competition. Poulin (1998*b*), on the other hand, suggests that these observations are due to a numerical effect rather than to a functional response of the parasites to competition. Paperna (1964) provided one of the best examples of competitive exclusion in his study of monogenean parasites on the gills of carp. Four species of *Dactylogyrus* were found and three of those species (*D. anchoratus, D. extensus*, and *D. minutus*) can apparently co-occur. The fourth species, *D. vastator*, causes damage to the gills of the fish, making them unsuitable for the other monogenean species and thus causing their exclusion. Ultimately, however, *D. vastator* damages the gills to a sufficient extent that it too can no longer survive. Interestingly, the now parasite-free gills heal gradually and can be recolonized. But not by *D. vastator*! A consequence of damaging the gills is that a specific immune response occurs in the fish, thereby permanently excluding *D. vastator* from the community. Other examples of interspecific interference competition in monogeneans are also known (Buchmann, 1988; Jackson *et al.*, 1998). Stock & Holmes (1987*a*) provide another good example of interspecific, interference competition. The cestode *Dioecocestus aspar*, found in the small intestine of red-necked grebes *Podiceps grisegena*, is very large (by several orders of magnitude) compared to other parasites found in the community. With its large mass, it dominates those infracommunities where it has colonized. The authors note that, whether due to the presence of the worm itself, or whether due to the pathological changes the worm induces in the gut, other smaller species (satellite species) are excluded from the community. Core species, which are also small, may persist in the community but their niches become altered.

14.5.6 Summary of niche restriction

Since the distribution of a parasite must equate to some measure of a parasite's niche, the evidence for niche restriction in parasites is unequivocal. Clearly, there is much debate on the mechanism(s) producing the phenomenon. Why this should be so, is not clear. As we have shown in earlier chapters, parasites are very diverse; it stands to reason then that parasite communities can be equally diverse. Is it reasonable to think that a single mechanism can explain the niche-occupation pattern seen in infracommunities? We think not; it is highly likely that many mechanisms, perhaps some yet to be identified, may be influential in determining infracommunity patterns.

14.6 The composition of parasite communities – species richness

Whether there is a pattern to the distribution of species in a community is the focus of our next section; here, we address factors attributed to driving species richness. Poulin (1998*b*), albeit in different words, and for a different reason, noted an important truism: there is an absolute maximum number of species available to any parasite infra-, component, or supracommunity. While seemingly trivial, Poulin's observation has important ramifications. What we actually see, in any community, will always be some subset of a theoretical maximum. Therefore, it is of considerable interest to look at species richness; stated differently, what determines the number of different parasite species in communities?

Before we begin looking at specific factors, the related ideas of supply and screens warrant discussion since they are, perhaps, the most important aspect of parasite community composition. The idea of 'supply' derives from the works of Connell (1985) and Gaines & Roughgarden (1985) and expressed as 'supply-side' ecology by Lewin (1986). Simply stated for parasites, supply means that a parasite, with any and all of its required hosts, is available in the system under study. Holmes (1987) was the first to propose, formally, the notion that the mere presence of a parasite in a system does not mean that just any host can be infected. Indeed, he notes, 'The parasite communities in individual hosts are non-random samples of the parasites available in the environment they occupy . . .'. In Fig. 14.9, we depict how supply and screens might influence species richness patterns. The Global Pool contains 15 parasites (each species is represented by a different letter) that can infect, potentially, one or more host species. Because of supply, hypothetical Regional Pool I will contain only nine species, whereas the second hypothetical Regional Pool will contain 10 species; i.e., supply determines what is actually available at the regional level. Some species are common to both regional pools but Regional Pool I can never have species B, F, H, I, K, O. Often, supply problems would equate to historical/zoogeographical events, e.g., a glaciation. Alternatively, anthropogenic events, e.g., introductions, may alter supply in a contemporary time frame. It is at the Local Pool level that screens become important. Screens determine what species are actually found in any particular component community (the smallest ellipses in the figure) and can take a variety of forms. For example, the screens may reflect biotic features such as phylogeny (including both descent and co-speciation coupled with host specificity), interactions between species, host immunology, or host food habits. Alternatively, screens may reflect abiotic features such as acid rain, episodic droughts, or extinctions. In Fig. 14.9, no screen effectively prevents any potential parasite from being available at Local Pool I; some may be important at the component community level as implied by the three arrows pointing to different component communities. However, some powerful screen(s) does exclude parasites C, L, and M from being available to hosts that draw their parasites from Local Pool II. Therefore, just as parasites B, F, H, I, K, and O were unavailable to Regional Pool I, parasites C, L, and M are, additionally, unavailable to Local Pool II. And, just as in Local Pool I, some other screen(s), implied by the three arrows, may be important in determining what parasites infect different component communities. Recognize too that screens need not, necessarily, be important in determining the species found in component communities; the occurrence of parasites in the component communities drawing their parasites from Local Pool I could simply reflect random colonization. Likewise, once the screen(s) excluding parasites C, L, and M is considered, the component communities drawing their parasites from Local Pool II might represent stochastic colonization. Bearing in mind the potential importance of supply and screens, we now turn to a consideration of some of the commonly proposed determinants of species richness. We attempt to separate the pro-

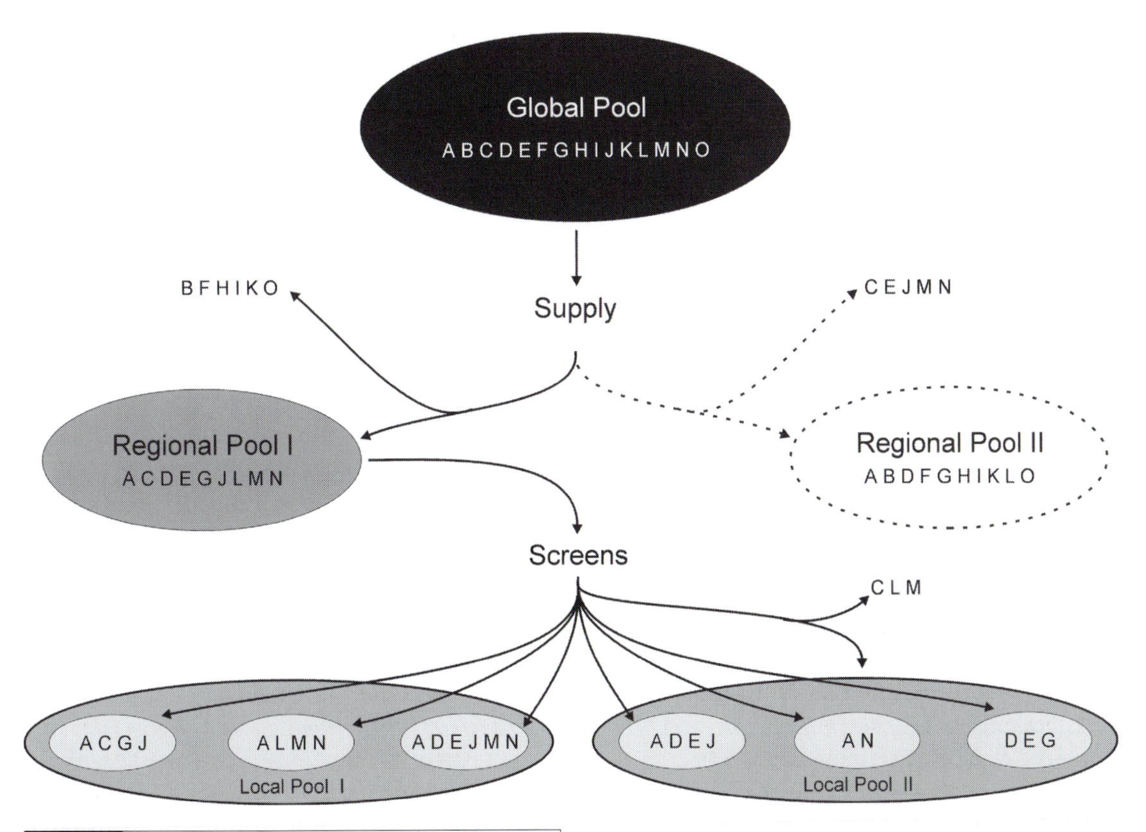

Fig. 14.9 The relationship between the presence of parasites in global, regional and local pools. The small ellipses within local pools represent component communities and the letters designate different species of parasites. Supply will determine the subset of parasites available to different regional pools whereas screens will determine what actually is available to local pools and, therefore, to component communities. See text for a more detailed explanation.

posed determinants into groupings, e.g., those attributable to the parasites, to the host, or to the phylogeny/ecology of the hosts based on how the majority of the authors have viewed their data. Although our presentation suggests discrete determinants, there is much overlap.

14.6.1 Parasite characteristics as determinants of species richness

Beyond parasites' adaptations for transmission and colonization (see Chapter 13), evidence for the actual contribution of parasites themselves to species richness is not common. Three attributes of parasites, competitive ability, facilitation, and population size, seem to have such potential in some host–parasite systems, at least some of the time.

Competitive exclusion of one parasite by another could determine species richness. Such a mechanism seems clear and unambiguous in some studies on digenean larvae in some gastropod hosts (Sousa, 1994). For other parasites, reports of competitive exclusion are very rare, as we noted under niche restriction mechanisms. Based on this, we would have to conclude that competitive exclusion, as a determinant of species richness in most helminth communities, is of limited importance.

In marked contrast to the negative effects of competition, the presence of one species of parasite can actually enhance the presence of a second species. This form of facilitation has been demonstrated, but only rarely (e.g., Ewing *et al.*, 1982). Clearly, it is not a major determinant of species richness.

A positive correlation between the prevalence and intensity of a parasite within a host species has been reported in a number of community studies. Indeed, this is an initial relationship that suggests the core/satellite dichotomy. Poulin & Rohde (1997) extended the observation of the relationship between prevalence and intensity within a host species by examining a very large data set (ectoparasites on 111 species of marine fishes). They too found that species richness and parasite abundance correlate strongly. What this actually means, at their level of analysis (e.g., across host species), is not clear. However, in what appears to be a logical extension of this idea, Poulin (1998*c*), using data on metazoan parasites (both endo- and ectoparasites) in, or on, Canadian freshwater fishes, found a negative correlation between the numbers of host species used by a parasite and that species' prevalence or intensity. He interpreted these data as suggesting a trade-off between how many host species a parasite can exploit and its fitness in those hosts. Poulin (1999) then considered the relationship of prevalence and intensity in the helminths of 158 bird species from Azerbaijan. In this analysis, he found a positive relationship between prevalence and abundance, i.e., exactly the opposite of what he had found previously in fishes and thus negating the idea of a trade-off, at least in bird parasites. His explanation for the latter observation was that 'Birds tend to harbour richer, denser and more diverse helminth communities than fish', as had been foreshadowed in the studies of Kennedy *et al.* (1986) and Bush *et al.* (1990).

14.6.2 Host traits as determinants of species richness

Various features of hosts themselves have been suggested to influence species richness. The most common of these include some measures of host size, host feeding habits, and host physiology/immunity. We will first consider, in some detail, the influence of the physical size of the host to emphasize how little we really know about some determinants of species richness; we then discuss, in less detail, the influence of the other two traits.

Interpreted broadly, host 'size' can be determined in three different ways. First is the actual physical size, usually length, or mass, of the host; second is the geographical range of the host; and third is host density. Here, we consider only physical size; the other two measures are examined in the following section. The implications of host size for species richness are several: (1) larger hosts provide more niches, (2) larger hosts are larger targets, (3) larger hosts eat more, and (4) larger hosts live longer. These ideas seem intuitive. However, where the hosts involved are fishes, studies need to be interpreted with extreme caution. The reasons are several. Big fish begin life as small fish and often show ontogenetic shifts in their diet. Certainly, organisms in other host groups begin smaller than their adult size but, often, they feed similarly throughout life. Most often, the studies attempting to elucidate a pattern between richness and fish size are meta-analyses and the measure used for fish size is usually the adult fish body size, despite the fact that many of the surveys used in the analyses are size-structured. Another serious shortcoming of using fishes is that fish size and fish age are covariates. In other words, although the authors may state that they found, or didn't find, a relationship with 'size' and species richness, their conclusions may actually reflect host age and species richness. According to Poulin (1998*b*), 'In particular, host-species body size and geographical range have proven good predictors of faunal richness.'

Aho & Bush (1993) used a meta-analysis to assess local versus regional determinants of parasite species richness in North American freshwater fishes. They found that both total adult parasite species richness and total parasite species richness, i.e., larval forms for which the fishes were acting as intermediate hosts were included, correlated significantly with host body size. Poulin (1995) also performed a meta-analysis using a much larger body of previously published data on a very diverse group of fishes and found that, of all vertebrate hosts, the only significant relationship was between fish size and endoparasite species richness. There was, however, no significant relationship between fish size and ectoparasite species richness. Poulin & Rohde (1997), in a study of 111 species of fishes where 'Almost all fish sampled were adults', also found no significant relationship between fish size and ectoparasites species richness. However, Morand *et al.* (1999) did find a significant relationship between host size and ectoparasites species richness. Perhaps an important feature, the last three studies were 'corrected' for host phylogeny (see Box 14.1).

Box 14.1 Phylogeny *versus* ecology or phylogeny *and* ecology: you be the judge

Often, when ecologists are confronted with evolutionary questions, they find it difficult to disentangle the manifestation of historical effects (phylogeny) from contemporary effects (ecology). This is not surprising – you are what you are because of your phylogeny, but your phylogeny has been shaped through an ecological context. Recently, this problem has been posed with reference to parasites and their hosts and the use of phylogenetic (or independent) contrasts has become a much-favored paradigm in some studies, most notably those of Gregory (1990), Gregory *et al.* (1991), and Poulin (1998*b* and references therein).

The observation that engenders the need for phylogenetic contrasts is that, because all species are related to some degree, they are not independent of each other. The word 'independent' here does not mean that species do not differ from one another; certainly they do or we would not consider them 'species'! Instead, independence is used in a statistical context and a very meaningful one at that. All statistical analyses have a number of underlying assumptions and, perhaps, none is more important than the assumption: *observations being tested must be independent of one another.*

Recognizing this, Felsenstein (1985) was the first to propose, formally, that comparative analyses, using related species as independent points, were statistically flawed. Because they are related, they cannot be 'regarded for statistical purposes as if drawn independently from the same distribution'. Stated differently, and if you know just a bit about statistics, using related species as independent variables will artificially inflate the degrees of freedom thereby increasing the probability of making a Type I error (rejecting the null hypothesis of no difference when, in fact, it is true). Felsenstein (1985) proposed a possible solution for the lack of independence based on the suggestion that each character is assumed to be evolving by a Brownian motion that is independent in each lineage. That model of character change, plus knowledge of the phylogeny of the organisms, and a bit of computer savvy, allows one, ultimately, to make comparisons using independent contrasts.

There are several problems with the use of phylogenetic contrasts, particularly when applied to parasites. Some are methodological, while others are philosophical. Here, we will highlight some of the methodological and philosophical problems and then we will ask you to judge.

For phylogenetic contrasts to be meaningful, one must have acceptable, robust phylogenies. At present, this is seldom the case. For example Poulin (1995) notes 'For most taxa of birds, mammals, and fish, the phylogeny has not been resolved down to the species level; therefore, I chose to carry out the analyses at the genus level.' This can also lead to the necessity of some important caveats such as, while the branch lengths in the phylogenetic tree for birds could be estimated, 'Accurate estimates of branch length could not be obtained for the phylogenies of mammals and fish . . ..'

A second problem with phylogenetic contrasts is that they lack statistical power. Although there appear to be many studies on the parasites of various host groups, when data are collapsed, as they must be in phylogenetic contrasts, the result is that there are often so few contrasts that the probability of making a Type II error (failing to reject the null hypothesis when it is really false) becomes unacceptably high. The following serves as an excellent example.

In a study to examine the 'time hypothesis' (for parasites – phylogenetically older hosts should have more parasites than phylogenetically younger hosts), Bush *et al.* (1990) performed a meta-analysis on parasite communities in 582 species of vertebrate hosts. In addition to refuting the time hypothesis for parasite communities in vertebrate host groupings, e.g., fishes, herptiles, birds, and mammals, they also noted that aquatic representatives within a host group had significantly more parasites than their terrestrial counterparts, e.g., aquatic birds had significantly richer parasite communities than did terrestrial birds. They concluded by suggesting that birds were the 'tropics' with respect to parasite species richness compared to other vertebrate hosts. Poulin (1995) re-examined some of these conclusions. For birds, an initial 98 different studies on host species, after correcting for different sampling efforts and for 'phylogenetic contrasts', was reduced to only six contrasts, and of these, five were positive, e.g., aquatic birds had higher species richness than terrestrial birds. ($\beta = 0.54$ meaning that the power of the test, $1 - \beta$, was only 0.46.) In fact, the only way to reject the null hypothesis of no difference in parasite species richness between aquatic and terrestrial birds would be for all six contrasts to have been positive. Fortunately, some authors, e.g., Poulin (1995, 1997*b*) recognize this problem and provide either sufficient data, or values for β, thus allowing their interpretations to be questioned.

A third methodological problem is the acceptance of the Brownian motion model of evolution. Felsenstein (1985), while proposing the 'solution' on the one hand, notes, on the other hand, 'All of the above has been predicated on the acceptance of the Brownian motion model as a realistic statistical model of character change. There are certainly many reasons for being skeptical of its validity.' The acceptance of the model is hotly debated today and it is fair to say that the debate will be lively, and will continue, for some time.

A final methodological problem is the parasites themselves. The impetus for Felsenstein's (1985) proposal seemed to be a number of papers, recent at that time, which used a comparative approach to assessing various character traits between species, e.g., brain weight on body weight, brain weight on body weight and basal metabolism, population density and body size, etc. The problem with using independent contrasts for parasites and their hosts is that, with very rare exception, parasites are not directly inheritable as are character states. They are not 'traits' in a traditional sense of the word. Even worse, parasites are living creatures and they too have a phylogeny! Currently, there exist no methods that can deal with both host and parasite phylogenies simultaneously. Thus, although the host contrasts are independent the parasites are not. And what then of statistical independence?

Once there are acceptable phylogenies, and these are becoming increasingly more common thanks to molecular techniques, once there is an acceptable model for evolution, once there are studies on various hosts species such that there will be more degrees of freedom for independent contrasts, and once there is a methodology that will allow for simultaneous evaluation of independent contrasts for both hosts and their parasites, it is certain that phylogenetic contrasts will prove very powerful tools for some questions. Here is where philosophical problems often arise. The first philosophic problem relates to the above methodological problem that parasites are not traits; although host genetics have, on occasion, been shown to determine the presence or absence of a parasite species, it is not a general phenomenon. Hosts, for the most part, are born without parasites; what ultimately colonizes them will be determined by a wide variety of factors of which phylogenetic *potential* is only one. Depending on the nature of the question(s) asked, species richness, *no matter how it is derived (descent, host-switching, sympatric speciation, etc.)*, may be inherently interesting. Finally, an underlying philosophical problem with phylogenetic contrasts is that it removes attention from the level of individual species to its closest ancestor. In other words, it obscures, a priori, that very phylogenetic event that makes a species different from its ancestor. How 'related' must that ancestor be? Indeed, species are related by genera, genera by families, and so on. We could reach the conclusion that everything is related to some glob in a primordial soup and, therefore, we might as well all go home and stop asking these dumb questions!

Figure 1 presents a modified summary of the actual data from the meta-analysis of Bush *et al.* (1990). Assume that your grade in this course will be based solely on turning in a parasite collection and that the more species of parasites in your collection, the higher will be your grade. You must select three hosts for necropsy, one host individual from each of the three groups, e.g., one herptile (either aquatic or terrestrial), one bird (either aquatic or terrestrial) and one mammal (either aquatic or terrestrial). Remember, Gregory *et al.* (1991) and Poulin (1995) conclude that there are no significant differences between aquatic hosts and their terrestrial counterparts. You be the judge.

Presumably, you all want to do well and you picked the aquatic member from each group. Why? Certainly not because of the arguments presented above! You probably made your choices based on the patterns you could see without regard for the statistical analyses of Bush *et al.* (1990), or Poulin (1995). Assuming you picked the aquatic hosts for each group does that make Bush *et al.* right and Poulin wrong? No! There is, currently, no 'correct' (in the statistical sense), robust methodology to address some questions that we ask of our data.

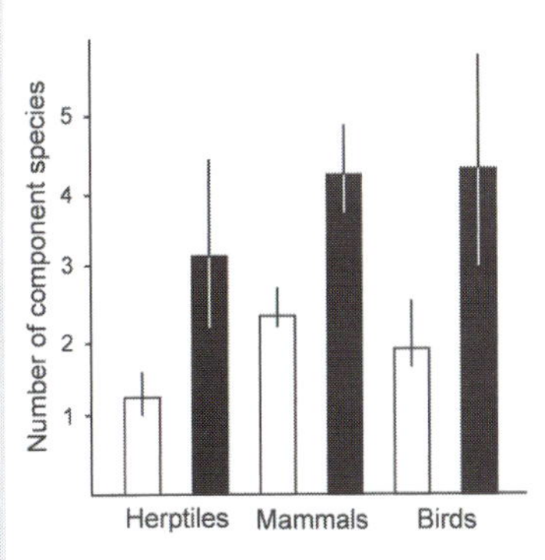

Fig. 1 Mean and asymmetric 95% confidence intervals for the number of parasites found in various terrestrial (white) and aquatic (black) vertebrates. 'Herptiles' include both reptiles and amphibians. Note that these are real data based on the actual study of parasites found in host species; they have not been 'corrected'! (Modified from Bush, A. O., Aho, J. M. & Kennedy, C. R. [1990] Ecological versus phylogenetic determinants of helminth parasite community richness. *Evolutionary Ecology*, **4**, 1–20. Reprinted with kind permission from Kluwer Academic Publishers.)

Although the same methodological problems associated with fish 'size' are not as prominent in birds and mammals, the results of phylogenetically 'corrected' tests for a relationship between mass and parasite species richness are equally problematic. For birds, Gregory *et al.* (1991) found a positive relationship between the host weight and the richness of digeneans and nematodes, but not cestodes, and Gregory *et al.* (1996) found an overall positive relationship between bird body mass and species richness. On the other hand, Poulin (1995) found no significant relationship between bird body mass and parasite species richness. Similarly, Gregory *et al.* (1996) found a significant relationship between mammal body mass and parasite species richness but neither Poulin (1995) nor Morand & Poulin (1998) found a significant relationship in their studies. Is host 'size' a good predictor of parasite species richness? Based on the data, whether phylogenetically 'corrected' or not, there are no clear answers.

For endoparasites, the notion that host feeding habits might influence parasite species richness also seems intuitive, largely because most endoparasites exploit food webs to colonize hosts. In a host–parasite sense, this means 'you have what you eat'. Consider, for example, a tadpole that feeds on plant material and an adult form, of the same species, which feeds on insects. It is common to find lung flukes *Haematoloechus* spp. in the adult frogs, but never their tadpoles. The reasons are obvious; tadpoles rely on gills rather than lungs so this is an obvious ontogenetic factor but, second, another ontogenetic factor is the change in diet. Since *Haematoloechus* spp. use insects as intermediate hosts, and tadpoles cannot eat insects, the parasite is not even available to them. Often, the most important aspect of host diet and parasite species richness seems to be related to the amount of animal material ingested by the host followed by the breadth of diet (see Kennedy *et al.*, 1986 and Bush, 1990 for some examples). In some studies, where parasites specific to certain intermediate hosts are found, the parasites themselves can tell us something about the food habits of the hosts. Still, although the diet of a host may be the single most important determinant of parasite species richness, a priori studies, incorporating a suitable null hypothesis, have not been done. In other words, most studies invoking the importance of host diet are inferential and *post hoc*. There are also many confounding variables. For example, we have already shown there exists a controversy about the importance of host size; an assumption was that larger hosts eat more. Attempting to explain the generally species-poor parasite communities in freshwater fishes versus the generally species-rich parasite communities in aquatic birds, Kennedy *et al.* (1986) identified five factors they felt were important in determining species richness: (1) complexity and/or physiology (endothermy/ectothermy) of the host gut, (2) host vagility, (3) a broad diet, (4) selective feeding on prey which serve as intermediate hosts for a wide variety of parasites, and (5) exposure to parasites with direct life cycles. Note that, although only (3) and (4) address feeding specifically, all but (5) are related, directly, to host feeding habits. In short, there are many surrogates for how the host diet might be important in determining species richness of parasite communities. We should also note that host diet might impact negatively on parasite species richness. For example, Huffman *et al.* (1996) suggest that leaf swallowing by chimpanzees may control the infection of some parasites, i.e., chimps are using herbal medicines.

Sadly, the implication that a host's physiology or immune status can determine species richness has received far too little attention. There are, however, some hints to their importance. As many hosts age, they may become less susceptible to a parasite even though they have never been exposed to the parasite, i.e., transferred, or acquired immunity, is not involved. For example, as dogs age, they become increasingly resistant to hookworm infection; a mature dog, even though never exposed to hookworms, may be completely refractory to an infection. Stress on a host often results in the production of corticosteroids and these can influence the presence or absence of some parasites. (See Esch *et al.*, 1975 for a more detailed discussion of parasitism and stress.) We have already discussed (Chapters 11 and 12) how immunity can influence the population of a parasite and a logical extension is that it can influence the richness of a parasite community by causing the complete, or temporal, elimination of a particular parasite species. Schad (1966) proposed that a parasite might induce an immune response in the host that could be directed towards another parasite species, but not itself. This might be likened to a form of interspecific competition mediated through the host's response.

14.6.3 Historical and ecological determinants of parasite species richness

We have noted, in Box 14.1, that it is sometimes difficult to disentangle historical events from ecological events. Nonetheless, history impacts on virtually every community. Kennedy & Bush (1992) investigated the occurrence of multiple congeners (often called 'species flocks') and confamilials from a large body of published literature. Although they were unable to identify a pattern in the distribution of these 'flocks', their mere existence is ample evidence of how important phylogeny may be in determining species richness. We have noted previously that many parasites show some degree of specificity, whether it is to a host species, a host genus, a host family, or to an even higher level of host taxon. Quite obviously, specificity can be a direct determinant of species richness. Poulin (1997*a*) proposes a novel idea based on his investigation of parasite species richness in Canadian freshwater fishes. His meta-analysis showed that the parasites found in species-rich communities occurred in fewer host species while the parasites found in depauperate communities occurred in many host species. In other words, rich communities contained specialists and generalists, whereas depauperate communities were comprised of generalists. The study by Herreras *et al.* (1997) on the parasites of harbor porpoises *Phocoena phocoena* shows this latter feature very well, as do almost all studies on marine mammals. Examples of the importance of phylogeny are so common that more than a general textbook would be required to do it justice. What is less common are insightful questions and novel ideas such as proposed by Poulin and noted above. Too often, investigators simply invoke an historical argument and leave it there with no further attempt to understand why this should be so. Seemingly, to some, history is enough. Many interesting questions remain to be addressed, however, some foreshadowed by earlier works, others yet to be identified. For example, it is clear that some parasite communities in closely related host species appear mostly to be descendents of some ancestral community. Bush *et al.* (1990), noting the importance of phylogeny, speculate that, although aquatic mammals might be expected to have richer communities than their terrestrial counterparts, when mammals secondarily return to marine environments, they would have depauperate communities. Several studies on marine mammals lend credence to that speculation, e.g., Wazura *et al.* (1986), Balbuena & Raga (1993), Herreras *et al.* (1997). Other phylogenetically isolated host groups might be expected to show similarly isolated communities of parasites and, although studies are rare, that by Aznar *et al.* (1998) on the parasites of the loggerhead sea turtle *Caretta caretta* is exemplary. Interesting is the notion that, if hosts are regarded as distinct species, with all of the implications that entails, what constrains parasites to retain similarity to an ancestral fauna? On the other hand, if the parasite communities diverge to some large extent between closely related host species, what is, or was, the driving force? It is this latter question that drives the idea that ecology can influence the diversity of parasite communities. The importance that phylogeny has on parasites, and parasitism, are further amplified in Chapters 15 and 16 and we turn now to a consideration of ecological determinants of species richness.

Perhaps surprisingly, some 'ecological' determinants of species richness are nothing more than surrogates for host feeding; a host factor that we have previously discussed. For example, several authors (Gregory, 1990; Poulin, 1997*b*) note that species richness increases with host geographic range (geographic range has also been considered as a measure of host 'size'). The implication is that by having extended geographical ranges, hosts may encounter more parasites. An embedded implication is that they will be exposed to more parasites because they will encounter a wider array of intermediate hosts. Bush (1990) studied the same host species on its breeding grounds in several freshwater environments and on its wintering grounds in several saltwater environments. No significant difference in species richness was found, but there was a radical change in the kinds of parasites, from mostly cestodes in freshwater to mostly digeneans in saltwater. He noted too that it was nothing more than host feeding patterns that produced what appear to be dramatic differences. Edwards & Bush (1989) showed that ecological events could override phylogenetic events in determining species richness in avocets *Recurvirostra americana*. Why would this happen? Under certain environmental conditions, avocets were forced to feed on amphipods infected with several species of parasite normally found only in lesser scaup ducks.

Some anthropogenic factors seem clearly related to feeding as well. For example, much of the early work on the ecology of parasite communities was based on studies of parasites in fish from different lakes. One of the earliest to address this was Wisniewski (1958) who examined hosts from eutrophic Lake Druzno and mesotrophic Lake Goldapivo in Poland. After sampling many thousands of potential vertebrate and invertebrate hosts, Wisniewski concluded that parasites were moving primarily from fishes, amphibians, and certain invertebrates, to birds. In other words, parasites were following food chains. Esch (1971) speculated that eutrophic systems were 'open' and subject to more opportunities for aquatic–terrestrial interaction such as may occur between piscivorous birds. Oligotrophic systems, on the other hand, were described as more 'closed', with less aquatic–terrestrial interaction and less opportunity for the transmission of parasites through piscine intermediate hosts to avian definitive hosts. The speculation was, therefore, essentially linked to the nature of predator–prey relationships; according to Esch (1971), this was the primary determinant of parasite community richness in oligotrophic versus eutrophic ecosystems. Marcogliese & Cone (1997) examined the parasite communities of eels taken from Nova Scotian (Canada) rivers with acidic and more neutral pHs. They observed lower species richness in rivers with reduced pH. They attributed these lower diversities to reduced complexities in food chains that would typically involve parasite transmission.

Ecological determinants of species richness need not be related directly to feeding, however. For example, the impact of anthropogenic factors may be independent from what a host eats. This is particularly true of digeneans that have molluscs as obligate first intermediate hosts. Curtis & Rau (1980) showed that the presence of strigeids in their second intermediate hosts, fish, was related to calcium ion concentrations in lakes. Lakes with low concentrations did not support snails, the obligate first intermediate host. Valtonen *et al.* (1997) concluded that the low prevalence of digeneans and glochidia in Lake Vatia, Finland was due to chemical pollutants, which caused a depauperate molluscan fauna; this would obviously impact directly on glochidia and indirectly on digeneans (because of the lack of first intermediate hosts). Following several years of reduced organochlorine input into the lake, the prevalences of digeneans and glochidia approached levels of unpolluted lakes. The same authors noted that the immune system, which may be important in regulating populations of monogeneans, was compromised in fish under pollution stress. Sentinel fish from unpolluted lakes introduced into a cage in Lake Vatia showed a rapid increase in populations of monogeneans but caged control fish in the unpolluted lake did not show a similar increase. The use of parasitological data to monitor environmental health (Kennedy, 1997; Overstreet, 1997) attests further to the importance that anthropogenic factors may have as a determinant to species richness.

Others have suggested that non-anthropogenic physicochemical properties may impact on species richness in fish parasite communities (e.g., Kennedy, 1978; Marcogliese & Cone, 1991). Rohde (1980, 1992 and references therein) suggested that latitudinal gradients could explain patterns in species richness of ectoparasites on marine fish. Poulin (1995) found no significant relationship between species richness in ectoparasite communities and latitude. However, Rohde (1994) noted 'Temperature appears to be the most important factor affecting the geographical range of marine parasites.' Poulin & Rohde (1997) did find a significant relationship between parasite species richness and temperature. When the data were analyzed independently for the Pacific versus the Atlantic Oceans, the relationship held for the former, but not for the latter. They attribute the relationship between richness and temperature in the Pacific fishes to the fact that evolution, and therefore **speciation**, is faster in the tropics. They attribute the lack of the relationship of temperature and richness in Atlantic fishes to the poor power of their test (see Box 14.1).

Other non-anthropogenic factors may also impact on species richness independent of feeding. Recently, Poulin & Morand (1999) have suggested that geographical distance might be important in determining species richness. They considered three sets of previously published data on parasite communities in fish (in each study, fish were collected from several lakes within the same region) and concluded that the physical distance between lakes was a significant determinant of species richness in two of the three studies. This seems to be a very good example where local processes (ecological) are more important than regional processes (historical). Aho & Bush (1993) also concluded that local processes were more important than regional processes in determining species richness in three species of bass and nine species of sunfish. The latter authors did not, however, consider geographical distance between individual sample lakes.

Epidemiologists have long been aware of a

positive relationship between parasite transmission and host density (also a measure of host 'size'). Such a relationship has not been investigated often for the richness of helminth parasites. Perhaps the first to do so were Moore *et al.* (1988). They suggest that they were looking at the importance of covey size in bobwhite quail *Colinus virginianus*, and parasite richness and intensity; however so many caveats were necessary to accept that the 'covey' was a meaningful parameter that their data are more easily thought of as large versus small groups of hosts. They found that the prevalence of some parasites was linked positively to large groups of hosts while others showed no relationships. Morand & Poulin (1998) examined parasites from 79 species of terrestrial mammals and found a significant relationship between parasite species richness and host density. They conclude that colonization by parasites is based on how many hosts there are in a particular region and, therefore, parasitism may be a cost of host density.

14.6.4 Summary of determinants of species richness in parasite communities

Interspecific interactions appear to be important determinants in some, but not all, larval digenean communities in snails. We have too little information on other intermediate host systems to draw even broad inference to possible mechanisms. For parasite species richness in vertebrate hosts, the situation is slightly clearer. Interspecific antagonisms are unlikely to be important determinants except, perhaps, in very isolated examples. Whether being 'bigger is better' with respect to parasite species richness is a 'can of worms'. Despite the bold statements by some authors attesting to the importance of host size and parasite species richness, nothing can be interpreted from the conflicting results (whether they be 'corrected' for phylogeny or not). Host feeding is clearly implicated as being important since most parasites exploit food webs. However, because there are so many confounding variates of host feeding, it is difficult to disentangle the actual processes. It seems that one feature is most likely to influence species richness *in general*. That feature, a feature of the parasites themselves, is host, habitat, or ecological specificity.

14.7 The structure of parasite communities

Much of what we have discussed above, e.g., niche restriction, specificity, species richness, contributes to the structure of communities at one level or another. Here we will conclude the chapter on parasite communities by looking at co-occurrences between parasite species, i.e., whether they form predictable patterns.

Ideally, as a first step to elucidating the structure of a community, we would like to know if the communities are, in fact, structured, or do they merely reflect random colonization by parasites? The debate continues and is unlikely to be resolved for some time. For example, among recent studies of monogenean parasites, there are studies invoking random colonization (Bagge & Valtonen, 1999), deterministic colonization (Gutiérrez & Martorelli, 1999), and both forms of colonization in the same community (Jackson *et al.*, 1998). Here is where an *appropriate* null model is desirable. Were such a model available, we could then test the observed patterns against a model that assumes random structure. Goater *et al.* (1987) tested the distribution of parasites in several species of sympatric salamanders against a Poisson distribution, which is a measure of randomness. Unfortunately, a Poisson distribution assumes that there is an equal probability for all parasites present to infect a host. As we have shown in section 14.6, this may, or may not, be true. An alternative is to use a null model based on the prevalences of the parasite species (Janovy *et al.*, 1995). Poulin (1998*b*) suggests that this approach offers a more appropriate test to determine randomness than does a Poisson distribution because it accounts for, at least in part, different prevalences and, thus, different probabilities of infection. Unfortunately this too suffers methodological problems that are equally as severe, and perhaps more severe, than are the assumptions of a Poisson variate. 'Prevalences' are not determined in a vacuum; often, the prevalence of a particular parasite is based on some extrinsic variable (section 14.6). What this means then, is that some form of 'structure' is embedded into the model and it is no longer 'random'. It seems safe to suggest that, at this point in our knowledge of parasite communities, there are no truly null models against which to test patterns.

14.7.1 Species co-occurrences

One of the more interesting observations made on communities of parasites is the significant co-occurrence seen in some parasites. By significant co-occurrence, we mean that two or more species of parasite occur in the same host individual more often than expected by chance. This observation itself is not new; indeed, ecologists working on free-living organisms have been analyzing recurrent groups for decades (e.g., Fager, 1957, Fager & McGowan, 1963). Further, however, many studies on parasites, where significant co-occurrence is detected, note an excess of positive co-occurrences between some species (e.g., Hobbs, 1980; Lotz & Font, 1985; Bush & Holmes, 1986*a*; Toft, 1986; Stock & Holmes, 1987*b*). Lotz & Font (1991, 1994) pursued this observation using their data sets on parasites in bats (Lotz & Font, 1985). These authors make several observations germane to parasite community structure. First, they show that the proportion of positive covariances is a function of the proportion of rare species, the proportion of common species, and the sample size of the hosts. If rare species dominate, there will be an excess of negative associations and if common species dominate, there will be an excess of positive associations. This notion, that a rare species can never be a frequent member of a common species' environment, was foreshadowed in the earlier works on free-living communities by Fager (1957). Second, they note that the excess of positive associations detected in their studies on bats disappeared after restricting the analyses to hosts in which both members of a species pair were present. They conclude from this that interspecific facilitation, e.g., that the presence of one species modifies the habitat in a way beneficial to a second species, was not important (see Gutiérrez & Martorelli, 1999, for an alternative viewpoint for monogeneans on fish gills). Finally, they emphasize that communities of helminths in individual bats are not random samples from a pool of potential colonists.

Bush *et al.* (1993) provide a possible clue as to how positive species' co-occurrences might happen in the many host–parasite systems relying on predator–prey interactions. They note that some intermediate hosts might be likened to 'source communities'; in other words, the community seen in a particular definitive host may simply mirror what it has ingested from an intermediate host. They caution that, if this pattern of association is transferred from source to target communities, trying to explain 'pattern' (interspecific interactions, co-occurrences of species) in definitive hosts may be futile. In other words, one must differentiate between transferred and post-transmission processes. Lotz *et al.* (1995) extended these ideas to a null model for transferred patterns. Noting that almost all helminth communities are recruitment-driven (new individual parasites are added from outside the host and not by birth within the definitive hosts) and spatially discontinuous (the hosts represent discrete patches of habitat), they show that source communities may well explain the excess of positive associations found in many target communities. Their model, looking at the dynamics of immigration and death processes indicates four things: (1) recruitment-driven target communities will grow to an asymptote, (2) the frequency distribution of recruits will determine the frequency distribution in the target communities, (3) interspecific associations in source communities will be transferred to target communities, but the magnitude of the associations will depend on survivorship in the target communities, and (4) the magnitude of a negative association will be less than the magnitude of a positive association in the target community if the two associations are of equal magnitude in the source community.

14.7.2 Hypotheses about parasite community structure

A number of hypotheses has been posited trying to explain observed patterns in communities of parasites. It is probably true that each of these hypotheses will explain some of the observed patterns, in some hosts, some of the time. Perhaps the only hypothesis that provides a *general* framework for considering parasite communities is the isolationist/interactive hypothesis of Holmes &

Price (1986). Isolationist communities are those characterized by parasite species with low transmission/colonization rates, the communities are non-equilibrial, and the parasite species are insensitive to the presence of other parasite species. Interactive communities are composed of parasite species with high transmission/colonization rates, they are equilibrial, and interspecific interactions occur, or have occurred.

There are some problems, even with this general hypothesis. For example, far too many authors use 'isolationist' simply as a synonym for 'depauperate' and 'interactive' as a synonym for 'rich'. They consider neither the implications of equilibrium nor the implication of transmission/colonization rates. Indeed, depauperate communities may well show either positive, or negative, interactions among species and species-rich communities may show either positive, or negative, or both, interactions among the species comprising the communities. Indeed, some species-rich communities such as that studied by Bush & Holmes (1986*a*, *b*) may show hybrid patterns when the properties of individual species are considered. For example, core species in the parasite communities of lesser scaup fit the predictions of an interactive community, but the randomly distributed satellite species, part of the same infracommunity, fit the predictions of an isolationist community.

The truth is, there really are no suitable hypotheses about communities of parasites. But, why should there be? In the early sections of this book, we emphasized the diversity of parasites, both in form and function. Is it realistic to expect that some general hypothesis that can account for patterns in the distribution of organisms from two kingdoms, and many, many phyla can ever be formulated, or meaningful? We think not. Even when we restrict our attention to some small subset of animal parasites, our current knowledge of the parasites themselves is too inadequate for general conclusions. What we need, from a community perspective, are more autecological studies on individual parasites (Bush & Aho, 1990). They are forthcoming but, until we have many more, we do not have a prayer of making meaningful generalizations!

References

Adamson, M. L. & Caira, J. N. (1994) Evolutionary factors influencing the nature of parasite specificity. *Parasitology* (Supplement), **109**, S85–S95.

Aho, J. M. & Bush, A. O. (1993) Community richness in parasites of some freshwater fishes from North America. In *Species Diversity in Ecological Communities*, ed. R. E. Ricklefs & D. Schluter, pp. 185–193. Chicago: University of Chicago Press.

Aznar, F. J., Balbuena, J. A., Bush, A. O. & Raga, J. A. (1997) Ontogenetic habitat selection by *Hadwenius pontoporiae* (Digenea: Campulidae) in the intestines of Franciscanas (Cetacea). *Journal of Parasitology*, **83**, 13–18.

Aznar, F. J., Badillo, F. J. & Raga, J. A. (1998) Gastrointestinal helminths of loggerhead turtles (*Caretta caretta*) from the western Mediterranean: constraints on community structure. *Journal of Parasitology*, **84**, 474–479.

Balbuena, J. A. & Raga, J. A. (1993) Intestinal helminth communities of the long-finned pilot whale (*Globicephala melas*) off the Faroe Islands. *Parasitology*, **106**, 327–333.

Bagge, A. M. & Valtonen, E. T. (1999) Development of monogenean communities on the gills of roach fry (*Rutilus rutilus*). *Parasitology*, **118**, 479–487.

Brooks, D. R. (1980) Allopatric speciation and non-interactive parasite community structure. *Systematic Zoology*, **29**, 192–203.

Buchmann, K. (1988) Interactions between the gill-parasitic monogeneans *Pseudodactylogyrus anguillae* and *P. bini* and the fish host *Anguilla anguilla*. *Bulletin of the European Association of Fish Pathologists*, **8**, 98–99.

Bush, A. O. (1990) Helminth communities in avian hosts: determinants of pattern. In *Parasite Communities: Patterns and Processes*, ed. G. W. Esch, A. O. Bush & J. M. Aho, pp. 197–232. London: Chapman & Hall.

Bush, A. O. & Aho, J. M. (1990) Concluding remarks. In *Parasite Communities: Patterns and Processes*, ed. G. W. Esch, A. O. Bush & J. M. Aho, pp. 321–325. London: Chapman & Hall.

Bush, A. O. & Holmes, J. C. (1983) Niche separation and the broken-stick model: use with multiple assemblages. *American Naturalist*, **122**, 849–855.

Bush, A. O. & Holmes, J. C. (1986*a*) Intestinal helminths of lesser scaup ducks: patterns of association. *Canadian Journal of Zoology*, **64**, 132–141.

Bush, A. O. & Holmes, J. C. (1986*b*) Intestinal helminths of lesser scaup ducks: an interactive community. *Canadian Journal of Zoology*, **64**, 142–152.

Bush, A. O. & Lotz, J. M. (2000) The ecology of 'crowding'. *Journal of Parasitology*, **86**, 212–213.

Bush, A. O., Aho, J. M. & Kennedy, C. R. (1990) Ecological versus phylogenetic determinants of helminth parasite community richness. *Evolutionary Ecology*, **4**, 1–20.

Bush, A. O., Heard, R. W. Jr & Overstreet, R. M. (1993) Intermediate hosts as source communities. *Canadian Journal of Zoology*, **71**, 1358–1363.

Bush, A. O., Lafferty, K. D., Lotz, J. M. & Shostak, A. W. (1997) Parasitology meets ecology on its own terms: Margolis *et al.* revisited. *Journal of Parasitology*, **83**, 565–583.

Caswell, H. (1978) Predator-mediated coexistence: a non-equilibrium model. *American Naturalist*, **112**, 127–154.

Connell, J. H. (1985) The consequences of variation in initial settlement vs. post-settlement mortality in rocky intertidal communities. *Journal of Experimental Marine Biology*, **93**, 11–45.

Courtney, C. H. & Forrester, D. J. (1973) Interspecific interactions between *Hymenolepis microstoma* (Cestoda) and *Heligmosomoides polygyrus* (Nematoda) in mice. *Journal of Parasitology*, **59**, 480–483.

Curtis, L. A. (1995) Growth, trematode parasitism, and longevity of a long-lived marine gastropod (*Ilyanassa obsoleta*). *Journal of the Marine Biological Association UK*, **75**, 913–915.

Curtis, L. A. (1997) *Ilyanassa obsoleta* (Gastropoda) as a host for trematodes in Delaware estuaries. *Journal of Parasitology*, **83**, 793–803.

Curtis, L. A. & Tanner, N. L. (1999). Trematode accumulation by the estuarine gastropod *Ilyanassa obsoleta*. *Journal of Parasitology*, **85**, 419–425.

Curtis, M. A. & Rau, M. E. (1980) The geographical distribution of diplostomiasis (Trematoda: Strigeidae) in fishes from northern Quebec, Canada, in relation to the calcium ion concentrations of lakes. *Canadian Journal of Zoology*, **58**, 1390–1394.

Edwards, D. D. & Bush, A. O. (1989) Helminth communities in avocets: importance of the compound community. *Journal of Parasitology*, **75**, 225–238.

Esch, G. W. (1971) Impact of ecological succession on the parasite fauna in centrarchids from oligotrophic and eutrophic ecosystems. *American Midland Naturalist*, **86**, 160–168.

Esch, G. W., Gibbons, J. W. & Bourque, J. E. (1975) An analysis of the relationship between stress and parasitism. *American Midland Naturalist*, **93**, 339–353.

Esch, G. W., Kennedy, C. R., Bush, A. O. & Aho, J. M. (1988) Patterns in helminth communities in freshwater fish in Great Britain: alternative strategies for colonization. *Parasitology*, **96**, 519–532.

Ewing, M. S., Ewing, S. A., Keener, M. S. & Mulholland, R. J. (1982) Mutualism among parasitic nematodes: a population model. *Ecological Modelling*, **15**, 353–366.

Fager, E. W. (1957) Determination and analysis of recurrent groups. *Ecology*, **38**, 586–595.

Fager, E. W. & McGowan, J. T. (1963) Zooplankton species groups in the North Pacific. *Science*, **140**, 453–460.

Felsenstein, J. (1985) Phylogenies and the comparative method. *American Naturalist*, **125**, 1–15.

Gaines, S. & Roughgarden, J. (1985) Larval settlement rate. *Proceedings of the National Academy of Sciences, US A*, **82**, 3707–3711.

Goater, C. P. & Bush, A. O. (1988) Intestinal helminth communities in long-billed curlews: the importance of congeneric host-specialists. *Holarctic Ecology*, **11**, 140–145.

Goater, T. M., Esch, G. W. & Bush, A. O. (1987) Helminth parasites of sympatric salamanders: ecological concepts at the infracommunity, component and compound community levels. *American Midland Naturalist*, **118**, 289–300.

Gregory, R. D. (1990) Parasites and host geographic range as illustrated by waterfowl. *Functional Ecology*, **4**, 645–654.

Gregory, R. D., Keymer, A. E. & Harvey, P. H. (1991) Life history, ecology and parasite community structure in Soviet birds. *Biological Journal of the Linnean Society*, **42**, 249–262.

Gregory, R. D., Keymer, A. E. & Harvey, P. H. (1996) Helminth parasite richness among vertebrates. *Biodiversity and Conservation*, **5**, 985–997.

Grutter, A. S. (1996) Parasite removal rate by the cleaner wrasse *Labroides dimidiatus*. *Marine Ecology Progress Series*, **130**, 61–70.

Gutiérrez, P. A. & Martorelli, S. R. (1999) The structure of the monogenean community on the gills of *Pimelodus maculatus* in Rio de la Plata (Argentina). *Parasitology*, **119**, 177–182.

Hanski, I. (1982) Dynamics of regional distribution: the core and satellite species hypothesis. *Oikos*, **38**, 210–221.

Hanski, I. (1991) Reply to Nee, Gregory, and May. *Oikos*, **62**, 88–89.

Herreras, M. V., Kaarstad, S. E., Balbuena, J. A., Kinze, C. C. & Raga, J. A. (1997) Helminth parasites of the digestive tract of the harbour porpoise *Phocoena phocoena* in Danish waters: a comparative geographical analysis. *Diseases of Aquatic Organisms*, **28**, 163–167.

Hobbs, R. P. (1980) Interspecific interactions among gastrointestinal helminths of pikas of North America. *American Midland Naturalist*, **103**, 15–25.

Holmes, J. C. (1961) Effects of concurrent infections on *Hymenolepis diminuta* (Cestoda) and *Moniliformis dubius* (Acanthocephala). I. General effects and comparison with crowding. *Journal of Parasitology*, **47**, 209–216.

Holmes, J. C. (1962*a*) Effects of concurrent infections on *Hymenolepis diminuta* (Cestoda) and *Moniliformis dubius* (Acanthocephala). II. Effects on growth. *Journal of Parasitology*, **48**, 87–96.

Holmes, J. C. (1962*b*) Effects of concurrent infections on *Hymenolepis diminuta* (Cestoda) and *Moniliformis dubius* (Acanthocephala). III. Effects in hamsters. *Journal of Parasitology*, **48**, 97–100.

Holmes, J. C. (1987) The structure of helminth communities. *International Journal for Parasitology*, **17**, 203–208.

Holmes, J. C. & Price, P. W. (1986) Communities of parasites. In *Community Ecology: Pattern and Process*, ed. D. J. Anderson & J. Kikkawa, pp. 187–213. Oxford: Blackwell Scientific Publications.

Huffman, M. A., Page, J. E., Sukhdeo, M. V. K., Gotoh, S., Kalunde, S., Chandrasiri, T. & Towers, G. H. N. (1996) Leaf-swallowing by chimpanzees: a behavioral adaptation for the control of strongyle nematode infections. *International Journal of Primatology*, **17**, 475–503.

Jackson, J. A., Tinsley, R. C. & Hinkel, H. H. (1998) Mutual exclusion of congeneric monogenean species in a space-limited habitat. *Parasitology*, **117**, 563–569.

Janovy, J. Jr, Clopton, R. E., Clopton, D. A., Snyder, S. D., Efting, A. & Krebs, L. (1995) Species density distributions as null models for ecologically significant interactions of parasite species in an assemblage. *Ecological Modelling*, **77**, 189–196.

Jokela, J. & Lively, C. M. (1995) Spatial variation in infection by digenetic trematodes in a population of freshwater snails. *Oecologia*, **103**, 509–517.

Kennedy, C. R. (1978) An analysis of the metazoan parasitocoenoses of brown trout *Salmo trutta* from British lakes. *Journal of Fish Biology*, **13**, 255–263.

Kennedy, C. R. (1997) Freshwater fish parasites and environmental quality: an overview and caution. *Parassitologia*, **39**, 249–254.

Kennedy, C. R. & Bush, A. O. (1992) Species richness in helminth communities: the importance of multiple congeners. *Parasitology*, **104**, 189–197.

Kennedy, C. R. & Bush, A. O. (1994) The relationship between pattern and scale in parasite communities: a stranger in a strange land. *Parasitology*, **109**, 187–196.

Kennedy, C. R., Bush, A. O. & Aho, J. M. (1986) Patterns in helminth communities: why are birds and fish different? *Parasitology*, **93**, 205–215.

Kuris, A. M. (1990) Guild structure of larval trematodes in molluscan hosts: prevalence, dominance and significance of competition. In *Parasite Communities: Patterns and Processes*, ed. G. W. Esch, A. O. Bush & J. M. Aho, pp. 69–100. London: Chapman & Hall.

Kuris, A. M. & Lafferty, K. D. (1994) Community structure: larval trematodes in snail hosts. *Annual Review of Ecology and Systematics*, **25**, 189–217.

Lang, B. Z. (1967) *Fasciola hepatica* and *Hymenolepis microstoma* in the laboratory mouse. *Journal of Parasitology*, **53**, 213–214.

Lewin, R. (1986) Supply-side ecology. *Science*, **234**, 25–27.

Lie, K. J., Basch, P. F. & Umathevy, T. (1966) Studies on Echinostomatidae (Trematoda) in Malaya. XII. Antagonism between two species of echinostome trematodes in the same lymnaeid snail. *Journal of Parasitology*, **52**, 454–457.

Lotz, J. M. & Font, W. F. (1985) Structure of enteric helminth communities in two populations of *Eptesicus fuscus* (Chiroptera). *Canadian Journal of Zoology*, **63**, 2969–2978.

Lotz, J. M. & Font, W. F. (1991) The role of positive and negative interspecific associations in the organization of communities of intestinal helminths of bats. *Parasitology*, **103**, 127–138.

Lotz, J. M. & Font, W. F. (1994) Excess positive associations in communities of intestinal helminths of bats: a refined null hypothesis and a test of the facilitation hypothesis. *Journal of Parasitology*, **80**, 398–413.

Lotz, J. M., Bush, A. O. & Font, W. F. (1995). Recruitment-driven, spatially discontinuous communities: a null model for transferred patterns in target communities of intestinal helminths. *Journal of Parasitology*, **81**, 12–24.

Marcogliese, D. J. & Cone, D. K. (1991) Importance of lake characteristics in structuring parasite communities of salmonids from insular Newfoundland. *Canadian Journal of Zoology*, **69**, 2962–2967.

Marcogliese, D. J. & Cone, D. K. (1997) Parasite communities as indicators of ecosystem stress. *Parassitologia*, **39**, 227–232.

Martin, D. R. (1969) Lecithodendriid trematodes from the bat *Peropteryx kappleri* in Colombia, including discussions of allometric growth and significance of ecological isolation. *Proceedings of the Helminthological Society of Washington*, **36**, 250–260.

Moore, J., Simberloff, D. & Freehling, M. (1988) Relationship between bobwhite quail social-group size and intestinal helminth parasitism. *American Naturalist*, **131**, 22–32.

Morand, S. & Poulin, R. (1998) Density, body mass and parasite species richness of terrestrial mammals. *Evolutionary Ecology*, **12**, 717–727.

Morand, S., Poulin, R., Rohde, K. & Hayward, C. (1999) Aggregation and species coexistence of ectoparasites of fishes. *International Journal for Parasitology*, **29**, 663–672.

Nee, S. R., Gregory, R. D. & May, R. M. (1991) Core and satellite species: theory and artefacts. *Oikos*, **62**, 83–87.

Overstreet, R. M. (1997) Parasitological data as monitors of environmental health. *Parassitologia*, **39**, 169–175.

Paperna, I. (1964) Competitive exclusion of *Dactylogyrus extensus* by *Dactylogyrus vastator* (Trematoda: Monogenea) on the gills of reared carp. *Journal of Parasitology*, **50**, 94–98.

Patrick, M. J. (1991) Distribution of enteric helminths in *Glaucomys volans* L. (Sciuridae): a test for competition. *Ecology*, **72**, 755–758.

Potts, G. W. (1973) Cleaning symbiosis among British fish with special reference to *Crenilabrus melops* (Labridae). *Journal of the Marine Biological Association UK*, **53**, 1–10.

Poulin, R. (1995) Phylogeny, ecology and the richness of parasite communities in vertebrates. *Ecological Monographs*, **65**, 283–302.

Poulin, R. (1997*a*) Parasite faunas of freshwater fish: the relationship between richness and the specificity of parasites. *International Journal for Parasitology*, **27**, 1091–1098.

Poulin, R. (1997*b*) Species richness of parasite assemblages: evolution and patterns. *Annual Review of Ecology and Systematics*, **28**, 341–358.

Poulin, R. (1998*a*) Comparison of three estimators of species richness in parasite component communities. *Journal of Parasitology*, **84**, 485–490.

Poulin, R. (1998*b*). *Evolutionary Ecology of Parasites: From Individuals to Communities*. London: Chapman & Hall.

Poulin, R. (1998*c*). Large-scale patterns of host use by parasites of freshwater fishes. *Ecology Letters*, **1**, 118–128.

Poulin, R. (1999) The intra- and interspecific relationships between abundance and distribution in helminth parasites of birds. *Journal of Animal Ecology*, **68**, 719–725.

Poulin, R. & Morand, S. (1999) Geographical distances and the similarity among parasite communities of conspecific host populations. *Parasitology*, **119**, 369–374.

Poulin, R. & Rohde, K. (1997) Comparing the richness of metazoan ectoparasite communities of marine fishes: controlling for host phylogeny. *Oecologia*, **110**, 278–283.

Price, P. W. (1980) *Evolutionary Biology of Parasites*. Princeton: Princeton University Press.

Rantanen, J. T., Valtonen, E. T. & Holopainen, I. J. (1998) Digenean parasites of the bivalve mollusc *Pisidium amnicum* in a small river in eastern Finland. *Diseases of Aquatic Organisms*, **33**, 201–208.

Richardson, D. J. & Nickol, B. B. (1999) Physiological attributes of the pyloric ceca and anterior intestine of green sunfish (*Lepomis cyanellus*) potentially influencing microhabitat specificity of *Leptorhynchoides thecatus* (Acanthocephala). *Comparative Biochemistry and Physiology Part A*, **122**, 375–384.

Rohde, K. (1979) A critical evaluation of intrinsic and extrinsic factors responsible for niche restriction in parasites. *American Naturalist*, **114**, 648–671.

Rohde, K. (1980) Diversity gradients of marine Monogenea in the Atlantic and Pacific Oceans. *Experientia*, **36**, 1368–1369.

Rohde, K. (1982) *Ecology of Marine Parasites*, St. Lucia: University of Queensland Press.

Rohde, K. (1992) Latitudinal gradients in species diversity: the search for the primary cause. *Oikos*, **65**, 514–527.

Rohde, K. (1993) *Ecology of Marine Parasites.* 2nd edn. Wallingford, UK: CAB International.

Rohde, K. (1994) Niche restriction in parasites: proximate and ultimate causes. *Parasitology* (Supplement), **109**, S69–S84.

Root, R. B. (1967) The niche exploitation pattern of the blue-green gnatcatcher. *Ecological Monographs*, **37**, 317–350.

Root, R. B. (1973) Organization of a plant–arthropod association in simple and diverse habitats: the fauna of collards (*Brassica oleracea*). *Ecological Monographs*, **43**, 95–124.

Rózsa, L., Reiczigel, J. & Majoros, G. (2000) Quantifying parasites in samples of hosts. *Journal of Parasitology*, **86**, 228–232.

Schad, G. A. (1966) Immunity, competition, and the natural regulation of parasite populations. *American Naturalist*, **100**, 359–364.

Shostak, A. W., Sharampaul, S. & Belosevic, M. (1993) Effects of source of metacercariae on experimental infection of *Zygocotyle lunata* (Digenea: Paramphistomidae) in CD-1 mice. *Journal of Parasitology*, **79**, 922–929.

Sogandares-Bernal, F. (1959) Digenetic trematodes of marine fishes from the Gulf of Panama and Bimini, British West Indies. *Tulane Studies in Zoology*, **7**, 69–117.

Sousa, W. P. (1990) Spatial scale and the processes structuring a guild of larval trematode parasites. In *Parasite Communities: Patterns and Processes*, ed. G. W. Esch, A. O. Bush & J. M. Aho, pp. 41–67. London: Chapman & Hall.

Sousa, W. P. (1993) Interspecific antagonism and species coexistence in a diverse guild of larval trematode parasites. *Ecological Monographs*, **63**, 103–128.

Sousa, W. P. (1994) Patterns and processes in communities of helminth parasites. *Trends in Ecology and Evolution*, **9**, 52–57.

Stock, T. M. & Holmes, J. C. (1987*a*) *Dioecocestus asper* (Cestoda: Dioecocestidae): an interference competitor in an enteric helminth community. *Journal of Parasitology*, **73**, 1116–1123.

Stock, T. M. & Holmes, J. C. (1987*b*) Host specificity and exchange of intestinal helminths among four species of grebes (Podicipedidae). *Canadian Journal of Zoology*, **65**, 669–676.

Stork, N. E. & Lyal, C. H. C. (1993) Extinction or 'co-extinction' rates? *Nature*, **366**, 307.

Sukhdeo, M. V. K. & Croll, N. A. (1981) The location of parasites within their hosts: bile and the site selection behavior of *Nematospiroides dubius*. *International Journal for Parasitology*, **11**, 157–162.

Sukhdeo, M. V. K. & Sukhdeo, S. C. (1994) Optimal habitat selection by helminths within the host environment. *Parasitology* (Supplement), **109**, S41–S55.

Toft, C. A. (1986) Communities of species with parasitic life-styles. In *Community Ecology*, ed. J. Diamond & T. J. Case, pp. 445–463. New York: Harper & Row.

Tompkins, D. M. & Clayton, D. H. (1999) Host resources govern the specificity of swiftlet lice: size matters. *Journal of Animal Ecology*, **68**, 489–500.

Valtonen, E. T., Holmes, J. C. & Koskivaara, M. (1997) Eutrophication, pollution, and fragmentation effects on parasite communities in roach (*Rutilus rutilus*) and perch (*Perca fluviatilis*) in four lakes in central Finland. *Canadian Journal of Fisheries and Aquatic Sciences*, **54**, 572–585.

Walther, B. A. & Morand, S. (1998) Comparative performance of species richness estimation methods. *Parasitology*, **116**, 395–405.

Wazura, K. W., Strong, J. T., Glenn, C. L. & Bush, A. O. (1986) Helminths of the beluga whale (*Delphinapterus leucas*) from the Mackenzie River Delta, Northwest Territories. *Journal of Wildlife Diseases*, **22**, 440–442.

Williams, H. H., McVicar, A. H. & Ralph, R. (1970) The alimentary canal of fish as an environment for helminth parasites. In *Aspects of Fish Parasitology*, ed. A. E. R. Taylor & R. Muller, pp. 43–77. Oxford: Blackwell Scientific Publications.

Wisniewski, W. L. (1958) Characterization of the parasitofauna of an eutrophic lake (parasitofauna of the biocoenosis of Druzno lake, part I). *Acta Parasitologica Polonica*, **6**, 1–64.

Zander, C. D. (1998) Ecology of host parasite relationships in the Baltic Sea. *Naturwissenschaften*, **85**, 426–436.

Zelmer, D. A. & Esch, G. W. (1999) Robust estimation of parasite component community richness. *Journal of Parasitology*, **85**, 592–594.

Chapter 15

Biogeographical aspects

Biogeography attempts to describe and explain the patterns in the geographical distribution of organisms. Most biogeographical studies attempt to identify and reconstruct the processes and events that account for present-day distributions. These studies rely on information from various disciplines such as systematics, ecology, evolution, paleontology, geology, climatology, and oceanography or limnology. In some cases, such as when dealing with strictly historical or ecological biogeographical studies, it is difficult to decide whether evolution, ecology, or biogeography is the driving discipline.

Biogeographic research embodies both historical and ecological approaches. **Historical biogeography** attempts to understand the sequences of origin, dispersal, and extinction of organisms to explain how geological events such as continental drift, glaciations, emergence and submergence of land masses, etc., influence the present patterns of distribution. Historical biogeography is divided into two branches, vicariant biogeography and dispersal biogeography. **Vicariant biogeography** attempts to explain species distributions based on the assumption that present-day distributions result from the fragmentation (vicariance) of the biota of a region into isolated populations due to the emergence of geographical barriers. **Dispersal biogeography** follows the simple premise that, from a center of origin, organisms disperse across pre-existing barriers by their own means. In the end, however, both **vicariance** and dispersal are likely to be responsible for the patterns of distribution seen in nature. **Ecological biogeography**, on the other hand, focuses mostly on extant organisms and attempts to account for their distribution patterns in terms of ecological processes occurring over short temporal and small spatial scales. The aim is to identify the processes that influence the distribution of species, maintain their diversity, and explain patterns of gradients in species richness and island colonization.

What about parasite biogeography? Fortunately, parasites have been in the forefront of much biogeographical research, mainly historical biogeography. Early in the development of the discipline (see Box 15.1), several authors noticed the close relationship that existed between hosts and parasites, and their geographical distributions. However, it was not until the advent of cladistic and phylogenetic analyses that an objective, analytical methodology became available to study historical biogeography allowing the field to move forward (Morrone & Crisci, 1995). Ecological biogeography, on the other hand, had a slow start. However, since the middle 1970s, a great deal of work by a handful of researchers has focused on parasites and created a solid foundation for further studies.

15.1 Factors affecting the geographical distribution of parasites

Different sets of factors and processes can shape the patterns of distribution of hosts and their parasites depending on the scale under consideration. On a global scale, distributions can be affected by the following: (1) The fragmentation and movement of land masses and the

Box 15.1 A brief record of historical parasite biogeography

The beginning of biogeography as a discipline can be traced back almost 200 years to the great explorer–naturalist, Alexander von Humboldt. His studies, beginning in 1805, were the first to conceptualize and quantify the primary role of climate in the distribution and morphology of plants around the world. The study of animal distribution (zoogeography) lagged behind phytogeography by several years. W. L. Sclater in 1858 made the first significant zoogeographic effort; at that time, he proposed a number of faunal regions based on the distribution of birds. Interestingly, the regions he created are very similar to the present-day classification of biogeographic realms, i.e., Nearctic, Palearctic, Oriental, Australian, Ethiopian, and Neotropical. At least 20 more papers were published before the turn of the century, marking the beginning of a prolific era in the biogeographical investigation of free-living organisms.

Biogeographic studies on parasites began in 1891 with the work of H. von Ihering, who recognized the significance of parasitic organisms in the zoogeography of host species. He was interested in the contribution of biological data to the question of land connections, especially among the landmasses in the Southern Hemisphere. Although he was not a parasitologist, he noted the presence of closely related species of *Temnocephala* (commensalistic turbellarians) on species of freshwater crustaceans on both sides of the Andes, and in New Zealand. He then presented evidence for an ancient biogeographic relationship between New Zealand and South America, arguing for the existence of a former land connection. Later, Metcalf (1929, 1940) studied the opalinid protozoans of leptodactylid frogs. He observed that the same genus of opalinid protozoans (*Zelleriella*) was present in frogs from South America and Australia, but was absent in frogs and toads from the Northern Hemisphere. He reasoned that leptodactylid frogs had not migrated into Eurasia because none of the Eurasian amphibians had *Zelleriella* as part of their parasite faunas. On the other hand, *Zelleriella* parasitized non-leptodactylid frogs in Brazil, such as *Bufo* and *Hyla*. He speculated, then, that these parasites had been acquired as a result of host capture from sympatric leptodactylid frogs. Metcalf postulated that leptodactylid frogs from South America and Australia were closely related based on the similarity of their opalinid protozoans. He also reviewed some previous studies on avian lice by V. Kellogg and S. Johnston, two investigators who also had studied the congruent distributions of host and parasites. These studies led Metcalf to suggest close relationships between certain species of European and North American birds, as well as between the South American rhea and the African ostrich, two species then thought to be unrelated. This approach to the study of relationships between groups of hosts based on the comparison of their parasite faunas and their distributions was referred to as the 'von Ihering method' of biogeography.

At this point, we should recall that the Theory of Continental Drift, although proposed by Wegener in 1912, did not gain acceptance until the 1960s. However, researchers such as Metcalf and Johnston embraced it early because their biogeographical findings agreed with the premises of Continental Drift. Unfortunately, the lack of support for the theory in the geological community at the time, and the fear of other scientists to embrace it and be stigmatized, tempered the success and expansion of their views.

Harold Manter (1940, 1955, 1963) made the next significant contribution. His interests were in the geographical distribution and affinities of digeneans along the Atlantic and Pacific coasts of Central America, of digeneans in marine fishes on a global scale, and on the biogeographical affinities of digeneans in freshwater fishes of South America. Although he did not embrace the Theory of Continental Drift until the 1960s, most of his biogeographic arguments required vicariant events that could not have occurred without Continental Drift. Manter's empirical approach kept biogeographical studies of parasites afloat until new methods were developed. Then, in the late 1970s, Brooks (for references see Brooks & McLennan, 1993) re-energized the field of historical biogeography. He combined the biogeographic concepts of his predecessors, the now widely accepted theory of Continental Drift, and Hennig's phylogenetic systematics, into a new research program with objective and repeatable methods for reconstructing and interpreting biogeographical and evolutionary patterns.

appearance of natural geographical barriers that divide and separate biotas (vicariance). For terrestrial organisms, the fragmentation of a land mass, the rising of a mountain, or the creation of a river, can be an effective barrier; for marine organisms, the emergence of a land isthmus, or even a change in the pattern of circulation of the ocean currents can fragment and restrict their distribution. (2) The ability of the organisms to move and increase their distributional range, regardless of vicariant events. This process includes both active and passive dispersal. In the case of parasites, many free-living stages disperse actively, whereas eggs, and the parasites inside a host, disperse passively over geographical areas. (3) The time frame within which vicariance and dispersal have occurred. Most geographical distributions are likely to be a consequence of both vicariant and dispersal events. The critical point in most instances is to determine if the disjunct distribution of a group of organisms (on both sides of a mountain range, or an ocean, for example) is because the barrier divided the original population (vicariance) or because a group of organisms was able to move and disperse across the barrier (dispersal).

Success in the colonization of new environments through dispersal is influenced by many abiotic and biotic factors that affect the organism's ability to cope and survive. Of the abiotic factors affecting the distribution of organisms, temperature and climatic conditions are, in many cases, the primary determinants. For many free-living organisms, competition, predation, and parasitism have been recognized as important biotic factors influencing their distributions; for parasites, however, the nature and biology of the host is probably the most important biotic factor influencing their distributions.

As scales decrease, the variables affecting host and parasite distributions become more specific. For example, pH, dissolved oxygen, the presence of calcium, water flow rates, substrate types, and depth, can affect host and parasite distributions in freshwater ecosystems; the ephemeral nature of a body of water also may play an important role. In terrestrial environments, soil type and composition, moisture content, vegetation, and altitude can be biogeographical determinants. In marine systems, temperature, salinity, depth, physiography of the substrata, and currents (as distinctive bodies of water) are important factors affecting distributions.

15.1.1 Freshwater and terrestrial systems

Even though terrestrial and freshwater systems seem to be rather discontinuous environments, they often are closely interrelated from a parasite's perspective. Significant numbers of 'terrestrial' parasites are dependent on freshwater systems. Many cestodes, most digeneans, and some acanthocephalans rely almost exclusively on aquatic or semi-aquatic intermediate hosts to complete at least one step of their life cycles. In these cases, the factors affecting the distribution and dispersal of parasites in terrestrial definitive hosts can be constrained by processes operating within aquatic systems.

Most cestodes in exclusively terrestrial animals such as lizards, some turtles, birds and mammals rely on predator–prey interaction in at least one stage of their life cycles. As a result, patterns of parasite distribution on a relatively local scale are determined by the presence of both intermediate and definitive hosts; these, in turn, are affected by local climatic and physiographic conditions. Ultimately, these patterns will be influenced by biotic factors related to the host such as sex, age, trophic status, vagility, and foraging areas.

Distributions of parasites with life cycles that do not require intermediate hosts, e.g., some nematodes and arthropods such as lice and fleas, depend, to a great extent, on specific environmental factors, mainly temperature and humidity. The latter factors are extremely significant for nematodes with terrestrial free-living stages since humidity conditions, temperature, and substrata types directly affect the survivorship and infectivity of their eggs or larvae. The same is true for the free-living stages of parasites with complex life cycles, i.e., those that require an intermediate host. The protostrongylid nematode *Elaphostrongylus rangiferi*, has a complex life cycle including a terrestrial snail as intermediate host. The adult nematode parasitizes reindeer *Rangifer tarandus*, and causes extensive neurotrophic disorders. Skorping (1982) investigated the effects of temperature and moisture on the infection of the snail host. He found that infection of the terrestrial snail was temperature-dependent, with higher rates of infection at higher temperatures. Infection was high when the snail was exposed to the nematode larvae in water and relatively low when soil was used as a substratum. Although this may sound like nonsense for a terrestrial snail–parasite system, the larvae require at least a thin film of water in order to move effectively. The higher the moisture content of the soil, the higher the infection rate of the snail by the nematode. Therefore, unless temperature and moisture conditions of the soil are appropriate, the life cycle of this and other species cannot be completed and the parasite cannot become permanently established in the area. Most nematodes become inactive between 5 °C and 15 °C and have a thermal optimum between 15 °C and 30 °C. A similar temperature-dependent rate of infection has been observed for digenean miracidia in aquatic environments.

Some of the factors affecting parasite distribution in freshwater systems have been described by Bailey & Margolis (1987) in their study of the parasite fauna of juvenile sockeye salmon *Oncorhynchus nerka*, in a series of lakes along the northwest coast of North America. They reported that the parasite fauna was similar among lakes of similar trophic status and occurring within the same biogeoclimatic zones. Their analysis also revealed that many parasites with similar modes of transmission or using the same intermediate hosts within a geographic range had the same distributions. This finding underscores the importance of both intermediate and definitive hosts in the geographic distribution of parasites.

The same study also revealed a substantial variation in the pattern of infection over time in relatively mesotrophic lakes and less variation in oligotrophic lakes. This finding lends support to the relationship between the allogenic–autogenic species concept and the trophic status of a particular lake (Esch *et al.*, 1988). In oligotrophic systems, most of the parasite species are autogenic, completing their life cycles within the confines of the system; this allows for some degree of 'predictability' from year to year in terms of species distributions. Eutrophic lakes, on the other hand, have an increased number of allogenic species whose life cycles are completed in birds and mammals that are not confined to the lakes. In these systems, the presence of many larval parasites in fishes will depend on the migration patterns and residency times of their 'ephemeral' definitive hosts. Once again, the distribution of intermediate and definitive hosts as a consequence of the trophic status of a lake has a significant impact in the distribution of parasites. Bailey & Margolis (1987) concluded 'evidently, geography influences the characteristics of the parasite fauna, but the trophic status of the lake and many biotic variables clearly have strong influences on the parasite faunas studied'.

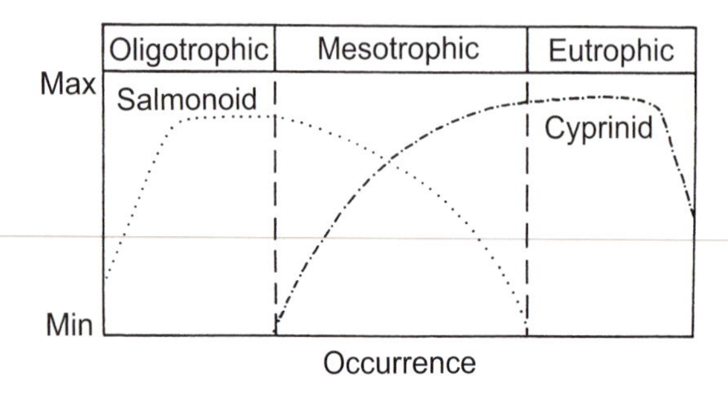

Fig. 15.1 Occurrence of salmonid and cyprinid fishes in relation to lake types. Salmonid fishes are dominant in oligotrophic waters, cyprinids in eutrophic lakes. Mesotrophic lakes contain mixed populations and have the greatest range of species. (From Chubb, 1970, with permission, Blackwell Scientific Publications.)

Chubb (1970) noted a similar pattern in his study of the parasite fauna in British freshwater fishes, which emphasized aspects of parasite diversity in terms of host distribution and lake trophic status. He observed that the distribution patterns of acanthocephalans, for example, were determined primarily by the presence and absence of isopod intermediate hosts. He reported that salmonid fishes and their parasites were dominant in oligotrophic lakes, whereas cyprinids and their parasites dominated eutrophic systems. Mesotrophic lakes possessed mixed host species populations and had a greater range of both host and parasite species (Fig. 15.1).

The significance of water flow and drainage systems was investigated by Arai & Mudry (1983) to determine the potential movement of parasites across the Continental Divide in North America because of a possible water diversion in northeastern British Columbia, Canada. They found 88 parasite species in 20 species of fishes from both the Arctic and the Pacific drainage basins. Twenty-six parasite species had disjunct distributions, occurring either in the Pacific or the Arctic basin even though most of their hosts were present in both drainage basins. This suggested a close association of the parasites with a given drainage system and indicated a restricted dispersal, against the flow, across the Continental Divide.

Unlike most endoparasites, ectoparasites are affected directly by the physical and chemical characteristics of the water. Ectoparasites of freshwater fishes, such as protozoans, monogeneans, and copepods, are affected by changes in salinity. Polyanski (1961) reported that some protozoans disappeared completely as salinity increased. The best examples for the effect of salinity on parasite distributions are provided by anadromous and catadromous fishes such as salmon (*Oncorhynchus*) and eels (*Anguilla*), respectively. Anadromous salmonids lose their freshwater ectoparasites soon after they reach marine waters, whereas catadromous eels lose their marine ectoparasites soon after entering freshwater. There are, however, some exceptions to this pattern. Among the copepods, for example, some species in the genus *Ergasilus* are euryhaline. Johnson & Rogers (1973) grouped the copepod species present in several Gulf of Mexico drainage systems into five categories according to their salinity tolerance; these included coastal, estuarine, coastal/inland, estuarine/inland, and inland species. Those with greater tolerance for salinity were the coastal/inland species, e.g., *Ergasilus versicolor* and *Ergasilus clupeidarum*, and the estuarine/inland species, e.g., *Ergasilus arthrosis*. Not surprisingly, these copepods are normally associated with fish hosts such as *Mugil cephalus* (striped mullet), *Alosa* spp., and *Derosoma* spp., which also have a high tolerance to salinity changes.

A determinant typically overlooked when dealing with digenean distributions in freshwater systems is the amount of calcium (hardness) in the water. This is an important limiting factor for the establishment and maintenance of molluscan communities. Because molluscs are obligate first intermediate hosts for most species of digeneans, the richness and abundance of molluscs constrain the occurrence and distribution of most digeneans. Water pH also plays an important role in the distribution of parasites, affecting both the parasite and the hosts. Low pH (acidic conditions), for example, interferes with calcium deposition in molluscs, making them extremely sensitive to acidification. Cone *et al.* (1993) studied the distribution of parasites of yellow eels *Anguilla rostrata*, in acid and near-neutral rivers in Nova Scotia, Canada. Eels from the river with near-neutral pH had the highest number of parasites, most of them specific to eels. Eels from the acidic river had fewer species of parasites, and the few species present were the same as those found in the eels from the near-neutral river. The missing species in the acidic river were all digeneans that use clams or snails as first intermediate hosts. A decline or elimination of these molluscs in the acidic river probably accounts for the local extinction of digeneans in eels.

15.1.2 Marine systems

The abiotic factors affecting the distribution of marine parasites are similar to those in freshwater systems, although temperature and salinity are of paramount importance because they characterize and define masses of water.

A now-classic example of the effect of temperature on parasite distribution in the ocean is that of the hemiurid digeneans *Lepidapedon sachion*, *Lepidapedon elongatus*, and *Derogenes varicus*. Manter (1934) found these digeneans in fishes collected at depths ranging from 270 to 550 meters off the coast of Florida (USA). However, in more northern latitudes, these species are abundantly distributed in fishes from much shallower waters. *Derogenes varicus* is truly cosmopolitan, with an almost complete lack of host specificity and a characteristic **amphitropical** distribution. It has been found in well over 100 species of fishes from near the Arctic to near the Antarctic, at a wide range of depths, and from coastal to bentho-demersal fishes. This type of amphitropical distribution has been reported many times for free-living benthic species that follow isothermal bands of water across the ocean floor. Cold water (4 °C) is much closer to the surface at the poles than in the tropics. These masses of polar water move from pole to pole across the Atlantic and Pacific Oceans following their submergence at the **Antarctic Convergence** (75° S), an oceanic frontal system delimiting the Antarctic from the sub-Antarctic. It is believed that the distribution of *D. varicus* in tropical waters is determined by its association with masses of cold water that move through the tropics at greater depths.

The significance of the Antarctic Convergence from a strictly oceanographic view has been widely acknowledged and so too has been its impact on the distribution of marine, or marine-dependent, organisms. Its importance in parasite distribution is also becoming more apparent. Hoberg (1986*a*) studied certain ecological and biogeographical characteristics of acanthocephalans in the genus *Corynosoma* in Antarctic seabirds. He examined various aspects of the host–parasite relationships and the geographical distribution of different species in piscine paratenic hosts, and in avian and mammalian definitive hosts. The study concluded that host–parasite coevolution probably had an important influence on the species composition of acanthocephalans in different hosts within historical times. However, the current, restricted distribution of *Corynosoma* spp. in their hosts, suggested that oceanographic factors such as the Antarctic Convergence could be limiting their range too. The Antarctic Convergence probably restrains the distribution of the first intermediate hosts (mostly amphipods) and piscine paratenic hosts, confining them to specific masses of water in the southern oceans.

The potential effect of depth on parasite distribution is not yet clear because of the confounding influences of other correlated variables such as temperature, barometric pressure, and relative abundance of food items. Although the effect of depth on the distribution of specific parasites has not been well studied, overall parasite diversity decreases significantly with depth. Because free-living organisms exhibit the same trend, changes in parasite diversity may be a consequence of changes in abundance and diversity of the free-living fauna and, consequently, of potential intermediate hosts (Campbell *et al.*, 1980); the same processes may, however, affect parasites in a more direct manner (see also section 15.3.3).

15.2 Patterns of distribution

The goal of historical biogeography is to reconstruct processes involved in the origin, dispersal, and extinction of taxa and biotas. Vicariance events and dispersal patterns that affect the host and the parasite, host specificity, host capture, coevolution, and co-accommodation are some of the elements commonly considered as significant in these processes. The relative importance of each one of these factors varies according to the parasite group and the hosts involved.

Many terrestrial and freshwater organisms are highly restricted in their distribution because of the powerful geographical barriers present in these environments. Land barriers between bodies of water, bodies of water isolating landmasses, mountain ranges, and the patchy distribution of favorable habitats clearly limit the dispersal of species. Depending on the nature of the hosts, of course, many of these barriers can be breached. Clearly, these barriers have less effect on many birds and migratory species than other organisms. Present and historical land discontinuities such as the current structure of the continents and their historical genesis through plate tectonics, i.e., Pangaea, Laurasia, and Gondwana, and the appearance and disappearance of land bridges between them, i.e., the Bering Strait, Isthmus of Panama, etc., are also critical factors affecting the distribution of both parasites and their hosts.

In the marine environment, patterns of distribution are affected by somewhat different physiographic factors. First, although the ocean is continuous and is constantly moving, it certainly is not homogeneous. There are very distinctive masses of water that normally have a distinctive fauna, at least in terms of their pelagic and planktonic species. Second, it is a three-dimensional realm and organisms are influenced by geographic location and depth. Although terrestrial and freshwater environments are also three-dimensional, the extreme depths and continuity of the marine habitat provide for a much more diverse array of conditions. Third, factors that, in the terrestrial environment constitute bridges or barriers, mean just the opposite for marine biotas. For example, a bridge was formed between North and South America (the Isthmus of Panama) 6 million ybp and facilitated the dispersal of many terrestrial and freshwater organisms between the two continents; it also, however, caused isolation of the Pacific and Atlantic marine biotas across the tropical belt. In contrast, the submergence of the Bering Strait significantly reduced faunal exchanges between Asian and North American terrestrial realms, whereas it created a connection that allowed for the exchange of biotas between the North Pacific and Arctic Oceans.

15.2.1 Patterns of parasite distribution in terrestrial and freshwater habitats

There have not been many comprehensive attempts to examine zoogeographical aspects of terrestrial parasites within the context of their host's distribution. Because most authors have recognized the parallel nature of host and parasite distributions, many of the studies have included reference to host–parasite relationships and coevolution, and have used parasites as 'biological markers' or indicators of possible host origins.

As discussed in Box 15.1, von Ihering (1891, 1902) and Metcalf (1929, 1940) were the first to assign paleobiological significance to the distribution of parasites and their hosts (respectively, *Temnocephala* in crustaceans and opalinids in frogs). Johnston (1912) reported that digeneans present in Australian frogs had their nearest morphological relatives in digeneans of anurans from South America, Africa, Asia, and Europe, not among digeneans living in other Australian vertebrates. At the time, highly complex and contrived explanations, some involving hypothetical land bridges of different types, were necessary to explain the nature of these rather 'spotty' distributions. However, with the development of modern geology and geophysics, the notion of a dynamic earth crust with moving continental plates, and of the oceans expanding and shrinking at oceanic ridges and subduction zones, the early explanations were greatly simplified. In doing so, a reliable account of the history of the continental masses was generated (Fig. 15.2). (For a detailed explanation of plate tectonics and Continental Drift, see Dietz & Holden, 1970). It is now widely accepted that the break-up of the ancestral continent of Pangaea about 150 million ybp led to geographical isolation (a vicariant event) and allopatric speciation of many organisms, including parasites.

Prudhoe & Bray (1982) provided a detailed account of the platyhelminth parasites in Amphibia worldwide. Although their emphasis was on systematics, they recognized that most genera of amphistome digeneans infecting recent amphibians, particularly frogs, have arisen through geographical isolation. Sey (1991), in his

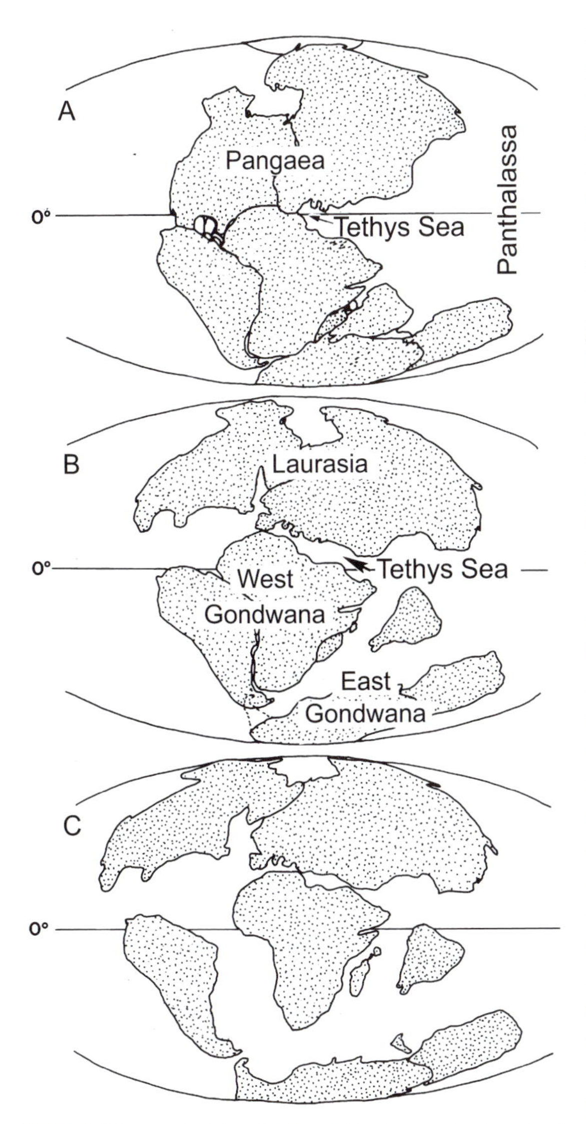

Fig. 15.2 Past configuration of the landmasses and oceans at different times based on actual views of plate tectonics and Continental Drift. (A) Supercontinent Pangaea in the Triassic, 200 million ybp. Panthalassa was the ancestral Pacific Ocean and the Tethys Sea the ancestral Mediterranean. (B) In the late Jurassic or early Cretaceous, after 65 million years of drift (135 million ybp) the Northern Pangaea became Laurasia; the Southern Pangaea, after becoming Gondwana, split into East Gondwana, West Gondwana. and India. The Atlantic and Indian Oceans are already opened. (C) At the end of the Cretaceous, after 135 million years of drift (65 million ybp), the Mediterranean is clearly recognizable; the Atlantic has become a major ocean separating South America from Africa, while Australia still remains attached to Antarctica.

monograph on amphistomes of all host groups, agreed with Prudhoe & Bray (1982) but he was able to characterize even further the extent of dispersal by the parasites. For example, amphistomes in non-mammalian hosts (mostly amphibians and reptiles) have intercontinental distributions that agree with the distribution of intermediate and definitive hosts. These distributions can best be explained in terms of the main continental disjunctions caused by continental drift (vicariance). The distribution of amphistomes in mammalian hosts, on the other hand, is much more variable. Although the role of continental drift is still recognizable in their distribution patterns, their present distribution has been determined mostly by the greater dispersal capabilities of their mammalian hosts.

A rather different pattern of distribution, with vicariance and dispersal but no speciation, is shown by the nematode *Rhabditis (Pelodera) orbitalis*. Adults and preadults of *Rhabditis orbitalis* are microbotrophic nematodes and live in the nests of rodents throughout the Holarctic. The 3rd stage larvae, however, are parasitic and enter the orbital fluid of rodents where they remain as parasites for several days. Schulte & Poinar (1991) completed reciprocal crosses between European and American strains, corroborating that all the populations belong to the same species. They argued that because *R. orbitalis* has a Holarctic distribution, the association with the rodents must have evolved when the continents were still connected by the now submerged Bering Strait. They attributed the lack of divergence between American and Eurasian populations to the similarity in ecological conditions present on the two continents. However, it is also possible that the time that has passed since the disappearance of the land connection between Eurasia and America has not been long enough to allow for genetic divergence.

Manter (1963) demonstrated the similarity in patterns of parasite distribution in freshwater and terrestrial parasites by studying the geographical affinities of digeneans in South American fishes. Intercontinental relationships normally included related genera present in fishes in South America, Africa, and India, all components of the Gondwana supercontinent. The different subfamilies of amphistome digeneans, for example, had the strongest pattern of Gondwanic distribution because they have an exclusive African–South American distribution. On the other hand, fishes from North America and Eurasia share at least eight genera and related species of digeneans not present in other areas, probably a consequence of their past history as components of the Laurasian supercontinental fauna. Studies like those of Manter, completed more than 35 years ago, provide an important baseline for the development of zoogeographical studies on parasitic organisms.

In the last 20 years, several investigators have combined biogeographic and phylogenetic studies of both hosts and parasites to examine evolutionary aspects of host–parasite relationships from an historical perspective. Such an approach was used to examine species of the digenean *Glypthelmins* parasitizing frogs (Brooks, 1977; Brooks & McLennan, 1993). These studies showed that *Glypthelmins* is a relatively ancient lineage that existed before the fragmentation of the supercontinent of Pangaea (see Fig. 15.2). The genus was apparently widespread throughout South America and Laurasia. The breakup of Pangaea first isolated the neotropical species from the Laurasian species. Later, when Laurasia split into Eurasia and North America, the remaining species were also isolated. When the distribution and phylogenies of the amphibian hosts of *Glypthelmins* are considered, hosts and parasites show a reasonable amount of congruence. However, it also becomes apparent that a significant degree of host switching occurred. Although it is not clear how the switching took place, detailed studies of the North American species suggest that changes in the life cycle of the digeneans are responsible for the switch. In the primitive (plesiomorphic) condition, the cercaria, with the aid of its penetration stylet, penetrates and encysts in the skin of frogs. Later, the frogs become infected when they ingest their epidermis along with the encysted metacercaria. In the North American species that exhibit host switching, the cercariae lack penetration stylets and, instead, they crawl into the nares of their hosts when they are still tadpoles. They then migrate to the abdominal cavity where they become unencysted metacercariae; later, they migrate into the intestine and mature when the tadpoles metamorphose into adults. The beauty of this change in life cycle becomes evident when the habitat of the new hosts is considered. The 'original' hosts (ranid frogs) spend a significant amount of time in and around the water and the aquatic cercariae have an extended window of opportunity for recruitment into the frog. The new hosts (hylid frogs), however, spend little time in water. Indeed, once they become adults, they only come to the water for a very short time to breed. In this case, the window of opportunity for infection to occur with the ancestral life cycle would be extremely narrow. But a derived condition, in which the tadpoles, and not the adults, are the targets of infection by the cercariae, greatly increases the chances of finding a host and completing the life cycle. In short, the biogeography and diversification of this group of digeneans appears to be a product of geographical events (vicariance), dispersal, host switches, and changes in their life cycle patterns.

Brooks (1979) also examined the evolutionary relationships between crocodiles and their digeneans. Again, based on phylogenetic and biogeographical analyses, he argued that the intercontinental relationships of the digeneans were coincident with the patterns of continental fragmentation since the Cretaceous; these relationships were also congruent with the biogeographic affinities of the crocodiles. The results provided a strong indication of coevolutionary events, but also biological evidence for the proposed patterns and timing in the fragmentation of the supercontinents.

Based on studies such as these, it is important to emphasize the relative contribution of Continental Drift, geophysical events, dispersal, coevolution, host–parasite associations, and local ecological conditions, in describing the observed patterns of distribution of the different groups.

15.2.2 Patterns of distribution in marine parasites

Much information is available about patterns of distribution for marine parasite species; importantly, substantial numbers have been studied on a more or less global scale in association with their hosts. The reasons for the difference in the number of studies between marine and terrestrial/freshwater systems are mostly related to economic factors. Marine organisms, from shellfish to fishes to whales, have been, and some still are, a significant and diverse source of protein and, therefore, an important component of the human diet almost worldwide. As a result, many studies have been undertaken to assess parasites as potential pathogens for marine organisms and the human consumer. Similarly, the realization that some parasites could be used as 'biological markers' for the identification of breeding stocks, foraging areas, or migratory routes, of many fish species also increased the funding for research directed toward the elucidation of these more 'practical' problems. From a purely academic perspective, however, these studies have provided valuable information about host–parasite coevolution and the origins of some host groups. Most of the work has focused on ecological biogeography and parasite distribution in relation to patterns of host specificity and distribution. Little biogeographical work has been done using analytical historical approaches but, fortunately, a significant amount of information can be extracted from evolutionary studies of host and parasites using **phylogenetic systematic** approaches.

HISTORICAL BIOGEOGRAPHY

Excellent examples of historical biogeography are the studies by Hoberg (1986*b*, 1992, 1995, 1997) on the cestodes of alcid birds (puffins, murres, auks, and their relatives) and pinniped mammals (seals, sea lions, walruses, and their relatives) in the Arctic. At the present time, it is common practice to attribute the low number of species present at high latitudes to widespread extinction episodes that occurred during the glacial epochs of the Pliocene and Pleistocene. In other words, the fauna that we see today at high latitudes would be the remnants of a pre-existing fauna that managed to survive the inclement environmental conditions. The cestode fauna of alcids and pinnipeds, however, challenges this generalized view. The available data show some interesting features. First, the cestode fauna of these two distinct host groups is depauperate. Second, the host and parasite phylogenies for alcids and their cestodes and pinnipeds and their cestodes are highly incongruent, indicating very little coevolution but extensive host switching with subsequent radiation. Third, host switches occurred mainly among hosts of similar trophic and ecological requirements. Fourth, the patterns of speciation of cestodes in alcids and in pinnipeds are congruent and synchronic. Based on these observations, it seems that the host–parasite assemblages in the Arctic are not remnants of a previous biota. Instead, they represent new assemblages structured by host switching, geographical colonization, and **speciation of peripheral isolates** driven by the changing geographic ranges of the hosts due to the dramatic fluctuations in the environment during the Pliocene–Pleistocene glaciations. Therefore, during periods of maximum glaciation, the sea level dropped, exposing the continental shelf and trapping populations of alcids, phocids, and their parasites in small areas, or refugia, throughout the high latitudes of the Holarctic region. These conditions favored cestode speciation through vicariance and host switching. Then, during the periodic retreat of the glaciers, the ocean level rose and allowed the confined species to disperse and extend their geographical range, increasing the chances of host switching even more. These events form the foundation for Hoberg's **marine refugium hypothesis**, which suggests that extreme environmental fluctuations during the Pliocene–Pleistocene glaciation promoted the diversification, not the extinction, of some organisms. The sequence of events that led to the biogeographic patterns seen in the cestodes of alcids (*Alcataenia* spp.) and pinnipeds (*Anophryocephalus* spp.) during the Pliocene and Pleistocene is illustrated in Fig. 15.3.

PARASITE DISTRIBUTION AND HOST SPECIFICITY

Broad geographic distributions of parasites in marine environments result from high degrees of ecological plasticity and tolerance of various environmental conditions, low host specificity for one or more hosts in the life cycle, widespread distributions of one or more hosts in the life cycle or, more likely, a combination of several of these factors.

The broad geographic distribution of the digenean *Derogenes varicus*, in association with masses

Fig. 15.3 Summary of historical biogeography for species of *Anophryocephalus* among phocids and otariids and for species of *Alcataenia* among alcids. Map shows the extent of exposed continental shelf (shaded area) during a reduction in sea level to −100 m during glacial maxima. Partitioning of the North Pacific and Arctic basin into regional (e.g., Arctic Basin, Pacific Basin, Okhotsk Sea) and insular refugia (Aleutian Islands and Kurile Islands) was influenced by fluctuations in sea level. (1) The postulated origin of *Anophryocephalus* in *Phoca* sp. and *Alcataenia* in larids (seagulls) was in the North Atlantic about 3.0–3.5 million ybp. (2) Range expansion occurred through the Arctic Basin and resulted in the development of early Holarctic distributions for hosts and parasites about 2.5–3.0 million ybp. (3) Initial entry to the North Pacific Basin through the Bering Strait occurred soon after the submergence of Beringia, about 2.5–3.0 million ybp. *Alcataenia* diversified following colonization of puffins, through sequential colonization and radiation in auklets, murres, and guillemots. Radiation of *Anophryocephalus* continued among species of *Phoca* in the North Pacific Basin. (4) Secondary Holarctic ranges were attained later by species of *Alcataenia* among murres and guillemots during the Quaternary period. (5) *Anophryocephalus* colonized otariid mammals less than 2.0 million ybp. (From Hoberg, 1992, with permission, *Journal of Parasitology*, **78**, 601–615.)

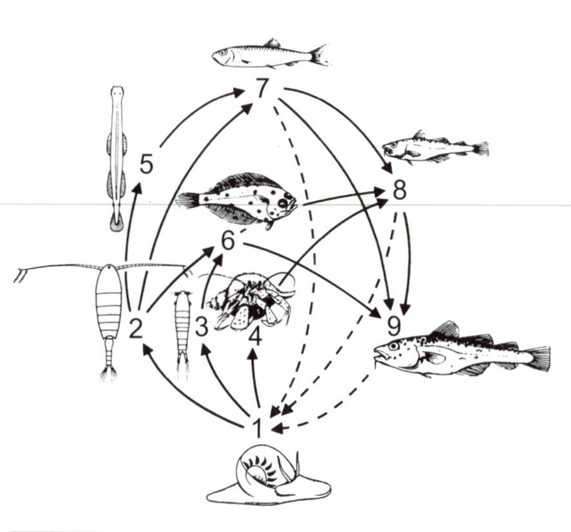

Fig. 15.4 The life cycle of the digenean *Derogenes varicus*, involving planktonic invertebrates: (1) *Natica* spp. (benthic mollusc); (2) calanoid copepod; (3) harpacticoid copepod; (4) hermit crab (benthic); (5) *Sagitta* spp.; (6) small fishes; (7) planktophagous fishes; (8) benthophagous and piscivorous fishes; (9) piscivorous fishes (large cod). (From Køie, M. (1979) On the morphology and life history of *Derogenes varicus* (Muller, 1784) Loos, 1901 (Trematoda, Hemiuridae). *Zeitschrift für Parasitenkunde*, **59**, 67–78. © Springer-Verlag.)

of cold water, has already been mentioned. There are, however, other features of some importance in this parasite's ubiquitous distribution, such as the almost complete lack of host specificity for both definitive and intermediate hosts. Køie (1979) examined its life cycle and determined that cercariae from naturally infected molluscs (*Natica* spp.) were infective to calanoid and harpacticoid copepods, as well as to benthic hermit crabs, where they develop into mesocercariae (Fig. 15.4). A diverse group of fishes with different feeding habits, from benthic to planktivorous, as well as chaetognaths (*Sagitta* spp.), become the next intermediate hosts upon ingestion of hosts carrying mesocercariae. Definitive hosts include fish species with diverse feeding habits, ranging from planktivory to piscivory in both the pelagic and benthic regions. Moreover, adult worms also can be transferred from host to host by predation of fishes infected with adult digeneans.

For parasites with complex life cycles, this type of host utilization seems to be the only possible way to achieve a global distribution. The nema-

Table 15.1 Latitudinal gradients in numbers of monogenean and digenean genera of commercial marine fishes

Biogeographical zone	Number of species of commercial fish	Number of genera[a]	
		Monogenea	Digenea
Arctic	approx. 20	15	12
North Boreal	40–50	20	50
South Boreal	>50	50	160
Tropical	100–120	95	300
North Antiboreal	40–50	40	80
South Antiboreal plus Antarctic	approx. 20	3	10

Note:
[a] Parasites of tropical fish are poorly known compared to fish from higher latitudes and, since 1969 when this table was composed, many new genera of Monogenea and Digenea have been described from tropical oceans.

Source: From Lebedev (1969).

tode *Anisakis simplex*, for example, also has a similar pattern of broad distribution, with an almost complete lack of host specificity for intermediate hosts, and with innumerable paratenic hosts.

15.3 Ecological aspects

Various aspects of parasite distribution have been examined throughout this chapter in terms of vicariance, dispersal, and the impact of evolutionary history. In this section, we focus on ecological factors that might influence the distribution of organisms producing spatial gradients of species richness. Some of the gradients that affect the distribution and abundance of species are latitude, altitude, salinity, depth, light intensity, and, although not a gradient, the frequency and intensity of physical disturbances.

Pianka (1983) and Brown (1988) reviewed a number of hypotheses directed at understanding patterns of diversity and distribution in free-living organisms. They emphasized that the various hypotheses that had been proposed are not necessarily exclusive; indeed, in most cases, a combination of several hypotheses provides the best explanation. The hypotheses considered include time since perturbation, climatic stability and predictability, productivity, habitat heterogeneity, interspecific interactions (competition and predation), ecological and evolutionary time, and rates of speciation and extinction. Similar patterns to those described for free-living organisms have been documented for parasites, and some of these hypotheses will be discussed next in the context of parasites.

15.3.1 Latitudinal gradients in species diversity and richness

The significant increase in diversity from the poles to the Equator is one of the more clear-cut patterns in ecological biogeography. Although the rate of increase in species richness towards the equator varies among taxonomic groups, the trend is consistent in species of high taxonomic groupings. In other words, if you consider the species in a given genus, the genus might not show an increase in species richness towards the tropics but, if you consider the species within an order, or class, these will show an increase in richness towards the tropics. The database available for parasites is far from comprehensive. Nonetheless, observations on several genera of Monogenea and Digenea in marine teleost fishes show the same trend. Rohde (1978*a*, 1993) summarized data from Lebedev (1969) (Table 15.1) and showed that the number of monogeneans and digeneans increases toward the tropics. He emphasized that this trend might be even stronger because the number of

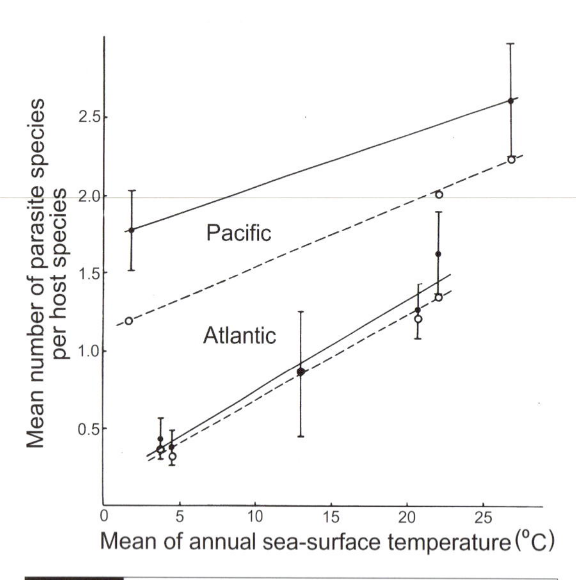

Fig. 15.5 Relative species richness (average number of parasites per host species) of monogenean gill parasites on teleost fish in the Pacific and Atlantic Oceans plotted against the approximate mean of annual sea-surface temperature at various localities. Open circles: total number of monogenean species / total number of host species examined at each locality; closed circles: idem, but monogenean species occurring in *x* host species counted *x* times. (Modified from Rohde, K. [1980] Diversity gradients of marine Monogenea in the Atlantic and Pacific Oceans. *Experientia*, **36**, 1369–1371.)

Monogenea and Digenea in the tropics is probably underestimated. The increase in number of parasite species toward the equator is also correlated, to some extent, with the increase in number of host species available. However, if the number of monogenean species/fish species (relative richness) is considered, the trend of increase towards the equator persists (Fig. 15.5) (Rohde, 1980). One of the few exceptions to this pattern is for parasites of marine mammals; in these hosts, species richness is greater in the cold waters of North Temperate Zones (Table 15.2). A partial explanation for this observation is the greater diversity of marine mammals in the Arctic regions when compared with the tropics. On the other hand, Hoberg's (1992) hypothesis of a Boreal marine refugium during the Pliocene and Pleistocene (see section 15.2.2) can probably explain the overall greater diversity of parasites in marine mammals at high latitudes.

One hypothesis commonly used to explain the latitudinal pattern of species richness relates to changes in niche breadth and the intensity of interspecific interactions. It has been argued, for example, that species in tropical latitudes are less limited by abiotic factors and more by biotic factors, with interspecific interactions being more important than intraspecific ones. Of the various parameters that can be used to define the niche of a parasite species, food requirements, host range, and microhabitat, seem to be the most important (Rohde, 1979).

Rohde (1978*b*, 1981, 1993) examined this question when he attempted to identify correlations between diversity gradients and possible gradients in the 'width' of the fundamental niches of parasites. Based on observations from several parasitological surveys, Rohde (1978*b*) indicated that the Digenea of marine fishes have more restricted host ranges in warm than in cold waters. Monogenea, on the other hand, have similar host ranges at all latitudes (Fig. 15.6). The only significant data set available to document microhabitat utilization in ectoparasites of marine fishes showed a lack of correlation between the number of microhabitats available and the number of parasites present (Fig. 15.7). Similarly, the ratio of monopisthocotylean to polyopisthocotylean monogeneans (a grouping based on the morphology of the opisthaptor that correlates with feeding strategy) calculated from many extensive surveys in different localities did not show any trend (Rohde, 1993). The rationale for the latter effort was that polyopisthocotylean monogeneans are obligatory blood feeders, whereas most monopisthocotyleans feed on mucus and epithelial cells and thus represent different feeding guilds. Based on these results, Rohde (1993) concluded that the fundamental niches of marine parasites were not affected significantly by the number of species in the community and that interspecific competition probably was not an important determinant in the dimension of the fundamental niche. Therefore, the increase in species richness toward the equator could not be explained in terms of narrower niches that could allow for a tighter packing of the species. The evident conclusion was that many empty niches were available in hosts from temperate and polar regions.

Table 15.2 | Helminth species of pinnipeds and cetaceans at different latitudes

	Digeneans	Cestodes	Nematodes	Acanthocephalans	Total
Arctic	6	7	9	3	25
Boreal	30	24	31	9	94
Tropical	7	2	18	4	31
Antiboreal	0	9	21	8	38
Antarctic	1	7	5	1	14

Source: With permission, from Rohde, K. (1993) *Ecology of Marine Parasites*, 2nd edn. Wallingford, UK: CAB International.

Climatic stability and predictability, another explanation commonly used to account for some latitudinal gradients, do not differ significantly between cold and warm seas (Rohde, 1989). Although temperature tends to be more constant in tropical and polar waters than in temperate areas, Rohde argued that the changes are not significant and noted that temperature fluctuations in tropical waters can be dramatic during tidal cycles. Spatial heterogeneity (at least for monogeneans) is not a significant factor in latitudinal gradients either, especially when the nature of the fish host is considered. For example, counts of gill filaments and surface area are quite similar among fishes at different latitudes. Productivity is known to be similar in cold and warm waters, and biotic interactions such as competition and predation also seem to be unimportant. Parasite competition, as discussed in terms of niche width, was not significant in the studied group (monogeneans), and predation, as seen in cleaning symbioses, for example, provided insufficient evidence because little is known about this association at higher latitudes, or about its ecological effects.

Rohde (1989) considered two other hypotheses, ecological time, which claims that species capable of colonizing certain habitats exist but have not had enough time to spread into them, and evolutionary time, which claims that species for certain habitats have not yet had time to evolve. He disregarded the importance of ecological time, arguing that temperate regions have a long, undisturbed history and probably have existed as long as tropical regions. The hypothesis related to evolutionary time, however, provided an acceptable explanation for latitudinal patterns in species diversity. Although the actual time available for evolution to occur has been the same at all latitudes, low latitudes with higher temperatures have had a longer 'effective' time for evolution to occur. This results from the acceleration of physiological processes and the shortening of generation times of the organisms with higher and more homogeneous temperatures. Presumably, this would have led to an increase in mutation rates and provided more morphological and physiological 'grist' for selection. Although the hypothesis of evolutionary time seems to work, the primary cause seems to be temperature as a determinant of evolutionary speed.

Finally, another pattern emerges on examination of data in Tables 15.1 and 15.2. There is much greater species richness in the Northern Hemisphere than in the Southern Hemisphere. Szidat (1961) and Parukhin (cited by Rohde, 1993) have made this observation previously for the parasites of marine fishes and mammals in the Atlantic and for the nematode fauna of some 200 fish species worldwide. It is possible, however, that this pattern is an artifact of the unequal amount of research carried out in the Northern and Southern Hemispheres.

15.3.2 Species richness in different oceans

A quantitative analysis of species richness between the Atlantic and Pacific Oceans suggests that the relative species richness of monogeneans is higher in the latter (Fig. 15.5) (Rohde, 1986). Parukhin (cited by Rohde, 1993) also reported that the number of nematode species was greater in the southern Indo-Pacific than in the southern Atlantic. Rohde (1993), citing data from Lebedev (1969), noted that **endemicity** of monogeneans and digeneans was greater in the Indo-Pacific

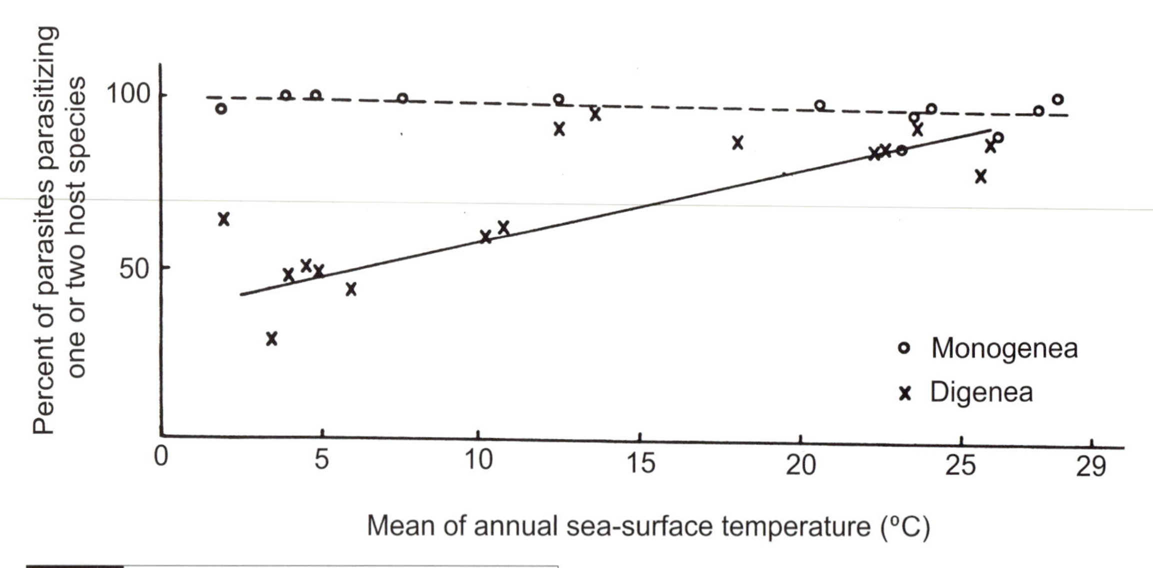

Fig. 15.6 Host specificity of marine Monogenea and Digenea at different latitudes as indicated by the percent of species parasitizing one or two host species, plotted against means of annual sea-surface temperature at various localities. The difference between both regressions is highly significant ($p < 0.001$). (Modified from Rohde, K. [1978] Latitudinal differences in host-specificity of marine Monogenea and Digenea. *Marine Biology*, **47**, 125–134. © Springer-Verlag.)

than in the Atlantic. Moreover, an analysis of the parasitological data from different latitudinal regions within and between oceans revealed that the differences between the two oceans were almost entirely due to a much greater number of species in the northern Pacific than in the northern Atlantic.

Rohde (1986, 1993) proposed that, again, an evolutionary-time hypothesis might best explain the differences in richness between the two oceans. Geological evidence and plate tectonics indicate that the Pacific Ocean has existed for the longest geological time, whereas the Atlantic began to form and expand only 150 million ybp with the break up of Pangaea (see Fig. 15.2). Although the pre-existing Pacific and the newly formed Atlantic have been in contact several times during their geological histories, the degree of endemicity of their biotas suggests that the exchange between them has been rather limited. An alternative explanation, based on ecological time, suggests that during the last glaciation event, the ice sheet that covered the Atlantic was significantly larger than the one in the Pacific, possibly leading to a higher extinction rate of Monogenea in the Atlantic than in the Pacific. This phenomenon would account for the higher richness of species in the Pacific Ocean.

Rohde (1986) also attempted to explain the higher species richness of the Pacific Ocean in terms of a species-area relationship as shown by island biogeography (to be discussed in section 15.3.4). For example, in the North Pacific the average number of Monogenea species per host species is more than twice that in the North Atlantic. It seems that the Pacific is much larger than the Atlantic, and consequently, a larger area translates into more species. However, after studying the richness of monogeneans in both oceans, in particular gyrodactylid monogeneans, and comparing the surface area available in both oceans by incorporating the complexity of the shoreline in a planimetric analysis, Rohde (1986) did not find a relationship between species richness and area. Indeed, the surface area available and complexity of the North Pacific is not greater than the North Atlantic as indicated by the length of the coastline and the number of islands. So, if the surface areas are similar, then the greater richness of the Pacific Ocean is more likely related to the older age of the North Pacific Ocean, once again, the evolutionary and ecological time hypothesis.

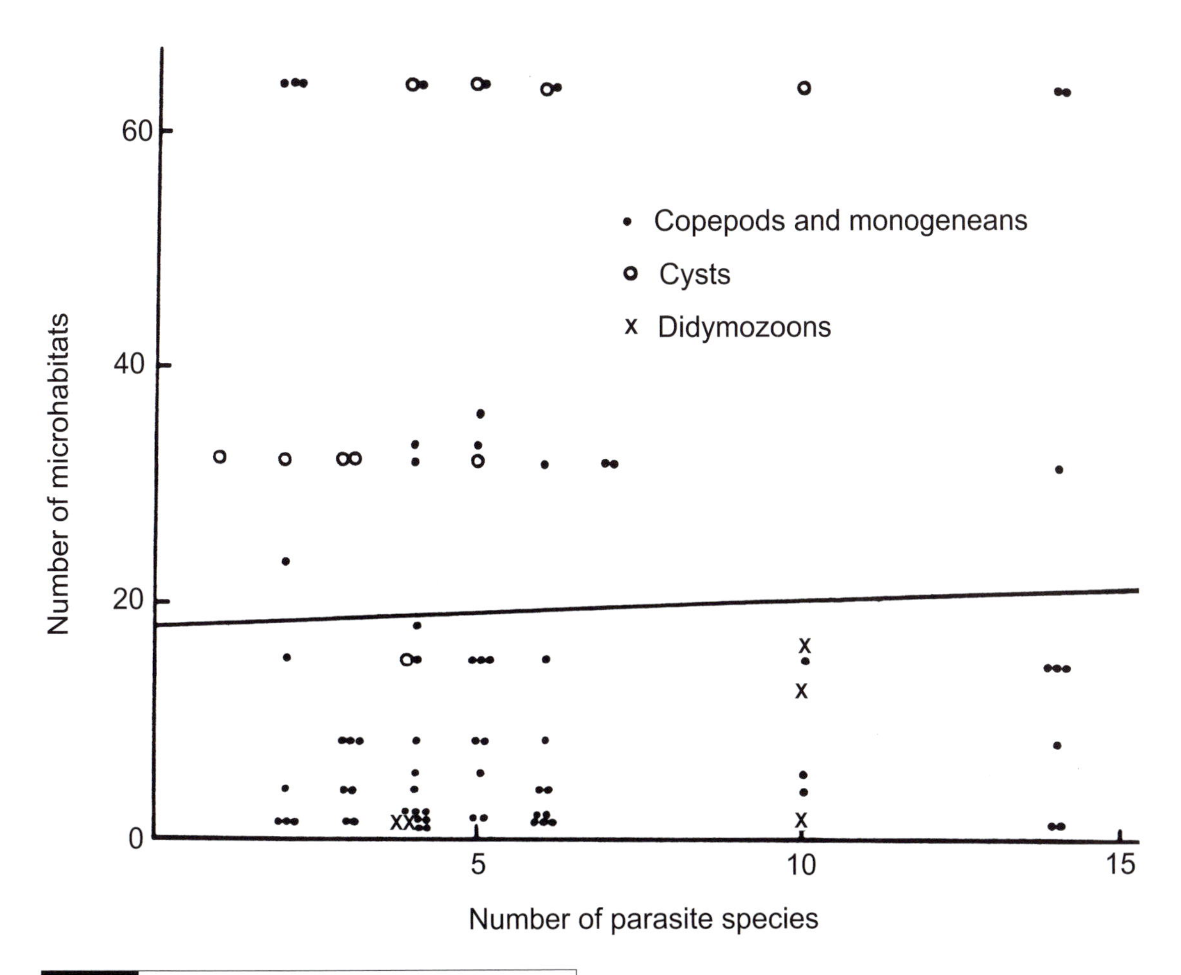

Fig. 15.7 Number of microhabitats per species of parasite on the gills and in the mouth cavity of species of fish with various numbers of parasite species. (Modified from Rohde, K. [1981] Niche width of parasites in species-rich and species-poor communities. *Experientia*, **37**, 359–361.)

15.3.3 Parasites and depth gradients

In both marine and freshwater environments, richness of free-living organisms generally decreases with increasing depth. The decrease in temperature and light, the increase in pressure, and the lack of seasonal change, are a few of the factors that may be responsible for establishing and maintaining this gradient. In marine environments, depth gradients must be considered within the context of bottom topography, e.g., continental shelf, slope, rise, and abyss. The three-dimensional distribution of organisms in the water column (epipelagic, mesopelagic, bathypelagic) also should be addressed since the biotic, physical, and chemical factors affecting each of these zones are different.

PARASITES IN PELAGIC FISHES

Pelagic fishes are distributed throughout the water column and are not associated with the substrate. They feed upon other pelagic organisms such as fishes, tunicates, crustaceans, chaetognaths, and dead organisms or organic material falling through the water column. A few investigators have addressed the nature of parasitism in meso- and bathypelagic fishes, including some species that migrate vertically (Collard, 1970; Noble, 1973; Gartner & Zwerner, 1989). In general, these studies have shown that the distribution of host biomass in the pelagic environment, and the longevity of the hosts, determines parasite richness. Host biomass decreases from the epipelagic realm towards the bathypelagic, so too does parasite richness. Organisms living in the mesopelagic region have adapted to maximize energy sources unavailable to bathypelagic fishes and that accounts, in part, for their greater parasite richness. These adaptations include daily vertical migrations to overlying waters where food (and thus, intermediate hosts) is more abundant. Other factors influencing parasite distribution and richness include the schooling behavior of mid-water fishes that provide concentrations of prey for piscivorous predators. The organically rich layer created by changes in water density at the permanent thermocline also is believed to attract populations of potential hosts (Campbell, 1983).

The findings of the few studies mentioned above indicate that mesopelagic fishes harbor many larval parasites common to pelagic prey that are peculiar to a given geographical region. Adult helminths (mostly monogeneans and acanthocephalans) are rare, and only a few digenean species with relatively low host specificity are present. Larval cestodes, but mainly larval nematodes, are dominant, probably due to their low host specificity. In contrast, bathypelagic fishes have the poorest parasite fauna, partly because host density is very low. Their diet includes mostly crustaceans and some bottom-dwelling invertebrates that apparently are not suitable intermediate hosts for parasites in deeper water. Nonetheless, bathypelagic fishes have a few parasite species that are highly specific for them and that are, therefore, well adapted to the extreme environmental conditions (Campbell, 1990).

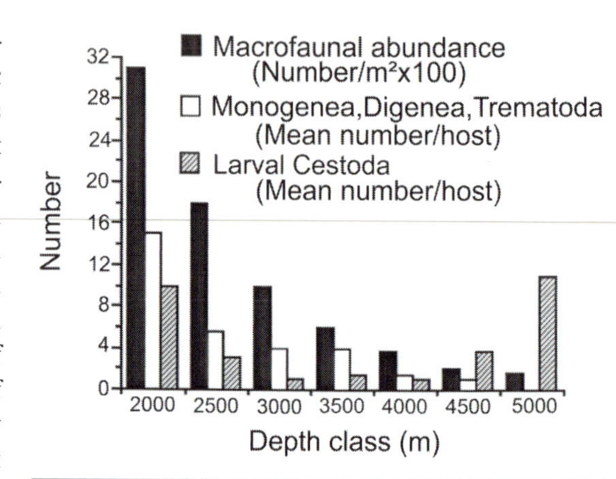

Fig. 15.8 Rates of infection by helminth parasites in *Coryphaenoides armatus* and overall abundance of macrofaunal invertebrates, arranged by depth. The macrofaunal values are for the Hudson Submarine Canyon. The number of *C. armatus* examined in each depth class, from shallow to deep are 17, 86, 85, 11, 10, 3, 1. (Modified from Campbell, R. A., Haedrich, R. L. & Munroe, T. A. [1980] Parasitism and ecological relationships among deep-sea benthic fishes. *Marine Biology*, **57**, 301–313. © Springer-Verlag.)

PARASITES IN BENTHIC FISHES

Bottom-dwelling fishes, at any depth, have greater parasite richness than pelagic fishes at similar depths. Among benthic fishes, however, parasite richness also decreases from the continental shelf into the abyss at various rates, depending on the feeding habits of the fish, the nature of the free-living species assemblages, and the nature of physical and biological oceanographic phenomena occurring in the upper layers. The studies by Manter (1934), Campbell *et al.* (1980), and Campbell (1983, 1990) summarize most of the information and provide some generalizations for the assemblage of parasites in deep benthic fishes of the northwest Atlantic. For example, parasite densities of deep benthic fishes are directly related to host population densities. Host specificity of monogeneans and digeneans is high. Monogeneans exhibit strict host specificity, most digeneans and cestodes have narrow host specificity, and acanthocephalans and nematodes show no host specificity. Overall, the parasite fauna of benthic fishes in deep water is less diverse than in shallow waters. The best evidence supporting this observation is shown by the parasites of the rattail *Coryphaenoides armatus* (Fig. 15.8). The decrease in parasite richness is paralleled by a decrease in macrofaunal abundance with depth and with changes in the consumption of certain prey, probably determined by their availability. Although it is not shown in Fig. 15.8, Campbell (1983) indicated that the abundance of parasites without intermediate hosts, particularly monogeneans, does not change significantly until they reach the lower slope and abyssal plains (4000–5000 m), where such parasites disappear.

Two factors seem to have a strong influence on the gradients of parasite richness with depth in both benthic and pelagic regions; one is the relative size of the deep-sea ecosystem and the other is the permanence of the host's association with particular invertebrate communities. Parasite richness is correlated with the richness of the invertebrate fauna that provides suitable intermediate hosts for the completion of complex life cycles. Similarly, low fish density and the low level of interaction among fishes in a population are probably major constraints for the establishment of parasites with direct life cycles. The differences between pelagic and benthic organisms at a given depth are attributed to differences in the food supply which, in turn, translates into parasite supply. Meso- and bathypelagic fishes are spread throughout a significant volume of water and their life spans are also relatively short, about 2 years. The brief life span means shorter 'windows' of exposure and the recruitment of fewer parasites. Deep-water benthic fishes, on the other hand, have significantly longer life spans and live in a more restricted two-dimensional zone with a concentrated and extensive food supply. Consequently, they are able to recruit parasites for a longer time and the richness and quantity of parasites to which they are exposed is greater.

15.3.4 Species–area relationship

It has long been noticed that the number of species increases systematically in samples from larger areas, but that the rate of increase in number of species decreases with progressively larger areas (MacArthur & Wilson, 1967). In other words, once a plateau in species richness has been reached, the number of species then remains essentially constant despite an increase in the area sampled. This pattern relies on the fact that habitat heterogeneity increases as surface area increases. In this sense then, area should be considered more as an indicator than as an explanation for species number.

The species–area relationship has been applied to parasites on a number of different scales. Rohde (1989) examined these relationships at three different levels using gill parasites of marine fishes. The gills of fishes seem to represent an ideal system for examining species–area relationships because the gills of various fishes are similar in structure and hosts can be studied both within, and between, latitudinal regions. Using fish length as an indicator of gill size, Rohde (1989) found that larger fishes with larger gills did not have more monogeneans than smaller fishes (Fig. 15.9). He also examined monogenean data from three species of *Scomber* (mackerel-like fishes) of similar size, but with different ranges of geographical distribution. Once again, he did not find a general correlation between host range and parasite species richness. He then considered copepod data for several scombrid genera and found a significant relationship between the number of copepod species and the number of geographical areas where the host genus occurred. More importantly, however, was a significant relationship between the number of copepod species and the number of host species in the genus. Indeed, this latter factor accounted for 93% of the variance in his regression analysis. Therefore, richness of the host group was the most important factor determining parasite richness, clearly a phylogenetic component.

Rohde (1989) concluded that 'area as indicated by size of fish, gills, geographical range of hosts, and geographical area occupied by a fauna (as in the monogeneans of the North Atlantic and North Pacific), is not predominant in determining species richness. The effect of area, if it exists at all, is probably overridden by other factors such as evolutionary time, richness of the group, etc.' Factors related to phylogeny and historical processes that affect the host can also override the effects of area.

Price & Clancy (1983) employed a similar approach to examine the relationship between parasite richness and host geographic range, size, and feeding habits of freshwater fishes in Great Britain. Unlike Rohde (1989), however, Price & Clancy (1983) found a significant correlation between geographic range of the host and the total number of parasites per host. Geographic range accounted for 68% of the total variation and feeding habits for an additional 5%. However, a re-examination of the data used by Price & Clancy (1983), after removing introduced species and controlling for the effect of sampling effort, showed no significant relationship between helminth species richness and host range (Guégan & Kennedy, 1993). These findings provide little

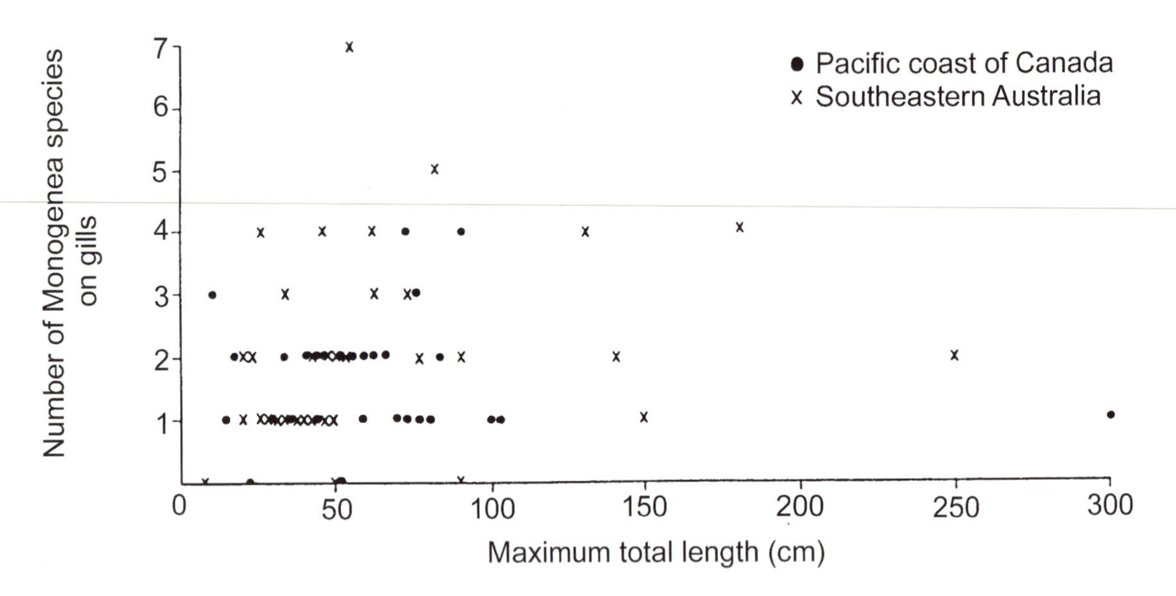

Fig. 15.9 Number of gill monogeneans in marine teleost fish of different lengths (total length) at two localities. (Modified from Rohde, 1989, with permission, *Evolutionary Theory*, **8**, 305–350.)

support for the idea that parasite species richness and host geographical range are related in terms of the Theory of Island Biogeography. Other studies also have shown that the theory is a poor predictor of parasite richness among freshwater fish species (Kennedy, 1978; Kennedy *et al.*, 1986). Moreover, the results obtained by Guégan & Kennedy (1993) support an alternative explanation in terms of the colonization time hypothesis, namely, parasite species richness is related to the time since the fish host colonized Great Britain.

The true nature of the relationship between the number of parasites and the geographic range of the host species is still subject to debate. It appears that many confounding factors may produce correlations between parasite species richness and host geographic range (Gregory, 1990). For example, there may be a causal relationship between these two variables. Host species with larger ranges may acquire more parasite species because larger areas will include the ranges of more parasite species. Also, it is possible that the relationship is a function of a correlation between the number of parasite species and the host range with another variable. For instance, host species with larger ranges may have been more heavily sampled and so, more parasite species have been identified (Kuris & Blaustein, 1977). Host species with larger ranges may be larger, providing more niches for more parasite species. It is also possible that host species with larger ranges may occur in higher population densities, or may be more gregarious, thereby facilitating increased rates of parasite transmission (Dogiel *et al.*, 1961; Price, 1980).

Another possibility, one that is currently receiving much attention, is that the effects of phylogeny may have a confounding effect on biological variables (Pagel & Harvey, 1988). A relationship between number of parasites and host range across all species may be a consequence of the host's phylogenetic history. In effect, host species that share common ancestry are likely to be the subject of similar evolutionary constraints. They are also likely to be morphologically similar and occupy similar niches. It would seem, therefore, that their parasite infracommunities might also be similar. As a case in point, Poulin (1995) studied the relative contribution of various ecological characteristics to the richness of parasite component communities using two methods. In one, each host species was considered to be independent of each other. In the other, the host data was analyzed controlling for phylogenetic effects. When using the former approach, a number of significant correlations between parasite richness and several ecological characteristics were found, but when the second approach, controlling for phylogenetic effects was used, some, but not all, of the correlations were no longer significant. In short, even though correlations between parasite richness and many ecological factors do exist, they are not necessarily due to ecological factors *per se* but possibly to the phylogenetic legacy of the hosts and/or parasites. It seems that whenever host–parasite interactions are studied from an ecological point of view, phylogenetic factors should not be ignored (see Box 14.1). (For an overview on the analysis of data using comparative methods, including phylogenetic contrasts, see Pagel & Harvey [1988], Harvey & Pagel [1991], and Garland *et al.* [1992].)

15.4 Applied aspects of biogeography

15.4.1 Parasites as 'biological markers'

As mentioned earlier in this chapter, one of the reasons for the great amount of parasitological data available from marine fishes is the economic value of the hosts. The efficient management of fisheries requires a complete understanding of the biology of the host. Sometimes, however, traditional fisheries' biology methods and techniques are inadequate to answer questions regarding observed patterns of biogeographical distribution. In such cases, parasitological studies may provide the appropriate answers. The use of parasites as tags in fisheries has some advantages over conventional tagging methods. For example, parasites used as tags do not interfere with the normal behavior of the host; they are cheaper, less labor intensive, and less time consuming because they do not require the catch–mark–release of the host. Currently, parasites are being used as markers in population studies to distinguish host stocks, to indicate migratory routes, to identify feeding and breeding areas, and to follow the recruitment pattern of juveniles into stocks.

Even though most hosts are likely to harbor parasites, not all parasites can be used as biological tags. In order to be used as a tag, a parasite should fulfill several general requirements. First, the parasite should not cause selective mortality or behavioral changes in the host. Second, the parasite used should have relatively high prevalence in the host population to facilitate the finding of tags when sampling a host population. Third, the levels of prevalence, intensity, or both should be significantly different in the different studied areas. Fourth, the parasite, or evidence of its past presence, should be detectable in the host throughout the time scale of the study. Fifth, the parasite should be easy to recognize. Sixth, the life cycle should preclude the possibility of inter-host transfer at sea. Normally, most ectoparasites (monogeneans, copepods, and isopods) are not useful markers because they are easily transmitted horizontally from host to host, regardless of the original area of infection. In addition, they may be more susceptible to changes in various abiotic parameters such as temperature and salinity than their internal counterparts. The best markers are endoparasites, in particular, larval stages which are almost invariably sequestered in parts of the body without direct access to the external environment and thus not subject to expulsion.

Parasites were first used as markers some 40 years ago. In the Pacific Northwest, offshore salmon fishing is an important industry. When the industry began in the early 1950s, it was not known if salmon from the Pacific and Asian rivers were intermixing in the ocean or if they remained as distinct populations. Similarly, the geographical origin of the salmon caught by the different countries was unknown. Since then, the migrations of salmon in the North Pacific have been well studied using parasites as markers. Margolis (1963) was able to distinguish between sockeye salmon *Oncorhynchus nerka* of Pacific and Asian origin based on two of their parasites. Sockeye salmon from Alaska were infected by the plerocercoid larva of the cestode *Triaenophorus crassus*. Infection was acquired in some of the lakes of western Alaska. Sockeye salmon from Kamchatka, on the other hand, were infected by the nematode *Dacnitis truttae*. Infection in this case was acquired by juvenile sockeye in the rivers of Kamchatka. Sampling of salmon at sea showed that salmon from both Alaska and Kamchatka disperse up to 2000 km from their area of origin (Fig. 15.10) overlapping in their distributions. On the other hand, it appears that schools of salmon from different

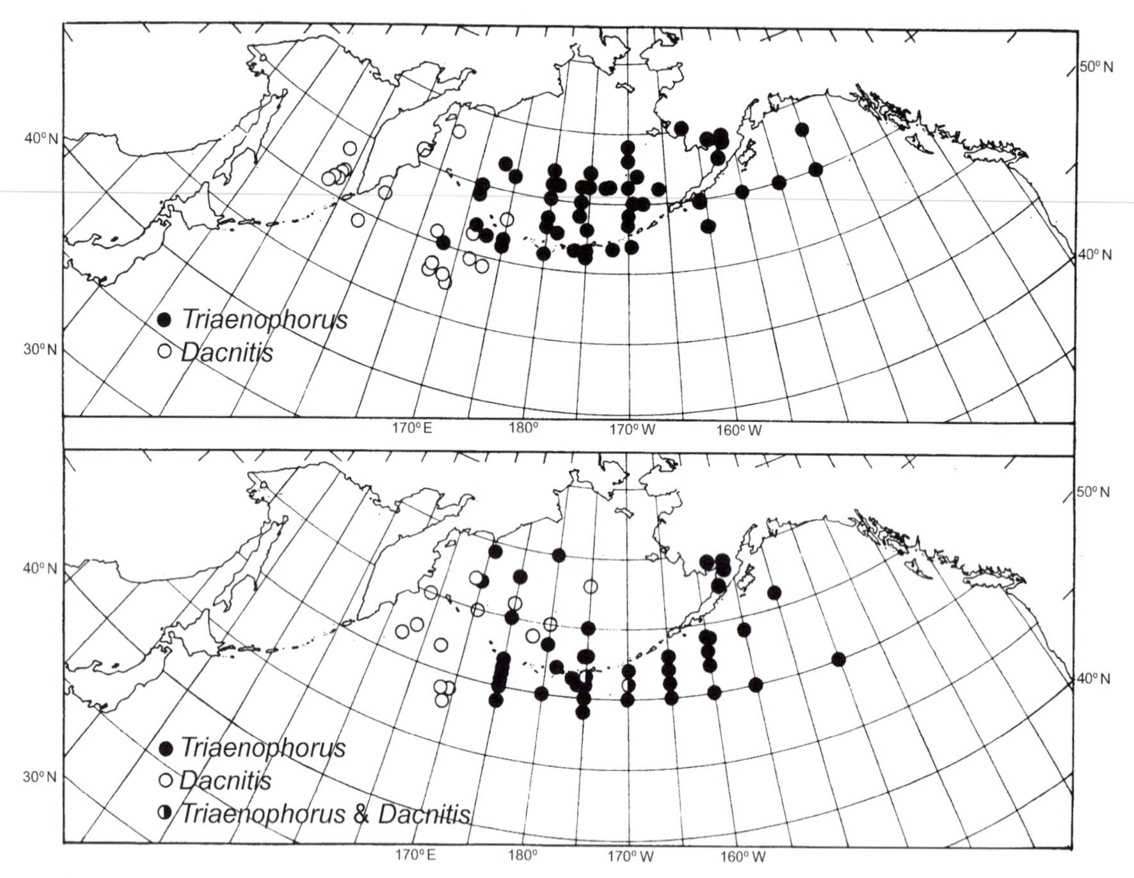

Fig. 15.10 Distribution of sockeye salmon *Oncorhynchus nerka*, infected with *Triaenophorus crassus* and *Dacnitis truttae* in the North Pacific. Top: maturing fish. Bottom: immature fish. (Modified from Margolis, 1965, with permission, *Journal of the Fisheries Research Board of Canada*, **22**, 1387–1395.)

streams in Alaska remain as distinctive schools in the sea. Margolis (1965) was able to recognize these different schools based on the different prevalences of infection of *Triaenophorus*, which are characteristic of different streams in western Alaska.

Similar studies on other salmonids have been equally successful at finding good markers and using them to determine the host's area of origin. Based on the nature of distinct parasite faunas, Hare & Burt (1976) were able to distinguish populations of the Atlantic salmon *Salmo salar*, from various tributaries in the Miramichi River system in eastern Canada. Recently, Margolis (1990) identified two digeneans as extremely good markers for populations of anadromous rainbow trout *Oncorhynchus mykiss*, originating in freshwaters of the Pacific coast of North America. Trout have a freshwater range that extends from California to Alaska, continuing to the eastern and western coasts of Kamchatka and the northern coast of the Okhotsk Sea. Metacercariae of *Nanophyetus salmincola* and adults of the digenean *Plagioporus showi* only occur in the North American freshwater systems, where they are acquired by juvenile fishes prior to their seaward migration. These parasites are able to survive in the trout throughout all its oceanic migration, an important requirement for parasites used as markers of anadromous fishes.

The presence of distinct, or characteristic, parasites in hosts of different origin is not always necessary in order to use them as markers. In some instances, the complete parasite assemblage can be used as a marker provided that parasite abundance differs among stocks, and that the appropriate mathematical tools are used to recognize numerical patterns in the parasite assemblage.

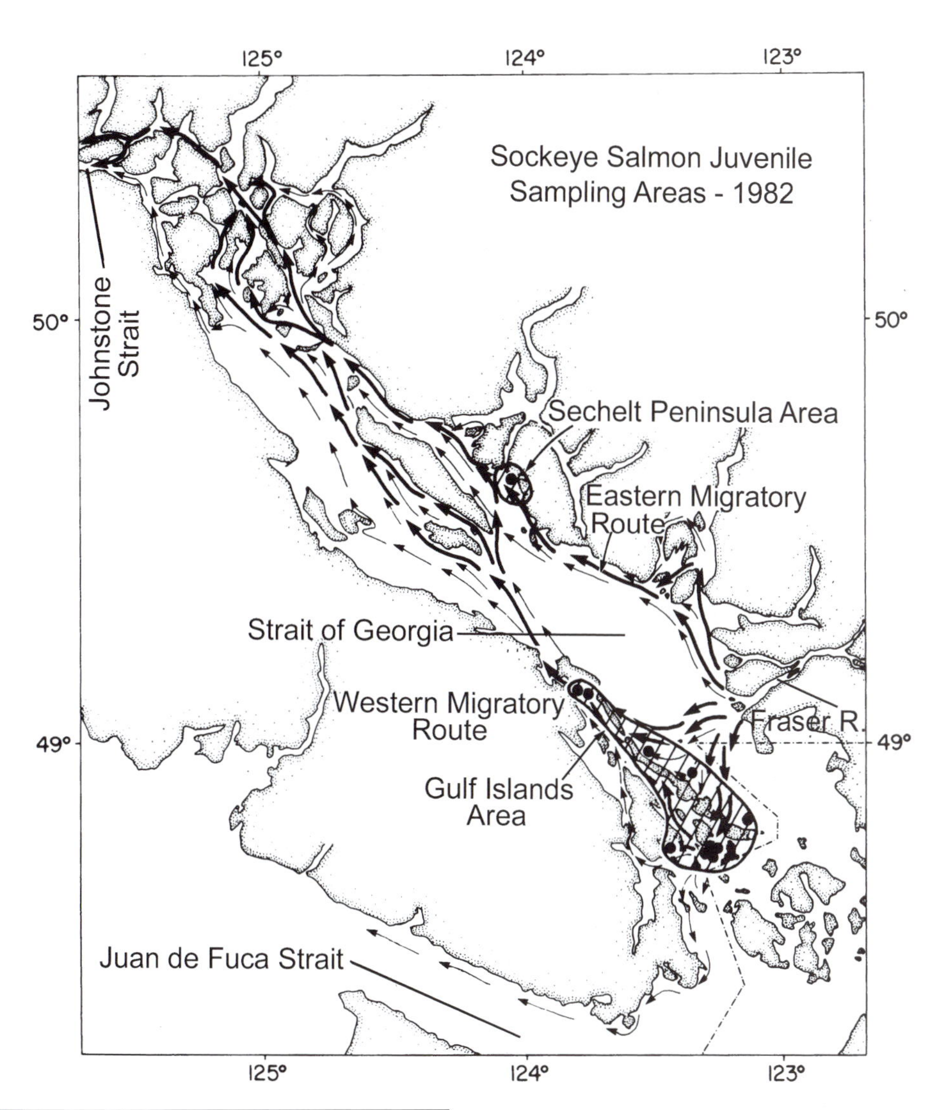

Fig. 15.11 Eastern and western migratory routes of sockeye salmon smolts *Oncorhynchus nerka*, in the Strait of Georgia, British Columbia. (Modified from Groot *et al.*, 1989, with permission *Canadian Journal of Zoology*, **67**, 1670–1678.)

Bailey *et al.* (1988) and Groot *et al.* (1989) adapted a multivariate maximum-likelihood mixture model used in fisheries to estimate the stock composition of a mixture of salmon stocks using parasitological data. They used this approach to find out if the different migratory routes of juvenile sockeye salmon were stock-specific or not. Soon after sockeye salmon *Oncorhynchus nerka* hatch in the gravel beds of the tributaries of the Fraser River system in British Columbia, Canada, they move into some of the many lakes in that system. Juvenile salmon spend one or more years in these nursery lakes before migrating to sea. When they reach the sea, they migrate north through the Strait of Georgia to the Pacific Ocean using two different routes (Fig. 15.11). One route turns north immediately after leaving the Fraser River estuary and travels along the mainland coast of Canada (eastern route). A second route (western route) crosses the Strait of Georgia towards Vancouver Island, turns north along the east side of the island and then diagonally towards the mainland where it joins the eastern migratory route. Nobody knew, however, if these different migratory routes were stock-specific, with salmon from a given nursery lake following a specific route. Because of the lack of parasites specific to the different nursery systems, Groot *et al.* (1989) used parasite assemblages and the multivariate maximum-likelihood mixture model to estimate the stock composition of the juvenile salmon along the two migratory routes. The results showed that most stocks of salmon from the Fraser system migrated north using both the eastern and western routes, regardless of their origin, meaning that there is no stock-specific route selection. They also showed that this model could be used in other systems where distinctive marker parasites are not present.

In marine fishes, most studies have focused on using parasites to separate stocks or to track stock movements between feeding and breeding areas. Larval anisakid nematodes have been widely employed as markers because they usually have longer life spans in their fish hosts than adult worms parasitizing the alimentary canal. Platt (1975, 1976) studied larvae of the nematodes *Anisakis* sp. and *Pseudoterranova decipiens*, in several populations of cod *Gadus morhua*, in the Arctic and northern Atlantic Oceans in the vicinity of Iceland and Greenland. He found significant differences in the parasite densities of these two species between the Arctic and North Atlantic cod populations and in the densities of *P. decipiens* between the two North Atlantic populations (Iceland and Greenland). Moreover, the differences in *P. decipiens* could be used to determine the geographical origin of cod spawning off the southern and western coasts of Iceland. Additional studies on cod from Norway and Canada showed, once again, the importance of larval nematodes as biological tags for stock recognition, although a few other parasites were also useful (Hemmingsen *et al.*, 1991; Khan & Tuck, 1995).

Herring, *Clupea harengus*, a strictly marine fish distributed widely in the cold and temperate waters of the Northern Hemisphere, has been extensively studied using parasites as markers. Sindermann (1957), for example, examined the parasite distribution and abundance in herring along the Atlantic coast of North America. The data suggested that there was little, if any, interchange between herring populations from the Gulf of Maine and the Gulf of Saint Lawrence. In addition, it appeared that immature herring in the eastern and western parts of the Gulf of Maine could be separate populations. McGladdery & Burt (1985) undertook a more comprehensive examination of the migration, feeding, and spawning behavior of herring. They found that six parasite species could be used as indicators of seasonal migration between the Bay of Fundy, the Nova Scotian Shelf, and the Gulf of Saint Lawrence. The biological characteristics of these parasites indicated that the seasonal variations in prevalence could be better explained by changes in composition of herring stocks rather than by changes in the component parasite communities within the same stock. The presence of a protozoan *Eimeria sardinae* permitted the separation of 'races' of herring spawning at different times of the year, an important piece of information when attempting to manage commercial fisheries. MacKenzie (1985) also was able to follow the migrations of herring in the North Sea and the Scotland Shelf by tracking the presence of the metacercariae of two digenean species and a larval cestode. All three were long-lived parasites acquired by herring in their nursing grounds where their prevalence varied significantly between the different locations.

Populations of Pacific herring *Clupea harengus pallasi*, along the California coast, have two major spawning grounds, Tomales Bay and San Francisco Bay (Fig. 15.12). In addition there is a third area, Monterey Bay, where little spawning occurs but that supports a modest summer fishery. Spawning lasts 1–3 weeks and no feeding occurs during this time. After spawning, the adults return to the open ocean. Juveniles remain in the bay for about a year, then leave and stay at sea for 2 years before returning to the bay to spawn for the first time. Moser (1990) and Moser & Hsieh (1992) examined the parasites of the Pacific herring to determine if the parasites could be used to separate and characterize the Tomales Bay and San Francisco Bay spawning stocks. They found consistent differences in the prevalence of one of the marker parasites between the two areas, suggesting that they are indeed different stocks and that they return to spawn in the same area. Moreover, the differences in infection of two groups of larval parasites between Tomales Bay and San Francisco, suggest that, while at sea, the Tomales Bay fish feed offshore whereas the San Francisco Bay fish feed

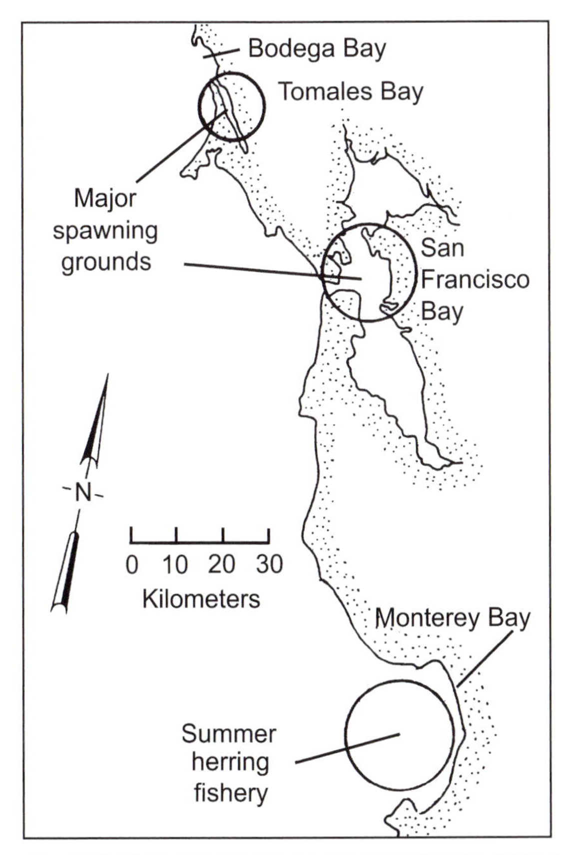

Fig. 15.12 Location of the spawning grounds and summer fishery of Pacific herring *Clupea harengus pallasi*, off central California. (From Spratt, 1981, California Department of Fish and Game, with permission from the author.)

inshore. How do we know? The definitive hosts of the more abundant larval parasites found in Tomales Bay are cetaceans common offshore, whereas the definitive hosts of the most abundant larval stages found in San Francisco Bay are pinnipeds, marine birds, and elasmobranchs that have more coastal distributions. Finally, based on parasite similarities, it seems that the non-spawning herring found in Monterey Bay during the summer and herring from San Francisco Bay belong to the same stock.

The use of parasites as biological markers is not restricted to fishes. For example, two digenean species, in addition to several population characteristics, were used to study the population discreteness of two species of shrimp (*Pandalus jordani* and *Pandalopsis dispar*) in British Columbia, Canada (Thompson & Margolis, 1987). The study found that the shrimp populations from two channels and one offshore area were distinct. The same result was obtained for both species of shrimp studied. Commercial fishing of cephalopods, mainly squid, is becoming more important worldwide and the need for biological markers has started to focus on parasites. Pascual & Hochberg (1996) reviewed the current status of this line of research and indicated that parasites already have been used to elucidate trophic interactions, zoogeographic patterns, ecological distributions, intraspecific groupings, stocks, seasonal migration patterns, and even sibling species of cephalopods throughout the world.

15.4.2 Historical centers of origin and host dispersal as indicated by parasites

Finding the geographical area where a given taxon evolved, and the dispersal routes of the derived taxa through time, has been one of the many applications of parasitological data on a global scale. It is assumed that hosts have a greater diversity of parasites in areas where they have occurred for a long time (their center of origin) than in areas they have more recently colonized. Similarly, the dispersal of hosts into new areas may lead to the loss of their 'original' parasites, or the acquisition of new ones, either by host capture or speciation of the 'original' species. Accordingly, parasites present in the area of origin should be regarded as ancestral to those present in the newly colonized locations.

Parasites also have been used to provide some insight into the origin of anadromus and catadromus fishes such as *Oncorhynchus* spp. and *Anguilla* spp., and of freshwater stingrays (Potamotrygonidae) in South American rivers. Margolis (1965) indicated that freshwater parasites had greater host specificity for Pacific salmonids (*Oncorhynchus* spp.) than marine parasites. Freshwater endoparasites included some with complex life cycles, suggesting a long association. There were only two marine parasites specific to salmon, a monogenean and a copepod, both with direct life cycles. These observations, in addition to other evidence, encouraged Margolis (1965) to conclude that Pacific salmonids had a freshwater origin and not a marine one as some ichthyologists had argued. Further studies on salmonids and their parasites using different taxonomic and spatial scales have helped to better understand the nature of the relationship between hosts and parasites over large geographic areas. A study of the parasite communities of *Oncorhynchus* in the geographical area in which the genus evolved (the heartland), and of *O. mykiss* throughout the world (where it has been introduced) uncovered some interesting patterns in addition to those originally mentioned by Margolis (Kennedy & Bush, 1994; see also section 14.4.1). In the heartland (western North America and eastern Siberia), *Oncorhynchus* is parasitized primarily by helminths that are specialists in salmonids and that seem to have had an evolutionary relationship with the host. The remaining species include broad generalists and non-salmonid specialists, representing an ecological association. When salmonids, in this case *O. mykiss*, have their geographical distribution increased by human introductions, their parasite fauna changes. As their distance from the heartland increases, they first lose their generic specialists (specific for *Oncorhynchus*), and then their salmonid specialists. They then acquire some generalist parasites, but their parasite communities become poorer and poorer, providing further evidence for the idea that, in general, hosts have a greater diversity of parasites in their area of origin.

Manter (1955) examined the data on parasites of *Anguilla* spp. from Europe, North America, Japan, and New Zealand, and suggested that *Anguilla* probably had a marine and, more specifically, a Pacific origin. He based his argument on the uniqueness of the Pacific parasites. Of nine species found in the Pacific Ocean, eight were specific for *Anguilla*; moreover, the only three genera of parasites specific for this host were also in the Pacific. Atlantic eels that occur in Europe and North America had a larger number of parasites than those in the Pacific, but these were mostly generalist digeneans common to other fishes. Further and more detailed studies on the parasites of *Anguilla* spp. have corroborated Manter's views of a Pacific origin but also shed some light on the mechanism of divergence of the Atlantic eels into a European and a North American species (Kennedy, 1995; Marcogliese & Cone, 1993, respectively). The species of Indo-Pacific eels from New Zealand and Australia harbor more parasites than the Atlantic eels from Europe or North America, even though the Atlantic region has been studied extensively. Also significant is that the Indo-Pacific eels harbor parasite specialists in a number of organs such as the heart, swim bladder, bladder, stomach, and body cavity, whereas in the Atlantic eels these organs are not parasitized or are inhabited by generalist parasites. It seems that the Atlantic and Indo-Pacific species of eels belong to two old, but separate and distinct groups, with the Indo-Pacific eels probably being the oldest (since they harbor the most parasites and have the larger proportion of eel specialists). The dominant specialist species of parasites in Indo-Pacific eels form a characteristic phylogenetic component, typical of parasite communities of a host in its area of origin (heartland).

Two general hypotheses (oceanic and vicariant) have been proposed to explain the divergence of European and North American eels from a common ancestor somewhere around the Atlantic Ocean (Avise *et al.*, 1990). In simple terms, the oceanic hypothesis argues that eels were native only to one side of the Atlantic, either Europe or North America, and that divergence occurred when the oceanic currents changed during the Pleistocene, carrying eel larvae to the new continent. The vicariant hypothesis, on the other hand, proposes that a single ancestral population existed across the North Atlantic, from Scandinavia to Greenland. This population was forced southwards by the Pleistocene glaciations and diverged into the European and American species. Parasitological evidence seems to supports the vicariant hypothesis (Marcogliese & Cone, 1993). Several species of freshwater parasites, some of them with complex life cycles, are present in both eel species. Because adult eels spawn in the ocean, dying soon after, and because eel larvae cannot acquire freshwater parasites at sea, these host-specific, freshwater parasites had to be present in a common geographic area before the eels diverged. Evidence from parasites, then, supports the scenario of a former single ancestral eel host, with a continuous distribution, which diverged into two groups by allopatric speciation, as a consequence of a glaciation event.

Brooks *et al.* (1981) and Brooks (1992) devised a clever methodology, combining biogeographic and phylogenetic information, to determine the evolutionary origins of the potamotrygonid stingrays present in the major river systems of eastern South America. It is believed that stingrays were secondary invaders of freshwater systems, being originally a marine group. Because all the drainage systems inhabited by potamotrygonids drain into the Atlantic side of South America, the common assumption was that their ancestor was an Atlantic marine or euryhaline stingray that dispersed into freshwater 3 to 5 million ybp. However, phylogenetic analyses of their parasites indicated that the parasites originated in the Pacific Ocean and that their closest relatives parasitize marine urolophid stingrays along the Pacific coast of South America. The phylogenetic analysis of the parasites, together with the geological history of the region, indicates that the ancestor of the potamotrygonid stingrays was a marine urolophid stingray present in the Pacific Ocean that was trapped in freshwater by the uplifting of the Andes Mountains, probably beginning in the early Cretaceous. Accordingly, during the Andean orogeny, the river drainages were altered, a new Continental Divide was established, and the rays became isolated from the ancestral group along the Pacific Coast of South America. Presently they are **relict** taxa in the Atlantic drainage system of South America. Recent phylogenetic studies of stingrays using only ichthyological data suggest that stingrays could have invaded freshwater habitats from the Caribbean Sea rather that the Pacific Ocean as suggested by previous studies (Lovejoy, 1997). However, studies of gnathostomatid nematodes of stingrays (Hoberg *et al.*, 1998) further confirm the view that the freshwater stingrays from rivers in South America are likely descendants of stingrays that inhabited the Pacific Ocean.

For many years, there has been an interesting controversy regarding the geographical origin and dispersal routes of species of the fish genus *Merluccius*, commonly known as hake, and close relatives of gadid fishes such as cod (*Gadus morhua*) and whiting (*Micromesistius merlangus*). Most species of hake are the object of an intense and extensive fishery throughout much of the world. Hakes are benthic-demersal fishes and their geographical distribution indicates that the different species inhabit the continental shelves and upper parts of the continental slopes (Fig. 15.13A). The question and the controversy regarding the geographical area in which *Merluccius* first evolved, and the subsequent routes of dispersion leading to their present distribution and speciation patterns, is an interesting example of how science works and how new approaches and collaboration between disciplines answers old questions (and, at the same time, creates new ones!). Two papers published in 1981 proposed a comprehensive view of the historical biogeography of hakes. One view (Inada, 1981) used osteological data of the host, whereas the other (Kabata & Ho, 1981) used parasitological data. Although they used different sets of biological information, both studies reached similar conclusions on most points. For example, both studies indicated that hakes originated in the eastern North Atlantic and dispersed southward following two routes, one along the west coast of Europe and the other along the east coast of North America. These studies also conceded that hakes entered the Pacific Ocean through the then-submerged Isthmus of Panama. The major discrepancy between the two studies lies in their account of the origin and dispersal of the Argentine hake *Merluccius hubbsi* and the subsequent origin of the New Zealand hake *M. australis*. The parasitological approach suggested that the Argentine hake originated from the western North Atlantic stock whereas the ichthyological data indicated that it originated from an eastern South Pacific stock that rounded Cape Horn to reach Argentina (see Fig. 15.13B). In the mean time, Fernández (1985) and Fernández & Durán (1985) studied the parasites of a Patagonic population of New Zealand hake occurring in southern Chile and corroborated the parasitological view of Kabata & Ho (1981) regarding the origin of Argentine and New Zealand hakes. Unfortunately, neither of these studies employed phylogenetic methods for either hosts or parasites similar to

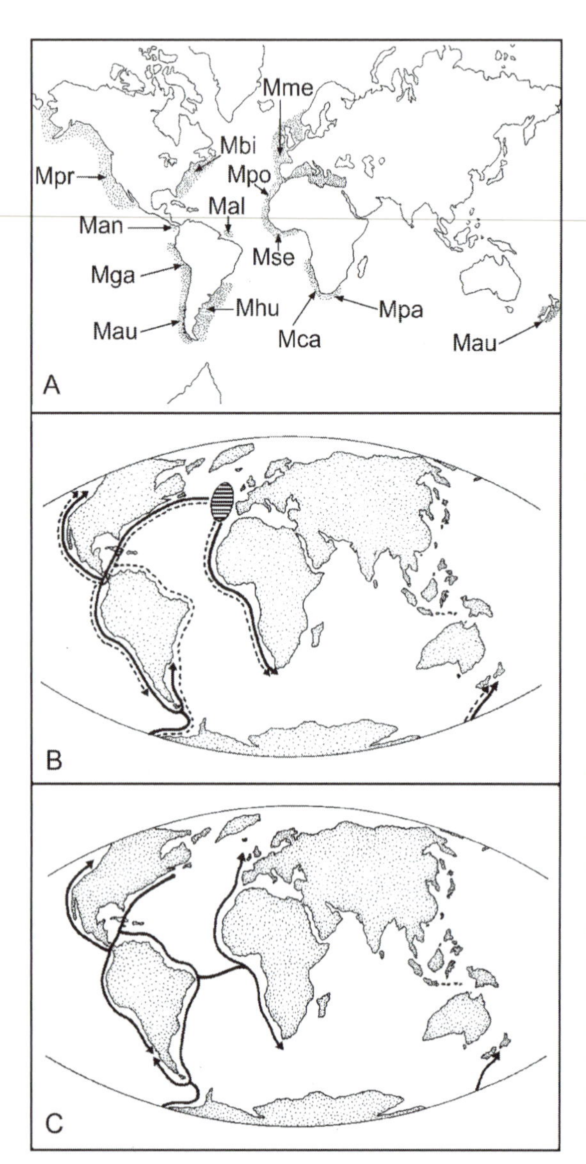

Fig. 15.13 Geographical distribution of hakes *Merluccius* spp., and proposed dispersal routes. (A) Geographical distribution of living species of hakes: Mal, *M. albidus*; Man, *M. angustimanus*; Mau, *M. australis*; Mbi, *M. bilinearis*, Mca, *M. capensis*; Mga, *M. gayi*; Mhu, *M. hubbsi*; Mme, *M. merluccius*; Mpa, *M. paradoxus*; Mpo, *M. polli*; Mpr, *M. productus*; Mse, *M. senegalensis*. (*Merluccius hernandezi*, found in the Gulf of California, was not included in the present analysis.) (B) Distribution and movements of hakes as proposed by Inada (1981) (—) and Kabata & Ho (1981) (---). (C) Distribution and movements of hakes as indicated by phylogenetic analysis of osteological characters. ((B, C) From Ho, 1989, *Fishery Bulletin, US*, **88**, 95–104, with permission from the author.)

the ones devised by Brooks *et al.* (1981) to study potamotrygonid stingrays.

Finally, Ho (1989) using the osteological data from Inada (1981) re-examined the relationships and biogeography of hakes using a phylogenetic approach. Some of his findings corroborated previous views, whereas others introduced some surprising new views. The phylogenetic analysis suggested that hakes originated in the western North Atlantic (Fig. 15.13C) and not in the eastern North Atlantic as suggested by the previous studies. Second, it suggested that hakes entered the Pacific over the submerged Isthmus of Panama *and* around the southern tip of South America, which corroborates the views proposed by the parasitological data. Third, it indicated that hakes migrated southward along the Atlantic coast of South America *and* then eastward across the mid-Atlantic towards Africa. This is a novel idea that had not been proposed before and that needs to be studied further in other bentho-demersal fishes. So, instead of putting an end to the current differences, the latest study confirms some of the previous views and closes some important gaps while at the same time raising new questions and opening new venues for research. Ho (1989) offers the best description of the current situation in his concluding remarks: 'Both the phylogenetic hypothesis and biogeographical model differ from the current views, but they are viewed as the better explanation of available data. They are presented here as a working model subject to modification as more exact information becomes available.'

References

Arai, H. P. & Mudry, D. R. (1983) Protozoan and metazoan parasites of fishes from the headwaters of the Parsnip and McGregor rivers, British Columbia: a study of possible parasite transfaunation. *Canadian Journal of Fisheries and Aquatic Sciences*, **40**, 1676–1684.

Avise, J. C., Nelson, W. F., Arnold, J., Koehn, R. K., William, G. C. & Thorsteinsson, V. (1990) The evolutionary genetic status of Icelandic eels. *Evolution*, **44**, 1254–1262.

Bailey, R. E. & Margolis, L. (1987) Comparison of parasite fauna of juvenile sockeye salmon (*Oncorhynchus nerka*) from southern British Columbia and Washington State lakes. *Canadian Journal of Zoology*, **64**, 420–431.

Bailey, R. E., Margolis, L. & Groot, C. (1988) Estimating stock composition of migrating juvenile Fraser River (British Columbia) sockeye salmon, *Oncorhynchus nerka*, using parasites as natural tags. *Canadian Journal of Fisheries and Aquatic Sciences*, **45**, 586–591.

Brooks, D. R. (1977) Evolutionary history of some plagiorchioid trematodes of anurans. *Systematic Zoology*, **26**, 277–289.

Brooks, D. R. (1979) Testing hypothesis of evolutionary relationships among parasites: the digeneans of crocodilians. *American Zoologist*, **19**, 1225–1238.

Brooks, D. R. (1992) Origins, diversification, and historical structure of the helminth fauna inhabiting neotropical freshwater stingrays (Potamotrygonidae). *Journal of Parasitology*, **78**, 588–595.

Brooks, D. R. & McLennan, D. A. (1993) *Parascript: Parasites and the Language of Evolution*. Washington, DC: Smithsonian Institution Press.

Brooks, D. R., Thorson, T. B. & Mayes, M. A. (1981) Freshwater stingrays (Potamotrygonidae) and their helminth parasites: testing hypothesis of evolution and coevolution. In *Advances in Cladistics: Proceedings of the First Meeting of the Willi Henig Society*, ed. V. A. Funk & D. R. Brooks, pp. 147–175. New York: New York Botanical Garden.

Brown, J. H. (1988) Species diversity. In *Analytical Biogeography*, ed. A. A. Myers & P. S. Giller, pp. 57–89. New York: Chapman & Hall.

Campbell, R. A. (1983) Parasitism in the deep sea. In *Deep Sea Biology*, vol. 8, *The Sea*, ed. G. T. Rowe, pp. 473–552. New York: John Wiley.

Campbell, R. A. (1990) Deep water parasites. *Annales de Parasitologie Humaine et Comparée*, **65**, Supplement 1, 65–68.

Campbell, R. A., Haedrich, R. L. & Munroe, T. A. (1980) Parasitism and ecological relationships among deep-sea benthic fishes. *Marine Biology*, **57**, 301–313.

Chubb, J. C. (1970) The parasite fauna of British freshwater fish. In *Symposia of the British Society for Parasitology*, vol. 8, *Aspects of Fish Parasitology*, ed. A. E. R. Taylor & R. Muller, pp. 119–144. Oxford: Blackwell Scientific Publications.

Collard, S. B. (1970) Some aspects of host-parasite relationships in mesopelagic fishes. In *A Symposium on Diseases of Fishes and Shellfishes*, ed. S. F. Snieszko, pp. 41–46. Washington, DC: American Fisheries Society.

Cone, D. K., Marcogliese, D. J. & Watt, W. D. (1993) Metazoan parasite communities of yellow eels (*Anguilla rostrata*) in acidic and limed rivers of Nova Scotia. *Canadian Journal of Zoology*, **71**, 177–184.

Dietz, R. S. & Holden, J. C. (1970) The break-up of Pangaea. *Scientific American*, **223**, 30–41.

Dogiel, V. A., Petrushevski, G. K. & Polyanski, Y. I. (eds.) (1961) *Parasitology of Fishes*. Edinburgh: Oliver & Boyd.

Esch, G. W., Kennedy, C. R., Bush, A. O. & Aho, J. M. (1988) Patterns in helminth communities in freshwater fish in Great Britain: alternative strategies for colonization. *Parasitology*, **96**, 519–532.

Fernández, J. (1985) Estudio parasitológico de *Merluccius australis* (Hutton, 1872) (Pisces: Merluccidae): aspectos sistemáticos, estadísticos y zoogeográficos. *Boletín de la Sociedad de Biología de Concepción*, **56**, 31–41.

Fernández, J. & Durán, L. (1985) *Aporocotyle australis* n. sp. (Digenea: Sanguinicolidae), parásito de *Merluccius australis* (Hutton 1872) en Chile y su relación con la filogenia de *Aporocotyle* Odhner, 1900 en *Merluccius* spp. *Revista Chilena de Historia Natural*, **58**, 121–126.

Garland, T. Jr, Harvey, P. H. & Ives, A. R. (1992) Procedures for the analysis of comparative data using phylogenetically independent contrasts. *Systematic Biology*, **41**, 18–32.

Gartner, J. V. & Zwerner, D. E. (1989) The parasite faunas of meso- and bathypelagic fishes of Norfolk Submarine Canyon, western North Atlantic. *Journal of Fish Biology*, **34**, 79–95.

Gregory, R. D. (1990) Parasites and host geographic range as illustrated by waterfowl. *Functional Ecology*, **4**, 645–654.

Groot, C., Bailey, R. E., Margolis, L. & Cooke, K. (1989) Migratory patterns of sockeye salmon (*Oncorhynchus nerka*) smolts in the Strait of Georgia, British Columbia, as determined by analysis of parasite assemblages. *Canadian Journal of Zoology*, **67**, 1670–1678.

Guégan, J.-F. & Kennedy, C. R. (1993) Maximum local helminth parasite community richness in British freshwater fish: a test of the colonization time hypothesis. *Parasitology*, **106**, 91–100.

Hare, G. M. & Burt, M. D. B. (1976) Parasites as potential biological tags of Atlantic salmon (*Salmo salar*) smolts in the Miramichi River System, New Brunswick. *Journal of the Fisheries Research Board of Canada*, **33**, 1139–1143.

Harvey, P. H. & Pagel, M. D. (1991) *The Comparative Method in Evolutionary Biology*. Oxford: Oxford University Press.

Hemmingsen, W., Lombardo, I. & MacKenzie, K. (1991) Parasites as biological tags for cod, *Gadus morhua* L., in northern Norway: a pilot study. *Fisheries Research*, **12**, 365–373.

Ho, J.-S. 1989. Phylogeny and biogeography of hakes (*Merluccius*; Teleostei): a cladistic analysis. *Fishery Bulletin, US*, **88**, 95–104.

Hoberg, E. P. (1986*a*) Aspects of ecology and biogeography of acanthocephala in Antarctic seabirds. *Annales de Parasitologie Humaine et Comparée*, **61**, 199–214.

Hoberg, E. P. (1986*b*) Evolution and historical biogeography of a parasite–host assemblage: *Alcataenia* spp. (Cyclophyllidea: Dilepididae) in Alcidae (Charadriiformes). *Canadian Journal of Zoology*, **65**, 997–1000.

Hoberg, E. P. (1992) Congruent and synchronic patterns in biogeography and speciation among seabirds, pinnipeds and cestodes. *Journal of Parasitology*, **78**, 601–615.

Hoberg, E. P. (1995) Historical biogeography and modes of speciation across high-latitude seas of the Holarctic: concepts for host–parasite coevolution among the Phocini (Phocidae) and Tetrabothriidae (Eucestoda). *Canadian Journal of Zoology*, **73**, 45–57.

Hoberg, E. P. (1997) Phylogeny and historical reconstruction: host–parasite systems as keystones in biogeography and ecology. In *Biodiversity II: Understanding and Protecting our Biological Resources*, ed. M. L. Reaka-Kudla, D. E. Wilson & E. O. Wilson, pp. 243–261. Washington, DC: Joseph Henry Press.

Hoberg, E. P., Brooks, D. R., Molina-Ureña, H. & Erbe, E. (1998) *Echinocephalus janzeni* n. sp. (Nematoda: Gnathostomatidae) in *Himantura pacifica* (Chondrichthyes: Myliobatiformes) from the Pacific Coast of Costa Rica and Mexico, with historical biogeographic analysis of the genus. *Journal of Parasitology*, **84**, 571–581.

Ihering, H. von (1891) On the ancient relations between New Zealand and South America. *Transactions and Proceedings of the New Zealand Institute*, **24**, 431–445.

Ihering, H. von (1902) Die Helminthen als Hilfsmittel der zoogeographischen Forschung. *Zoologischer Anzeiger*, **26**, 42–51.

Inada, T. (1981) Studies on Merluccid fishes. *Bulletin of the Far Seas Fishery Research Laboratory*, **18**, 1–172.

Johnston, S. J. (1912) On some trematode parasites of Australian frogs. *Proceedings of the Linnean Society of New South Wales*, **37**, 285–362.

Johnson, S. K. & Rogers, W. A. (1973) Distribution of the genus *Ergasilus* in several Gulf of Mexico drainage basins. *Agricultural Experiment Station Auburn University*, Bulletin No. **445**, 74 pp.

Kabata, Z. & Ho, J.-S. (1981) The origin and dispersal of hake (Genus *Merluccius*: Pisces: Teleostei) as indicated by its copepod parasites. *Oceanography and Marine Biology Annual Review*, **19**, 381–404.

Kennedy, C. R. (1978) The parasite fauna of resident char *Salvelinus alpinus* from Arctic Islands, with special reference to Bear Island. *Journal of Fish Biology*, **13**, 457–466.

Kennedy, C. R. (1995) Richness and diversity of macroparasite communities in tropical eels *Anguilla reinhardtii* in Queensland, Australia. *Parasitology*, **111**, 233–245.

Kennedy, C. R. & Bush, A. O. (1994) The relationship between pattern and scale in parasite communities: a stranger in a strange land. *Parasitology*, **109**, 187–196.

Kennedy, C. R., Laffoley, D. d'A., Bishop, G., Jones, P. & Taylor, M. (1986) Communities of parasites of freshwater fish of Jersey, Channel Islands. *Journal of Fish Biology*, **29**, 215–226.

Khan, R. A. & Tuck, C. (1995) Parasites as biological indicators of stocks of Atlantic cod (*Gadus morhua*) off Newfoundland, Canada. *Canadian Journal of Fisheries and Aquatic Sciences*, **52**, Supplement 1, 195–201.

Køie, M. (1979) On the morphology and life history of *Derogenes varicus* (Müller, 1784) Loos, 1901 (Trematoda, Hemiuridae). *Zeitschrift für Parasitenkunde*, **59**, 67–78.

Kuris, A. M. & Blaustein, A. R. (1977) Ectoparasitic mites on rodents: application of the island biogeography theory? *Science*, **195**, 596–598.

Lebedev, B. I. (1969) Basic regularities in the distribution of monogeneans and trematodes of marine fishes in the world ocean. *Zoologicheskii Zhurnal*, **48**, 41–50. [In Russian.]

Lovejoy, N. R. (1997) Stingrays, parasites, and neotropical biogeography: a closer look at Brooks *et al.*'s hypothesis concerning the origins of neotropical freshwater rays (Potamotrygonidae). *Systematic Biology*, **46**, 218–230.

MacArthur, R. H. & Wilson, E. O. (1967) *The Theory of Island Biogeography*. Princeton: Princeton University Press.

MacKenzie, K. (1985) Relationships between the herring, *Clupea harengus* L., and its parasites. *Advances in Marine Biology*, **24**, 263–319.

Manter, H. W. (1934) Some digenetic trematodes from the deep-water fish at Tortugas, Florida. *Carnegie Institute Publication no. 435, Papers from Tortugas Laboratory*, **27**, 257–345.

Manter, H. W. (1940) The geographical distribution of digenetic trematodes of marine fishes in the tropical American Pacific. *Allan Hancock Pacific Expeditions*, **2**, 531–547.

Manter, H. W. (1955) The zoogeography of trematodes of marine fishes. *Experimental Parasitology*, **4**, 62–86.

Manter, H. W. (1963) The zoogeographical affinities of trematodes of South American freshwater fishes. *Systematic Zoology*, **12**, 45–70.

Marcogliese, D. J. & Cone, D. K. (1993) What metazoan parasites tell us about the evolution of American and European eels. *Evolution*, **47**, 1632–1635.

Margolis, L. (1963) Parasites as indicators of the geographical origin of sockeye salmon, *Oncorhynchus nerka* (Walbaum) occurring in the North Pacific Ocean and adjacent seas. *Bulletin of the International North Pacific Fish Community*, **11**, 101–156.

Margolis, L. (1965) Parasites as an auxiliary source of information about the biology of Pacific salmons (genus *Oncorhynchus*). *Journal of the Fisheries Research Board of Canada*, **22**, 1387–1395.

Margolis, L. (1990) Trematodes as population markers for North Atlantic steelhead trout. *Bulletin de la Société Française de Parasitologie*, **8**, Supplement 2, 735.

McGladdery, S. E. & Burt, M. D. (1985) Potential of parasites for use as biological indicators of migration, feeding and spawning behaviour of northwestern Atlantic herring (*Clupea harengus*). *Canadian Journal of Fisheries and Aquatic Research*, **42**, 1957–1968.

Metcalf, M. (1929) Parasites and the aid they give in problems of taxonomy, geographical distribution, and paleogeography. *Smithsonian Miscellanea Collection*, **81**, 1–36.

Metcalf, M. (1940) Further studies on the opalinid ciliate infusorians and their hosts. *Proceedings of the United States Museum*, **87**, 465–634.

Morrone, J. J. & Crisci, J. V. (1995) Historical biogeography: Introduction to methods. *Annual Review of Ecology and Systematics*, **26**, 373–401.

Moser, M. (1990) Biological tags for stock separation in Pacific Herring (*Clupea harengus pallasi* Valenciennes) and the possible effect of 'El Niño' currents on parasites. *Proceedings of the International Herring Symposium, October 1990, Anchorage, Alaska*, pp. 245–254.

Moser, M. & Hsieh, J. (1992) Biological tags for stock separation in Pacific Herring *Clupea harengus pallasi* in California. *Journal of Parasitology*, **78**, 54–60.

Noble, E. R. (1973) Parasites and fishes in a deep sea environment. *Advances in Marine Biology*, **11**, 121–195.

Pagel, M. D. & Harvey, P. H. (1988) Recent developments in the analysis of comparative data. *Quarterly Review of Biology*, **64**, 413–440.

Pascual, S. & Hochberg, F. G. (1996) Marine parasites as biological tags of cephalopod hosts. *Parasitology Today*, **12**, 324–327.

Pianka, E. R. (1983) *Evolutionary Ecology*, 3rd edn. New York: Harper & Row.

Platt, N. E. (1975) Infestation of cod (*Gadus morhua* L.) with larvae of codworm (*Terranova decipiens* Krabbe) and herringworm, *Anisakis* sp. (Nematoda: Ascaridata), in North Atlantic and Arctic waters. *Journal of Applied Ecology*, **12**, 437–450.

Platt, N. E. (1976) Codworm – a possible biological indicator of the degree of mixing of Greenland and Iceland cod stocks. *Journal du Conseil International pour l'Exploration de la Mer*, **37**, 41–45.

Polyanski, Y. I. (1961) Ecology of parasites of marine fishes. In *Parasitology of Fishes*, ed. V. A. Dogiel, G. K. Petrushevsky & Y. I. Polyanski, pp. 48–83. Edinburgh: Oliver & Boyd.

Poulin, R. (1995) Phylogeny, ecology, and the richness of parasite communities in vertebrates. *Ecological Monographs*, **65**, 283–302.

Price, P. W. (1980) *Evolutionary Biology of Parasites*. Princeton: Princeton University Press.

Price, P. W. & Clancy, K. M. (1983) Patterns in number of helminth parasite species in freshwater fishes. *Journal of Parasitology*, **69**, 449–454.

Prudhoe, S. & Bray, R. A. (1982) *Platyhelminth Parasites of the Amphibia*. Oxford: Oxford University Press for British Museum (Natural History).

Rohde, K. (1978*a*) Latitudinal gradients in species diversity and their causes. II. Marine parasitological evidence for a time hypothesis. *Biologisches Zentralblatt*, **97**, 405–418.

Rohde, K. (1978*b*) Latitudinal differences in host-specificity of marine Monogenea and Digenea. *Marine Biology*, **47**, 125–134.

Rohde, K. (1979) A critical evaluation of intrinsic and extrinsic factors responsible for niche restriction in parasites. *American Naturalist*, **114**, 648–671.

Rohde, K. (1980) Diversity gradients of marine Monogenea in the Atlantic and Pacific Oceans. *Experientia*, **36**, 1369–1371.

Rohde, K. (1981) Niche width of parasites in species-rich and species-poor communities. *Experientia*, **37**, 359–361.

Rohde, K. (1986) Differences in species diversity of Monogenea between the Pacific and Atlantic oceans. *Hydrobiologia*, **137**, 21–28.

Rohde, K. (1989) Simple ecological systems, simple solutions to complex problems? *Evolutionary Theory*, **8**, 305–350.

Rohde, K. (1993) *Ecology of Marine Parasites*, 2nd edn. Wallingford, UK: CAB International.

Schulte, F. & Poinar, G. O. (1991) On the geographical distribution and parasitism of *Rhabditis (Pelodera) orbitalis* (Nematoda: Rhabditidae). *Journal of the Helminthological Society of Washington*, **58**, 82–84.

Sey, O. (1991) *CRC Handbook of the Zoology of Amphistomes*. Boca Raton: CRC Press.

Sindermann, C. J. (1957) Diseases of fishes of the western North Atlantic. V. Parasites as indicators of herring movements. *Maine Department of Sea and Shore Fisheries Research Bulletin*, **27**, 1–30.

Skorping, A. (1982) *Elaphostrongylus rangiferi*: influence of temperature, substrate, and larval age on the infection rate in the intermediate snail host, *Arianta arbustorum*. *Experimental Parasitology*, **54**, 222–228.

Szidat, L. (1961) Versuch einer Zoogeographie des SudAtlantik mit Hilfe von Leitparasiten der Meeresfische. *Parasitologische Schriftenreihe*, **13**, 1–98.

Thompson, A. B. & Margolis, L. (1987) Determination of population discreteness in two species of shrimp, *Pandalus jordani* and *Pandalopsis dispar*, from coastal British Columbia using parasite tags and other population characteristics. *Canadian Journal of Fisheries and Aquatic Sciences*, **44**, 982–989.

Chapter 16

Evolutionary aspects

Biological evolution is the process by which there is a change in **gene frequency** within a population from one generation to the next. The changes in gene frequency may be reflected in individual organisms through changes in morphological, behavioral, physiological, or ecological traits; these changes may be readily apparent, subtle and not always easy to evaluate, or not apparent at all. Unless these changes are transmitted from generation to generation, they cannot be considered within the framework of biological evolution.

Evolution, as for many other phenomena already described in this book, is a scale-dependent process. In general terms, **microevolution** is concerned with evolution as it occurs below the species level, whereas **macroevolution** addresses evolutionary questions at the species level and above. The temporal and spatial frames within which evolution can be studied at these two levels are very different, and so are the methodologies employed.

In more specific terms, microevolution is the systematic change in the frequencies of homologous alleles, chromosome segments, or entire chromosomes in a population. Macroevolution, on the other hand, involves phenotypic and genotypic changes of greater magnitude than those found in microevolution; these result in the evolution of characteristics distinguishing major groups such as genera, families, orders, etc. Microevolutionary changes occur in real or ecological time, whereas macroevolutionary developments take place over geological, or evolutionary, time. In evolutionary terms, all individuals within a species sharing a common gene pool form a population (also called a local population or **deme**) (but see Chapter 10 for population concepts as applied to parasites). The individual species becomes the line of separation between micro- and macroevolutionary phenomena because genetic discontinuities between species are absolute. This means that genetic change in one species cannot be transferred to another because, for the most part, sexually reproducing organisms of different species cannot exchange genes due to reproductive barriers (the possibility of horizontal gene transfer between species, however, will be considered in section 16.4.3). In effect, most species are independent evolutionary units. There are exceptions, however, and most of these are symbiotic in character; they involve mutuals, commensals, parasites, *and* their partners.

16.1 Microevolution

In sexually reproducing organisms, a **species** can be defined as a group of potentially interbreeding individuals that are reproductively isolated from other groups of organisms and that produce viable offspring. Members of a species, however, usually are not homogeneously distributed in space. Free-living organisms, e.g., hosts, are characteristically subdivided into smaller groups due to environmental patchiness. Geographic areas having favorable habitats are interspersed with unfavorable ones, reducing (or even eliminating) potential interbreeding among individuals occupying the different areas. Behavioral variations (schooling, flocking, herding, colony formation or, in contrast, territoriality) may also enhance the effects of spatial heterogeneity on reproduction and gene flow among host populations of a given species. Locally interbreeding individuals of geographically isolated, or otherwise structured, groups are referred to as local populations or demes. The significance of the spatial distribution of populations rests with the idea that it is within these demes where changes in gene frequencies are most likely to occur and, therefore, where evolution takes place. Because the various selection forces operating on these demes may not be the same across the geographic range of a species, the local population must be considered as the basic unit of evolution. On the other hand, the precise geographic boundaries for these local populations are not always fixed or clear. For example, migration of individuals into and out of a geographical area will reduce considerably the reproductive and genetic independence of populations.

The patterns of population structure associated with parasitic organisms are far more complex than those of their hosts. In part, this is related to several of the functional characteristics of parasitism itself. For example, parasites occur as potentially reproducing infrapopulations in a definitive host, within which effective gene flow will take place for all, or part, of each sexually reproductive cycle. These infrapopulations may then disappear, only to reappear as new and genetically distinct infrapopulations in the next, sexually reproductive generation. The genetic structure of the next infrapopulation, then, will be constrained by the population structure of the potential host and how, when, and where parasites are recruited. Unlike free-living organisms, infrapopulations of many metazoan parasite species increase in numbers only through immigration (recruitment), and not as a result of natality within or on the host. In each reproductive cycle, parasite infrapopulations may or may not be the same, depending on the life history of the parasite. If parasite generations do not overlap, the genetic makeup of the newly assembled infrapopulation in a given host individual could be completely different from the previous one. If there is overlap in parasite generations, the new genetic make up of the reproductive infrapopulation will depend on the genetic makeup of the immigrants and residual 'residents'.

The recruitment or immigration of parasites into an infrapopulation will be influenced by the vagility of the host and factors limiting infrapopulation density in a host. Obviously, there is no direct gene flow among parasites of a given species at the component or suprapopulation level. Intuitively, one might also predict that the aggregated distribution of parasite infrapopulations could contribute uniquely in some way to the population genetics of parasites. Given that most hosts themselves have aggregated distributions, perhaps the superimposition of parasite aggregation on the aggregation of the host might affect gene flow and enhance drift via the constant renewal of potential founder effects with each new reproductive cycle. On the other hand, it is also possible that within an aggregated host–parasite system, the infective stages are 'pooled' and redistributed each generation, which will likely swamp any within-host drift. Unfortunately, very little is known about this topic. The notion of selection pressure imposed by the host's immune responsiveness is also an important aspect of the population genetics of a parasite that needs to be addressed.

Price (1977, 1980) acknowledged the idea that populations were the basic units of evolution and emphasized that their population structure was a major factor in the evolutionary biology of parasitic organisms. On theoretical grounds, he argued that, because parasites are adapted to exploit small and discontinuous environments (hosts) in a matrix of inhospitable environments, they should have small and homogeneous populations with little gene flow between them. However, if host variability in time and space is considered, together with the specialized resource exploitation shown by the parasites, then three additional predictions are possible. First, if the environment (host) is stable in time and variable in space, then the local parasite population should be monomorphic and specialized, with several geographic races. But, if the environment is uniform in space and variable in time, the parasite population should be polymorphic, with several specialized types present on a **cline** within a geographical scale. Finally, when the host is variable in both space and time, Price (1980) predicted the formation of both geographic races and polymorphisms in the parasite population.

For evolution to occur a population must have some degree of genetic variability; this will enhance the probability of changes in gene frequencies over time. Heritable variation in a population can arise through **mutation** and genetic **recombination** (but see section 16.4.3 for the possibility of horizontal gene transfer). Genetic drift, gene flow (migration), and natural selection are, on the other hand, the processes by which these variations are enhanced, reduced, or eliminated from a population. Although the mechanisms of reproduction do not affect the allelic frequencies of a population, they have to be considered because they influence genotype frequencies.

At the microevolutionary level, genetic drift and gene flow are the forces driving the genetic variability of populations. Genetic drift is particularly important in the microevolution of small populations. The amount of genetic drift in a deme is determined only by the number of individuals contributing genes to the next generation (the effective population size, N_e) and not by the total number of individuals in the deme (Wright, 1931). In a deme (or infrapopulation of parasites) a small effective population size likely means that not many of the characteristic allele frequencies and genetic variability of the population will be transmitted to the next generation. With a small effective population size, the heterozygosity of the population will be reduced due to inbreeding, and allelic frequencies will change randomly due to sampling variance (genetic drift). Gene flow, the movement of genes within and between local populations, is generally accomplished by migration of individuals among these populations. The amount of gene flow among local populations, together with other population parameters, affects the rate of evolutionary change of these populations. Low gene flow among local populations increases their chance of evolving independently. High levels of gene flow, on the other hand, tend to homogenize the different local populations and, for evolutionary purposes, they all become one evolutionary unit.

The amount of genetic variability in a local population, its genetic structure, and potential for gene flow, are important parameters needed to study microevolutionary changes in local parasite populations. Genetic variation within a population frequently has been studied by gel electrophoresis of proteins. Molecular approaches using the polymerase chain reaction (PCR) and direct sequencing of DNA polymorphisms with techniques such as DNA fingerprinting of minisatellite regions and randomly amplified polymorphic DNA (RAPD), are contributing enormously to the study of genetic variation at the basic DNA level. Genetic variation in populations using protein electrophoresis is commonly quantified using **polymorphism** (*P*), the proportion of polymorphic loci in a population, and **heterozygosity** (*H*), the average frequency of heterozygous individuals per locus. Polymorphism and heterozygosity can also be estimated at the nucleotide level, where P_{nuc} is the proportion of polymorphic nucleotide sites, and H_{nuc} is the number of nucleotide substitutions per site.

Unfortunately, observed allelic differences among different populations, or populations in different geographic areas, cannot be used directly to characterize gene flow among these populations. Instead, indirect methods such as the use of *F*-statistics (F_{st}, F_{it}, F_{is}) as developed by Wright (1951), the tracking of rare or unique alleles among populations or demes, or the analysis of nodes and branch lengths of a gene tree using nucleotide sequence data, need to be applied (Nadler, 1995). *F*-statistics estimated from genetic data are particularly useful because they allow characterization of the breeding structure within populations of parasites and reveal the potential for interpopulation differentiation by genetic drift for neutral alleles. Although enzyme and DNA studies have provided a great deal of information about genetic variation, attempts to link this variation to selection forces have not been successful (Grant, 1994). For example, it is not known if the enzymatic or DNA variation that has been measured matters to the organism, because in most cases, it has been attributed no function.

16.1.1 Genetic variability and component population structure of parasites

The nature of the local random mating population, or deme, remains as one of the major problems to be solved in studies on the population genetics of parasites. Although intuitively, and for practical purposes, the infrapopulation constitutes the random mating population, its genetic structure may be highly variable depending on the host's and the parasite's ecology and basic biology.

Most parasite component populations have levels of genetic variation that are similar to those of free-living invertebrates (Tables 16.1, 16.2). Unfortunately, the available data regarding geographically isolated infra- and component populations of parasites are not extensive and, in many cases, the analysis of the data is incomplete. For example, the low levels of genetic variability found in a single population of parasites such as *Schistosoma japonicum*, *S. mekongi*, or *Paragonimus westermani* (Table 16.2) do not mean that the species lack genetic variability, only that the populations studied lack variability. Other populations of these species need to be studied before a definite conclusion can be reached regarding the genetic variability of the species.

Even though several studies address the genetic variability of parasite populations, not all of them address issues such as population structure or the amount of gene flow. The genetic structure of a population can be defined as the distribution of genetic variation among individuals sampled over different spatial scales (Anderson *et al.*, 1998). The genetic structure of a population is determined by the effective size of the population and the rate of gene flow among populations. Nadler (1995) identified a number of ecological and natural-history factors that may influence the population genetic structure of animal parasites (Table 16.3). Many of these have been confirmed by experimental studies.

Studies of enzyme polymorphism and genetic variability in the nematode *Ascaris suum* provided one of the first insights into the nature of variation in allele frequencies in relation to the extent of geographical separation (Leslie *et al.*, 1982). Levels of variation in *A. suum* in the localities studied (Iowa and New Jersey, USA) were in the lower range for other parasites (see Table 16.2). Based on genotypic frequency expectations, worms within each locality seemed to be members of a single component population, apparently sharing a common gene pool. Comparisons between the Iowa and New Jersey component populations (separated by a distance of nearly 2400 km) did not reveal marked differences. It was assumed, then, that gene flow between the two locations had been historically high, or that cessation of the gene flow may have been a recent event. Unfortunately, this pioneer study analyzed the parasite data based on geographical area (the component population level) and not based on individual pig hosts (the infrapopulation level). Subsequent studies into the genetic structure of *Ascaris suum* at three spatial scales (individual host, farm, and geographic regions) using both protein electrophoresis and RAPD PCR-based markers showed significant patterns of genetic structure for infrapopulations within a farm, for pooled infrapopulations within a farm, and for geographic areas (Michigan, Illinois, Indiana, USA) (Nadler *et al.*, 1995). For example, 91% of the total estimated genetic diversity was found within infrapopulations and 9% was partitioned among infrapopulations within a farm. On a larger geographical scale, the genetic diversity of all pooled infrapopulations within a region accounted for 92% of the diversity and only 8% was partitioned among the geographical areas. The data showed excess homozygosity for infrapopulations of *A. suum*, as indicated by high inbreeding coefficients, and significant genetic isolation among farms within geographical regions. In short, in contrast with the study of Leslie *et al.* (1982), Nadler *et al.* (1995) found low gene flow between close populations on a geographic scale, and also genetic isolation among the infrapopulations within geographic regions. It is likely that the genetic differentiation among infrapopulations has been promoted by genetic drift as a consequence of their small effective population size and possible **founder effects**. A study by Anderson & Jaenike (1997) on *Ascaris* spp. in Guatemala showed similar results, with subdivided populations. In their study, however, only 2.4% of the genetic variation of *Ascaris* was partitioned between geographic areas. Interestingly, they suggest that although low gene flow may explain the patterns found, the variation also could be explained by retention of ancient polymorphisms.

Studies of mitochondrial DNA in populations from different geographical regions of the trichostrongylid nematodes *Ostertagia ostertagi* and *Haemonchus placei* (from cattle), *H. contortus* and

Table 16.1 Heterozygosity (*H*) and polymorphism (*P*) and their coefficient of correlation (*r*) for various organisms

	H			*P*			
	Number of species	Mean	SD	Number of species	Mean	SD	*r*(*H*, *P*)
Coelenterata	5	0.140	0.042	5	0.481	0.191	0.840 ns
Vermes	6	0.072	0.079	6	0.289	0.222	0.949**
Mollusca	46	0.148	0.170	44	0.468	0.287	0.764***
Crustacea	122	0.082	0.082	119	0.313	0.224	0.879***
Insecta (except Drosophila)	122	0.089	0.060	130	0.351	0.187	0.753***
Drosophila	34	0.123	0.053	39	0.480	0.143	0.552***
Pisces	183	0.051	0.035	200	0.209	0.137	0.845***
Amphibia	61	0.067	0.058	73	0.254	0.151	0.735***
Reptilia	75	0.083	0.119	84	0.256	0.148	0.814***
Aves	46	0.051	0.029	56	0.302	0.143	0.497***
Mammalia	184	0.0.41	0.035	181	0.191	0.137	0.821***
Average for major groups							
Plants	56	0.075	0.069	75	0.295	0.251	0.842***
Invertebrates	361	0.100	0.091	371	0.375	0.219	0.769***
Vertebrates	551	0.054	0.059	596	0.226	0.146	0.792***

Note:
Levels of significance: ** = $p<0.01$, *** = $p<0.001$, ns = $p>0.05$ (not significant).

Source: Modified from Nevo, E., Beiles, A. & Ben-Shlomo, R. (1984) The evolutionary significance of genetic diversity: ecological, demographic and life history correlates. In *Evolutionary Dynamics of Genetic Diversity, Lecture Notes in Biomathematics*, Vol. 53, ed. G. S. Mani, pp. 13–213. © Springer-Verlag.

Table 16.2 Estimates of genetic variability for parasites

Species	Locality	Number of loci Surveyed	P^a	H^a	Reference[b]
Digenea					
Paragonimus westermani	Mie (1), Japan	18	0.12	0.035	Agatsuma & Habe (1985*b*)
Paragonimus westermani	Mie (2), Japan	18	0.06	0.033	Agatsuma & Habe (1985*b*)
Paragonimus westermani	Ohita, Japan	18	0.0	0.0	Agatsuma & Habe (1985*b*)
Schistosoma mansoni	(mean of 22 strains) various localities	18	0.13	0.04	Fletcher *et al.* (1981)
Schistosoma japonicum	Leyte, Phillippines	18	0.0	0.0	Woodruff *et al.* (1987)
Schistosoma mekongi	Laso	15	0.0	0.0	Woodruff *et al.* (1987)
Halipegus occidualis	North Carolina, USA	8	0.125	0.05	Goater *et al.* (1990)
Fascioloides magna[c]	Tennessee, USA(1)	22	0.643	0.09	Lydeard *et al.* (1989)
Fascioloides magna[c]	Tennessee, USA(2)	22	0.429	0.05	Lydeard *et al.* (1989)
Fascioloides magna[c]	South Carolina, USA(3)	22	0.429	0.13	Lydeard *et al.* (1989)
Fascioloides magna[c]	South Carolina, USA(4)	22	0.429	0.10	Lydeard *et al.* (1989)
Cestoda					
Progamotaenia festiva	Australia, various localities (species complex)	16	n/a[d]	0.03[e]	Baverstock *et al.* (1985)
Echinococcus granulosus	Australia, mainland	20	0.15	0.02	Lymbery & Thompson (1988)
Echinococcus granulosus	Tasmania	20	0.15	0.06	Lymbery & Thompson (1988)
Nematoda					
Contracaecum sp. 'I' (larvae)	Mexico	11	0.54	0.141	Vrijenhoek (1978)
Contracaecum sp. 'II' (larvae)	Mexico	11	0.54	0.193	Vrijenhoek (1978)
Contracaecum osculatum 'B'	n/a[d]	21	0.62	0.10	Bullini *et al.* (1986)
Contracaecum rudolphii 'A'	n/a[d]	21	0.57	0.17	Bullini *et al.* (1986)
Anisakis simplex (larvae)	North Atlantic Ocean	22	0.50	0.21	Nascetti *et al.* (1986)
Anisakis pegreffii (larvae)	Mediterranean Sea	22	0.32	0.12	Nascetti *et al.* (1986)
Anisakis physeteris (larvae)	n/a[d]	22	0.50	0.11	Bullini *et al.* (1986)
Phocascaris cystophorae	n/a[d]	21	0.24	0.10	Bullini *et al.* (1986)
Ascaris suum	Iowa, USA	38	0.21	0.066	Leslie *et al.* (1982)
Ascaris suum	New Jersey, USA	38	0.17	0.053	Leslie *et al.* (1982)
Ascaris suum	n/a[d]	24	0.17	0.03	Bullini *et al.* (1986)
Ascaris lumbricoides	n/a[d]	24	0.25	0.02	Bullini *et al.* (1986)

Parascaris equorum	Louisiana, USA	18	0.22	0.085	Nadler (1986)
Parascaris equorum	Central–eastern Europe	27	0.03	0.008	Bullini *et al.* (1978)
Parascaris equorum	n/a[d]	28	0.07	0.02	Bullini *et al.* (1986)
Parascaris univalens	Central–eastern Europe	27	0.03	0.015	Bullini *et al.* (1978)
Parascaris univalens	n/a[d]	28	0.11	0.03	Bullini *et al.* (1986)
Neoascaris vitulorum	n/a[d]	18	0.11	0.04	Bullini *et al.* (1986)
Toxocara canis	n/a[d]	18	0.33	0.10	Bullini *et al.* (1986)
Toxocara canis	Louisiana, USA	18	0.33	0.135	Nadler (1986)
Toxocara cati	Louisiana, USA	18	0.38	0.137	Nadler (1986)
Toxocara cati	n/a[d]	18	0.17	0.05	Bullini *et al.* (1986)
Toxascaris leonina	n/a[d]	18	0.11	0.02	Bullini *et al.* (1986)
Baylisascaris transfuga	n/a[d]	18	0.17	0.05	Bullini *et al.* (1986)
Arthropoda					
Glossina m. morsitans	Rekometjie, Zimbabwe	11–13	0.38	0.171	Gooding & Jordan (1986)
Glossina m. centralis	Nalusanga, Zambia	11–13	0.36	0.147	Gooding (1989)
Glossina m. centralis	Keembe, Zambia	11–13	0.18	0.093	Gooding (1989)
Glossina pallidipes	Lambwe, Kenya	11–13	0.17	0.092	Agatsuma & Otieno (1988)
Glossina pallidipes	Nguruman, Kenya	11–13	0.17	0.088	Agatsuma & Otieno (1988)
Glossina pallidipes	Lambwe, Kenya	11–13	0.33	0.153	Tarimo Nesbit *et al.* (1990)
Glossina pallidipes	Nguruman, Kenya	11–13	0.25	0.091	Tarimo Nesbit *et al.* (1990)
Glossina longipennis	Nguruman, Kenya	6	0.33	0.045	Tarimo Nesbit (1991)

Notes:

[a] Criteria for defining polymorphism (*P*) and estimating average heterozygosity (*H*) vary by author.

[b] References for helminths can be found in Nadler (1990); references for arthropods in Gooding (1992).

[c] (1) Shelby Forest Wildlife Management Area; (2) Reelfoot National Wildlife Refuge; (3) Savannah River Plant; (4) Webb Wildlife Center.

[d] n/a: not available.

[e] Mean for species studied.

Sources: Modified from Nadler (1990) and Gooding (1992). From Nadler (1990), by permission, *International Journal for Parasitology*, **20**, 11–29. Reprinted from Gooding, R.H. (1992) Genetic variation in tsetse flies and implications for trypanosomiasis. *Parasitology Today*, **8**, 92–95. Copyright © 1992, with permission from Elsevier Science.

Table 16.3 Examples of ecological and natural-history factors that may influence the population genetic structure of animal parasites

Factors increasing genetic structure	Factors reducing genetic structure
Sedentary definitive host or extreme morbidity of all infected hosts	Highly vagile hosts (definitive, intermediate, paratenic) or vectors
Life cycle includes a large number of specific obligate hosts	Persistent (long-lived) life-cycle stages in environment or definitive host
Suitable parasite niches patchily distributed in space or time	Low definitive host specificity/many reservoir hosts
Small effective size for parasite population	Uniform distribution of parasites among hosts
Parasite predominantly self-fertilizing	Life history with frequent metapopulation extinction followed by reestablishment
Physical contact between definitive hosts required for transmission	

Source: From Nadler (1995), with permission, *Journal of Parasitology*, **81**, 650–669.

Teladorsagia circumcincta (from sheep), and *Mazamastrongylus odocoilei* (from deer), revealed a different kind of genetic structure (Blouin *et al.*, 1992, 1995). The parasites of sheep and cattle had very high within-population diversities, with a pattern of high gene flow and low genetic differentiation among populations. The parasites of deer also had high within-population diversity, but genetic differentiation among populations was high, with populations much more subdivided than those of the domestic animals. The pattern observed in the deer parasite is one of isolation-by-distance among the populations, meaning that gene flow in the parasites is a function of host movement. The high gene flow and low differentiation observed in the nematodes of domestic animals is probably mediated by the long-distance transport of cattle throughout the USA, the spread of contaminated manure, or by other husbandry practices that increase the gene flow of the parasites. A similar pattern of high genetic diversity within populations, and low differentiation among populations, was observed in *Schistosoma haematobium* in different rivers in Zimbabwe using RAPD markers. The similarity between different populations of *S. haematobium* was more related to their straight-line geographical distance than to their proximity within a river system. It appears that dispersal by the definitive host (probably humans), together with the longevity of the parasite in this host, may be the most important mechanisms favoring increased gene flow among populations, which results in low differentiation among populations (Davies *et al.*, 1999). In short, host movement seems to be of paramount importance for the genetic structure of parasite component populations across geographical areas. Parasite populations in hosts with high vagility exhibit less genetic structure than parasites populations in less mobile hosts.

High host vagility obviously facilitates gene flow among parasite populations. Similarly, parasites with high dispersal capabilities (through the use of multiple hosts in the life cycle, for example) also should exhibit less genetic structure and differentiation across geographical areas. Paggi *et al.* (1991) and Nascetti *et al.* (1993) used protein electrophoretic data to show that populations of sibling species of the ascaridoid nematodes *Contracaecum* and *Pseudoterranova*, using seal definitive hosts and invertebrates and fishes as intermediate and paratenic hosts, have low genetic structure across extensive geographic ranges of the Boreal regions. Nematodes were obtained from definitive and intermediate hosts from over 20 localities from the eastern and western Atlantic Ocean as far north and west as the Barents Sea. Nascetti *et al.* (1993) hypothesized that the migration of the seal definitive hosts, in combination with the transport of larvae in intermediate and paratenic hosts, facilitates the gene flow over extensive geographic areas such as the eastern and western Atlantic Ocean.

The relationship between high parasite dispersion or host vagility and low genetic structure of parasite populations is not restricted to helminths. Studies of genetic variability among different populations of ticks such as *Amblyomma americanum, Ornithodoros erraticus,* and *O. senrai* in the USA, and among different populations of six species of reptilian ticks in Australia, revealed a pattern with low levels of genetic variability within, as well as between, different geographic populations (Wallis & Miller, 1983; Bull *et al.*, 1984; Hilburn & Sattler, 1986*a*, *b*). In the studies of ticks from the United States, both Nei's genetic distance and *F*-statistics were used to estimate the degree of inbreeding and divergence within and between purported populations. In the case of *A. americanum*, nine geographic populations were examined. Some of them exhibited quantitative and qualitative differences that argued in favor of a small degree of genetic structuring. However, the differences were not sufficiently large to affect the measurements of genetic relatedness. Because the degree to which populations diverge genetically is primarily determined by the amount of gene flow between them, even infrequent exchange between populations will prevent extensive genetic divergence. Since ticks are relatively sedentary when they are not on a host, the rate of gene exchange will depend on host vagility. The low host specificity and the number of hosts in the life cycle also increase the amount of gene flow between geographical populations. In the case of *A. americanum,* each tick attaches and feeds on three hosts during its lifetime but, each time the ticks molt or lay eggs, they leave the host and remain on the ground. This behavior enhances its chances of capturing a different individual or host species each time it reattaches, thereby increasing its chances of dispersal and gene flow.

The significance of parasite reproductive strategies and their impact on the genetic structure of parasite populations has not been fully studied. Research on the deer fluke *Fascioloides magna*, and the mostly self-fertilizing cestode *Echinococcus granulosus*, however, has provided many insights into the importance of reproductive strategies.

Lydeard *et al.* (1989) and Mulvey *et al.* (1991) studied the genetic structure of four geographically separated component populations of the liver fluke *Fascioloides magna*, in white-tailed deer *Odocoileus virginianus*, in the southeastern USA. Levels of genetic variability, including heterozygosity and polymorphism, in the different locations (two in Tennessee and two in South Carolina), were comparable to those reported for other species of parasites (Table 16.2). Populations of *F. magna* among localities within a single state showed low differentiation, suggesting that some gene flow may occur among these local, in-state, populations. Comparison of populations from South Carolina and Tennessee, on the other hand, showed high levels of differentiation, suggesting an isolation-by-distance effect. Also, a deficiency of heterozygous individuals was observed within localities. This might suggest extensive inbreeding or a high degree of parasite relatedness in the deer, possibly as a consequence of the asexual reproduction (polyembryony) that takes place in the snail intermediate host. For example, large numbers of genetically identical cercariae are released from an infected snail and encyst as metacercariae on vegetation within a narrowly defined geographic area. As individual deer or small groups of deer browse, their chances of acquiring clumps of identical metacercariae derived from a single, genetically distinct, adult would be considerably enhanced. Indeed, Mulvey *et al.* (1991) found that deer tend to be infected with parasites of the same genotype

The taxonomy of species in the cestode *Echinococcus* is currently in a state of flux because molecular genetics studies have shown that many previously defined intraspecific variants, or strains, deserve species status (Bowles *et al.*, 1995; Thompson *et al.*, 1995). Studies of *Echinococcus* have also shed some light on the genetic structure of its different populations. The levels of genetic variation (heterozygosity and polymorphism) found in *E. granulosus* in mainland Australia and Tasmania (Table. 16.2) are relatively low when compared to other parasites. This could be explained, in part, by the relatively recent colonization of mainland Australia and Tasmania by the parasite or by its reproductive strategy of self-fertilization. The genetic structure of plant and invertebrate species that reproduce by self-fertilization, like *E. granulosus*, is characterized by low genetic variation within populations and high variation between populations. However, most genetic variation in *E. granulosus* occurs within local populations with very little evidence of genetic variation between populations. Although self-fertilization is ubiquitous in this genus, it appears that genetic differentiation between populations is prevented by the dispersal of the parasite over wide geographical areas with its intermediate and definitive hosts, and probably by occasional outcrossing of the worms (Thompson & Lymbery, 1996; Lymbery *et al.*, 1997).

Unlike most previous studies where parasite populations show low genetic structure, studies of chewing lice (*Geomydoecus actuosi*) from pocket gophers in the USA, and lice in the *Heterodoxus octoseriatus* group from rock-wallabies in Australia, indicate that lice populations are highly structured with high differentiation among populations (Nadler *et al.*, 1990; Barker *et al.*, 1991). Both groups of lice showed low polymorphism and heterozygosity probably as a consequence of population bottlenecks, founder effects, and inbreeding. Both the genetic structure of the populations and their genetic variability seem to be strongly constrained by the natural history of the host and lice. In both studies, the social organization of the host and the fragmented habitat they inhabit reduces contact and limits transmission of the parasite which can only occur by direct contact between the hosts.

Even though hosts seem to have a significant impact on the genetic structure of parasite populations, very few studies have examined or compared the genetic structure of both host *and* parasite populations. In the study of deer and *Fascioloides magna* discussed previously (Mulvey *et al.*, 1991), the genetic distances between the deer populations were not concordant with the genetic distances between the parasite populations or with their geographic distribution, probably due to the mobility of deer. Dybdahl & Lively (1996), however, studying the population structure of the snail *Potamopyrgus antipodarum* and its digenean parasite *Microphallus* sp., in several lakes in New Zealand, found just the opposite. The snail populations were highly structured among the lakes, but not the parasite populations, suggesting higher levels of gene flow for the parasite (likely due to their dispersal by waterfowl and wading birds that serve as definitive hosts). Nonetheless, host and parasite genetic distances were correlated with each other and with the distances between the lakes. These results indicate that there is more gene flow among parasite populations than among host populations, but that both host and parasite have similar patterns of dispersal, creating a situation where lakes with similar host populations also have similar parasite populations. It is likely that aquatic birds moving between adjacent lakes disperse both hosts and parasites, but parasite dispersal in the birds (the definitive host) is more frequent than the dispersal of the snails.

The study by Davies *et al.* (1999) on the population genetics of the blood fluke *Schistosoma haematobiun* and its snail host *Bulinus globosus* from rivers in Zimbabwe also shows that the snail host has higher levels of population structure than its digenean parasite. The structure and differentiation of the snail populations could result from high levels of genetic drift. These snail populations are under strong seasonal variations, undergoing periodic bottlenecks. These factors increase genetic drift and genetic differentiation. On the other hand, the low genetic structure of the parasite populations suggests that the vagility of the parasite in the definitive host is higher than the vagility of the snail intermediate host. The relatively long life span of the parasite in the definitive host and its low specificity for the definitive host result in high levels of gene flow among parasite populations because of the homogenizing effects of these factors on a geographic scale.

The genetic structure of parasite populations seems to depend on a number of biological and ecological factors not only related to the parasite but also related to the host. This adds another dimension to the already complex problem of genetic variability and structure in parasite populations. Although available data show some patterns and trends, it is also evident that each case study is slightly different from others and that many more studies are needed to assess the differential impact of the various natural history parameters on the genetic structure of parasite populations.

16.1.2 Correlates of genetic variability

It is evident from Tables 16.1 and 16.2 that different species and component populations have varying degrees of polymorphism and heterozygosity. Because of these variations, two important questions arise. What mechanisms account for the differences and how are these differences maintained within parasite component populations? Two hypotheses have been proposed to explain the differing degrees of polymorphism and heterozygosity. One is based on a neutralist perspective and the other on a selectionist approach. The neutralist hypothesis is rooted in the observation that there is a relatively constant rate of molecular evolution in homologous molecules from different species' lineages because most of the allelic changes that take place over time are selectively neutral. Both hypotheses agree that new alleles appear by mutation. However, the neutralists argue that if alternate alleles have identical fitness, then variations in allelic frequencies will occur only by accidental change from generation to generation (genetic drift). Genetic drift, then, will be affected by changes in the effective infrapopulation sizes due to founder effects, bottlenecks, and departures from 1:1 sex ratios (Roughgarden, 1979; Nadler, 1987). The neutralist hypothesis recognizes, however, that deleterious mutations can be eliminated by natural selection. The selectionist hypothesis assumes that the changes in gene frequencies are adaptive, and not neutral. Consequently, adaptive variations will increase gradually in frequency over time at the expense of less adaptive ones. The selectionist hypothesis, to some extent, accounts for the diversity of organisms because it promotes their adaptation to new and different life styles. Although there has been a long-term debate between the two schools of thought, in reality they are not mutually exclusive. Nadler (1990) urged that the controversy should, more properly, be focused on what proportion of the observed variation is maintained by one mechanism or the other. The controversy has since subsided, with selectionists conceding that much molecular variation is close enough to being neutral and neutralists admitting that natural selection can be extremely sensitive in influencing molecular evolution (Powell, 1994).

According to the selectionist viewpoint, genetic variability among species and populations is often positively correlated with levels of ecological heterogeneity. Species or populations from 'heterogeneous' environments should have increased genetic variability in response to diverse selection pressures. Similarly, in homogeneous environments, or in organisms that experience the environment as homogeneous, genetic variability should be lower because a specialized genotype is more likely to be selected. Data from free-living organisms seem to support this hypothesis (Selander & Kaufman, 1973; Powell, 1975; Nevo, 1978; Nevo *et al.*, 1984). For example, geographically widespread generalists occupying heterogeneous habitats have greater genetic variability than geographically restricted specialists with narrow niches.

Bullini *et al.* (1986) attempted to compare the genetic structure of several ascaridoid nematodes with single- and multiple-host life cycles (Table 16.4). Their working hypothesis was based on the observed correlation between the genetic variation of free-living organisms and the degree of environmental heterogeneity they experienced. The evidence indicated that species whose life cycles were completed in a single homeothermic host had lower genetic variability (polymorphism and heterozygosity) than those species using both poikilothermic and homeothermic hosts to complete their life cycles (Table 16.4). They argued that, in species with more than one host in their life cycle, a given allele might have greater fitness in one stage of the life cycle whereas another allele may function optimally in another stage. In a multiple-host life cycle, then, heterozygosity would be favored. Bullini *et al.* (1986) also argued that ascaridoid nematode species with life cycles that were completed exclusively in homeothermic hosts were buffered against environmental variation and did not require genetic flexibility in order to cope with environmental heterogeneity. It was implicit in their argument that the number and kind of hosts were the source of heterogeneity. On the other hand, those species using both

Table 16.4 Parameters of genetic variability in ascaridoid nematodes that need one homeothermic host (upper group) or several hosts, both poikilothermic and homeothermic (lower group), to complete their life cycle

	N_g (average number of genes sampled per locus)	N_l (number of loci studied)	H_c (expected mean heterozygosity per locus)		P (proportion of polymorphic loci; 1% criterion)		A (mean number of alleles per locus)[a]	
Ascaris lumbricoides	92	24	0.02	[0.04][b]	0.25	[0.33][b]	1.38	[1.58][b]
Ascaris suum	445	24	0.03	[0.05]	0.17	[0.25]	1.29	[1.50]
Parascaris univalens	447	28	0.03	[0.08]	0.11	[0.25]	1.18	[1.50]
Parascaris equorum	131	28	0.02	[0.04]	0.07	[0.17]	1.11	[1.33]
Neoascaris vitulorum	253	18	0.04	[0.04]	0.11	[0.08]	1.11	[1.17]
Toxocara canis	157	18	0.10	[0.13]	0.33	[0.42]	1.50	[1.58]
Toxocara cati	44	18	0.05	[0.09]	0.17	[0.25]	1.22	[1.42]
Toxascaris leonina	46	18	0.02	[0.04]	0.11	[0.17]	1.11	[1.25]
Baylisascaris transfuga[c]	10	18	0.05	[0.09]	0.17	[0.25]	1.17	[1.33]
Average ± standard error	0.04 ± 0.01	[0.06 ± 0.01]	0.16 ± 0.03	[0.24 ± 0.04]	1.24 ± 0.05	[1.42 ± 0.05]		
Anisakis simplex A	272	22	0.12	[0.16]	0.64	[0.67]	2.32	[2.83]
Anisakis simplex B	251	22	0.21	[0.19]	0.64	[0.58]	2.60	[2.67]
Anisakis physeteris	84	22	0.11	[0.14]	0.50	[0.67]	1.95	[2.50]
Phocascaris cystophorae[c]	12	21	0.10	[0.17]	0.24	[0.42]	1.33	[1.58]
Contracaecum osculatum A	78	21	0.12	[0.10]	0.48	[0.50]	2.14	[2.33]
Contracaecum osculatum B	169	21	0.10	[0.15]	0.62	[0.75]	2.71	[3.17]
Contracaecum rudolphii A	87	21	0.17	[0.18]	0.57	[0.67]	2.24	[2.58]
Contracaecum rudolphii B	65	21	0.21	[0.19]	0.62	[0.67]	2.33	[2.67]
Average ± standard error			0.15 ± 0.02	[0.16 ± 0.01]	0.58 ± 0.02	[0.64 ± 0.03]	2.33 ± 0.10	[2.68 ± 0.10]
Student's t			5.61	[5.57]	10.29	[8.31]	10.23	[11.44]
Probability			<0.001	[<0.001]	<0.001	[<0.001]	<0.001	[<0.001]

Notes:

[a] Alleles with frequency <0.01 were not considered for computation.

[b] The values in brackets were calculated on the 12 loci shared by all the species studied.

[c] Not considered in computing the averages, because of the small sample size.

Source: From Bullini *et al.* (1986), with permission, *Evolution*, **40**, 437–440.

poikilothermic and homeothermic hosts were subjected to greater environmental variability and thus benefited from increased genetic flexibility. They also suggested that nematodes with low host specificity as larvae were subjected to extreme environmental heterogeneity because of the many species of fishes acting as potential paratenic hosts and should also exhibit a greater degree of genetic variability. Values of genetic variability of three sibling species of the ascaridoid *Contracaecum* that require both poikilothermic and homeothermic hosts to complete their life cycle also fall in the range observed by Bullini *et al.* (1986) (Nascetti *et al.*, 1993). The three sibling species of *Contracaecum*, then, provide further evidence for the positive correlation between genetic variability and environmental heterogeneity, supporting the hypothesis that natural selection may play a major role in this phenomenon.

Although these conclusions are intuitively logical, they may not be totally correct. It is possible that the experimental design and analyses of Bullini *et al.* (1986) compromised their results. For example, to consider the number of hosts as a measure of habitat heterogeneity may be an inaccurate assumption. Even parasites with a one-host life cycle may encounter a high degree of environmental heterogeneity if they are required to migrate extensively through different organs before reaching the final site of infection in a host. Habitat heterogeneity in these cases can be related to variations in temperature, pH, osmolarity, concentrations of dissolved oxygen, digestive enzymes, immunological responsiveness, and so on. An additional problem in their analysis was that all the species with a one-host life cycle were terrestrial, while the multiple-host species were all marine. Nevo *et al.* (1984) showed that among free-living invertebrates, species from marine environments had a higher degree of genetic variability than species from terrestrial environments. It is therefore possible that the differences observed by Bullini *et al.* (1986) could have been due to the quality of the habitat rather than the nature of the life cycle. These alternative explanations have not been tested.

Anderson *et al.* (1998) attempted to examine Bullini *et al.*'s (1986) hypothesis by comparing heterozygosity levels in nematodes with direct and indirect life cycles in four families of nematodes (Fig. 16.1). The data, however, do not provide strong support for, or against, Bullini *et al.*'s (1986) views. For example, filaroids with indirect life cycles have lower heterozygosity than strongylids and trichostrongylids with direct life cycles. Also, the direct life cycle strongylids and trichostrongylids show levels of variation overlapping both direct and indirect life cycle ascarids. However, within the ascaridoids, species with indirect life cycles tend to have higher heterozygosity levels than species with direct life cycles, providing a slight association between heterozygosity and environmental heterogeneity in terms of life-cycle complexity. Clearly, further data and more rigorous comparative tests are needed to investigate the selectionist view that genetic variability correlates with environmental heterogeneity.

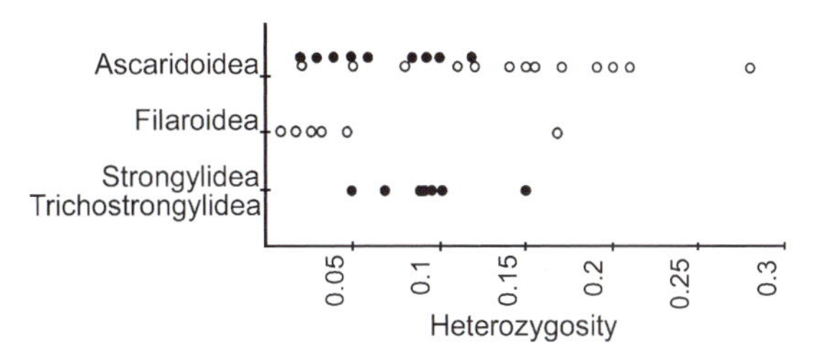

Fig. 16.1 Heterozygosity of nematode species with direct and indirect life cycles. Open circles represent species with indirect life cycles; filled circles represent species with direct life cycles. Each point represents one nematode species. In those species in which heterozygosity has been measured several times, the point represents the average value. (Modified from Anderson *et al.*, 1998, with permission, *Advances in Parasitology*, **41**, 219–283.)

If the correlation between enzyme polymorphism and environmental heterogeneity is used as an argument to explain the maintenance of enzyme polymorphisms in populations by means of natural selection, then comparative studies on the biochemical properties of different allozymes (under different temperatures, pH, etc.) are necessary to support such arguments. These studies should generate a better understanding of how selection operates. On the other hand, the hypothesis of increased genetic heterozygosity and polymorphism can also be tested in terms of the neutralistic theory of molecular evolution. Thus,

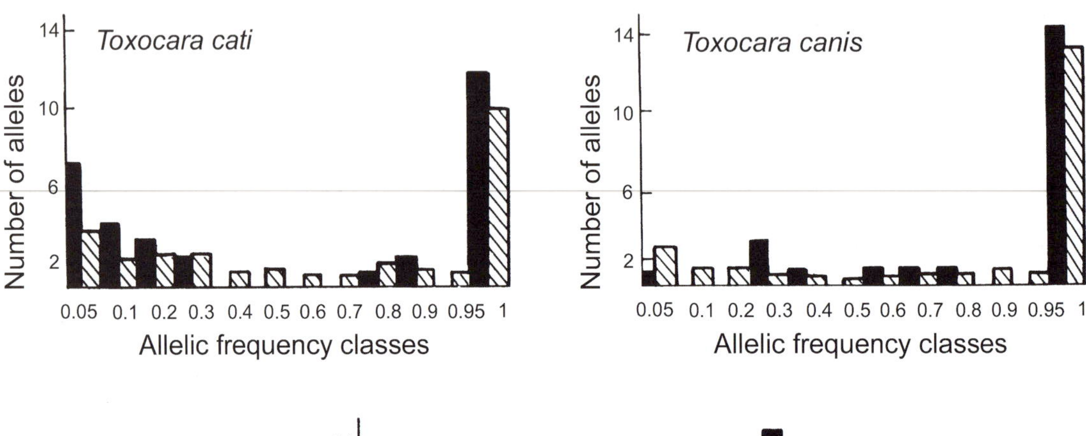

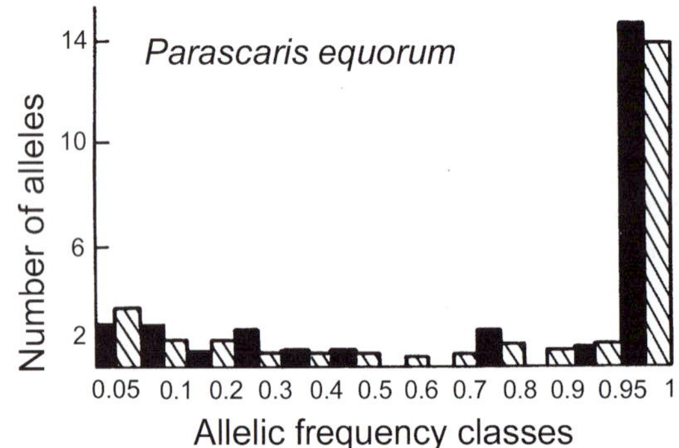

Fig. 16.2 Observed (solid) and expected (hatched) distributions of number of alleles by frequency for three species of ascaridoid nematodes. Observed distributions based on protein electrophoretic studies of individuals (Nadler, 1986). Expected distributions were calculated using the infinite allele–constant mutation rate model. (From Nadler, 1990, with permission, *International Journal for Parasitology*, **20**, 11–29.)

if heterozygosity is selected for, then allelic frequency distributions should depart from random expectations (the neutralist argument) to maximize heterosis (hybrid vigor). Nadler (1990) took this approach and compared the allelic frequency distributions in three nematode species (*Toxocara canis, T. cati,* and *Parascaris equorum*) against the infinite allele–constant mutation model that predicts expected frequencies according to the neutralistic theory (Fig. 16.2). No differences were found between the observed and expected distributions of allele frequencies in these species, indicating that selection for heterozygosity had not occurred. The observed allelic frequencies, then, may have resulted from random genetic events as predicted by the neutralist theory.

16.2 Evolutionary biology of parasites

16.2.1 Speciation in parasites

The formation of two or more species from a previous species is called speciation. This is a simple and straightforward definition that relies on the 'species concept'. Over the years, many authors have proposed many definitions for what constitutes a species, but none of the definitions seems to fit all organisms perfectly. The most accepted definition defines a species as all members of a group of populations that interbreed, or can potentially interbreed, with each other under natural conditions, and whose offspring are viable. In addition, the members are bound together by a unique common ancestor. One obvious problem with this definition is that it does not apply to asexual organisms. Even though the biological concept of species is the most widely used in a theoretical framework, from a purely practical perspective, the taxonomic concept of species is the most commonly utilized. A taxonomic species is a basic taxonomic category to which individual specimens are assigned based mainly on morphological characters, in fact, they are sometimes referred to as morphospecies. The taxonomic species often, but not always, corresponds to the biological species. The proper identification of a species is not an easy task in many cases, in particular when dealing with parasites. Without going into many details, identification of a species based on morphological characters must consider the intrinsic phenotypic variability of the organisms and the influences of the environment. When dealing with parasites, the hosts add another dimension to the problem of morphological variability. Host-induced variability in parasites has been widely documented (Blankespoor, 1974; Thompson, 1982; Downes, 1990, and references within).

Speciation, i.e., the origin of new species, requires that the gene flow between two populations of a given species be somehow interrupted. Without gene flow, the populations may become genetically different because of their adaptation to different local conditions and also because of genetic drift. As the populations become more and more genetically different, reproductive isolating mechanisms may appear thus completing the process of speciation. Over time, even if the original barrier that interrupted gene flow disappears, the resultant populations will continue to be genetically isolated. An important outcome of the speciation process is that the resulting **sister species** are each other's closest relatives.

Many models of speciation have been proposed. Some are widely supported by evidence, whereas others have been developed for, and applied to, very specific conditions and organisms. The three most common types include allopatric speciation type I, or vicariant speciation; allopatric speciation type II, or speciation by peripheral isolates; and sympatric speciation (Brooks & McLennan, 1991, 1993; for other models see Futuyma, 1986). **Vicariant speciation** occurs when an ancestral species is geographically separated into two or more isolated populations that become reproductively isolated and genetically divergent. The rate of change in these populations will depend on their genetic variability and the rate at which evolutionary novelties (mutations) originate. In speciation by peripheral isolates (also known as **peripatric speciation**), a new species arises from an usually small, isolated population, on the periphery of the larger ancestral population. Founder effects near the ancestral population are particularly significant in this type of speciation. Because the partition of the ancestral population in terms of size is very asymmetrical, the ancestral species does not exhibit evolutionary change associated with the vicariant event, only the peripheral population does. **Sympatric speciation** occurs when one or more new species arise without geographical separation of the populations. Gene flow is not eliminated by factors extrinsic to the populations, e.g., geographical barriers, as in the allopatric models, but by intrinsic biological processes such as hybridization, ecological partitioning, evolution of asexual or parthenogenetic reproduction, or a change in mate recognition systems.

Where do parasites stand in terms of species recognition and speciation? The identification of parasite species is based largely, as in free-living organisms, on morphological characters. One of the major problems with this approach has been recognition of host-induced parasite variability. Another problem has been the potential existence of **sibling species**, i.e., morphologically similar but reproductively isolated species. Host-induced parasite variability can be recognized to some extent by extensive sampling that may show clines in morphological variation, by experimental completion of their life cycles, or by genetic analysis. Sibling species among many parasites can, however, only be recognized through genetic analysis. Behavioral and physiological differences also help with the identification of sibling species, but such studies are not feasible in most groups of parasites. Protein electrophoresis, restricted fragment length polymorphism (RFLP), and DNA sequencing, have been successfully used to recognize sibling species as well as morphologically variable species, and to test their reproductive isolation in the field (for references see Nascetti *et al.*, 1993; Zahler *et al.*, 1995).

In terms of evolution and speciation, Price (1980) proposed that the evolution of new parasite–host relationships is driven predominantly by sympatric speciation via colonization of new hosts. These new hosts do not need to be related to the original hosts. Price (1980) argued that the small and relatively discrete nature of parasite demes (infrapopulations of reproductive individuals) facilitates sympatric speciation. Demes are important because they serve as units that can be reinforced by inbreeding to such an extent that they produce incipient species. Factors such as vagility, dispersion, habitat patchiness, mate recognition, and social structure affect inbreeding in free-living organisms. In theory, the more these factors tend to enhance inbreeding and inhibit immigration in individual demes, the more cohesive the deme will be and the greater the likelihood that incipient speciation will occur (Brooks & McLennan, 1993). Most of Price's arguments were based on models of sympatric speciation developed earlier by Bush (1969, 1974, 1975) for phytophagous insects showing a pattern of differentiation into host races. According to Bush (1974, 1975), the establishment of new sympatric host races could require only minor genetic changes involving only two alleles at each of two loci. One locus would control host selection and the other would control survival in the host. An example of this type of sympatric speciation in progress is described by Théron & Combes (1995) for the digenean *Schistosoma mansoni* on the Caribbean island of Guadeloupe. *Schistosoma mansoni*, normally a human parasite, has undergone a definitive host shift toward a murine host, with populations from both hosts showing microevolutionary changes in morphology, behavior, and allozyme frequencies. This is clearly a case of differentiation into sympatric host races because both populations have undergone microevolutionary changes.

Brooks & McLennan (1993) take a more realistic approach towards the importance of sympatric speciation for parasitic organisms, acknowledging that it may work only in some specific cases. They argue that parasite demes are not permanent enough in time, and lack the spatio-temporal continuity needed to allow for inbreeding and isolation to occur. In their view, in many cases, parasite demes are reassembled each generation by the relatively random immigration of larval stages from a much larger gene pool than the gene pool represented by the members of the original deme. This larger gene pool could be the parasite component population or suprapopulation. Brooks & McLennan (1993) conclude that rather than promoting sympatric speciation, the demic structure of many parasite species seems to increase the influence of immigration and decrease the cohesion of the deme.

There are several groups of parasites, however, whose demes are not ephemeral and are continuous in time. Parasites that autoinfect long-lived hosts can produce more than one generation of parasites in, or on, the same host organism, increasing the potential that any differences appearing within the deme could be maintained by inbreeding. In organisms with direct life cycles, and the potential for autoinfection of the host, such as monogeneans and oxyurid nematodes (pinworms), sympatric speciation is likely to occur and play a significant role in their evolution. Sympatric speciation in these organisms can be recognized by the presence of sister parasite species in the same host species which differ in some newly acquired ecological or genetic characteristics that could have produced independent species evolution (Brooks & McLennan, 1993). Among the Monogenea, it is common to find several congeneric species parasitizing the same fish species, raising questions about their possible origin. Phylogenetic studies of monogeneans in piranhas in the Amazon (Van Every & Kritsky, 1992) and in minnows in Africa (Guégan & Agnèse, 1991), suggest that sympatric speciation may indeed have occurred. Brooks & McLennan (1993) also suggest that sympatric speciation might have played a significant role in the evolution of species of *Spirorchis*, digeneans parasitic in the circulatory system of freshwater turtles, because many species occur in the same host species but occupy different sites.

Price (1980) considered host switching a possible route for sympatric speciation. Brooks & McLennan (1993) disagree with Price's view and consider host switching a case of peripheral–isolates speciation, where the new host may not represent a new resource for the parasite, but represents a new geographical distribution. They argue that if the host is treated as a new geographical area for the parasite, it creates the potential for speciation by the effects of geographical isolation on gene flow, even if the host *per se* (as a new environment) does not provoke an adaptive response by the parasite. Once the switch has occurred, the amount of gene flow between the ancestral parasite population and the host switchers will depend on the dispersing capabilities of the organisms and the extent of sympatry between the old and new host.

The difference between Price's and Brooks & McLennan's host switching is subtle and depends on whether sympatry and allopatry are evaluated from the host or the parasite point of view. Brooks & MacLennan (1993) seem to look at it from the parasites' point of view . . . as do the parasites. If a new host is considered to be a new resource, but the new parasite population still inhabits the same geographic area as the original parasite population, then it is sympatric speciation according to Price (1980). But if a host is considered to be a new geographic distribution for the parasite, even though the new parasite population is still within the same physical geographic area of the original parasite population, then it is speciation by peripheral isolates according to Brooks & McLennan (1993).

Vicariant speciation (allopatric speciation type I) is the most widely accepted mechanism of speciation and probably the most common in both free-living organisms and parasites. For vicariant speciation to occur, the parasite populations must be geographically separated into two or more isolated populations, effectively severing gene flow. As in the allopatric isolation by peripheral isolates, the rate of speciation will depend on the genetic variability of the ancestral population and the rate of origin of evolutionary novelties in the subdivided populations (Brooks & McLennan, 1991). Such isolation is likely to arise when the host population is geographically divided and the parasites with it. When the host population becomes divided and isolated, three things might happen. Either the host and the parasite may speciate, the parasite may speciate but not the host, or the host may speciate but not the parasite. These possible outcomes have great relevance for phylogenetic studies of host–parasite evolution. Moreover, vicariance biogeography, as discussed in Chapter 15, relies on this mode of speciation to detect episodes of parallel biological and geological evolution.

Is it possible to identify the types of speciation behind the evolution of present-day parasites? For some of them, the answer is probably yes. Brooks & McLennan (1993) performed a detailed analysis of a group of intestinal digeneans in the genus *Telorchis*, parasitic in reptiles, mostly freshwater turtles, based on a previous study by Macdonald & Brooks (1989). The new analysis combined a phylogenetic study of *Telorchis* with geographical, host, and morphological correlates, of phylogenetic diversification. The results indicate that the diversification of *Telorchis* was probably due to several events of vicariance speciation and speciation by peripheral isolates as a result of host switching. No evidence of sympatric speciation was found.

16.2.2 Some evolutionary misconceptions about parasites

Parasitism is one of the most common life styles on Earth and practically every animal species harbors some type of parasite. Parasitism, however, evolved independently many times in different groups of organisms. As a result, what we call parasites is just an artificial assemblage of very diverse organisms with a common lifestyle. From the beginnings of the Darwinian evolutionary era, researchers have attempted to formulate some general rules about the evolution of these organisms, trying to define the evolutionary paths that they have followed. In the process however, some ideas have become widely accepted as truths without the benefit of a careful analysis. Recently, Brooks & McLennan (1993) and Poulin (1995, 1996) have re-examined some of these general rules and assumptions shedding new light on the true nature of their evolutionary significance.

A common perception is that parasites are simple and degenerate, and that they evolve towards reduced structural complexity. This view of extreme simplification comes solely from comparing the morphology/physiology of parasites with that of their hosts. To determine whether parasites really evolve towards reduced complexity, parasites should be compared with their closest free-living relatives. Unfortunately, this kind of comparison has not been done. Nonetheless, Brooks & McLennan (1993) attempted to determine if parasites have become secondarily simplified, by comparing different characters and their degree of change within the Platyhelminthes, taking into account both character losses and character innovations (Table 16.5). Their analysis indicates that, of 1882 character changes found in the Platyhelminthes, only 10.6% represent secondary losses; in other words, most of the evolutionary changes among the platyhelminths (≈90%) involve character innovation rather than character loss. Further analysis of the proportion of lost characters against the proportion of change shows that digeneans tend to lose male characters, eucestodes tend to lose larval characters and preserve adult characters, whereas monogeneans show no significant trends. Based on this partial analysis of the major platyhelminth groups, it seems that character loss is very low and there is no general pattern in the nature of the character that is lost.

A second evolutionary myth regarding parasites is that they evolve toward smaller body sizes. As before, this view comes from comparing parasites with their hosts and not with their closest free-living relatives or closest free-living ancestors. Poulin (1995, 1996, 1998) compared the body size of parasitic and free-living nematodes, copepods, amphipods, and isopods. Nematodes parasitic in invertebrates were about the same size as related free-living species, whereas nematode parasites of vertebrates were much larger. In copepods, the transition from free-living to parasitic in invertebrate hosts resulted in a small and significant increase in body size; the change from parasites of invertebrate hosts to parasites of fishes was accompanied by an even larger increase in body size. In amphipods and isopods, however, the transition from free-living to parasitic resulted in diminution of body size, which could be related to the relative size of some of their hosts. So far, not enough critical studies have been pursued to establish clear patterns. However, the partial evidence presented by Poulin (1995, 1996, and references within) suggests that the size of parasitic organisms tends to either increase, or to be as variable, as the size of free-living organisms (Poulin, 1996). Moreover, it is possible that ectoparasites and endoparasites might experience different selective pressures, with the body size of endoparasites being more constrained due to space limitations. In free-living invertebrates, body size is usually correlated with fecundity. Larger bodies typically mean higher fecundity (Sibly & Calow, 1986). It is likely, then, that natural selection may favor increases in body size if fecundity becomes a critical factor for transmission success (Poulin, 1995).

Yet another commonly held assumption suggests that parasites evolve toward higher fecundity. Arguments for this trend state that parasites need to maximize fecundity to compensate for the heavy losses suffered during transmission from host to host (Price, 1974). This argument is probably flawed because it implies that free-living organisms do not try to maximize reproductive success. A second argument suggests that higher fecundity is an evolutionary response to the nutrient-rich environment provided by the host (Calow, 1983, and references within). Poulin (1995) concedes that this argument may be appropriate for many parasites but warns against generalizations. The available data indicate that, within groups such as nematodes, cestodes, and copepods, fecundity ranges widely from extreme highs to extreme lows. Other groups such as monogeneans are more consistent and fecundity is low overall. Unfortunately, there are no critical analyses deciphering the evolutionary nature of fecundity in parasites. Whenever such studies are undertaken, phylogenetic factors, as well as the evolutionary trade-offs of fecundity in the parasite's life history, need to be considered.

A number of other myths regarding the evolutionary biology of parasites, including topics such as the relationship between host specificity and speciation, parasite plasticity, adaptation, and the evolutionary nature of life cycles, are discussed in much more detail by Brooks & McLennan (1993) and Poulin (1995).

16.3 Evolution of host–parasite associations

One of the most challenging areas in evolutionary biology is the study of how interspecific interactions have influenced the rates of evolution and the patterns of adaptive radiation of the organisms involved. This is one of the most complex aspects of

Table 16.5 Comparison of character loss for four different character categories among major groups of parasitic Platyhelminthes

	Amount of character loss within each clade [a]				
Character category	Cercomeria	Digenea	Monogenea	Eucestoda	Total [b]
Male	4/20 (20%)	31/130 (23%)	27/188 (14%)	5/74 (5%)	67/427 (16%)
Female	3/26 (11%)	6/230 (3%)	14/98 (14%)	2/109 (2%)	25/493 (5%)
Adult nonreproductive	4/56 (7%)	31/222 (14%)	22/263 (8%)	6/128 (5%)	66/691 (9%)
Larval	8/55 (14%)	20/157 (13%)	10/38 (26%)	7/19 (37%)	45/271 (17%)
Total	19/157 (12%)	88/739 (12%)	73/587 (12%)	20/330 (6%)	200/1882 (10.6%)

Notes:

[a] Numerator denotes number of losses; denominator denotes number of derived characters in that category.

[b] Information from the Aspidobothrea, Gyrocotylidea and Amphilinidea included in totals.

Source: From Brooks & McLennan (1993), *Parascript: Parasites and the Language of Evolution*, Smithsonian Institution Press, with permission by the authors.

evolutionary biology because, to a great extent, it requires the synthesis of two incomplete theories, i.e., the genetic theory of evolution and the ecological theory of community structure (Futuyma, 1986). There are many types of associations in which reciprocal evolutionary interactions occur between the members of the association, i.e., prey–predator, plant–pathogen, plant–pollinator, host–parasite, etc. This reciprocal evolutionary interaction is broadly termed coevolution.

16.3.1 Models of host–parasite coevolution

The term coevolution was developed by Mode (1958), but it was the study by Ehrlich & Raven (1964) on the evolutionary influence that plants and herbivorous insects have had on each other that set the stage for current coevolutionary studies. Since its introduction, the term has been widely applied, and presently there is no agreement on its exact definition. For example, Roughgarden (1976) defined it as the type of evolution in which the fitness of each genotype depends on the population densities and genetic composition of the species itself and the species with which it interacts. A more specific definition provided by Janzen (1980) requires that each interacting species change its genetic structure in response to a genetic change in its partner. Thompson (1994) prefers a short definition instead and defines coevolution as reciprocal evolutionary change in interacting species.

Parasitologists and non-parasitologists have had little overlap in their approaches to studying the coevolution of interacting species; the former have focussed on the phylogenetic aspects of the interaction, and the latter on the ecology and population genetics of the interacting species. The microevolutionary approach used by non-parasitologists studying species interactions postulates the existence of three main coevolutionary models, i.e., allopatric cospeciation, resource tracking, and the evolutionary arms race. Brooks & McLennan (1991, 1993), however, have tried to merge both approaches, providing macroevolutionary patterns for the microevolutionary models of coevolution in an attempt to increase the understanding of the two approaches and produce a more robust theory of coevolution.

The model of **allopatric cospeciation** is based on the assumption that parasites and hosts are simply sharing space and energy. When gene flow between two host populations is interrupted by a vicariant event, so too is the gene flow of the parasites, resulting in allopatric speciation of both the host and the parasite populations. A reconstruction of the host and parasite phylogenies should reveal congruent branching patterns (Fig. 16.3) (Brooks & McLennan, 1991, 1993). The timing of speciation for the host and the parasite, however, is not necessarily the same, and cospeciation can be synchronous or delayed (Hafner & Nadler, 1988, 1990). In **synchronous cospeciation**, the host and parasite speciate simultaneously and their lineages show similar degrees of evolutionary divergence. In **delayed cospeciation**, one of the members of the association lags behind the other. If parasite speciation lags behind host speciation, the evolutionary divergence of the parasite should be less than that of the host. Conversely, if host speciation lags behind parasite speciation, the evolutionary divergence of the host should be less than that of the parasite. From a phylogenetic point of view, delayed cospeciation can still be seen as a congruent pattern of phylogenetic branching, assuming that the group that lags behind speciates before the group that speciated first speciates again.

The model of **resource tracking** is based on the ecological concept that hosts are patches of necessary resources that parasites track through evolutionary time. The parasite evolves in response to a change in the resource offered by the host. If a host changes the resource sought by the parasite, then the parasite will likely undergo evolutionary change to be able to exploit the new resource. The change in the parasite, however, will lag in time behind the change in the resource offered. From a phylogenetic point of view, resource tracking can produce congruent or incongruent phylogenetic patterns for the host and parasite (Brooks & McLennan, 1991, 1993). If the resource that the parasites are tracking is widespread or shows patterns of convergence throughout the host group, the phylogenetic trees of hosts and parasites will not show congru-

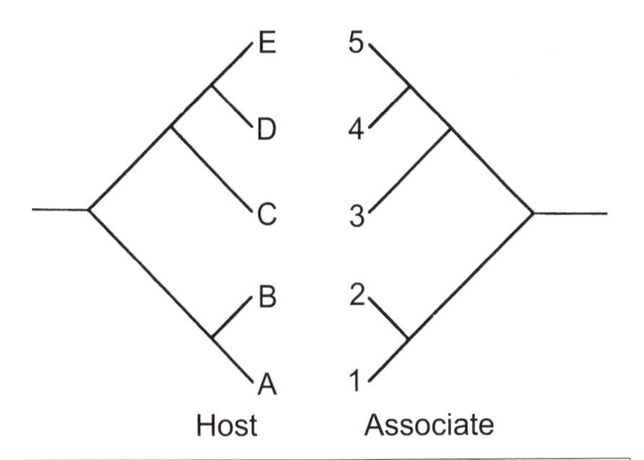

Fig. 16.3 Possible macroevolutionary outcome of the allopatric model of coevolution. Complete congruence between the phylogeny for the hosts (taxa represented by letters) and the phylogeny for their parasites (taxa represented by numbers) is due to simultaneous cospeciation. (Modified from Brooks, D. R. & McLennan, D. A. [1991] *Phylogeny, Ecology, and Behavior: A Research Program in Comparative Biology.* © 1991 University of Chicago Press. Reprinted with the permission of the University of Chicago Press.)

ence. This model, known as the **sequential colonization model** (Jermy, 1984), was proposed to explain the coevolution of phytophagous insects and plants. On the other hand, if the resource that the parasites are tracking is restricted to the host clade and has evolved within the clade, then host and parasite phylogenies are likely to be congruent. This model is also known as **phylogenetic tracking**. Brooks & McLennan (1993) proposed an evolutionary sequence of events to illustrate how this later process may occur: 'a new host species evolves, characterized in part, by an evolutionarily modified form of the required resource. This new species of host is then colonized by individuals of the parasite species associated with the ancestral host. Some of these species adapt to the new form of the resource, eventually producing, in their turn, a new species of parasite.' This scenario, however, requires that the ancestral host species remains unchanged after the speciation event that produced descendants with the new resource because the ancestral parasite species needs the ancestral resource to support the population through the colonization process. This situation is possible in cases of sympatric speciation or allopatric speciation by peripheral isolates (Brooks & McLennan, 1991, 1993).

A third model, the so-called **evolutionary arms race**, represents a more stringent view of coevolution, one requiring mutual adaptive responses between hosts and parasites. Although the model was originally proposed for plant–pathogen systems, it also might apply to parasites. In parasitological terms, the model assumes that parasites and hosts are constantly evolving in an aggressive manner toward each other. Selection on the parasite is always for greater exploitation of the host and selection on the host is always for a more efficient exclusion of the parasite. Evolutionary arms races are assumed to occur within the context of the **gene-for-gene hypothesis**, which proposes that for every gene causing resistance in a host, there is a corresponding (matching) gene for avirulence in the parasite. According to this hypothesis, the occurrence of resistance depends both on the presence of a gene for resistance in the host and a corresponding gene for avirulence in the parasite. Resistance in the host and avirulence in the parasite are the dominant genes (Table 16.6). Although the gene-for-gene hypothesis seems to explain common interactions between plants and pathogens, it is not clear to what extent the hypothesis may work in other systems (Thompson, 1994).

The possible sequence of events for the development of an arms race requires the following steps. First, the parasites reduce the fitness of their hosts, and the hosts acquire defense mechanisms against the parasites through mutation and/or recombination. Hosts with this novel defense mechanism have increased fitness and the trait spreads through the population (probably a new host species). Eventually, a new mutant and/or recombinant appears in the parasite population that is able to overcome the new host defense. If this counterdefense increases the fitness of the parasite population, it spreads through the parasite population (probably a new parasite species) and the new parasite population is able to parasitize the previously protected host group. The cycle would then repeat.

From a macroevolutionary perspective, evolutionary arms races can produce congruent or

Table 16.6 Expected resistance and susceptibility of a host to various parasite genotypes in the context of the gene-for-gene hypothesis.

	Host genotype[b]		
Parasite genotype[a]	*RR*	*Rr*	*rr*
AA	Host resistant	Host resistant	Host susceptible
Aa	Host resistant	Host resistant	Host susceptible
aa	Host susceptible	Host susceptible	Host susceptible

Notes:

[a] *A*, dominant parasite gene conferring avirulence; *a*, recessive parasite gene conferring virulence.

[b] *R*, dominant host gene conferring resistance to the parasite; *r*, recessive host gene conferring susceptibility.

incongruent phylogenies between hosts and parasites (Fig. 16.4) (Brooks & McLennan, 1991, 1993). For example, in systems were the defense and counterdefense mechanisms arise in a short evolutionary time, these traits will probably appear at the same point in the phylogeny of host and parasite, producing congruent phylogenetic patterns (Fig. 16.4A). However, if the time frame in which the traits for defense and counterdefense originate is longer than the time between the host speciation events, the macroevolutionary pattern will show the parasite group missing from most members of the host clade that possessed the defense trait (Fig. 16.4B). Partial incongruence between host and parasite phylogenies also can be found when some parasites with the counterdefense trait colonize relatively more ancestral members of the host clade (Brooks & McLennan, 1991, 1993). The macroevolutionary outcomes of the arms race model are highly variable, ranging from strict cospeciation to extensive host switching.

The macroevolutionary consequences of the microevolutionary models of coevolution show considerable overlap. However, if the relevant information about host and parasite phylogenies and the characters involved are available, then it is possible to identify the proper microevolutionary model responsible for the macroevolutionary pattern observed. Because coevolving systems are complex and dynamic evolutionary units, it is likely that the history of a given host–parasite association results from the differential effects of several of these microevolutionary processes.

16.3.2 Host–parasite coevolution: case histories

Cospeciation is the process where one species speciates in response to the speciation of another. Cospeciation, then, is the phylogenetic component of coevolution. Historically, many verbal accounts have been proposed about scenarios where host and parasites coevolved. However, the lack of rigor and testing has limited their value. With the advent of phylogenetic systematics, phylogenies can be contrasted and tested in a more rigorous way and the process of coevolution can be studied by comparing the phylogenies of hosts and parasites. If these phylogenies are congruent, then cospeciation probably occurred; if they are not, alternative explanations, e.g., host switches, extinctions, etc. may explain the discrepancies. Despite the progress made in the field in the last 25 years, too few studies have constructed, contrasted, and tested the phylogenies of host and parasites in a methodical fashion sufficient to provide a solid foundation for this area of evolution.

Nonetheless, the available data provide important insights into how the evolution of hosts and parasites might occur. Brooks & Glen (1982) examined the coevolution of primates and their pinworm nematodes (*Enterobius* spp.). The analysis of 13 species of *Enterobius* supported the notion that pinworms and primates have coevolved. The relationship between *E. vermicularis* and humans was in conflict, however, suggesting that the presence of pinworms in humans was the result of host switching or that man has been misclassified among the great apes. Additional studies by Glen

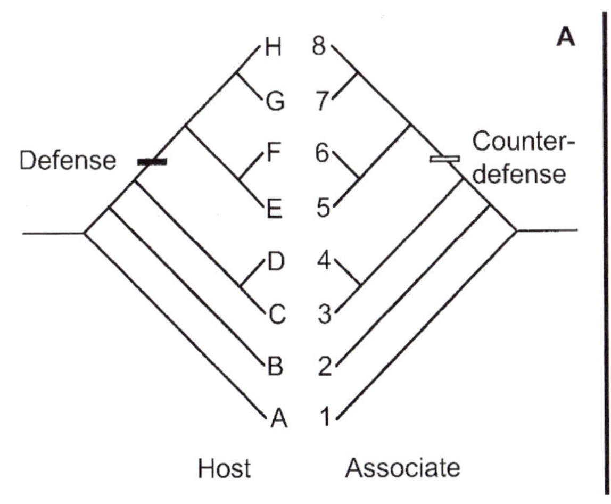

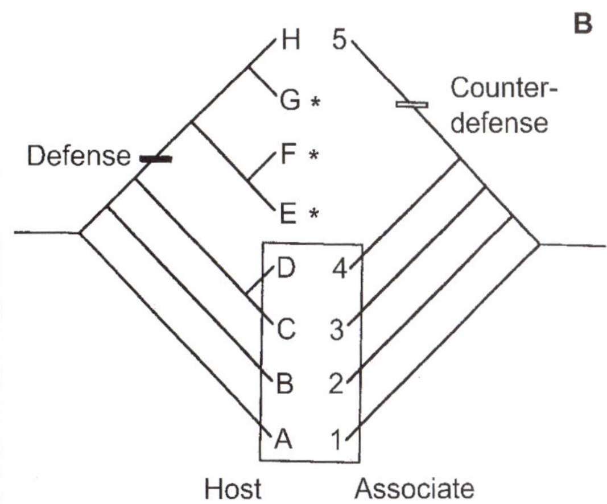

Fig. 16.4 Possible macroevolutionary outcomes of the evolutionary arms race model of coevolution. (A) Host and parasite phylogenies are congruent because traits for defense and counterdefense appear at the same point in the common phylogenies. (B) Host and parasite phylogenies are congruent (boxed area) up to the point at which the defense trait appears; if the origination of a counterdefense lags behind the origination of a defense while the host continues to speciate, the parasites without the counterdefense will not be able to parasitize the new host species (asterisks) until the counterdefense is acquired. Once the counterdefense appears, the host and parasite phylogenies rejoin (host H – parasite 5). (Modified from Brooks, D. R. & McLennan, D. A. [1991] *Phylogeny, Ecology, and Behavior: A Research Program in Comparative Biology.* © 1991 University of Chicago Press. Reprinted with the permission of the University of Chicago Press.)

& Brooks (1986) included a comprehensive phylogenetic analysis of both pinworms (*Enterobius* spp.) and strongylids (*Oesophagostomum* spp.) in hominoid primates. The analysis showed the same pattern of congruence among all groups except for two postulated cases of host switching in the great apes, one involving the hookworm *Oesophagostomum blanchardi*, and the other the pinworm *Enterobius vermicularis*. The latter was involved in a host switch between humans and gibbons.

The phylogenetic relationship between ancyrocephalid monogeneans *Ligictaluridus* spp., and their hosts, ictalurid fishes, was investigated using a traditional cladistic approach (Klassen & Beverley-Burton, 1987). The host and parasite cladograms were partially congruent, leading to the conclusion that ⅔ of these host–parasite associations were coevolutionary. Most of the evolutionary divergence in *Ligictaluridus* occurred in association with the ancestral host group, with two cases of host switching. Interestingly, at least one of the host switches occurred between hosts with the same habitat preference. In a similar study of ancyrocephalid monogeneans parasitic on the gills of basses and sunfishes (centrarchid fishes in the genera *Micropterus* and *Lepomis*), Klassen & Beverley-Burton (1988) found that the phylogenetic relationships of basses and their monogeneans were practically identical. However, the phylogenetic relationship between the species of sunfishes and their monogeneans was unclear, with many cases of host transfer. This lack of phylogenetic relationship was attributed to the extensive hybridization that occurs among sunfishes, which probably facilitated the many host transfers that occurred during the evolutionary history of these monogeneans.

Coevolutionary studies between seabirds and cestodes, and seabirds and chewing lice, showed two very different coevolutionary patterns. Hoberg *et al.* (1997) studied the relationship between marine birds in the family Alcidae and their cestodes in the genus *Alcataenia*. Because the hosts are a monophyletic group and the cestodes are relatively host specific, it was originally thought that their relationship involved a high degree of association by descent that would produce congruent host–parasite phylogenies. However, comparison of their phylogenies indicates minimum congruence. Further analysis suggests that the alcids and cestodes have a complex evolutionary relationship in which host switching and sequential colonization driven by continued compression of host populations into an Arctic refugium (see section 15.2.2), and not cospeciation, are the driving mechanisms. A completely different scenario is presented by Paterson & Gray (1997) for the coevolution of seabirds and their chewing lice. The host and parasite phylogenetic trees show a high degree of congruence suggesting a history of cospeciation between birds and lice. There are only two episodes where events other than cospeciation may have been involved. The natural history of chewing lice provides some powerful arguments for such intense cospeciation with their hosts. Lice are ectoparasites that rely on the host's body temperature and humidity for survival and reproduction, and consequently, they can survive away from their host only for a short time. Their life cycle is usually completed on one individual host and transmission usually occurs from parents to offspring and between mates, minimizing the chances for interspecific transmission and probable host switching. Even in the bird colonies, or during foraging, there is little physical contact between birds, hindering the chances of further interspecific transmission of lice (Paterson & Gray, 1997).

Probably one of the best-known animal host–parasite systems in terms of their coevolutionary relationships is the pocket gopher–chewing lice system (Hafner & Nadler, 1988, 1990; Page, 1993, 1996; Hafner *et al.*, 1994; Huelsenbeck & Rannala, 1997; Huelsenbeck *et al.*, 1997). The study began when Hafner & Nadler (1988) analyzed the phylogenetic relationships between eight species of pocket gophers (Geomyidae) and 10 species of ectoparasitic chewing lice (Trichodectidae) using biochemical data for both host and parasites. Host and parasite phylogenetic trees were topologically identical and highly congruent (Fig. 16.5), suggesting a long historical association. Sister taxa and branching sequences above the species level were also congruent. Host switches apparently occurred three times (see asterisks in Fig. 16.5) and, in all cases, the new hosts were sympatric. Moreover, the rates of evolution and the timing of the cospeciation events in gophers and lice indicated that the rate of genetic change within each group (standardized by generation time) had been approximately constant through time and that lice and pocket gophers may have evolved at about the same rate (Hafner & Nadler, 1990). The authors attributed the high degree of phylogenetic association and the similarity in rates of genetic divergence in host and parasite lineages to a strong coevolutionary coupling of genetic change between hosts and parasites. Later, Hafner *et al.* (1994) increased the data base to 15 species of gophers and 17 species of lice and used DNA sequences of the gene for mitochondrial cytochrome oxidase I from gophers and lice to generate phylogenetic trees. Although host and parasite phylogenetic trees were not exactly the same, the degree of similarity was statistically significant and their evolutionary rates were correlated. Also, Hafner *et al.* (1994) were able to determine that the rate of evolutionary change was approximately three times higher in lice than in their hosts. Further analysis of the pocket gopher–lice system confirms that although the host and parasite trees are not exactly the same, they provide evidence for cospeciation (Fig. 16.6) (Huelsenbeck & Rannala, 1997; Huelsenbeck *et al.*, 1997). Analysis of just a section of the proposed phylogenies (the top five gopher and lice species in Fig. 16.6) indicate that these species have truly cospeciated and that speciation time for hosts and parasites is identical. The differences in the topology of the remainder of the tree are probably due to host switching by the lice, by persistence of multiple ancestral louse lineages, or both.

Coevolutionary studies between hosts and parasites, following appropriate protocols to provide unambiguous information about their macro- and microevolutionary patterns of coevolution, are scarce. However, the database of phylogenetic studies of parasites is becoming larger (for references see Brooks & McLennan, 1993) and should facilitate the construction of more accurate host–parasite phylogenies.

16.3.3 The outcome of host–parasite interactions

The outcome of coevolutionary interactions between host and parasites has been widely debated. Until the 1960s, the paradigm for parasitologists was that, over evolutionary time, the interactions between host and parasites almost invariably evolved towards commensalism. In this

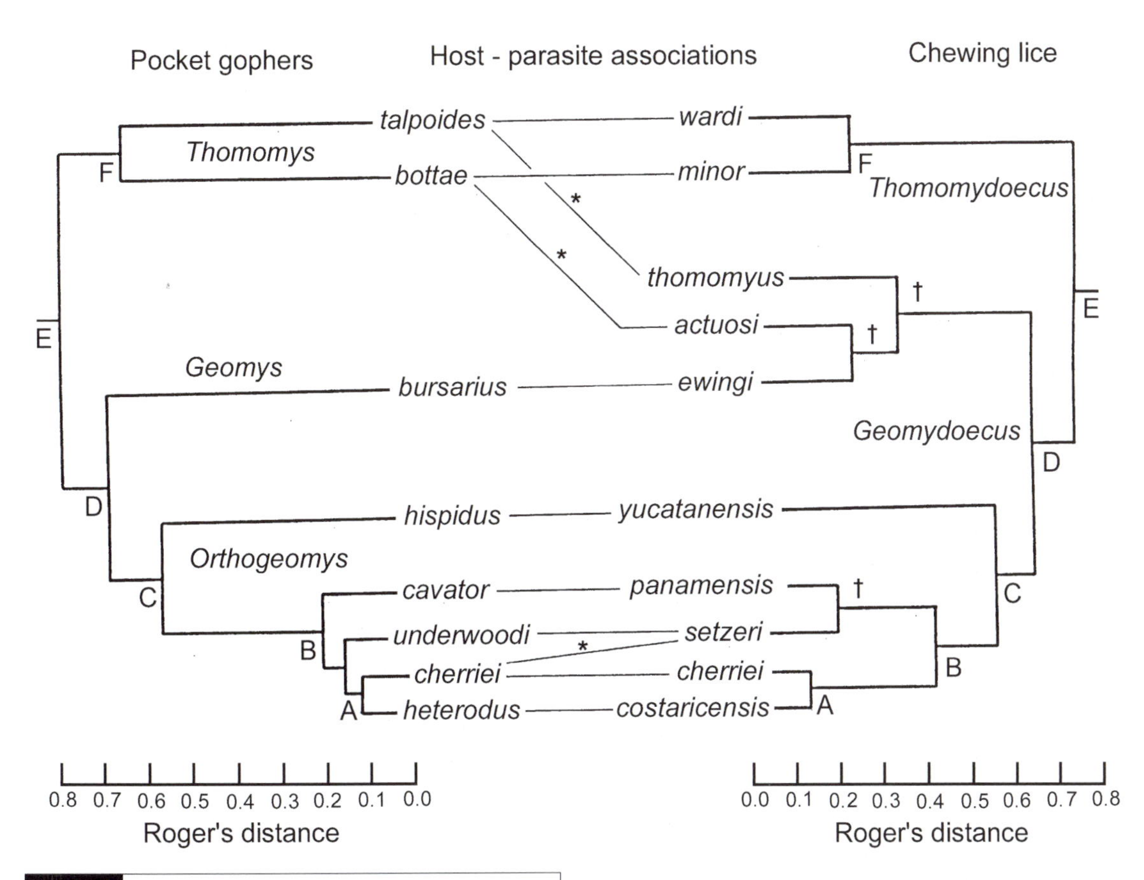

Fig. 16.5 Phylogenetic relationships between eight species of pocket gophers and their chewing lice (10 species) based on standard gel electrophoresis procedures. Lines connecting extant species indicate host–parasite associations. Asterisks indicate instances in which host switching probably took place. Daggers identify nodes on the louse tree that are absent on the pocket gopher tree. (Modified from Hafner, M. S. & Nadler, S. A. [1988] Phylogenetic trees support the coevolution of parasites and their hosts. *Nature*, **332**, 258–259. © 1988 Macmillan Magazines Ltd.)

context, present-day host–parasite associations in which the parasite is highly pathogenic were viewed as relatively new in evolutionary or even in ecological time. This view means that these antagonistic host–parasite relationships are either on their way to the extinction of one or both participants or are evolving towards a milder interaction. Over time, the gradual synthesis of ecological, evolutionary, and population genetics studies has changed these views. Even though many parasitologists remain loyal to the directional view of host–parasite interactions evolving exclusively towards commensalism, the paradigm has been replaced with the view that all possible outcomes, from commensalistic to highly antagonistic, are possible (Thompson, 1994).

The classic paradigm of host–parasite interactions evolving exclusively towards a benign outcome is based on the assumption that, all else being equal, a parasite that inflicts little damage would represent an advantage both to the individual host and to the parasite. Most of the evidence supporting this notion comes from observations that parasites are often less pathogenic toward their customary hosts than toward hosts newly introduced into their range and that newly evolved pathogens that are highly pathogenic gradually become more benign. In East Africa, for example, indigenous ruminants suffer mild infections, with insignificant morbidity, due to the hemoflagellate *Trypanosoma brucei*; introduced domestic ruminants, that have been bred in the area for a long time, develop more serious

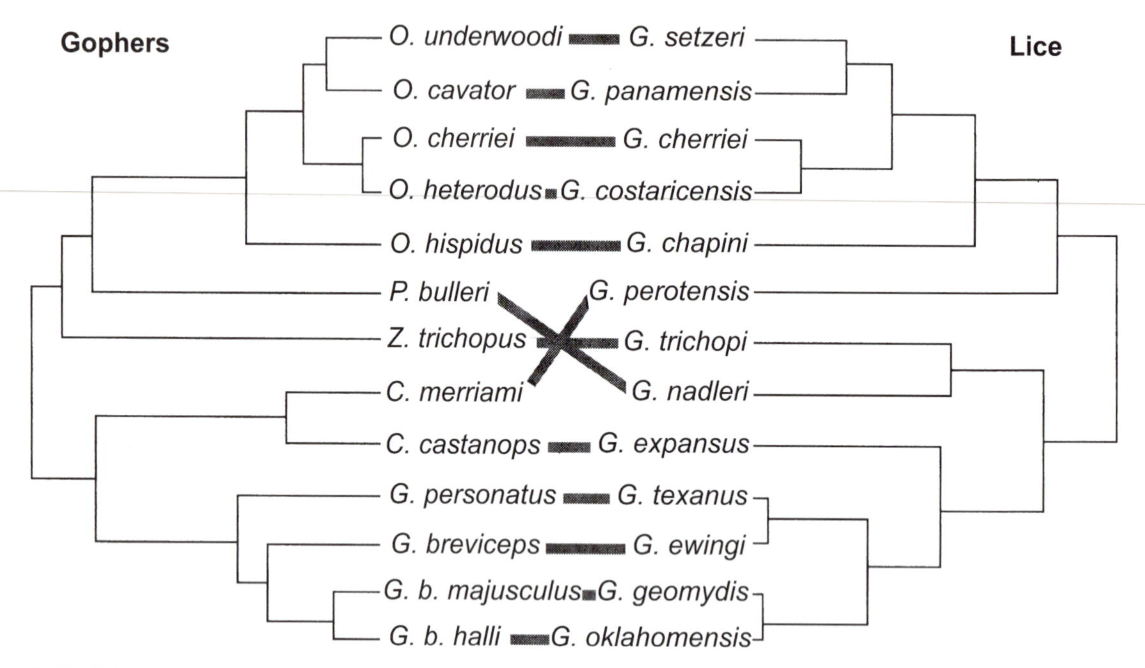

Fig. 16.6 Phylogenetic relationships between 13 species of pocket gopher and their parasitic lice using DNA sequence data from Nadler *et al.* (1994). Lines connecting extant species indicate host–parasite associations. The top five gopher and lice species indicate true cospeciation with synchronous speciation for hosts and parasites. The differences in the topology of the remainder of the tree are probably due to host switching by the lice or persistence of multiple ancestral louse lineages, or both. (The pocket gopher species are in the genera *Orthogeomys*, *Pappogeomys*, *Zygogeomys*, *Cratogeomys* and *Geomys*; *G.b.* stands for *Geomys bursarius*; the species of lice belong to the genus *Geomydoecus*.) (Reprinted with permission from Huelsenbeck, J. P. & Rannala, B. [1997] Phylogenetic methods come of age: testing hypotheses in an evolutionary context. *Science*, **276**, 227–232. © 1997 American Association for the Advancement of Science.)

symptoms and newly imported animals suffer infections that are typically fatal (Allison, 1982). Also, wild rats in cities that have had recent plague epidemics by *Yersinia pestis* show higher survival after challenge infections with the plague bacillus than rats from cities without a recent history of plague. Similarly, Holmes (1983) observed that parasitic infections appear to exert a more efficient regulatory role among newly introduced plants or animals, or when the parasites are introduced into new geographic localities.

Introduced parasites (or new host–parasite associations), however, are not always more harmful than adapted parasites. Experimental evidence shows that the opposite is also true; novel parasites seem to be less harmful, less infectious, and less fit than the same parasites infecting the hosts to which they are adapted (for references see Ebert & Herre, 1996). In spite of this evidence and current theories of the evolution of pathogenicity (to be discussed later), there are some cases, e.g., *T. brucei* and ruminants in East Africa, where novel associations, nonetheless, have had highly antagonistic outcomes. The explanation for the observed high pathogenicity of some new associations may be a simple one. It is likely that these cases are exceptions that have been recorded because of their extreme effects, whereas numerous other novel host–parasite associations have remained unnoticed because they have been benign (Read, 1994; Ebert & Herre, 1996). It is also possible that what we see as high virulence in some parasite introductions is simply an ecological response to a novel ecological interaction of two unadjusted populations, and not some evolutionary response to the interaction.

Current views indicate that the outcome of host–parasite interactions lies somewhere on a continuum, from antagonism to mutualism, depending on the natural history of the organisms involved, with emphasis on the mode of parasite transmission and reproduction (Anderson & May, 1982; May & Anderson, 1990; Ewald, 1995). Some of the prevailing ideas argue that different levels of parasite reproduction and pathogenicity depend on the extent to which the costs and benefits of extensive reproduction vary among different parasite species. For example, because hosts are temporary islands for parasites, effective transmission from one host to another is essential for the survival of parasite genes. Extensive parasite reproduction within a host will increase the number of parasites available to reach other hosts. If this occurs, host reproduction may be adversely affected and immobilization or death of the host would reduce the chances of the parasites to infect other hosts. On the other hand, if parasites are transmitted by vectors, such as biting arthropods, then immobilization of the host should increase the chances of being bitten, and extensive parasite reproduction may be beneficial because it increases the probability that a vector will obtain an infective dose (Ewald, 1983, 1994, 1995). In short, a reduction in host mobility should be especially costly for parasites that rely on host mobility for transmission; these parasites, then, should have lower levels of virulence. However, if transmission of the parasite is less affected by host mobility, the parasite's fitness should not be greatly affected by the immobilization of the host, and natural selection should favor a higher level of host exploitation and greater virulence (Ewald, 1995). The relationship between host mobility and virulence for different types of parasites is depicted in Fig. 16.7. Some of these interactions are discussed next. One problem with Ewald's (1995) views regarding parasite reproduction, degree of virulence, and host mobility is that it assumes a direct relationship between parasite reproduction and pathogenicity to the host, a view that has not been confirmed.

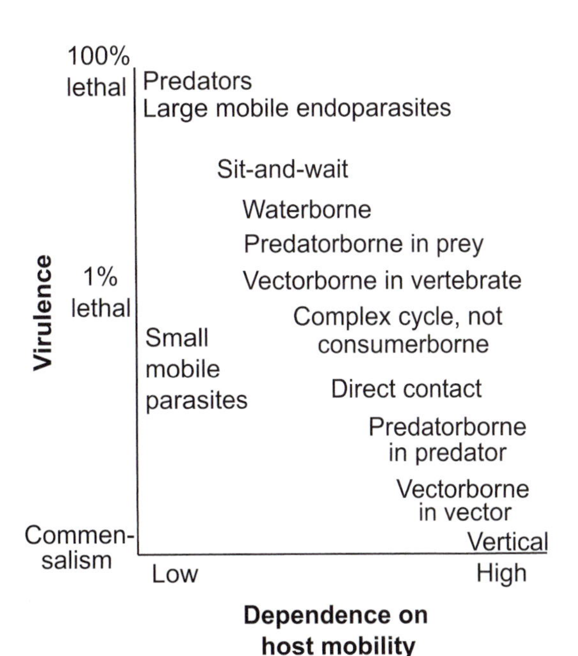

Fig. 16.7 Schematic representation of the relationship between host mobility and virulence from different categories of parasites showing the potential spectrum of harmfulness. The location of each category is meant to represent the central tendency of widely variable associations that form a continuum from commensalism to extreme harm. (From Ewald, 1995, with permission, *Journal of Parasitology*, **81**, 659–669.)

Ewald (1983) supported his views about virulence with data from human diseases. Using mortality rates as an estimate of pathogenicity, it was shown that diseases transmitted by insect vectors were significantly more lethal than those transmitted by other means. Similarly, by using immobilization of the host as an indicator of pathogenicity, vector-transmitted diseases were also significantly more lethal. For example, vector-borne parasites such as *Plasmodium* and *Leishmania* should benefit from host immobility, by reducing vector-avoidance behaviors of the host, whereas directly transmitted pathogens such as *Chlamydia* should minimize virulence because immobility of the host reduces transmission rates.

Not all parasites that rely on vectors for transmission show the same pattern, however. Variability related to the natural history of the host–parasite system can significantly change the outcome of the interaction. Mock & Gill (1984) studied the dynamics of the relationship between the red-spotted newt *Notophthalmus viridescens* and the trypanosome *Trypanosoma diemyctyli*. Long-term studies of trypanosome-infected hosts indicated that the parasites did not increase mortality or affect the reproductive ability of the host; on the contrary, the trypanosomes were rather benign. A look at the biology of newts shows that they have a semiaquatic life cycle and that at the end of each summer, juvenile newts leave the ponds where the adults live and go into the surrounding forest for about 6 years. Aquatic leeches are the vectors for this trypanosome and, as a consequence, trypanosomes can only be transmitted when the newts are in the pond. It seems that the interval of time between potential transmission events (the years in the forest) introduced a strong selection pressure against harmful trypanosomes. Newts parasitized with pathogenic trypanosomes probably die before returning to the pond to pass the infection around. Only newts parasitized with 'benign' trypanosomes survive long enough to be able to return to the pond and, with them, their parasites.

Generally, in terms of costs and benefits, parasites that are transmitted through invertebrate vectors can afford being pathogenic to their vertebrate hosts because weakness or immobilization of these hosts does not affect transmission of the parasite. The parasite, however, cannot afford being pathogenic to its vector host because the vector must remain mobile in order for transmission to proceed. Vector-borne parasites, then, should evolve toward mildness in their vectors.

An interesting aspect related to host immobilization, vectors, and transmission, but with a twist, is the degree of virulence of metazoan parasites in intermediate and definitive hosts. Macroparasites traditionally have been ignored or assumed to be simply benign, and have not been included into the greater picture of evolutionary outcomes. Ewald (1994, 1995) proposed that the relationship between macroparasites and their hosts was just a special case of the relationship between virulence and transmission with vectors. A vector (or micropredator) such as a mosquito can be envisioned as becoming larger and larger until it becomes a predator, instead of a micropredator. In this scenario, the vertebrate host of the vector-transmitted parasite becomes a host prey, whereas the larger and larger vector becomes the predator host. As in vector-borne parasites, exploitation of the prey host with the resulting pathogenicity should have a low fitness cost for the parasite because the parasite still can be transmitted successfully from a moribund or debilitated prey host by means of the predator. Immobilization or morbidity of the host may actually increase transmission because a weakened prey is at a higher risk of predation.

The prediction that prey hosts are usually adversely affected by their helminth parasites whereas predator hosts are rarely affected was tested based on literature surveys of the damage inflicted by various predator-borne parasites to their intermediate and definitive hosts (Table 16.7) (Ewald, 1995). The data seem to support the prediction that parasites are more pathogenic to prey hosts than to predator hosts. If the life cycle involves more than two hosts (prey and predator host), the parasite affects, adversely, all the hosts that become prey and remains less pathogenic to the final predator host (definitive host). Another prediction that seems to have support from the data is that definitive hosts should be more severely affected by parasites that are not predator-borne, that is parasites that actively search for their hosts, such as schistosomes. Because the data supporting many of the views presented by Ewald (1983, 1994, 1995) are fragmentary, and criticisms have been voiced regarding his assumptions, many of his ideas need to be subjected to rigorous testing. Only then can a clear vision of the relationship between parasite reproduction, parasite pathogenicity, host mobility, and parasite transmission emerge.

Vertical transmission is another aspect considered in the general theory of the evolution of virulence. Vertical transmission of parasites (from host parent to host offspring) should favor evolution toward benignness because the survival of the parasite is tied to the survival of the host. It also relates to the concept of mobility because the infected host must remain mobile and active enough to survive in the community. Low transmission rates and vertical transmission favor parasite genotypes that are less detrimental to host survival, whereas high transmission rates and horizontal transmission (between unrelated hosts) favor parasite genotypes that maximize their growth rate in the host. Experimental studies from viruses to multicellular parasites seem to support this idea. Experimental manipulation of three strains of *Escherichia coli* and several

Table 16.7 Severity of predator-borne helminths in relation to their mode of transmission

Parasite	Trophic status of the host during transmission[a]			References[c]
	Prey host[b]	Prey/predator host[b]	Predator host[b]	
Cestoda				
Hymenolepis diminuta	Beetles (S)		Rodents (m)	Keymer (1980)
Hymenolepis citelli	Beetles (S)		Rodents	Schom *et al.* (1981)
Echinococcus granulosus	Ungulates (S)		Canids (M)	Soulsby (1965); Leiby & Dyer (1971); Olsen (1974)
Echinococcus multilocularis	Rodents (s)		Carnivora (M)	Soulsby (1965); Leiby & Dyer (1971)
Monordotaenia taxidiensis	Ground squirrels (S)		Badgers	
Taenia multiceps	Lagomorphs (S), sheep (S)		Canids (M)	Boughton (1932); Soulsby (1965); Leiby & Dyer (1971)
Taenia mustelae	Rodents (S)		Mustelids (M)	Soulsby (1965); Leiby & Dyer (1971)
Taenia saginata	Cows (s)		Humans (m)	Soulsby (1965)
Taenia twichelli	Rodents (S)		Wolverine (M)	Soulsby (1965); Leiby & Dyer (1971)
Taenia pisiformis	Lagomorphs (s)		Carnivora (M)	Soulsby (1965); Leiby & Dyer (1971)
Taenia solium	Pigs (s)		Humans (m)	Soulsby (1965)
Diphyllobothrium spp.	Cyclops (S)	Fish (S)	Carnivora (m)	Haderlie (1953); Davis & Libke (1971); Griffiths (1973)
Spirometra spp.	Cyclops	Fish (S)	Carnivora (M)	Soulsby (1965); Ribelin & Migaki (1975)
Ligula intestinalis	Cyclops	Perch (s)	Gulls	Pitt & Grundman (1957)
Schistocephalus solidus	Copepods	Fish (S)	Birds	Haderlie (1953); Ribelin & Migaki (1975); McPhail & Peacock (1983)
Bothriocephalus acheilognathi	Cyclops		Mosquitofish (S)	Granath & Esch (1983)
Triaenophorus crassus	Cyclops (S)	Fish (S)	Fish	Watson & Price (1960); Rosen & Dick (1983*a*, 1983*b*)
Triaenophorus nodulosus	Cyclops	Fish (S)	Fish	Lawler (1969)
Archigetes iowensis	Oligochaetes (S)		Carp	Calentine (1962, 1964)
Corallobothrium spp.	Cyclops	Sunfish	Catfish (M)	Harms (1960)
Digenea				
Bucephalus sp.	Fish (S)		Fish	Olsen (1974)
Diplostomum baeri	Sticklebacks (S)		Mallard	Olsen (1974)

Table 16.7 *(cont.)*

Parasite	Trophic status of the host during transmission[a]			References[c]
	Prey host[b]	Prey/predator host[b]	Predator host[b]	
Diplostomum spathaceum	Fish (S)		Birds	Crowden & Broom (1980)
Halipegus occidualis	Ostracods (S)		Odonates to frogs	Wetzel & Esch (1995)
Nanophyetus salmincola	Fish		Mammals	Millemann & Knapp (1970); Butler & Milleman (1971)
Paragonimus kellicotti	Crayfish		Mink (m)	Davis & Libke (1971)
Uvulifer ambloplites	Fish (S)		Kingfisher	Olsen (1974); Lemly & Esch (1984)
Nematoda				
Trichinella spiralis	Mice (S)		Mammals	Olsen (1974); Rau (1983)
Ascaris columnaris	Mice (S)		Raccoons	Tiner (1954)
Acanthocephala				
Neoechinorhynchus cylindratus	Ostracods (S)		Fish	Olsen (1974)
Acanthocephalus dirus	Isopods (S)		Fish	Seidenburg (1973)
Plagyorhynchus cylindraceus	Isopods (s)		Birds	Moore (1983)

Notes:

[a] Prey/predator refers to hosts that acquire their parasites by eating the infected host and transmit the parasite to the next host in the cycle when eaten by the next host.

[b] S, severe (can cause death); s, severe non-lethal effects such as castration); M, mild (essentially asymptomatic) m, moderate to mild effects (inflammation or slightly reduced mobility).

[c] A complete listing of references can be found in Ewald (1995).

Source: From Ewald, 1995, with permission, *Journal of Parasitology*, **81**, 659–669.

variants of their parasitic bacteriophage viruses showed that when the interaction was maintained in culture in a way that minimized horizontal transfer of the viruses, natural selection favored the variants that were most benevolent to the hosts. But, when the cultures allowed horizontal spread of the viruses, the advantage of the benevolent variants was lost (Bull *et al.*, 1991; Bull & Molineaux, 1992). Clayton & Tompkins (1994) studied the relationship between virulence and the mechanism of transmission of two parasites of birds, vertically transmitted lice and horizontally transmitted mites. Their results show that virulence correlates with the presumed rates of vertical and horizontal transmission according to the proposed model.

The same kind of interaction seems to occur between fig wasps and their nematode parasites (Herre, 1993). The relationship between fig wasps and nematodes is worth describing because it shows what the current paradigm about the possible outcome of host–parasite interactions is all about. The continuum from low virulence to high virulence is dependent on the transmission dynamics. Each species of fig wasp pollinates only one species of fig. The number of female wasps that enters each fig is variable depending on the species of wasp. After one or more gravid female wasps enter each inflorescence (that eventually becomes the fig), they pollinate the flowers, lay their eggs, and die within that fig. The offspring develop within the seeds, hatch as adults, mate, and then the gravid female flies away to lay eggs in another fig inflorescence. The study by Herre (1993) included 11 species of fig wasps found in the vicinity of the Panama Canal, where each species of wasp was parasitized by a different species of nematode. The life cycle of the nematodes is closely connected with that of the fig wasps. Nematodes lay their eggs within the figs and the eggs hatch at the time the adult female fig wasps emerge. The nematodes eventually penetrate the wasp's body cavity and are transported with the female as she flies in search of another inflorescence to lay her eggs. The nematodes grow, eventually consume the adult female, emerge from her body, and start a new cycle by laying their eggs in the fig. If only one female wasp enters a fig, the nematodes from that single female only have access to that female's offspring, and the transmission is vertical. However, if several wasps oviposit in the same fig, the nematodes can parasitize the offspring of several wasps and transmission can be horizontal. Herre's (1993) study found that among the 11 species of nematodes, there was an almost linear relationship between virulence and the number of female wasps ovipositing. In figs with several females, horizontal transmission was high and virulence was high, favoring nematodes that are more destructive to their hosts. In figs with fewer females, vertical transmission was high and virulence was low, favoring nematodes that have no effect on the number of offspring produced by their host. In short, across species, increased parasite virulence correlates with increased horizontal transmission, and low virulence correlates with vertical transmission.

16.4 Parasite influence on the evolutionary biology of the host

The evolution of host–parasite interactions is not restricted to coevolutionary considerations. Outside the boundaries of historical association, or strict phylogeny, there is a wide array of host–parasite interactions that deserve attention from an evolutionary perspective. These include the possible role of parasites in the origin and evolution of sex, in host sexual selection, and in host genetic polymorphisms among others.

16.4.1 Parasites and the evolution of sex

From an evolutionary point of view, sex is a bad idea. Sex is a costly process in terms of the energy required to produce sex organs (and gametes), find mates, and participate in courtship and mating. Moreover, all things being equal, organisms that reproduce asexually are twice as efficient as sexual conspecifics in propagating their genes. In populations with a 1:1 sex ratio, sex is expensive for females. Females are better off if they can reproduce asexually so that they can pass their complete genome intact to their offspring, without being diluted by the father's genes. In addition, males seldom invest as much as females in reproduction and rearing of the offspring, placing an even heavier burden on the female. Other than for more whimsical reasons, why do organisms insist on having sex? One important advantage of sexual reproduction is genetic recombination, which constantly stirs and mixes the gene pool, resulting in a greater genetic diversity among the offspring of a given individual. Sexual reproduction ensures genetic variability.

Several hypotheses have been proposed to explain the maintenance of sex in populations, but most fall into two classes. Some hypotheses attribute the maintenance of sex to the existence of environmental heterogeneity in time or space. In such cases, variability will ensure that at least some genotypes should be successful in variable environments, through space or time. Other hypotheses attribute the maintenance of sex to the operation of frequency-dependent selection against common 'host' genotypes resulting from interspecific interactions, such as host–parasite interactions. This particular notion, known as the **Red Queen Hypothesis** (nothing more, really, than a more specific type of 'evolutionary arms race'), proposes that sex and recombination are favored in hosts because they produce rare phenotypes, which are expected to have a greater chance of escaping parasites. The parasites, then, would be capable of tracking these rare genotypes as they become common, initiating a frequency-dependent selection process (Jaenike, 1978; Hamilton, 1980; Bell & Maynard-Smith, 1987). Thus, sexually produced offspring should have new genotypes conferring resistance and should benefit from increased resistance to coadapted parasites. Offspring produced asexually, however, usually have the same combination of resistance genes as their parents and may suffer increased levels of exploitation by parasites (Ladle, 1992). Some genetic models and model simulations support the view that parasites are a likely driving force behind the maintenance of sex (Hamilton, 1980; Hamilton *et al.*, 1990; Howard & Lively, 1994; Lively & Howard, 1997). The Red Queen Hypothesis, in a simplified form, predicts continued evolutionary change (of host and parasite) under conditions of constancy in the physical environment. The name of the hypothesis comes aptly from the character in Lewis Carroll's book *Through the Looking Glass* who says that you must keep running just to stay in the same place.

If sex is an adaptation against parasites, it should be common where parasites are common; conversely, parthenogenesis should be common where parasites are rare. In New Zealand, populations of the dioecious freshwater snail *Potamopyrgus antipodarum* are formed by sexual and parthenogenetic individuals, providing a valuable model for testing the Red Queen Hypothesis regarding parasites. Lively (1992) found a positive correlation between the frequency of sexual reproduction in parthenogenetic snails and the degree of infection by digeneans. In populations where parasites were rare, parthenogenetic females had numerically replaced sexual females, whereas in populations where parasites were common, sexual females persisted. In other words, sex was rare when parasites were absent, but widespread when parasites were common. Lively *et al.* (1990) reached a similar conclusion while investigating parasitism in coexisting sexual and hybrid parthenogenetic forms of the topminnow *Poeciliopsis monarcha*, in Mexico. Clones of the parthenogenetic form accumulated parasites at a much faster rate than the sexual form. However, when the sexual populations of topminnows were inbred, the antiparasitic effect of sexual reproduction disappeared because their genetic diversity was extremely low. Interestingly, if other individuals entered the inbred population of topminnows, parasitism declined because genetic diversity increased. A study by Moritz *et al.* (1991) on the lizard *Heteronotia binoei* also indicated that hybrid parthenogenetic races are more prone to infection by mites than their asexual relatives. In these two last examples, however, the parthenogenetic populations were hybrids and it is possible that this factor alone might have affected the susceptibility to parasites (Ladle, 1992).

Although these examples provide evidence for the prediction that sexual reproduction should be common when parasites are present, and asexual reproduction when parasites are rare, no clear evidence existed for the prediction that frequency-dependent selection occurred, in which rare genotypes are favored and common genotypes are selected against. Dybdahl & Lively (1998) were able to gather evidence by studying clonal lineages of the snail *Potamopyrgus antipodarum* and its digenean parasites through a period of 5 years in a lake in New Zealand. They identified a number of genetic markers specific to each snail clone but, because these snails only reproduce asexually, the genetic markers remain unchanged within each snail clone through time. With these genetic markers, Dybdahl & Lively (1998) identified rare and common clones of snails and tracked their frequencies, and their parasites, over time. Over the 5-year period, the snail clones showed the characteristic oscillation of frequency-dependent selection, with cycling of host clones and time-lagged correlated responses by the parasites. For example, one of the infected common clones was driven down and replaced in the lake by what was, originally, a rare clone; this was also, over time, driven down and replaced. These clones were then experimentally exposed to the common parasite. In the laboratory, the host clones that had been rare during the previous 5 years were significantly less infected by the parasite when compared to the common clones. This study provides clear evidence for two aspects of the Red Queen Hypothesis regarding parasites and their role in the maintenance of sex. First, it shows that rare host genotypes are more likely to escape infection by parasites, and second, that host–parasite interactions can produce frequency-dependent oscillations in the host population.

Sometimes, the question pondered by a host is not whether to have sex or not but, what kind of sex to have. Many hosts are hermaphrodites and have the potential for outcrossing, i.e., exchange gametes with other individuals, or selfing, i.e., self-fertilization. Self-fertilization increases the inbreeding of the population, decreasing its genetic variability and increasing the genetic correlation, or genetic similarity, between parents and offspring. This situation is analogous to parthenogenesis and should lead to higher susceptibility to parasites. Outcrossing, or cross-fertilization, on the other hand, makes for an outbred population, with higher genetic variability and low genetic correlation between parents and offspring. The snail *Bulinus truncatus* is an hermaphrodite that can reproduce both by self-fertilization and cross-fertilization. Schrag *et al.* (1994) studied a number of biotic and abiotic ecological correlates of male outcrossing ability in Nigerian populations of *B. truncatus*. Their study showed that outcrossing correlated positively with the prevalence of the most abundant digenean, and also with indices of digenean diversity that incorporated prevalence and richness, but not with any of the other biotic or abiotic correlates analyzed. There was, however, a weak correlation between outcrossing and water conductivity, but because the relationship between snail biology, water chemistry, and habitat ecology are not well understood, the meaning of this latter correlation awaits further study.

16.4.2 Parasites and sexual selection

Sexual selection, the selection by one sex of variations within the opposite sex, frequently acts on traits that are phenotypically expressed in one sex only. Sexual selection results in sexual dimorphisms, for example, size differences between the sexes, differences in coloration, and presence of antlers. Although the idea, first proposed by Darwin (1871), that secondary sexual characters such as the long tails and bright plumage of birds evolved because females use them as cues in mate choice is widely accepted, the question of why females should prefer males with those exaggerated characters is still debated.

Hamilton & Zuk (1982) drew attention to the potential role of parasites in sexual selection. They proposed that secondary sexual traits such as the bright colors and elaborate courtship displays of many animal species allowed females to assess a potential reproductive partner's ability to resist parasites. By mating with such males, females would assure the passage of parasite resistance genes to their offspring. Hamilton & Zuk's hypothesis was based on three assumptions. First, the display of secondary sexual traits of individual animals is related to their general health and vigor; second, hosts coevolve with their parasites, and as a result, they constantly generate heritable resistance to parasites; and, third, parasites have a negative effect on host viability. The available evidence seems to support these assumptions.

There are three basic models of parasite-mediated sexual selection that can lead to adaptive changes in mate choice. First, individuals may avoid mating with individuals infected with contact-transmitted parasites, such as lice and various sexually transmitted diseases, which can be transmitted to the mate or offspring. Second, in species in which males provide parental care, females should choose healthy and vigorous mates to help care for the offspring. Third, if parasite resistance is heritable, females should mate with resistant males to obtain resistant genes for their offspring (for reviews of these models see Hamilton, 1990; Møller, 1990; Read, 1990). The last model, where there is genetically based resistance, is an example of a 'good genes' model of sexual selection because female choice leads to more viable offspring with increased fitness.

If females are choosing their mates based on secondary sexual characters, they should cue on male traits that honestly indicate genetic resistance because there is no advantage in choosing males that can masquerade their diseases. The mechanism by which the female chooses may be based on the **handicap principle** (Zahavi, 1975) which assumes that the cost of a signal guarantees its reliability, and that cheating is prevented because the cost of a unit of display is greater for low-quality than for high-quality individuals. The development of secondary sexual characters, then, is assumed to be costly, and these traits are even costlier for poor-quality males than for high-quality males. Only high-quality individuals, with superior viability genes, should be able to develop costly secondary sexual characters and still be able to maintain their viability. If poor-quality individuals attempt to cheat by developing a costly display, it will be at the expense of other functions, e.g., by diverting resources, reducing their overall viability and survival (Iwasa *et al.*, 1991). A special version of the handicap principle is the **immunocompetence handicap hypothesis** which suggests that secondary sexual characters develop in response to androgens or other compounds that increase the expression of secondary sexual traits but that reduce the ability of the immune system to fight parasites (Folstad & Karter, 1992). In other words, the immunocompetence handicap will allow high-quality males to develop secondary sexual traits while compromising its immune system. Despite the compromised immune system, these males still remain viable and relatively resistant to parasites. Saino & Møller (1994) studied the relationship between testosterone levels, secondary sexual characters, and parasitism in the barn swallow *Hirundo rustica*. They found that tail length, a secondary sexual character involved in sexual selection, was positively correlated to testosterone levels, whereas the prevalence and intensity of ectoparasites were unrelated to testosterone levels. Intensity of ectoparasites, however, was negatively related to the tail length of male barn swallows; also, unmated males, had higher intensities of ectoparasites than mated males. The absence of a correlation between parasite load and testosterone levels corroborates the prediction that male barn swallows with high levels of testosterone and large secondary sexual characters are reliably signaling their resistance to parasites. In other words, despite the high levels of androgens that are suppressing the immune system barn swallows are still able to avoid parasites.

Hamilton & Zuk's (1982) hypothesis for the evolution of secondary sexual characters, whereby these characters allow females to assess a potential partner's ability to resist parasites, makes a number of intra- and interspecific predictions. These predictions can be summarized as follows: (1) there is heritable variation in parasite resistance; (2) there is a negative relationship between parasite load and host viability; (3) the expression of secondary sexual characters is negatively correlated with variation in parasite burden over the range of infection intensities that affect host fitness in natural populations; (4) females should prefer those males which are more resistant to the parasites infecting the population at the time; and (5) there should be a positive correlation across species between the parasite burden experienced by individuals in a species and the level of complexity of their secondary sexual characters. In other words, in species that are particularly vulnerable to parasitism, sexual selection should favor greater development of 'health-certifying' traits, allowing females a more accurate assessment of the parasite load of a male.

The first four predictions can be tested intraspecifically, within a single host species, whereas the last prediction requires testing across a number of different species. In the past 10 to 15 years there has been a proliferation of studies attempting to test these premises. Whereas the lack of comparative studies in more host species does not permit definitive conclusions regarding the interspecific prediction of Hamilton & Zuk's (1982) hypothesis, studies dealing with the intraspecific predictions have been more supportive. Thus, the predictions that an expression of sexual ornamentation will reflect a certain level of parasitism, that host resistance to parasites is heritable, that females choose males with fewer parasites, and that parasites negatively affect host fitness have been supported in most of the studies in which these questions were investigated (see Møller, 1990 for a review; also Buchholz, 1995; Polak & Markow, 1995; Rosenqvist & Johansson, 1995; Taylor *et al.*, 1998).

Møller (1990) and Read (1990) provided a comprehensive review of the experimental evidence for parasite-driven host sexual selection. In contrast to the evidence available corroborating the intraspecific predictions, comparative studies testing the interspecific prediction (prediction number 5) have been equivocal, mainly because methodological problems compromise the results. For example, parasite prevalence may be affected by sampling bias, the reliability of host displays is not always assessed objectively (in most cases the scoring is subjective), host species are normally used as independent observations despite the fact that they are phylogenetically related, and environmental correlates may influence the extent of the display and cause correlations due to ecological conditions, not parasite signaling.

Hamilton & Zuk (1982) originally supported their hypothesis with observations that a positive association exists between the prevalence of blood infections (*Leucocytozoon*, *Haemoproteus*, *Plasmodium*, *Trypanosoma*, *Toxoplasma*, and nematode microfilariae) and three types of host displays (male brightness, female brightness, and song complexity) in passerine bird species (the interspecific prediction). A re-analysis of those data (Read, 1988), controlling for the variables mentioned in the previous paragraph, confirmed some of the Hamilton & Zuk (1982) findings and ruled out the possibility that the correlation could have been produced by taxonomic association of the hosts. A similar study conducted across families of British and Irish freshwater fishes (Ward, 1988, 1989) showed that the degree of sexual dimorphism was positively associated with parasite loads after controlling for various confounding variables. However, criticisms of Ward's conclusions indicate that the correlation in his data was to a large extent due to uncontrolled ecological variables. Chandler & Cabana (1991), for example, found that **dichromatism** in fishes was not correlated with the mean number of parasite species per individual host, but with the total number of parasite genera per host species, which can become an extremely biased measurement due to unequal study efforts. Indeed, Chandler & Cabana (1991) showed that the correlation between dichromatism and parasite genera per host species disappears when study effort is controlled statistically. A number of confounding variables are likely to exist in any particular study of the interspecific prediction of the Hamilton–Zuk hypothesis. For this reason, the best support for the prediction will be provided by positive results from a wider range of host and parasite taxa, because it is unlikely that the same confounding variable will exist in all the host–parasite systems (Read, 1990). It would be interesting, for example, to examine sexual selection where the females are the more brightly colored gender, and, subsequent to egg laying, the male assumes the role of rearing the young. Such a scenario exists in birds known as phalaropes (*Phalaropus* spp.)

Møller & Saino (1994) offered additional evidence for the interspecific prediction based on the immunocompetence handicap hypothesis. They reasoned that, if differences in secondary sexual characters are the result of parasite-mediated sexual selection, then species of birds that are sexually dichromatic should have higher parasite burdens and, therefore, require a greater immune response. In immunological terms, this means larger spleens, a larger bursa of Fabricius (an organ essential for the production of antibodies in young birds but that regresses before sexual maturity), and higher leucocyte counts. Their results comparing monochromatic and dichromatic birds indicate that sexual dichromatism was indeed associated with larger spleens and a bursa of Fabricius and higher leucocyte counts. In short, intensity of sexual selection and immune defense seem to be related interspecifically.

DEVELOPMENTAL STABILITY AND FLUCTUATING ASYMMETRY

Under ideal conditions, a developing embryo should attain perfect bilateral symmetry because both sides of a symmetrical trait are the products of the same genotype. However, environmental perturbations during development affect the developmental stability and may result in deviations from perfect symmetry in traits that are usually bilaterally symmetrical. **Fluctuating asymmetry** measures these small, random deviations from perfect bilateral symmetry in a given morphological trait. The factors that affect developmental stability are relatively well known. They include both environmental, e.g., climatic conditions, food deficiency, pesticides, parasitism, and genetic factors such as inbreeding and hybridization. Recent reviews indicate that there is, in general, a negative relationship between developmental instability and various fitness components such as growth, longevity, and reproduction (Parsons, 1990; Møller, 1996, 1997; Polak, 1997a). In other words, symmetrical individuals tend to have higher fecundity, better survival, and faster growth, than more asymmetrical individuals.

Fluctuating asymmetry also may be important in sexual selection. Studies of male mating success in scorpionflies (*Panorpa japonica* and *P. vulgaris*) and dung flies (*Scathophaga stercoraria*) suggest that symmetrical males are more successful in male–male competitive interactions for mates (Thornhill, 1992; Thornhill & Sauer, 1992; Liggett *et al*., 1993). Similarly, studies of female choice in barn swallows (*Hirundo rustica*) and zebra finches (*Taeniopygia guttata*) show that females respond to fluctuating asymmetry and prefer males with the most symmetrical ornaments (Møller, 1992*a*, 1993, 1994; Swaddle & Cuthill, 1994*a*, *b*). Other studies involving damselflies, crickets, fruitflies, houseflies, ladybird beetles, midges, and oribus (a vertebrate) also have found that fluctuating asymmetry affects mating success. However, the mechanisms involved (intrasexual competition, mate choice, etc.) have not been elucidated (for references see Polak, 1997*a*).

Thornhill (1992) discussed the idea that pathogens, including parasites, may be involved in producing fluctuating asymmetry. Parasites may cause fluctuating asymmetry by competing with their hosts for available nutrients during the host development, impairing host metabolism, development, growth rate, etc. In addition, the parasite may be responsible for specific pathological effects on parts of the host body that eventually may lead to fluctuating asymmetry. If parasites increase the fluctuating asymmetry of their hosts, they may ultimately affect the mating success of their host. If symmetry is an indication of 'good genes', females could use symmetry to assess the quality of potential mates. Moreover, if parasitism increases fluctuating asymmetry, then sexual selection could favor parasite-free males and this may lead to the evolution of resistance to parasites (Polak, 1997*b*). So far, studies on barn swallows (*Hirundo rustica*), reindeer (*Rangifer tarandus*), and *Drosophila nigrospiracula* provide evidence that parasites increase the fluctuating asymmetry of their hosts (Møller, 1992*b*; Polak, 1993; Folstad *et al*., 1996).

Parasites do not always affect the symmetry of the host harboring the parasites. Sometimes the offspring of the parasitized host are the ones affected. Polak (1997*b*), for example, has shown that female *Drosophila nigrospiracula* parasitized by the hematophagous mite *Macrocheles subbadius*, produced sons with significantly higher asymmetry than did mite-free females. In spite of their asymmetry, however, the mating success of the male offspring was similar to that of more symmetrical males.

Although there is some evidence that fluctuating asymmetry has a negative effect on mating success, and that parasites increase fluctuating asymmetry in their hosts, the direct effect of parasites on mating success due to the induced fluctuating asymmetry has not been definitely demonstrated. Very few studies have evaluated the connection between fluctuating asymmetry–mating success and fluctuating asymmetry–parasitism simultaneously in a single host population (Polak, 1997*a*). Additional studies in this rather new research area are needed to determine the potential impact of parasites in developmental stability and their true role in host sexual selection.

16.4.3 Parasites, host genetic polymorphisms, and evolutionary novelty

Parasites are gaining recognition as important evolutionary agents beyond the intensely studied areas of sex evolution and host sexual selection. Aspects of the evolutionary biology of the hosts that are, or may be, affected by parasites include the vast gene polymorphisms seen in both human and mouse histocompatibility systems, the polymorphisms of blood proteins, the phenomenon of reproductive isolation by symbiont incompatibility, evolutionary change by the acquisition of symbionts, and potential gene transfer.

HOST GENETIC POLYMORPHISMS

Haldane (1949, 1954), while studying human blood groups, proposed that parasites could be a determining factor in the maintenance of certain blood-group polymorphisms. He argued that if parasites were successful because they could not be distinguished antigenically from self by the host, then the success of a single parasite genotype would be reduced in a host population that was variable for blood protein antigens. Parasites would tend to evolve adaptations to attack the more common host genotypes and, as a result, the rare host genotypes would be at a selective advantage. This process would result in a continuous, frequency-dependent selection of both hosts and parasites as each responded to the influences of the other.

Increasing evidence indicates that host genetic factors play a major role in determining susceptibility to many infectious and parasitic diseases. Likewise, there has been extensive debate concerning the evolutionary mechanisms that have maintained the unique diversity of the **major histocompatibility complex** (MHC). The MHC is a multi-gene family that produces cell-surface glycoproteins that serve to identify 'self' from 'non-self'. Its most striking characteristic, however, is the high level of polymorphism found at certain MHC loci, which are among the most polymorphic loci known in living organisms. The important role of MHC molecules in immune responses against pathogens led to the proposal that the diversity of MHC molecules evolved as a result of host–parasite interactions (Snell, 1968; Doherty & Zinkernagel, 1975). Current views indicate that some form of balancing selection such as overdominant selection (heterozygote advantage) may be responsible for most of the MHC polymorphism because of the advantage of such diversity in enhancing an individual's ability to bind a wide array of pathogens. Frequency-dependent selection, however, has also been considered as a mechanism because it may explain the frequency of some genotypes under appropriate conditions (Hill, 1996; Hughes *et al.*, 1997). Recent proposals suggest that the parasites which have influenced MHC polymorphisms the most are those with a long history of coevolution with their hosts, whereas parasites that switched hosts recently, in the last 10 000 years or so, have had little effect on MHC polymorphism (Klein & O'Huigin, 1997).

Opposite views, however, regarding the role of parasites on the polymorphism of the MHC system are held by Potts *et al.* (1997). Their studies indicate that mating preferences in mice are dependent on the genetic makeup of the MHC of the mating pair, favoring matings that would produce progeny with heterozygous MHC system. In mice, products of the MHC are eliminated with urine, and this confers specific odors to the potential mates. Mating is based on the recognition of olfactory signals that provide specific information about the MHC genotypes, favoring matings that increase polymorphism and avoid genome-wide inbreeding. Obviously, the debate is not settled and further research is needed to clarify the significance of parasites in the evolution of MHC diversity.

SYMBIONTS AND REPRODUCTIVE ISOLATION

Symbionts are ubiquitous in all organisms. Most insects harbor bacterial symbionts that can cause reproductive incompatibility between host populations because of incompatibility between the symbionts. The bacteria *Wolbachia* are symbionts common to a substantial number of arthropods. They are transmitted vertically (and sometimes horizontally) causing reproductive incompatibility in the host. Crosses between males and females infected with *Wolbachia* are unaffected and the result is viable offspring. However, if mating occurs between infected males and uninfected females, the bacteria causes cytoplasmic incompatibility by disrupting the incorporation of male chromosomes in the fertilized eggs, and development ceases. If the hosts are treated with antibiotics that kill the bacteria, however, the incompatibility ceases and reproductive isolation disappears. In some cases, infection results in hosts that become parthenogenetic because the symbionts prevent segregation of chromosomes in unfertilized eggs, or, in others, the presence of symbionts changes the sex ratio biasing it toward production of female progeny (Werren *et al.*, 1986; Stouthamer & Luck, 1993). From an evolutionary point of view, if reproductive isolation between populations with incompatible bacteria is maintained through time, speciation may occur (Thompson, 1994).

Three different types of symbiont-induced host speciation have been proposed based on the reduced viability or fertility induced in hosts (Thompson, 1987). In one interaction, a cytoplasmically inherited symbiont reduces the viability or fertility of hybrids in crosses between incompatible populations, and speciation occurs by hybrid inferiority. It is also possible that the symbiont may cause sterility in males in some populations, favoring the evolution of parthenogenetic sibling species. In a third interaction, a non-cytoplasmically inherited parasite such as *Plasmodium* sp. may have a more detrimental effect in some populations than others, again favoring speciation by selecting against hybrid organisms.

Studies of host–endosymbiont interactions are relatively new and, although little is known about how these interactions evolve, some evidence indicates that they may evolve quickly. Ebbert (1991) studied the interaction between endosymbionts and several populations of *Drosophila willistoni* by multiple crossing combinations. The outcomes varied both with the host and the bacterial lines crossed, and in some cases, it appeared that coevolution had occurred in some of the laboratory cultures. Rapid evolution has been found in other systems too. A laboratory strain of *Amoeba proteus* became accidentally infected with virulent bacteria that killed most of the amoeba. Some *A. proteus* survived, however, and within years became totally dependent on the bacteria. After two decades in the laboratory, the dependence on the bacteria continues to the extent that removal of the bacteria results in loss of cell viability; reinsertion of the bacteria restores the viability of *A. proteus* (Jeon, 1972, 1991).

Although few examples of endosymbiont-induced changes in host populations are available, they have proven that these changes are possible and that they can occur very rapidly (Thompson, 1994). The ultimate change, however, occurs when the outcome is speciation, or even new forms of life as has been suggested by the endosymbiotic theory of eukaryotic cells proposed by Margulis and colleagues (Margulis & Bermudes, 1985; Margulis & Fester, 1991). Evidence supporting this theory has been accumulating slowly since the original proposal that eukaryotic cells descended from symbiotic unions of simpler prokaryotic cells. In addition to the widely accepted prokaryotic origin of mitochondria and chloroplasts, a freshwater cryptomonad alga is thought to have originated from the fusion of two eukaryotes, a heterotrophic eukaryote and a red alga containing chloroplasts. More recently, Kohler *et al.* (1997) found that parasitic apicomplexan protozoans acquired a plastid by secondary endosymbiosis, probably from an alga. Although its function is not yet known, the fact that it has been retained through evolutionary history suggests that it might be essential.

GENE TRANSFER

General evolutionary theory interprets evolution as the product of variation generated by random mutation and recombination, with natural selection acting on the variation produced. In recent years, however, the idea of **horizontal gene transfer** also has received attention. Horizontal gene transfer is defined as the non-sexual transfer of genetic information between the genomes of organisms of different species (Kidwell, 1993). This phenomenon is known to occur mainly among prokaryotic organisms, but transfers between prokaryotes and eukaryotes and between eukaryotes appear to have occurred too (for references see Kidwell, 1993). The mechanisms for horizontal gene transfer between prokaryotes include transformation, conjugation, and transduction. In eukaryotes, however, a vector is normally required to transport DNA sequences between reproductively isolated species. Various viruses and possibly a mite have been identified as effective vectors.

The idea of gene transfer in host–parasite systems, introduced by Howell (1985), was based primarily on the ability of some parasites to produce host-like molecules (antigens) and thus to avoid recognition by the host. Such phenomena would certainly be possible if parasites have genomes containing DNA sequences that are identical, or closely related, to those of their hosts (DNA homology). This kind of homology could be the product of either long-term conservation, or a more recent adaptive change through mutation, recombination, and selection. Howell (1985) argued that these DNA homologies also might be the product of direct incorporation of host genetic material into the parasite genome. This process of interspecies gene transfer could be a two-way flow from host to parasite and vice versa. Incorporation of parasite genetic material by the host may lead to tolerance of the parasite if the information is expressed and recognized by the host as 'self'. On the other hand, incorporation of host DNA into the parasite genome may lead to the expression of host macromolecules by the parasite, facilitating the evasion of the immune response. Possible mechanisms for such two-way flow require the presence of retroviruses, RNA viruses that, upon infection of a host cell, can copy their viral RNA into DNA by means of the enzyme reverse transcriptase. This DNA then enters the cell nucleus and integrates into the host DNA where it is transcribed and its products are translated into proteins. In these terms, the potential of the virus to act as a vehicle for exchange of genetic material between hosts and parasites is based on its ability to transduce DNA, cross species boundaries, and enter the germ lines of their hosts.

At the present time, the evidence for this phenomenon between hosts and parasites is scarce. Although virus or virus-like particles similar to RNA viruses have been observed in several parasite groups, from protozoans to nematodes (for references see Howell, 1985), there is no direct evidence of gene exchange. Circumstantial evidence, for example, includes the similarity between human and *Ascaris* collagens (Michaeli *et al.*, 1972), the ability of *Fasciola hepatica* to synthesize human blood group substances (Ben-Ismail *et al.*, 1982), the numerous instances of antigen-sharing by host and parasite (Capron *et al.*, 1968; Damian, 1979), and the presence of mouse DNA in one of its parasites, the digenean *Schistosoma japonicum* (Iwamura *et al.*, 1991). In the latter case, host DNA was apparently integrated into the genome of the adult parasites and was detectable in their eggs.

Although the focus of this section is on the potential horizontal gene transfer between hosts and parasites, there is a relatively well-studied case in which a semiparasitic mite accomplishes gene transfer between two species of *Drosophila* with the potential for host evolutionary change (Houck *et al.*, 1991; Houck, 1994). The mite *Proctolaelaps regalis* feeds on *Drosophila* eggs by rapidly pinching and piercing a number of nearby eggs with its chelicerae. This feeding behavior provides potential for transfer of cellular inclusions, including DNA, from one host to another as successive eggs are pierced quickly and material for one is accidentally transferred to the next. Interestingly, this behavior simulates the method of egg microinjection used by researchers for intra- and interspecific transfer of genes in *Drosophila*.

In summary, horizontal gene transfer between hosts and parasites is a potential source of genetic variation upon which natural selection can operate, in addition to mutation and recombination. Even if the frequency of this phenomenon is low, Rohde (1990) suggests that it might have a profound impact on the phylogeny of many groups. In this context, phylogenetic systematics, based strictly on molecular data, would not be reliable in its entirety because of the possible transfer of genes between unrelated species. In his view, mosaic evolution (with gene transfer acting as a source of isolated evolutionary units) might explain, for example, the conflicting evidence regarding the relationships of some parasitic taxa.

References

Allison, A. C. (1982) Coevolution between hosts and infectious disease agents and its effects on virulence. In *Population Biology of Infectious Diseases*, ed. R. M. Anderson & R. M. May, pp. 245–268. New York: Springer-Verlag.

Anderson, R. M. & May, R. M. (1982) Coevolution of parasites and hosts. *Parasitology*, **85**, 411–426.

Anderson, T. J. C. & Jaenike, J. (1997) Host specificity, evolutionary relationships and macrogeographic differentiation among *Ascaris* populations from humans and pigs. *Parasitology*, **112**, 325–342.

Anderson, T. J. C., Blouin, M. S. & Beech, R. N. (1998) Population biology of parasitic nematodes: applications of genetic markers. *Advances in Parasitology*, **41**, 219–283.

Barker, S. C., Briscoe, D. A., Close, R. L. & Dallas, P. (1991) Genetic variation in the *Heterodoxus octoseriatus* group (Phthiraptera): a test of Price's model of parasite evolution. *International Journal for Parasitology*, **21**, 555–563.

Bell, G. & Maynard Smith, J. (1987) Short-term selection for recombination among mutually antagonistic species. *Nature*, **328**, 66–68.

Ben-Ismail, R., Mulet-Clamagirand, C., Carme, B. & Gentillini, M. (1982) Biosynthesis of A, H and Lewis blood group determinants in *Fasciola hepatica*. *Journal of Parasitology*, **68**, 402–407.

Blankespoor, H. D. (1974) Host-induced variation in *Plagiorchis noblei* Park, 1936 (Plagiorchiidae: Trematoda). *American Midland Naturalist*, **92**, 415–433.

Blouin, M. S., Dame, J. B., Tarrant C. A. & Courtney, C. H. (1992) Unusual population genetics of a parasitic nematode: mtDNA variation within and among populations. *Evolution*, **46**, 470–476.

Blouin, M. S., Yowell, C. A., Courtney, C. H. & Dame, J. B. (1995) Host movement and the genetic structure of populations of parasitic nematodes. *Genetics*, **141**, 1007–1014.

Bowles, J., Blair, D. & McManus, D. P. (1995) A molecular phylogeny of the genus *Echinococcus*. *Parasitology*, **110**, 317–328.

Brooks, D. R. & Glen, D. R. (1982) Pinworms and primates: a case study in coevolution. *Proceedings of the Helminthological Society of Washington*, **49**, 76–85.

Brooks, D. R. & McLennan, D. A. (1991) *Phylogeny, Ecology, and Behavior: A Research Program in Comparative Biology.* Chicago: University of Chicago Press.

Brooks, D. R. & McLennan, D. A. (1993) *Parascript: Parasites and the Language of Evolution*. Washington, DC: Smithsonian Institution Press.

Buchholz, R. (1995) Female choice, parasite load and male ornamentation in wild turkeys. *Animal Behavior*, **50**, 929–943.

Bull, C. M., Andrews, R. H. & Adams, M. (1984) Patterns of genetic variation in a group of parasites, the Australian reptile ticks. *Heredity*, **53**, 509–525.

Bull, J. J. & Molineaux, I. J. (1992) Molecular genetics of adaptation in an experimental model of cooperation. *Evolution*, **46**, 882–895.

Bull, J. J., Molineaux, I. J. & Rice, W. R. (1991) Selection of benevolence in a host–parasite system. *Evolution*, **45**, 875–882.

Bullini, L., Nascetti, G., Paggi, L., Orecchia, P., Mattiucci, S. & Berland, B. (1986) Genetic variation of ascaridoid worms with different life cycles. *Evolution*, **40**, 437–440.

Bush, G. L. (1969) Sympatric host race formation and speciation in frugivorous flies of the genus *Rhagoletis* (Diptera, Tephritidae). *Evolution*, **23**, 237–251.

Bush, G. L. (1974) The mechanism of sympatric host race formation in the true fruit flies (Tephritidae). In *Genetic Mechanisms of Speciation in Insects*, ed. M. J. D. White, pp. 3–23. Brookvale: Australia and New Zealand Book Co.

Bush, G. L. (1975) Sympatric speciation in phytophagous parasitic insects. In *Evolutionary Strategies of Parasitic Insects and Mites*, ed. P. W. Price, pp. 187–206. New York: Plenum Press.

Calow, P. (1983) Patterns and paradox in parasite reproduction. *Parasitology*, **86**, 197–207.

Capron, A., Biguet, J., Vernes, A. & Afchain, D. (1968) Structure antigénique des helminthes: aspectes immunologiques des relations hôte–parasite. *Pathologie et Biologie*, Paris, **16**, 121–138.

Chakraborty, R., Fuerst, P.A. & Nei, M. (1980) Statistical studies on protein polymorphism in natural populations. III. Distribution of allele frequencies and the number of alleles per locus. *Genetics*, **94**, 1039–1063.

Chandler, M. & Cabana, G. (1991) Sexual dichromatism in North American freshwater fish: do parasites play a role? *Oikos*, **60**, 322–328.

Clayton, D. H. & Tompkins, D. M. (1994) Ectoparasite virulence is linked to mode of transmission. *Proceedings of the Royal Society of London, B*, **256**, 211–217.

Damian, R. T. (1979) Molecular mimicry in biology adaptation. In *Host–Parasite Interfaces*, ed. B. B. Nickol, pp. 103–126. New York: Academic Press.

Darwin, C. (1871) *The Descent of Man, and Selection in Relation to Sex*. London: John Murray.

Davies, C. M., Webster, J. P., Krüger, O., Munatsi, A., Ndamba, J. & Woolhouse, M. E. J. (1999) Host–-parasite population genetics: a cross-sectional comparison of *Bulinus globosus* and *Schistosoma haematobium*. *Parasitology*, **119**, 295–302.

Doherty, P. C. & Zinkernagel, R. (1975) Enhanced immunologic surveillance in mice heterozygous at the H-2 gene complex. *Nature*, **256**, 50–52.

Downes, B. J. (1990) Host-induced morphology in mites: implications for host-parasite coevolution. *Systematic Zoology*, **39**, 162–168.

Dybdahl, M. F. & Lively, C. M. (1996) The geography of coevolution: comparative population structures for a snail and its trematode parasite. *Evolution*, **50**, 2264–2275.

Dybdahl, M. F. & Lively, C. M. (1998) Host–parasite coevolution: evidence for rare advantage and time lagged selection in a natural population. *Evolution*, **52**, 1057–1066.

Ebbert, M. A. (1991) The interaction phenotype in the *Drosophila willistoni*–spiroplasma symbiosis. *Evolution*, **45**, 971–988.

Ebert, D. & Herre, E. A. (1996) The evolution of parasitic diseases. *Parasitology Today*, **12**, 96–101.

Ehrlich, P. R. & Raven, P. H. (1964) Butterflies and plants: a study in coevolution. *Evolution*, **18**, 586–608.

Ewald, P. W. (1983) Host–parasite relations, vectors, and the evolution of disease severity. *Annual Review of Ecology and Systematics*, **14**, 465–485.

Ewald, P. W. (1994) *Evolution of Infectious Disease*. New York: Oxford University Press.

Ewald, P. W. (1995) The evolution of virulence: a unifying link between parasitology and ecology. *Journal of Parasitology*, **81**, 659–669.

Folstad, I. & Karter, A. J. (1992) Parasites, bright males, and the immunocompetence handicap. *American Naturalist*, **139**, 603–622.

Folstad, I., Arneberg, P. & Karter, A. J. (1996) Antlers and parasites. *Oecologia*, **105**, 556–558.

Futuyma, D. J. (1986) *Evolutionary Biology*, 2nd edn. Sunderland, MA: Sinauer Associates.

Glen, D. R. & Brooks, D. R. (1986) Parasitological evidence pertaining to the phylogeny of the hominoid primates. *Biological Journal of the Linnean Society*, **27**, 331–354.

Gooding, R. H. (1992) Genetic variation in tsetse flies and implications for trypanosomiasis. *Parasitology Today*, **8**, 92–95.

Grant, W. N. (1994) Genetic variation in parasitic nematodes and its implications. *International Journal of Parasitology*, **24**, 821–830.

Guégan, J.-F. & Agnèse, J.-F. (1991) Parasite evolutionary events inferred from host phylogeny: the case of *Labeo* species (Teleostei, Cyprinidae) and their dactylogyrid parasites (Monogenea, Dactylogyridae). *Canadian Journal of Zoology*, **69**, 595–603.

Hafner, M. S. & Nadler, S. A. (1988) Phylogenetic trees support the coevolution of parasites and their hosts. *Nature*, **332**, 258–259.

Hafner, M. S. & Nadler, S. A. (1990) Cospeciation in host–parasite assemblages: comparative analysis of rates of evolution and timing of cospeciation events. *Systematic Zoology*, **39**, 192–204.

Hafner, M. S., Sudman, P. D., Villablanca, F. X., Spradling, T. A., Demastes, G. W. & Nadler, S. A. (1994) Disparate rates of molecular evolution in cospeciating hosts and parasites. *Science*, **265**, 1087–1090.

Haldane, J. B. S. (1949) Disease and evolution. *La Ricerca Scientifica* (Supplement), **19**, 68–76.

Haldane, J. B. S. (1954) The statics of evolution. In *Evolution as a Process*, ed. J. Huxley, A. C. Hardy & E. B. Ford, pp. 109–121. London: Allen & Unwin.

Hamilton, W. D. (1980) Sex versus non-sex versus parasite. *Oikos*, **35**, 282–290.

Hamilton, W. D. (1990) Mate choice near or far. *American Zoologist*, **30**, 341–352.

Hamilton, W. D. & Zuk, M. (1982) Heritable true fitness and bright birds: a role for parasites? *Science*, **218**, 384–387.

Hamilton, W. D., Axelrod, R. & Tanese, R. (1990) Sexual reproduction as an adaptation to resist parasites (a review). *Proceedings of the National Academy of Sciences, USA*, **87**, 3566–3573.

Herre, E. A. (1993) Population structure and the evolution of virulence in nematode parasites of fig wasps. *Science*, **259**, 1442–1445.

Hilburn, L. R. & Sattler, P. W. (1986*a*) Electrophoretically detectable protein variation in natural populations of the lone star tick, *Amblyomma americanum* (Acari: Ixodidae). *Heredity*, **56**, 67–74.

Hilburn, L. R. & Sattler, P. W. (1986*b*) Are tick populations really less variable and should they be? *Heredity*, **57**, 113–117.

Hill, A. V. S. (1996) Genetic susceptibility to malaria and other infectious diseases: from the MHC to the whole genome. *Parasitology* (Supplement), **112**, S75–S84.

Hoberg, E. P., Brooks, D. R. & Siegel-Causey, D. (1997) Host–parasite co-speciation: history, principles, and prospects. In *Host–Parasite Evolution: General Principles and Avian Models*, ed. D. H. Clayton & J. Moore, pp. 212–235. Oxford: Oxford University Press.

Holmes, J. C. (1983) Evolutionary relationships between parasitic helminths and their hosts. In *Coevolution*, ed. D. J. Futuyma & M. Slatkin, pp. 161–185. Sunderland, MA: Sinauer Associates.

Houck, M. A. (1994) Mites as potential horizontal transfer vectors of eukaryotic mobile genes: *Proctolaelaps regalis* as a model. In *Mites, Ecological and Evolutionary Analyses of Life-History Patterns*, ed. M. A. Houck, pp. 45–69. New York: Chapman & Hall.

Houck, M. A., Clark, J. B., Peterson, K. R. & Kidwell, M. G. (1991) Possible horizontal transfer of *Drosophila* genes by the mite *Proctolaelaps regalis*. *Science*, **253**, 1125–1129.

Howard, R. S. & Lively, C. M. (1994) Parasitism, mutation accumulation and the maintenance of sex. *Nature*, **367**, 554–557.

Howell, M. J. (1985) Gene exchange between hosts and parasites. *International Journal for Parasitology*, **15**, 597–600.

Huelsenbeck, J. P. & Rannala, B. (1997) Phylogenetic methods come of age: testing hypotheses in an evolutionary context. *Science*, **276**, 227–232.

Huelsenbeck, J. P., Rannala, B. & Yang, Z. (1997) Statistical tests of host–parasite cospeciation. *Evolution*, **51**, 410–419.

Hughes, A. L., Hughes, M. K., Howell, C. Y. & Nei, M. (1997) Natural selection at the class II major histocompatibility complex loci of mammals. In *Infection, Polymorphism and Evolution*, ed. W. D. Hamilton & J. C. Howard, pp. 89–97. London: Chapman & Hall for the Royal Society.

Iwamura, Y., Irie, Y., Kominami, R., Nara, T. & Yasuraoka, K. (1991) Existence of host-related DNA sequences in the schistosome genome. *Parasitology*, **102**, 397–403.

Iwasa, Y., Pomiankowski, A. & Nee, S. (1991) The evolution of costly mate preferences. II. The 'handicap' principle. *Evolution*, **45**, 1431–1442.

Jaenike, J. (1978) An hypothesis to account for the maintenance of sex within populations. *Evolutionary Theory*, **3**, 191–194.

Janzen, D. H. (1980) When is it coevolution? *Evolution*, **34**, 611–612.

Jeon, K. W. (1972) Development of cellular dependence of infective organisms: microsurgical studies in amoebas. *Science*, **176**, 1122–1123.

Jeon, K. W. (1991) Amoeba and x-bacteria: symbiont acquisition and possible species change. In *Symbiosis as a Source of Evolutionary Innovation: Speciation and Morphogenesis*, ed. L. Margulis & R. Fester, pp. 118–131. Cambridge: MIT Press.

Jermy, T. (1984) Evolution of insect/host plant relationships. *American Naturalist*, **124**, 609–630.

Kidwell, M. G. (1993) Lateral transfer in natural populations of eukaryotes. *Annual Review of Genetics*, **27**, 235–256.

Klassen, G. H. & Beverley-Burton, M. (1987) Phylogenetic relationships of *Ligictaluridus* spp. (Monogenea: Ancyrocephalidae) and their ictalurid (Siluriformes) hosts: an hypothesis. *Canadian Journal of Zoology*, **54**, 84–90.

Klassen, G. H. & Beverley-Burton, M. (1988) North American freshwater ancyrocephalids (Monogenea) with articulating haptoral bars: host–parasite coevolution. *Systematic Zoology*, **37**, 179–189.

Klein, J. & O'Huigin, C. (1997) MHC polymorphism and parasites. In *Infection, Polymorphism and Evolution*, ed. W. D. Hamilton & J. C. Howard, pp. 81–88. London: Chapman & Hall for the Royal Society.

Kohler, S., Delwiche, C. F., Denny, P. W., Tilney, L. G., Webster, P., Wilson, R. J. M., Palmer, J. D. & Roos, D. S. (1997) A plastid of probably green algal origin in apicomplexan parasites. *Science*, **275**, 1485–1489.

Ladle, R. J. (1992) Parasites and sex: catching the Red Queen. *Trends in Ecology and Evolution*, **7**, 405–408.

Leslie, J. F., Cain, G. D., Meffe, G. K. & Vrijenhoek, R. (1982) Enzyme polymorphism in *Ascaris suum* (Nematoda). *Journal of Parasitology*, **68**, 576–587.

Liggett, A. C., Harvey, I. F. & Manning, J. T. (1993) Fluctuating asymmetry in *Scathophaga stercoraria* L.: successful males are more symmetrical. *Animal Behavior*, **45**, 1041–1043.

Lively, C. M. (1992) Parthenogenesis in a freshwater snail: reproductive assurance versus parasitic release. *Evolution*, **46**, 907–913.

Lively, C. M. & Howard, R. S. (1997) Selection by parasites for clonal diversity and mixed mating. In *Infection, Polymorphism and Evolution*, ed. W. D. Hamilton & J. C. Howard, pp. 1–11. London: Chapman & Hall for the Royal Society.

Lively, C. M., Craddoci, C. & Vrijenhoek, R. C. (1990) Red Queen Hypothesis supported by parasitism in clonal and sexual fish. *Nature*, **344**, 864–866.

Lydeard, C., Mulvey, M., Aho, J. M. & Kennedy, P. K. (1989) Genetic variability among natural populations of the liver fluke *Fascioloides magna* in white-tailed deer, *Odocoileus virginianus*. *Canadian Journal of Zoology*, **67**, 2021–2025.

Lymbery, A. J., Constantine, C. C. & Thompson, R. C. A. (1997) Self-fertilization without genomic or population structuring in a parasitic tapeworm. *Evolution*, **51**, 289–294.

Macdonald, C. A. & Brooks, D. R. (1989) Revision and phylogenetic analysis of the North American species of *Telorchis* Luhe, 1899 (Cercomeria: Trematoda: Digenea: Telorchiidae). *Canadian Journal of Zoology*, **67**, 2301–2320.

Margulis, L. & Bermudes, D. (1985) Symbiosis as a mechanism of evolution: status of the cell symbiosis theory. *Symbiosis*, **1**, 101–124.

Margulis, L. & Fester, R. (1991) *Symbiosis as a Source of Evolutionary Innovation: Speciation and Morphogenesis*. Cambridge: MIT Press.

May, R. M. & Anderson, R. M. (1990) Parasite–host coevolution. *Parasitology* (Supplement), **100**, S89–S101.

Michaeli, D., Senyk, G., Maoz, A. & Fuchs, S. (1972) *Ascaris* cuticle collagen and mammalian collagens: cell mediated and humoral immunity relationships. *Journal of Immunology*, **109**, 103–109.

Mock, B. A. & Gill, D. E. (1984) The infrapopulation dynamics of trypanosomes in red-spotted newts. *Parasitology*, **88**, 267–282.

Mode, C. J. (1958) A mathematical model for the co-evolution of obligate parasites and their hosts. *Evolution*, **12**, 158–165.

Møller, A. P. (1990) Parasites and sexual selection: current status of the Hamilton and Zuk hypothesis. *Journal of Evolutionary Biology*, **3**, 319–328.

Møller, A. P. (1992*a*) Female swallow preference for symmetrical male sexual ornaments. *Nature*, **357**, 238–240.

Møller, A. P. (1992*b*) Parasites differentially increase the degree of fluctuating asymmetry in secondary sexual characters. *Journal of Evolutionary Biology*, **5**, 691–699.

Møller, A. P. (1993) Female preference for apparently symmetrical male sexual ornaments in the barn swallow *Hirundo rustica*. *Behavioral Ecology and Sociobiology*, **32**, 371–376.

Møller, A. P. (1994) Sexual selection in the barn swallow (*Hirundo rustica*). IV. Patterns of fluctuating asymmetry and selection against asymmetry. *Evolution*, **48**, 658–670.

Møller, A. P. (1996) Developmental stability of flowers, embryo abortion, and developmental selection in plants. *Proceedings of the Royal Society of London, B.*, **263**, 53–56.

Møller, A. P. (1997) Developmental stability and fitness: a review. *American Naturalist*, **149**, 916–932.

Møller, A. P. & Saino, N. (1994) Parasites, immunology of hosts, and host sexual selection. *Journal of Parasitology*, **80**, 850–858.

Moritz, C., McCallum, H., Donnellan, S. & Roberts, J. D. (1991) Parasite loads in parthenogenetic and sexual lizards (*Heteronotia binoei*): support for the Red Queen Hypothesis. *Proceedings of the Royal Society of London, B*, **244**, 145–149.

Mulvey, M., Aho, J. M., Lydeard, C., Leberg, P. L. & Smith, M. H. (1991) Comparative population genetic structure of a parasite (*Fascioloides magna*) and its definitive host. *Evolution*, **45**, 1628–1640.

Nadler, S. A. (1987) Genetic variability in endoparasitic helminths. *Parasitology Today*, **3**, 154–155.

Nadler, S. A. (1990) Molecular approaches to studying helminth population genetics and phylogeny. *International Journal for Parasitology*, **20**, 11–29.

Nadler, S. A. (1995) Microevolution and the genetic structure of parasite populations. *Journal of Parasitology*, **81**, 395–403.

Nadler, S. A., Hafner, M. S., Hafner, J. C. & Hafner, D. J. (1990) Genetic differentiation among chewing louse populations (Mallophaga: Trichodectidae) in a pocket gopher contact zone (Rodentia: Geomyidae). *Evolution*, **44**, 942–951.

Nadler, S. A., Lindquist, R. L., & Near, T. J. (1995) Genetic structure of midwestern *Ascaris suum* populations: a comparison of isoenzyme and RAPD markers. *Journal of Parasitology*, **81**, 385–394.

Nascetti, G., Cianchi, R., Mattiucci, S., D'Amelio, S., Orecchia, P., Paggi, L., Brattey, J., Berland, B., Smith, J. W. & Bullini, L. (1993) Three sibling species within *Contracaecum osculatum* (Nematoda, Ascaridida, Ascaridoidea) from the Atlantic Arctic–Boreal region: reproductive isolation and host preferences. *International Journal for Parasitology*, **23**, 105–120.

Nevo, E. (1978) Genetic variation in natural populations: patterns and theory. *Theoretical Population Biology*, **13**, 121–177.

Nevo, E., Beiles, A. & Ben-Shlomo, R. (1984) The evolutionary significance of genetic diversity: ecological, demographic and life history correlates. In *Evolutionary Dynamic of Genetic Diversity, Lecture Notes in Biomathematics*, vol. 53, ed. G. S. Mani, pp. 13–213. Berlin: Springer-Verlag.

Page, R. D. M. (1993) Genes, organisms, and areas: the problem of multiple lineages. *Systematic Zoology*, **42**, 77–84.

Page, R. D. M. (1996) Temporal congruence revisited: comparison of mitochondrial DNA sequence divergence in cospeciating pocket gophers and their chewing lice. *Systematic Biology*, **46**, 151–167.

Paggi, L., Nascetti, G., Cianchi, R., Orecchia, P., Mattiucci, S., D'Amelio, S., Berland, B., Brattey, J., Smith, J. W. & Bullini, L. (1991) Genetic evidence for three species within *Pseudoterranova decipiens* (Nematoda, Ascaridida, Ascaridoidea) in the North Atlantic and Norwegian and Barents Seas. *International Journal for Parasitology*, **21**, 195–212.

Parsons, P. A. (1990) Fluctuating asymmetry: an epigenetic measure of stress. *Biological Reviews*, **65**, 131–145.

Paterson, A. M. & Gray, R. D. (1997) Host–parasite co-speciation, host switching, and missing the boat. In *Host–Parasite Evolution: General Principles and Avian Models*, ed. D. H. Clayton & J. Moore, pp. 236–250. Oxford: Oxford University Press.

Polak, M. (1993) Parasites increase fluctuating asymmetry of male *Drosophila nigrospiracula*: implications for sexual selection. *Genetica*, **89**, 255–265.

Polak, M. (1997*a*) Parasites, fluctuating asymmetry, and sexual selection. In *Parasites and Pathogens: Effects on Host Hormones and Behavior*, ed. N. E. Beckage, pp. 246–276. New York: Chapman & Hall.

Polak, M. (1997*b*) Ectoparasitism in mother causes higher positional fluctuating asymmetry in their sons: implications for sexual selection. *American Naturalist*, **149**, 955–974.

Polak, M. & Markow, T. A. (1995) Effect of ectoparasitic mites on sexual selection in a Sonoran Desert fruit fly. *Evolution*, **49**, 660–669.

Potts, W. K., Manning, C. J. & Wakeland, E. K. (1997) The role of infectious disease, inbreeding and mating preferences in maintaining MHC genetic diversity: an experimental test. In *Infection, Polymorphism and Evolution*, ed. W. D. Hamilton & J. C. Howard, pp. 99–108. London: Chapman & Hall for the Royal Society.

Poulin, R. (1995) Evolution of parasite life history traits: myths and reality. *Parasitology Today*, **11**, 342–345.

Poulin, R. (1996) The evolution of life history strategies in parasitic animals. *Advances in Parasitology*, **37**, 107–134.

Poulin, R. (1998) *Evolutionary Ecology of Parasites: From Individuals to Communities*. London: Chapman & Hall.

Powell, J. R. (1975) Protein variation in natural populations of animals. *Evolutionary Biology*, **8**, 79–119.

Powell, J. R. (1994) Molecular techniques in population genetics: a brief history. In *Molecular Ecology and Evolution: Approaches and Solutions*, ed. B. Schierwater, B. Streit, G. P. Wagner & R. DeSalle, pp. 131–156. Basel: Birkhauser Verlag.

Price, P. W. (1974) Strategies for egg production. *Evolution*, **28**, 76–84.

Price, P. W. (1977) General concepts on the evolutionary biology of parasites. *Evolution*, **31**, 405–420.

Price, P. W. (1980) *Evolutionary Biology of Parasites*. Princeton: Princeton University Press.

Read, A. F. (1988) Sexual selection and the role of parasites. *Trends in Ecology and Evolution*, **3**, 97–102.

Read, A. F. (1990) Parasites and the evolution of host sexual behavior. In *Parasitism and Host Behaviour*, ed. C. J. Barnard & J. M. Behnke, pp. 117–157. London: Taylor & Francis.

Read, A. F. (1994) The evolution of virulence. *Trends in Microbiology*, **2**, 73–76.

Rohde, K. (1990). Phylogeny of Platyhelminthes, with special reference to parasitic groups. *International Journal for Parasitology*, **20**, 979–1007.

Rosenqvist, G. & Johansson, K. (1995) Male avoidance of parasitized females explained by direct benefits in a pipefish. *Animal Behavior*, **49**, 1039–1045.

Roughgarden, J. (1976) Resource partitioning among competing species – a coevolutionary approach. *Theoretical Population Biology*, **9**, 388–424.

Roughgarden, J. (1979) *Theory of Population Genetics and Evolutionary Ecology: An Introduction*. New York: Macmillan.

Saino, N. & Møller, A. P. (1994) Secondary sexual characters, parasites and testosterone in the barn swallow, *Hirundo rustica*. *Animal Behavior*, **48**, 1325–1333.

Schrag, S. J., Mooers, A. O., Ndifon, G. T. & Read, A. F. (1994) Ecological correlates of male outcrossing ability in a simultaneous hermaphrodite snail. *American Naturalist*, **143**, 636–655.

Selander, R. K. & Kaufman, D. W. (1973) Genic variability and strategies of adaptation in animals. *Proceedings of the National Academy of Science, USA*, **70**, 1875–1877.

Sibly, R. M. & Calow, P. (1986) *Physiological Ecology of Animals: An Evolutionary Approach*. Oxford: Blackwell Scientific Publications.

Snell, G. D. (1968) The H-2 locus of the mouse: observations and speculations concerning its comparative genetics and its polymorphism. *Folia Biologica*, **14**, 335–358.

Stouthamer, R. & Luck, R. F. (1993) Influence of microbe-associated parthenogenesis on the fecundity of *Trichogramma deion* and *T. petiosum*. *Entomologia Experimentalis et Applicata*, **67**, 183–192.

Swaddle, J. P. & Cuthill, C. (1994*a*) Female zebra finches prefer males with symmetric chest plumage. *Proceedings of the Royal Society of London, B*, **258**, 267–271.

Swaddle, J. P. & Cuthill, C. (1994*b*) Preference for symmetric males by female zebra finches. *Nature*, **367**, 165–166.

Taylor, M. I., Turner, G. F., Robinson, R. L. & Stauffer, J. R. Jr (1998) Sexual selection, parasites and bower height skew in a bower-building cichlid fish. *Animal Behaviour*, **56**, 379–384.

Théron, A. & Combes, C. (1995) Asynchrony of infection

timing, habitat preference, and sympatric speciation of schistosome parasites. *Evolution*, **49**, 372–375.

Thompson, J. N. (1987) Symbiont-induced speciation. *Biological Journal of the Linnean Society*, **32**, 385–393.

Thompson, J. N. (1994) *The Coevolutionary Process*. Chicago: University of Chicago Press.

Thompson, R. C. A. (1982) Intraspecific variation and parasite epidemiology. In *Parasites – Their World and Ours*, ed. D. F. Mettrick & S. S. Desser, pp. 369–378. New York: Elsevier.

Thompson, R. C. A. & Lymbery, A. J. (1996) Genetic variability in parasites and host–parasite interactions. *Parasitology* (Supplement), **112**, S7–S22.

Thompson, R. C. A., Lymbery, A. J. & Constantine, C. C. (1995) Variation in *Echinococcus*: towards a taxonomic revision of the genus. *Advances in Parasitology*, **35**, 145–176.

Thornhill, R. (1992) Fluctuating asymmetry and the mating system of the Japanese scorpionfly, *Panorpa japonica*. *Animal Behavior*, **44**, 867–879.

Thornhill, R. & Sauer, P. (1992) Genetic sire effects on the fighting ability of sons and daughters and mating success of sons in a scorpionfly. *Animal Behavior*, **43**, 225–264.

Van Every, L. R. & Kritsky, D. C. (1992) Neotropical Monogenoidea. 18. *Anacanthorus* Mizelle and Price, 1965 (Dactylogyridae, Anacanthorinae) of piranha (Characoidea, Serrasalmidae) from the Central Amazon, their phylogeny, and aspects of host–parasite coevolution. *Proceedings of the Helminthological Society of Washington*, **59**, 52–75.

Wallis, G. P. & Miller, B. R. (1983) Electrophoretic analysis of the ticks *Ornithodoros* (*Pavlocskyella*) *erraticus* and *O.* (*P.*) *sonrai* (Acari: Argasidae). *Journal of Medical Entomology*, **20**, 570–571.

Ward, P. I. (1988) Sexual dichromatism and parasitism in British and Irish freshwater fish. *Animal Behavior*, **36**, 1210–1215.

Ward, P. I. (1989) Sexual showiness and parasitism in freshwater fish: combined data from several isolated water systems. *Oikos*, **55**, 428–429.

Werren, J. H., Skinner, S. W. & Huger, A. M. (1986) Male-killing bacteria in a parasitic wasp. *Science*, **231**, 990–992.

Wright, S. (1931) Evolution in Mendelian populations. *Genetics*, **16**, 97–159.

Wright, S. (1951) The genetical structure of populations. *Annals of Eugenics*, **15**, 323–354.

Zahavi, A. (1975) Mate selection – a selection for a handicap. *Journal of Theoretical Biology*, **53**, 205–214.

Zahler, M., Gothe, R. & Rinder, H. (1995) Genetic evidence against a morphologically suggestive conspecificity of *Dermacentor reticulatus* and *D. marginatus* (Acari: Ixodidae). *International Journal for Parasitology*, **25**, 1413–1419.

ADDITIONAL REFERENCES

Clayton, D. H. & Moore, J. (eds.) (1997) *Host–Parasite Evolution: General Principles and Avian Models*. Oxford: Oxford University Press.

Futuyma, D. J. & Slatkin, M. (eds.) (1983) *Coevolution*. Sunderland, MA: Sinauer Associates.

Otte, D. & Endler, J. A. (eds.) (1989) *Speciation and its Consequences*. Sunderland, MA: Sinauer Associates.

Glossary

acanthella: Larval stage of acanthocephalans; it develops between the acanthor and cystacanth stages.

acanthor: Acanthocephalan larva that hatches from the egg; it is the first larval stage.

acetabula: The sucker-like attachment structures on the scolex of some cestodes; they resemble the suckers on digeneans.

acetabulum: Another name for the ventral sucker on a digenean. It is a holdfast only.

acidosis: A physiological state in which the pH of the blood drops into the acid range. A serious metabolic disturbance leading to pathology.

aclid organ: Spinose structure located at the anterior end of the acanthor used to penetrate and migrate through the internal organs of the intermediate host.

active transmission: Movement of a parasite to a host by its own means. The parasite uses stored energy in the process.

aerobic: Organisms that require free oxygen for their respiration.

aesthetascs: A type of chemoreceptor usually found in the antennule of crustaceans that functions in food, mate, and probably host recognition.

agglutinate: The clumping together of particles by antibody binding to antigens on the surface of the particle.

aggregated distribution: A term that refers to the manner in which a parasite component population is distributed within a given host population; the variance is greater than the mean.

allergic: Allergies are symptomatic reactions to antigens present in the environment. The result of interactions between antigens and T-cells or antibody produced by an earlier primary exposure to the antigen.

allogenic: Organisms whose life cycles are completed in hosts not confined to a discrete system (usually aquatic).

allopatric cospeciation: Cospeciation of host and parasites due to a vicariant event.

alternation of generations: Life cycle in which sexual individuals alternate with asexual individuals.

ametabolous: Development without metamorphosis in which the newly hatched larva is a miniature of the adult and growth occurs by successive molts with no significant morphological changes.

amino acid: A carboxylic ($-COOH$) acid with one or more amine ($-NH_2$) groups. Amino acids are the basic building blocks used for the synthesis of peptides and proteins.

amphid: Sensory organ in each side of the anterior end of nematodes.

amphistome: The body form of a digenean in which the acetabulum is situated at the posterior end.

amphitropical: Distributed on both sides of the tropics.

anabolic: Metabolism that leads to the synthesis of macromolecules that are important energy stores or structural materials.

anadromous: Fishes that spend the adult phase of their life cycles in salt water but move up streams and rivers to spawn.

anaerobic: Organisms that cannot survive in the presence of free oxygen. Facultative anaerobes can live with or without oxygen.

anaphylactic shock: An allergic reaction due to binding of an antigen to IgE antibody that is bound to mast cells. This leads to the release of inflammatory mediators with circulatory collapse and possible suffocation due to swelling of the trachea.

anapolysis: In tapeworms, the release of a proglottid from a strobila after the eggs are shed.

anemia: A deficiency of red blood cells and hemoglobin in the blood.

annuli: Superficial, false segmentation of pentastomids.

anoxia: Oxygen depletion, often used in reference to depletion at the local tissue level.

Antarctic Convergence: An oceanic frontal system, located around the Antarctic Continent at 75° S that delimits the Antarctic from the sub-Antarctic. Cold masses of water from the Antarctic area submerge at the Antarctic Convergence and move north at greater depths.

antennae: Second pair of appendages in Crustacea.

antennules: The most anterior pair of appendages in Crustacea. Homologous to the antennae of insects.

anthropozoonotic: A human infectious disease than can be transmitted to a variety of reservoir hosts.

antibody: Large-molecular-weight plasma protein that is produced in response to immunization by an antigen.

antigen: A large-molecular-weight molecule that elicits a specific immune response.

apical complex: A complex of anteriorly located organelles characteristic in some life cycle stages of all apicomplexan species.

apolysis: In tapeworms, the release of an egg-containing (gravid) proglottid from a strobila.

arrest: Some nematodes, in resistant hosts, have 3rd stage, unsheathed larvae that persist in the tissues of the host. In female hosts, these larvae may be mobilized for transmammary or transplacental transmission.

autecology: That aspect of ecology dealing with structure and function of individual organisms or species.

autogenic: Organisms that complete their life cycles within the confines of a system (usually aquatic).

axostyle: A prominent, rod-like, supporting ultrastructure composed of microtubules arranged in ribbons or sheets and usually extending from the anterior to the posterior end of the cell. Found in the Zoomastigophorea.

bacillus: A rod-shaped bacterial cell.

basal body: The same as kinetosome or centriole.

B-cell: One of the two major classes of lymphocytes. B-cells, after activation by an antigen, differentiate into cells that produce antibody to that antigen.

binary fission: The simple division of one cell into two daughter cells. Before division, the organelles of the original cell are duplicated.

biogeography: Science that attempts to describe and explain the patterns in the geographical distribution of organisms.

blastocyst: That portion of a larval tapeworm into which the body may be withdrawn.

bothria: A grooved attachment device on the scolex of certain cestodes, e.g., pseudophyllideans.

bradyzoite: A slowly growing merozoite. The term is generally used to describe some stages in the life cycle of certain coccidian sporozoa, e.g., *Toxoplasma*.

buccal cone: Section of the mouthpart of mites and ticks formed by the hypostome and labrum.

bulla: Non-living structure secreted by the head and maxillary glands of lernaeopodid copepods used to anchor the copepod to the host. The maxillae are permanently attached to this structure.

bursa: In some male nematodes and all male acanthocephalans the posterior part of the body is drawn out into a fleshy appendage (retractable in acanthocephalans, permanent in nematodes) that is used to grasp the female's body during copulation.

cachexia: A clinical condition in which there is a depletion of blood proteins, and fat deposits. Tissue proteins are eventually catabolized with the ultimate possibility of severe tissue atrophy and muscle wasting. There is a progressive weight loss.

cadre: Sclerotized mouth lining of pentastomids.

calotte: Anterior end of a dicyemid mesozoan.

capitulum: The anterior region of the two basic body parts of mites and ticks. It carries the mouthparts and feeding appendages.

carapace: Structure formed by the extension of the dorsal sclerites of the head in many Crustacea, usually covering and/or fusing with one or more thoracic somites.

carbohydrate: Organic compounds consisting of a chain of carbon atoms to which hydrogen and oxygen atoms are attached in a 2:1 ratio. Carbohydrates include sugars, glycogen, cellulose, etc.

catabolism: Metabolism involved in the breakdown of organic molecules, usually leading to the formation of ATP needed for anabolic pathways.

catadromous: Fishes that spend most of their life cycle in fresh water but spawn in salt water.

catecholamine: Amine (—NH_2) containing compounds derived enzymatically from the amino acids phenylalanine and tyrosine. They include compounds such as epinephrine, norepinephrine, dopamine (DOPA), etc., which are important hormones and neurotransmitters.

cement glands: Glands in a male acanthocephalan that produce secretions to seal the female reproductive tract after copulation.

cephalization: The concentration of vital functions into the anterior end of the body, or head.

cercaria: An immature digenean, usually free swimming, produced by a sporocyst or a redia.

cercarial dermatitis: A localized hypersensitivity reaction in response to the penetration of avian and mammalian cercariae in the skin of humans. Also called 'swimmer's itch'.

cercomer: An embryonic structure on cestode procercoids on which occur the oncosphere's hooks.

chalimus: Specialized parasitic copepodid.

chelicerae: The anterior-most pair of appendages in chelicerate arthropods; used in feeding.

chemotactic: A chemical substance that induces host white blood cells or parasites to move towards, or away from, it.

chigger: Common name of mites in the family Trombiculidae.

chloride cells: Specialized epidermal ion-transporting cells found in pentastomids.

cilia: Short, hair-like, cylindrical organelles consisting of nine pairs of outer and two central microtubules arising from an internal base, the kinetosome. It is covered by the cell membrane. The cilium is usually involved in locomotion but may also be important in feeding or attachment to a substrate.

cline: Gradual change of a character in a population or species that parallels or correlates with a gradient in the environment.

clitellum: Glandular area on the body of oligochaetes and hirudineans that secretes a cocoon to enclose the eggs.

cloaca: The structure into which opens some combination of the terminal gut (i.e., anus), genital ducts, and excretory ducts.

clotting: A complex biochemical process in which red blood cells and platelets are clumped during the formation of a matrix of serum proteins.

clumped: Same as aggregated distribution.

cnidocyte: Specialized cell of cnidarians that produces a nematocyst. Also called a cnidocil.

coelomocyte: In nematodes, cells of unknown function found in the pseudocoel.

coenurus: A cysticercus-type cestode larval stage in possession of many scolices that remain attached to the bladder wall; always found in a mammalian intermediate host.

coevolution: Reciprocal evolutionary change in interacting species.

colonization: Establishment of a parasite species where none of that species was previously. An uninfected host is colonized when it gets an infection.

community: A group of populations of different species occupying a discrete area.

competition: The negative interaction between different individuals, or populations, involved in securing a mutually limiting resource, e.g., space, nutrients, etc.

complement: Plasma proteins that can be activated by antigen–antibody interactions, or by spontaneously interacting with molecules on the surface of a pathogen. Complement proteins aid in the direct killing of the pathogen or in the removal of the pathogen by phagocytes.

component population: All of the individuals of a specified life-history phase at a particular place and time.

concomitant immunity: Resistance to reinfection of a host by a specific parasite when the host is currently infected with that parasite.

conjugation: Reciprocal fertilization; a form of sexual reproduction (exchange of micronuclei) that occurs between members of different mating types in ciliates.

contractile vacuole: A cytoplasmic organelle involved in osmoregulation. It pulsates with a regular frequency emptying dissolved wastes and water to the exterior of the cell through a pore on the cell's surface.

copepodid: Juvenile stage that follows the naupliar stage in copepods.

copulatory bursa: See **bursa**.

copulatory cap: Cement-like cap applied to the genital opening of the female by a male acanthocephalan following copulation.

coracidium: The free-swimming, ciliated larva of certain cestodes.

corona: Specialized ciliated area found in the anterior end of rotifers. Also called a 'wheel organ'.

cospeciation: Speciation by one species in response to the speciation of another.

cotylocidium: The free-living, partially ciliated, larval stage of an aspidobothrean platyhelminth.

coxal glands: The excretory organs of chelicerates. A sac, a tubule, and an opening on the coxa form the gland.

cryptogonochoristic sexuality: Condition in which the sexes are separate, but hidden, giving the appearance of hermaphroditism, as in rhizocephalan cirripedes.

ctenidia: Series of stiff, peg-like spines on the head and body of many fleas.

cuticulin: Protein component of the arthropod cuticle or exoskeleton.

cypris or **cyprid**: Larval stage that follows the nauplius stage in cirripedians.

cyst: An inactive, dormant stage in the life cycle of many protozoans. It is considered more environmentally resistant than other stages in the life cycle and is often the stage important in transmission of the parasite.

cystacanth: Juvenile stage of acanthocephalans that is infective to the definitive host.

cysticercoid: A cestode larval stage in possession of a non-invaginated scolex and a 'tail' to which may be attached the embryonic hooks; almost always found in invertebrate intermediate hosts.

cysticercus: A cestode larval stage possessing an invaginated scolex and a bladder; always found in a mammalian intermediate host.

cytokines: Cytokines are chemical messengers (proteins) made by cells that influence the behavior of other cell types. Cytokines produced by lymphocytes are also referred to as lymphokines or interleukins (IL-1, 2, etc.).

cyton: The body of the syncytial tegument of parasitic

flatworms containing the nuclei and other organelles.

cytopharynx: A tubular passageway going from the cytostome into the interior cytoplasm of ciliates. Normally, food vacuoles are formed at the lower end of the cytopharynx.

cytoproct: In ciliates, it is equivalent to the cell's anus. Normally, a permanent opening in the cell surface at the posterior end of the cell through which solid wastes are excreted.

cytostome: In ciliates, the cell's mouth; usually a permanent opening through which food passes into the inner cytoplasm via the cytopharnyx.

decomposer: Any heterotrophic organism at the bottom of the food chain that breaks down dead and decaying material of organic origin. Decomposers keep the world clean of organic material and allow for the recycling of elements such as carbon.

definitive host: That host in a parasite's life cycle in which the parasite reaches sexual maturity. Also called a 'final' host.

deirid: Sensory papilla on either side near the anterior end of some nematodes.

delayed cospeciation: Process of change in which the speciation of one of the members of the association is delayed and lags behind the speciation of the other.

deme: A local population of a species; usually a small, panmictic population.

dendritic cell: Mononuclear phagocytic cells that originate from bone marrow precursors. They are present in a variety of lymphoid and non-lymphoid tissues and are important in antigen presentation to T- and B-cells. They are named according to their tissue location, e.g., dendritic cells in the skin are called Langerhans cells.

density-dependent: Attributes such as growth, fecundity, etc. that are affected by population density.

density-independent: Attributes such as growth, fecundity, etc. that are unaffected by population density.

dermotropic: Host white blood cells, or parasites, that have a predilection for host dermal tissue.

detritivore: An animal that feeds on dead and decaying organic material.

developmental arrest: A form of diapause generally seen in 3rd stage larvae of certain nematodes. (See **arrest**.)

diapause: A brief interruption, genetically programmed, in the developmental progression of an organism.

dichromatism: The presence of two color patterns, one in females and the other in males.

didelphic: A condition found in some female nematodes where most reproductive structures (e.g., ovaries, oviducts, uteri) are doubled.

diporpa: A stage in the life cycle of *Diplozoon*, a bizarre ectoparasitic monogenean.

direct life cycle: The life cycle of a parasite where only one host is required for completion.

dispersal: The process of breaking up from an assembled state and moving away from a source.

dispersal biogeography: Branch of historical biogeography that attempts to explain the distribution of organisms based on the premise that the organisms themselves disperse across barriers by their own means.

distome: The body form of a digenean with two suckers, one anteriorly (the oral sucker), and another located mid-ventrally (the ventral sucker or acetabulum).

dorsal organ: Glandular structure characteristic of pentastomid larva. It secretes a mucoid substance that flows through the facette and becomes part of the first protective egg layer surrounding the larva.

drift (= genetic drift): Random changes in the frequencies of two or more alleles or genotypes within a population.

ecdysis: Shedding of the exoskeleton during the life cycle of some invertebrates. Often under hormonal control.

ecological biogeography: Branch of biogeography that attempts to understand the distribution patterns of organisms in terms of ecological processes occurring over short temporal and spatial scales.

ecosystem: The sum total of a biotic community and potentially interacting abiotic factors within a predefined region of interest.

ecsoma: A retractile 'tail' of some hemiurid flukes that functions as a feeding organ.

ectoparasite: A parasitic organism that lives on the surface of its host.

endemic (= endemicity): Confined to a particular geographical region.

endodyogeny: Formation of two daughter cells, each surrounded by its own membrane, while inside the mother cell. Similar to endopolyogeny but only two daughter cells are produced.

endoparasite: A parasitic organism that lives inside its host.

endopolyogeny: Formation of multiple daughter cells,

each surrounded by its own membrane, while inside the mother cell.

endotoxin: Bacterial toxins that are only released following bacterial cell damage.

enzyme: Normally refers to protein molecules that act as chemical catalysts in biological systems.

eosinophil: Polymorphonuclear cells containing eosinophilic granules. They have receptors for IgE and are prominent in allergic reactions and in some parasitic infections.

epidemiology: A term used to describe the biology of a disease process affecting humans.

epigametic sex determination: Type of sex determination in an individual in which the environment determines the sex of the individual.

epitope: A small segment of a large macromolecule that activates the immune response and ultimately interacts with specific antibody or T-cells.

epizootic: When a large number of non-human animals in a particular area or region is infected by a disease. When the disease circulates among humans, it is an epidemic.

epizootiology: A term used to describe the biology of a disease process affecting animals other than humans.

erythropoiesis: The formation of red blood cells.

eukaryotic: Cells with a nuclear membrane, mitotic spindle, a variety of internal organelles, structurally complex flagella or cilia, etc. Eukaryotes include the algae, protozoa, fungi, and all higher animal and plant cells.

euryhaline: Organisms with a wide tolerance to varying levels of salinity.

eutely: Cell or nuclear constancy.

evolutionary arms race: Model of coevolution in which there are mutual adaptive responses between the organisms involved. (See **Red Queen Hypothesis**.)

exploitative competition: The interaction between two species or individuals that is mediated indirectly through the use of a common resource which is in limited supply.

externa: The external reproductive structure of the female rhizocephalan body.

extrusome: A general term used to describe a variety of membrane-bound structures located beneath the pellicle and which can be released to the cell's exterior upon appropriate stimulation.

facette: Funnel-shaped opening in the inner membrane of the egg through which secretions, produced by the dorsal organ of the larva of pentastomes, flow.

facultative parasite: A parasite in which the adult stages may be free-living or use a host, usually depending on the vagaries of certain environmental conditions.

fatty acids: Long-chain hydrocarbons, a component of lipids. Fatty acids can be either saturated, or unsaturated.

flagellum: A long, thread-like structure that has nine pairs of outer and two central microtubules. Essentially, a long cilium. The flagellum, or flagella, can be used in locomotion and feeding. The eukaryotic flagellum is not structurally homologous to the prokaryotic flagellum.

fluctuating asymmetry: Small, random deviations from perfect bilateral symmetry in a given morphological trait.

founder effect: A type of genetic drift that occurs as the result of the founding of a population by a small number of individuals carrying just a fraction of the total genetic variation of the source population.

fundamental niche: The maximum possible distribution of an organism in an *n*-dimensional hypervolume. Often, limiting factors, whether abiotic or biotic, limit the distribution to a subset of the fundamental niche. (See **realized niche**.)

gametogony: The portion of the life cycle during which gametes are formed.

gasterostome: The body form of a digenean in which the oral sucker is located toward the mid-ventral area and the acetabulum is absent.

gene flow: The movement of alleles into, or out of, a population.

gene-for-gene hypothesis: Hypothesis proposing that, for each gene causing resistance in a host, there is a matching gene coding for avirulence in a parasite.

gene frequency: The proportion of a specific gene in a population.

genetic drift: Random changes in the frequencies of two or more alleles or genotypes within a population.

glochidium: Larval stage of some freshwater bivalves parasitic on the gills of fishes.

glycocalyx: A fuzzy, secreted layer external to the cell membrane and covering the cell surface (surface coat). The composition of the glycocalyx will depend upon the species of protozoan. It can consist of a variety of different macromolecules, e.g., polysaccharides and glycoproteins.

glycogen: A large polysaccharide synthesized from glucose. It is one of the main energy storage products in most animal cells. After hydrolysis, it is converted to glucose that can be used for the formation of ATP.

gnathosoma: The anterior region of the two basic body parts of mites and ticks. It carries the mouthparts and feeding appendages.

granuloma: A site of chronic inflammation usually caused by the continued presence of any infectious agent that results in tissue necrosis. The site contains numerous macrophages, T-cells, and fibroblasts. It may also include B-cells, neutrophils, and eosinophils.

gubernaculum: An accessory, cuticularized structure found in some male nematodes having spicules. It guides the spicules out through the cloaca.

guild: A group of organisms that exploit resources in a similar fashion. The nature of the exploitation and the resources used will determine the identity of the guild. The taxonomy of the organisms is irrelevant to the guild concept.

gynecophoric canal: The external grooved surface of male schistosomes in which the female resides *in copula*.

haemozoin: A malaria pigment, a by-product of hemoglobin metabolism that accumulates in the host's reticuloendothelial system during a malaria infection.

Haller's organ: A large set of different types of receptive sensilla clumped together on the first tarsi of ticks; it includes mecano-, thermo-, hygro- and chemoreceptors.

halteres: Vestigial pair of wings in dipterans; used for balance and flight stability.

handicap principle: In sexual selection, the development of secondary sexual characters is costly; the cost of producing the trait guarantees its reliability and prevents cheating because the cost of a unit of display is greater for low-quality individuals than for high-quality individuals.

haplodiploidy: A sex-determining system where unfertilized eggs give rise to haploid males and fertilized eggs give rise to diploid females.

hemimetabolous: Development with simple, or incomplete metamorphosis, in which the newly hatched larva resembles the adult, but lacks wings, gonads, and external genitalia; the larvae are called nymphs.

hemoglobinemia: A deficiency of hemoglobin in the blood.

hemoglobinuria: The presence of hemoglobin in the urine. In malaria infections, this is due to severe hemolysis associated with high parasitemias.

hemolysis: The lysis of red blood cells.

heterogonic: In contrast to some rhabditid nematodes having homogonic life cycles, these nematodes produce free-living, dioecious adults. Larvae from free-living females are infective.

heterozygosity: Average frequency of heterozygous individuals for a given gene locus in a population.

hexacanth: Also called an oncosphere; a Eucestoda embryo having three pairs of hooks.

hirudin: Anticoagulant produced by blood-sucking leeches to aid during their feeding.

historical biogeography: Branch of biogeography that attempts to understand the sequences of origin, dispersal, and extinction of organisms.

HIV: The human immunodeficiency virus. A retrovirus that infects CD-4 T-cells, and is the cause of the acquired immunodeficiency syndrome (AIDS).

holometabolous: Development with complete metamorphosis in which the immature stages (larva and pupa) are quite different from the adults.

holostome: The body form of a digenean which appears to be divided into two parts, a larger anterior region in which are located the suckers, and a smaller posterior region in which the genitalia are located.

homeostasis: The maintenance of a stable internal physiological environment or equilibrium in an organism.

homogonic: In some rhabditid nematodes, the organisms are protandric hermaphrodites (the individual is first a male, then a female) and there is no free-living adult generation.

horizontal gene transfer: Non-sexual transfer of genetic information between the genomes of different species.

hydatidosis: The disease associated with infection by hydatid cysts, larval stages of the cestode *Echinococcus*.

hydrogenosome: A single, membrane-bound cytoplasmic organelle characterized by containing hydrogenases. The hydrogenosome is found in the anaerobic protozoa, such as the trichomonads. It is involved in maintaining the cell's oxidation–reduction balance by the formation of H_2.

hyperparasite: A parasite that uses another parasite as a host.

hypersensitivity: Symptomatic reactions that are produced upon re-exposure to an antigen. Hypersensitivity is an inflammatory reaction resulting from antigen–antibody interactions (immediate hypersensitivity) or antigen-T–cell interactions (delayed-type hypersensitivity). A granuloma is the ultimate result of a localized, delayed-type, hypersensitivity reaction.

hypobiosis: When an unusually large number of 3rd or 4th stage nematode larvae remain in the tissues extending the prepatent period. Hypobiosis may be an important process allowing the parasites to survive periods when environmental conditions are unfavorable for larvae.

hypoglycemia: A physiological condition in which the blood glucose level is reduced.

hypopharynx: Tongue-like lobe arising from the floor of the mouth in insects.

hypostome: A portion of the mouthpart of mites and ticks formed by the fusion of the coxae of the pedipalps.

idiosoma: The posterior part of the two basic body parts of mites and ticks. It bears the legs and most internal organs.

immunocompetence handicap hypothesis: In sexual selection, secondary sexual characters develop in response to hormones that increase the expression of these traits but that reduce the ability of the immune system to fight parasites. This allows superior males to remain viable and relatively parasite-free despite their compromised immune system.

immunoglobulin: Large-molecular-weight plasma protein produced by B-cells. Immunoglobulins can be divided into five classes based upon differences in their heavy chains.

indirect life cycle: A parasite's life cycle in which two or more hosts are required for completion.

indolamine: Amine-containing compounds enzymatically derived from the amino acid tryptophan. The indolamines include such compounds as melatonin and serotonin that have important hormonal and neurotransmitter activity.

inflammation: The local accumulation of fluids, plasma proteins, and white blood cells produced as the result of an infection, an immune response, or physical injury.

infrapopulation: All of the organisms of a single species within a single host at a particular time.

infusoriform larvae: Ciliated larva produced by an infusorigen within a dicyemid mesozoan.

infusorigen: Mass of reproductive cells within a rhombogen, a developmental stage of dicyemids.

interactive site segregation: Refers to the specialization, or segregation, of niches by two species in which the realized niche of one, or both, is reduced by the presence of the second species.

interference competition: Direct confrontation between organisms. Can lead to competitive exclusion.

interleukins: A term for cytokines produced by leukocytes (See **cytokines**).

intermediate host: That host in a parasite's life cycle required by the parasite to complete its life cycle and in which some morphological change or development occurs.

interna: The internal nutrient-absorbing structure of the female rhizocephalan body.

ionocytes: Specialized, epidermal ion-transporting cells found in pentastomids.

Katayama fever: A high fever caused by *Schistosoma japonicum* during the acute phase of the disease.

kentrogon: Specialized larval stage in the rhizocephalan crustaceans that follows the cypris stage. This is the stage that attaches to the host.

kinete: A motile zygote, similar to the *Plasmodium* ookinete stage, occurs during the life cycle of the piroplasms. They are formed from the fusion of two gametes. Kinetes invade cells and, intracellularly, they form sporozoites.

kinetid: Repeating structural units of the ciliate cortex (also some flagellates). It consists basically of a kinetosome(s), the cilia, membranes in the area, as well as microtubules, microfilaments, etc. Sometimes included are extrusomes and other organelles.

kinetosome: A complex, subpellicular ultrastructure consisting of nine sets of three microtubules forming a tubular cylinder plus a core containing microfilaments. Distally, the kinetosome is continuous with the peripheral microtubules of the cilium or flagellum.

kinins: Family of straight-chain polypeptides generated by the enzymatic hydrolysis of plasma precursors. Kinins are potent vasodilators of most vessels in the body except for vasoconstrictors of the pulmonary bed. They also increase vascular permeability and promote leukocyte migration through small vessels into tissue spaces.

koilin lining: The hard lining comprising the grinding surface inside a bird's gizzard.

***K*-selection:** Selection that favors maintenance of the individual and production of few offspring. Tendency towards low fecundity, low mortality, longer life span and greater complexity of the individuals.

labium: Mouthpart of insects.

labrum: Plate forming an upper lip, or anterior closure, in the mouth in insects.

lacunar system: System of canals in the body wall of acanthocephalans. It functions as a circulatory system.

lamphredin: Anticoagulant produced by blood-sucking lampreys to aid during their feeding.
landscape epidemiology: The processes of population biology associated with spatial components of a parasite's transmission and colonization.
larval migrans: The tissue migration of a larval helminth usually in an abnormal host.
larval stem nematogen: Ciliated larva that is the earliest developmental stage of dicyemids found in juvenile cephalopods.
Laurer's canal: A non-functional tube connecting the surface of some digeneans to the seminal receptacle; it may be a vestigial vagina.
lemniscus: Sac-like, paired structure found attached to the inner margin of the neck of acanthocephalans.
ligament sac: Envelope surrounding the gonads and accessory organs in acanthocephalans.
lipids: A heterogeneous class of organic compounds, e.g., fats, waxes, phospholipids. They are important components of the cell membrane. Lipids are poorly soluble or are insoluble in water.
long slender stage: A rapidly dividing stage in the life cycle of the African trypanosomes in the mammalian host.
lycophore: The decacanth embryo of a cestodarian.
lymphocyte: Cells involved in the immune response that are recognized by specific surface antigens. There are two main classes of lymphocytes; the B-lymphocytes and the T-lymphocytes. (See **T-** and **B-cells**).
lymphokine: Cytokines produced by lymphocytes. (See also **cytokines** and **interleukins**.)
lysosome: A membrane-bound cytoplasmic organelle in eukaryotic cells that contains a variety of hydrolytic enzymes capable of digesting all types of bioorganic compounds.

macroevolution: Evolution above the species level, brought about by great phenotypic changes.
macrogamete: The larger of paired gametes, usually not flagellated, and generally considered the female gamete.
macroparasite: A parasite that is usually visible with the naked eye, e.g., helminths, arthropods.
macrophage: Large, migrating mononuclear phagocytic cells derived from bone marrow precursors. Important in innate immunity and as antigen presenting cells.
major histocompatibility complex (MHC): A cluster of genes in mammalian chromosomes, which encode the MHC proteins (or MHC antigens). There are two classes of MHC molecules. The class I proteins that present antigenic epitopes to CD-8, T-cells, and class II proteins that present antigenic peptides to the CD-4, T-cells. The MHC gene cluster is highly polymorphic with a number of alleles at different loci; it serves to identify 'self' from 'non-self'.
malaise: A term used to describe vague bodily discomfort. It may include feeling tired, loss of appetite, headache, nausea, etc.
Malpighian tubules: Elongated, blind, excretory tubes that arise near the anterior end of the hindgut and extend into the body cavity in many arthropods.
manca: Well-developed larval stage (juvenile) of some isopods.
mandibles: Paired appendages of arthropods used primarily for feeding.
mange: Dermatitis caused by several species of mites.
marine refugium hypothesis: Proposes that extreme environmental fluctuations during the Pliocene–Pleistocene glaciation promoted the diversification, and not the extinction, of some organisms.
marsupium: Specialized, ventral brooding chamber of isopods.
mast cell: Large cells, found in connective tissue throughout the body, that have an important role in allergic reactions. Mast cells have a high-affinity Fc receptor for IgE, as well as large granules that contain vasoactive amines. Interactions between the antigens and the Fc-bound IgE lead to the release of the vasoactive mediators and local or systemic allergic reactions.
mastax: Jaw-like structure found in the pharynx of rotifers forming a crushing organ.
matricidal endotoky: In some rhabditid nematodes with heterogonic cycles, and in many free-living rhabditid nematodes, the female produces a few large eggs that hatch and develop to sheathed 3rd stage larvae. These larvae consume her internal organs before breaking through her cuticle to become free-living.
maxillae: Paired appendages of Crustacea used for feeding.
maxilliped: Appendage of Crustacea used usually for feeding but also for attachment in some parasitic forms.
maxillules: Paired appendages of Crustacea used primarily for feeding.
mean: The sum of all observations divided by the number of observations.
Mehlis' gland: Specialized cells surrounding the ootype and which apparently play a role in the formation of the eggshell.

meningoencephalitis: Disease of the brain and spinal cord that involves the meninges, often with a lymphocytic response in the spinal fluid.

merogony: The portion of the protozoan life cycle involved in the formation of merozoites from a precursor cell.

merozoite: The final life cycle stage in the process of merogony. It develops into either a new meront following invasion of a new host cell or, at the appropriate time and place, into either a micro- or a macrogamete.

mesocercaria: A life-cycle stage between a cercaria and a metacercaria in which there is no developmental change.

mesoparasite: Parasites intermediate between ecto- and endoparasites. It generally applies to copepods, where part of the copepod body remains outside the host, and the remaining part, usually an anchoring device, reaches deep into the body cavity of the host.

meta-analysis: Analyses that make use of existing published data, often from multiple studies, and often including host–parasite checklists for hypothesis testing.

metacercaria: A developmental stage of digeneans between a cercaria and an adult; usually sequestered within a cyst in a second intermediate host.

metacyclic: The final stage of development of parasitic kinetoplastid flagellates during the invertebrate (or vector) phase of their life cycle. The metacyclic forms are infective for their vertebrate host.

metamerism: The division of the body into segments (metameres).

metamorphosis: A marked change in body form during development.

metazoan: Animals having more than one cell and having at least two tissue layers. In the strictest sense, the term also excludes the possession of choanocytes thus excluding the phylum Porifera.

metraterm: A muscular terminus of the uterus; present in some digeneans.

microaerophilic: Environment with limited (or only traces of) free oxygen available.

microevolution: Evolution at, or below, the species level brought about by changes in the gene frequencies of natural populations.

microgamete: The smaller of paired gametes in protozoans, often flagellated, and generally considered the male gamete.

microparasite: A parasite that usually requires a microscope to be seen, e.g., viruses, bacteria, protozoans.

microtriches: The finger-like projections on the tegument of a tapeworm or digenean.

miracidium: The ciliated, free-swimming larval stage of a digenean that emerges from an egg.

modular iteration: The process involved in forming a tapeworm strobila.

monostome: The body form of a digenean in which there is only an oral sucker; the acetabulum is absent.

monoxenous: A parasite having a single taxonomic species of host.

monozoic: The form of a tapeworm in which modular iteration, or strobilization, does not occur.

morbidity: A term to describe illness, or sickness, in animals.

multilocular hydatid cyst: A cysticercus-type cestode larval stage in possession of many scolices (= protoscolices) produced in so-called 'brood capsules' that float freely in the fluid of the parental cyst; asexual budding is both exogenous and endogenous.

multiple fission: The simultaneous division of a cell into many daughter cells.

mutation: An inheritable change in the DNA sequence of a chromosome; any spontaneous, heritable change in the genotype.

mycetome: Specialized organ of some insects that bears mutualistic bacteria.

myiasis: Infection caused by fly larvae (maggots).

myocyton: The non-contractile body of the nematode muscle.

natural selection: A process of interaction between organisms and their environment that results in the differential rate of production of different phenotypes in the population. The differential contribution of genotypes to the gene pool of the next generation under natural conditions.

nauplius: The earliest larval stage of crustaceans.

negative binomial model: The negative binomial model is given by the expression $(p-q)^k$ where $q=1+p$ and k is a positive exponent. The model is frequently used to describe population frequency distributions that are clumped or aggregated.

nematocyst: A threadlike stinging organelle formed within cnidocytes and characteristic of cnidarians.

nematogen: Stage in the life cycle of a dicyemid.

neotenic plerocercoid: The stages of several pseudophyllidean cestodes that become sexually mature as larvae.

nephrocytes: Specialized cells capable of picking up and accumulating waste products.

neutralize: The ability of an antibody molecule to inhibit the infectivity of an infectious agent, the toxicity of some molecules, or the enzymatic activity of an enzyme.

neutrophil: These are cells with a multilobed nucleus and neutrophilic granules. They are a major class of white blood cell in the peripheral blood of mammals and play a key role in innate immunity. They are phagocytic cells and are, therefore, involved in the killing of extracellular parasites.

nucleic acids: Large polymers of nucleotides arranged in unbranched chains. There are two classes of nucleic acids, deoxyribonucleic acid (DNA) and ribonucleic acid (RNA). They are similar in composition except that in RNA, ribose replaces deoxyribose as the sugar portion of the nucleotide, and also that the pyrimidine uracil replaces thymine.

nymphs: Juvenile instars of insects with hemimetabolous development, of mites and ticks, and of pentastomids.

obligate parasite: An adult parasite that requires a host for survival and perpetuation.

oncomiracidium: The ciliated, free-living larval stage of monogeneans.

oncosphere: The hexacanth embryo of the Eucestoda.

oocyst: An encysted zygote in the sporozoan life cycle. The zygote sporulates (sporogony) to form sporozoites.

ootype: A dilation of the female oviduct in platyhelminths, usually surrounded by the Mehlis' gland.

opisthaptor: A large attachment device located on the ventral surface at the posterior end of monogeneans.

osmoconformers: Organisms that do not osmoregulate; most nematodes are thought to be osmoconformers.

ostia: Slit-like opening in the heart of insects to allow the passage of blood into the heart.

ovijector: A highly muscular vagina through which the eggs are ejected forcibly by muscular contraction.

ovoviviparous: Eggs develop within the female body but there is no nutrition provided to the eggs by the female.

oxidation: Chemical reaction involving the loss of electrons or hydrogen by an atom or molecule.

pansporoblast: Myxosporidian sporoblast that gives rise to more than one spore.

paramphistomiasis: A disease produced by paramphistome digeneans, usually in ruminants.

parapodia: Appendages present on the body segments of polychaetes used for locomotion, tube building, or to create water currents.

paratenic (= transport) host: A host in which development does not occur, but which may serve to bridge an ecological, or trophic, gap in a parasite's life cycle.

paraxial rod (= paraflagellar rod): An electron-dense ribbon or rod consisting of a three-dimensional fibrillar lattice associated with the flagellum, often for most of its length. It is contained within the flagellar membrane, and is present in a number of the parasitic flagellates.

parenchyma: Mass of mesenchymal cells filling spaces between the organ systems, muscles, or epithelia.

paroxysm: A severe, sudden attack; a fit, or convulsion, frequently associated with malaria.

passive transmission: Movement of a parasite from one host to another without expenditure of energy by the parasite.

pathogen: Organisms that cause disease when they infect a host.

pedipalps: The second pair of appendages in chelicerate arthropods.

peptides: Two or more amino acids linked together to form small molecules. Peptides consist of less than 100 amino acids; if they contain more, they are called polypeptides or proteins.

pereiopod: Thoracic appendage of crustaceans.

peripatric speciation: Speciation that results from the isolation of a small population on the periphery of a larger ancestral population.

peritrophic membrane: A chitinous, non-cellular membrane that surrounds the food in the midgut of most insects; it usually pulls loose from the midgut, remains around the food, and passes out with the feces.

phagocytosis: The process of engulfing and internalizing particulate matter by cells. *In vivo*, both neutrophils and macrophages are important phagocytic cells. The sacrodinian, or ameboid, protozoa are generally phagocytic.

phagolysosomal vacuole: Consists of the phagosome fused with a lysosomal vacuole, important in the destruction and killing of engulfed pathogens.

phasmid: Sensory pit on either side near the end of the tail of some nematodes.

phyllidea: Also called bothridia; leaf-like structures associated with the scolices of some cestodes.

phylogenetic systematics: An analytical method used to infer evolutionary histories.

phylogenetic tracking: Model of resource tracking in which the resource that the parasites are tracking is restricted to a host clade; this results in congruent host and parasite phylogenies.
pinocytosis: The intake of fluid by a cell through the formation of small, fluid-filled vacuoles derived from the cell membrane.
planula larva: Larval stage of cnidarians.
plasmodium: Multinucleated, amoeboid stage of orthonectids.
platelet: A small, round, granular body present in the blood of mammals. It is derived from a bone marrow cell and, in the blood, is involved in the clotting process. Platelets also contain histamine and serotonin that can be released during an immediate hypersensitivity reaction. Platelets have a complement receptor on their surface and, therefore, are involved in immune adherence.
pleopods: Abdominal appendages of the Crustacea.
plerocercoid: A second larval stage (after the procercoid) in the life cycle of certain cestodes, e.g., Pseudophyllidea; a solid larva possessing a scolex resembling that of the adult.
plerocercoid growth factor: A low-molecular-weight agent produced by plerocercoids of *Spirometra mansonoides*; the agent mimics the effects produced by vertebrate growth hormone.
poecilogyny: In some oxyurid nematodes, this is a process where two different forms of females are produced. One type produces thin-shelled eggs that may play a role in autoinfection. The other type of female produces thick-shelled eggs that pass out with the feces and play a role in colonizing new hosts.
polar capsule: Compartment bearing the polar filaments in myxozoans.
polar filament: Thread-like organelles present in myxozoans similar to the nematocysts of cnidarians.
polyembryony: The formation of two or more 'embryos' from a single zygote.
polymorphism: The proportion of polymorphic loci in a population.
polyphyletic: Species currently assigned to a single taxonomic group evolved from not one, but a number of different taxonomic groups.
polysaccharide: A carbohydrate composed of numerous simple sugars or monosaccharide units joined into long chains. Glycogen, cellulose, and starch are all examples of polysaccharides.
polyzoic: A tapeworm strobila in which proglottid formation is distinct.
praniza: Parasitic larva of gnathiid isopods. It parasitizes fishes and feeds on blood.
precipitate: The interaction between antigen and antibody resulting in the formation of very large complexes that ultimately come out of solution as a precipitate. The measurement of the amount of precipitate can be used in a serological assay for determining the amount of antibody or antigen present in a solution.
predator: An organism that kills, and consumes, many prey during its lifetime.
primary larva: First larval stage of pentastomids. It remains inside the egg and excysts after the egg is ingested by an intermediate host.
proboscis: In acanthocephalans, the anterior retractile organ bearing hooks and used for attachment to the host. In nemerteans, a protrusible structure used in food capture and defense.
procercoid: A small, sausage-shaped larva of certain cestodes, usually parasitic in the hemocoel of microcrustaceans and usually in possession of three pairs of embryonic hooks located on the cercomer.
proctodaeum: The part of the digestive tract that arises as a posterior invagination of the ectoderm and becomes the anus in the adult.
procyclic: A developmental stage in the life cycle of parasitic kinetoplastid flagellates in the invertebrate host. It is most often found in the midgut of the host.
proglottid: The so-called 'segment' in a tapeworm's strobila.
prohaptor: A large attachment organ located on the surface at the anterior end of monogeneans.
prokaryotic: Cells lacking a nuclear membrane, large complex chromosomes, a mitotic or meiotic spindle, Golgi, or other organelles of the more complex eukaryotic cells. Prokaryotes include the bacteria.
protandrous hermaphroditism: Type of hermaphroditism in which the male gonads develop first; the female organs develop later.
protandry: Development of the male reproductive system and sperm production first, followed by development of the female reproductive system.
protein: A complex macromolecule composed of one or more polypeptide chains, each chain made up of amino acids joined by peptide bonds. Proteins are composed of more than 100 amino acids.
protogyny: Development of the female reproductive system and ova first, followed by the development of the male reproductive system in a hermaphrodite.
protonephridium: The unit of structure and function in the osmoregulatory system of the Platyhelminthes.

pseudocoelomate: Triploblastic animals with a body cavity incompletely lined by mesoderm, i.e., not lined by a peritoneum.
pseudocyst: A cyst-like structure, often containing numerous protozoan parasites, that is formed by host cells, not by the parasites themselves.
pseudopodia: Temporary retractile cytoplasmic protrusion (a false-foot), which is characteristic of amoeboid protozoa. The pseudopodium functions in locomotion, feeding, or both.
purine: A component of nucleic acids. The parent compound of the nitrogen-containing bases adenine and guanine.
pustule: Any small, fluid-filled swelling, such as a pimple or blister.
pyrimidine: A component of nucleic acids. The parent compound of the nitrogen-containing bases cytosine, thymine, or uracil.

random distribution (= Poisson): A term used to describe the distribution of a population of organisms in which the variance equals the mean.
realized niche: The actual distribution of an organism in an *n*-dimensional hypervolume.
recombination: The formation of new gene combinations; in eukaryotes it may be accomplished by new associations of chromosomes produced during sexual reproduction or crossing-over.
recruitment: The acquisition of an infective stage or propagule.
Red Queen Hypothesis: Predicts continued, reciprocal, evolutionary change in an assemblage of interacting organisms, allowing them to maintain their level of interaction. (See **evolutionary arms race**).
redia: An intramolluscan developmental stage produced by a digenean sporocyst.
reduction: A chemical reaction involving the gain of electrons or hydrogen atoms by a molecule.
regular distribution (= uniform): A term used to describe the distribution of a population of organisms in which the variance is less than the mean.
regulation: The mechanism(s) involved in controlling the size of a population.
relict: A fauna that has been isolated from its parent fauna for a long time.
renette cell: Ventral gland cells unique to the Nematoda that open to the exterior via a midventral pore. Thought to be associated with the excretory system.
reservoir host: Host in which a parasite can mature and reproduce but is not considered the normal host. Reservoir hosts help maintain a parasite when the normal hosts are not available.
resource tracking: Model in which hosts are necessary resources that parasites track through evolutionary time.
rete system: Highly branched system of tubules located on the muscular region of acanthocephalans. It appears to help with the contraction stimuli.
rhombogen: Stage in the life cycle of dicyemids.
rostellum: A muscular organ located at the tip of a scolex and which may or may not be armed with spines or hooks; used in attachment to the surface of the host's intestine.
***r*-selection:** Selection which favors production of many offspring at the expense of maintaining the individual.

Saefftigen's pouch: Internal muscular sac near the posterior end of male acanthocephalans; it aids in the functioning of the copulatory bursa.
Salivaria: One taxonomic section of the genus *Trypanosoma*.
salvage pathway: Biochemical pathways that use preformed molecular intermediates, present in the organism's environment, for the synthesis of macromolecules.
sarcocyst: The last generation meront of the sporozoan *Sarcocystis* spp. It is found in the muscle tissue of the intermediate host.
sarcoptic mange: Allergic condition caused by the burrowing of mites of the genus *Sarcoptes* under the skin. **Scabies** is another term for this condition.
schistosomule: The migratory stage of schistosomes in the vertebrate host that follows penetration of cercariae and shedding of its tail, but preceding maturation to the adult.
schizogony: A general term referring to the formation of daughter cells by multiple fission. Its use is restricted to sporozoan groups. If the final products of division are merozoites, the process is referred to as merogony, if sporozoites, sporogony, and if gametes, gametogony.
sclerotized: A structure where protein molecules are linked, or tanned, by quinones resulting in hardening of the structure.
scolex: The muscular attachment organ located at the anterior end of all cestodes.
scutum: Sclerotized dorsal plate of ticks.
selective site segregation: Refers to site selection that is non-interactive, implying the absence of current interspecific interactions.

seminal receptacle: A structure associated with the female reproductive system where sperm from another individual is stored prior to being used for fertilization.

seminal vesicle: A structure associated with the male reproductive system where sperm, produced by the individual, is stored.

sequential colonization model: Model of resource tracking in which the resource that the parasites are tracking is widespread; this results in poor congruence of host and parasite phylogenies.

sex reversal: Capability of some organisms to change their sexual condition from male to female, or from female to male.

sexual dimorphism: Found in those dioecious organisms where there is a consistent external morphological difference between males and females.

sexual selection: Selection by one sex based on variations within the opposite sex.

short stumpy stage: Refers to a morphological stage in the life cycle of the African trypanosome in the mammalian host. It is a non-dividing stage, physiologically pre-adapted for transmission to the insect vector.

sibling species: Species that are difficult or impossible to distinguish by morphological characters.

sister species: Species that are each other's closest relatives.

sparganosis: A form of larval migrans in humans produced by plerocercoids of several pseudophyllidean cestodes, primarily *Spirometra*.

speciation: The division of one species into two or more distinct species; the process by which new species are formed.

speciation of peripheral isolates: Speciation that results from the isolation of a small population on the periphery of a larger ancestral population.

species: A group of organisms that actually (or potentially) interbreed in nature and are reproductively isolated from all other such groups.

species diversity: A derived value that relates both the number of species, and their abundances, in a prescribed area of interest.

species richness: The actual number of different species in a prescribed area of interest.

spermatophore: A specially formed capsule or package containing sperm that is placed in or on the body of the female.

spicules: Cuticularized structures found in some male nematodes that, during copulation, are inserted into the female vulva keeping it open against the strong hydrostatic pressure.

spiracles: External openings of the tracheal system to allow gas exchange with the environment.

spirochaete: A motile, spiral-shaped bacterial cell.

splenomegaly: A clinical condition often the result of a parasitic infection in which the spleen becomes enlarged.

spore: A very general term referring to a variety of different resistant and transmissible stages in the life cycle of different taxonomic groups of protozoans. It is being replaced by the more specific term **oocyst**, especially by students of the Apicomplexa.

sporocyst: An intramolluscan, asexual developmental stage of digeneans **or** a stage present in the life cycle of most coccidians. The latter refers to a cyst that contains sporozoites that is formed within the oocyst.

sporogony: A type of schizogony in which a zygote undergoes a single meiotic division followed by mitotic divisions. This results in the formation of sporocysts and sporozoites during the life cycle of many apicomplexans.

sporoplasm: Amoeboid portion of a myxozoan spore that is infective to the next host.

sporozoite: The motile, infective stage often present within a cyst or shell produced during sporogony.

stem nematogen: Stage in the life cycle of dicyemids that produces vermiform embryos, asexually, within its axial cells.

stenohaline: Having a narrow tolerance to changing salt concentrations. Usually applies to those aquatic organisms whose distribution is limited by too high salinity.

stercoraria: One taxonomic section of the genus *Trypanosoma*. It includes trypanosomes in which the infective metacyclic stages are found in the hindgut–rectum of the vector and are released with feces onto the skin of the mammalian host.

sterile immunity: A form of immunity that is permanent.

stichosome: A glandular esophagus with one to three rows of individual cells (stichocytes). These stichocytes are large and have a single pore communicating with the lumen of the esophagus.

stomodaeum: The part of the digestive tract that arises as an anterior invagination of the ectoderm and becomes the mouth in the adult.

strobila: The ribbon-like arrangement of the body of cestodes beginning immediately posterior to the scolex.

strobilization: The process of forming a tapeworm strobila.

strobilocercus: A cestode larval stage, possessing a fully developed scolex and a strobila-like neck to which is attached a bladder; always found in a vertebrate intermediate host, usually a rodent.
stylets: Modified mouthparts forming a needle-like structure used by sucking insects **or** the tooth-like structures found on the anterior of cercariae and used to aid in penetration of the host.
stylostome: Specialized hardened, tube-like structure used for feeding by chiggers.
sugar: Any monosaccharide, e.g., glucose or dissacharide, e.g., dextrose.
suprapopulation: A population that includes all of the developmental phases of a species at a particular place and time.
swimmer's itch: A localized hypersensitivity reaction to the accidental penetration of avian or mammalian cercariae into the skin of humans.
sympatric speciation: Speciation without the geographical isolation of populations; populations become isolated by intrinsic factors.
synchronous cospeciation: Host and parasites speciate simultaneously; their lineages show similar degrees of evolutionary divergence.
syncytial: The term refers to a single cell possessing many nuclei; the tegument of parasitic flatworms is said to be syncytial, with nuclei scattered in cytons extending below the surface of the tegument.
synecology: A term used to describe the ecology of a group of organisms of different species.
synlophe: Cuticular ridges found in the trichostrongyles.
syzygy: The side-by-side or end-to-end association of gamonts prior to the formation of gametocysts and gametes. Generally used in reference to the gregarines.

tachyzoites: This term generally refers to the fast-developing merozoites that are produced in aggregate during the acute phase of a *Toxoplasma* infection.
tagma: Distinct region of the body of arthropods formed by the fusion of segments (metameres).
tagmatism: The specialization of body segments into functional groups to form different regions of the body.
T-cell: Thymus-dependent (T) lymphocytes are lymphocytes that require the thymus for their ultimate maturation to functional T-cells.
tegument: The superficial covering of multicellular organisms.
telson: The posterior part of the last abdominal segment in the Crustacea.
tetrathyridium: A solid-bodied larval stage of the cestode *Mesocestoides* that apparently, in some species, is capable of asexual reproduction via binary fission.
thrombosis: The coagulation of blood in a blood vessel or in the heart leading to partial or complete blockage of blood flow.
tomite: A small, free-swimming, non-feeding stage in the life cycle of a number of parasitic ciliates. Usually, it has recently emerged with a number of other tomites from a cyst in which rapid division has taken place.
tracheae: A system of branching tubules through which gaseous exchange is effected.
transmammary transmission: A method of transmitting infective stages of some parasites during nursing.
triactinomyxon: Stage in the life cycle of a myxozoan. It produces characteristic triactinomyxon spores.
tribocytic organ: A secretory structure used in the release of proteolytic enzymes on the external surface of strigeid digeneans.
trichogon: Male larva of a rhizocephalan cirripede that penetrates the female externa and implants itself in a male receptacle.
trophozoite: The active, feeding stage in the life cycle of protozoans and myxozoans.

undulating membrane: In parasitic flagellates, the extension of the body's surface membrane that is joined for part, or all, of its length with the membrane that surrounds the flagellar substructure and the axoneme. In the ciliates it is an organelle lying within the buccal cavity.
unilocular hydatid cyst: A cysticercus-type cestode larval stage in possession of many scolices (= protoscolices) produced in so-called 'brood capsules' that float freely in the fluid of the parental cyst; all asexual reproduction is endogenous.
unsaturated: Refers to long-chain hydrocarbons that contain one or more double-bonded carbon atoms.
uropod: One of the terminal pair of abdominal appendages in Crustacea.
uterine bell: Egg-sorting structure found in female acanthocephalans; it allows fully developed eggs to leave the body while retaining immature ones.

variance: A mathematical term used to describe the distribution, or dispersion, around the mean, or some other measure of central tendency.
vector: A micropredator that transmits a parasite from one host to the next. Development in the vector may, or may not, occur.

vermiform embryo: Stage in the life cycle of dicyemids that attaches to the kidney of the host.

vicariance: Distribution of organisms due to geographical isolation and fragmentation of landmasses.

vicariant biogeography: Branch of historical biogeography that attempts to explain the distribution of organisms based on the assumption that their present-day distribution results from fragmentation and isolation due to the emergence of geographical barriers.

vicariant speciation: Speciation that results from the geographical isolation and separation of an ancestral population into two or more isolated populations.

viscerotropic: Host white blood cells or parasites that have a predilection for visceral tissue sites.

vitellaria: Glandular cells that contribute to the formation of eggshells and yolk.

ybp: Years before present.

zoonosis: A disease normally cycling through animals which, under certain conditions, can be transmitted to humans.

zygocyst: A synonym of the term oocyst.

zygote: The diploid (2*N*) stage of an organism resulting from the fusion of gametes.

zymodeme: A strain of organisms differentiated from another strain based on electrophoretic patterns.

Index